Genetics and Human Variation

By

William L. Daniel

University of Illinois

Copyright © 1982, 1986, 1991
Stipes Publishing Co.

ISBN 0-87563-385-4

Published by
STIPES PUBLISHING COMPANY
10-12 Chester Street
Champaign, Illinois 61820

I would like to express my gratitude to my friends and family for their support and patience during preparation of this text, to Ms. Lauren Goralski, Susan Lansky-Shafer, and Mr. Richard Green for the karyotypes used in the text, and to the medical students, human genetics students, and biology instructors who provided many useful suggestions for improvement of an earlier text from which the current edition evolved.

Preface

This text has been written for use by upper division undergraduate students, for graduate students in the biological sciences, and for medical students studying the basic medical sciences. It is assumed that students have completed an introductory course in genetics, although some introductory material has been included in the text. A sampling of problems is provided at the end of each chapter with solutions appended to the text. Each chapter also contains a bibliography as a guide for further reading relevant to special topics that may be of interest to the student. The following journals frequently publish articles in various areas of human genetics.

Acta Geneticae Medicae et Gemellologiae	Gene
Advances in Human Genetics	Genomics
American Journal of Human Genetics	Human Genetics
American Journal of Medical Genetics	Human Heredity
Annals of Human Genetics	Japanese Journal of Human Genetics
Annual Review of Biochemistry	Journal de Genetique Humaine
Annual Review of Genetics	Journal of Inherited Metabolic Disease
Biochemical Genetics	Molecular and Cellular Biology
Cancer Genetics and Cytogenetics	Nature
Cell	New England Journal of Medicine
Clinical Genetics	Nucleic Acids Research
Cytogenetics and Cell Genetics	Proceedings of the National Academy of Sciences (U.S.A.)
Developmental Genetics	Progress in Medical Genetics
Federation Proceedings	Science

Table of Contents

Chapter 1

Introduction

GENETICS AND HUMAN HEALTH

Awareness of the importance of genetics in human health has markedly increased during recent years. The terms gene and chromosome appear frequently in the press; and patients and families are confronting health care professionals with questions concerning risks of occurrence and recurrence of genetic diseases and their underlying mechanisms. These demands underscore the need for better training of health care professionals in the fundamentals of medical genetics.

Genetic abnormalities affect a surprisingly large segment of the human population. About 15% of diagnosed pregnancies abort during the first trimester, and approximately 40% of these aborted fetuses possess abnormal chromosome constitutions. As many as one in four human pregnancies may terminate at earlier stages, and a higher percentage of these embryos appear to have abnormal chromosome complements. Liveborn infants run a risk of 4-8% for major genetic handicaps. A typical series of 1000 newborn would include about 5 babies with chromosome anomalies, 20 to 30 infants who have or who will develop complications caused by major genes, and 20 to 40 infants who have major malformations with multifactorial causes. Effects of certain abnormal genes may not appear until late childhood, adolescence, or adulthood. Examples of late onset diseases possessing significant genetic components include Huntington's disease, cancers, heart disease, diabetes, emphysema, mental illness, and epilepsy.

Patients with genetic disorders tend to be admitted to hospitals more frequently than those with non-genetic diseases. Patient management often demands the coordinated efforts of the family physician, nurse, dietician, physical therapist, teachers, and a variety of clinical specialists. The burden of the disease upon both the patient and the family often precipitates problems that must be addressed by psychiatrists and social workers. Increased use of screening programs for detection and treatment of certain relatively common genetic diseases, more extensive application of cytogenetics and molecular genetics for the diagnosis of cancers, and increased reliance of the medicolegal community upon protein and molecular markers for demonstration of relationship and source of criminal evidence have broadened the numbers of health care professionals and others involved to some degree with human genetics.

There is also a critical need for education of the public with regard to human genetics. Legal, ethical, and moral issues arising from pregnancy intervention, genetic engineering, delivery of health care, and hazards of pollutants are among many impacting upon both individual families and society. An informed public is essential for effectively dealing with these difficult issues and the problems associated with them.

THE CELLULAR BASIS OF BIOLOGICAL ORGANIZATION

Normal human function requires the careful integration of the processes occurring within each of the body's component parts. Organs such as the brain, lungs, heart, and liver comprise a macroscopic level of body organization. Each organ is composed of tissues, including various combinations of muscle, epithelium, bone and connective tissue. Each tissue consists of an ordered array of cells which perform certain tasks characteristic of the cell type. This complexity of body organization is determined by both the intrinsic genetic machinery within cells and the environments in which they develop. Although the same genes occur in almost all cells, their patterns of expression often vary among diferent cell types. As a consequence an alteration of a gene that interrupts its normal expression will often produce somewhat different effects in various cell types.

Robert Hooke (1665) initially described the microscopic structure of cork and proposed the term "cells" for the chambers he observed. These chambers actually were the dried remains of cells, and Van Leeuwenhoek (1632-1723) is credited with the discovery of living cells corresponding to what are presently classified as either bacteria or protozoa. During the nineteenth century, the observations of Dutrochet, Schwann, Schleiden and Virchow established the tenets of the **Cell Theory**: 1) All living organisms consist of one or more cells; 2) These cells are capable of independent existence; and 3) All cells arise from preexisting cells. Virchow is also credited with emphasizing that cells are dynamic

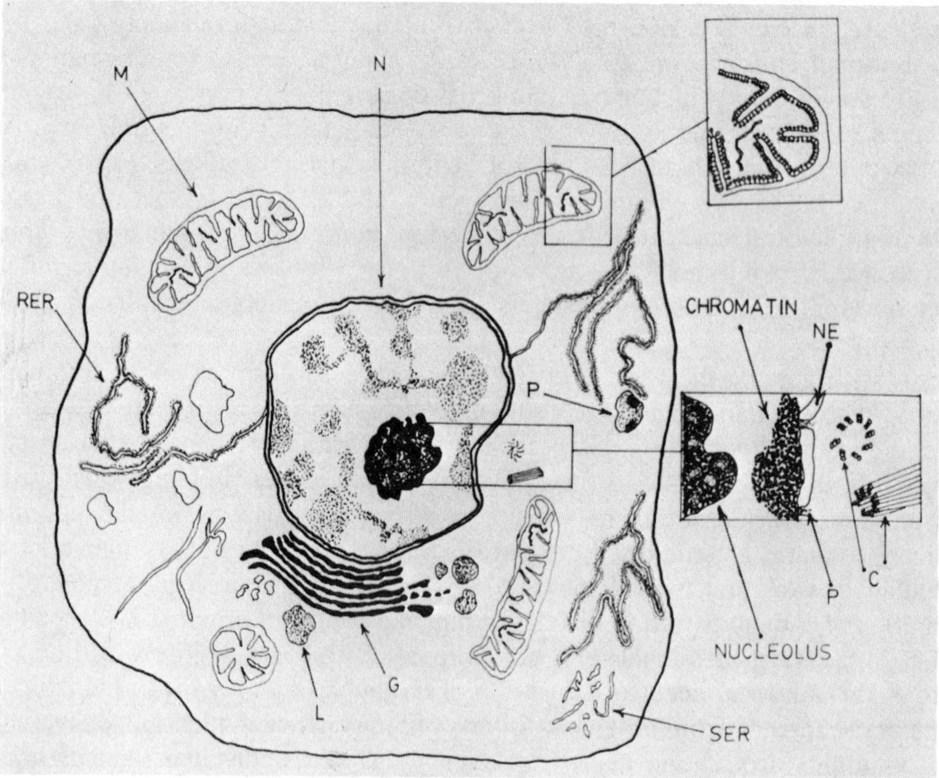

Fig. 1-1. Schematic diagram of a human cell. N=nucleus; M=mitochondrion; RER=rough endoplasmic reticulum; SER= smooth endoplasmic reticulum; L=lysosome; G=Golgi apparatus; P=peroxisome. Inset: NE=nuclear envelope; P=pore.

functional entities, not merely structural elements of living systems. He stressed that the integrity of an individual is derived from the properties of his or her individual cells, and that disease stems from abnormal cellular function.

COMPONENTS OF LIVING CELLS

A typical cell is illustrated in Fig. 1-1. The **cell membrane** is a lipid bilayer that encloses cellular contents and serves as a selective barrier to free exchange of chemicals between the cell and its environment. The lipids are arranged such that their charged (polar) regions project toward the outer and inner membrane surfaces, while their uncharged (nonpolar) portions are sequestered in the core of the membrane. Some of these lipid molecules have carbohydrate chains attached to their polar regions, which project into the aqueous layer contacting the membrane. These glycolipids are especially enriched on the surfaces of nerve membranes and on red blood cell surfaces. Proteins are also found imbedded in the cell membrane. Some of these proteins are anchored to the inner or outer membrane surface, while others transect the membrane one or more times. The cell membrane is a dynamic structure with its protein and glycolipid components moving from one place to another in the lipid bilayer. Certain membrane proteins serve as receptors that participate in the internalization of molecules such as lipoproteins. Others relay signals from the exterior to the interior of the cell. Still others may aggregate to form pores or channels through which small molecules may enter and exit the cell.

The **nucleus** is a large spherical structure that is bounded by a double membrane, the **nuclear envelope.** A large number of pores penetrate the nuclear envelope, through which the nucleus communicates with the cytoplasm. The nucleus contains the majority of the cell's genetic information which is stored in **chromatin.** The composition and function of chromatin will be discussed in a later chapter. The **nucleolus** is a spherical body encompassed by the nucleus which is the site of manufacture of several components of **ribosomes.** Ribosomes participate in the synthesis of proteins. The immature components of ribosomes appear as granular structures in the nucleolus. Ribosomes consist of both proteins and nucleic acids called RNA. The ribosomal proteins are made in the cytoplasm and imported into the nucleolus where they are assembled with ribosomal RNAs to form the large and small ribosomal subunits. These subunits are exported to the cytoplasm where they participate as a complex in the synthesis of proteins.

Protoplasm is the living substance of the cell and consists of a number of small structures or organelles suspended in a chemical soup containing carbohydrates, lipids, proteins, various salts, other small molecules, and water. Protoplasm consists of two general portions: the **nucleoplasm** enclosed by the nuclear envelope and the **cytoplasm** which fills the space between the nuclear envelope and cell membrane. Four thousand to 5000 proteins and many other biomolecules are present in the cytoplasm. The cytoplasm should be viewed as a highly organized compartment that serves as a complex factory to produce the energy, metabolites, and other needs for the life of the cell.

Organelles found in the cytoplasm of a typical human cell include mitochondria, the Golgi apparatus, lysosomes, peroxisomes, and the endoplasmic reticulum. The **endoplasmic reticulum (ER)** is an extensive network that extends throughout the cell. The ER is the site of synthesis of proteins and lipids that become incorporated in the cell membrane or other organelles and of proteins that are secreted from the cell. The **rough endoplasmic reticulum (RER)** is studded with ribosomes, while the **smooth endoplasmic reticulum (SER)** lacks ribosomes on its surface. The endoplasmic reticulum is continuous with both the outer membrane of the nuclear envelope and with the cell membrane. The flow of proteins synthesized on ribosomes of the RER is from the ribosome to the lumen of the RER, where they are chemically modified and then exported to the Golgi apparatus for further processing prior to incorporation into other organelles or secretion from the cell. These events will be discussed more fully in a later chapter.

Following export from the RER, proteins enter a series of stacked vesicles, the **Golgi apparatus,** where they are chemically modified. The movement of proteins from the RER to the Golgi and from the proximal (closest to the RER) to the distal regions of the Golgi occurs by enclosing the proteins in small vesicles which bud from the membrane of one compartment and fuse with membranes of more distal regions of the Golgi. Chemical modifications occurring within the Golgi include glycosylation (addition of carbohydrate) to proteins, cleavage of peptides from proteins, and modification of amino acids within proteins. These alterations affect the three dimensional structure, stability, and function of the proteins as well as targeting them to various sites within the cell or for release from the cell.

Lysosomes contain a large variety of enzymes that catalyze the beakdown of complex molecules, viruses, and bacteria. These enzymes are synthesized on the RER and modified by the Golgi apparatus. The lysosomes appear to bud from the distal Golgi region and are capable of fusion with other lysosomes or with vacuoles containing ingested particles and/or proteins from the exterior of the cell. The lysosomes release their enzymes into the vacuoles, digesting their contents. Lysosomal enzymes are "trapped" within lysosomes by "limited proteolysis", a process that trims small peptides from the enzymes, making it difficult for them to exit the lysosomal membrane. Lysosomes are an adaptation that greatly increases the concentration of these enzymes, many of which are trace proteins (1/5000 to 1/12000 protein molecules) within the cell. The additional flexibility of lysosomal fusion also increases the ability of the cell to produce combinations of enzymes that are most appropriate for the stepwise degradation of large molecules. In addition to breaking down materials ingested by the cell, lysosomes also participate in the dynamic remodelling of cellular architecture that occurs throughout the life of the cell.

Mitochondria are the "power plants" of living cells. Mitochondria are similar to prokaryotic cells and are believed to have been incorporated into eukaryotic cells where they evolved symbiotically with these cells, becoming specialized for energy production. Genes are found not only in the nucleus, but also in mitochondria. Human mitochondrial DNA contains 37 genes, far to few for the structural and functional needs of the mitochondrion. A larger number of nuclear genes encode proteins that are incorporated into mitochondria. Mitochondrial enzymes may be entirely encoded by mitochondrial genes, entirely encoded by nuclear genes, or may contain nuclear-encoded subunits and mitochondrial-encoded subunits. Mitochondria of sperm are sequestered in a region behind the sperm head, a portion of the sperm that enters the egg but decays after fertilization. Consequently, all of the embryo's mitochondria are derived from the egg; and traits determined by mitochondrial genes display maternal inheritance. These traits occur in both sexes, but are transmitted to the next generation by the mother and not the father. In principle, all mitochondria present in the world today are derived from a single female ancestor (the "Eve Principle"). The "Eve Principle" and other unusual aspects of mitochondrial genes and their expression will be addressed in subsequent chapters.

Peroxisomes are lysosome-sized vesicles that arise by budding from the ER. The enzymes found in peroxisomes are imported from the **cytosol,** the liquid portion of the cytoplasm. These enzymes are specialized for executing chemical reactions that use molecular oxygen. One function of these enzymes includes the detoxification of peroxides and other compounds that are capable of harming the cell.

CHEMICAL COMPONENTS OF PROTOPLASM

Protoplasm contains a variety of molecules that perform a large number of vital functions. An overview of the nature and roles of several types of these molecules will be provided in the following section.

PROTEINS

Proteins are macromolecules which consist of **amino acids.** Genes specify twenty of these amino acids, while others are known to be formed by modification of these twenty amino acids after the protein is synthesized. A simple (**monomeric** = 1 subunit) protein consists of a chain of 30 to several hundred amino acids that are folded into a three dimensional structure placing the water-repelling (**hydrophobic**) amino acids in the core and the water-attracting amino acids (**hydrophilic**) on the surface of the protein. Proteins may contain from approximately 30 amino acids to more than 1000 and have one to several (**multimeric**) subunits. Examples of proteins include insulin (41 amino acids), hemoglobins (O_2-transporting protein of the blood which contains two α-globin chains (141 amino acids each) and two β-globin chains (146 amino acids each)), and IgG (an antibody consisting of two light chains and two heavy chains possessing a total of 1064 amino acids). Proteins may serve as structural elements (collagens), transport molecules (hemoglobins), hormones (insulin), receptors (LDL-receptor), as a defense against disease (IgG), and as facilitators of chemical reactions (enzymes).

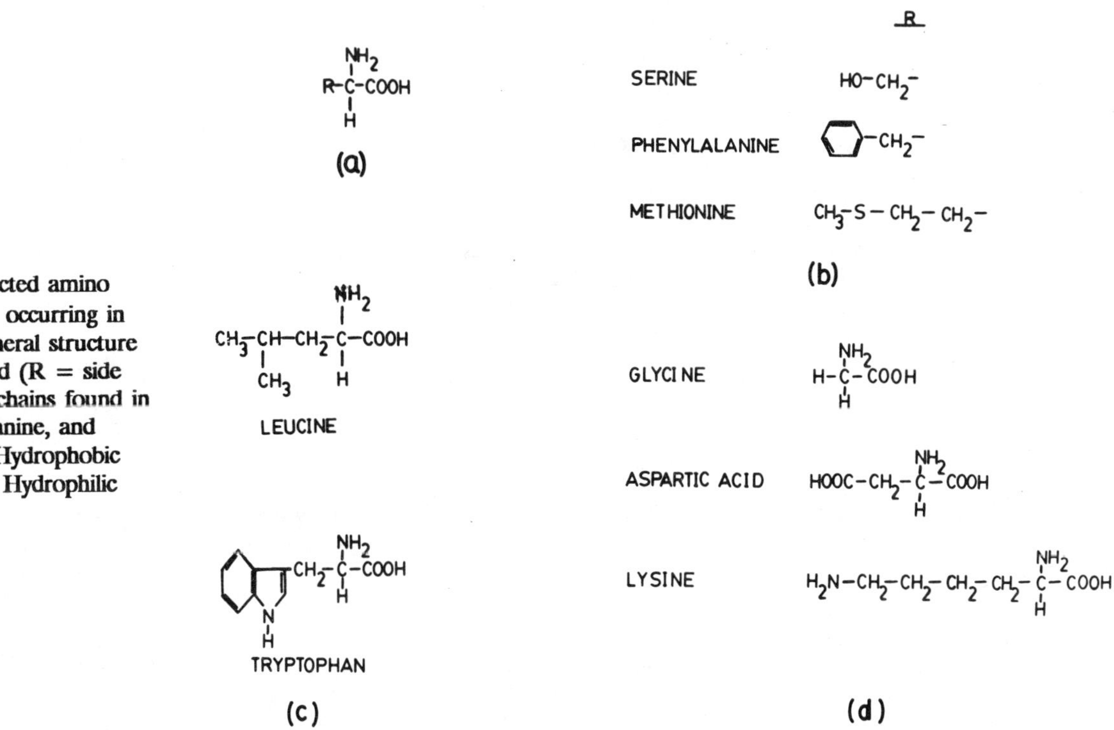

Fig. 1-2. Selected amino acids commonly occurring in proteins. a) General structure of an amino acid (R = side chain); b) Side chains found in serine, phenylalanine, and methionine; c) Hydrophobic amino acids; d) Hydrophilic amino acids.

Representative amino acids are displayed in Fig. 1-2. Each amino acid has one or more basic groups (e.g. an amino group = $-NH_2$) and one or more acidic groups (e.g. a carboxyl group = -COOH) (Fig. 1-2a). The net charge of the amino acid depends upon the relative number of acidic and basic groups it contains and the pH of the medium in which the amino acid occurs. The effects of pH upon the net charge of glycine are presented in (Fig. 1-3). Glycine is positively charged in acid media, neutral at physiological pH, and negative in basic (alkaline) environments. The net charge of a protein reflects the sum of the charges of its surface amino acids and the pH of their environment. The intracellular environment is approximately neutral, and amino acids may be positively charged (lysine), negatively charged (aspartic acid), or neutral (glycine). Amino acids with long carbon side chains or rings tend to be hydrophobic (Fig. 1-2c), while those with small side chains or groups on their side chains that are charged at physiological pH tend to be hydrophilic (Fig. 1-2d).

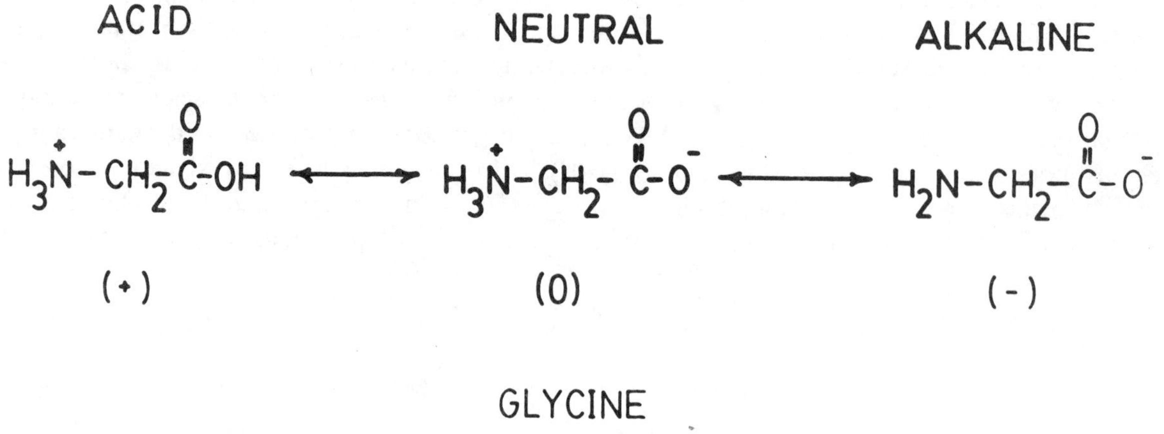

ACID	NEUTRAL	ALKALINE
$H_3\overset{+}{N}-CH_2-\overset{O}{\overset{\|}{C}}-OH$	$H_3\overset{+}{N}-CH_2-\overset{O}{\overset{\|}{C}}-O^-$	$H_2N-CH_2-\overset{O}{\overset{\|}{C}}-O^-$
(+)	(0)	(−)

GLYCINE

Fig. 1-3. Behavior of glycine in acidic, neutral, and alkaline (basic) solutions. The net charge is given in parentheses below the amino acid. Notice how glycine has a net charge of 0 at physiological pH in most cells and body fluids. However, in stomach secretions which are strongly acidic, glycine would have a net charge of +1.

Polypeptides are linear arrays of amino acids which are linked together by peptide bonds connecting the amino group of one amino acid with the carboxyl group of the preceding amino acid in the chain (Fig. 1-5). One end of each polypeptide has a free amino group, and the other contains a free carboxyl group. Proteins may contain from one to many polypeptide subunits, each of which has folded into the most energetically stable conformation, in which the polypeptides develop a hydrophobic core and hydrophilic surface. This structure adapts the protein to its largely aqueous environment. Both the over-all shape of the protein and its net surface charge strongly influence its behavior and interactions with other proteins. This section has addressed the general structure of proteins that are found in the cytosol and other aqueous solutions. Proteins that are anchored in or which

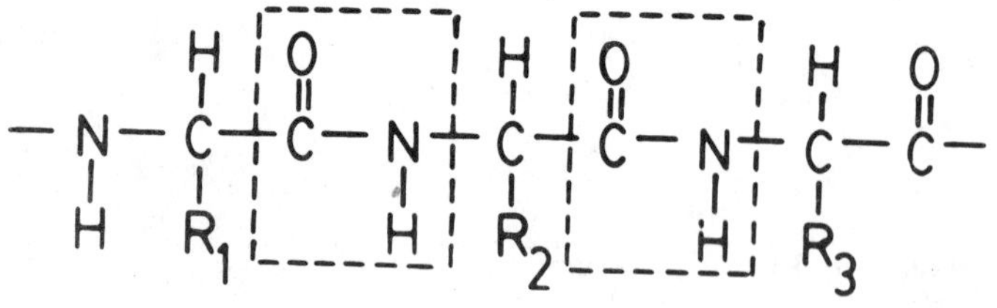

Fig. 1-4. Peptide bonds linking three amino acids of a polypeptide. R_1, R_2, and R_3 are side chains.

transect membranes contain hydrophilic and hydrophobic domains. The hydrophilic portions are localized at the membrane surfaces, while the hydrophobic domains interface with lipids and other hydrophobic protein regions in the central regions of membranes.

Enzymes are proteins that catalyze (facilitate) chemical reactions by reducing the amount of energy required for the reaction to take place. Enzymes that participate in the assembly of molecules promote interactions between appropriate chemical groups of the molecules and formation of new bonds by bringing the reactants (**substrates**) closer together (Fig. 1-5a). The region of the enzyme in immediate contact with the substrates and products is called the **active site.** Enzymes that degrade larger molecules into their constituent parts interact with the substrate to destabilize existing bonds between components of the large molecule, causing their separation into smaller products (Fig. 1-5b).

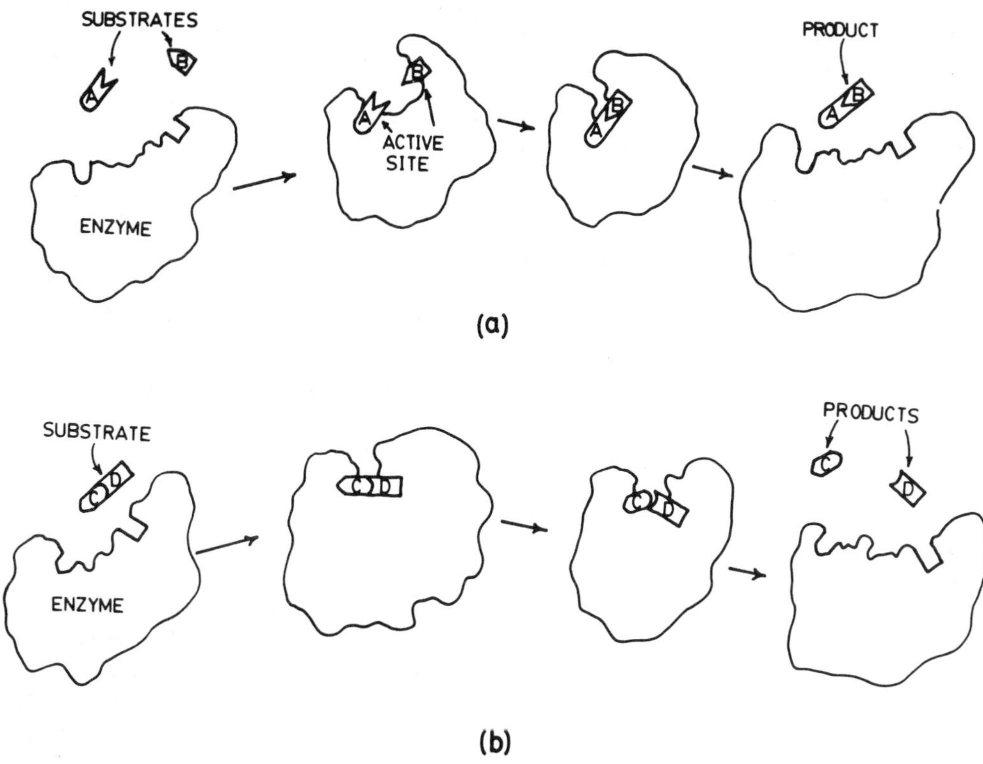

Fig. 1-5. Mechanisms of enzyme action in synthetic (a) and degradative (b) reactions. In both cases, the substrates bind to the active site where they are chemically modified by the enzyme to form the respective products.

Enzymes often utilize **cofactors** to increase the efficiency of enzyme-catalyzed reactions. Cofactors include ions of copper, zinc, and molybdenum and certain organic molecules such as pyridoxal phosphate (Fig. 1-6a). Pyridoxal phosphate is a **coenzyme,** a molecule that participates in but is not permanently changed during the chemical reaction. Coenzymes are synthesized from **vitamins,** compounds that cannot be synthesized by our cells and must be included in our diets. For example, vitamin B_6 (Fig. 1-6b) is phosphorylated to produce pyridoxal phosphate. Both metal ions and coenzymes are integral components of functional enzymes, binding to locations in the enzyme that are near the active site.

Enzymes, like many other proteins, vary considerably in size. Monomeric enzymes contain a single polypeptide chain.   Many enzymes are dimeric, possessing either two identical subunits (homodimer) or two dissimilar polypeptides (heterodimeric). Lactate dehydrogenase, an important enzyme occurring in a variety of different tissues is a tetrameric enzyme.   Several enzymes may be tightly associated in a complex.   For example, the pyruvate dehydrogenase complex contains three different enzymes and possesses more than 40 subunits and several coenzymes.

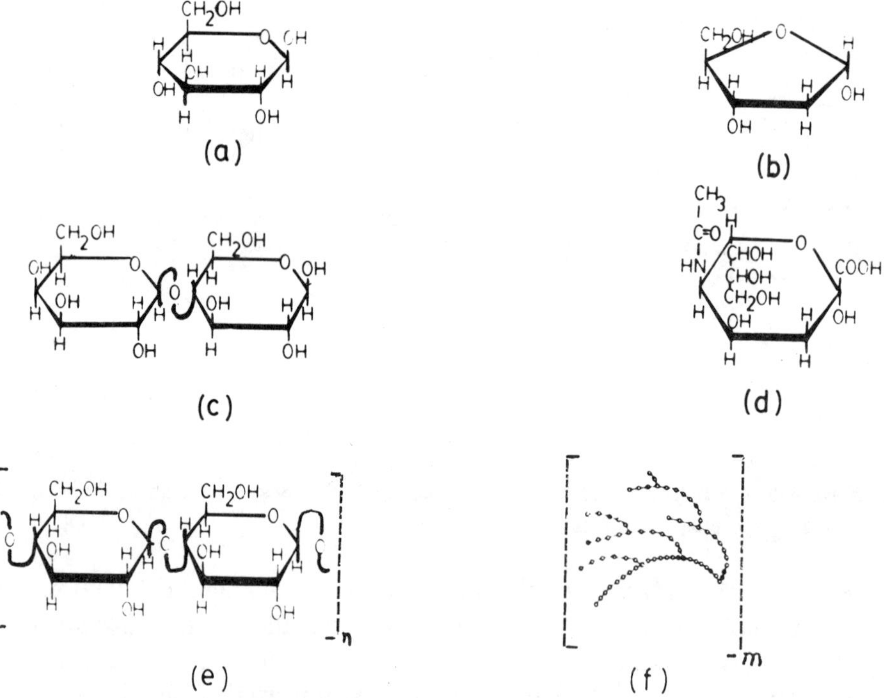

Fig. 1-6. Pyridoxal phosphate (a), a coenzyme, and its vitamin, vitamin B_6.

Fig. 1-7. Carbohydrates occurring in eukaryotic cells. a) Glucose; b) Deoxyribose; c) Lactose; d) Sialic Acid; e) Cellobiose; f) Glycogen. n and m designate integers.

CARBOHYDRATES

Carbohydrates contain carbon, hydrogen, and oxygen and generally have the formula $(CH_2O)_n$, where n is an integer.   Sugars are a common form of carbohydrate and vary in size from simple sugars (**monosaccharides**) to very complex **polysaccharides**, containing hundreds to thousands of subunits (Fig. 1-7).   Carbohydrates serve a variety of purposes in the cell. Glucose functions as an immediate energy source. Deoxyribose is a component of DNA. Lactose, a disaccharide, is a common milk sugar. Sialic acid is an important constituent of glycoproteins, increasing their solubility. Polysaccharides include cellobiose, a component of plant cellulose, and glycogen, an important energy store in animals.

LIPIDS

Lipids comprise a class of diverse molecules that can be extracted from tissues using detergents or solvents such as ether, certain alcohols, and various petroleum products. Lipids are found in cell membranes, (glycolipids, phosphatides, and steroids), are carried in chylomicrons and by lipoprotein particles in the blood (cholesterol and triglycerides), are stored as inclusions in fat cells (triglycerides), are important components of steroid hormones and visual pigments (retinoids), and may act as developmental signals (retinoids). Several lipids are illustrated in Fig. 1-8. These molecules possess relatively long hydro-

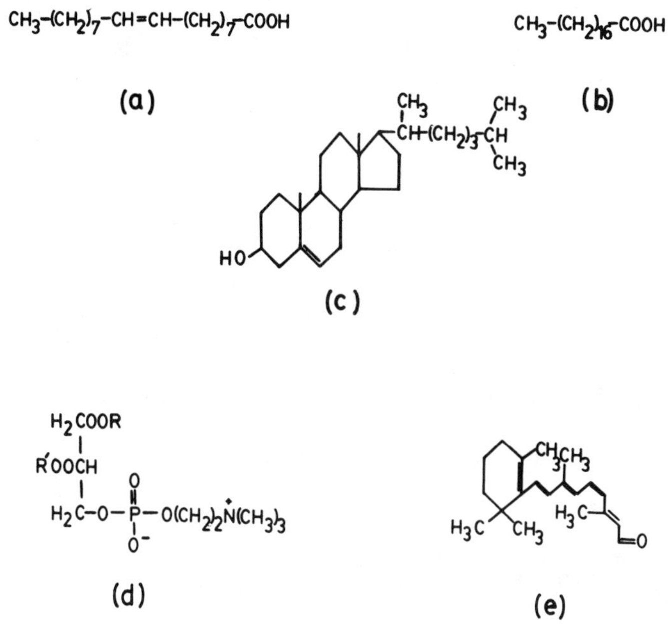

Fig. 1-8. Representative lipids. a) Oleic acid, an unsaturated fatty acid; b) Stearic acid, a saturated fatty acid; c) cholesterol, a steroid; d) phosphatidyl choline, a phosphoglyceride; e) cis-retinene, a component of visual pigments.

carbon chains, multiple rings, or both. Phosphoglycerides are interesting molecules that are major components of membranes. These molecules consist of a head that has a strong affinity for water and a hydrophobic tail composed of fatty acids. They are oriented in cell membranes such that the polar head faces the aqueous interface of the membrane and the tail projects into the interior of the membrane.

NUCLEIC ACIDS

Nucleotides (Fig. 1-9) have a variety of functions in the cell. Nucleotide triphosphates such as ATP serve as energy sources, with hydrolysis of their phosphodiester bonds (arrows) releasing energy required for biosynthetic reactions, transport, movement, and other physiological processes requiring energy. Nucleotides contain a planar base and a pentose (5-carbon sugar).

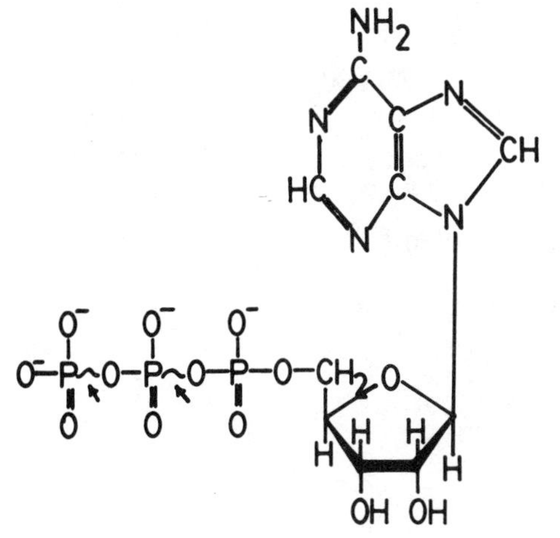

Fig. 1-9. Adenosine triphosphate. This high energy nucleotide is generated in large quantities in mitochondria under high oxygen conditions and in lesser amounts in the cytosol. Cleavage of the phosphodiester bonds (arrows) releases the stored energy.

9

The nucleotide illustrated in Figure 1-9 contains the base, adenine, which has two unsaturated rings and the pentose, ribose. Nucleotide triphosphates can be enzymatically coupled to form polynucleotides. **RNA** (Ribonucleic acid) is a polynucleotide possessing from about 75 to thousands of ribonucleotides linked to one another by a sugar-phosphate backbone. **Deoxyribonucleic acid, DNA**, consists of two polynucleotide chains that are interconnected by base pairs. The base pairs form the steps, and the sugar phosphate backbones of each polynucleotide chain form the sides of the DNA which assumes a three dimensional shape resembling a spiral staircase. These important nucleic acids will be discussed in more detail in the following chapter.

METABOLISM

The synthesis of complex molecules from smaller carbohydrates, lipids, and amino acids occurs through a series of **anabolic** chemical reactions that are enzymatically catalyzed and that consume substantial amounts of energy. **Catabolic** reactions convert large molecules into smaller metabolites that are either recycled or excreted from the cell as waste. Energy is often produced at specific points in the sequence of catabolic reactions and is used for synthesis of other substances or for other energy-requiring cellular processes. **Metabolism** encompasses the sum of anabolic and catabolic reactions occurring within a given cell.

The number of chemical reactions taking place in the typical human cell is unknown. However, Escherichia coli, an intestinal bacterium, conducts approximately 2500 metabolic reactions during its lifetime. Human nuclei contain about 100,000 different protein-encoding genes, suggesting that the number of chemical reactions executed by human cells during the lifetime of an individual greatly exceeds those occurring in E. coli.

Synthesis of complex molecules requires relatively enormous amounts of energy. Enzymes provide a partial solution to this energy dilemma by reducing the quantity required for individual reactions. Metabolic pathways provide a second solution by breaking the biosynthetic process into many small steps, each step requiring a small proportion of the total energy necessary for construction of the macromolecule. A **metabolic pathway** is a sequence of related anabolic or catabolic reactions. Cells generally utilize separate pathways for the assembly and breakdown of macromolecules. The synthesis of a complex molecule may be compared to an automobile assembly line where each of the component parts is installed in a specific order until the entire structure is achieved. The catabolism of a complex substance may be viewed as an assembly line working in reverse.

A hypothetical metabolic pathway is presented in Fig. 1-10. Compound A has four metabolic fates (end-products): E, K, L, and N. Branch points at 1 and 2 provide an opportunity for external factors to influence the flow of metabolites through the pathway. Depending upon cellular needs, compound A will be predominantly converted to one of the four end products. However, a change in the intracellular state could alter the flow pattern such that A is preferentially converted to another of the four end products. This metabolic plasticity permits the cell to readily adapt to the demands of its changing environment.

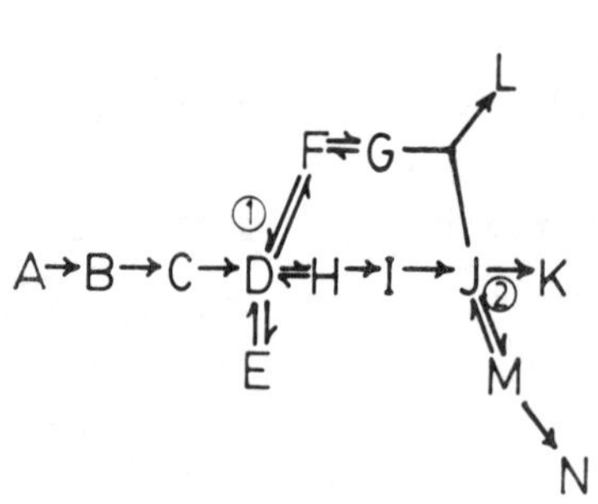

Fig. 1-10. A hypothetical metabolic pathway. Double arrows indicate that the reaction is reversible.

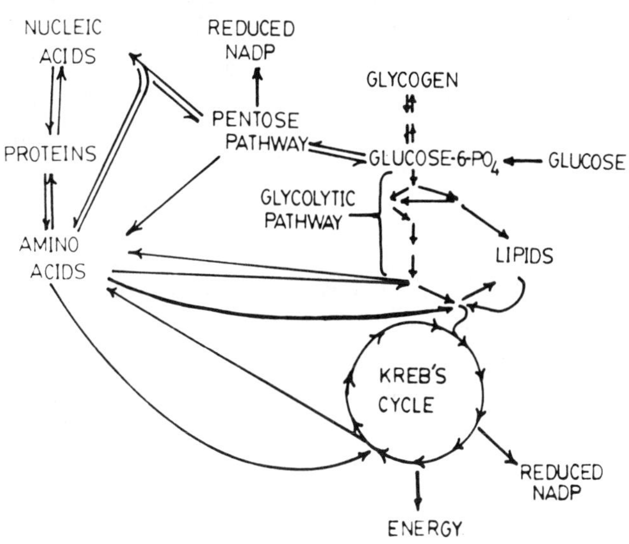

Fig. 1-11. Metabolic pathways connecting lipids, carbohydrates, and proteins. Double arrows represent reversible reaction sequences.

Metabolic pathways are intricately interconnected (Fig. 1-11). The diagram provides a highly simplified view of the interconversion of carbohydrates, lipids, and proteins and their constituent amino acids. Glycogen is the usual form of stored energy, although proteins and fats are also potential energy sources. Glycogen is converted to glucose-6-PO_4 which subsequently enters the glycolytic pathway and Kreb's Cycle. Both of these pathways generate energy in the form of ATP (see Fig. 1-9). Proteins and lipids are less desirable energy sources because byproducts are produced that interfere with brain function. Severe stages of diabetes mellitus are associated with coma and other nervous system complications caused by these harmful byproducts. The pentose pathway is also of central importance to cellular metabolism. It generates the carbon skeletons required for certain amino acids and for carbohydrates that are components of nucleic acids. The pentose pathway is also a major producer of reduced NADP, a coenzyme that is essential for many biosynthetic steps. Red blood cells lack mitochondria; therefore, they must rely upon energy derived from the glycolytic cycle for their metabolism and upon reduced NADP from the pentose pathway for stability of their cell membranes.

Flow of metabolites through the pathways presented in Fig. 11 is influenced by the cell state. Certain enzymes that are especially sensitive to regulation usually catalyze the step that commits a molecule to entry into a given pathway.

SUMMARY

The cell is a complex element of all living systems. Cells contain protoplasm composed of specialized organelles that are suspended in a heterogeneous soup of diverse chemicals. These organelles include the nucleus, mitochondria, Golgi apparatus, lysosomes, and peroxisomes. The nucleus contains the vast majority of genes. Human mitochondria possess a few genes, and generate most of the cell's energy. Polypeptides are synthesized by ribosomes, a portion of which are anchored to the endoplasmic reticulum (ER). Polypeptides synthesized by ribosomes anchored to the ER enter the ER lumen, where carbohydrate is added, and subsequently traverse the compartments of the Golgi apparatus. During transit, the polypeptides are further modified into functional proteins and targeted for secretion from the cell or enclosure within lysosomes. Lysosomes arise from the Golgi apparatus and contain degradative enzymes that participate in the digestion of intra-and extracellular debris. Peroxisomes detoxify peroxides produced as intermediates or byproducts during cellular metabolism.

The cytoplasm is situated between the nuclear envelope and cell membrane. The liquid portion, or cytosol, is composed of a large number of chemicals of varying size and complexity dissoleved or suspended in water. Proteins, carbohydrates, lipids and RNAs are three general classes of molecules found in the cytosol. These molecules contribute to cell structure and function in a variety of ways. Enzymes are special types of protein that catalyze the chemical reactions transforming cytoplasmic and nucleoplasmic molecules from one type to another. These chemical reactions occur in sequences called metabolic pathways. The cell's pathways are interconnected in an intricate network that is sensitive to changes in the intra- and extracellular environment and which responds to these changes by altering the flow of metabolites to meet cellular needs.

PROBLEMS

1. Diagram a typical cell and label the cellular components. List the functions of the following organelles: nucleus, nucleolus, mitochondrion, lysosome, Golgi apparatus, and peroxisome.

2. Why does hemoglobin display different rates of migration in an electrical field under neutral, alkaline and acidic conditions?

3. How do cells attempt to solve their energy crisis?

4. A patient has unusually low levels of enzyme B activity. Suggest ways in which this might occur.

5. Why do cells have membranes?

6. Patients who possess an unusual form of a gene often present with a number of clinical problems. Can you suggest an explanation for this multiplicity of gene effects?

7. Name four general types of molecules whose components are derived from glucose.

8. A patient lacks the enzyme that converts H to I in Fig. 10. Predict the consequences of this metabolic block.

9. Individuals who lack nucleoli most likely die during early development. Why?

10. Cells often use separate metabolic pathways for synthesis and degradation of specific molecules. Can you suggest a reason for this?

GLOSSARY OF TERMS

Active Site the region of an enzyme that interacts with the substrate.

Amino Acid building blocks of proteins that may be acidic, basic, or neutral, depending upon their environment. Amino acids derive their name from the occurrence of at least one amino and one acidic group in their structure.

Anabolic energy-requiring reactions leading to the construction of complex molecules.

Carbohydrates molecules which contain carbon, hydrogen, and oxygen, usually in the ratio of 1:2:1.

Catabolism that portion of metabolism that involves the degradation of molecules. Energy is often released by these reactions.

Catalyze to assist a chemical reaction without being consumed by that reaction.

Cell membrane a thin network of lipid and protein molecules which surrounds the cytoplasm of the cell.

Cell theory all living organisms are composed of cells that are capable of independent existence and which arise from preexisting cells.

Chromatin the genetic substance occurring within the nucleus.

Chromosome a linear aggregate of genes.

Coenzyme a molecule associated with an enzyme and which is required for enzyme function.

Cofactor any molecule which binds to an enzyme, facilitating its action. This term is generally applied to all such molecules, while **coenzyme** is restricted to organic molecules.

Cytoplasm that portion of the protoplasm which is situated between the nuclear envelope and the cell membrane.

Endoplasmic reticulum a membranous structure located in the cytoplasm. The **rough** ER is studded with ribosomes, while the **smooth** ER lacks ribosomes. The ER is interconnected with both the nuclear envelope and the cell membrane. Peroxisomes arise from the ER, and certain proteins are synthesized by the RER.

Enzymes proteins that catalyze chemical reactions in the cell.

Gene DNA sequence that specifies the structure of a polypeptide or non-protein encoding RNA.

Glycolytic cycle a sequence of chemical reactions converting glucose to smaller carbohydrates and producing energy.

Golgi apparatus cytoplasmic organelle which participates in the maturation of polypeptides and proteins and their targeting to intracellular and extracellular sites.

Hydrophilic water attracting.

Hydrophobic water repelling.

Kreb's cycle the mitochondrial metabolic pathway generating large amounts of chemical energy in the form of ATP.

Lipid hydrophobic molecules possessing long hydrocarbon chains, ringed structures or both.

Lysosomes vesicles containing degradative enzymes.

Metabolic pathway a sequence of related chemical reactions involved with synthesis or degradation of molecules.

Metabolism anabolism + catabolism.

Mitochondrion cytoplasmic organelle devoted to the production of energy.

Monomer a protein or enzyme consisting of one polypeptide chain.

Monosaccharide a simple sugar or carbohydrate.

Multimer a protein or enzyme consisting of two or more polypeptides,

Nuclear envelope double membrane surrounding the nucleus.

Nucleic acids polymers of nucleotides complexed with proteins.
  DNA deoxyribonucleic acid, the genetic substance of genes and chromosomes.
  RNA ribonucleic acid- nucleic acid found both in nuclei and cytoplasm.

Nucleotide building block of nucleic acids and a form of chemical energy (ATP).
  Nucleotides contain a base, a monosaccharide, and phosphate(s).

Nucleolus spherical particle within the nucleus that is involved with the synthesis and processing of ribosomal RNA.

Nucleoplasm that portion of the protoplasm bounded by the nuclear envelope.

Nucleus organelle that contains the genetic substance.

Pentose pathway metabolic pathway which generates carbon skeletons for certain amino acids, monosaccharides occurring in nucleotides, and reduced NADP for biosynthetic reactions.

Peroxisome cytoplasmic organelle that contains enzymes which degrade peroxides.

Polymer a large molecule consisting of many identical or similar subunits.

Polypeptide a polymer of amino acids.

Polysaccharide a polymer of monosaccharides.

Protein molecule consisting of one or more polypeptides.

Protoplasm living substance consisting of organelles suspended in an aqueous solution of chemicals.

Ribosomes structures composed of RNAs and proteins that participates in protein synthesis.

Substrate the chemical acted upon by an enzyme.

Vesicles small sacs containing fluid.

Vitamin organic molecules required in the diet that are metabolized to coenzymes by cells. Cells cannot synthesize vitamins.

ADDITIONAL LEARNING RESOURCES

Alberts, B., D. Bray, J. Lewis, M. Raff, K. Roberts, and J.D. Watson. <u>Molecular Biology of the Cell</u>, 2nd ed. New York: Garland Publishing, Inc. 1989.

Chapter 2

Gene Structure and Expression

The complexity of the cell and the intricacy of the chemical processes occurring within it place rigorous demands upon the mechanisms which coordinate and control those processes. **Genes**, the basic units of heredity, play a major role in this control. This chapter is devoted to a discussion of the gene, its composition, and the means by which it accomplishes its effects.

CHEMICAL COMPOSITION OF A GENE

Molecules that serve as genes must satisfy the following requirements: 1) the molecule must be capable of information storage; 2) it must be able to reproduce itself accurately; 3) the molecule must be stable and yet capable of change; and 4) there must be a mechanism for efficient retrieval of stored information, transport of that information to appropriate sites, and translation of the information into molecules required for cell life.

The vast majority of our genes are found in the nucleus. Chromatin, the genetic substance, consists of protein, DNA, and a small amount of RNA. Studies of bacteria and viruses have led to the identification of DNA as the genetic material in most cells. Avery, MacLeod, and McCarty performed a series of experiments in 1944 that proved DNA was the agent responsible for the transmission of bacterial traits (1). <u>Pneumococcus</u> types I and II breed true. That is, type I bacteria produce only type I bacteria, and type II bacteria produce only type II bacteria. When DNA extracted from killed type I pneumococci was introduced to live type II cells, some of the type II cells were <u>transformed</u> to type I pneumococci.

Hershey and Chase reported results of experiments in 1952 that demonstrated that DNA was the genetic material of certain viruses (2). The virus was known to consist of a DNA core surrounded by a protein coat. The DNA was labelled with radioactive phosphorus, and the protein was labelled with radioactive sulfur. These radioisotopes have different decay properties, permitting the scientists to monitor the behavior of both the DNA and the protein. During an infection, the virus attaches to the bacterial surface and injects its genetic material into the bacterial cell. The viral genetic substance then utilizes the bacterial cell's machinery to make new viruses. Some of the new viruses will contain some of the labelled genetic material from the infecting viruses. Hershey and Chase allowed the virus to infect bacteria, and then used a Waring blender to dislodge the viral components remaining on the bacterial surfaces. The radioactive phosphorus (present in the DNA) and <u>not</u> the radioactive sulfur appeared in the subsequent generation of viruses, implicating the DNA as the viral genetic material. Subsequent experiments by a large number of investigators have demonstrated that the majority of living organisms including humans utilize DNA as their genetic material. Some organisms such as the RNA viruses use RNA as their genetic material. However, these viruses form a DNA intermediate prior to making new viruses. These viruses and one of their enzymes, **reverse transcriptase**, will be discussed in later sections of this text. Reverse transcriptase is an enzyme that can make DNA using a RNA molecule as a

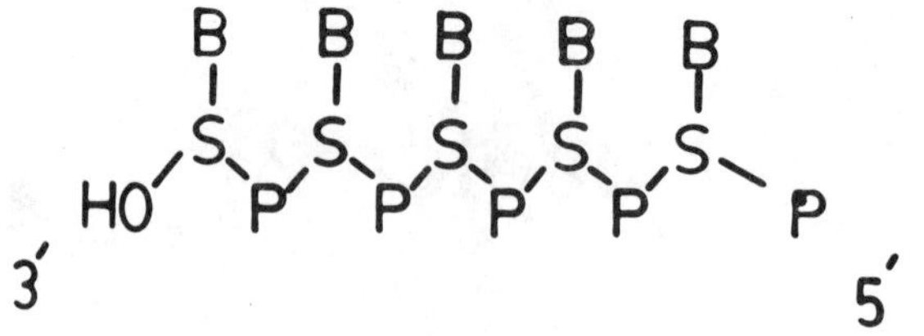

Fig. 2-1. Nucleic acid polymer. The macromolecule extends from the 3' hydroxyl (OH) group to the 5' phosphate (P) group. S= a 5-carbon sugar; B= base.

template or mold on which to build the new DNA molecule.

DNA and RNA are macromolecules which possess a sugar-phosphate backbone and a series of bases that project from the backbone (Fig. 2-1). DNA contains the sugar, deoxyribose, while ribose occurs in RNA. Adenine, guanine, and cytosine are bases that occur in both DNA and RNA. A fourth base, thymine, is also found in DNA; however, this base is replaced by uracil in RNA. The five bases found in nucleic acids are illustrated in Fig. 2-2.

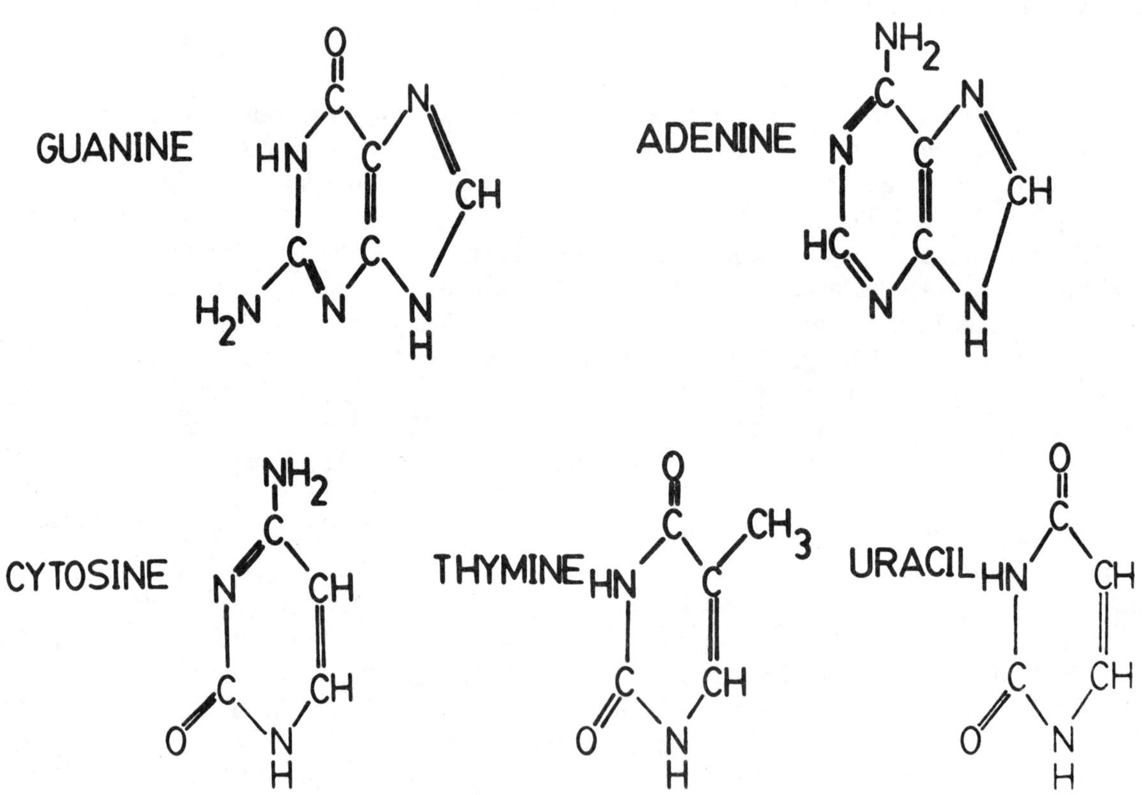

Fig. 2-2. Bases occurring in DNA and RNA. Guanine and adenine contain two rings and are designated purines. Cytosine (occurring in both nucleic acids), thymine (in DNA), and uracil (in RNA) contain a single ring and are pyrimidines.

Watson and Crick (3) proposed a structure for DNA in which two strands form a double helix. The sugar-phosphate backbones of the DNA strands are located on the outside of the double-helical structure, and the bases of one DNA strand project to the inside and pair with the complementary bases

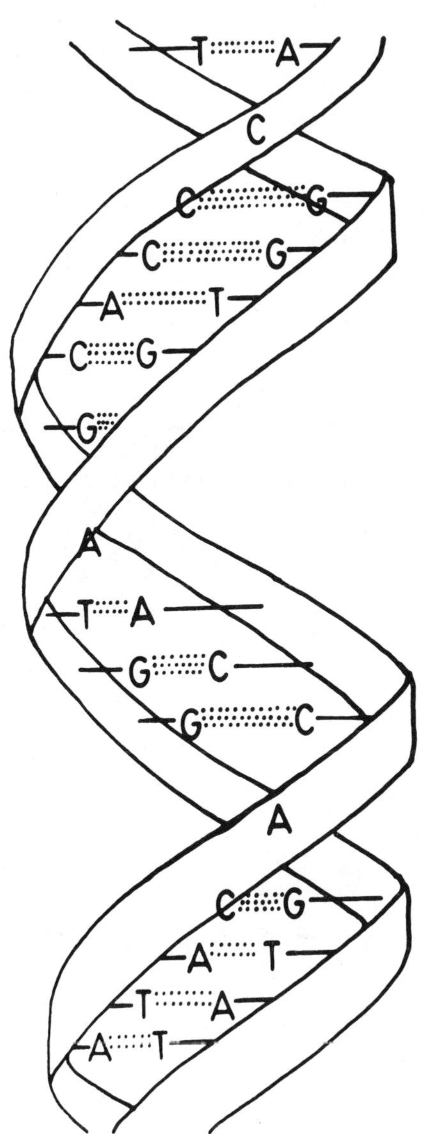

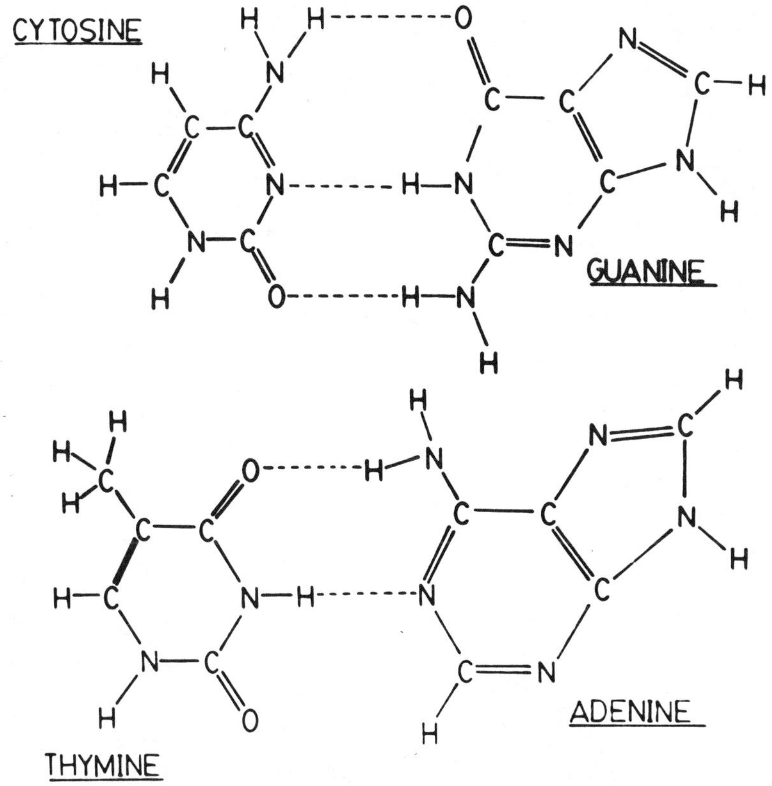

Fig. 2-3. Double-helical structure of DNA. The rungs of the ladder are formed by the purine-pyrimidine base pairs. The sugar-phosphate backbones constitute the sides of the ladder. Dotted lines denote hydrogen bonds. A= adenine; G= guanine; C= cytosine; and T= thymine.

Fig. 2-4. Pairing properties of purines and pyrimidines. The hydrogen bonds are indicated by dashed lines.

of the other DNA strand (Fig. 2-3). The base pairs consist of one purine (A,G) and one pyrimidine (T,C). Adenine pairs with thymine, and guanine pairs with cytosine (Fig. 2-4). Note that guanine-cytosine pairs are linked by <u>three</u> hydrogen bonds, and adenine-thymine pairs by <u>two</u> bonds. Consequently, DNA which is enriched in guanine-cytosine pairs tends to be more resistant to forces such as heat that denature or break apart the double helix. The two strands of DNA are antiparallel (Fig. 2-5); that is, the sugar-phosphate backbone of one chain begins with a 5' phosphate (top strand) and ends with a 3' hydroxyl group, while its partner strand runs from 3' hydroxyl to 5' phosphate group.

The sequence of the purines and pyrimidines of one strand is complementary to that of the other. This complementarity satisfies one of the requirements for molecules serving as the genetic material: the ability to accurately reproduce itself. The details of this process will be discussed in the next section. The sequence of bases also satisfies the second requirement for the genetic material: information storage. The bases serve as the letters of the genetic alphabet. They occur in varying combinations and lengths, providing more than enough capacity for the information that must be stored. For example, if the four DNA bases form three-letter code words, 4^3 or 64 different words are possible. This number is more than enough to specify the 20 amino acids commonly occurring

19

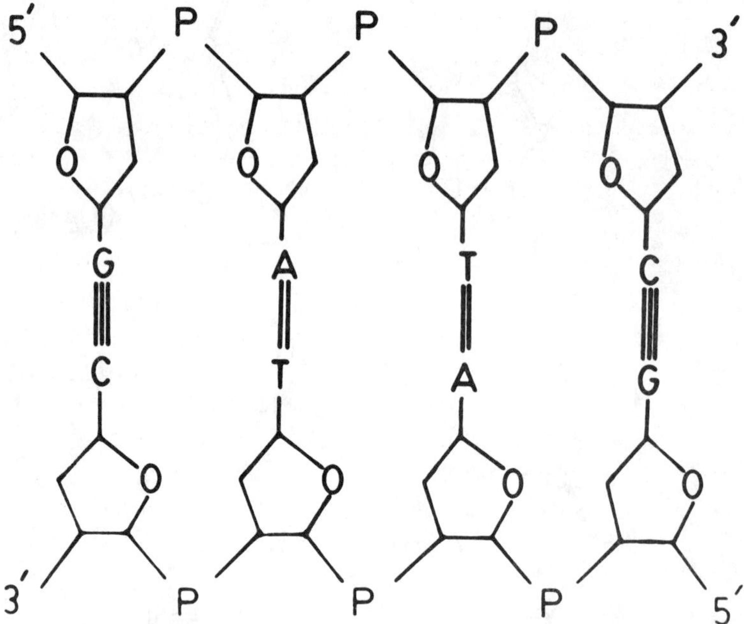

Fig. 2-5. Two antiparallel sequences of DNA. Notice the complementarity of the purines and pyrimidines. 3' and 5' refer to the carbons of deoxyribose possessing the hydroxyl and phosphate groups participating in the phosphosphodiester bonds.

in the newly synthesized polypeptides. The relationships among genes, the genetic code, and proteins will be discussed later in this chapter.

REPLICATION OF DNA

Replication is the process by which DNA forms an exact copy of itself. Watson and Crick (4)

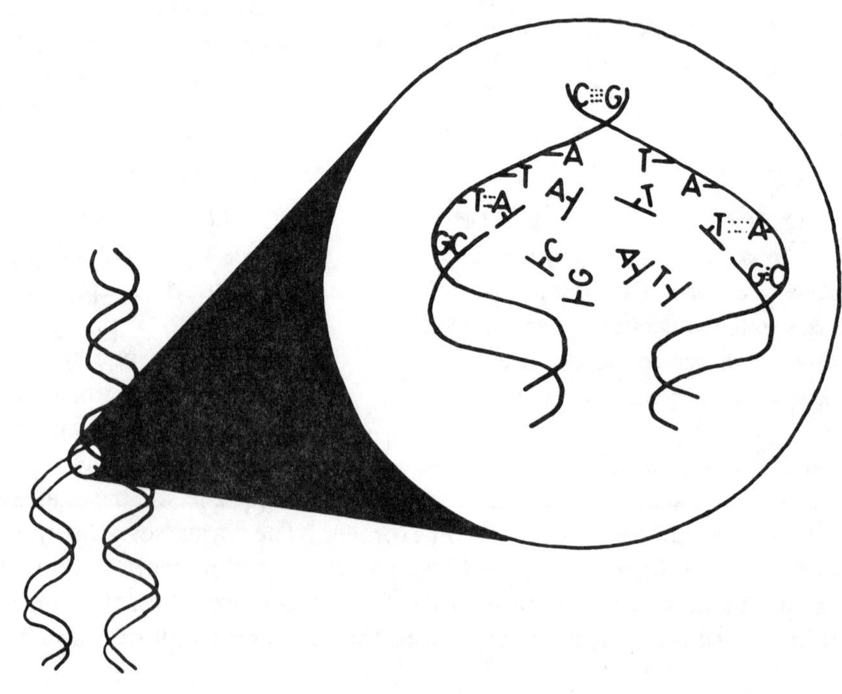

Fig 2-6. Semiconservative model for DNA replication. The deoxyribonucleotides are ordered according to specific base pairing relations (G-C and A-T) and linked enzymatically to form the new DNA strand.

proposed a mechanism for DNA replication that has subsequently been verified by an abundance of experimental data. The strands of the DNA double helix separate at the origin of replication and progressively unwind as replication proceeds (Fig. 2-6). Bases of entering nucleotides (base + sugar + phosphate) pair with appropriate bases of the old strand, and the adjacent new nucleotides are joined by phosphodiester bonds to form the new DNA strand. The specificity of the base pairing process assures that a complementary strand is formed for each strand of the double helix, resulting in the production of two daughter double helices that are exact copies of the parental double helix.

Replication is **semiconservative.** The old strands remain intact during replication, and each new double helix consists of one new and one old DNA polymer. Meselson and Stahl (5) demonstrated the semiconservative nature of DNA replication by growing bacteria in heavy nitrogen (N^{15}). As the bacteria multiplied in this medium, the heavy nitrogen was incorporated into the bases of their DNA. The bacteria were then transferred to a new medium which contained ordinary nitrogen (N^{14}). The bacteria were permitted to complete two replications in this medium. Bacteria were removed after the first and second round of replication, and their DNA was extracted and centrifuged in a cesium chloride gradient. The latter procedure bands the DNA in a region of the cesium chloride gradient that has a density equal to that of the DNA. The results of these experiments are presented in Fig. 2-7. The DNA from cells

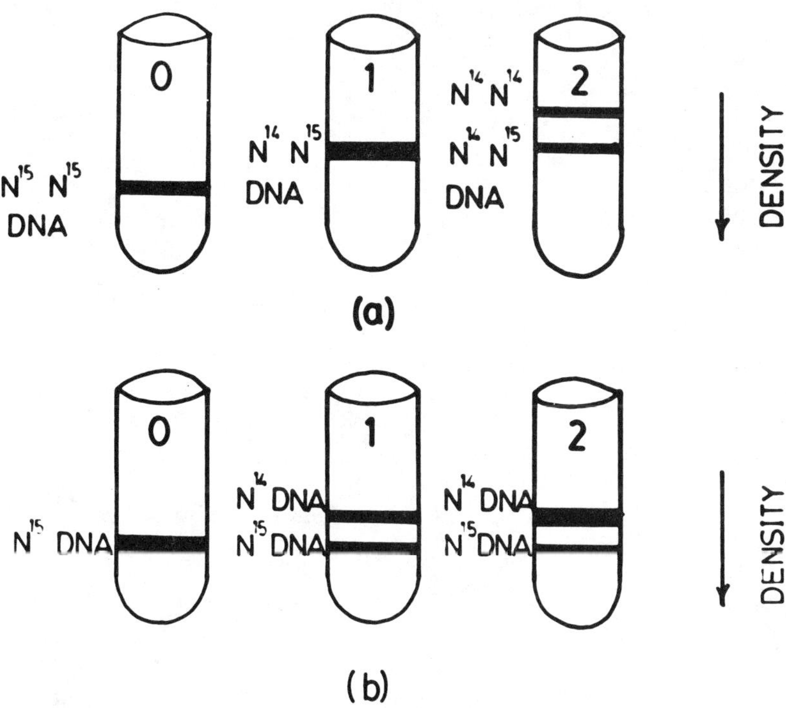

Fig. 2-7. Schematic representation of the Meselson and Stahl experiments. a) Density-gradient centrifugation of intact DNA from E. coli cells after 0, 1, and 2 replications in the N^{14} medium. b) Centrifugation of heat-denatured DNA from cells at comparable stages of replication.

that had completed one round of replication in the N^{14} medium banded at a lower density than DNA from cells that had been freshly isolated from the N^{15} medium. DNA isolated from cells that had completed two rounds of replication in the N^{14} medium banded in two locations: one corresponding to the hybrid band and another which occurred at a lower density and corresponded to DNA containing only

N^{14}. Heat treatment separates (**denatures**) the strands of the DNA double helix. Centrifugation profiles of denatured DNA from cells completing one and two rounds of replication in the N^{14} medium displayed two bands (Fig. 2-7b) that occurred in approximately 1:1 and 3:1 (N^{14}:N^{15}) ratios, respectively. The best interpretation of these experiments is that DNA replication occurs by a semiconservative mechanism.

RNA

Ribonucleic acid (**RNA**) is a single-stranded polynucleotide containing guanine, cytosine, adenine, and uracil as its principal bases and possessing ribose as its sugar moiety. Other bases, such as inosine, are occasionally found in RNAs. Several types of RNA have been identified in human cells: messenger RNA, ribosomal RNA, transfer RNA, and small nuclear RNA. RNA also occurs as an integral component of chromosomes. The function of the chromosomal RNA has not been established; however, it may serve a structural role (see Chapter 4).

Messenger RNA (**mRNA**) serves as a template for protein synthesis. During **transcription**, the synthesis of an RNA molecule using a DNA template, a relatively large RNA is formed that is subsequently processed to make the mRNA. The mRNA exits the nucleus where its is **translated** by ribosomes to form the polypeptide specified by the gene. The mRNA contains a linear sequence of bases that is translated into the amino acid sequence of the polypeptide. The coding portion of the mRNA consists of base triplets, called **codons**, that stipulate the type of amino acid that should be placed at the corresponding position in the polypeptide. This role of mRNA satisfies a third requirement for the genetic material: the delivery of the stored information to a site where it can be utilized.

Transfer RNA (tRNA) shuttles amino acids to the sites of protein synthesis. tRNA is enzymatically coupled to specific amino acids and serves as an adapter molecule to place these amino acids in their appropriate location in the polypeptide being synthesized. The ordering of the amino acids is accomplished by base pairing between the bases in the **anticodon** of the tRNA and the codon of the mRNA (Fig. 2-8). After the tRNA has completed its task, it is cleaved from the amino acid, and the amino acid is coupled to the adjacent amino acid by

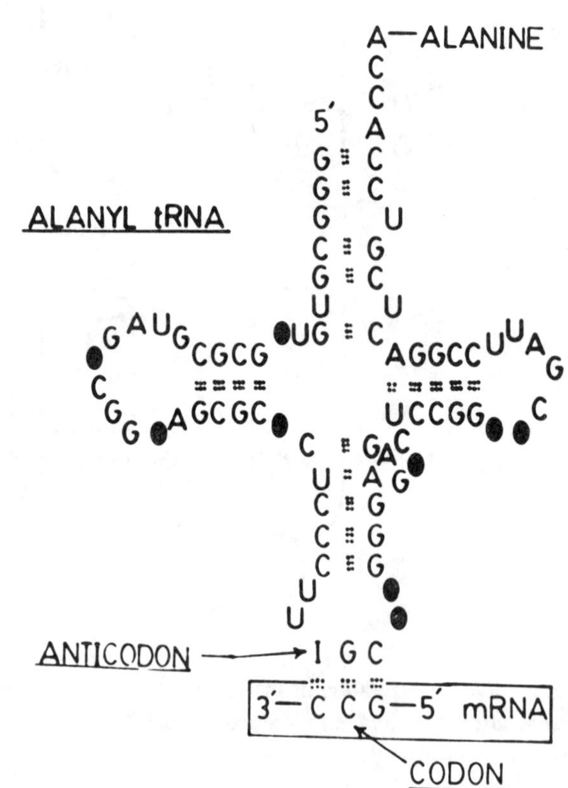

Fig. 2-8. Structure of a representative tRNA. ● = unusual bases; I = inosine. The anticodon, IGC, of the tRNA binds by base pairing to the codon, CCG, of the mRNA.

22

a peptide bond. The tRNA is recycled to pick up another amino acid of the same type.

Ribosomal RNA (**rRNA**) is a structural component of the ribosome. Ribosomes consist of approximately equal portions of rRNA and protein. Ribosomes contain a large and small subunit, each possessing different rRNAs. 18S rRNAs are found in the small subunit, while 28S rRNAs occur in the large subunit. "S" is the symbol for the sedimentation coefficient, a value that is influenced by both the size and shape of the molecule, that is determined by centrifugation. Tissue homogenates are centrifuged at various speeds to pellet cell components. For example, a lysosome has a sedimentation coefficient of 9400S and can be pelleted by medium speed centrifugation (20,000 times gravity for about 20 minutes), while ribosomes are 80S and are pelleted after approximately 3 hours at 150,000 times gravity. Two other rRNAs are found in the large subunit: 5S and 5.8S. The 5.8S, 18S, and 28S rRNAs are formed by processing of a large 45S RNA (Fig. 2-9). This large RNA is transcribed from many tandemly arranged

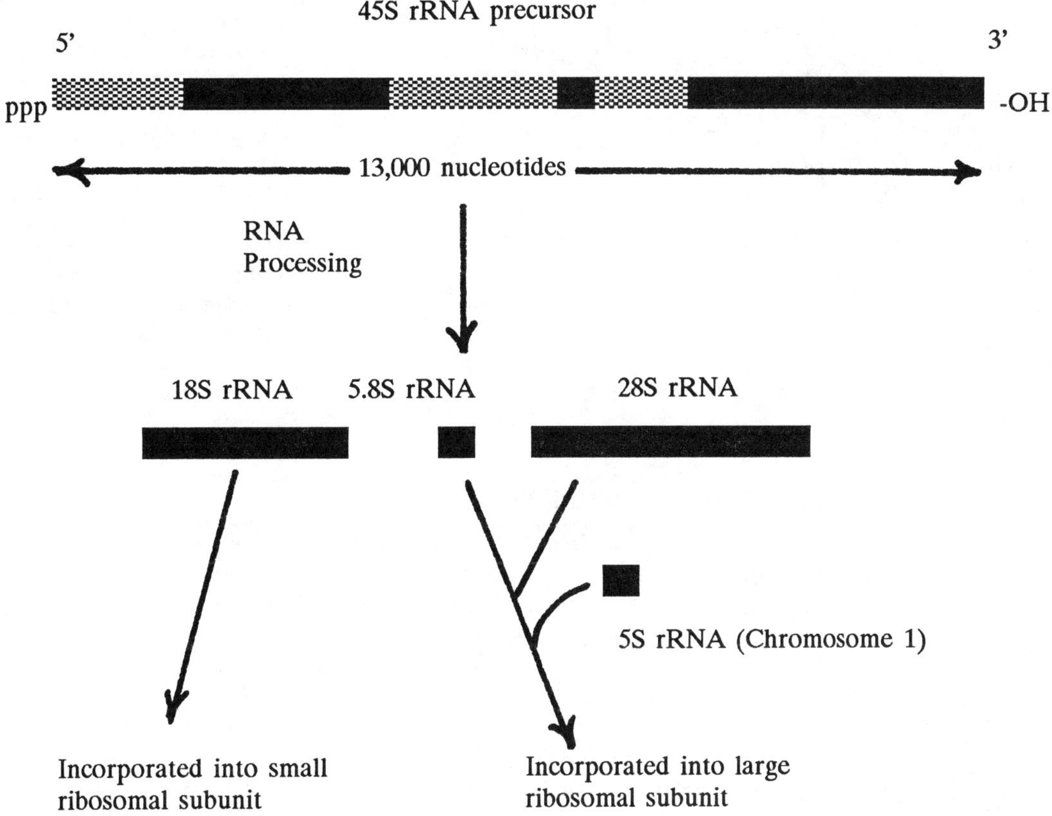

Fig. 2-9. 45S rRNA precursor processing. This precursor is transcribed from one of many tandemly repeated genes that are localized in the nucleolar organizer regions of chromosomes 13, 14, 15, 21, and 22. The light-colored sequences are excised and degraded during RNA processing in the nucleolus. 5S RNA is derived from another gene cluster that has been mapped to chromosome 1.

genes that are located on the short arms of chromosomes 13-15, 21, and 22. The regions carrying the rRNA genes are called **nucleolar organizing regions**. Transcription and processing of the RNAs occurs in the nucleolus which is formed by condensation of the nucleolar organizing regions from each of the five chromosome pairs. 5S ribosomal RNA is transcribed from a cluster of genes located on chromosome

1, enters the nucleolus, and associates with the large subunit of the ribosome. The function of the ribosome is to assist with the translation of the genetic message contained in the mRNA by "reading" this message and aligning the tRNA-amino acid complexes with the appropriate codons of the mRNA. Both rRNAs and their associated proteins assist the attachment of both the mRNA and tRNA to the ribosome. Other regions of rRNA may interact with ribosomal proteins to generate the three-dimensional structure of the ribosome. The translational unit consisting of tRNA, mRNA, and the ribosome is schematically represented in Fig. 2-10.

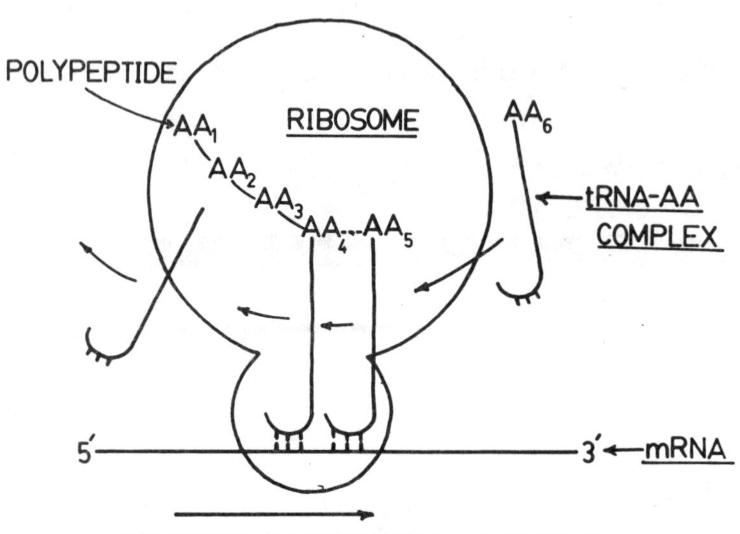

Fig. 2-10. Translational unit involved with the synthesis of the polypeptide. The ribosome reads from the 5' to the 3' end of the mRNA, bringing the tRNA-amino acid complex into alignment with the appropriate codon. The growing polypeptide is transferred to this tRNA by forming a peptide bond with the new amino acid (AA_5), and the old tRNA is released. The tRNA carrying the polypeptide is transferred to the left-hand binding site. This process is repeated until a termination codon is encountered in the mRNA.

Small nuclear RNA (**snRNA**) refers to RNA molecules about 250 nucleotides in length. These snRNAs are often found in a nuclear structure called the **spliceosome** which removes non-coding sequences (**introns**) from the initial RNA transcript formed from protein-encoding genes. The spliceosome possesses both snRNA and several proteins that participate in the excision of the intron and the joining of the appropriate coding sequences (**exons**) flanking the intron.

THE GENETIC CODE

The nature of the genetic code was defined by a series of experiments that were performed during the late 1950's and early 1960's. Nirenberg and Matthei (6) demonstrated the direction of protein synthesis in a cell-free system by template RNA extracted from <u>E. coli</u> and by a synthetic RNA, polyuridylic acid. Furthermore, the latter RNA supported the incorporation of C^{14}-labelled phenylalanine into protein, suggesting that a code word for phenylalanine was U^n, where n was some integer. Extension of this method by Ochoa and his colleagues (for example, see Speyer <u>et al.</u> (7)) led to the delineation of the genetic code and its properties. Crick <u>et al.</u> (8) summarized the characteristics of the code: 1) the code consists of triplets of bases; 2) the code is commaless (the sequence of bases is read continuously from a fixed point; 3) the code is non-overlapping (that is, the sequence UAGCGA is read as UAG CGA and not as UAG AGC GCG CGA); and 4) the code is degenerate (an amino acid may be encoded by two or more triplets). The genetic code was subsequently found to contain three terminators (periods): UAA, UGA, and UAG. The first AUG occurring in an mRNA is used as a start signal for translation of the mRNA. methionyl-tRNA binds to the first AUG codon, and the mRNA is translated continuously from that point to the first terminator. The methionine is cleaved from the protein shortly after initiation of translation, and does not appear in the mature protein. The complete series of codons is presented in Table 2-1.

Table 2-1. The Genetic Code.

First Base (5' end)	Second Base				Third Base (3' end)
	U	C	A	G	
U	Phe	Ser	Tyr	Cys	U
	Phe	Ser	Tyr	Cys	C
	Leu	Ser	Term	Term	A
	Leu	Ser	Term	Trp	G
C	Leu	Pro	His	Arg	U
	Leu	Pro	His	Arg	C
	Leu	Pro	Gln	Arg	A
	Leu	Pro	Gln	Arg	G
A	Ile	Thr	Asn	Ser	U
	Ile	Thr	Asn	Ser	C
	Ile	Thr	Lys	Arg	A
	Met	Thr	Lys	Arg	G
G	Val	Ala	Asp	Gly	U
	Val	Ala	Asp	Gly	C
	Val	Ala	Glu	Gly	A
	Val	Ala	Glu	Gly	G

Note: To obtain the codon for methionine (Met), read the first base at the left, the second at the top, and the third at the right (AUG).
Ala=alanine; Arg=arginine; Asn=asparagine; Asp=aspartic acid; Cys=cysteine; Gln= glutamine; Glu–glutamic acid; Gly=glycine; His=histidine; Ile=isoleucine; Leu–leucine; Lys=lysine; Met=methionine; Phe=phenylalanine; Pro=proline; Ser=serine; Thr=threonine; Trp=tryptophan; Tyr=tyrosine; Val=valine; Term=terminator.

TRANSCRIPTION

Transcription is the process by which the base sequence of <u>one</u> (the "master" strand) of the two strands of the gene is used as the template for the synthesis of a complementary RNA. This complementary RNA is then processed to form the mRNA which directs protein synthesis in the cytosol. Protein encoding genes are transcribed by an enzyme complex that includes RNA Polymerase II. Two other RNA polymerases have been described that transcribe rRNA precursors (RNA Polymerase I) and tRNA precursors (RNA Polymerase III). The transcription complex enters the chromosomal region of the gene upstream from its first exon and moves to the gene's **promoter**, a region several hundred nucleotides in extent which contains regulatory sequences governing the amplitude, timing, and location of the expression of that gene. The complex unwinds the double helix, and RNA Polymerase II synthesizes the complementary RNA starting at the "cap" site and continuing through the exons and introns to a termination sequence located downstream from the last exon of the gene. The cap site is a single nucleotide that is positioned about 50 nucleotides upstream from the first exon of the gene. This site and other key structures associated with the promoter and with downstream regions will be discussed in a later section of this chapter. The ribonucleotides are positioned according to their base-pairing properties (A-U, G-C) and connected by phosphodiester bonds (Fig. 2-11). Notice that RNA Polymerase II constructs the phosphodiester bond from the 3' hydroxyl group of the growing RNA to the 5' phosphate of the nucleotide triphosphate being linked to the RNA. This manner of bond formation requires reading of the master strand of the gene in a 3' to 5' direction and the formation of the RNA from its 5' end to its 3' end (Fig. 2-12). As the RNA is formed, it disengages from the DNA; and the DNA double helix reforms. The RNA product of transcription is generally much larger than the mRNA that enters the cytosol and must be subjected to a number of processing steps before it assumes its functional form.

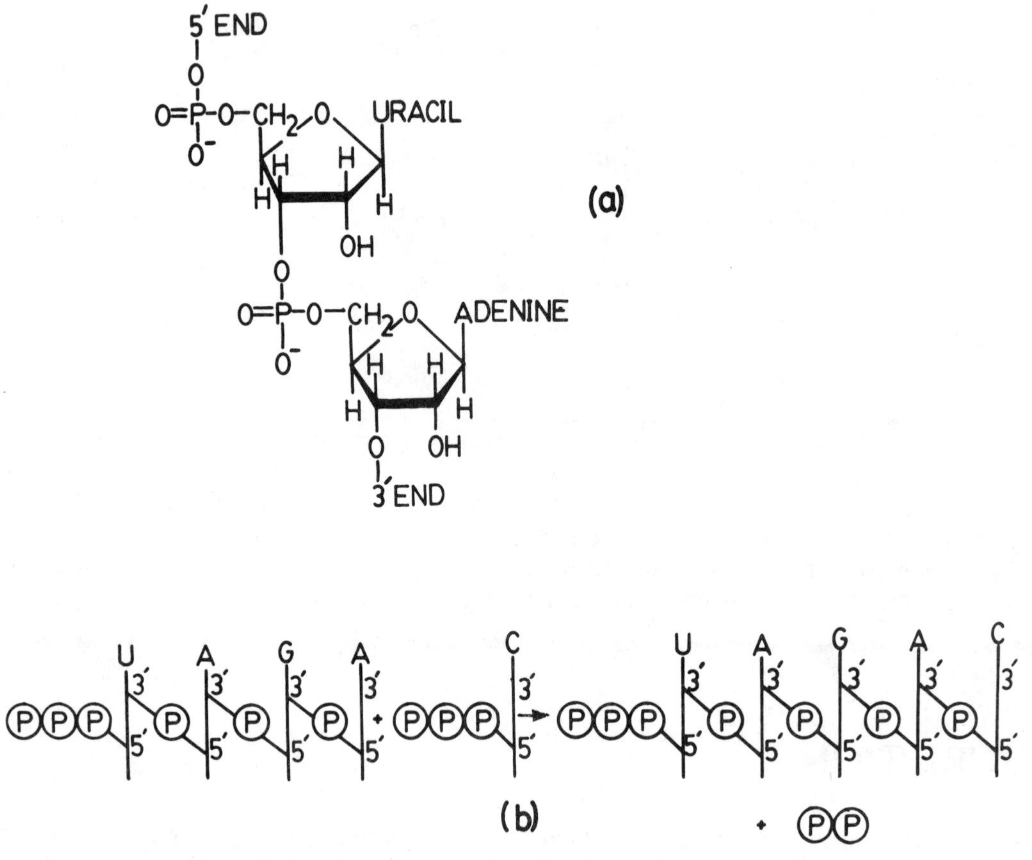

Fig. 2-11. Synthesis of RNA from DNA templates. a) Directionality of bonding in the RNA; b) synthesis of RNA by RNA Polymerase II. P=phosphate.

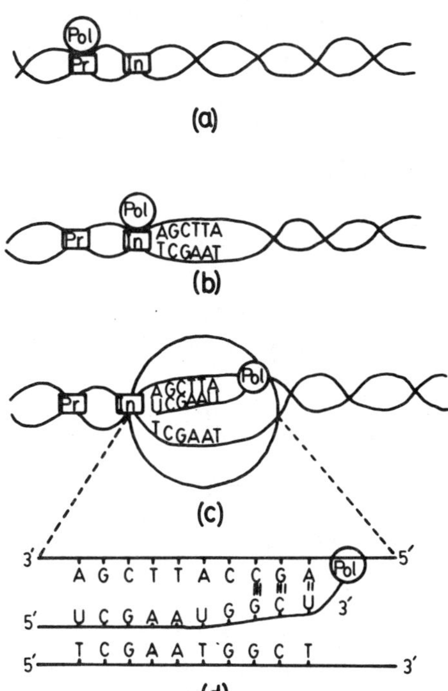

Fig. 2-12. Sequential steps in transcription. a) Binding to the promoter (Pr) site; b) Translocation to the cap (In) site and unwinding of DNA; c) Synthesis of RNA by RNA polymerase (Pol); d) Direction of reading of DNA and synthesis of RNA.

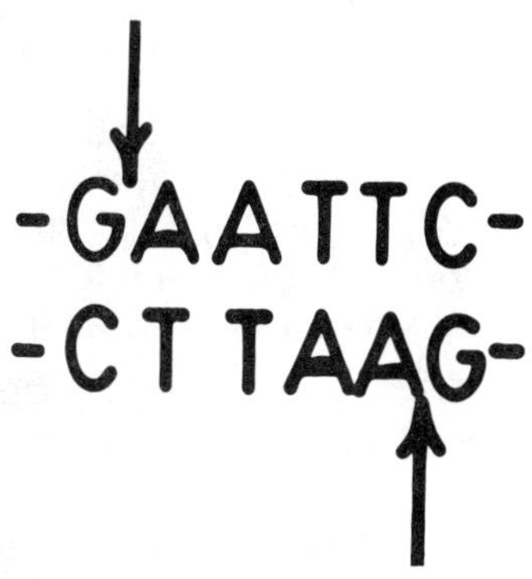

Fig. 2-13. Cleavage sites (arrows) for the restriction endonuclease, EcoRI.

MOLECULAR ANALYSIS OF GENE STRUCTURE

Construction of gene libraries. Libraries of human genes may be constructed using all of the DNA or only that portion of the **genome** that is actively expressed at a given time and in a specific tissue. The term, **genome**, refers to the total DNA occurring in an individual cell. **Genomic libraries** can be made by extracting the total DNA from a cell and digesting it with a restriction enzyme such as EcoRI. Restriction enzymes cut DNA at sites which possess special sequences (Fig. 2-13). Wherever the indicated sequence occurs along the chromosome, EcoRI will cut the double helix, forming fragments (called **restriction fragments**) that are of variable length, depending upon the distances between the restriction sites. The pieces of DNA are packaged in a cloning "vector" or **plasmid**. Plasmids are circular DNAs that are self-replicating and which were first isolated from bacteria where they imparted antibiotic resistance to their hosts. Plasmids have been exploited for construction of of gene libraries because genes within the libraries can be amplified in number independent of the bacterial host's replication controls. Many different plasmid vectors have been developed for these purposes. Plasmids that can be used for smaller DNA fragments (0-5kb) include pGEM and Lambda$_{gt}$11. The structure of pGEM is presented in Figure 2-14. pGEM contains the Amp^r gene which allows its host bacterium to grow in the presence of the antibiotic, ampicillin. This plasmid also contains the lacZ gene which encodes a β-galactosidase. The lacZ gene possesses an EcoRI site flanked by two promoters, SP6 and T7. The DNA fragment is inserted at the EcoRI site and, following incorporation of the "**recombinant**" plasmid in bacteria, can be transcribed from either direction of the insert. This capability greatly facilitates expression of the inserted piece because the researcher does not have to be concerned about the possibility that the insert might enter the plasmid "backwards", preventing its transcription. pGEM is called an "**expression vector**" because it allows the transcription of its insert in the bacterial host. Another plasmid that has often been used for construction of libraries is pBR322. Unfortunately, this vector does not have a promoter for transcription of the inserted DNAs, and the inserts must be transferred (**subcloned**) to an expression vector such

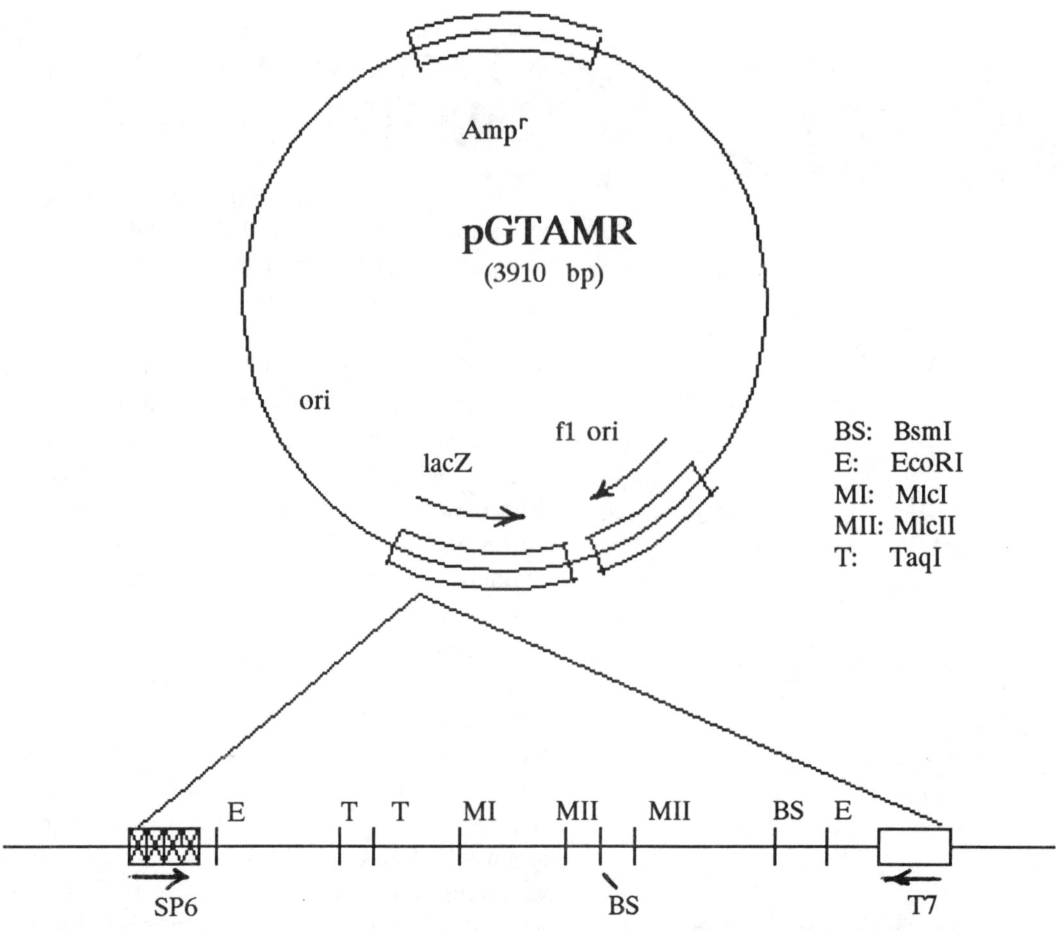

Fig. 2-14. The structure of pGEM - 3Zf(-) vector containg a DNA insert (expanded at bottom of figure). pGEM contains the Amp^r gene for ampicillin-resistance and the lacZ gene which encodes β-galactosidase. The foreign DNA has been inserted in an EcoRI site contained within the lacZ sequence. The EcoRI site is flanked by two strong promoters, SP6 and T7. Either promoter can support transcription of the insert when in a basterial host. Arrows indicate the direction of transcription.

as pGEM for this purpose. Lambda$_{gt}$11 is a popular expression vector for gene libraries. This vector was developed from lambda phage and contains the lacZ gene. If the DNA is inserted "in phase" with the lacZ gene, it will be transcribed into a hybrid mRNA containing both the β-galactosidase and insert codons and translated into a "fusion protein" consisting of β-galactosidase fused to the protein encoded by the insert. DNA sequences up to 4.8kb in length can be packaged in Lambda$_{gt}$11; however, larger DNA fragments approximating 38-54kb in size can be packaged in **cosmid** libraries. A cosmid is formed by introducing cos sites at both ends of the inserted foreign DNA (cos = cohesive site). The cos sites are derived from lambda phage DNA. Lambda DNA is linear, containing single-stranded ends. When these ends join to form circular DNA, the joining region is designated the cos site. Insertion of the 38-54kb DNA fragment between the cos sites forms a vector sufficiently large to be packaged into a lambda phage particle. The infective particle can now inject the cosmid DNA into a bacterial host, where the DNA can be replicated.

Libraries can also be formed exclusively from genes that are being expressed in a given tissue. mRNA is polyadenylated. This property can be exploited for both isolation of mRNA and for synthesis of DNA that is complementary to the mRNA sequence. RNA is extracted from cells and passed through an oligo-dT column. The latter contains groups that consist of strings of thymidine residues. These residues project from the column and bind to the polyA tails of the mRNAs, retaining them on the column while the other RNAs pass through. The mRNA can then be recovered from the column and used as a template for an enzyme, reverse transcriptase, which can synthesize the complementary DNA (**cDNA**) using the mRNA and an oligo-dT "primer" (Fig. 2-15). The primer pairs with the polyA tail of the

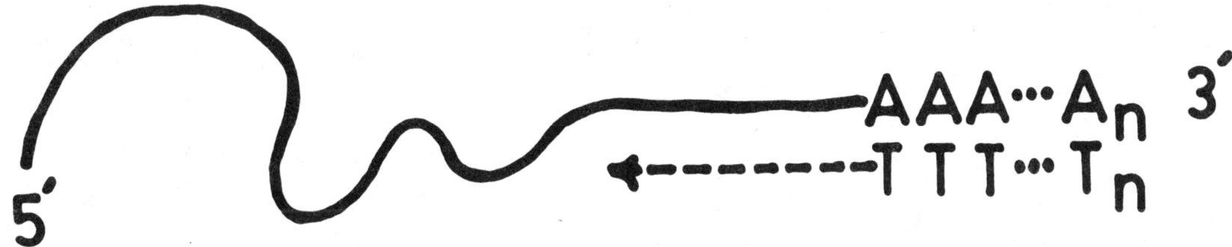

Fig. 2-15. Formation of cDNA from an mRNA template by reverse transcriptase. An oligo-dT primer, generally 16 to 18 nucleotides in length, pairs with a sequence of adenylate residues in the polyA tail and serves as a focus for extension by reverse transcriptase which uses the remainder of the mRNA as a template.

mRNA and serves as an initiation point for the cDNA. The cDNA is then made into a duplex by using a second polymerase and inserted into the appropriate vector. cDNAs are usually much smaller than genomic fragments and can be packaged into vectors such as pBR322 and lambda$_{gt}$11; however, the cDNAs lack promoter sequences, introns, and downstream regions important for normal expression of genes. Therefore, genomic libraries must be screened to isolate the genomic counterparts of the cDNAs in order to provide information regarding the structures of these sequences.

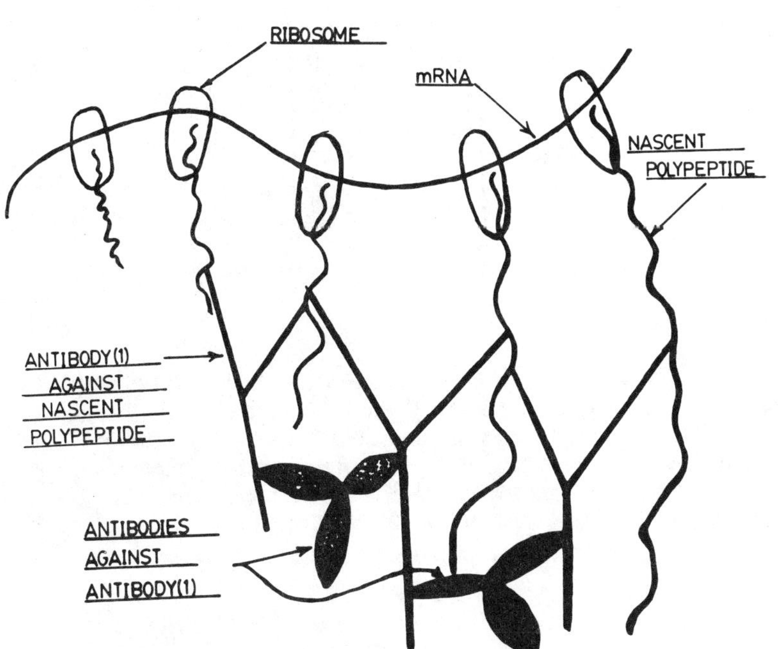

If the gene of interest is expressed as an abundant message (more than 2% of the total mRNAs synthesized by the cell), the mRNA may be isolated by immuno-precipitation of polysomes (Fig. 2-16). **Polysomes** are formed when several ribosomes translate the mRNA in series, a common occurrence in the cytosol. An antibody is raised against the polypeptide, and allowed to react with the polypeptides projecting from the polysome. A second antibody directed against the first antibody is then used to form a large aggregate. When the polysome aggregates are passed through a

Fig. 2-16. Isolation of mRNA by double antibody precipitation of polysomes.

Protein A-Sepharose column, the antibody in the complex binds to the Protein A; and the aggregate is retained on the column, while all other proteins and RNAs pass through. The complex is eluted, and dissociated. The mRNA can then be isolated by oligo-dT-cellulose chromatography as described above.

Once the DNA has been packaged into a suitable library, the library can be screened for the gene of interest. An example of a screen of a genomic library for a specific gene is presented in Figure 2-17. Bacteria that have been approved for use in recombinant DNA research are infected with the recombinant vectors and grown in a culture medium. As the bacteria grow and multiply, the recombinant molecules multiply in their cytoplasms, amplifying the human DNA segments they carry. The bacteria

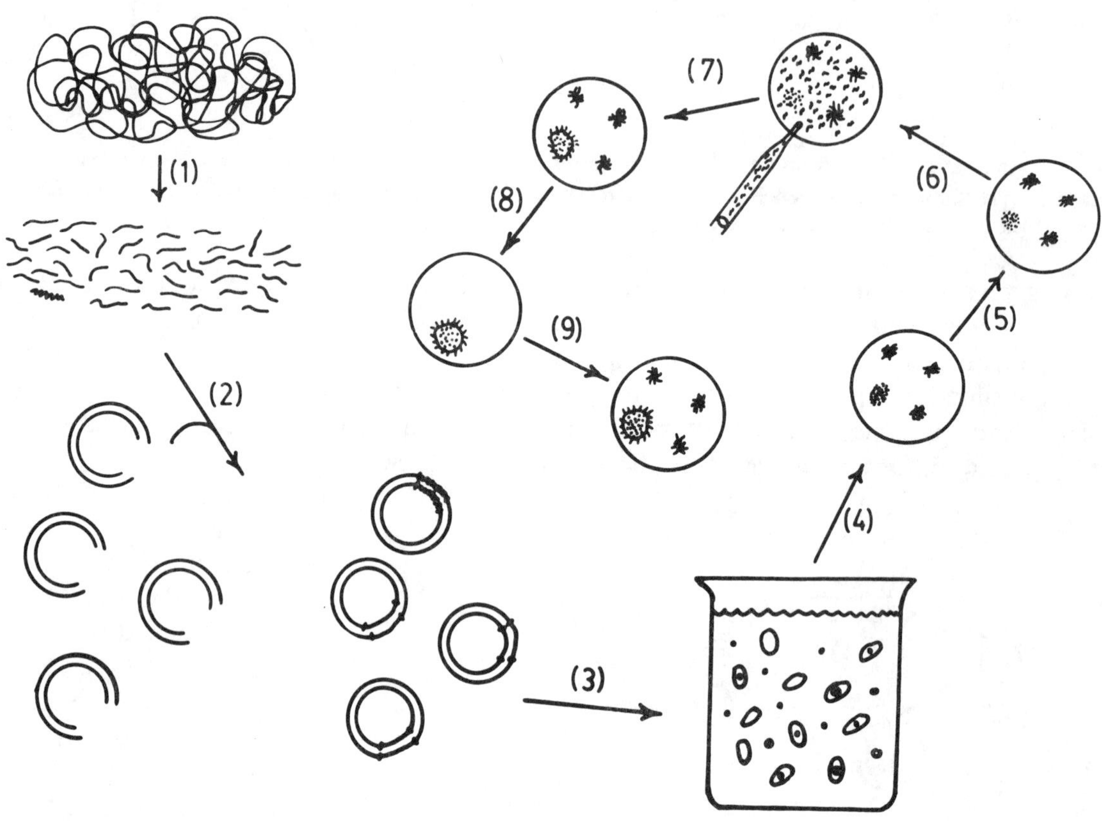

Fig. 2-17. Cloning of sequences from total nuclear DNA. Nuclear DNA is extracted and fragmented by treatment with a restriction endonuclease (1). The fragments are packaged in a cloning vector (2) and grown in bacteria (3). The bacteria are thinly plated to obtain clones (colonies derived from single cells) and transferred to nitrocellulose filters (5). The filter is flooded with a radioactive DNA or RNA probe which contains a sequence complementary to a portion of the gene of interest and washed to remove unhybridized probe (6). The probe binds to the clone containing the sequence desired (7), exposing a photographic emulsion overlay at that site on the filter (8). This permits identification of the clone on the original plate that contains the desired DNA sequence (9). Cells from this clone are cultured to obtain many copies of the gene for analytical purposes.

are plated, and the clones containing the gene of interest are identified as described in the figure legend. Clones containing the gene or gene segments are picked from the plate and mass cultured to obtain sufficient DNA for analysis.

Recent progress in DNA biochemistry has led to the development of a new mechanism for amplifying gene sequences. This procedure, called the **polymerase chain reaction**, is now being used extensively in both basic research and clinical diagnosis of genetic disorders. The steps involved with the polymerase chain reaction or PCR are illustrated in Figure 2-18. The basic PCR reaction utilizes the series: thermal unwinding of the DNA double helix, primer-DNA annealing, and primer extension. The

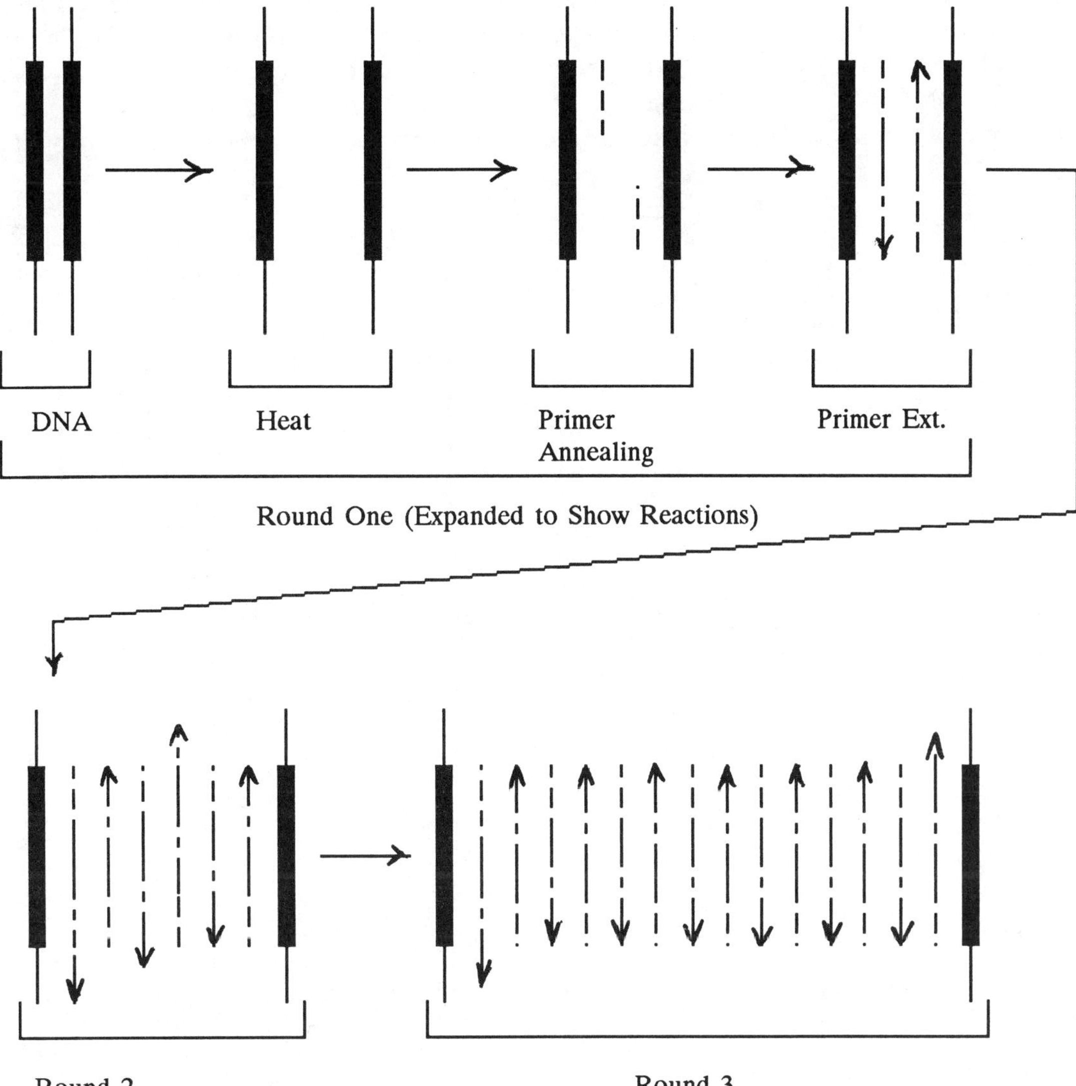

Fig. 2-18. The Polymerase Chain Reaction (PCR). The DNA is heated to denature (unwind) the double helix, and a short primer containing a sequence from the gene being investigated is allowed to pair (anneal) with a complementary sequence in the gene. The primer is then extended by a DNA polymerase. This process is repeated rapidly for approximately 30 cycles under conditions which favor replication of short stretches of DNA, namely those from the gene of interest.

PCR is repeated for 30 cycles in a medium containing a vast excess of the primer and the four deoxynucleotides. These conditions favor replication of only the gene of interest and not other genomic DNAs.

Separation and Identification of DNA and RNA Fragments. All of the procedures described above depend upon identification of the nucleic acid sequence being investigated and separation of that sequence from others of similar type. Most separation techniques exploit electrophoresis of the nucleic acids in agarose gels that have large pore sizes. The nucleic acids are negatively charged under the conditions utilized and migrate toward the positive pole. The distance migrated depends upon both their size and shape. Double-stranded DNA segments behave as cylinders within the gel. The principal factor influencing their migration is their size, with the smaller fragments migrating the longer distances. RNA molecules are much more irregular with respect to shape, such that both shape and size influence their separation. Gels used for separating RNA fragments generally have larger pore sizes than gels employed for resolution of DNAs. A typical experiment for separating DNA fragments is illustrated in Figure 2-19. For example, a sample of the restricted genomic DNA presented in Figure 2-17 is applied to the

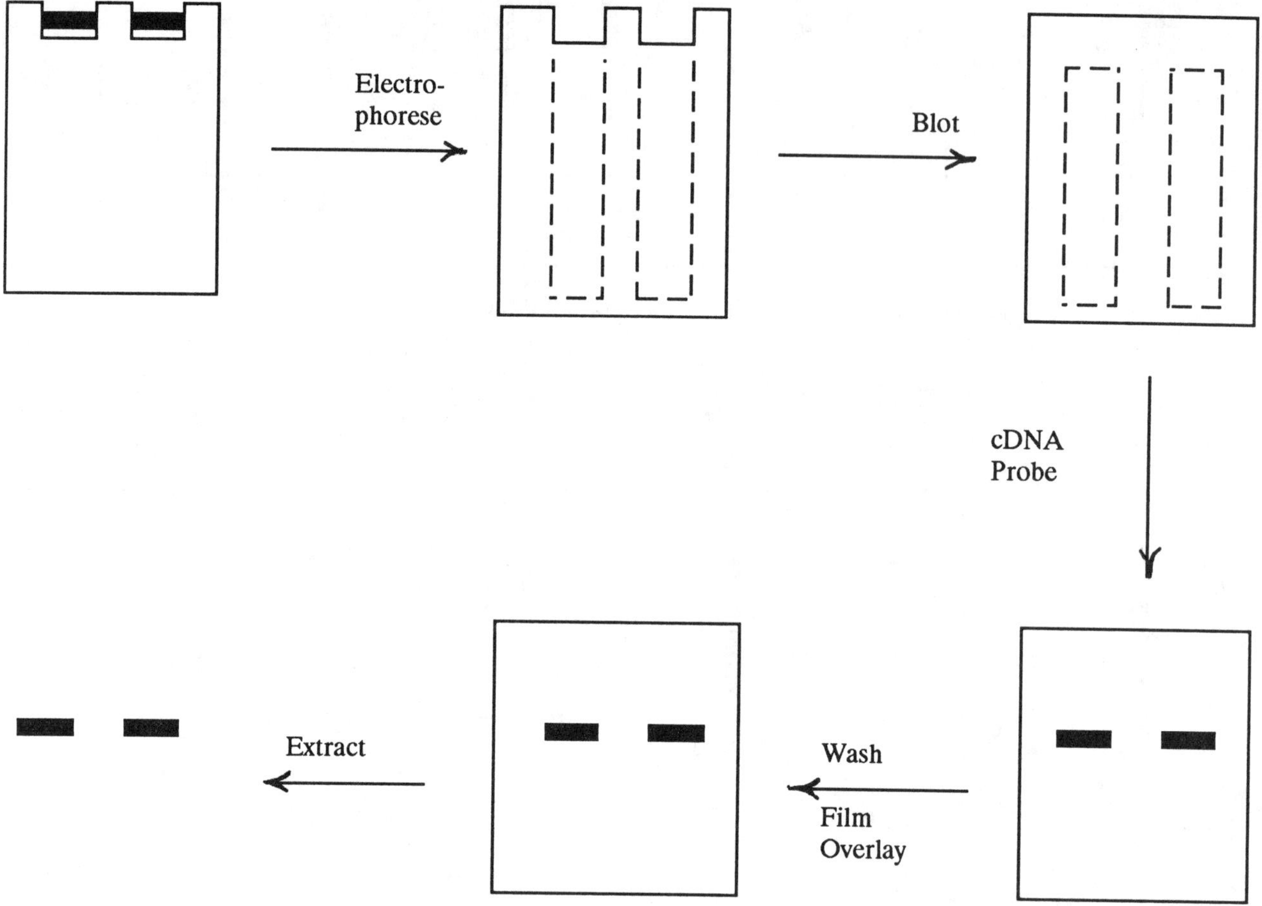

Fig. 2-19. Southern Blotting. DNA is restricted with an endonuclease and layered into wells at the top of the gel. The fragments are separated in an electrical field according to size and then denatured. The denatured DNA is transferred to a nitrocellulose filter and annealed with a radiolabelled cDNA probe. Non-hybridized probe is washed from the filter, and the filter is overlayed with X-ray film. The decaying radioisotope in the probe exposes the film at the site of hybridization. The DNA corresponding to the X-ray image is located and excised for further characterization.

well at the top of the gel and electrophoresed for several hours. The DNA fragments form a smear

5'-³²P-C<u>A</u>GGC<u>A</u>TG<u>A</u>CTGG-3'

5'-³²P-CAGGCATG CTGG-3'
5'-³²P-CAGGC TGACTGG-3'
5'-³²P-C GGCATGACTGG-3'

(a)

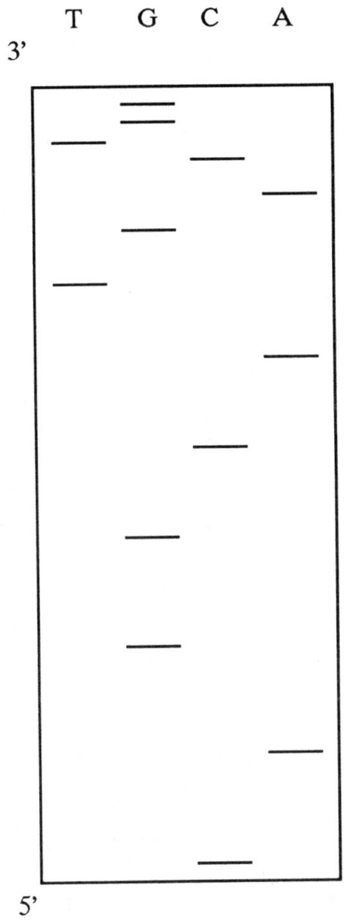

    T   G   C   A

3'

5'

(b)

Fig. 2-20. Sequencing of DNA. The DNA is denatured, and divided into four samples. Each sample is treated with a different chemical which cleaves at A, C, G, or T and generates a different series of fragments (a). The fragments are separated by electrophoresis, generating four different ladders (b). The sequence is inferred by reading the gel from the bottom.

according to their relative sizes that extends from the well toward the bottom of the gel. The DNA in the gel is denatured, and transferred to a nitrocellulose filter. The filter is flooded with a cDNA probe labelled with radioactive phosphate and containing a sequence complementary to the DNA of interest. The probe and the gene are allowed to anneal, and excess probe is washed from the filter. Autoradiography is then employed to locate the DNA that has hybridized to the probe. The corresponding DNA band can then be cut out and used for further experiments. This type of blotting procedure (DNA hybridized to cDNA probe) is called a Southern blot, named after the scientist who developed the procedure. Several variations of the Southern blot have been made, among them are the northern blot and the western blot. For the northern blot, RNAs are separated by electrophoresis and hybridized with cDNA probes. Western blots are used to locate the products of expressed cloned genes. The protein products are separated electrophoretically and transferred to nitrocellulose. The filter is flooded with a radiolabelled antibody or antibody coupled to a fluorescent dye, and the protein and antibody are allowed to form a complex. Uncomplexed antibody is washed from the filter, and the filter is either subjected to autoradiography or viewed under fluorescent light. Bands that are identified correspond to the desired protein-antibody complex.

Sequencing Nucleic Acids. Two methods are commonly used to sequence DNA. One of these utilizes random chemical cleavage of a DNA sequence that has been labelled at the 5' end with radioactive phosphate. The labelled DNA is divided into four portions. The first aliquot is treated with a chemical that cleaves the DNA strand wherever an adenine occurs, generating the series of fragments illustrated in Figure 2-20a and in the A track of the gel in Figure 2-20b. The second sample is digested with a second chemical that cleaves the sequence at cytosine residues, producing the fragments illustrated in track C of Figure 2-20b, and so on. The smallest fragments are located at the bottom of the gel; therefore, the gel is read from bottom to top generating the sequence of the DNA strand.

A second method utilizes a chain termination procedure. The DNA is denatured, and the 3'---->5' strand is used to synthesize a radioactive DNA using a pool of A, T, G, and C in which a small number of one of the bases lacks a 3'-OH group. Wherever this "defective" base is added to the new strand of DNA, synthesis stops, generating a fragment. The fragments are separated by electrophoresis, producing the ladders shown in Figure 2-21. This method is referred to as the "dideoxy sequencing procedure" and has been automated, greatly simplifying the sequencing task.

GENE STRUCTURE

Rapid progress toward determining the structure of mammalian genes has occurred during recent years. The majority of human genes, perhaps more than 80%, occur as two copies per nucleus. Many of these genes are expressed in most tissues at a low level and throughout the life of the individual. These genes are often referred to as "**housekeeping**" genes because their products appear to carry out "routine" functions in the cell. A second class of genes may occur as two to several copies per nucleus (Table 2-2) and are expressed at high levels, only in certain cell types, and usually only during specific periods in the life of the individual. These genes are referred to as "**specialized**" genes. The promoters of specialized and housekeeping genes are quite different, so each class will be treated separately during the following discussion of gene structure.

Most human genes are much larger than the sizes required for specifying the amino acid sequences of the proteins they encode. For example, the housekeeping gene, **HEXA**, which encodes the alpha chain precursor of the hexosaminidase A enzyme, spans about 40 kilobases (kb) of DNA, while the amount of DNA required for the amino acid sequence of the alpha chain precursor is 1587 bases. The noncoding regions include introns that vary considerably in size and upstream and downstream regions which often contain regulatory sites.

The human protein, hemoglobin, plays an important role in respiration. The hemoglobin molecule contains two α-like chains, two β-like chains, and four molecules of a multiringed structure called heme. The actual polypeptides occurring in the hemoglobin molecule vary with the stage of development. Furthermore, only erythroid cells synthesize significant amounts of globin. The globin genes display a high degree of similarity in their structures, with each globin gene possessing three exons and two introns (Fig. 2-22). A schematic representation of the human β-globin gene is presented in Figure 2-22(a) and the sequences of specific sites are highlighted in Figure 2-22(b). The latter panel displays the sequence of the complementary strand of DNA, since its sequence approximates that of the RNA transcibed from the master or sense strand of the gene. The globin gene promoter contains several elements that are found in specialized gene promoters. These include several proximal promoter sites (CCAAT and TATA or ATA) and a sequence that specifies the cap site (TTGCTT*AC;* denotes site of capped base in RNA). The distal promoter region of several of the Left: globin genes contains sites that influence not only the expression of the gene to which they are attached, but also other globin

Fig. 2-21. Ladder from DNA Sequencing by the Dideoxy Method. The sequence is derived from insert from Fig. 2-14. from SP6 promoter; Right: from T7 promoter. Lanes: ACGTAGTC in both.

34

Table 2-2. A Sampling of Specialized Genes.

Protein Class	Number of Genes per Nucleus	Expressing Cell	Chromosomal Location
Globins			
α-like	Eight	Erythroid	16
β-like	Ten	Erythroid	11
Collagens	>26	Chondrocyte	1,2,6,7,12,13,17,21
Carbonic Anhydrase	Eight	Varies according to gene type	8, 16
HLA	>30	Class I-general	6
		ClassII-immune system	6
Immunoglobulins	Thousands	B-cell lineage	2, 14, 22

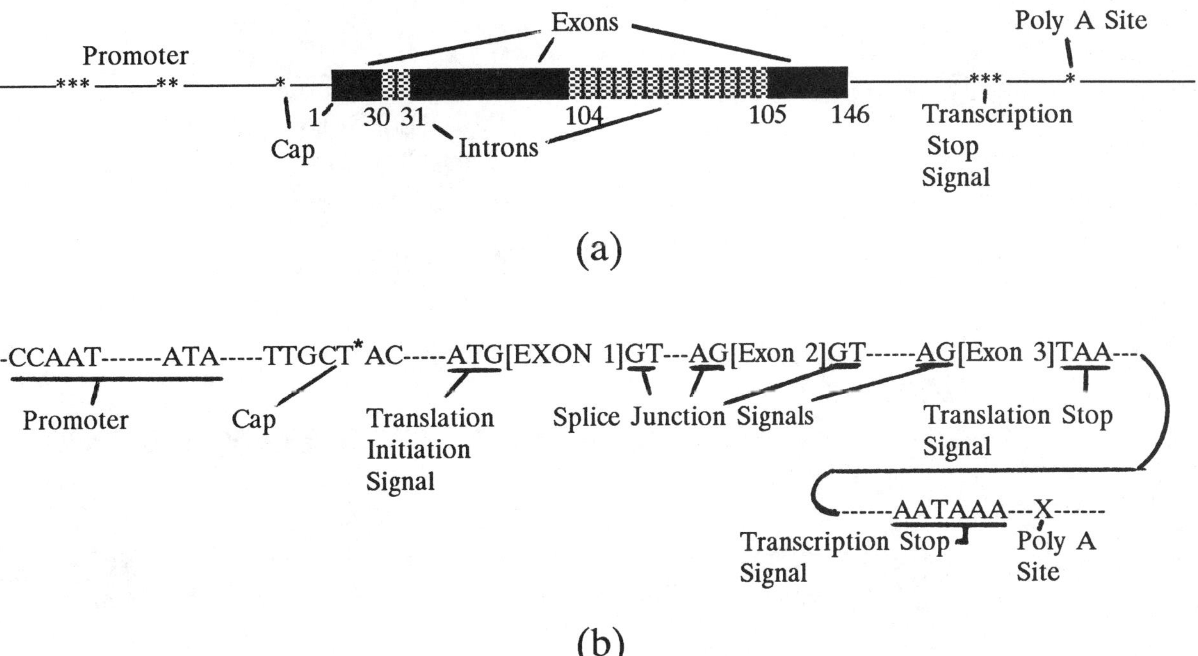

Fig. 2-22. Structure of the human β-globin gene. a) schematic diagram of the entire gene (numbers at exon boundaries refer to corresponding amino acids in the β-globin polypeptide); b) sequences of the promoter, intron/exon junctions, and downstream regions.

genes within the cluster. These sites will be discussed in the following chapter. Immediately adjacent to the first exon is the triplet that serves as the initiation signal for translation. The corresponding codon in the mRNA would be AUG which specifies methionine in the nascent polypeptide. This amino acid serves a translation initiation function, but does not appear in the mature protein. The intron/exon junctions contain characteristic sequences that signal the spliceosome to cut and join the exons at that particular point. You will notice that the first and second introns of the β-globin gene begin with GT and end with AG. Mutations affecting any of these four bases will influence gene expression by preventing normal splicing of the exons to form the mature mRNA. The β-globin gene has a translational stop signal (TAA) immediately following the triplet specifying the 146th amino acid and a transcriptional signal (AATAAA) 107 nucleotides further downstream. Transcription normally proceeds beyond this site; however, the RNA is cut 17 nucleotides downstream and polyadenylated (Poly A site). RNA transcripts that are not cut at the appropriate site are not polyadenylated and are rapidly degraded prior to exiting the nucleus.

The entire structure (distal and proximal promoter regions, exons, introns, and downstream regions) is referred to as a **transcriptional unit.** Genetic diseases can often be traced to lack of or non-functional proteins. The aberrant gene expression responsible for a genetic disease may be caused by

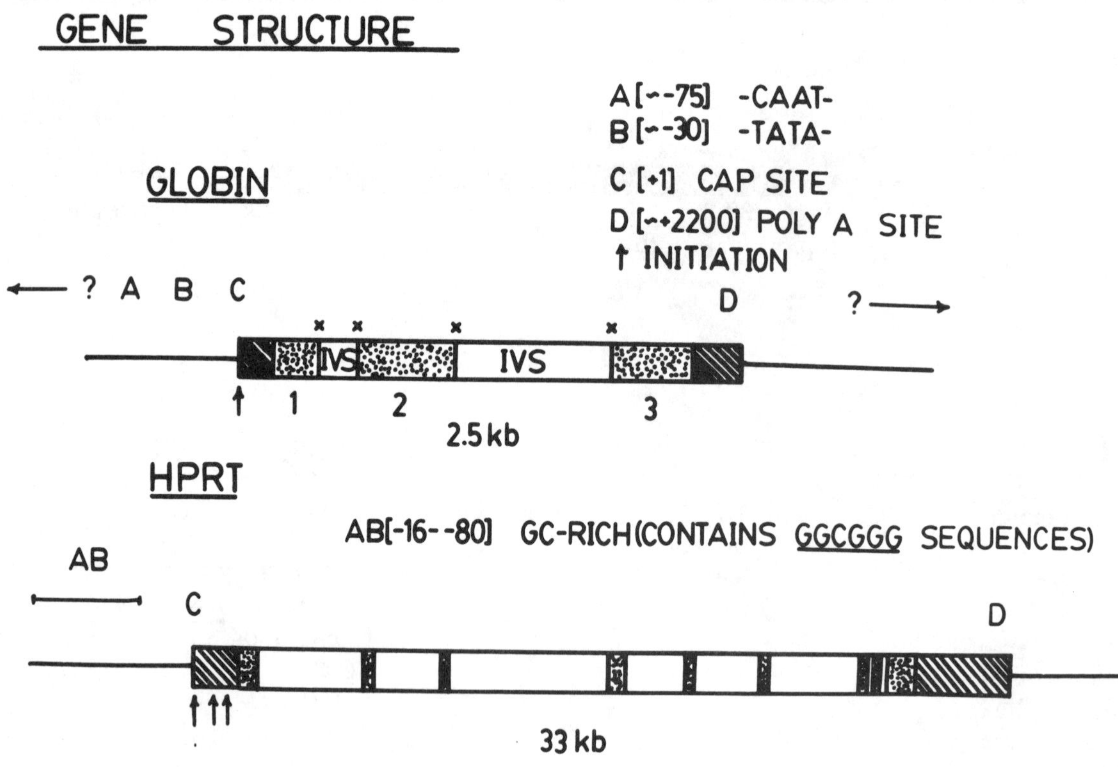

Fig. 2-23. Comparison of the structures of the globin and HPRT genes. The exons of the HPRT gene are dispersed over 33kb of DNA. The promoters of the two gene types are strikingly different. HPRT, a housekeeping gene, has a GC-rich promoter which lacks CCAAT and TATA promoter elements. Upstream and downstream noncoding regions are hatched, and exons are stippled. Nucleotides are numbered from the cap site. IVS = intervening sequence = intron.

any one of several mutations affecting non-coding regions (promoter, intron/exon junctions, downstream elements) or coding regions (exons).

The promoters of housekeeping genes lack CCAAT and TATA boxes and tend to be GC-rich (Fig. 2-23). The HPRT gene, the gene encoding hypoxanthine phosphoribosyltransferase, is situated near the distal end of the long arm of the X chromosome. The gene spans 33kb and includes nine exons and eight introns. The enzyme catalyzes steps in a purine "salvage" pathway that converts the free bases,guanine and hypoxanthine, that are formed from degradation of nucleotides and nucleic acids into the nucleotides GMP and IMP, respectively. Salvage pathways tend to be more economical with respect to energy consumption than pathways that produce the same nucleotides de novo by a longer route. The proximal promoter of the HPRT gene contains multiple copies of the sequence GGCGGG. These sequences contain CpG doublets that tend to be unmethylated in active genes and methylated in non-transcribing genes. These "CpG islands" are most common among housekeeping genes, but are also found in a number of specialized genes. The mechanisms involved with expression of GC-rich promoters is poorly understood at present.

RNA Processing

A large pool of RNA exists within the nucleus that contains sequences which are similar to those appearing in the cytoplasmic mRNA population. This large pool is called **hnRNA**, or heterogeneous nuclear RNA. About 95% of hnRNA is rapidly degraded and does not contribute to the mRNA population; however, some of the hnRNA molecules are processed into mRNAs which then enter the cytoplasm where they are translated. A typical human cell contains a few mRNAs that occur in more than 10,000 copies per cell, several hundred mRNAs that are present in about 500 copies per cell, and more than 10,000 mRNAs that occur in less than 20 copies per cell. RNA processing has received considerable attention during the last few years and serves as a key point at which gene expression can be regulated.

Capping. An enzyme that is part of the transcription complex caps the 5' end of the hnRNA formed by RNA polymerase II at a very early stage of transcription. The capping process involves several steps which include addition of a guanine to the 5' nucleotide, methylation of this guanine, and methylation of the adjacent nucleotide on its ribose moiety. The structure of a typical cap is displayed in Figure 2-24. The cap is essential in two respects: 1) it participates in the formation of the initiation complex during translation and 2) it protects the 5' end of the RNA from degradation. Uncapped hnRNAs are cleared from the nucleus very rapidly and rarely contribute to the cytoplasmic mRNA pool.

$$_3HC-GpppNpNp--CH_3$$

Fig. 2-24. Cap Structure. The guanine nucleotide is methylated at the 7 position and linked by a triphosphate group to the adjacent nucleotide (N = A, T, G, or C; p = phosphate). The adjacent nucleotide is also methylated on its ribose ring.

Splicing. Introns are removed and the flanking exons are joined by a complex of snRNPs (small nuclear ribonucleo-proteins) called the spliceosome. The splicing of a hnRNA to form a mRNA molecule is illustrated in Figure 2-25. The U1 snRNP attaches to the 5' exon/intron junction, and U2 snRNP binds to an adenine residue at an internal site. This base "attacks" the 5' junction, resulting in cleavage, and formation of a lariat. Cleavage at the 3' intron/exon junction and joining of the exons then occur. All of the reactions involved with spicing have not been elucidated; however, the spliceosome is about as large as a ribosome, raising the possibility that the process is complex and requires many proteins and snRNAs.

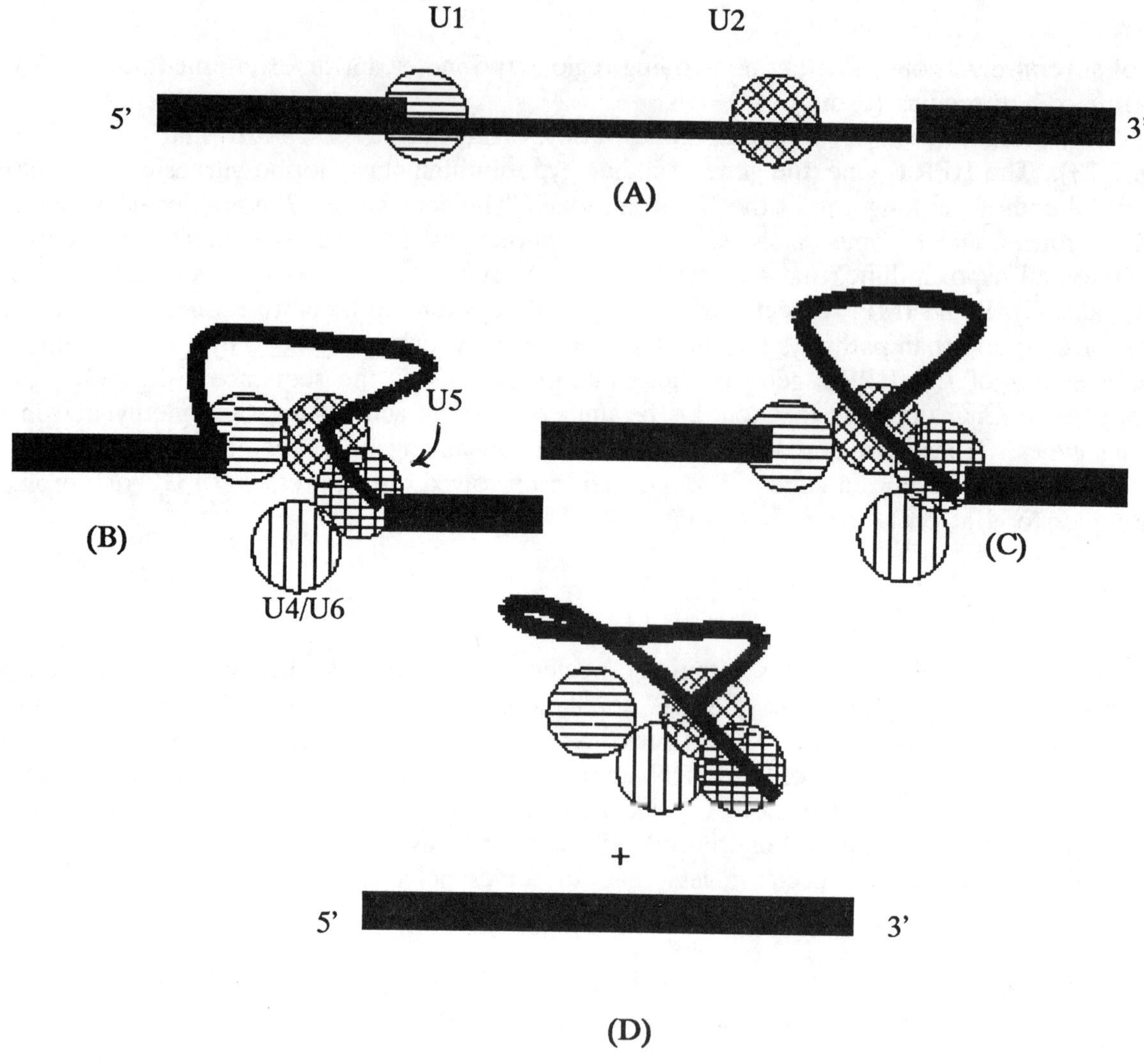

Fig. 2-25. Splicing of hnRNA to form mRNA. A) U1 snRNP binds to the 5' exon/intron junction, and U2 snRNP binds to a sequence containing an A residue. B) U4, U6, and U5 snRNPs attach, and the adenine attacks the 5' splice junction. C) The junction is cleaved and a branched linkage is formed with the adenine, producing a lariat. D) The 3' intron/exon junction is cleaved, and the exons are spliced together. The intron lariat is eventually degraded.

Different mRNAs can be produced by combining exons in varying combinations. In this way a gene can specify several products having either similar or quite distinct functions. An example of this phenomenon is provided by the myosin contractile protein. Myosin consists of light and heavy chains encoded by different genes. However, two light chains, MLC 1_f (MLC = myosin light chain) and MLC 3_f are encoded by the same gene (Fig. 2-26). The MLC gene contains two promoters. The MLC 1_f and MLC 3_f hnRNAs are transcribed from different promoters and processed such that exons 1, 4, and 5-9 are spliced into the MLC 1_f mRNA, and exons 2, 3, and 5-9 are included in the MLC 3_f mRNA. Alternative mRNA splicing is common among human genes, increasing the diversity of products that are possible using a finite number of genes. Another example of alternative splicing will be discussed in a later chapter in relation to antibody targeting. Antibody heavy chain genes contain "secretory" and "membrane" exons following the other exons in the cluster. All exons of the cluster are transcribed into the hnRNA. The early splicing pattern removes the secretory exon and splices in the membrane exons. The mRNA is then translated into a protein that is anchored in the cell membrane where it serves as a receptor. After antigen exposure, the splicing pattern changes such that the secretory exon is spliced in and the membrane exons are excised during the formation of the mRNA. The protein translated from this mRNA is secreted from the cell and attacks the antigen involved.

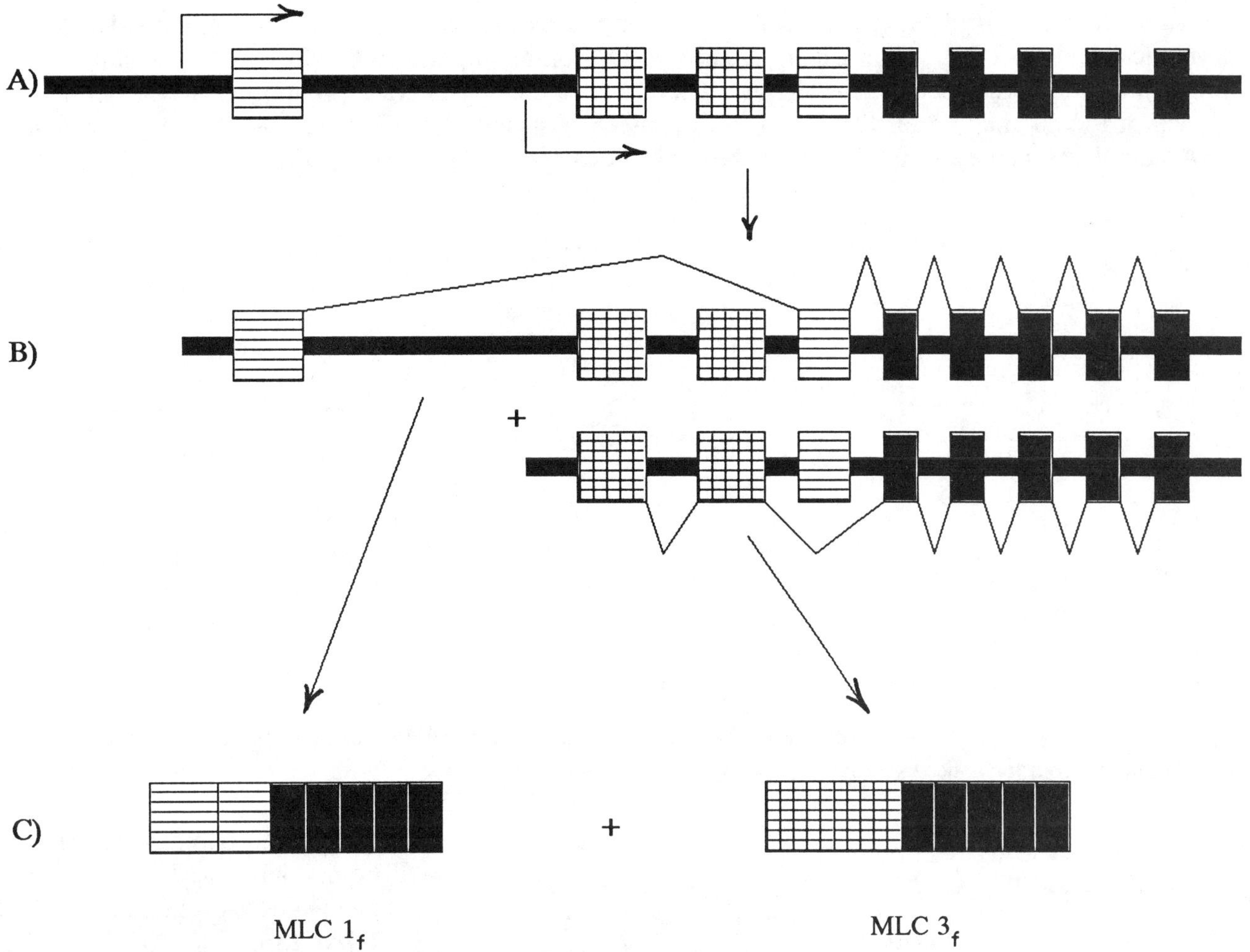

Fig. 2-26. Alternate splicing of a hnRNA to form two different mRNAs. A) The gene encoding myosin light chain 1$_f$/3$_f$. The arrows indicate transcription from two promoters within the same gene. B) hnRNAs transcribed from the MLC 1$_f$/3$_f$ gene. Exons participating in the alternative splicing are connected by thin lines. C) MLC 1$_f$ and MLC 3$_f$ mRNAs.

Polyadenylation. The sequence AATAAA signals RNA polymerase II to cease transcribing several bases downstream and a second enzyme within the polymerase complex begins the addition of approximately 200 adenylate residues to the end of the hnRNA. This polyA tail stabilizes the RNA, protecting it from degradation from the 3' end by nucleases. hnRNAs lacking polyA tails are degraded rapidly and rarely participate in translation. The polyA tail is included on the mRNA derived from the hnRNA. After the mRNA reaches the cytosol, the polyA tail can be shortened or lengthened by different enzymes. The relative success of the lengthening vs. shortening of the polyA determines the half-life of the message.

Translation

     Translation refers to the mechanism through which the base sequence of the mRNA orders the amino acid sequence of a polypeptide. This process requires mediation by an adapter molecule, tRNA, since amino acids cannot attach directly to the bases of the mRNA with the degree of specificty necessary for synthesis of functional proteins. Amino acids are coupled to the appropriate tRNA enzymatically (Fig. 2-27). For example, the tRNA containing the **anticodon**, AAA, is linked with phenylalanine to form phenylalanyl-tRNA. There is one loading enzyme (aminoacyl-tRNA synthetase) for each of the 20 amino acids that occur in the nascent polypeptides. Following loading, the aminoacyl-tRNA complexes move to the site of synthesis of the polypeptide where they are aligned with the

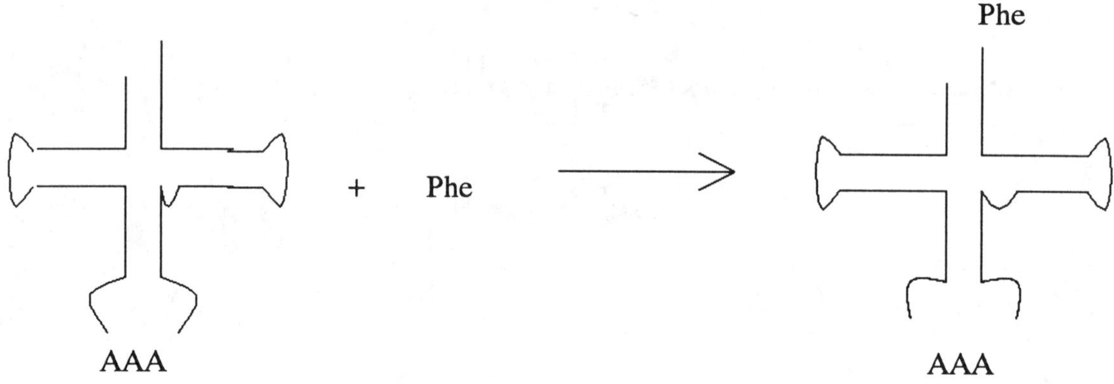

Fig. 2-27. Loading of tRNA with phenylalanine (Phe). A specific phenylalanyl-tRNA synthetase couples Phe to the tRNA at the end opposite from the anticodon, AAA.

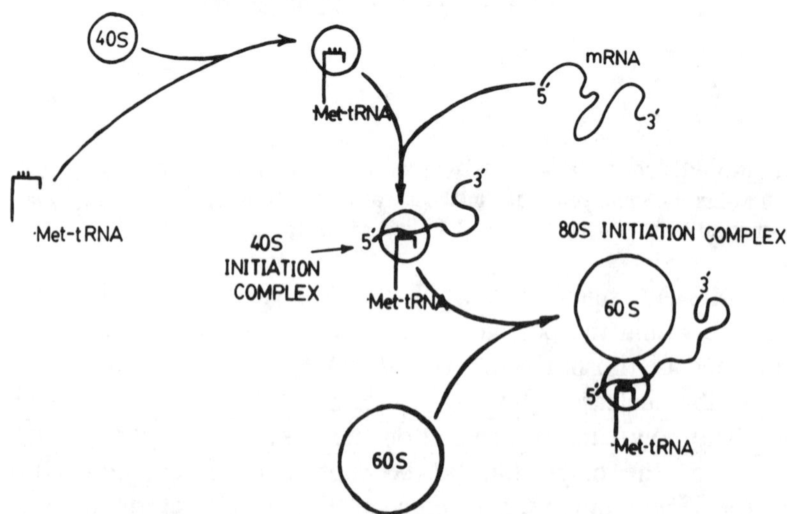

appropriate codon in the mRNA. The first aminoacyl-tRNA to bind with the mRNA is Met-tRNA. This tRNA forms a 40S initiation complex with the small ribosomal subunit and the mRNA (Fig. 2-28). The Met-tRNA anticodon binds to the first AUG codon occurring in the mRNA. The

Fig. 2-28. Formation of the translational initiation complex. Initiation occurs in two steps with the formation of the 40S complex followed by the binding of the large ribosomal subunit to construct the 80S complex.

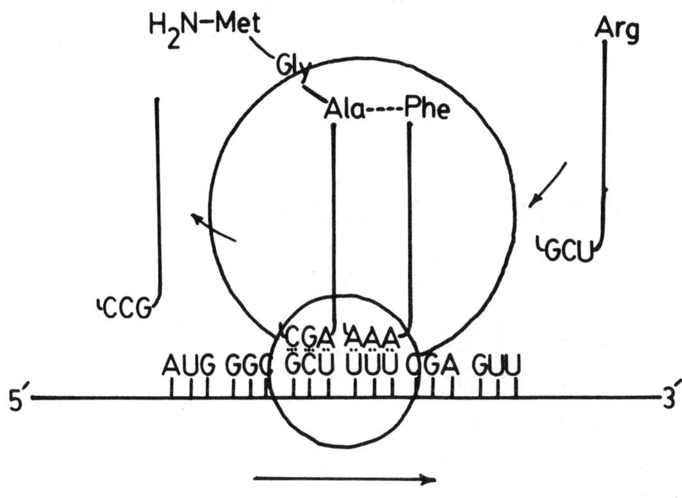

Fig. 2-29. Synthesis of a polypeptide.

large ribosomal subunit binds to complete initiation, forming an 80S initiation complex. Each ribosome may have as many as three attachment points for the aminoacyl-tRNAs. The initial site is called the A-site (aminoacyl-tRNA binding site). When the next aminoacyl-tRNA enters, the first moves to the P-site (peptidyl-tRNA binding site); and the new aminoacyl-tRNA occupies the A-site. As the third aminoacyl-tRNA enters the A-site, the first translocates to an exit site, and the second moves to the P-site. The amino acid attached to the tRNA at the A-site is linked to the growing peptide that is attached to the tRNA at the P-site by a peptidyl transferase. Arrival of the fourth aminoacyl-tRNA results in the exiting of the first tRNA from the ribosome and the successive translocations of the others to the next site available. This process is repeated many times until the stop codon is encountered by the ribosome. Five to ten ribosomes may be attached to a single mRNA, depending upon its length, at any one time forming a **polysome**. The ribosome possesses many proteins which participate either directly or indirectly in protein synthesis. These proteins facilitate interactions among the various components of the translational machinery, cleave the initial methionine from the N-terminal end of the polypeptide, and assist elongation of the peptide. The reaction catalyzed by the peptidyl transferase is presented in Figure 2-30.

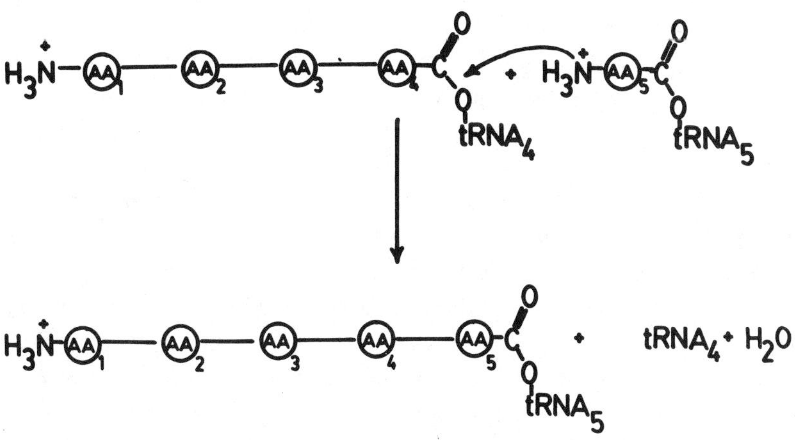

Fig. 2-30. Formation of a peptide bond between adjacent amino acids. Sequential formation of these bonds produces an amino acid polymer (polypeptide). Peptide bond formation requires participation of the enzyme, peptidyl transferase, and other proteins that are components of the ribosome.

The relationship between the base sequence of the master strand of the DNA and the amino acid sequence of the polypeptide is illustrated in Figure 2-31. An abridged polypeptide is shown for clarity. Notice that the N-terminal methionine is no longer present.

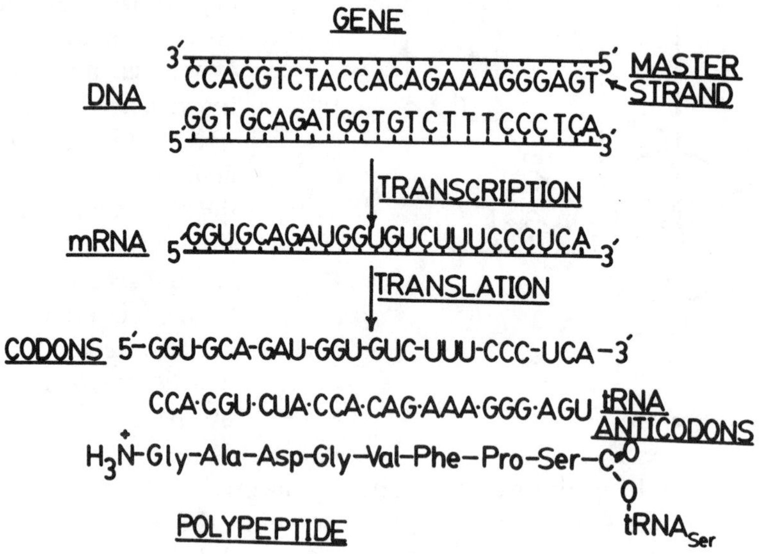

Fig. 2-31. Mechanism by which the base sequence of the coding strand of the gene determines the amino acid sequence of the polypeptide.

Posttranslational Modification

During synthesis the polypeptide begins to fold to adopt the most energetically acceptable conformation. This folding places the hydrophobic amino acids in the core of the protein and the hydrophilic amino acids on the surface. The folding generally occurs without enzyme intervention; however, numerous other modifications occur as a consequence of enzyme action. These modifications include formation of amino acid derivatives, addition of carbohydrate and/or lipid, and sizing of the protein to its functional form.

The genetic code contains information for specification of twenty amino acids; however, more than 140 appear in mature proteins. Phosphorylation and sulfation of tyrosine, serines, and threonines are commonly observed and most likely play a role in regulation of the proteins in which they occur, their stability, or both. Proline is hydroxylated during the synthesis of collagen, and modified lysine residues also appear in this protein group. Other modifications include the formation of disulfide bridges between nearby cysteine residues to form cystine, the oxidation of homocysteine to homocystine, and the methylation or acetylation of a number of other proteins.

Cleavage of primary translation products to form smaller, biologically active proteins is commonly observed. Some of the better known examples include the conversion of the proinsulins to insulin, the limited proteolysis of pancreatic proenzymes to their active counterparts, and the series of changes that occur in clotting factors during the clotting "cascade". The formation of an active protein from an inactive precursor provides an additional level of control of the expression of the respective genes involved. In many cases, this control tends to be protective, preventing autodigestion of the acinar cells of the pancreas by the enzymes they synthesize and inappropriate clotting of the blood.

Cell membranes have polar surfaces and hydrophobic internal regions. Proteins that are components of membranes must accommodate both environments. Membrane-spanning proteins often contain alternating hydrophilic and hydrophobic domains. These proteins adopt an orientation in which the hydrophilic domains are situated at the external surfaces of the membrane and the hydrophobic domains interdigitate with lipids in the membrane core. This biphasic structure of the protein is augmented by the addition of carbohydrate to the hydrophilic domains and lipid to the hydrophobic domains. Membrane receptors provide good examples of this type of structure.

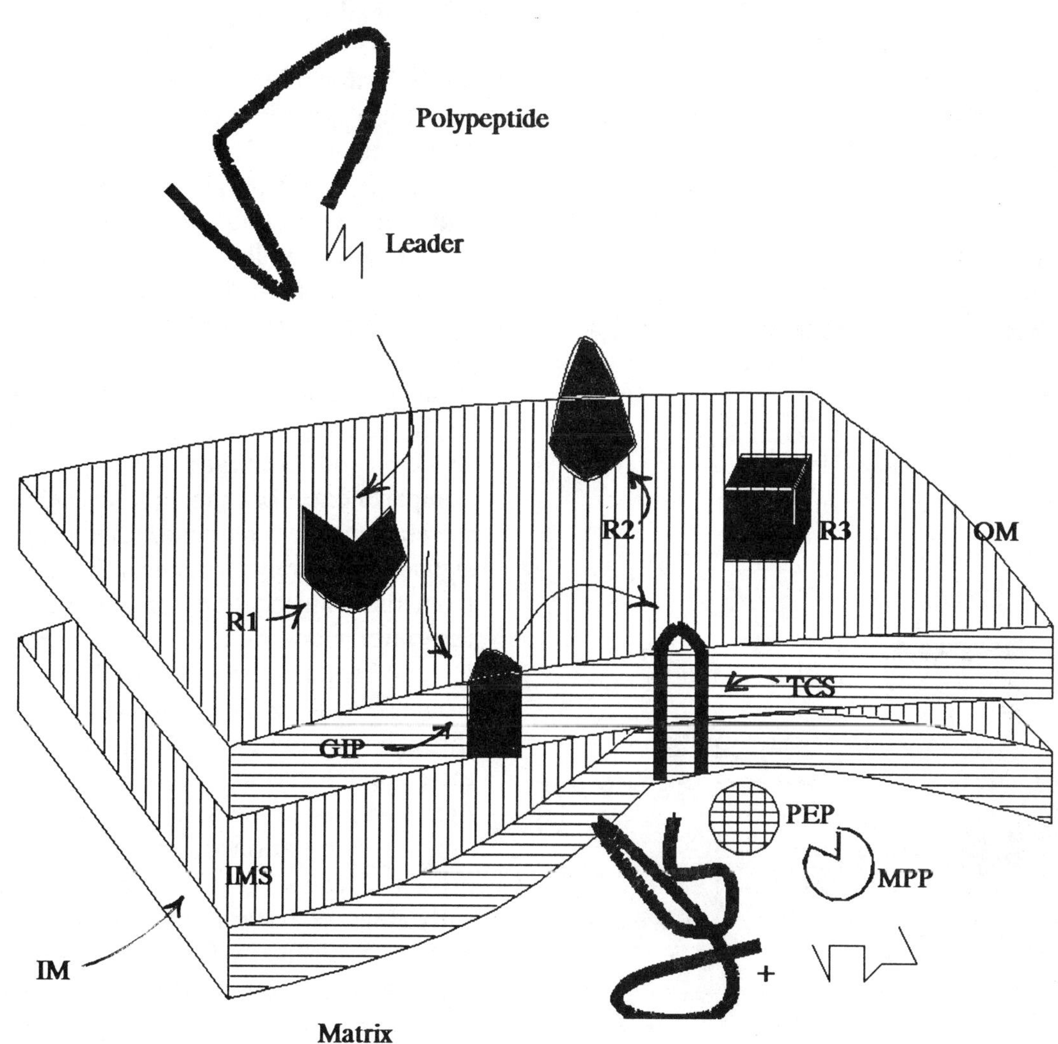

Fig. 2-32. Model for insertion of a nuclear-encoded protein into the mitochondrion. The polypeptide containing a leader sequence docks with a receptor (R1) and is transferred to a general insertion protein (GIP). The GIP assists the insertion of the precursor polypeptide (leader sequence first) into the outer membrane. An unfolding protein converts the conformation of the polypeptide to a shape suitable for passage through the membrane. Some proteins are sorted at this step and channeled to the outer membrane (OM). Others are sent through the channel formed at translocation contact sites (TCS) into the matrix where the joint action of processing-enhancing protein (PEP) and matrix processing peptidase (MPP) cleave off the leader sequence. The protein will either stay in the matrix or be sorted for inclusion in the intermembrane space (IMS) or inner membrane (IM).

Protein Targeting. Nascent polypeptides that are destined for cellular organelles or for secretion usually contain leader sequences at their N-terminal ends. These leader sequences are generally about 20-30 amino acids in length and contain the signals that are essential for entry into the appropriate organelle. Two examples of protein targeting will be presented here. Mitochondrial DNA contains insufficient information to supply all of the proteins that contribute to its structure and function. The majority of mitochondrial proteins are synthesized from mRNAs derived from nuclear genes. These proteins are synthesized by cytosolic ribosomes and are targeted for either the outer mitochondrial membrane, the intermembrane space, the inner membrane, or the matrix of the mitochondrion. The pathway between the site of synthesis of the protein and its final destination is complicated and varies among proteins according to their final location (Fig. 2-32). Nuclear-encoded mitochondrial proteins are synthesized as precursor proteins that often contain an N-terminal leader sequence that targets them to the mitochondrion. This N-terminal sequence interacts with a specific receptor (R1, R2, R3...(Fig. 2-32)) on the outer membrane surface of the mitochondrion and are inserted into the membrane by a general insertion protein. The leader sequence apparently enters first, with the remainder of the protein unfolding to permit entry through translocation contact sites. Following arrival in the matrix, the leader is cleaved from the protein by matrix processing peptidase. A second protein, processing-enhancing protein facilitates the cleavage reaction. One or more additional proteins refold the nuclear-encoded mitochondrial protein and facilitate sorting to the various compartments of the mitochondrion. Some nuclear-encoded mitochondrial proteins contain signals (stop-transfer signals) which divert them from the translocation contact site pathway into the outer membrane of the mitochondrion.

Proteins that are destined for secretion from the cell or for inclusion in lysosomes pass through a series of compartments before reaching their final location. These proteins contain a signal sequence possessing an ER recognition signal. Translation of mRNAs encoding these proteins begins on cytosolic ribosomes. At an early stage of translation, a signal recognition particle (SRP;Fig. 2-33) binds to both the ribosome and the signal peptide (SP). The SRP transfers the entire translational complex to the ER, where the SRP binds to a specific receptor (SRPR). Note how this transfer is occurring during the translation of the mRNA. The polypeptide is inserted into the ER lumen through a pore that is adjacent to the SRP receptor. The signal peptide is believed to remain within the membrane,

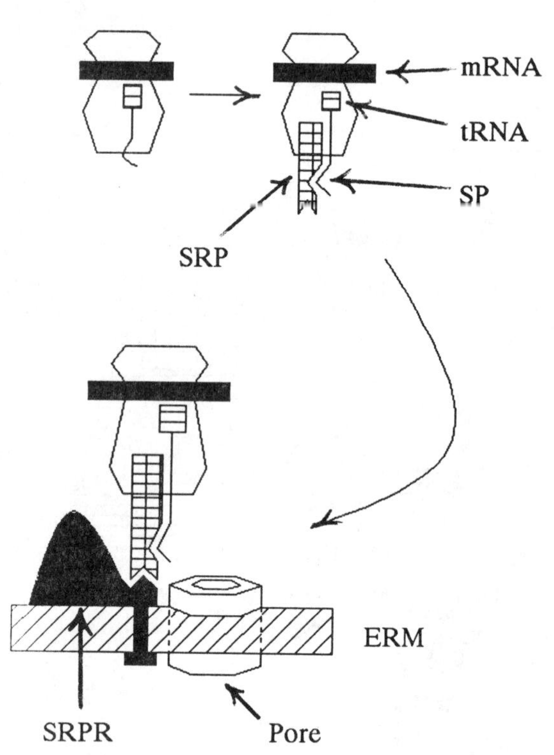

Fig. 2-33. Targeting of proteins to the ER membrane. The signal recognition particle (SRP) transfers the translational complex to a receptor (SRPR), and the polypeptide is inserted through an adjacent pore into the ER lumen.

forming a U-shaped structure with the N-terminus of the signal peptide pointing toward the cytosolic surface of the ER, and the remainder of the polypeptide is driven by an energy-requiring process from the site of synthesis on the ribosome into the lumen of the ER. Following insertion, the signal peptide is cleaved from the polypeptide by a membrane associated enzyme, signal peptidase. The "mature" protein is now glycosylated as described below and transferred by small saccules to the Golgi apparatus where they are further modified.

Some polypeptides targeted to the ER contain not only a signal peptide, but also one or more stop-transfer signals. If the polypeptide contains one stop-transfer signal, it will be inserted up to that point and anchored within the membrane. This results in a membrane-spanning protein with its N-terminus projecting into the ER lumen and its C-terminus extending into the cytosol. Polypeptides containing two or more stop-transfer sequences may span the ER membrane several times.

The Golgi apparatus is arranged as a series of stacked compartments that have been histochemically demonstrated to have different functions. The Golgi stack may be divided into cis (proximal to ER), medial, and trans regions (Fig. 2-34). While the protein destined for secretion or lysosomes is within the ER lumen, a high mannose oligosaccharide (Fig. 2-35) is attached to certain arginine residues in the protein, and these are designated "N-linked oligosaccharides". As the protein moves to the cis region of the Golgi, some of the oligosaccharides are phosphorylated. One of these phosphorylated mannose moieties is an important marker for a receptor-mediated targeting of enzymes to lysosomes. As the proteins enter the medial Golgi, N-acetylglucosamine (GlcNAc) is added to the N-linked oligosaccharide chains. Some of the oligosaccharides (those which were not phosphorylated) have a number of mannose residues removed in the medial compartment. The latter proteins are further modified in the trans Golgi by addition of Galactose (Gal) and sialic acid (SA). These modified carbohydrates are termed "complex" oligosaccharides, while their unmodified and phosphorylated counterparts are called "high mannose" oligosaccharides. The different N-linked oligosaccharides influence the sorting of proteins such that some are targeted to secretory vesicles for secretion, to lysosomes, or sent directly to the cell membrane. Proteins may also be glycosylated on the side chains of certain serine and threonine residues. These "O-linked" oligosaccharides are also modified during the transit of the protein through the Golgi.

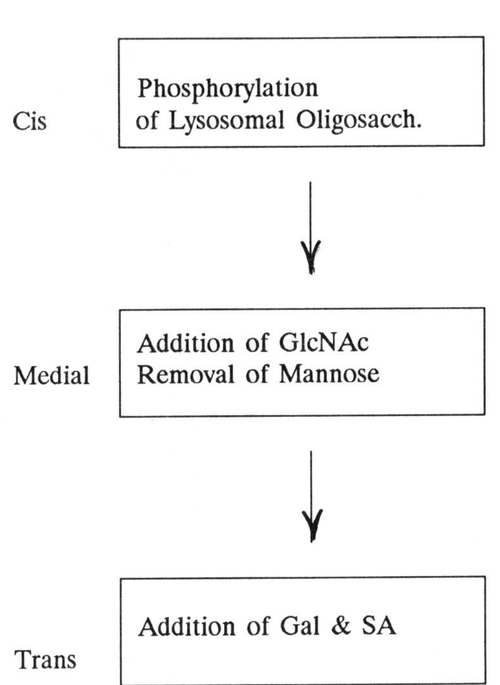

Fig. 2-34. Sequential action of Golgi processing enzymes. The Golgi cysternae are organized chemically into subregions: cis (proximal to ER), medial, and trans. Enzymes in these regions modify the oligosaccharides as indicated. GlcNAc = N-acetylglucosamine; Gal = Galactose; SA = Sialic acid.

There is a significant amount of intercellular exchange of proteins and enzymes. For example, some enzymes are exported from the cell, bound to specific receptors on the adjacent cell surface, and internalized by receptor-mediated endocytosis. This process requires the presence of an exposed mannose-6-phosphate residue on the N-linked oligosaccharide of the protein involved. Following internalization, the protein is enclosed in a vesicle that is chemically modified and eventually fuses with lysosomes of the cell. Therefore, some lysosomes contain not only enzymes from the cell in which they occur, but also enzymes synthesized by neighboring cells. This enzyme exchange is essential for normal cellular metabolism, because persons who have a mutation inactivating a processing enzyme develop a lethal disorder characterized by the lack of a large group

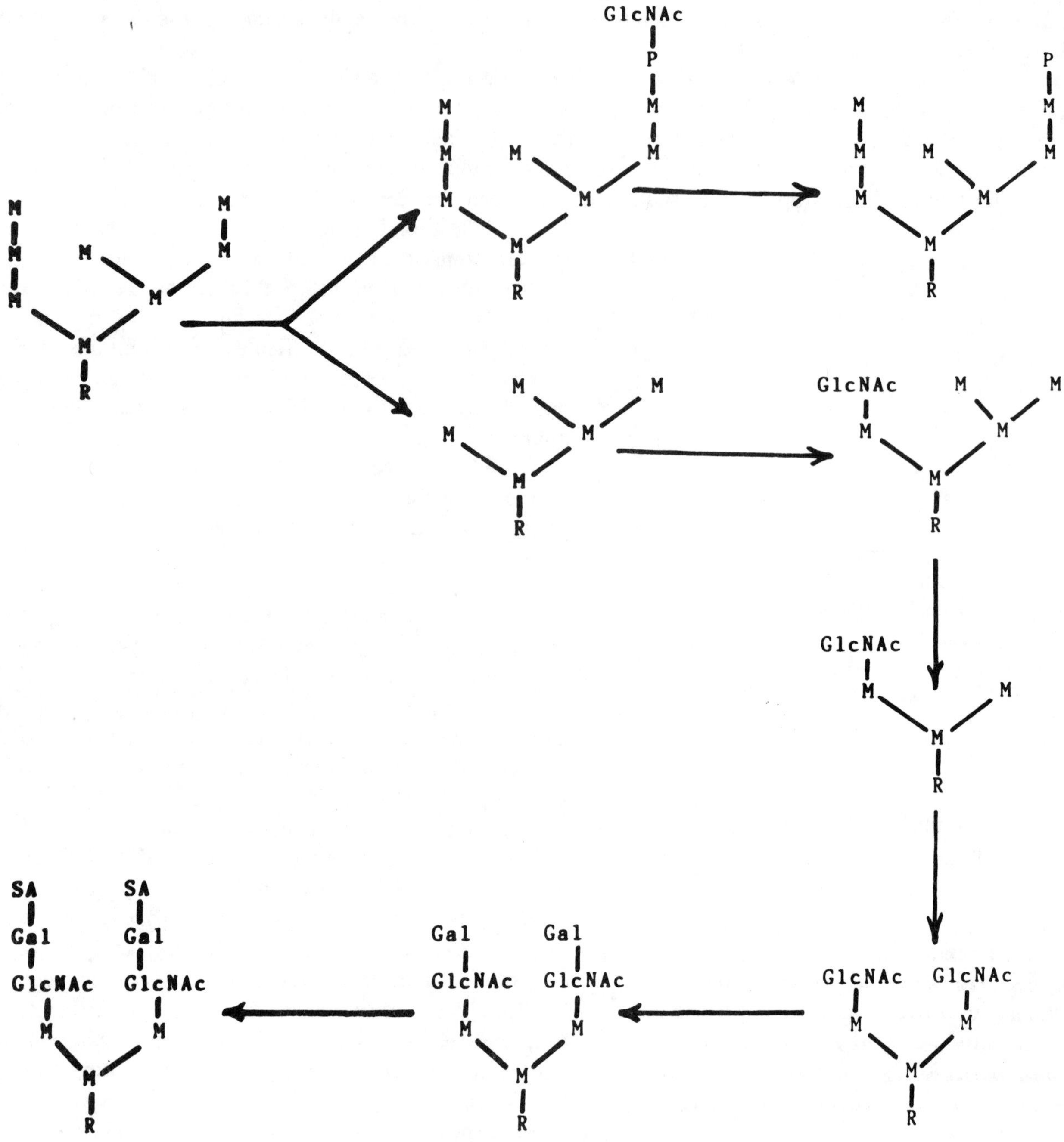

Fig. 2-35. Synthesis of the high mannose (upper right) and complex (lower left) N-linked oligosaccharides. Carbohydrate abbreviations are identical to those in Fig. 2-34. R = remainder of molecule.

of lysosomal hydrolases and the accumulation of their respective substrates within lysosomes. The inter-cellular exchange of enzymes is schematically represented in Fig. 2-36.

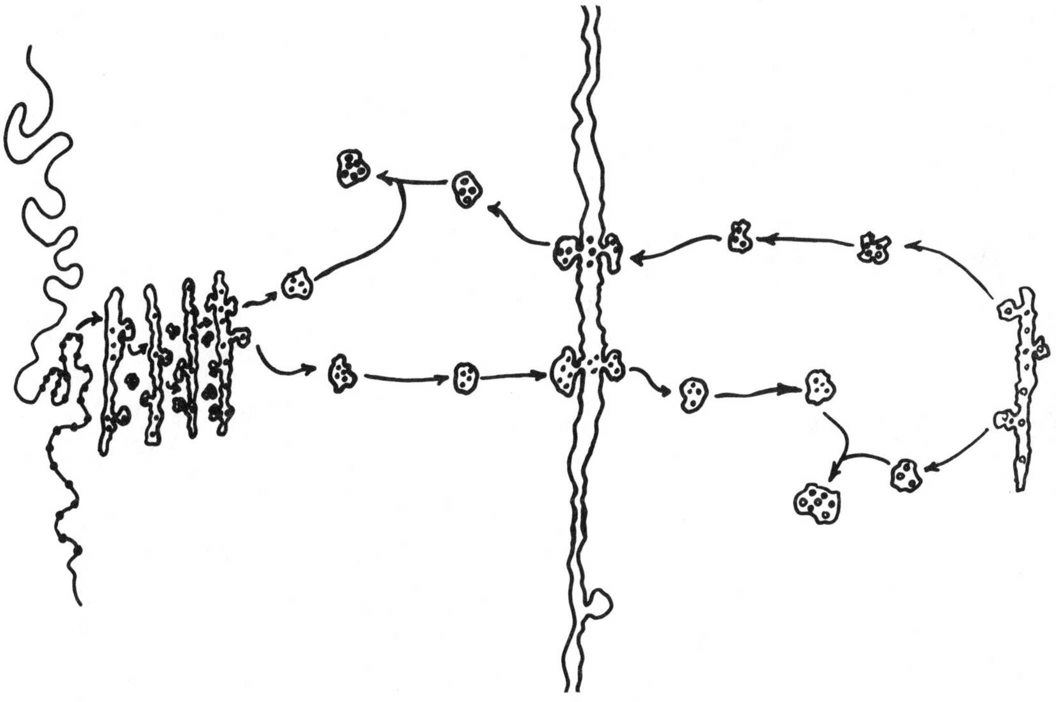

Fig. 2-36   Intercellular exchange of lysosomal enzymes. These enzymes are targeted to the endoplasmic reticulum (left), where a high mannose oligosaccharide is appended to certain arginine residues. These oligosaccharides are phosphorylated in the cis Golgi region and N-acetylglucosamine residues are added in the medial compartment. Transport between cisternae of the Golgi is accomplished by means of small vesicles. Enzymes containing phosphorylated high mannose oligosaccharides are either incorporated in lysosomes that bud from the trans Golgi or are exported from the cell. The exported enzymes are taken up by a receptor-mediated mechanism by the neighboring cell (right). Uptake by both lysosomes and neighboring cells is dependent upon the presence of an exposed mannose-6-phosphate on the N-linked oligosaccharide. Once the enzymes have been internalized by the neighboring cell, they are enclosed in vesicles that fuse with lysosomes already present, producing a secondary lysosome containing a mixture of enzymes derived from the host cell and those that were internalized by receptor-mediated endocytosis.

Proteins may also be trimmed (limited proteolysis) to smaller sizes during their movement through the Golgi apparatus, once they reach their final destination, or both. Hexosaminidase A is a lysosomal enzyme that contains two different types of subunit: α and β. The α and β subunits are encoded by different genes: **HEXA** and **HEXB**. Both of these genes have very similar structures with 14 exons and 13 introns that span about 35-40kb of DNA. The mRNAs from each gene are translated into a polypeptide containing more than 500 amino acids, and these proteins are inserted into the ER lumen where the signal peptide is cleaved from the molecule. While in the ER lumen, the remainders of the α and β polypeptides form a heterodimer ($\alpha\beta$ = pre-hexosaminidase A) and a homodimer ($\beta\beta$ = pre-hexosaminidase B). The present discussion will be confined to the processing of pre-hexosaminidase A. The heterodimer is N-glycosylated at sites on both the α and β polypeptides. The respective oligosaccharides are phosphorylated and N-acetylglucosamine is added during the passage of the pre-hexosaminidase A through the Golgi apparatus. Pre-hexosaminidase A enters the lysosome by the receptor-mediated process and is subsequently "trapped" within the lysosome by two events:
limited proteolysis and partial degradation of the oligosaccharides. The α polypeptide is cut internally and near the C-terminus, losing 66 amino acids and producing two fragments that are connected by disulfide bonds (Fig. 2-37). Similarly, the β polypeptide is cleaved internally at two locations and near the C-terminus generating three fragments. These five subunits are assembled into the mature active

47

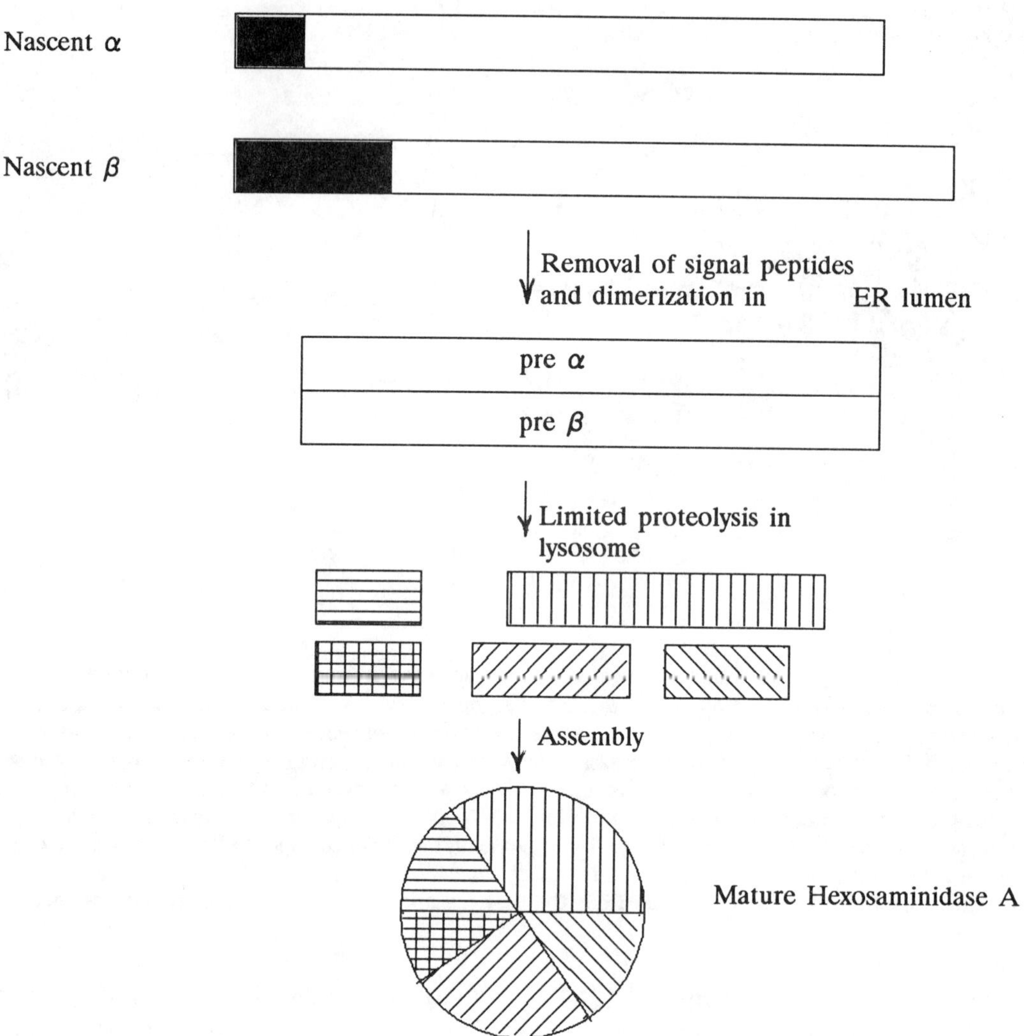

Fig. 2-37. Proteolytic processing of the *α* and *β* polypeptides of Hexosaminidase A. The *α* and *β* mRNAs are encoded by different genes, HEXA and HEXB, respectively. The nascent polypeptides enter the ER lumen, and their signal peptides are removed. The prepolypeptides associate, forming a heterodimer. Certain arginyl residues of both polypeptides are glycosylated with a high mannose oligosaccharide, and the heterodimers enter the Golgi apparatus. Mannose phosphorylation occurs in the cis Golgi, and N-acetylglucosamine moieties are added in the medial compartment. The glycosylated heterodimer enters the lysosome by a receptor-mediated mechanism. The pre *α* polypeptide is cleaved internally and near the C-terminus, forming two fragments. The pre *β* polypeptide is cut at two internal sites and near the N- and C-termini to form three segments. The *α* and *β* segments assemble and are linked by disulfide bonds to form mature Hexosaminidase A. This enzyme is entrapped within the lysosome by partial degradation of its oligosaccharide side chains.

form of hexosaminidase A. An overview of the expression of lysosomal enzymes is illustrated in Fig. 2-38. Note how the expression of a lysosomal enzyme is not only subject to mutations directly affecting its transcriptional unit, but also those that affect expression of the mannose-6-phosphate receptor, Golgi

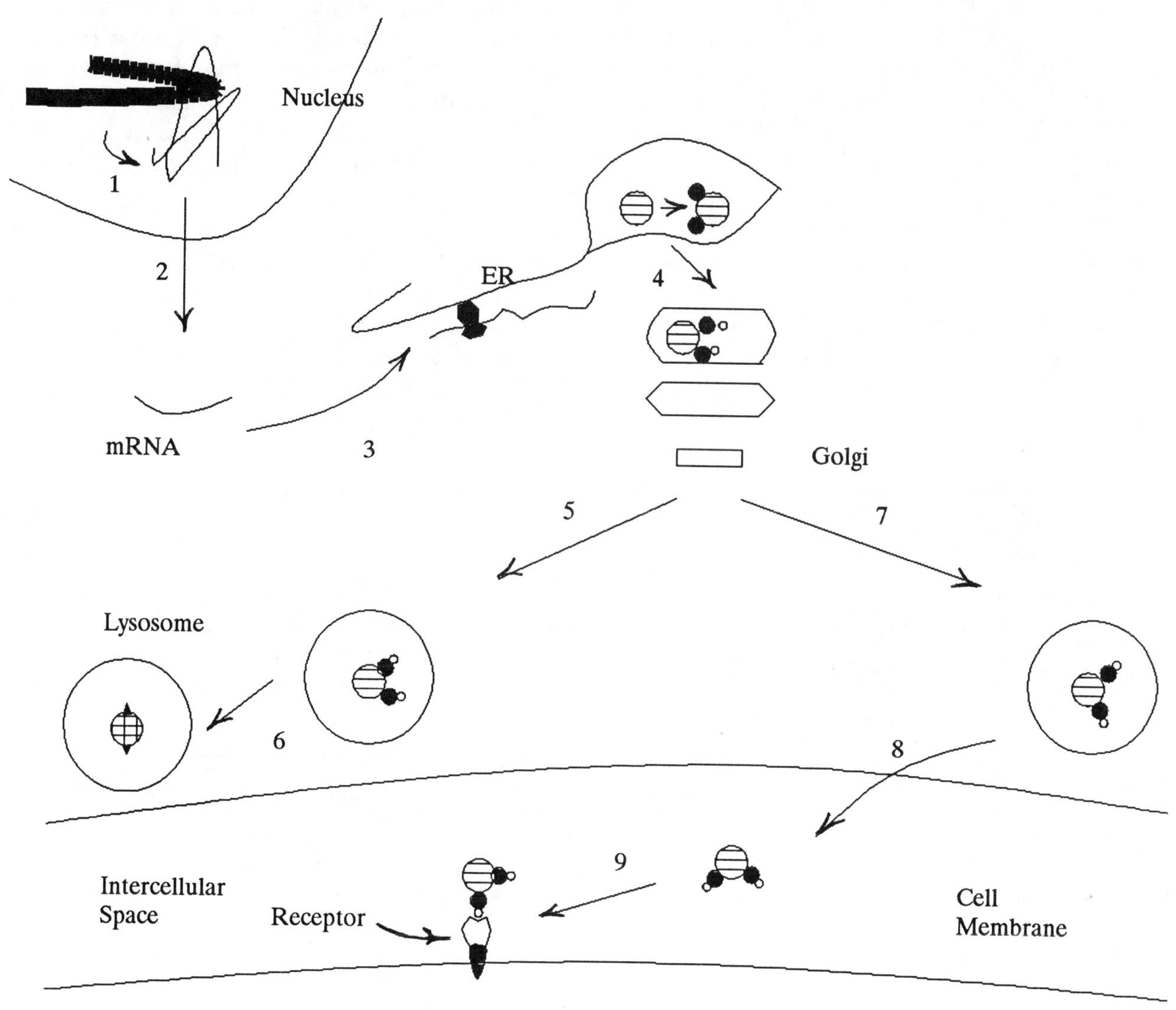

Fig. 2-38. Processing of lysosomal enzymes. 1) Transcription; 2) Processing of hnRNA to form mRNA; 3) Translation of mRNA, insertion of polypeptide into ER lumen, and cleavage of signal peptide; 4) Folding of polypeptide and addition of high mannose oligosaccharides and transport to Golgi apparatus where phosphorylation of mannose and further glycosylation occurs; 5) Transport to lysosome; 6) Entrapment by limited proteolysis and modification of oligosaccharides; 7) Transport to secretory vesicle; 8) Secretion into intercellular space; 9) Receptor-mediated endocytosis into neighboring cell for eventual inclusion in lysosomes.

processing enzymes, and lysosomal processing enzymes.

Several products derived from one gene by post-translational modification. A number of systems have been described where a single gene specifies a nascent polypeptide that is subsequently processed into two or more functional products. Several peptide hormones are derived from larger polypeptide precursors. The precursors appear to have a sequence of five to seven amino acids located at the sites of cleavage. A pair of basic amino acids is present in the core of this sequence: Arg-Arg, Lys-Arg, or Lys-Lys. The sequence appears to generate a β-turn at the surface of the precursor which is recognized in a specific manner by processing enzymes called maturases. Pro-opiomelanocortin is determined by a gene expressed in the pituitary gland. The processing of pro-opiomelanocortin varies among different regions of the pituitary (Fig. 2-39). Processing of the prohormone in the anterior pituitary favors the production of β-LPH and ACTH, while α-MSH and β-endorphin are preferentially produced in the intermediate lobe of the pituitary. One possible explanation for the differential processing pattern may be that the processing enzymes recognizing the appropriate surface features at the sites of cleavage may occur in one region of the pituitary but not another. ACTH is an important hormone that stimulates increased production of steroid hormones by the adrenal cortex. β-endorphin acts as an analgesic, diminishing sensitivity to or appreciation of pain. Data derived from animal experiments suggest that the various forms of MSH influence memory.

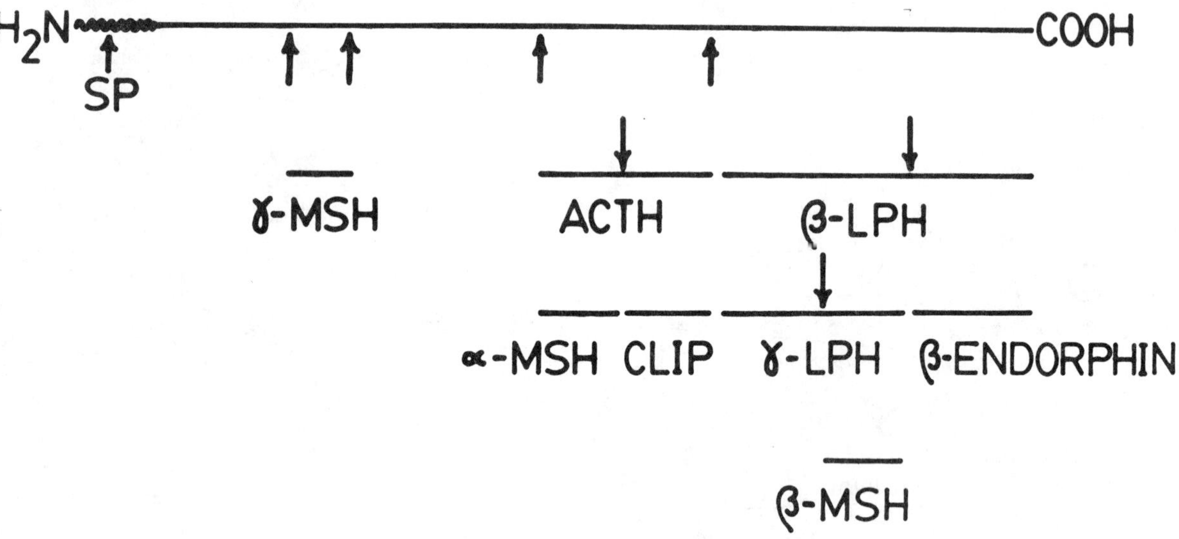

Fig. 2-39. Processing of pro-opiomelanocortin. Arrows indicate the sites of cleavage. SP = signal peptide; MSH = melanocyte-stimulating hormone; LPH = lipotropin; ACTH = adrenocorticotropic hormone; CLIP = corticotropin-like intermediate-lobe peptide. (Modified from Krieger and Martin (9)).

Protein Clearance. Protein clearance refers to the disappearance of a protein from the cell as a consequence of either degradation or secretion. Proteins that have "destabilizing" amino acids near their N-terminal ends tend to have short half-lives (time necessary for half of the protein molecules to disappear from the cell). These destabilizing amino acids (including His, Arg, Lys) "tag" a protein for early destruction by the **ubiquitin** pathway (Fig. 2-40). Ubiquitin is a small protein that occurs in both the nucleus and cytosol which appears to have a number of different functions. One is the "conditioning" of a protein for degradation. An ubiquinating enzyme complex binds to the N-terminus of a protein containing a destabilizing amino acid and couples several ubiquitin molecules in series to the ϵ amino group of a nearby lysine, and then disengages from the protein. The ubiquitin chain attracts the attention of a large protease which attacks the protein, degrading it to its constituent amino acids and releasing the ubiquitin molecules for reuse. Cells contain a large 19-20S particle called a prosome or proteosome. This particle consists of a complex of proteases that degrade cellular proteins and may include the large protease that degrades ubiquinated proteins. Two other mechanisms are known which target proteins for degradation. Phosphorylation of certain amino acids in proteins renders them unstable. Furthermore, multimeric proteins may contain arginine residues near contact points between subunits. Deamidation of these arginines changes the conformation of the proteins, breaking them apart and targeting them for degradation. By contrast, stabilization of proteins is often accompanied by acetylation of their amino acids.

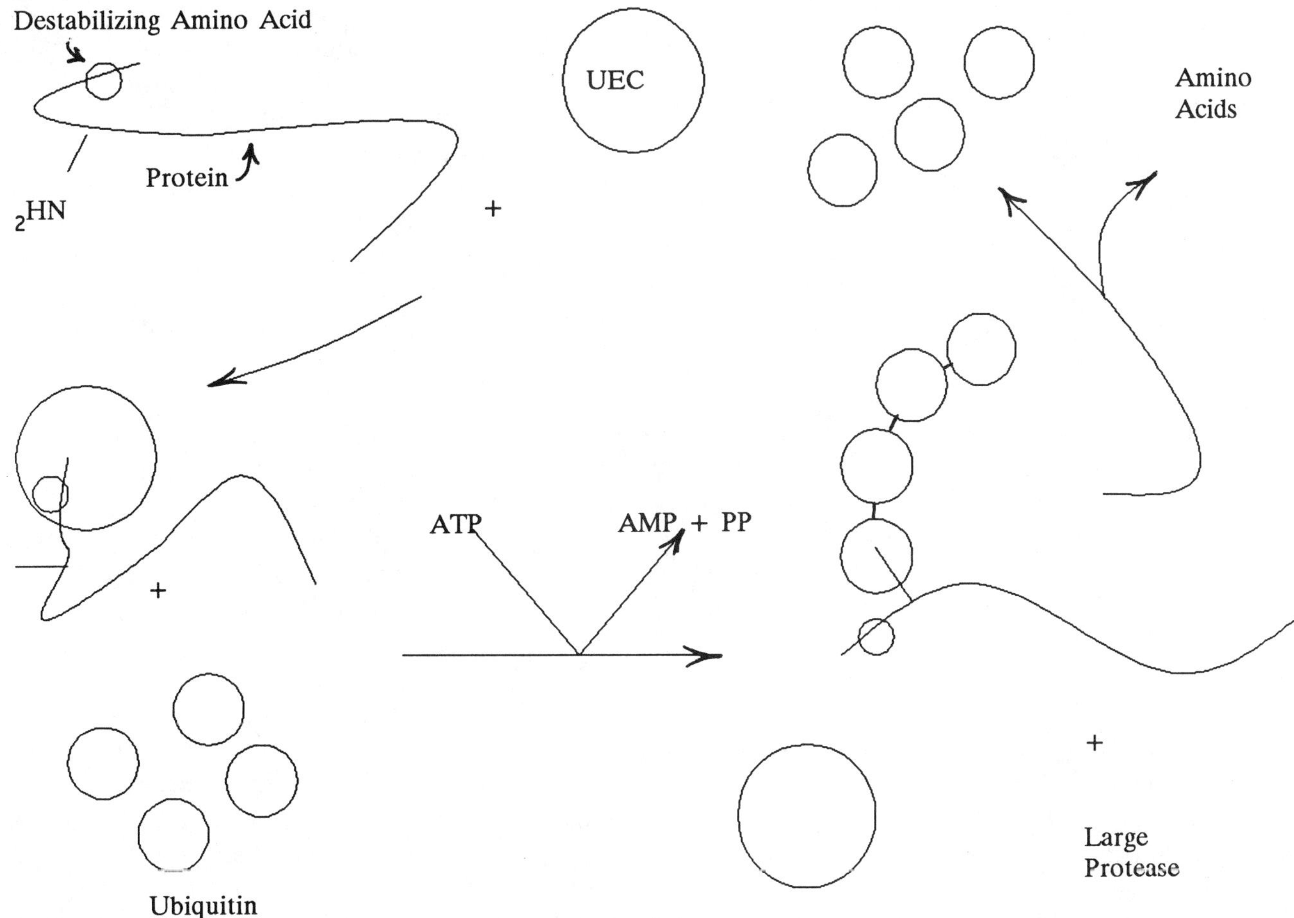

Fig. 2-40. Ubiquitin pathway for protein degradation. Proteins containing a destabilizing amino acid near their N-termini are bound by a ubiquinating enzyme complex (UEC). This complex adds several ubiquitin molecules to the ϵ amino group of a nearby lysine and disengages. The ubiquinated protein is attacked by a large protease that degrades the protein to amino acids and releases the ubiquitins for reuse.

Protein realization. The level of functional protein in a cell at a given time (protein realization) is subject to influences at a variety of levels (Fig. 2-41). The interconversion of a gene between active and inactive transcriptional states is affected by both higher order chromosome structure and proteins that bind to its promoter elements. The hnRNA transcribed from the gene may be degraded to nucleotides or processed into a functional mRNA. The mRNA may be degraded or translated into a polypeptide. This polypeptide may be processed to a functional form or prematurely destroyed by proteases. The functional protein may be secreted, targeted to an intracellular location,

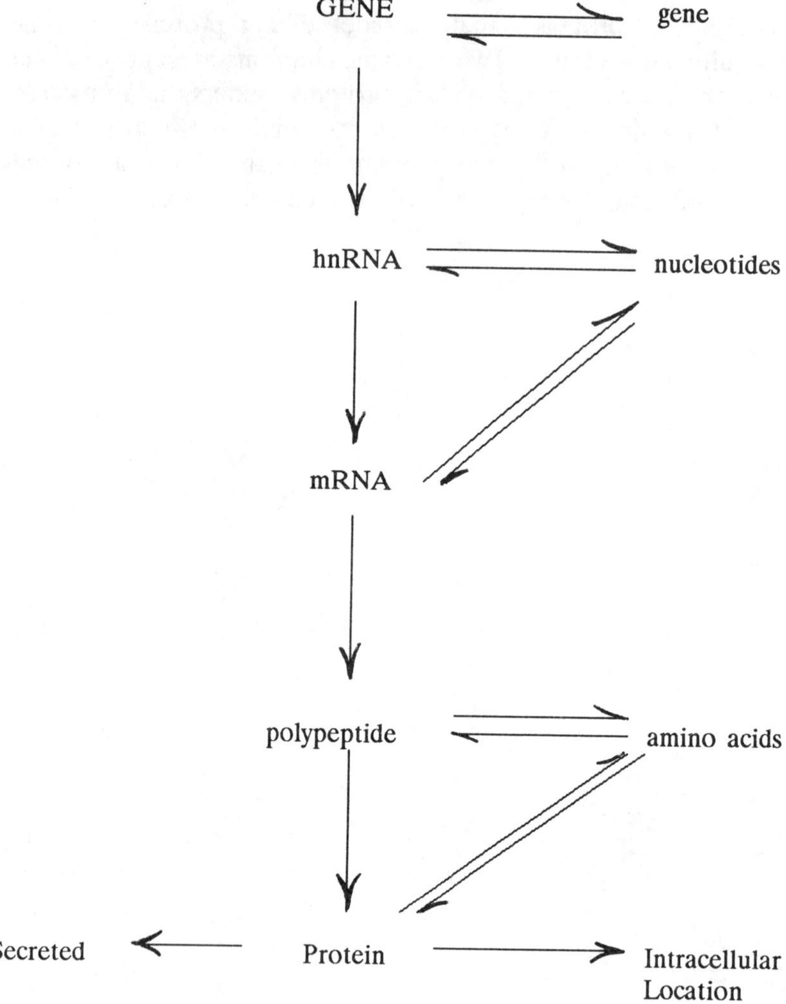

Fig. 2-41. Protein realization. The figure illustrates the many steps in gene expression between transcription and the final product. Factors acting at any one or a combination of points along this path can profoundly affect the level of the protein encoded by the gene.

or degraded. The sum of the balances between these processes determines the intracellular level of the protein.

MUTATION

Genes are subject to rare changes in gene structure. **Mutation** is defined as a heritable change in gene structure. A mutation may affect any site within the transcriptional unit. A common mutation replaces one base with another (Fig. 2-42). Suppose the second base of the exon triplet, **AAC**, is replaced by a **G**. This changes the sense of the triplet such that the mRNA codon **UUG** which encoded leucine is replaced by **UCG**. The protein determined by that gene will have the normal leucine at that position replaced by serine. Serine and leucine have quite different properties, and the protein containing the serine will most likely function less efficiently than the normal protein.

Alternatively, the second base may be replaced by T. This alteration generates a termination codon in the mRNA at that site. This premature terminator will result in the production of an unusually short polypeptide that will have abnormal properties and most likely will be rapidly degraded. A base substitution at the third position of the DNA triplet results in the formation of a synonymous codon and no amino acid substitution in the protein. This would prove to be a silent mutation and would only be detected by sequencing of the gene or its mRNA. This latter base substitution illustrates the protective effect of the degeneracy of the genetic code.

Mutations in non-coding regions of the transcriptional unit can profoundly affect its level of expression. Alterations of elements in the distal promoter may disturb the tissue pattern of expression, the timing of expression, or the response of the gene to hormones. Changes in proximal promoter regions may permanently silence the gene or lead to an inappropriately high level of expression. Mutations involving splice junctions prevent removal of the intron or uncover cryptic splice sites, resulting in unusual proteins. If splicing does not occur, the hnRNA is usually degraded and does not exit the nucleus. Mutations converting the translational stop signal to a triplet encoding an amino acid causes the synthesis of unusually large proteins that are rapidly degraded. Mutations in the AATAA sequence

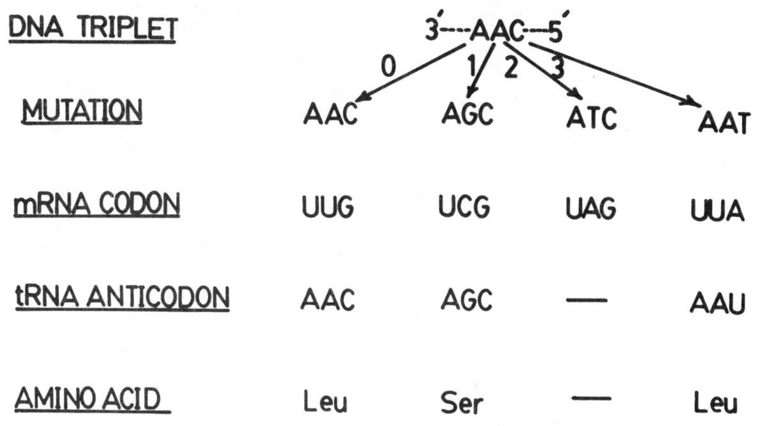

Fig. 2-42. Effects of base substitutions upon protein structure. Mutations affecting the first two bases tend to be more harmful than those changing the third base of the triplet.

result in runon transcription, generating a large hnRNA that is not properly polyadenylated. These hnRNAs are rapidly degraded. Examples of these mutations will be provided in a later chapter in relation to thalassemia, an inherited blood disorder.

GENE REGULATION

This section will focus upon factors that regulate the expression of genes at the transcriptional level; however, as mentioned previously, gene expression may be influenced by factors acting at many levels. Hormones may increase or decrease gene expression by acting at the transcriptional level and one or more post-transcriptional points. Examples of these systems will be provided where appropriate.

Regulation of transcription appears to involve both specific recognition sites in the DNA (**response elements**) and DNA-binding proteins. Response elements are most commonly located in the promoters of protein-encoding genes; however, they may occasionally be found within introns or among downstream sequences. DNA response elements are generally relatively short sequences of fewer than 30 base pairs. Examples include enhancers, hormone response elements, tissue-specific regulatory sites, and metalloprotein-binding sites.

DNA-binding proteins may be present in all cell types, or may display a tissue- or developmental stage-specific pattern of distribution. These proteins appear to have two general types of interactive sites. One or more domains of the DNA binding protein interacts directly with some aspect of the DNA, while other domains are essential for important interactions among two-or more DNA-binding proteins.

Families of DNA-binding proteins. Many DNA-binding proteins have been isolated and sequenced. These appear to fall into groups of structurally related proteins, called protein families, that share particular types of domains.

Helix-turn-helix motif. One large group of DNA-binding proteins that is widely distributed among prokaryotes and eukaryotes possesses a characteristic structure involving two α-helical domains separated by a loop (Fig. 2-43). The helical regions bind in the major groove of the DNA double helix through a variety of contacts between amino acids in the protein and bases within the groove and between the amino acids and the phosphate backbones of the double helix. Together these contacts impart sufficient specificity to direct the binding proteins to the proper sites near the gene being regulated. Examples of human DNA-binding proteins that contain the helix-turn-helix motif include protein E12, an enhancer-binding protein, and MyoD, a regulatory protein involved with muscle differentiation. These regulatory proteins are believed to interact, such that contacts between the enhancer protein and MyoD bring the enhancer element into close proximity of the gene being regulated, increasing the frequency of transcription of the gene controlled by the promoter.

Zinc finger motif. Another type of protein structure that is found in certain DNA-binding

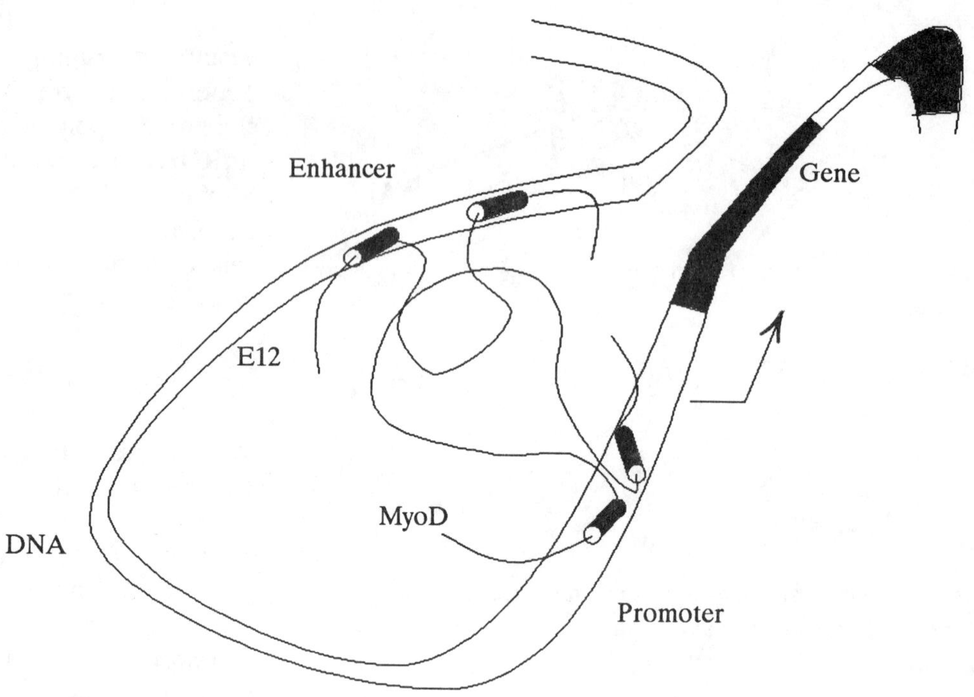

Fig. 2-43. **Helix-turn-Helix regulatory proteins.** E12, an enhancer-binding protein, and MyoD, a regulatory protein involved with muscle differentiation, appear to interact to bring the enhancer element close to the promoter. This increases expression of the gene controlled by that promoter. Amino acids within the helical domains interact with both base pairs in the major groove of the DNA and the sugar-phosphate backbones of the DNA double helix to bind specifically with the enhancer and promoter elements, respectively.

proteins is the zinc finger (Fig. 2-44). The zinc finger is a loop of amino acids that is formed by a complex including a zinc ion bonded to either two cysteine and two histidine residues or four cysteines (glucocorticoid receptor; Fig. 2-44). The glucocorticoid receptor is a globular protein consisting of two identical subunits. Each subunit possesses two zinc fingers and two α-helical regions. One α-helix of each subunit and its adjacent zinc finger constitute a DNA-binding domain that lodges in the major groove of the DNA. The other α-helix and associated zinc finger appear to interact with the corresponding α-helix and zinc finger of the partner subunit to form the dimeric receptor complex. Therefore, each glucocorticoid receptor contains two DNA-binding domains that bind to the glucocorticoid response element (GRE) in alternate turns of the DNA double helix. The GRE has the consensus sequence: 5'-NNGGTACANNNTGTTCTNN-3', where N is any base. Note that the element consists of two hexamers separated by three bases. The hexamers or "half-sites" appear to be located in alternate turns of the DNA double helix. The two DNA-binding domains of the glucocorticoid receptor interact with the two glucocorticoid response element half sites by contacting two AT base pairs at the center of each half-site. Not only do amino acids in the DNA-binding region connect with these AT base pairs, but adjacent amino acids in the DNA-binding region also interact with nearby bases that occur in the GRE but not other hormone response elements. This selectivity permits regulation of the appropriate genes by the receptor and its attached hormone.

   Leucine zipper motif. A third structural motif found in many DNA-binding proteins provides an important means for linking DNA-binding proteins that cooperate to regulate gene transcription. Leucine zippers are formed by sequences of about five leucine residues that interdigitate with the corresponding sequence of the second regulatory protein of the complex. Each protein has a basic amino

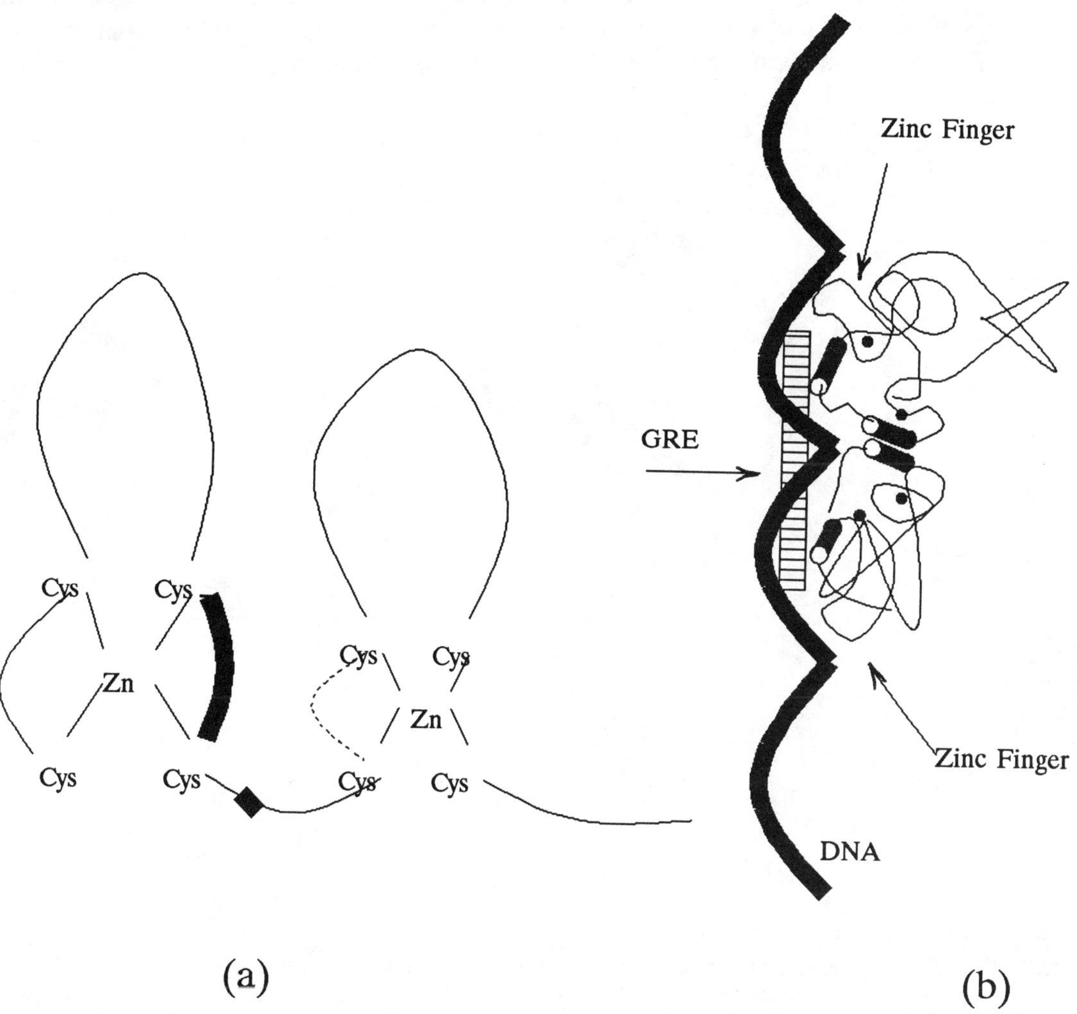

Fig. 2-44. Glucocorticoid receptor. The glucocorticoid receptor contains two zinc fingers per subunit (A). The thick lines indicate amino acids participating in response element discrimination, and the dashed line represents amino acids that are located at key intersubunit contacts. b) The glucocorticoid receptor is a homodimer. One zinc finger and adjacent α-helix of each subunit enter the major groove of alternate turns of the DNA double helix and bind with base pairs in the glucocorticoid response element (GRE). The remaining α-helix and zinc finger of each subunit are situated between the subunits where they particpate in protein-protein interactions.

acid domain near the leucine zipper which makes contact with a response element in the DNA (Fig. 2-45). The proto-oncogenes c-fos and c-jun encode DNA-binding proteins that regulate transcription of genes possessing an AP-1 response element in their promoter regions. Osteocalcin is a bone protein produced by osteoblasts (bone forming cells). Osteocalcin participates in the mineralization of bone, and its synthesis is subject to enhancement by both Vitamin D and Vitamin A metabolites. The osteocalcin promoter has both an AP-1 response element and a vitamin response element closely associated within its promoter region. Sequencing studies have demonstarted that the vitamin response element for both retinoic acid (Vitamin A metabolite) and Vitamin D_3 contains the AP-1 site in the center of the element.

When the c-fos/c-jun dimer is bound to this AP-1 site, the synthesis of osteocalcin is shut off, and the gene does not respond to either of the two vitamins. Therefore, the c-fos/c-jun heterodimer functions as a negative regulator in this system, preventing bone development and remodelling. The retinoic acid

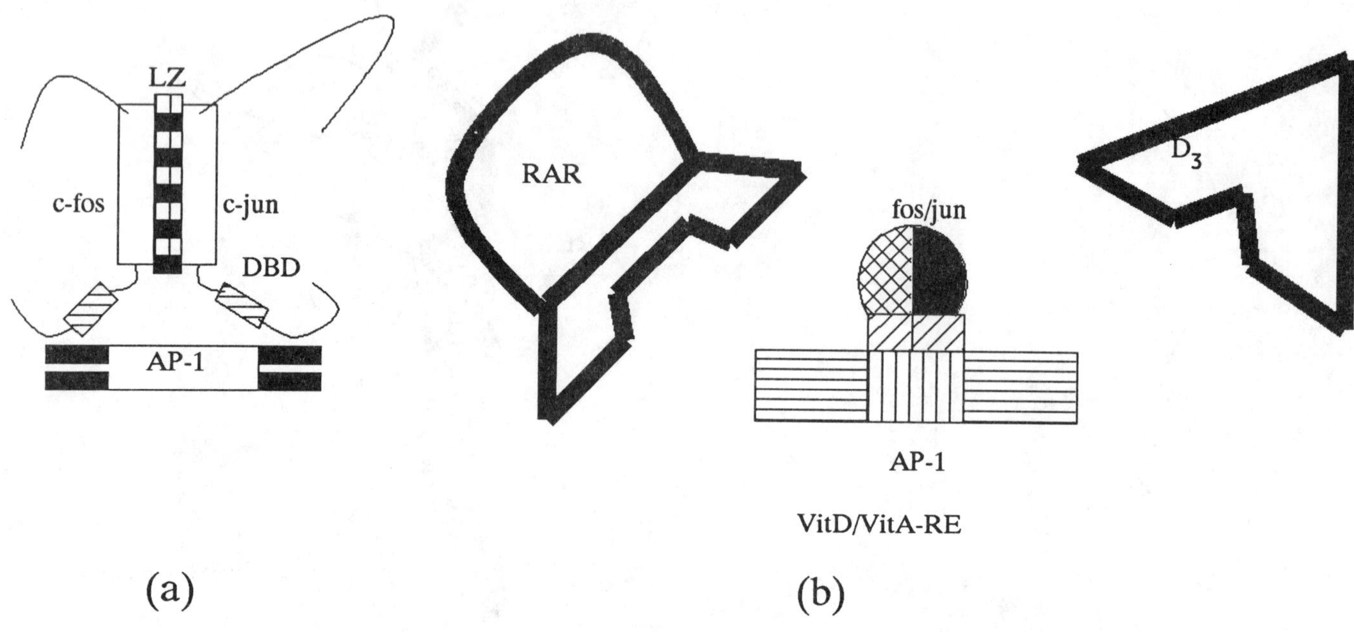

Fig. 2-45. Leucine zippers. a) The regulatory proteins, c-fos and c-jun, contain a run of five leucines which interdigitate to form a "leucine zipper" (LZ). This protein-protein interaction domain is near a domain containing basic amino acids (DBD = DNA-binding domain) and binds to the a consensus sequence (TGACTCA) called the AP-1 response element. b) The osteocalcin gene promoter contains an AP-1 site nested within the vitamin D/Vitamin A response element (VitD/VitA-RE). When the c-fos/c-jun heterodimer occupies the AP-1 site, the retinoic acid receptor (RAR) and the Vitamin D$_3$ receptor cannot bind to the VitD/VitA-RE, and the osteocalcin gene is silenced and will not respond to either of the vitamin-derived hormones.

receptor and Vitamin D$_3$ receptors are shown as separate molecules in Figure 2-45. Several other regulatory protein families have been described whose members share special DNA-binding or protein-protein interaction domains. It is likely that the number of motifs and protein families will continue to increase as research progresses in this highly active field.

       Mechanisms of hormone amplification of gene expression. Steroid hormones are important regulators of gene expression in human cells. These hormones accomplish their effects through similar pathways and may increase gene expression at both transcriptional and post-transcriptional levels. For example, testosterone and its metabolite, dihydrotestosterone, bind to the same cytoplasmic receptor (Fig. 2-46). The binding of the hormone to the receptor releases it from an inactive state. The hormone-receptor complex may translocate to the nucleus and bind to androgen response elements, amplifying transcription from the adjacent structural gene. Lin and Ohno (10) have demonstrated that androgen receptor also has a high affinity for mRNA, raising the possibility that the hormone-androgen receptor may also influence gene expression at the mRNA level.

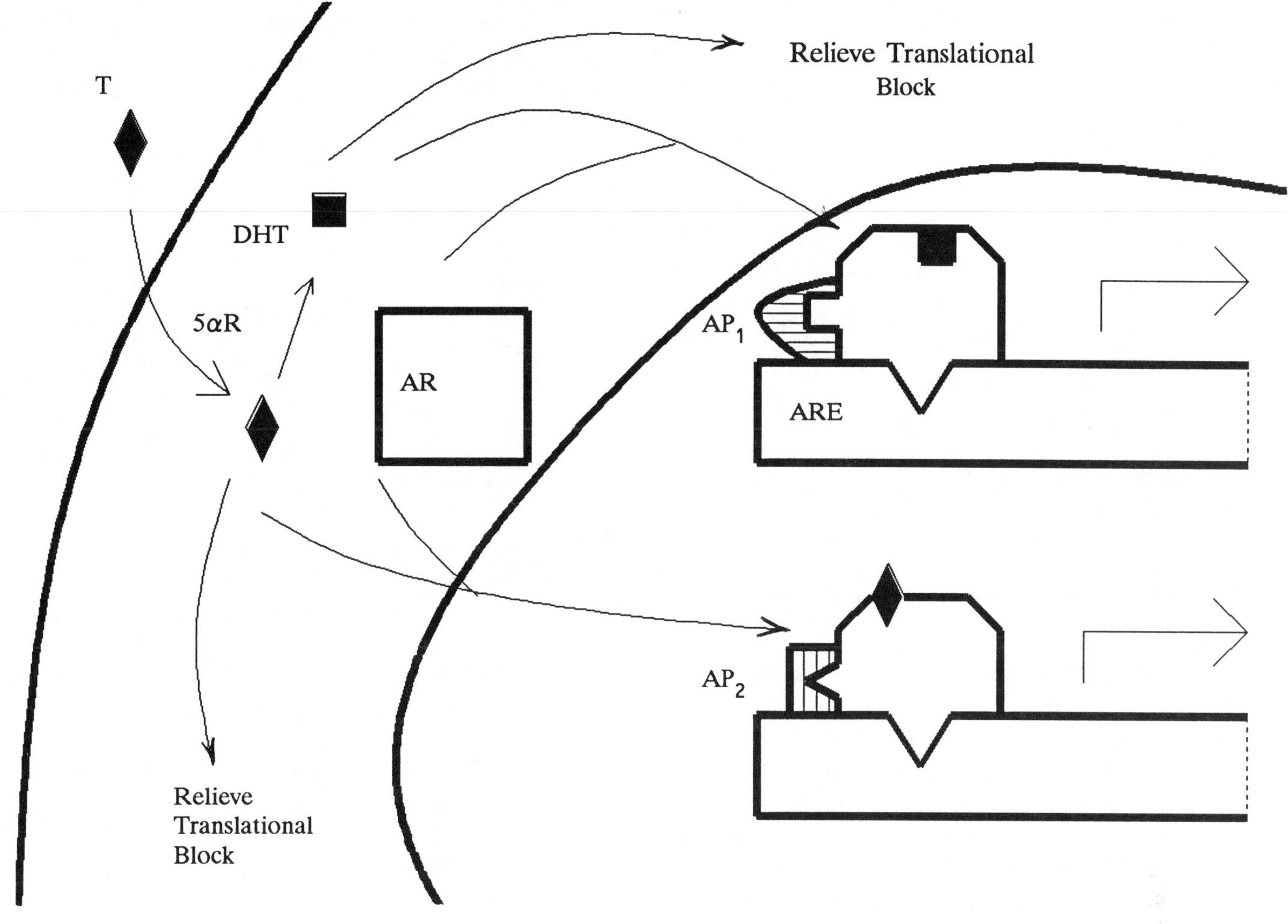

Fig. 2-46. Model for induction of gene expression by androgens. Testosterone (T) penetrates the cell and enters one of three pathways. Testosterone may be converted by 5-α-reductase (5αR) to dihydrotestosterone (DHT). DHT may relieve a translational block, increasing translation of certain mRNAs. DHT may bind to the androgen receptor (AR), activating the receptor which enters the nucleus and interacts with both the androgen response element (ARE) and an accessory protein AP$_1$. Transcription of the adjacent gene is subsequently amplified. The DHT-AR complex may also be the active factor involved with enhanced translation of existing mRNA in the cytosol. Unmodified testosterone may also relieve translational blocks of certain mRNAs, either directly or in combination with the AR. Recent evidence suggests that the latter may be the active factor. The T-AR complex may translocate to the nucleus and bind to an ARE and a different accessory protein (AP$_2$), activating the adjacent gene. Note that the hormone response element for both androgen-AR complexes is the same, and that this model suggests that the different arrays of genes activated by the two hormones are determined in part by protein-protein interactions.

Other steroid hormones also influence genes through a receptor-mediated mechanism. Estrogen receptors are intranuclear, and the hormone enters the nucleus where it activates the receptor, facilitating its binding with estrogen receptor elements within the promoters of estrogen-responsive genes. The hormone response elements for steroid hormones possess consensus sequences that indicate that these response elements can be grouped into families (Table 2-2). The glucocorticoid receptor (up regulation), androgen receptor, and progesterone receptor share a common consensus response element. The differential effects of these hormones upon the expression of various genes must be determined by

Table 2-2. Consensus Hormone Response Elements.

Hormone	Response Element
Androgen	GGTACANNNTGTTCT
Progesterone	.
Glucocorticoid	.
(up reg.)	
Estrogen	AGGTCANNNTGACCT
Ecdysone	AGGGTTNNNTGCACT
Thyroid Hormones	TCAGGTCA-TGACCTGA
Retinoic Acid	.
Glucocorticoid	ATYACNNNNTGATCW
(down reg.)	

Modified from Beato (11).

bases surrounding the consensus sequence and by interactions of the receptors with other DNA-binding proteins attached to the genes' promoters.

Steroid hormone receptors also appear to share a number of structural features. The N-terminus of each receptor has been shown by mutation experiments to be essential for full biological activity of the receptor and displays extensive amino acid variation from one receptor to another. This "modulator" region ranges from 50 to 500 amino acids in length and, in the case of the estrogen receptor, has also been demonstrated to play a role in the cell specificity of receptor action. The DNA-binding region tends to be more highly conserved among different receptors and contains zinc finger sequences. The hormone binding domain is located at or near the C-terminus of steroid receptors. Steroid receptors are dimeric and bind to their response elements in a manner that is similar to the one illustrated for the glucocorticoid receptor (Fig. 2-44). Genes for hormone receptors are often much larger than the sequences necessary for specification of receptor sequences. For example, the human androgen receptor is more than 90kb in size, and its coding sequences account for about 3%. The human androgen receptor gene contains 8 exons. The first encodes the modulator region, and exons 2 and 3 encode the DNA-binding domain. The last five exons determine the hormone-binding domain of the receptor. At least four reports have appeared describing mutations in the human androgen receptor gene that result in androgen insensitivity and partial to complete feminization of the 46,XY individuals possessing the mutations. All of these mutations have affected the hormone-binding domain; however, more are anticipated where other functional domains of the gene may be involved.

SUMMARY

Genes are the fundamental units of heredity and consist of DNA. DNA contains the bases adenine, guanine, cytosine, and thymine, the sugar deoxyribose, and phosphate. DNA occurs as a double strand that is coiled into a helical shape. Replication or reproduction of the double helix requires relaxation and separation of the strands. Each parental strand serves as the template for a complementary daughter strand whose base sequence is ordered by a DNA polymerase using the base sequence of the parental strand as a guide. Base pairing is governed by the rules: adenine pairs with thymine and guanine pairs with cytosine. Replication is semiconservative. Each daughter double helix consists of an intact parental strand and an intact new chain of DNA.

Recently developed analytical methods have permitted study of the fine structure of human genes. Genes are generally much larger than the sizes required to specify the amino acid sequences of the proteins they encode. Most genes appear to be interrupted by noncoding sequences called introns. Junctions between introns and the coding sequences (exons) they separate are highly conserved among different genes and serve as signals for a splicing complex that processes the hnRNA into functional mRNA. Genes also have extensive promoter regions that include transcriptional signals determining the amplitude, timing, and cell pattern of gene expression. Promoters of specialized genes contain two proximal elements (TATA and CCAAT) that are missing from promoters of housekeeping genes. Housekeeping genes and some specialized genes have CpG islands associated with their promoters. Differential methylation of these CpG islands often distinguishes active or transcriptionally silent genes. The cap site, a sequence that specifies a methylated base in the RNA transcribed from the gene, is the first nucleotide appearing in the mRNA. This RNA modification facilitates initiation of translation and protects the message from nuclease digestion from its 5' end. The first TAC triplet specifies an AUG codon in the mRNA which serves as an instruction to place methionine at the N-terminal end of the nascent polypeptide. This amino acid is removed from the polypeptide during an early phase of its synthesis. Downstream sequences that influence gene expression include a translational stop triplet at the 3' end of the last exon and the AATAA sequence which specifies the site of termination of transcription and polyadenylation of the mRNA.

Transcription is the process by which RNA is formed from a DNA template. RNA is composed of the sugar ribose, phosphate, and the four bases: uracil, guanine, adenine, and cytosine. Uracil pairs like thymine. Only one DNA strand (master or coding strand) serves as the template for RNA synthesis. The initial RNA product, hnRNA, is capped, spliced, and polyadenylated to form the mRNA. These alterations are collectively referred to as RNA processing.

The base sequence of the mRNA is read in groups of three bases. A tRNA bearing the appropriate anticodon is paired by the ribosome with the right codon in the mRNA by following the RNA pairing rules. The codons of the mRNA are read continuously from the 5' to the 3' end of the mRNA, and the amino acids carried by the tRNAs are linked in succession by peptide bonds to form the nascent polypeptide. This polypeptide is then processed into a mature protein by a variety of post-translational modifications. Proteins are targeted for specific intra- and extra-cellular sites by a leader or signal sequence at their N-terminal ends. Once the proteins have reached their correct destination, the signal sequence is cleaved from the molecule. Other post-translational modifications include glycosylation, addition of lipids, and sizing of the polypeptide to the dimensions and composition of the active protein.

Proteins differ with respect to their half-lives within the cell. They may be secreted along well defined paths or targeted for degradation. Proteins having destabilizing amino acids near their N-terminus are ubiquinated and degraded by a large protease to their constituent amino acids. Other modifications that increase a protein's susceptibility to degradation include phosphorylation and deamidation.

Regulation of gene expression may occur at any one or a combination of points ranging from transcription to degradation. The level of functional protein in a cell at any one time is called protein realization. Transcriptional controls involve the interaction of DNA-binding proteins with DNA response elements in or near promoters and interactions with other DNA-binding proteins in their vicinity. Steroid hormone induction or repression of gene expression involves complexing of the hormone with a hormone receptor that subsequently binds with specific hormone response elements in the promoters of responsive genes. Hormone response elements can be grouped into families whose sequences are recognized by a small number of steroid hormone-receptor complexes. Steroid hormone receptors also display similar structural motifs. The general structure of these receptors includes a variable N-terminal sequence that influences the tissue specificity and biological response to the hormone, a DNA-binding region characterized by two or more zinc finger motifs, and a C-terminal hormone-binding domain.

PROBLEMS

1. A segment of a gene possesses the following bases: 3'-**AAGTCGAAA**-5'. Predict the amino acid sequence of the polypeptide encoded by this sequence of bases.

2-12. Match the following with their appropriate functions in the cell:

2. mRNA	a. chain terminators
3. cap structure	b. structural component of ribosomes
4. tRNA	c. contains degradative enzymes
5. ribosomes	d. initiator of protein synthesis
6. **UGA, UAA, UAG**	e. site of translation
7. hnRNA	f. site of posttranslational modification of proteins
8. rRNA	g. contains precursors of mRNA
9. methionyl-tRNA	h. aggregate of several ribosomes, mRNA and tRNAs
10. polysome	i. facilitates initiation of translation
11. lysosome	j. orders sequence of amino acids in proteins
12. Golgi apparatus	k. shuttles amino acids to ribosomes

13. Consider the N-terminal sequences of the following polypeptides:

$$\text{Normal: } H_2N\text{-Phe-Arg-Ser-}$$

$$\text{Mutant: } H_2N\text{-Phe-Glu-Arg-}$$

Diagram the event(s) (two or less) responsible for the generation of the mutant gene.

14. A protein occurs in both membrane-bound and membrane-free (soluble) forms. The two proteins differ at their C-terminal ends. The membrane-bound form has a short hydrophobic sequence, while its soluble counterpart has a hydrophilic sequence at the same location. Both proteins are known to be encoded by the same gene. Experiments have revealed that the two forms of the protein are derived from two different mRNA species which have different 3' termini. Provide an explanation for the origin of the two protein forms.

15. Human erythrocytes synthesize Hb A_1 ($\alpha_2\beta_2$) and Hb A_2 ($\alpha_2\delta_2$) in an approximate 97:3 ratio. Both forms of adult hemoglobin occur in all red cells, and neither is made from the other. Levels of α, β, and δ mRNA have similar half-lives and occur in the proportions expected on the basis of the prevalence of the globin chains they encode. How would you account for the observed levels of the two adult hemoglobins?

16. A patient has a severe disease that has been attributed to low levels of enzyme "A". The level of this enzyme is about 1% of that occurring in cells of unrelated persons lacking this disease. How would you account for the low levels of enzyme found in this patient's cells?

17. Diagram the mRNA sequence, tRNA anticodons, and amino acid sequence determined by the DNA base sequence: 3'-GCATTACCC-5'.

18. A patient has a severe anemia caused by low activities of red cell carbonic anhydrase, an enzyme
essential for normal red cell metabolism. An immunological test revealed that normal quantities
of enzyme protein are present. Furthermore, the enzyme protein is of normal size and has a normal
carbohydrate content. This patient most likely has a mutation in

a) the structural gene for carbonic anhydrase
b) a processing gene for carbonic anhydrase
c) a regulatory gene affecting the rate of synthesis of the enzyme
d) a regulatory gene influencing the rate of degradation of the enzyme
e) a gene controlling secretion of carbonic anhydrase by red cells

19. The non-α chains of human adult hemoglobins occur in close array on the same chromosome. An
unusal hemoglobin, Hb Lepore, contains a hybrid non-α chain: N-terminal half = N-terminal δ
sequence and C-terminal half = C-terminal β sequence. This fusion protein is believed to be encoded
by a gene that arose by a mispairing of the β and δ genes during meiotic division, breakage in the two
genes, exchange of their respective parts, and fusion of the segments to form the hybrid gene. What
does the order of the β and δ sequences in the fusion globin suggest regarding the order of the
respective genes on the chromosome?

20. NIH has smiled upon you and given you $100,000 to isolate and sequence the X-linked gene
encoding superzymase. How would you accomplish this task?

21. Cellulose acetate electrophoresis separates proteins on the basis of their surface charge. Three
phenotypes were observed in a family:

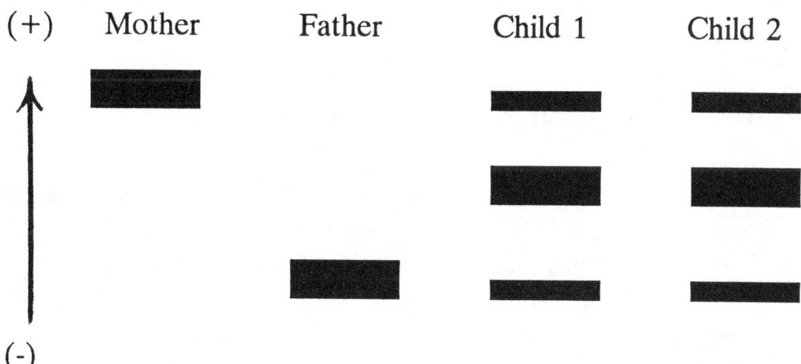

The mother's isozyme pattern is the standard for the population. What do the isozyme profiles of
these family members suggest regarding the structure of this enzyme?

22. Suggest an experiment that would demonstrate that the new protein you have isolated is a DNA-
binding protein that binds to the promoter for gene **A**. Hint: You have cloned the promoter by
inserting the promoter between the two EcoRI sites of the plasmid, pGEM, and have adequate

promoter DNA to test. Take advantage of two properties of DNA sequences complexed with their DNA-binding proteins.

GLOSSARY OF TERMS

Anticodon - triplet of bases in the tRNA that pairs with the appropriate codon of the mRNA.

Blotting - transfer of nucleic acids from a gel to another support, such as nitrocellulose, for the purpose of hybridization with labelled or fluorescent probes.
-Southern: DNA probed with cDNA.
-northern: RNA probed with cDNA.
-western: proteins probed with antibody against protein of interest.

Cap - array of bases, one of which is methylated, at the 5' terminus of the mRNA. The cap facilitates initiation of translation.

cDNA - complementary DNA synthesized using a mRNA template using reverse transcriptase.

Chromosome - linear array of genes and their associated proteins. Each chromosome consists of one continuous DNA double helix.

Cloning - method employed for isolating and amplifying a gene.

Cloning vector - modified virus or plasmid employed for packaging the DNA to be cloned. These vectors replicate independently from the host chromosome.

Codon - triplet of bases in the mRNA that specifies the amino acid to be placed at the corresponding site in the polypeptide.

Conformation - the three-dimensional structure of a molecule.

Cosmid - cloning vector with lambda cohesive ends; can be used for packaging larger DNA molecules.

CpG island - area within or near promoters that influences transcription. Methylated CpG islands are associated with transcriptionally silent genes.

DNA - deoxyribonucleic acid; contains adenine, guanine, cytosine, uracil, deoxyribose, and phosphate.

DNA sequencing - method of determining the primary structure of DNA. The DNA is cut into fragments, and the fragments are separated by gel electrophoresis, producing a ladder. The ladder is read from bottom to top to decifer the base sequence.

Elongation factor - enzyme or other protein that participates in the growth of the polypeptide during translation.

Enhancer - DNA sequence that participates in amplification of gene transcription in a time- and cell-specific manner.

Exon - coding sequence of a gene.

Genetic code - sixty-four codons specifying the 20 amino acids found in newly synthesized proteins and three terminators.

Genome - all of the DNA in the nucleus, mitochondrion, or chloroplast of an organism.

Helix-turn-helix motif - structural region of a DNA-binding protein in which there is a loop separating two alpha-helical regions.

Hormone receptor - a protein that mediates the effects of hormones on DNA transcription.

Hormone response element - short sequence of DNA that provides a DNA-binding site within the promoter for the hormone receptor complex.

Housekeeping gene - a gene expressed at low levels in all tissues and at all times.

Hypersensitive site - region upstream from active or potentially active genes that have a relaxed structure and are susceptible to digestions by certain nucleases.

Initiation complex - complex containing Met-tRNA, mRNA, and small ribosomal subunit (40S) or these components plus the large ribosomal subunit (80S).

Initiation factors - proteins which assist the formation and operation of the initiation complexes.

Intron - noncoding sequence of a gene that is situated between two exons.

Leucine zipper motif - an interactive site consisting of a series of leucines which promotes dimerization of DNA-binding proteins.

Libraries - incorporation of sets of DNA in plasmids or modified viral vectors for future screening for genes of interest.
    genomic - library formed from the total DNA of the cell.
    cDNA - library formed from the expressed genes of a particular cell or tissue type.

Limited proteolysis - partial digestion of proteins by endoproteases and/or exoproteases, trimming them to active forms and/or entrapping them in organelles.

Mutation - a heritable change in the DNA.

Plasmid - a circular, self-replicating DNA molecule found in bacteria.

Polymerase chain reaction - a method for rapid amplification of gene sequences lying between two primers that flank the sequence of interest.

Polysome - an aggregate of several ribosomes translating different regions of the same mRNA.

Posttranslational modification - processing of a polypeptide following its synthesis.
  Capping - methylation at the 5' end of the mRNA which protects the 5' end from nuclease attack and facilitates initiation,
  Polyadenylation - the addition of about 200 adenyl residues to stabilize the 3' end of the mRNA.
  Splicing - removal of introns and joining of their flanking exons to make a continuous mRNA coding region.
  Spliceosome - the complex of proteins and RNAs performing the splicing reaction.

Promoter - Upstream region of the transcriptional unit or gene that controls its level and timing of expression.

Prosome (proteosome) - large complex of proteases which degrades proteins.

Protein realization - the level of functional protein present in a cell at a given time.

Protein targeting - mechanism for directing the movement of a protein to its appropriate intra- or extra-cellular site.

Purine - DNA or RNA base possessing two rings.

Pyrimidine - DNA or RNA base having one ring.

Recombinant DNA - DNA molecule artificially constructed from two or more DNA molecules.

Replication - the process forming two new DNA double helices using the old DNA as a template.
  Semiconservative - the new DNA molecules consist of one new strand and the old template strand of DNA.

Restriction endonuclease - an enzyme that cuts DNA at prescribed sites to form DNA fragments.

Reverse transcriptase - enzyme that synthesizes a cDNA using a mRNA template.

RNA - nucleic acid containing ribose, phosphate, adenine, guanine, uracil, and cytosine. Some RNAs possess additional bases such as inosine.
  hnRNA - heterogeneous nuclear RNA; volatile RNA pool that includes nascent transcripts from genes.
  mRNA - messenger RNA; processed transcript from protein encoding genes.
  rRNA - ribosomal RNA; RNAs contributing to structure and function of the ribosomes.
  snRNA - small nuclear RNA; these RNAs are found in the spliceosome and participate in RNA processing.

RNA editing - replacement of one or more bases in the RNA following transcription.

RNA processing - capping, splicing, and polyadenylation of the nascent RNA to form mature mRNA. Trimming of other RNA species to form functional RNAs (eg. rRNA precursor).

Transcription -the process by which the master or coding strand of the DNA is used to direct the formation of an RNA complementary to that DNA sequence.

Transcriptional unit - the sequence of DNA that includes both coding and noncoding sequences that participates in the regulation and expression of a gene and the determination of the sequence of the protein it encodes (promoter + exons + introns + downstream sites).

Transformation - incorporation of foreign DNA into a cell, usually by natural means, resulting in genetic modification of its function or appearance.

Translation - the conversion of the codon sequence of a mRNA into the amino acid sequence of a polypeptide.

Ubiquitin - a protein that assists with the degradation of proteins.

Zinc finger motif - a structure found in DNA-binding proteins that functions in recognition of the DNA-binding site and with interactions between subunits of the binding protein.

BIBLIOGRAPHY

1. Avery, O.T., C.M. MacLeod, and M. McCarty. 1944. Studies on the chemical nature of the substance inducing transformation of pneumococcal types. J. Exptl. Med. 79:137-156.

2. Hershey, A.D. and M. Chase. 1952. Independent functions of viral protein and nucleic acid in growth of bacteriophage. J. Gen. Physiol. 36:39-56.

3. Watson, J.D. and F.R.C. Crick. 1953a. A structure for deoxyribose nucleic acid. Nature 171:737-738.

4. Watson, J.D. and F.R.C. Crick. 1953b. Genetical implications of the structure of deoxyribonucleic acid. Nature 171:964-967.

5. Meselson, M. and F.W. Stahl. 1958. The replication of DNA in _Escherichia coli_. Proc. Natl. Acad. Sci. (USA) 47:671-682.

6. Nirenberg, M.W. and J.H. Matthaei. 1961. The dependence of cell-free protein synthesis in _E. coli_ upon naturally occurring or synthetic polyribonucleotides. Proc. Natl. Acad. Sci. (USA) 47:1588-1602.

7. Speyer, J.F., P. Langyel, C. Basilio, and S. Ochoa. 1962. Synthetic polynucleotides and the amino acid code. IV. Proc. Natl. Acad. Sci. (USA). 48:441-448.

8. Crick, F.R.C., L. Barnett, S. Brenner, and R.J. Watts-Tobin. 1961. General nature of the genetic code for proteins. Nature 192:1227-1232.

9. Krieger, D.T. and J.B. Martin. 1981. Brain peptides. N. Eng. J. Med. 304:876-885.

10. Lin, S. and S. Ohno. 1982. The interactions of androgen receptor with poly(A)-containing RNA and polyribonucleotides. Eur. J. Biochem. 124:283-287.

11. Beato, M. 1989. Gene regulation by steroid hormones. Cell 56:335-344.

ADDITIONAL LEARNING RESOURCES

1. Alberts, B.M. 1990. Recipes for replication. Nature 346:514-515.

2. Alberts, B., D. Bray, J. Lewis, M. Raff, K. Roberts, and J.D. Watson. <u>Molecular Biology of the Cell</u>, 2nd ed., New York, NY, Garland, 1989.

3. Beato, M. 1989. Gene regulation by steroid hormones. Cell 56:335-344.

4. Harris, R.B. 1989. Processing of pro-hormone precursor proteins. Arch. Biochem. Biophys. 275: 315-333.

5. Jones, N. 1990. Transcriptional regulation by dimerization: Two sides to an incestuous relationship. Cell 61:9-11.

6. Lewin, B. 1990. Commitment and activation at Pol II promoters: A tail of protein-protein interactions. Cell 61:1161-1164.

7. Pfanner, N. and W. Neupert. 1990. The mitochondrial import apparatus. Annu. Rev. Biochem. 59: 331-353.

8. Ptahsne, M. and A.A.F. Gann. 1990. Activators and targets. Nature 346:329-331.

9. Smith A.W.J., J.G. Patton, and B. Nadal-Ginard. 1989. Alternative splicing in the control of gene expression. Annu. Rev. Genet. 23:527-577.

Chapter 3

Chromosome Structure and Replication and the Cell Cycle

During the late nineteenth century, biologists described the process of cell division, noting that cell division was preceded by nuclear division. Shortly before this period, cytological stains were developed and utilized for studies of cell structures, and several researchers described structures in dividing cells which were rod-like and which had an affinity for these stains. The rod-like structures were called **chromosomes** (colored bodies). Sutton (1) and Boveri (2) independently noted the relationship between the behavior of chromosomes and inherited traits and proposed that the chromosomes were the bearers of the hereditary information. Development of new cytological methods permitted the discovery of 46 chromosomes in human cells (3), and the identification of the first chromosomal abnormality, Down syndrome, soon followed (4).

The average human cell contains about three billion base pairs of DNA and exceeds one meter in length. The packaging of this large amount of DNA and its partitioning among daughter cells present major problems which the cell has solved by packaging the DNA into chromosomes and developing cell division mechanisms that facilitate their appropriate distribution to the daughter cells. Although the average chromosome number for human cells is 46, some cells lack chromosomes (red blood cells), some have 23 (gametes), and others may have as many as 138 (hepatocytes) or 184 chromosomes (certain brain cells). The latter cell types exhibit high metabolic rates, and their high chromosome numbers provide a means for increasing gene dosage to meet these unusual metabolic needs.

CHROMOSOME STRUCTURE

Chromosomes consist of approximately equal amounts of DNA and **histones** (basic proteins) and smaller amounts of nonhistone proteins and RNA. The proportions of these chromosomal constituents appear to vary during the cell cycle. For example, metaphase chromosomes contain an unusually large proportion of RNA compared to interphase chromosomes. Their high RNA content most likely reflects contamination with nucleolar RNA following disintegration of the nucleolus during prophase.

DNA. The DNA of a typical chromosome is present as one uninterrupted double helix. This was demonstrated by observing the effects of proteases, RNases, and DNases upon chromosome structure. Neither proteases nor RNases cleaved chromosomes into smaller fragments; however, DNases successfully digested about 50% of the chromosomal material. These experiments were interpreted to suggest that 1) protein or RNA "linkers" between DNA double helices do not occur in chromosomes, and 2) that about 50% of the DNA in chromosomes must be complexed with histones to form a nuclease-resistant structure. This complex, subsequently called a **nucleosome**, will be described in a later section.

DNA occurring in eukaryotic chromosomes is heterogeneously distributed with respect to its base composition. Human chromosomes possess AT-rich chromosome regions and other areas that are GC-rich. Ag^+ preferentially binds to AT-rich DNA. When DNA extracted from human cells is spun at high speed in tubes containing gradients of cesium sulfate and Ag^+ or Hg^{+2} ions, the DNA bands in one major and four minor fractions. The DNA occurring in the minor fractions is called **satellite DNA**. The four satellite fractions collectively account for about 6% of the total DNA found in human chromosomes. Main band DNA has a GC content approximating 40%, while satellites I and II have GC compositions of 26% and 36%, respectively.

DNA possesses sequences that occur in widely varying multiplicities. When DNA is cut into short lengths, denatured by heating, and then cooled to promote reannealing, the DNA reforms a double helix at a rate commensurate with its base composition. DNA that consists of largely unique sequences of bases, such as those found in protein-encoding genes, will pair slowly ($C_ot > 100$, where C_o = initial concentration of the DNA, and t = time in seconds). Moderately repetitious DNA containing intermediate numbers of DNA segments, each consisting of nearly identical base sequences, will reanneal at an intermediate rate (C_ot = 1 to 100). Highly repetitive DNA anneals rapidly with a C_ot ranging from 10^{-1} to 10^{-4}. As the numbers of repeated base sequence increase, more opportunites for base pairing are provided, facilitating the reannealing process.

About 20% of all human DNA is moderately to highly repititious and tends to be located in centromeric regions of human chromosomes. The **centromere** is a primary constriction on mitotic or meiotic chromosomes which provides an attachment site for microtubules connecting the chromosome to the pole of the spindle. An additional 15-20% of human DNA displays moderate to intermediate repetition of a common sequence and largely occurs in non-centromeric regions. Much of the moderately repetitive DNA tends to be interspersed among "islands" of unique gene sequences and thoughout the chromosome arms. The majority of human DNA sequences occur in two to a few copies per nucleus.

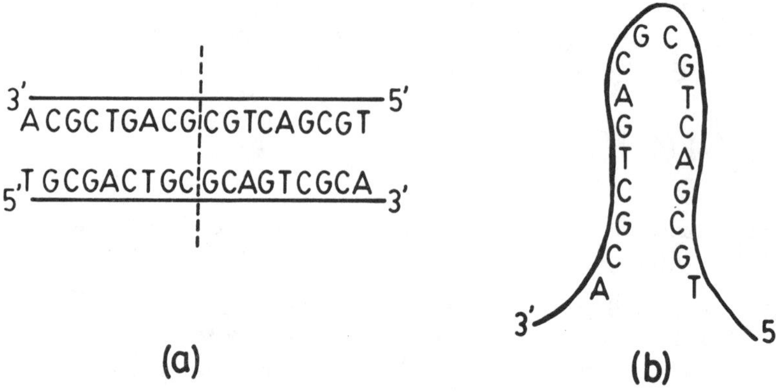

Fig. 3-1. Structure of a palindrome. Palindromes (a) can fold back upon themselves to form hairpin loops (b).

A special type of DNA reanneals very rapidly. This DNA, termed "snapback" DNA, occurs in reverse repeats known as palindromes (Fig. 3-1). Notice that the sense of the base pairs on either side of the central plane is the same regardless of which way the message is read. The base composition of palindromes does not differ significantly from that of DNA in general. Palindromes are generally 300 to 1200 bases in extent and often occur in clusters of two to four. Thousands of these clusters appear to be distributed throughout the human genome. Palindromes are capable of forming hairpin loops in which the two halves of the reverse repeats pair with themselves.

The relationships between satellite DNAs and repetitive sequences are not straightforward. Obviously only a minority of repetitive DNA occurs in the satellite fraction, since the latter only accounts for about 6% of the total genomic DNA. The chromosomal distribution of the four satellite DNAs varies and will be discussed below.

Histones. Histones are relatively small, basic nucleoproteins (molecular weight ca. 20,000) and occur in five general types in human cells: H1, H2a, H2b, H3, and H4. Histones are widely distributed among eukaryotes with H3 and H4 displaying a high degree of evolutionary conservatism. That is, the sequences of these proteins are very similar among distantly related species. This conservatism suggests that the majority of amino acids in these proteins are essential for protein-protein or protein-DNA contacts that are important for normal chromosome structure. Both H2a and H2b appear to be more variable; however, much of this variability is confined to their N-terminal ends. Histones H2a, H2b, H3, and H4 possess positively charged N-terminal ends that are especially appropriate for binding with the negatively charged phosphate groups of the DNA backbones. The remainder of these histone molecules have a globular structure. Histone H1 exhibits the greatest amount of amino acid sequence variability, not only differing among species, but also among different cell types within a species, indicating that a single species has several genes encoding the H1 molecule. Actually, each of the four histone genes occur in several copies that are clustered in various combinations on as few as two human chromosomes. Histones are highly basic. H1 has a high lysine content, while H2a and H2b are moderately lysine-rich. H3 and H4 possess a high arginine content.

Histones H2a, H2b, H3, and H4 form the core of a structure called a **nucleosome** (Fig. 3-2).

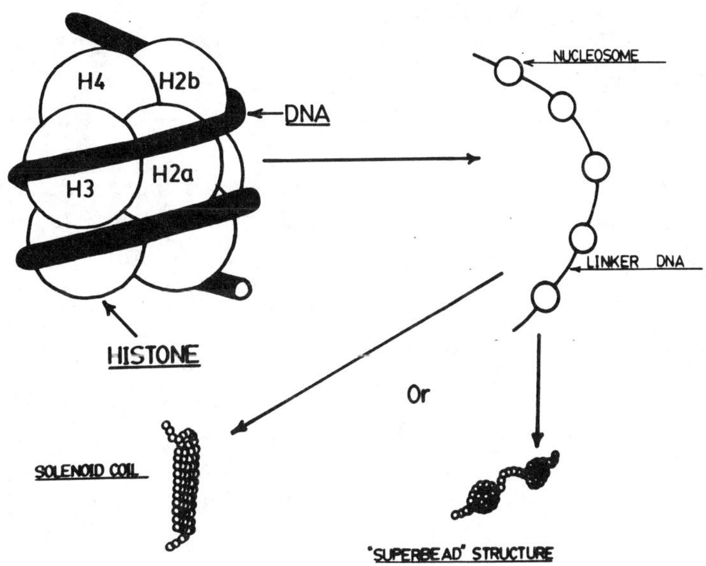

Fig. 3-2. Structure of a chromosome fiber.

Each nucleosome consists of an octamer formed from two copies of each of the four core histones surrounded by a double turn of the DNA. The basic N-terminal ends of the histones interact with the phosphates of the DNA, imparting stability to the structure. Nucleosomes are connected by linker DNA that is free of histone. Histone H1 bonds from the exit point of the DNA from one nucleosome to the entry point of the double helix to the adjacent nucleosome. Contraction of the histone pulls the nucleosomes together, contributing about a 7:1 compaction ratio for the chromatin. Coiling of this chain of nucleosomes generates a solenoidal coil which folds to form the fundamental interphase chromatin fiber, 240-400 angstroms in diameter. The linker DNA is susceptible to digestion by certain nucleases, while DNA in the nucleosome is resistant to digestion by these enzymes.

Histones are synthesized in the cytoplasm during S-phase, the period of DNA synthesis. The histones associate with the replicating DNA in a conservative fashion, that is, individual nucleosomes contain entirely newly synthesized histones or old histones from the parent double helix (see below). Histones may be modified by phosphorylation, acetylation, methylation, and ribosylation. Phosphorylation of serine and threonine residues of histones H1 and H3 is associated with the early stages of condensation of the extended interphase chromosomes into compact mitotic and meiotic chromosomes. Acetylation of certain lysines of histones H1, H2a, and H4 is often found in actively transcribing chromosome regions or regions capable of transcription and appears to be partially responsible for a more relaxed chromosome structure that permits access of the transcription complex to the promoters of genes located in these areas. Methylation of lysines in H3 and H4 and histidines in H1 has been described, appears to occur during late S-phase and G_2 of cell division, and is irreversible. Such methylation could affect DNA-histone interactions and could contribute to the more condensed chromosome structure found in regions possessing inactive genes. Five to ten percent of nucleosomes contain an A24 complex. This complex is formed by the binding of ubiquitin to H2A. A24 complexes are enriched in non-transcribing regions, suggesting a role in gene inactivation.

Non-histone proteins. Most non-histone proteins can be removed from chromatin by treatment with 0.35M NaCl, indicating that they are not integral components of chromosome structure. This group includes DNA- and RNA-polymerases, other replication and transcription enzymes, processing proteins, transferases, and kinases, all of which play an important role in gene replication and expression. About 70% of these proteins are generally distributed among different cell types and are detectable at many developmental stages. The remainder display some degree of tissue and temporal specificity. The latter group of proteins is believed to play a significant role in tissue-specific and stage-specific expression of genes.

A group of small (8,000 to 26,000 MW) proteins with rapid electrophoretic mobilities also occurs in the non-histone protein fraction. These proteins have been labelled **HMG (high mobility group) proteins.** Although the precise functions of these proteins remain unknown, they have been observed to be associated with nucleosomes. Ubiquitin (HMG-20) was mentioned above and may participate in gene inactivation. Other HMG proteins, including HMG-14 and HMG-17, are bound to actively transcribing nucleosomes. DNase digestion studies of specific genes whose nucleosomes are either complexed with HMG-14 and HMG-17 or free of these proteins suggest that HMG-14 and HMG-17 may participate in the relaxation of chromosome structure, permitting expression of genes in these regions. Weisbrod <u>et al.</u> (5) employed sensitivity to DNase I digestion as an indicator of gene activity. For example, DNase I digests adult globin genes in red cell precursors where they are expressed, but not in hepatocytes where they are transcriptionally silent. Chromatin was isolated from β-globin-synthesizing cells, and HMG-14 and HMG-17 were removed with 0.35M NaCl. The HMG-depleted chromatin was divided into two portions. One portion (A) was reconstituted with HMG-14 and HMG-17, and the other (B) was not. Both (A) and (B) were treated with DNase I and subsequently with a restriction endonuclease to produce DNA fragments. The fragments were separated by gel electrophoresis and probed with a β-globin cDNA to identify fragments containing the β-globin gene. The β-globin gene in the HMG-free DNA (B) proved to be intact; whereas, no fragments corresponding to the β-globin gene were observed in the HMG-reconstituted DNA (A). Therefore, HMG-14 and HMG-17 restored DNase I sensitivity to the β-globin gene, suggesting that they may be involved with maintaining a chromosome conformation that allows accessibility to the enzyme and, by extrapolation, to the transcription complex.

Hypersensitive sites are DNA sequences that display exquisite sensitivity to DNase I digestion. These sites are usually located upstream from chromosome areas that are actively transcribing or capable of transcription. Hypersensitive sites are nucleosome-poor, and are believed to provide "windows" for access of the transcriptional complex to the genes nearby.

Chromosomal domains and isochores.   Interphase chromosomes are approximately 250 angstroms in diameter and are organized into loops that average 100kb in length. These loops or **chromosomal domains** (Fig. 3-2) are anchored to the **nuclear matrix** or **nuclear scaffold**, a substructure of the nucleus that imparts an ordered three-dimensional structure to interphase chromatin. The

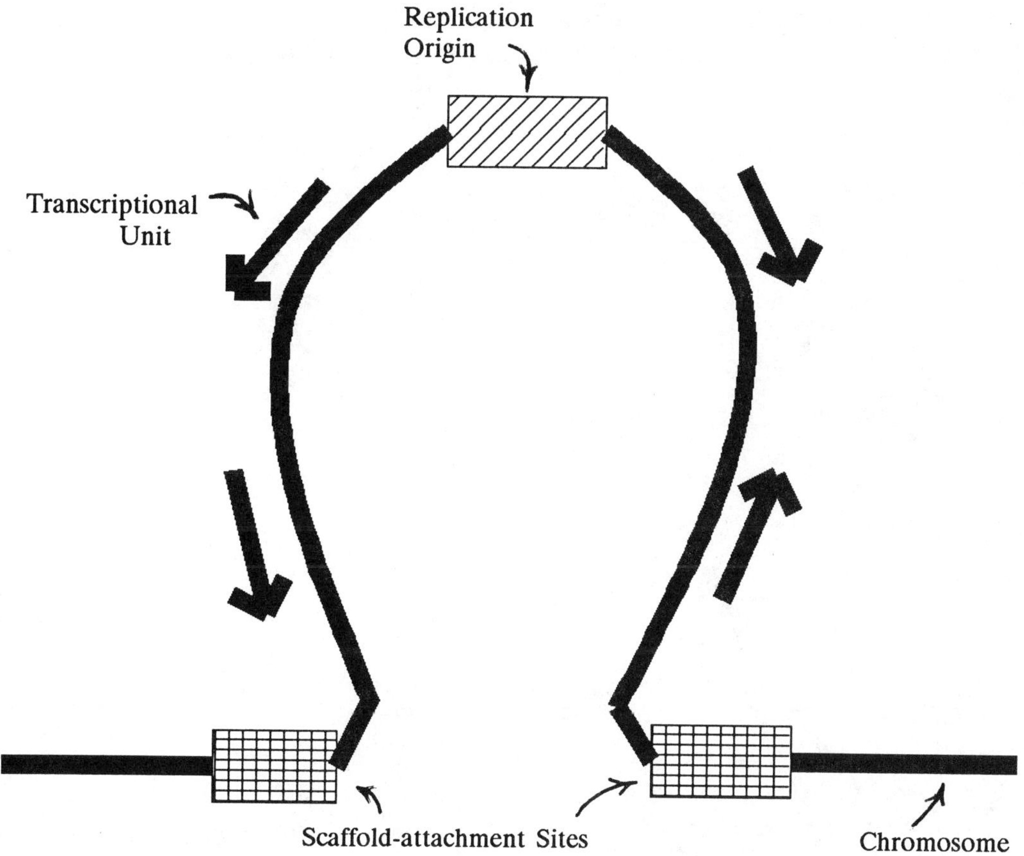

Fig. 3-2. Chromosomal domain. The 250 angstrom interphase chromosome is folded into loops that are anchored to the nuclear scaffold. Each chromosomal domain contains a central replication origin and from one to several transcriptional units. The chromosomal domains are attached to the scaffold at both ends of the loop. The chromosomal domain provides a means for regulating the timing of replication of genes and possibly for coordinate regulation of their expression.

size of the chromosomal domain closely corresponds to the average size of replicating chromosome regions. Therefore, the chromosomal domain is viewed as the structural entity of the chromosome that corresponds to the **replicon,** the unit of chromosome replication. The chromosomal domain also appears to correspond to the average size of co-expressed genes or transcriptional units. The transcriptional units in Figure 3-2 are represented by arrows that are oriented in the direction of their transcription. There is a strong relationship between the timing of gene replication and the active or inactive transcriptional state of a gene. Expressed genes of a tissue replicate early, while those that are silent in that tissue replicate late during S-phase. Future work may demonstrate that groups of active genes may be clustered in one domain, while inactive genes are clustered in another.

  Isochores are defined as long stretches of DNA that share very similar base compositions. Isochores may be GC-rich or AT-rich. Active genes tend to be imbedded in GC-rich isochores, while late-replicating, non-functional genes are located within or near AT-rich isochores.

  Transition from the interphase to the mitotic or meiotic chromosome. Interphase chromatin consists of an extended 250 angstrom fiber that is folded into loops or domains (Fig. 3-2). These loops appear to be clustered into rosettes (Fig. 3-3). As the cell enters mitosis or meiosis, further compac-

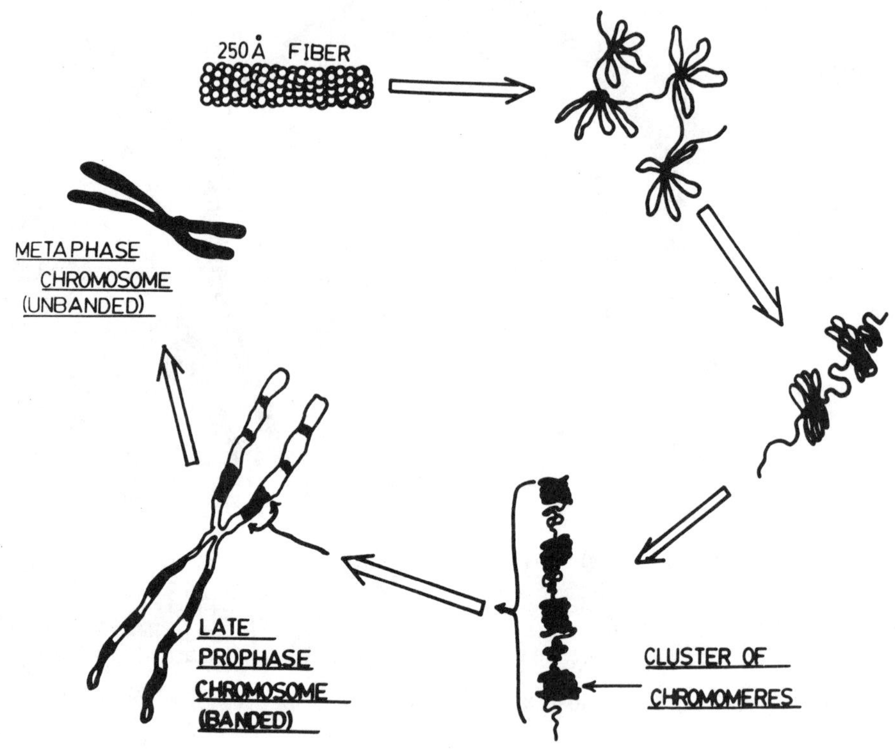

Fig. 3-3. Higher order chromosome structure.

tion of the chromosome fiber occurs by coiling and folding to produce the chromomeric structure of prophase chromosomes (most apparent during meiosis). These chromosomes continue to condense, successively forming the late prophase-prometaphase chromosomes which contain a characteristic "banded" fine structure and metaphase chromosomes in which the sub-bands have fused to produce larger bands.

The banding pattern of chromosomes can be visualized by using appropriate staining methods that will be described in a later section. The fully condensed metaphase chromosome is the final product of the compaction process (upper left of Fig. 3-3).

Heterochromatin and euchromatin. Heterochromatin has long been associated with genetic inactivation. Heterochromatin tends to be highly condensed and is late-replicating. By contrast, euchromatin has a more relaxed structure and contains actively-transcribing genes. Classical genetic experiments using Drosophila demonstrated that relocating genes from a euchromatic region to a heterochromatic region resulted in the transcriptional silencing of those genes. Conversely, placement of heterochromatin within or close to a euchromatic region inactivated genes in the euchromatic region in gradient-like fashion. That is, euchromatic genes closest to the heterochromatin were completely inactivated in all cells, while those more distant from the heterochromatin were partially inactive. The latter genes either were expressed at lower than normal levels or were expressed in some cells and not others within the same tissue. This pattern of inactivation suggests that inactivation most likely occurs by a spreading mechanism involving, as yet, unidentified chemicals. Nucleosomes are found in both euchromatic and heterochromatic chromosome regions; however, enzyme digestion studies indicate that the conformation of the nucleosomes is most likely different in the two regions. DNaseI rapidly degrades accessible DNA to an acid-soluble form. Structural genes for hemoglobins are readily digested by DNaseI in expressing cells, but are resistant to DNaseI in adult hepatocytes which do not express hemoglobin. By contrast, ovalbumin, a liver-expressed gene, is digested by DNaseI in hepatocytes but not red cell precursors. As discussed above, certain HMG proteins are differentially distributed among active and inactive genes, and both the acetylation and phosphorylation patterns of histones differ between actively transcribing and transcriptionally silent chromosome sectors. All of these factors may influence the transcriptional state of a gene.

Heterochromatin has been classified as **constitutive** or **facultative.** Constitutive heterochromatin contains sequences that are not expressed in RNA during the life of the organism and is found in centromeric regions or may be interspersed among euchromatic regions along the chromosome arms. The latter type of constitutive heterochromatin is often called **intercalary** heterochromatin, is AT-rich, and contributes to the chromomeric structure of meiotic pachytene chromosomes. Centromeric heterochromatin largely consists of satellite DNA and may be AT-rich, GC-rich, or intermediate, depending upon the particular satellite predominating in the centromeric region of a particular chromosome. Most euchromatic areas tend to be GC-rich. This uneven distribution of GC and AT base pairs provides a chemical foundation for the chromosome banding procedures described in a subsequent section of this chapter.

Facultative heterochromatin contains gene sequences that may be active during one stage of development of an organism and not another. The best known example of facultative heterochromatin in human cells is provided by the inactive X chromosome. **X-inactivation** is the process that silences many of the genes on the long arm of one X chromosome in female somatic cells (Fig. 3-4). There appears to be an "inactivation center" located in the medial portion of the long arm of each X chromosome. The center may contain a gene that determines an inactivating substance. The gene functions on only one of the two X chromosomes in female embryonic somatic cells. The substance spreads in both directions from the inactivation center, shutting down transcription of genes it encounters. The centromere apparently retards the spread of the inactivating substance, since genes located on the distal third of the short arm may not be inactivated. X-inactivation is a random process in cells giving rise to the female embryo. Either the maternal or paternal X chromosome may be inactivated in a particular cell. Furthermore, inactivation in a particular cell line is permanent, such that the same X is inactivated in all progeny cells. X-inactivation produces a **mosaic** expression pattern for X-linked genes.

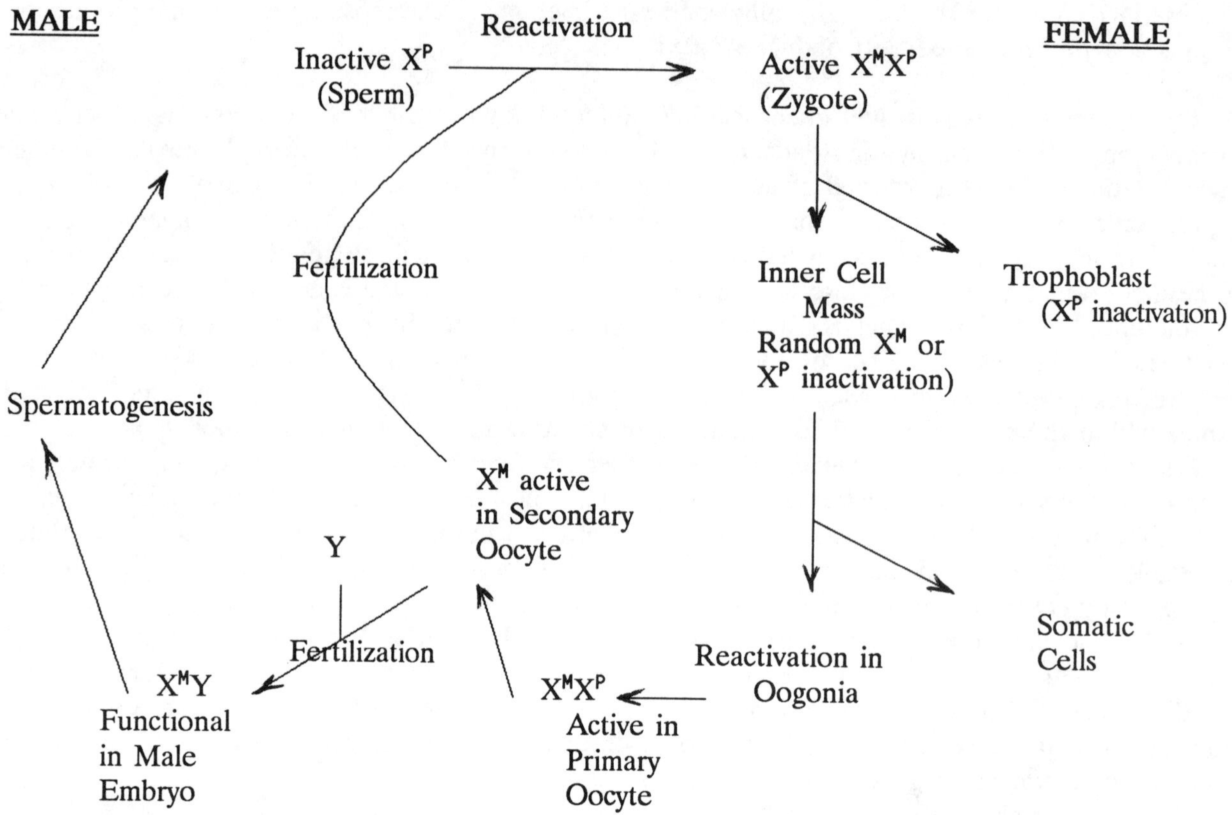

Fig. 3-4. Inactivation and reactivation of the X chromosome in human cells. A cyclic inactivation and reactivation of the human X chromosome occurs in human cells. Female zygotes possess two functional X chromosomes. The paternally-derived X chromosome (X^P) is inactivated in trophoblast cells which contribute to the placenta and fetal membranes. The inner cell mass, cells giving rise to the embryo proper, display random inactivation of the X chromosome, where either the maternally-derived X (X^M) or X^P may be inactivated in a single cell and all of its descendants. The inactive X is reactivated as oogonia enter meiosis, such that both X chromosomes are functional in the primary oocyte. Male embryos contain a single (maternally-derived: X^M) functional X chromosome. This X chromosome is inactivated during spermatogenesis and reactivated during fertilization.

Most female tissues contain a mixture of cells, approximating a 50:50 distribution, which expresses the array of X-linked alleles derived from the individual's father or those from her mother (Fig. 3-5). Since inactivation is a random event, the percentage of cells expressing the X-linked alleles from a given parent follows a Gaussian or Normal Distribution. Occasional women are encountered who, by chance, preferentially express X-linked genes from either their mother or their father (eg. a 90:10 split). Consequences of this unusual expression pattern will be discussed in relation to X-linked inheritance.

The inactive X chromosome replicates late during S-phase and is manifested as a **Barr Body** in somatic cells (Fig. 3-6). The Barr body is a mass of heterochromatin adjacent to the nuclear membrane and may be used to estimate the number of X chromosomes in cells (Table 3-1). In diploid cells (two complete chromosome sets, where 23 chromosomes constitute a normal chromosome set), the number of Barr bodies is equivalent to the number of X chromosomes minus one. Cells containing more than two chromosome sets generally follow the relation:

$$B = X - (P/2)$$

where B is the number of Barr bodies, X is the number of X chromosomes, and P is the number of chromosome sets. For example a 92,XXXX cell would have 4-(4/2) = 2 Barr bodies. In this example, the total number of chromosomes is given, followed by the sex chromosome complement. This relation holds for even-numbered polyploids; however, triploid cells (three chromosome sets) do not follow the rule. 69, XXX cells may have one or two Barr bodies, and 69,XXY cells usually lack Barr bodies.

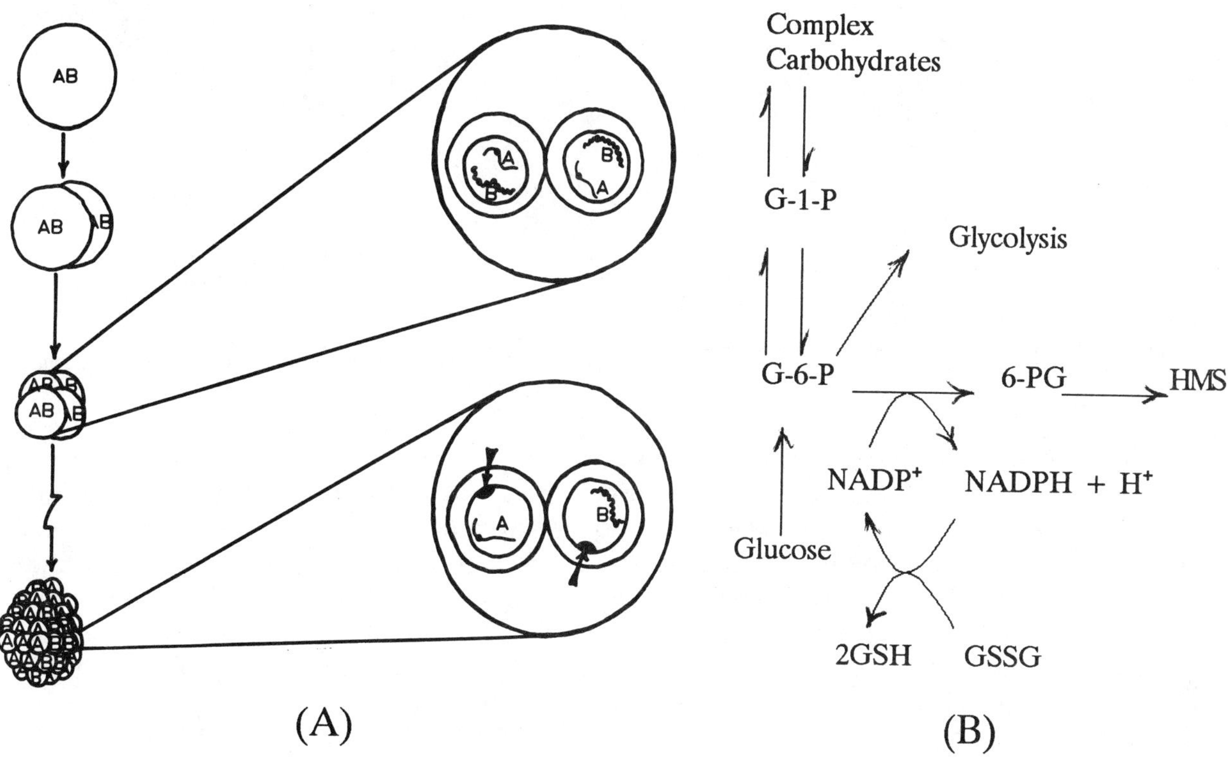

Fig. 3-5. Mosaicism produced by random X-inactivation of Glucose-6-phosphate dehydrogenase (G6PD). A) X-inactivation of the G6PD alleles, A and B, during pre-implantation development. Arrows indicate the Barr body, and smooth and wavy lines distinguish the functional maternal and paternal X chromosomes, respectively. The G6PD-A and G6PD-B alleles determine a fast or slow migrating form of the enzyme, respectively. Any one cell will synthesize only one of the two possible isozymes following X-inactivation. B) G6PD catalyzes the conversion of glucose-6-phosphate (G-6-P) to 6-phosphogluconate (6-PG). This enzyme is particularly important for the generation of reduced nicotinamide adenine dinucleotide phosphate (NADPH) and reduced glutathione (GSH) which is necessary for integrity of the red cell membrane. Deficiencies of the enzyme can lead to anemia. Other abbreviations: G-1-P = glucose-1-phosphate; HMS = hexose monophosphate shunt; NADP⁺ and GSSG = oxidized forms of nicotinamide adenine dinucleotide phosphate and glutathione, respectively.

The Y chromosome can also be identified in the nuclei of interphase cells by using an appropriate fluorescent stain and viewing the cells with a fluorescence microscope. The Y appears as a bright dot called a Y body. There is one Y body for each Y chromosome (Table 3-1). The term, **sex chromatin**, traditionally has been used to refer to the Barr body, but, more recently, has been broadened to include the Y body.

Table 3-1. Interphase Sex Chromatin Pattern and
Sex Chromosome Complements.

Chromosome Complement	Barr Bodies	Y Bodies
45,X	0	0
46,XY	0	1
46,XX	1	0
47,XXY	1	1
47,XXX	2	0
47,XYY	0	2
48,XXYY	1	2
48,XXXX	3	0
49,XXXYY	2	2
92,XXYY	0	2
92,XXXX	2	0

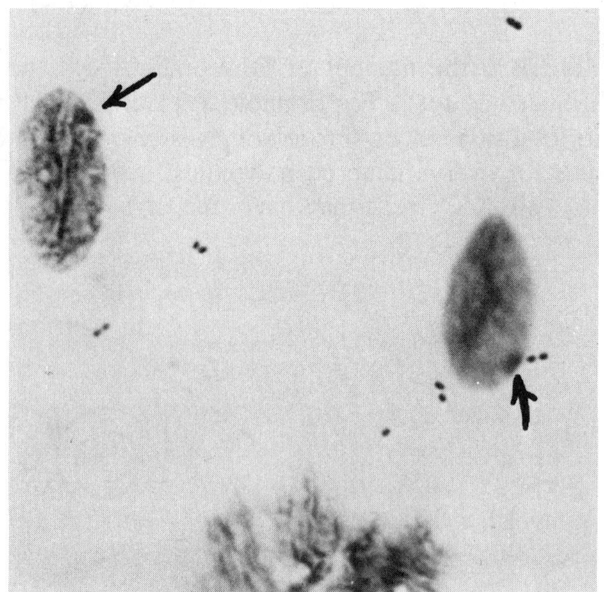

Fig. 3-6. Interphase manifestations of X-inactivation. Barr bodies (arrows).

The mechanism of X-inactivation remains unknown. The process is most likely complicated, involving at least two steps: an initial inactivating event and a consolidation event that stabilizes the inactivation. Marsupials such as the kangaroo also display X-inactivation; however, inactivation is nonrandom, with the paternally-derived X chromosome involved in all cells. A second difference between eutherian and metatherian mammals concerns the pattern of gene expression on the inactivated chromosome. Paternal X-linked alleles at some loci are inactivated in all cells, while other paternal X-linked alleles may display partial activity in some tissues and not others (6). DNA methylation, particularly of CpG islands upstream from housekeeping genes, has been proposed as a mechanism for X-inactivation. However, inactivation of the paternal **G6PD** allele in marsupials appears to occur in the absence of DNA methylation. Furthermore, studies of female mouse embryos have revealed that methylation of sites in the **Hprt** allele on the inactive X chromosome does not occur until several days after the inactivation event has occurred (7). These results and others indicate that methylation is important in maintaining X-inactivation but is not involved with the inactivation event itself. Driscoll and Migeon (8) studied the methylation patterns of five housekeeping and two specialized X-linked genes in human fetal ovarian cells. Methylation-sensitive enzymes were used to digest the DNA in germ cells from a series of stages ranging from entry into meiosis through cessation of DNA synthesis. These enzymes will cut DNA at unmethylated restriction sites, but not those that are methylated. The restriction enzyme, MspI, cuts at sites containing the sequence CCGG. Thirty-two CCGG sites that were known to be located in CpG islands near the seven genes and 25 other CCGG sites located elsewhere within these genes were examined. The patterns obtained from digestion of the ovarian cells were compared with those from male meiotic germ cells and with those from male and female somatic cells. CpG islands were not methylated in fetal ovarian germ cells nor adult testicular germ cells. The authors interpreted this observation as a possible explanation for the reversibility of X-inactivation in germ cells described above. Another important result was the discovery that the 25 non-clustered CCGG sequences were similarly methylated in male germ cells and female somatic cells **but not in fetal ovarian germ cells**. This sex difference in methylation pattern may be partially responsible for parental imprinting of X-linked genes that has been observed in some systems (see Fragile X syndrome).

Recent evidence (9) suggests that a simple diffusion model may not be adequate for explaining the pattern of X-inactivation observed. Figure 3-7 illustrates the distribution of loci that have been found to be functional or non-functional on the X-chromosome. Somatic cell hybrids were constructed using both a selection system that favored retention of the inactive X chromosome and no selection system. This experimental procedure involves fusion of mouse and human cells to form dikaryons which contain both human and mouse nuclei. The nuclei fuse, and cells are derived which possess one copy of each human and one copy of each mouse chromosome. As the hybrid cells divide, the human chromosomes are lost one by one, producing an array of progeny that differ from one another with respect to the types of human chromosomes present. Clones of cells are established from individual hybrid cells, and their biochemistry and genetics can be studied. By making the cultures dependent upon a product encoded by a gene located on a particular chromosome, clones may be recovered that contain that chromosome and the mouse chromosomes present in the original hybrid cells. Drs. Brown and Willard prepared hybrids from mouse and human female cells. The hybrids were selected to produce clones containing either an active or inactive human X chromosome. RNA was extracted from each type of clone, and cDNAs corresponding to these transcripts were prepared. The cDNAs were amplified by PCR, and probed with gene sequences from the human X chromosome. The results of these studies indicated 1) that active and inactivated genes are inter-

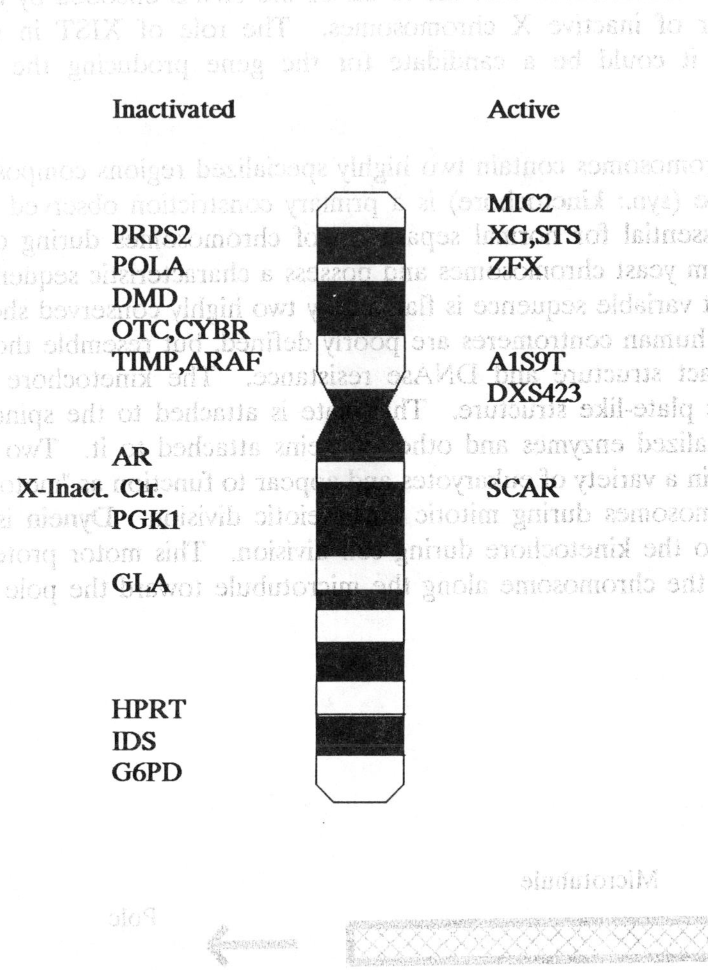

Fig. 3-7. Expression of genes on the inactive human X chromosome. Inactivated genes are listed at the left, and functional genes are are presented at the right of the X chromosome. The X-inactivation center is located between the AR and PGK1 genes on the proximal long arm.

spersed on the short arm of the inactive X chromosome, and 2) that an active locus (SCAR) is very close to the X-inactivation center on the long arm. Other functional loci are anticipated farther down the long arm of the inactive X chromosome. These results suggest that there may be a fundamental difference between the promoters or nearby regions of genes that are susceptible to X-inactivation and those which escape X-inactivation, and that X-inactivation is not simply determined by the distance of a gene from the X-inactivation center nor its position relative to the centromere.

Drs. Brown and Willard have also isolated a gene that is only expressed in human females and is only expressed in somatic cell hybrids containing the **inactive X chromosome**. This gene, called XIST (X_i - specific transcripts), contains five exons and maps to the region of the X-inactivation center. The gene is transcribed and processed into five different mRNAs by alternative splicing. Studies of persons with multiple X chromosomes have demonstrated that the levels of the RNAs encoded by this gene increase in proportion to the number of inactive X chromosomes. The role of XIST in the inactivation process is unknown; however, it could be a candidate for the gene producing the X-inactivating substance.

Centromeres and Telomeres. Chromosomes contain two highly specialized regions composed of unusual base sequences. The **centromere** (syn.: **kinetochore**) is a primary constriction observed on mitotic and meiotic chromosomes that is essential for normal separation of chromosomes during cell division. Centromeres have been cloned from yeast chromosomes and possess a characteristic sequence in which an AT-rich core of uniform size but variable sequence is flanked by two highly conserved short stretches of DNA. The DNA sequences of human centromeres are poorly defined, but resemble those of yeast with respect to their highly compact structure and DNAse resistance. The kinetochore of eukaryotic chromosomes has a characteristic plate-like structure. This plate is attached to the spindle fibers or microtubules and has several specialized enzymes and other proteins attached to it. Two of these enzymes have been extensively studied in a variety of eukaryotes and appear to function as "motors" that are responsible for movement of chromosomes during mitotic and meiotic division. **Dynein** is a multimeric enzyme that has been localized to the kinetochore during cell division. This motor protein appears to generate the major force pulling the chromosome along the microtubule toward the pole of

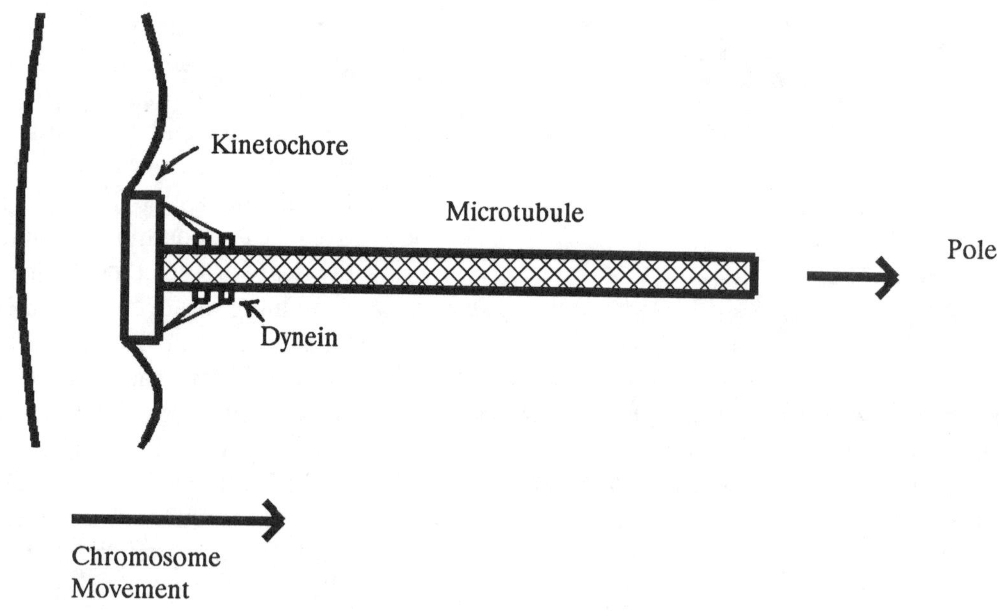

Fig. 3-8. Chromosome movement along the microtubule during cell division.

of the dividing cell during anaphase (Fig. 3-8). The energy for this movement is derived from the hydrolysis of ATP by dynein. As the microtubule is pulled toward the kinetochore, the microtubule is degraded, accounting for the shortening of the spindle fibers during this phase of cell division. The second motor protein, **kinesin**, may function at an earlier stage of cell division. During prometaphase, microtubules appear to grow out of the centrosomes, and become attached to the kinetochores of the chromosomes. Chromosomes have been observed to congregate at the poles during early prometaphase, and then move to the metaphase plane. This movement from the pole to the metaphase plane is believed to be facilitated by kinesin. Like dynein, kynesin obtains the energy required for its action from ATP.

     Telomeres are protein-DNA complexes located at the ends of the chromosome arms. Telomeres prevent fusion of chromosome ends and are often associated with the nuclear membrane and with one another. All organisms appear to have a consensus sequence which has the general formula, T_nG_m, where G and T are the predominant bases on the DNA strand running 5' to 3' toward the telomere. Human telomeres possess from 2 to 20 kilobases, primarily consisting of tandem repeats of the consensus sequence 5'-TTAGGG-3' with copies of TTGGGG and TGAGGG occasionally interspersed among them (Fig. 3-9). A proportion of these telomeric sequences are lost during each cycle of cell division as a consequence of a difficulty with their replication. However, addition of repeats following replication of chromosomes can be accomplished using a ribonucleoprotein, called telomerase. Telomerase has a RNA segment that contains a sequence complimentray to about 1½ copies of the G-rich strand repeat. This RNA appears to act as a template for the formation of additional copies of the repeat at the end of the chromosome. Therefore, telomeres should be considered to be dynamic entities

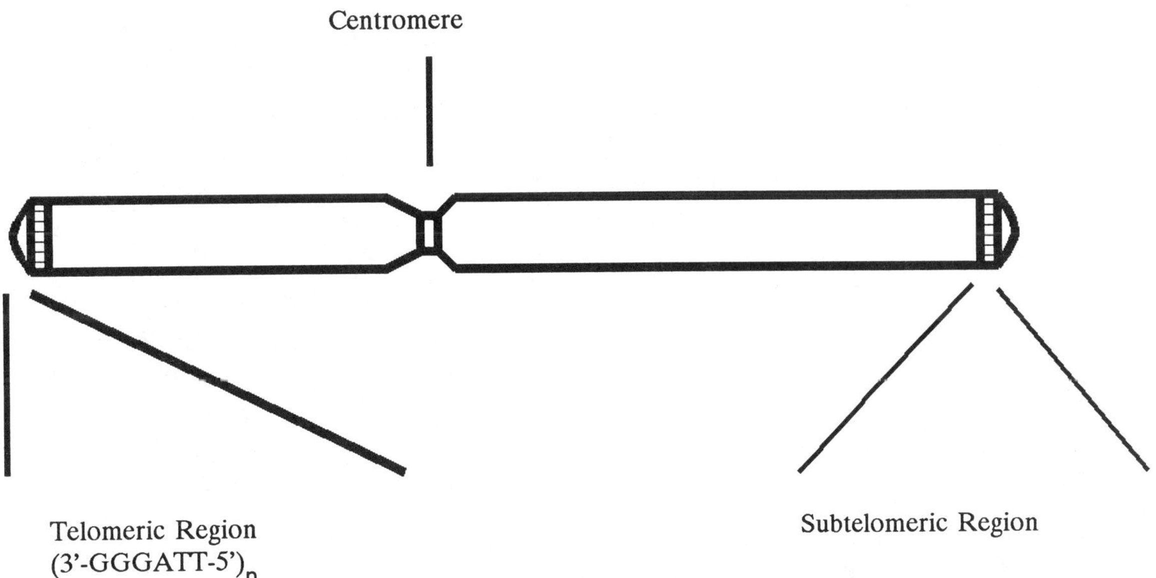

Fig. 3-9. Telomeric and subtelomeric regions of a human chromosome. The telomeric region consists of tandemly repeated sequences with the consensus: Telomere-3'-GGGATT-5'-Centromere. Occasional runs of 3'-GGGGTT-5' and 3'-GGGAGT-5' are nested in the more general consensus sequence. The subtelomeric zone, or telomere-associated DNA region, displays both inter-chromosomal and inter-individual variation. Tandemly repeated DNA sequences also occur in the subtelomeric zone and may be similar to or distinct from those present in telomeres, depending upon the chromosome involved.

whose length reflects a balance between the loss of sequences occurring during cell division and their replacement by telomerase. Recent studies have indicated that aging is associated with a successive shortening of the telomeric sequences of human chromosomes, and cancer chromosomes also have relatively short telomeric segments. In the latter case, progressive loss of telomeres could lead to fusion of the ends of two chromosomes, producing a product with two centromeres. Dicentric chromosomes are highly unstable and promote chromosome breakage and associated loss and/or rearrangement of chromosome segments. These changes are frequently observed in cancer cells as they progress from relatively benign to malignant states. Subtelomeric areas of human chromosomes also possess repeated DNA sequences. Some chromosomes appear to have repeats of the same consensus sequence found in telomeres nested in non-telomeric DNA in their subtelomeric regions. These nests of telomeric DNA may have been introduced to the subtelomeric regions by "shuffling" of telomeric DNA among different chromosomes, possibly by recombination. In other cases, the subtelomeric areas have repeats of sequences not occurring in telomeres. Some of these sequences are unique to certain chromosome types, while others may be shared among two or more different chromosomes. Both intercellular and inter-individual variation of subtelomeric DNA from the same chromosome, eg. chromosome 7, has also been reported.

Cell Cycle

The cell cycle encompasses the time for one cell generation and varies in length from one cell type to another, with the nutritional state of the cell, and with the developmental stage of the organ containing the cell. Most actively dividing cells require approximately 24 hours to progress through one cell cycle. The cell cycle has been divided into four stages: G_1, S, G_2, and M (Fig. 3-10). The length of each stage varies among cell types and often among different cells within the same tissue. Most cells spend the vast majority of a cycle in the **Interphase** period which includes the G_1, S, and G_2 stages. Following completion of cell division, the metabolic processes of the cell are resumed (G_1). Cytoplasmic signals related to nutritional state, size of the cell, and other, often ill-defined, factors influence the cell to enter S-phase. Replication of chromatin occurs during S-phase, with euchromatin replicating during early S-phase and heterochromatin completing its replication during late S-phase. A short interval (G_2) exists between completion of replication and initiation of cell division. Note

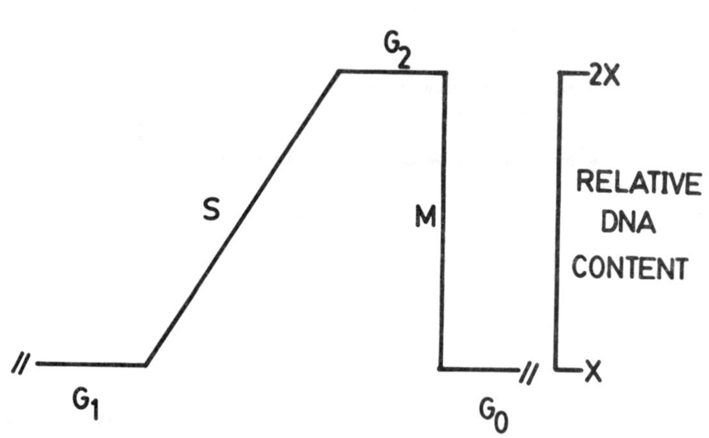

Fig. 3-10. The cell cycle. G_1 = preDNA synthetic phase; S = period of chromatin replication; G_2 = postDNA synthetic phase; M = division phase.

how a G_2 cell has twice as much DNA as a G_1 cell. Many adult cells typically do not divide, remaining in G_0 for many years. Repair of damage to the organs in which they are found is accomplished by a small subpopulation of cells that retain their ability to divide. Certain brain and hepatic cells may undergo two or more rounds of replication without cell division. These cells contain three or four times the usual amount of DNA (6N and 8N, respectively). The term **endoreduplication** refers to the replication of chromosomes without subsequent nuclear and cytoplasmic division. Fusion of cells followed by fusion of their nuclei could also produce cells with polyploid (4N, 6N, 8N) nuclei. Whether one or both of these mechanisms is involved in the generation of the polyploid brain and liver cells is currently unknown.

Nuclear division and partitioning of chromosomes between daughter nuclei occur during M-phase of the period of mitotic (meiotic) division. Synthesis of proteins and nucleic acids largely ceases during M-phase, and extensive chromosome condensation is observed during this time. Cytoplasmic division (**cytokinesis**) immediately follows M-phase, and the nuclear and cytoplasmic membranes are reconstructed. Each of the daughter cells formed by mitotic division contains the same number of chromosomes and genes as the parental G_1 cell from which they originated. During the partitioning of the cytoplasm, cytoplasmic organelles, including mitochondria, are also divided among the daughter cells. The partitioning of these mitochondria occurs at random. If a mutation is present in some of the mitochondria, the number of mutant mitochondria may vary among progeny cells, leading to a mosaic expression of the effects of the mutation.

Control of cell division appears to be mediated by cytoplasmic factors. When cells in M-phase are fused with cells in G_1- or G_2-phase, the chromosomes from the G_1 or G_2 cell are induced to condense as they would during M-phase. Studies of mitosis in yeast cells have identified a number of factors that influence the cell cycle (Fig. 3-11). At least three check points are known in the cycle.

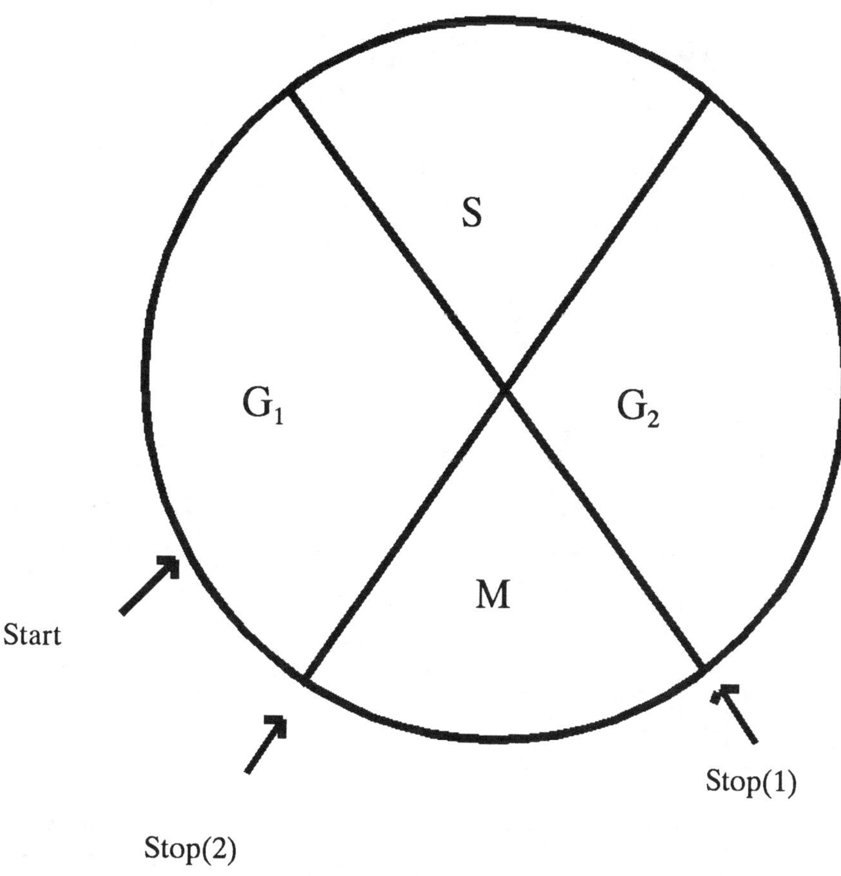

Fig. 3-11. Regulation of the yeast mitotic cell cycle. See text for explanation.

Commencing of the cell through G_1- to S-phase requires sufficient size and adequate nutrients. If either condition is not met, the cell will not enter S-phase. A stop control occurs near the end of G_2. At this point DNA synthesis is checked to make sure it has been completed. Exit from M-phase requires an assessment of the adequacy of mitotic spindle function.

Several factors have been identified recently that appear to regulate progression through the cell cycle. Many of these regulatory proteins are widely distributed among eukaryotes and presumably play similar regulatory roles. M-phase kinase (MPF; Note: M-phase kinase was first described as a maturation promotional factor that would induce arrested Xenopus eggs to complete their development. The abbreviation for the M-phase kinase is derived from this factor.) consists of two proteins, cyclin and p34. MPF phosphorylates histone H1, initiating condensation of chromosomes into the compact rods observed during metaphase. Activation of MPF occurs by phosphorylation of its p34 component near the onset of M-phase (Fig. 3-12). The enzyme is inactivated by the combined effects of cyclin degradation and p34 dephosphorylation. Both changes occur at the end of M-phase. This phosphorylation-dephosphorylation cycle is catalyzed by a family of kinases and a family of phosphatases that are being

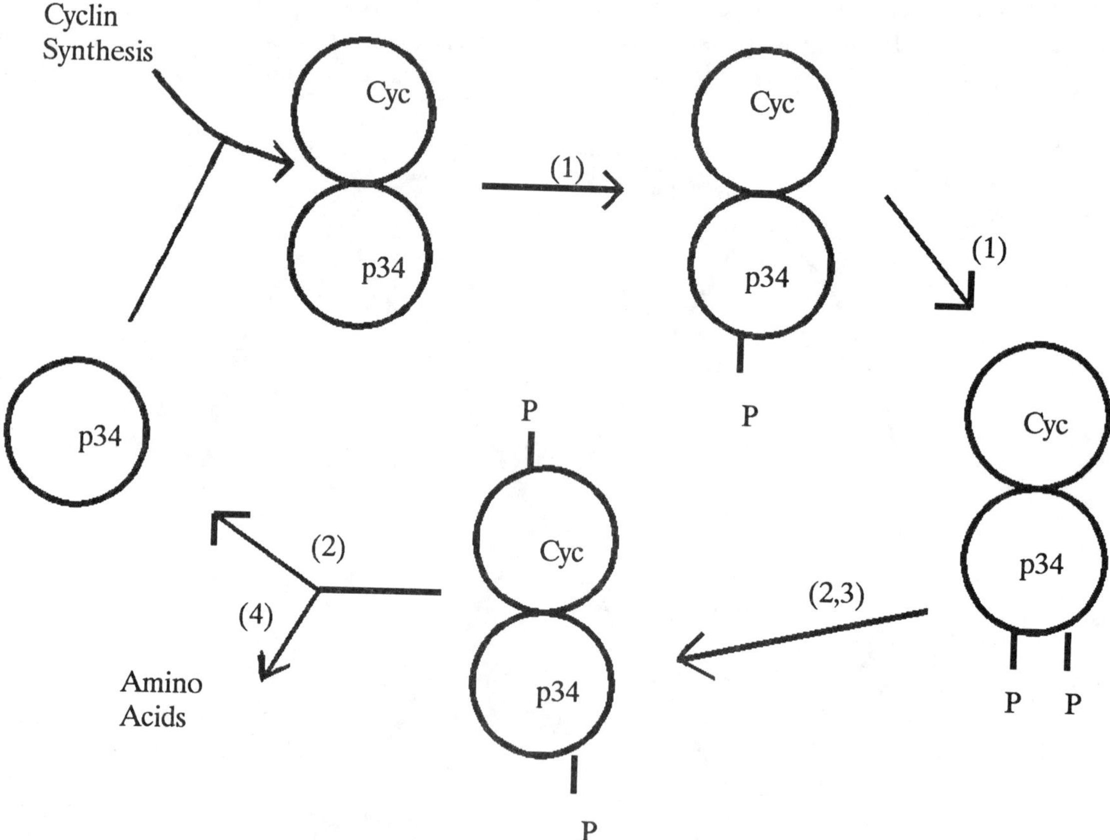

Fig. 3-12. Activation and inactivation of M-phase kinase during the cell cycle. Cyclin a component of M-phase kinase is synthesized during G_1 and combines with p34 to form M-phase kinase. M-phase kinase is phosphorylated on its p34 subunit (1), activating the enzyme complex. M-phase kinase phosphorylates histone H1, triggering the condensation of chromosomes to their compact metaphase structure. Toward the end of M-phase, p34 is dephosphorylated (2) and cyclin is phosphorylated (3). Phosphorylation of cyclin targets the subunit for degradation (4), inactivating M-phase kinase.

appreciated as important regulatory factors during cell division. Additional events occurring during the cell cycle may also be dependent upon the actions of these enzymes. A second factor, S-phase promoting factor (SPPF) is activated near the end of G_1 and induces the synthesis of DNA and histones. SPPF is inactivated at some time during S-phase, and, as its level falls, DNA synthesis declines. The roles of MPF and SPPF are illustrated in relation to the cell cycle in Figure 3-13.

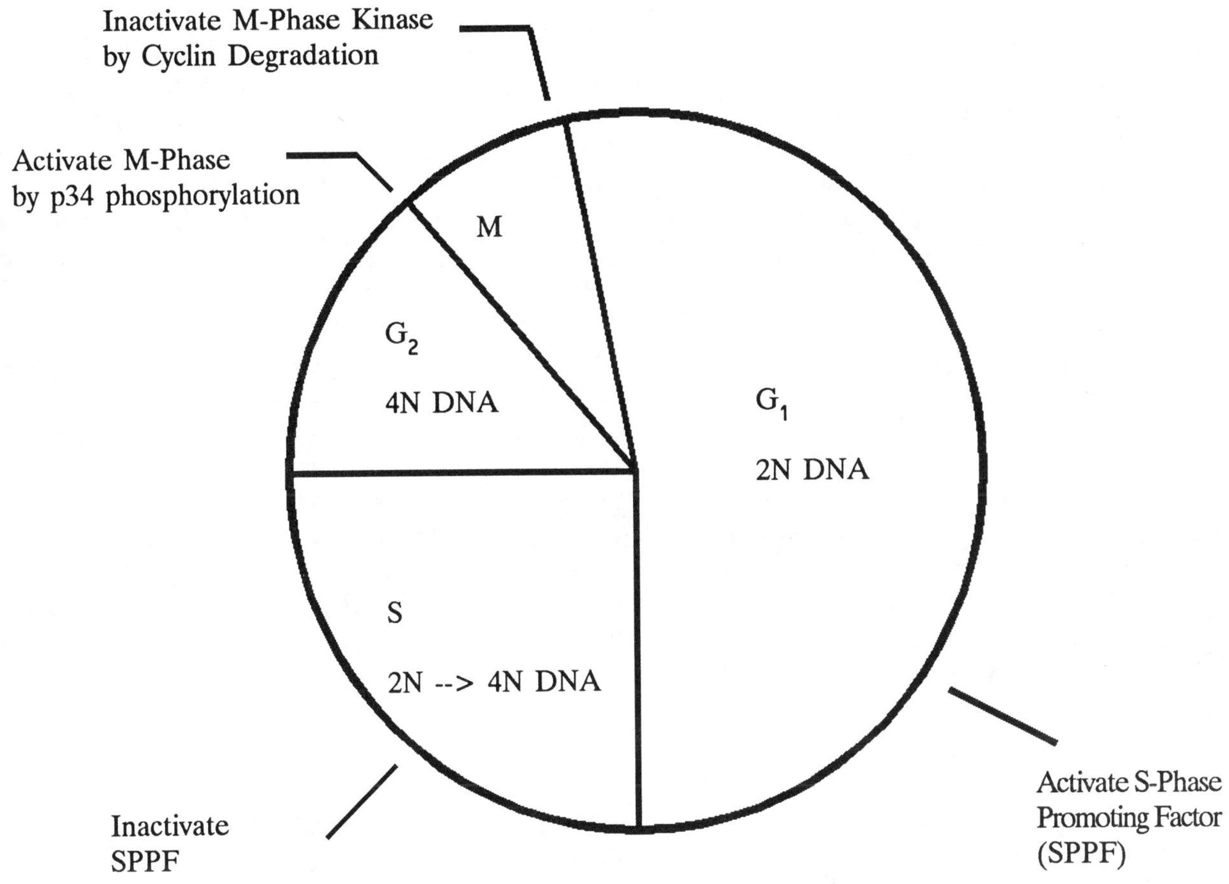

Fig. 3-13. Integration of molecular and cytological events during the cell cycle. Cells complete cell growth and carry on metabolism during G_1-phase. Near the end of G_1, SPPF is activated, inducing DNA and histone synthesis. All chromosomes and genes are replicated during S-phase, increasing the DNA content to twice the amount present in a G_1 cell. SPPF is inactivated during S-phase, resulting in a decline of DNA synthesis and progression of cells to G_2. Near the end of G_2-phase, cyclin levels reach a peak, combining with p34 to form M-phase kinase. This kinase is activated by phosphorylation of its p34 subunit and catalyzes the phosphorylation of histone H1, resulting in chromosome condensation as the cell enters M-phase. The majority of gene transcription ceases as chromosome structure becomes more compact. The chromosomes and genes are apportioned among the two daughter cells, restoring the G_1 chromosome and gene numbers. M-phase kinase is inactivated near the end of M-phase as a consequence of cyclin degradation and dephosphorylation of p34. The chromosomes elongate, forming the threadlike structures present in G_1 cells; and transcription of genes occurs, reactivating the metabolic processes of the cell.

Other events are also occurring during the transition from G_2- to M-phase, and many of these activities appear to involve phosphorylation and dephosphorylation of nuclear proteins. Many of these changes remain to be elucidated; however, breakdown of the nuclear membrane is known to depend, in part, upon the phosphorylation of lamins and several other proteins found in the nuclear envelope. The phosphorylation of these proteins precedes disassembly of the nuclear envelope during M-phase.

Chromatin Replication

Replication of DNA and construction of new chromosomes occurs during S-phase of the cell cycle. The large size and complex structure of human chromosomes require additional processes compared to prokaryotic chromosome replication; however, the two systems share a number of features in common. DNA replication is semiconservative, bidirectional and discontinuous in both systems. Each human chromosome exhibits many replicons along its length, with each replicon encompassing about 100kb of DNA and its associated proteins. Replicons that are close together tend to replicate at similar times during S-phase; however, chromosome regions that are far apart, as well as different chromosomes, may replicate at different times during S-phase. As described earlier, heterochromatic regions of chromosomes complete their replication later than euchromatic regions. Histone synthesis is coordinated with DNA synthesis, providing one of the few examples of proteins that are produced during S-phase. Chromatin replication includes several steps: 1) relaxation of higher order structure, 2) unwinding and stabilization of the DNA double helix, 3) priming with a short RNA segment, 4) synthesis of DNA using the RNA primer as an initiation point, 5) removal of the RNA and gap filling with DNA, 6) ligation of the DNA segments, and 7) restoration of higher order chromosome structure.

Relaxation of chromosome structure. A typical human chromosome present in an interphase (G_1) nucleus consists of a 30nm fiber folded into a series of looped domains that are anchored to the nuclear scaffold. The fiber has a packed nucleosome structure that must be relaxed to permit replication. This relaxation occurs by unknown mechanisms that elongate the chromosome, permitting entry of the replication complex. Yeast chromosomes contain a special sequence at origins of replication. Such sequences have not been observed in human chromosomes. Replication appears to initiate at sites that are located between nucleosomes, with the replication fork progressing in both directions into the nucleosomes on either side (Fig. 3-14). The histones present in the nucleosome do not leave the DNA,

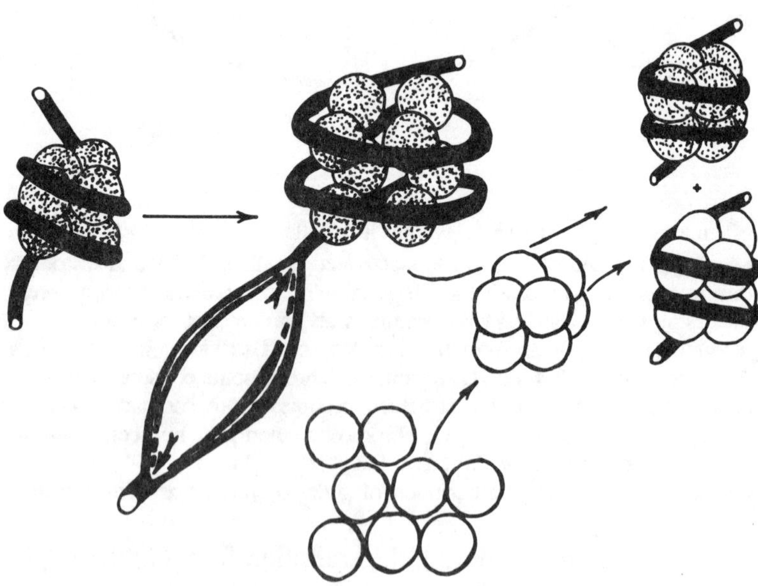

Fig. 3-14. Relaxation of nucleosome structure during chromosome replication.

but adopt a temporary relaxed structure which allows entry of the replicating enzymes (10). Several models have been proposed to explain the conformational changes occurring within nucleosomes during replication; however, the particulars of this phenomenon remain unclear.

Chromatin replication is bidirectional. Studies of both prokaryotic and eukaryotic replicons have revealed that replication proceeds in both directions from the replication origin (Fig. 3-15). Human chromatin often contains several clustered replicons in various stages of growth. Replication "bubbles" appear, enlarge, and merge with adjacent bubbles, until the replication of the entire chromosome has been completed. This bidirectionality of replication created an apparent paradox. DNA polymerases are capable of replicating DNA in one direction only by coupling a free 3' hydroxyl to a 5' phosphate group, such that the direction of chain growth would be anticipated to proceed from 5' to 3'. The bubbles appeared to be growing in both 5' to 3' and 3' to 5' directions. The apparent paradox was solved by recognizing that DNA was synthesized as a continuous strand when growth was in the "correct" direction (5'-->3') and discontinuous when overall growth was "against the grain" (3' to 5'). Discontinuous growth required multiple priming and DNA elongation events, producing a series of DNA fragments called **Okazaki** fragments (Fig. 3-16). Growth of each Okazaki fragment occurred in the correct 5' to 3' direction.

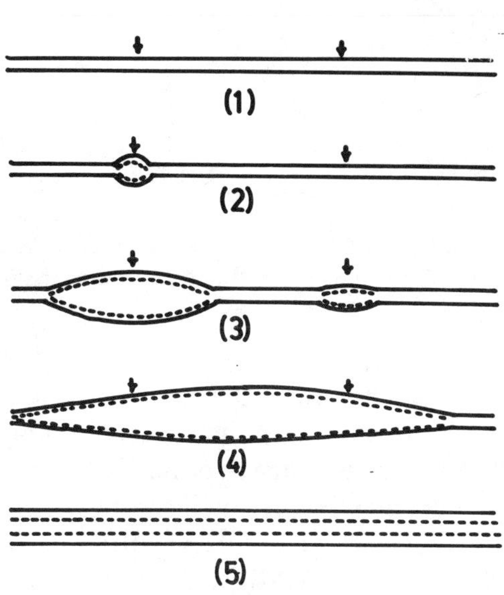

Fig. 3-15. Bidirectional replication of DNA. The double helix is represented in an unwound state. 1) Initiation sites for DNA replication (arrows). 2) Strand separation at first replication site and replication of DNA in both directions. 3) Bidirectional growth of first bubble and initiation of replication at second site. 4) Fusion of adjacent replication bubbles. 5) Daughter double helices consisting of one new and one old strand.

DNA polymerases. Several DNA polymerases have been found in human cells (Table 3-2) and have been renamed by a recent conference (11). DNA polymerase α participates with a primase in the initiation of DNA synthesis and synthesizes the **lagging** strand of new DNA (the strand that is formed as Okazaki fragments). DNA polymerase β is a small enzyme whose function is unknown. DNA polymerase γ is encoded by nuclear genes and catalyzes replication of mitochondrial DNA. Synthesis of the **leading** strand of new DNA (continuous growth in 5' to 3' direction) is accomplished by DNA polymerase δ. Repair of DNA damage appears to be effected by DNA polymerase ϵ. Preliminary evidence suggests that further heterogeneity may exist within one or more of these categories; however, for the purpose of the following discussion, each class will be assumed to contain one enzyme.

DNA Replication. The DNA animal virus, SV40, has been used as a model system for the in vitro analysis of eukaryotic DNA replication (12). Many aspects of the replication of this virus appear to be shared by human DNA replication; however, the description of human DNA replication provided in this section should be considered tentative until verified by appropriate experiments with human chromosomal DNA.

Replication of SV40 DNA begins with the binding of the T antigen protein, topoisomerase I,

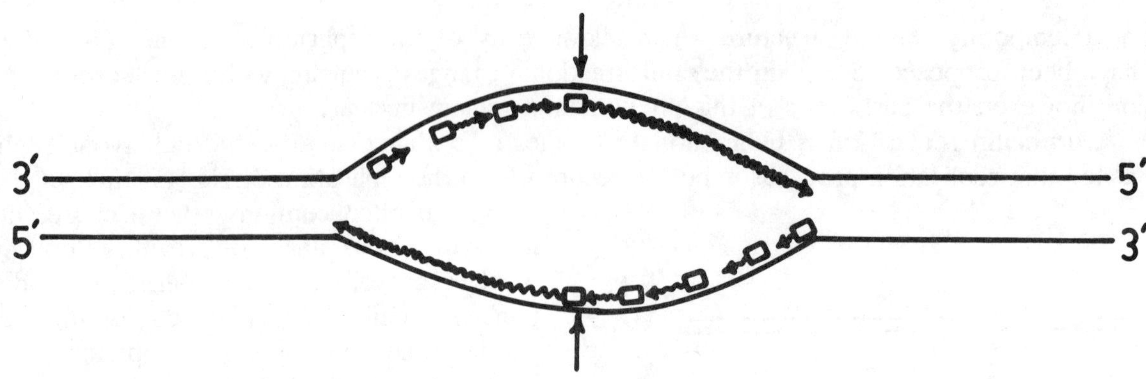

Fig. 3-16. Discontinuous DNA synthesis. The replication origin is indicated by an arrow, and blocks represent RNA primers. New DNA strands are symbolized by wavy lines. Notice how DNA is made as one continuous strand when new chain growth is in the 5' to 3' direction and as Okazaki fragments when <u>overall</u> growth is in the opposite direction.

Table 3-2. Human DNA Polymerases.

Type	Function
α	DNA synthesis: initiation & lagging strand
β	Function unknown
γ	Nuclear-encoded; mitochondrial DNA synthesis
δ	DNA synthesis: leading strand
ε	DNA repair

and helicase to the DNA. These enzymes collaborate in the unwinding of the double helix and formation of single-stranded DNA. This single-stranded DNA is stabilized by a protective protein called RF-1. DNA polymerase α, its associated primase, and a protein known as RF-C attach to each stabilized strand, one complex on either side of the origin. The primase lays down a short RNA sequence of about nine bases, and DNA polymerase α begins to synthesize DNA using the end of the RNA primer as a focus of initiation. The RF-C protein appears to facilitate interaction of the primase and polymerase with the DNA template. As the new DNA is formed, the protective protein is dislodged from the DNA. After synthesizing an Okazaki fragment, the RF-C/Polα complex moves to a new site closer to the 3' end of the DNA molecule serving as a template and repeats the process. The original Okazaki fragment is then elongated by a second complex containing DNA polymerase δ and PCNA, a protein that functions like RF-C. After replication of the virus is completed, the RNA primers are excised; and the gaps are filled with DNA complementary to the template DNA in the same regions. New DNA segments are joined by a DNA ligase, and the DNA is rewound, possibly by topoisomerase II. This sequence of events is summarized in Figure 3-17.

Many of the proteins participating in SV40 replication have been isolated from human cells. These include Polα, Polδ, primase, helicase, ligase, one or more single strand-binding proteins, an endo-nuclease, and topoisomerase II. These enzymes appear to occur in a large complex that is associated with the nuclear scaffold. One view of human DNA replication hypothesizes that the origin of replication of a chromosomal domain transiently attaches to the nuclear scaffold in association with the replication complex. As replication progresses, the replication forks remain associated with the scaffold and the daughter DNA helices are released as twin loops (Fig. 3-18). This hypothesis has recently been supported by the discovery that newly synthesized DNA is associated with the nuclear scaffold and that this nuclear fraction is enriched with replication forks (13).

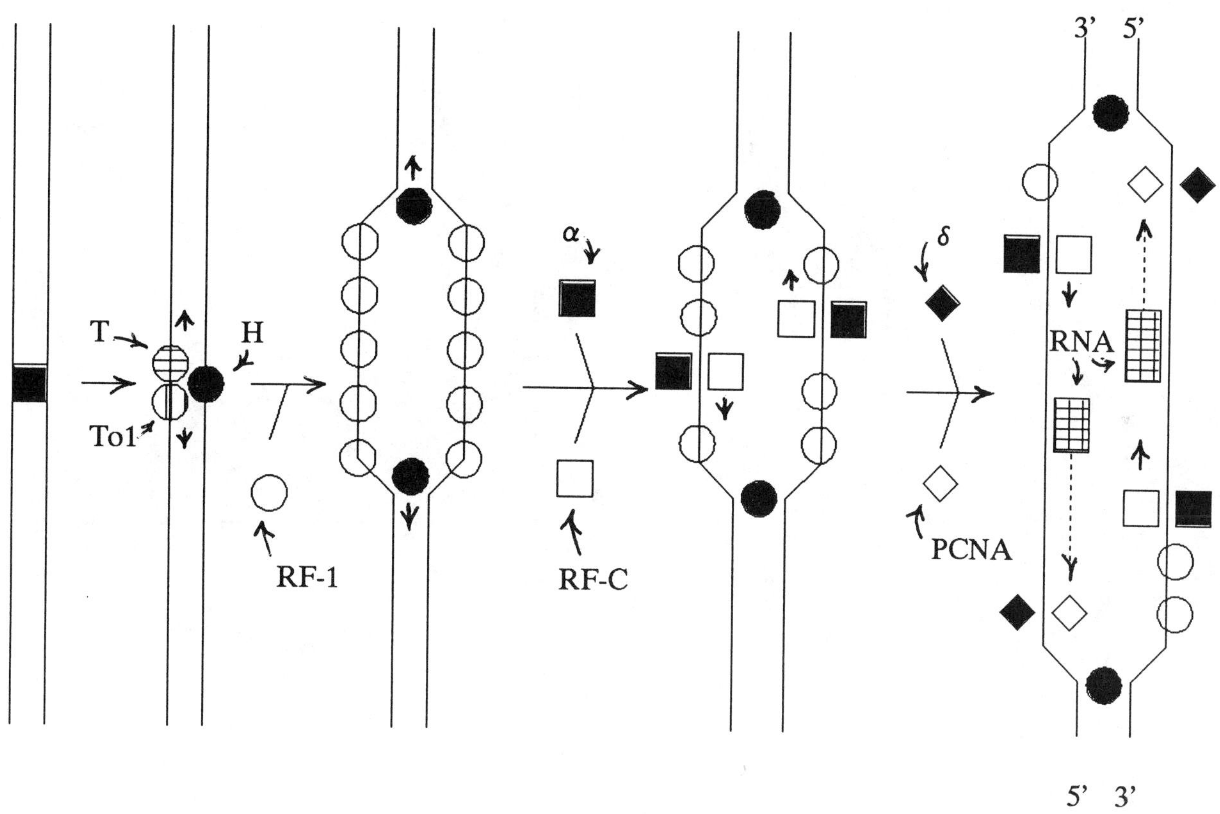

Fig. 3-17. Model for replication of the SV-40 virus. The origin of replication is represented by a shaded box in the double helix at the far left of the figure. 1) T-antigen (T), helicase (H), and topoisomerase I (To1) bind at the origin of replication and begin to unwind the double helix. 2) RF-1 protects the single-stranded DNA, and helicase continues to unwind the double helix in both directions. 3) DNA polymerase α, primase (both represented by α), and RF-C attach to each single strand on either side of the origin. Primase synthesizes a short RNA segment, and DNA polymerase α forms an Okazaki fragment. 4) The Polα/primase/RF-C complex translocates toward the 3' end of each template strand and begins to synthesize a second Okazaki fragment. DNA polymerase δ (δ) and PCNA continue to elongate the first Okazaki fragment on each strand. 5) RNA primers are excised by an endonuclease, and the gaps are filled with DNA. 6) Ligase joins the new DNA segments, and topoisomerase 2 rewinds the double helix.

     Segregation of histones among daughter nucleosomes occurs by a conservative mechanism. Histones are synthesized during S-phase on cytoplasmic polyribosomes and enter the nucleus where they associate with the replicating DNA in a conservative fashion (Fig. 3-19). That is, nucleosomes on a given daughter double helix contain all old histones or all new histones. One model for histone segregation proposes that old histones are incorporated into nucleosomes associated with those portions of the DNA

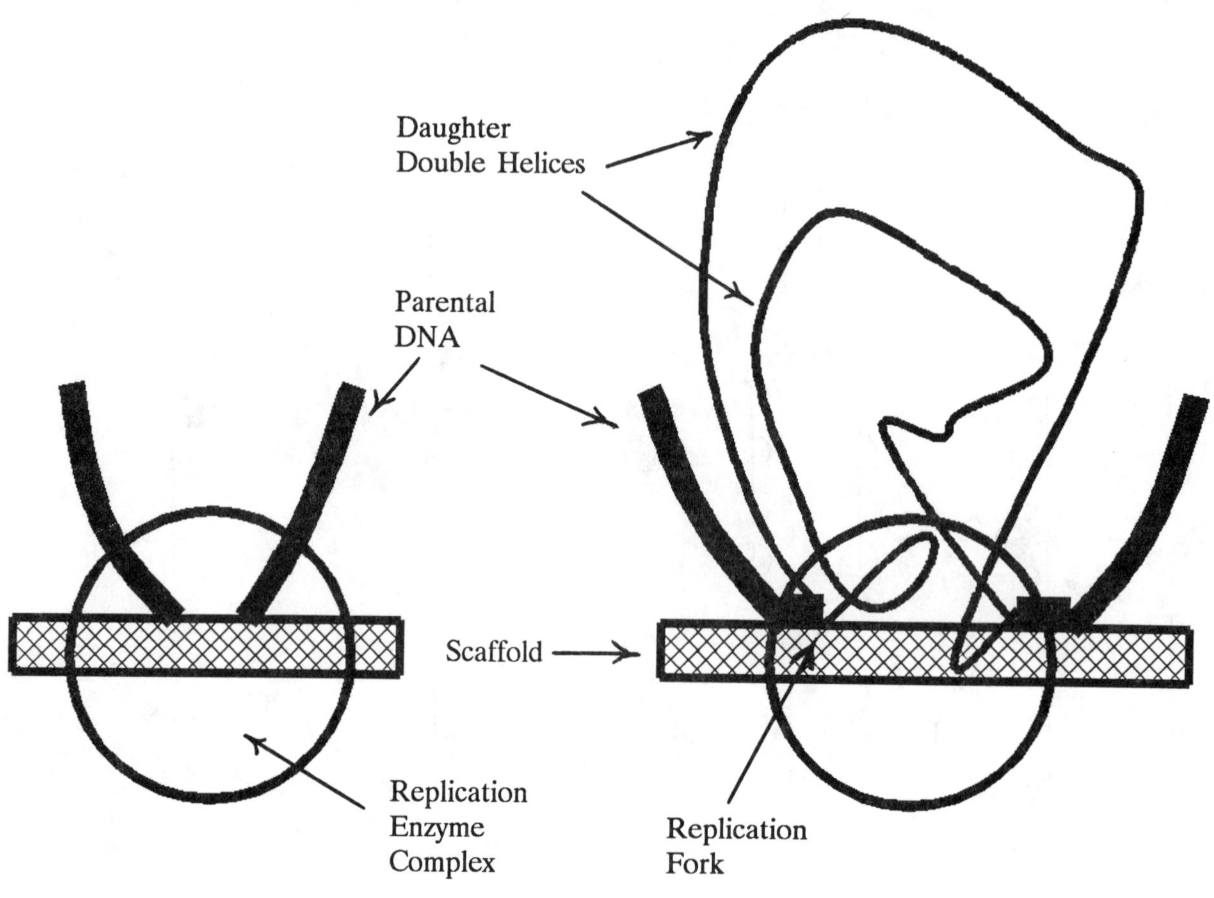

Fig. 3-18. Possible configuration for replication of DNA on the nuclear scaffold. Left: The DNA domain attaches to the scaffold at the replication origin. Right: DNA of the domain is reeled in from either side, with the replication forks remaining attached to the scaffold; and the daughter helices exit the scaffold between the replication forks.

that were formed by continuous DNA synthesis (catalyzed by Polδ), and new histones assemble into nucleosomes attached to the lagging strand. Nucleosomes in the vicinity of replicating DNA have a relaxed conformation; however, this loose assembly is quickly restored to the more compact nucleosomes observed in non-replicating areas. Maturation of nucleosomes includes alteration of their spacing along the DNA, chemical modification of histones, and a general compaction of their structure which denies access to DNases. This maturational process is completed by the time the replication fork has moved 5-40kb beyond a given nucleosome.

Mitosis

Mitosis is the process by which a single cell produces two daughter cells, each having a genetic complement identical to that of the parental cell. Mitotic division is preceded by chromosome replication. Each cell entering G_2 of the cell cycle now has twice the amount of DNA and twice the number of

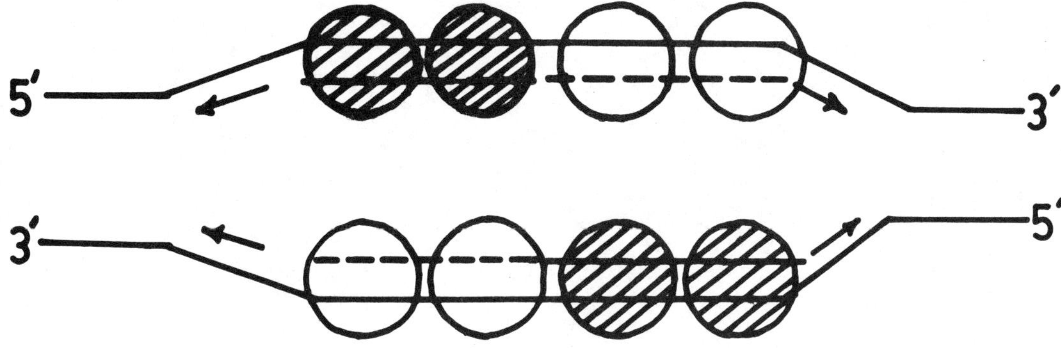

Fig. 3-19. Segregation of nucleosomes between daughter chromosomes behind the replication fork. Open circles represent nucleosomes containing new histones, while hatched circles are nucleosomes with old histones. The new nucleosomes are shown associating with the lagging strand composed of Okazaki fragments, while old nucleosomes are found on the leading or continuously-synthesized DNA.

genes and chromosomes present in the G_1 cell. Shortening of the chromosomes signals the onset of cell division, and this period of chromosome condensation is referred to as **prophase** (Fig. 3-20). As the chromosomes thicken, their double structure becomes evident. Each half of a mitotic chromosome is called a **chromatid.** Each chromatid contains a complete DNA double helix and its associated proteins and RNA, and will ultimately become a functional chromosome in each daughter cell produced by mitotic division. The two chromatids of each mitotic chromosome are held together at the centromere. During prophase the centriole, a structure adjacent to the nucleus, divides and migrates to opposite poles of the dividing cell. The nuclear envelope disappears, and nuclear structure breaks down.

The mitotic chromosomes line up in the central or equatorial plane of the cell during **metaphase.** The elements of the **mitotic spindle**, the centrioles and spindle fibers, are now visible. The mitotic spindle has evolved to assure equal distribution of the chromatids of each mitotic chromosome to the respective daughter cells. The spindle fibers are microtubules consisting of proteins that extend from the centriole to the centromere of each chromosome or from centriole to centriole.

Separation of the chromatids of each mitotic chromosome now occurs, and each chromatid begins to migrate to opposite poles of the dividing cell. The force responsible for chromatid movement appears to be generated by motor proteins associated with the kinetochore or centromere of each chromosome. As the motor proteins pull the chromatids toward the poles, the microtubules are degraded to their constituent amino acids, contributing to the overall shortening of the spindle fibers observed during this period of mitosis. The stage of mitosis during which chromatids migrate toward the poles of the dividing cell is called **anaphase.**

When daughter chromosomes arrive at the poles, the cells enter **telophase.** The cytoplasm divides in the region of the equatorial plane (**cytokinesis**), and the nuclear envelope is formed, isolating the nucleoplasm of each daughter cell from the cytoplasm. Each progeny cell reconstitutes its nuclear structure, and the chromosomes uncoil and become indistinct. The offspring of the parental cell are now complete, and metabolic activity resumes. The major purpose of mitosis is to insure that each daughter cell receives the same genes and chromosomes present in the cell from which they are derived. This type of cell division transforms the zygote into the mature individual who possesses more than 10^{12} cells, each cell possessing an identical genetic composition.

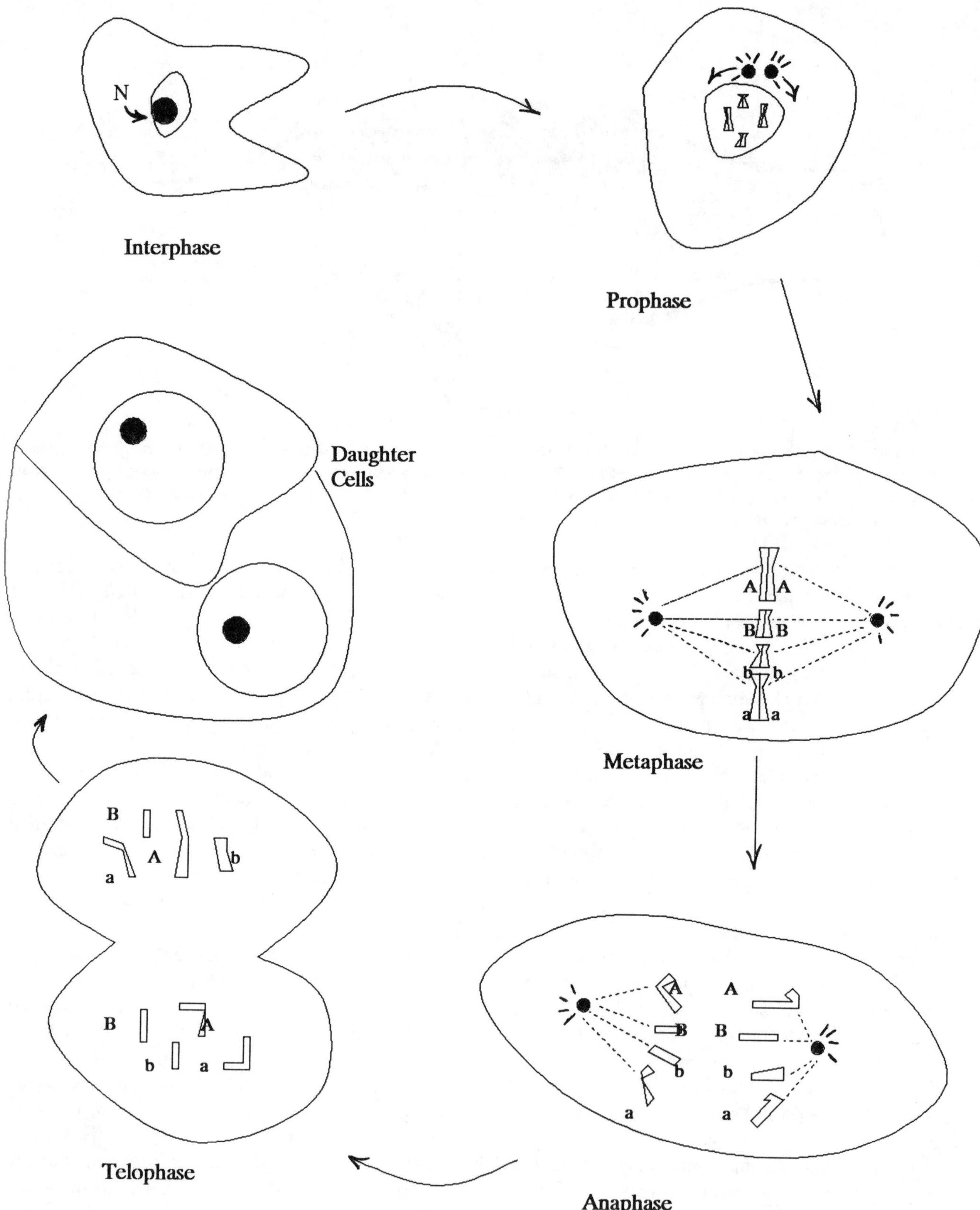

Fig. 3-20. Mitotic Division. A, a, B, and b are alternate forms (alleles) of the genes A and B, respectively. N = nucleolus. The centrioles are displayed above the prophase nucleus and at the left and right of the metaphase and anaphase chromosomes.

The effect of mitotic division may be summarized:

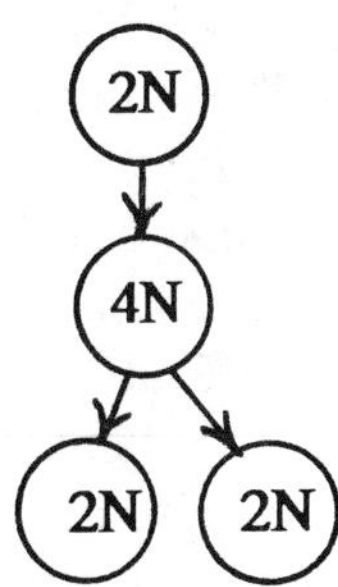

The letter **N** represents the **haploid** DNA complement or one set of chromosomes and associated genes. As indicated, each cell typically possesses two complete sets of chromosomes and two copies of each gene (excepting sex-linked genes where males have only one copy of each X-linked gene and one copy of each Y-linked gene).

Two types of chromosomes are presented in Figure 3-20. The larger chromosomes carry the gene **A** or its **allele, a.** The smaller chromosome carries **B** or **b.** The centromeres of the chromosomes carrying **A** or **a** are **submetacentric,** or displaced slightly toward one end of the chromosome. The chromosomes carrying **B** or **b** have centromeres close to one end and are **acrocentric** chromosomes. In general, human chromosomes may have centrally placed centromeres (**metacentric**), may be submetacentric, or may be acrocentric (Fig. 3-21). The chromosomes shown in the figure are derived from

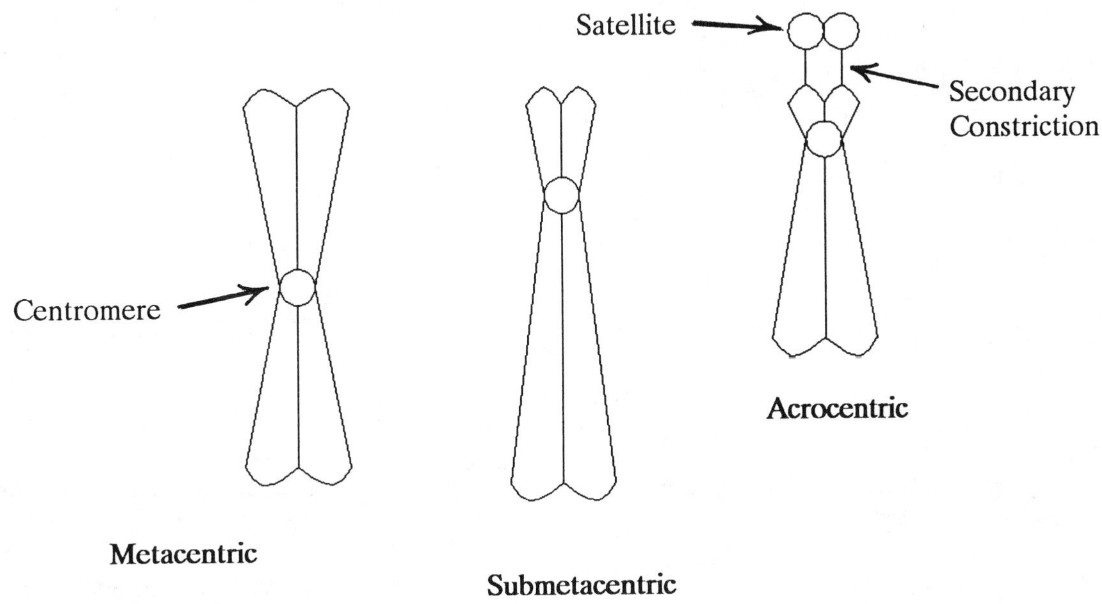

Fig. 3-21. Metacentric, submetacentric, and acrocentric metaphase chromosomes. Each chromosome possesses two chromatids (left and right halves). A chromatid contains one DNA double helix and will become a functional chromosome in the daughter cells. Acrocentric chromosomes are frequently satellited. The stalk, or secondary constriction possesses rRNA genes.

metaphase cells, and each metaphase chromosome contains two chromatids. Chromosomes of the same pair (eg. the chromosomes carrying the **A** and **a** alleles) are said to be **homologous** chromosomes. Note how the homologous chromosomes are located at different places in the metaphase plane. **A** and **B** are situated on chromosomes that are members of different pairs, or **heterologous** (**non-homologous**) chromosomes. Notice how alleles occupy the **same** relative position on homologous chromosomes. The site of a gene on a chromosome is called a **locus** (plural: **loci**).

Several important points are illustrated in Figure 3-20. 1) Homologous chromosomes do not associate in the equatorial plane at metaphase. 2) Daughter cells receive one maternally-derived and one paternally-derived chromosome of each pair. 3) Each cell inherits one maternal allele and one paternal allele of each gene. 4) The genetic content of each progeny cell is identical to that of its parent. This fundamental property of mitotic division is responsible for the faithful transmission of the genetic information present in the fertilized egg to all of our somatic cells.

Preparation of Human Mitotic Chromosomes

Human mitotic chromosomes (Fig. 3-22) are usually isolated from cultured cells derived from peripheral white blood cells (specifically T-lymphocytes), skin, or bone marrow. T-lymphocytes are most commonly used when a baby, child, or adult is suspected of having a chromosome abnormality that

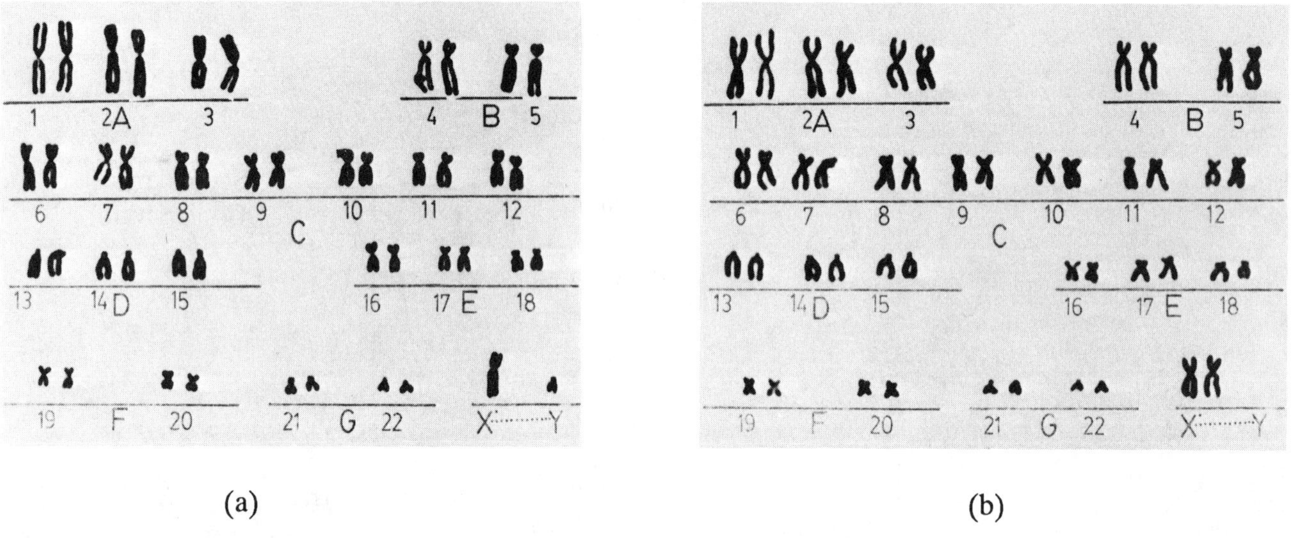

(a)   (b)

Fig. 3-22. Normal male (a) and female (b) karyotypes. The chromosomes have been stained with Giemsa without pre-treatment. (Courtesy S. Shafer)

involves most or all somatic cells. When both normal and abnormal cells are found in white blood cells, skin fibroblasts may be studied to provide an average estimate of the proportion of body cells possessing the normal and abnormal chromosome complements, respectively. Leukemias arise in bone marrow cells. Chromosomal changes associated with cancer are restricted to cancer cells. Therefore, only the leukemic cells would display the defect, making bone marrow the most reliable source of cells, particularly during the early stages of the leukemias. Prenatal diagnosis may be undertaken when a fetus is at relatively high risk for a chromosomal disorder. The baby is cushioned by the amniotic fluid which contains cells derived from the skin, respiratory, gastrointestinal, and genitourinary systems of the baby. These cells

92

are obtained by a process known as **amniocentesis** which will be described in a later chapter. At earlier stages of pregnancy, cells may be obtained from **chorionic villi**, tissue that does not become a part of the baby, but which possesses the same genes and chromosomes found in the developing infant. Either amniocytes or chorionic villus cells can be used to determine whether a fetus has an abnormal or normal chromosome complement.

Processing of cells for chromosome analysis varies with the source of the cells; however, the following procedures are often used in each case. White blood cells are grown for about 70 hours in a medium containing a mitotic stimulant, antibiotics, and appropriate nutrients. The cells are treated with a spindle poison such as colcemid that breaks down the spindle, arresting the cells at metaphase. The cells are harvested and placed in a hypotonic solution which swells the cells and permits adequate separation of the chromosomes. The cells are "fixed" (treated to stop the swelling), flattened on glass slides, dried, and processed for staining. The staining methods are described below.

Trypsin-Giemsa Banding. This is the most common procedure utilized for human chromosome analysis. The chromosomes are briefly treated with a protease called trypsin and then stained with Giemsa, a DNA-specific dye mixture. This staining protocol most intensely stains heterochromatic regions. Therefore, the dark bands identify non-expressed chromosome regions, while the light bands contain genes that were transcribed during the previous G_1-phase (Fig. 3-23). A fluorescent method,

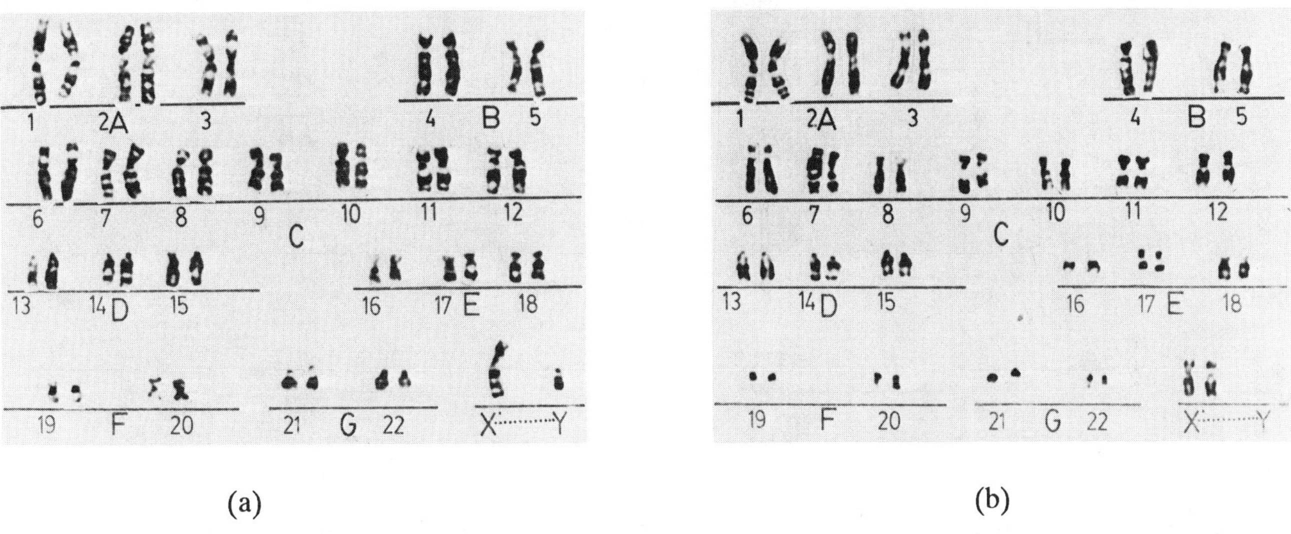

(a)　　　　　　　　　　　　　　　　　　　　(b)

Fig. 3-23. Normal male (a) and female (b) karyotypes of trypsin-Giemsa banded chromosomes. Note the greater detail of chromosome structure. (Courtesy S. Shafer).

Q-banding, utilizes a dye known as quinacrine-HCl to stain chromosomes. AT-rich areas stain brightly with this stain, whil GC-rich regions quench the fluorescence. Since heterochromatin is frequently AT-rich, Q-banding also preferentially stains heterochromatic regions of the chromosomes.

Reverse Banding. When chromosomes are treated with heat and/or alkaline salt solutions and then stained with Giemsa, a pattern complementary to that of TG-banding is obtained. The dark bands now represent active gene regions, while the light bands contain heterochromatin. R-banding is used when structural modifications of chromosomes involve euchromatic areas, because better detail is obtained, permitting a more precise determination of the sites involved in the alteration.

93

Centromeric Banding. One of the most severe procedures utilized for pretreatment of chromosomes involves heating of slides in alkali solution. This treatment promotes denaturation of DNA and depurination. This depurination is most severe in highly repetitious, highly condensed chromatin. The areas that stain most intensely following this type of pretreatment include centromeric heterochromatin and certain other regions such as the distal half of the Y chromosome. C-banding is used when questions regarding the centromeric region, the presence of Y chromosomes, or consequences of rearrangements involving C-banded areas arise. The effects of these staining procedures upon chromosomes 1 and 13 and the X and Y chromosomes are illustrated in Figure 3-24.

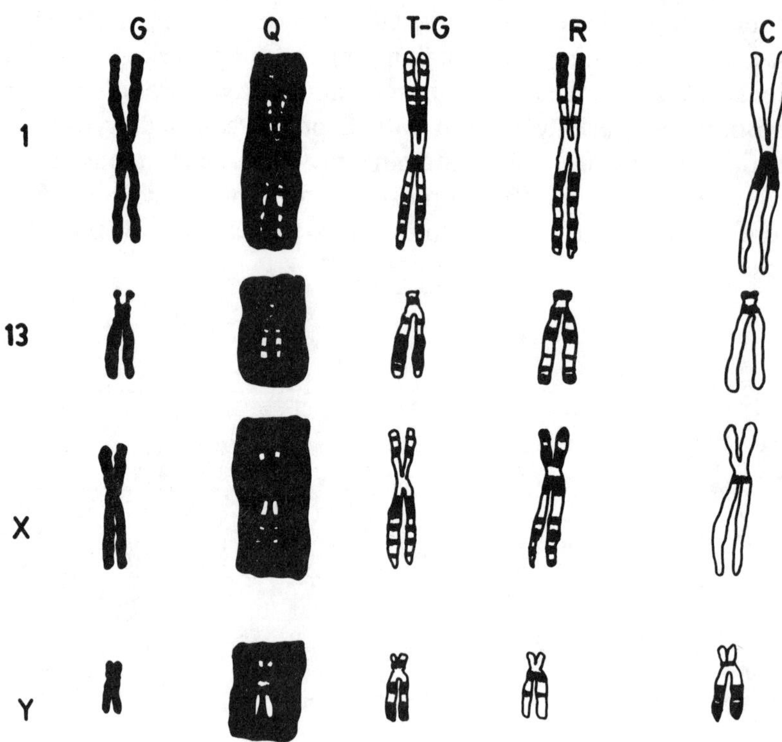

Fig. 3-24. Differential staining of human metaphase chromosomes. G = general Giemsa staining procedure; Q = Quinacrine-banding method; T-G = trypsin-Giemsa banding method; R = reverse banding method; and C = centromeric heterochromatin method.

Prometaphase chromosome banding. When small deletions or subtle chromosomal rearrangements are suspected, long chromosomes are advantageous for their detection. Cells are partially synchronized (forced to enter mitosis at the same time) by adding methotrexate, increasing the proportion of cells that are preparing to enter metaphase (prometaphase). Chromosomes are less condensed during prometaphase, such that single bands present during metaphase appear as two or more sub-bands during prometaphase permitting high resolution analysis of chromosome structure. Reducing the level of colchemid also enhances chromosome length.

Several other staining procedures, including both fluorescent and visible wavelength stains, have been developed for use in special circumstances. **Silver staining** of nucleolar organizers, regions of the chromosomes containing rRNA genes is often employed when attempting to distinguish between normal satellites and small pathological rearrangements affecting the satellite regions. Semiconservative replication results in the production of sister chromatids (chromatids of the same mitotic chromosome) that differ with respect to their relative content of new DNA after two rounds of replication. After the second round of replication, one chromatid will have a double helix that contains one new and one old DNA chain, while the sister chromatid contains an entirely new DNA molecule. If 5-bromodeoxyuridine (BrdU) is added to a cell culture, the BrdU replaces thymidine in the replicating DNA. Cells that are permitted to complete two DNA replications in the dark (to avoid photolysis of BrdU-containing strands) will possess chromosomes that contain twice as much BrdU in one chromatid compared to its sister chromatid. BrdU quenches fluorescence, and fluorescent dyes such as Hoescht 33258 will stain sister chromatids differently. Furthermore, sequential staining with Hoescht 33258 and Giemsa permits demonstration of sister-chromatid exchanges under visible light conditions (Fig. 3-25). This method of chromosome staining is called **Sister Chromatid Staining** or **Harlequin Staining**. Rapid progress is being made with the development of probes for identification of specific chromosomes or chromosome regions. For example, a series of biotinylated probes have been designed which display chromosome-specific binding. Use of these probes and fluorescent microscopy permits identification of small derivative chromosomes and chromosome fragments produced by chromosome rearrangements. A useful gene-specific probe recently was employed to detect a chromosome rearrangement in interphase chromatin that is associated with chronic myeloid leukemia (CML). CML arises in cells that have sustained a translocation (exchange of chromosome segments between heterologous chromosomes) disrupting the **c-abl** gene on chromosome 9 and a **bcr** gene on chromosome 22. A biotinylated probe for the **c-abl** gene and a probe for the **bcr** gene that was laballed with

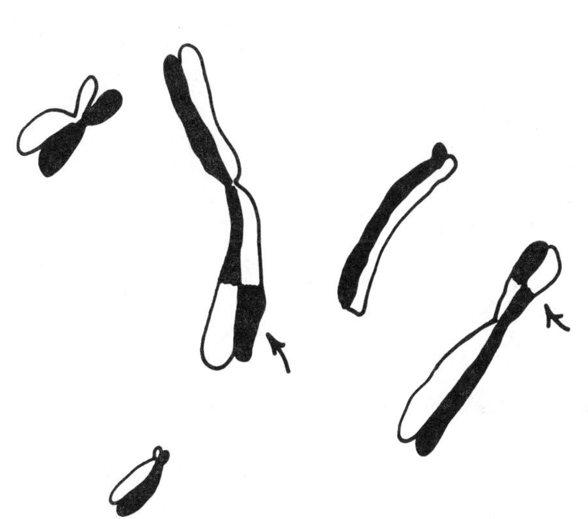

Fig. 3-25. **Sister chromatid exchange revealed by culture of cells in BrdU-containing medium followed by staining with Hoechst 33258 and Giemsa.**

digoxigenin. The **c-abl** probe was detected with the fluorochrome, Texas red, and the **bcr** probe was identified with fluorescein isothyocyanate, another fluorochrome that displays green fluorescence. When interphase cells from CML patients were hybridized with the probes and treated with the fluorochromes, the intact **c-abl** gene appeared as a red dot, and the intact **bcr** gene fluoresced green. The **bcr-abl** fusion gene formed as a consequence of the translocation appeared yellow (14). This procedure which utilizes hybridization of a probe with a specific gene sequence of a chromosome is called <u>in situ</u> hybridization.

The Normal Human Karyotype and Chromosome Nomenclature

Human mitotic chromosomes are arranged according to a system defined by the Denver (15) and Paris (16) conferences and several others that have been held since the latter meeting. This arrangement is called a **karyotype**. Most human nuclei have 46 chromosomes that can be arranged in 23 pairs (Fig. 3-23). Two of these chromosomes are heteromorphic, the X and Y chromosomes, and are referred to as **sex chromosomes**, since they participate in sex determination. The remaining 22 pairs are called **autosomes**. Males have 22 pairs of autosomes and one X and one Y chromosome. Females also have 22 autosomal pairs; however, the Y chromosome is replaced by an additional X chromosome. The autosomes are apportioned among seven groups (A-G). Each of the D- and G-group chromosomes are acrocentric and possess satellites adjacent to tandemly repeated rRNA gene sequences.

Karyotypes are labelled according to a convention that places the total number of chromosomes first, followed by the sex chromosome complement and the abnormality found, if any. Abbreviations commonly used for these labels are presented in Table 3-3. For example, normal male and female

Table 3-3. Symbols Commonly Used to Describe Human Karyotypes.

p = short arm	p- = deletion affecting short arm
q = long arm	t = translocation chromosome
/ separates cell lines with different karyotypes	inv = inversion
-21 = entire chromosome 21 is missing	i = isochromosome
r = ring chromosome	+13 = an extra chromosome 13 is present
q+ = additional segment attached to long arm	der = derivative chromosome

karyotypes would be designated 46,XY and 46,XX, respectively. Additional examples of karyotype labels are presented in Table 3-4. Detailed descriptions of the structural and numerical alterations displayed and their clinical consequences will be given in the next chapter.

Human Sex Determination and Sex Differentiation

Determination of human sex and differentiation of the male and female reproductive systems and secondary sexual characteristics are displayed in Figure 3-26. The fetal gonads (reproductive organs) at 5-6 weeks of gestation are bipotential - they are capable of developing into either testes or ovaries. The fetus also has two paired duct systems, the mullerian ducts and the mesonephric (Wolffian) ducts. The commitment to ovarian or testicular development occurs through the action of the product of a testis-determining factor encoded by a gene located on the short arm of the Y chromosome, **TDF** (Fig. 3-27). If a 7-week fetus has a normal Y chromosome carrying **TDF**, the factor encoded by this gene activates one or more genes that participate in development of the testis. Testis-determining factor is believed to be a DNA-binding protein that interacts with the promoters of the genes directing testicular development, inducing their expression. The resulting fetal testis produces two hormones, **testosterone** and **mullerian inhibitory factor (MIF)**. Testosterone perpetuates differentiation of the Wolffian ducts into the male accessory duct system, including the vas deferens and epididymis. MIF inhibits further development of the mullerian ducts which form the fallopian tubes, uterus and a part of the vagina.

Table 3-4. Selected Examples of Karyotype Labelling.

47,XY,+21	Male with Down syndrome, a condition caused by the presence of three copies of chromosome 21.
46,XX/45,X	A woman who is mosaic for Turner syndrome, a syndrome associated with lack of one sex chromosome. This woman has two cell lines, one normal and the other Turner's.
46,XX,t(9;22)(q34;q11)	This woman has CML. Her leukemic cells carry the translocation described earlier, where breaks occurred in band 34 of the long arm of chromosome 9 and band 11 of the long arm of chromosome 22. The segments distal to the break points exchanged places and fused to chromosomes 9 and 22 respectively. This translocation is diagrammed in the next chapter.
46,XXqi	This woman has one normal X chromosome and one isoX chromosome consisting of identical arms corresponding to the long arm of the normal X.

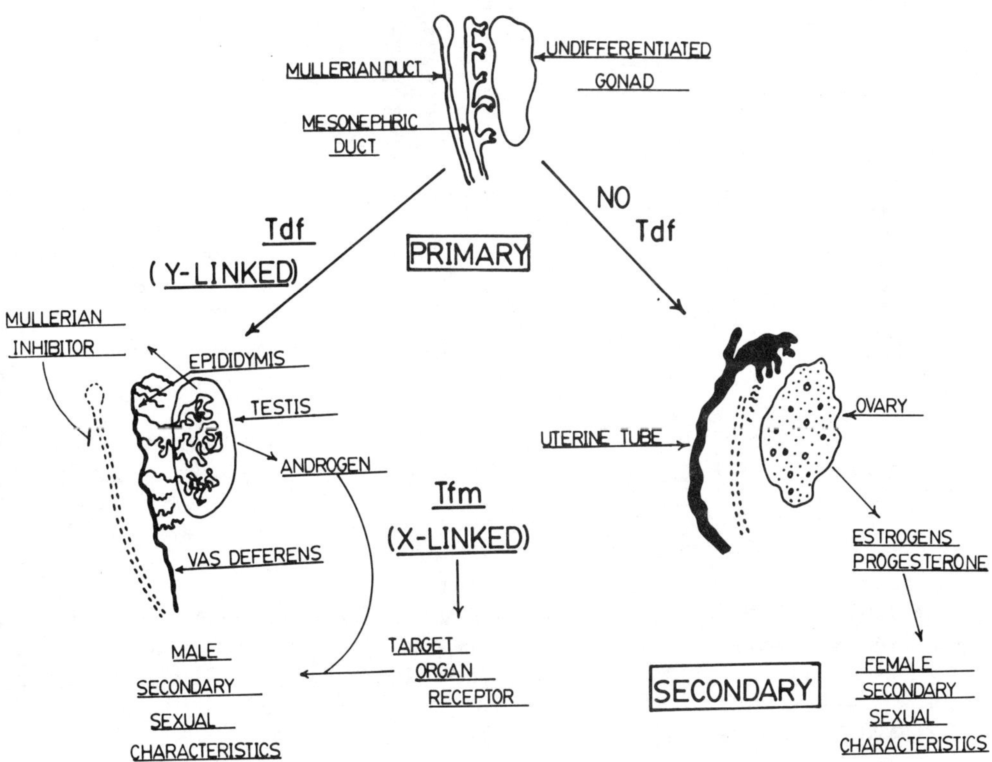

Fig. 3-26. Human sex determination and development of the male and female reproductive systems.

Fetuses who have a normal 46,XX karyotype lack a Y chromosome and its associated **TDF** gene. In the absence of testis-determining factor, the testicular differentiation genes are not activated, and an alternative cascade of differentiation events takes place that promote ovarian development. These events are complex and most certainly mediated by genes located on the X and on various autosomes. Very little is known regarding the early genetic and molecular changes that are responsible for ovarian differentiation. The ovary produces neither testosterone nor MIF. In the absence of testosterone, the Wolffian ducts gradually disappear, preventing appearance of the structures they form in males. Lack of MIF permits continued differentiation of the mullerian ducts into the fallopian tubes, uterus and upper third of the vagina.

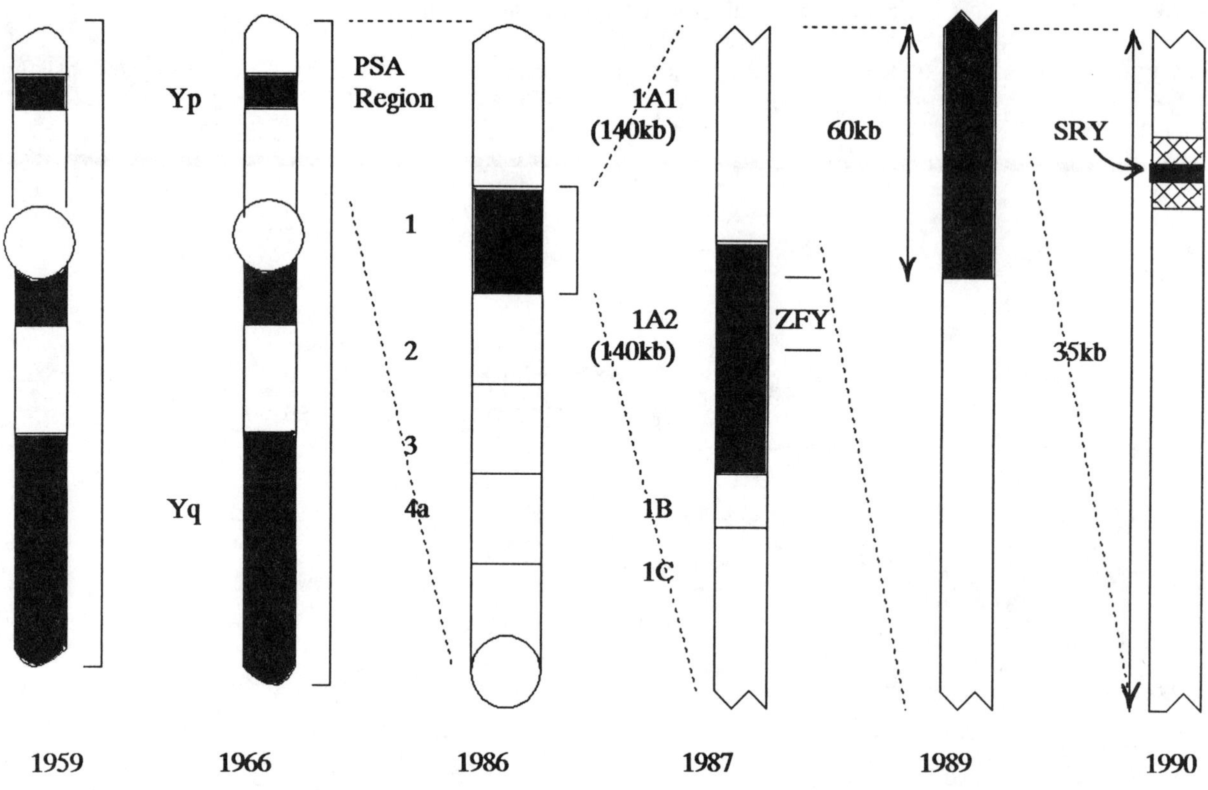

Fig. 3-27. The search for TDF. The Y chromosome was identified as the sex-determining chromosome in humans and mice in 1959, and the sex-determining region was localized to the short arm in 1966. Advent of molecular probes and studies of Y-X and Y-autosome translocations have rapidly confined the region to smaller segments of Yp. PSA = pseudoautosomal region that pairs and recombines with a similar region on the X-chromosome during meiosis; ZFY = zinc-finger Y; SRY = sex-determining region of Y. (Modified from McLaren (17))

The quest for the male sex-determining gene. Improved cytogenetic methods led to the identification of the Y chromosome as the bearer of the male sex-determining region in 1959, and about seven years later it was found that the short arm of the Y was sufficient for male sex determination. Candidate genes for **TDF** must meet several requirements: 1) they must be present in all individuals who have testes and absent in all persons who lack them; 2) the gene must be Y-linked and located in the sex-determining region; 3) the gene must be evolutionarily conserved, at least among mammals; and 4) the gene should have sequences that encode DNA-binding motifs. The first Y-linked candidate gene was discovered when tissue from male purebred mice was transplanted to female mice from the same inbred strain. The females rejected the male tissue. The only genetic difference between the donor and recipient of the grafted skin could be attributed to the Y chromosome that was not present in the female mice. The Y-linked gene became known as H-Y (Histocompatibility-Y), and it was soon identified in a wide variety of mammals, including humans. Almost without exception, the H-Y antigen was only present in males with testes and absent in persons lacking testes. However, the exceptions proved to be the demise of H-Y. Mice were observed who had testes but who lacked H-Y antigen. The human H-Y gene was subsequently mapped to the proximal long arm of the Y chromosome, a region that is not involved with sex-determination. H-Y may be required for normal spermatogenesis.

The second candidate gene proved even more exciting. A 46,XX **male** patient was discovered who lacked all of the Y except for sequences from region 1 of the short arm. This region was cloned in its entirety by Dr. David Page and his associates, and a gene was soon found that had several zinc fingers and satisfied the requirements for a DNA-binding protein capable of regulating transcription. The gene was evolutionarily preserved and was found in the exceptional patients and mice who lacked H-Y. This gene was called **ZFY (Zinc finger-Y)**. Normal men had ZFY and women lacked the gene. All was well until three XX men and an intersexual person with testicular tissue were found who did not possess the ZFY gene. Although the three men lacked ZFY they possessed sequences from the adjacent region, 1A1. Data were also emerging from murine genetics which also suggested that ZFY was not the testis-determining gene. ZFY transcripts were found in several tissues that had nothing to do with sex determination, and ZFY appeared to be expressed at the wrong time for testis-determination in mice. This gene and the homologous genes, **ZFX and ZFA**, which are very similar to ZFY but located on the X chromosome and on an autosome may be involved with some later step in testicular development.

The latest candidate gene for **TDF** has been detected in a 35kb subregion of 1A1. The mouse homolog of the gene, **SRY (sex-determining region of the Y)** is expressed at the time of sex-determination in the bipotential mouse gonad and contains a highly conserved DNA-binding motif that has been identified in a wide range of organisms extending from yeast to humans. Further research is necessary to establish **SRY** as the long sought **TDF**.

The role of hormones in sex differentiation. Several hormones are known to play important roles in male and female sexual development. The early actions of MIF and testosterone were described above. Testosterone and its derivative, dihydrotestosterone were discussed in Chapter 2. Both of these hormones utilize the same androgen receptor and identical or very similar hormone response elements to regulate gene transcription. Testosterone is essential for normal spermatogenesis. It not only induces the initiation of spermatogenesis, but also participates in the feedback regulation of the release of pituitary hormones (gonadotropins) that control testosterone synthesis and release as well as other aspects of male fertility. Dihydrotestosterone is essential for the formation of the prostate gland and works synergistically with testosterone to promote other male secondary sexual characteristics such as hair pattern and growth, enlargement of the male genitalia, muscle development, and depth of voice. At least some of these androgen effects depend upon induction of different genes by testosterone and dihydrotestosterone. Figure 2-46 in the previous chapter proposes one model for this discrimination. The conformation of the androgen receptor is modified in different ways by the two hormones, affecting their binding to the androgen response element and nearby accessory proteins. This model requires experimental confirmation. Both men and women possess the androgen receptor gene and express the protein it determines. Female pubic and axillary hair are androgen-induced and require the androgen receptor for their normal appearance. The female adrenal gland is the usual site of synthesis of the androgens produced by women, and these hormones are synthesized at a much lower rate than characteristic of male testosterone and dihydrotestosterone biosynthesis.

Estrogens and progesterone are copiously synthesized by the ovary at puberty, triggering the onset of female secondary sexual differentiation. Both hormones utilize a receptor-mediated mechanism for gene activation. For example, estrogens enter the nucleus and bind to the estrogen receptor. The complex binds to the estrogen response element within the promoters of estrogen-responsive genes, influencing their transcription. Normal estrogen and progesterone action are required for breast development, full development of the female external genitalia, normal adolescent growth, feminine body proportions, menstruation, and pregnancy. Estrogen and progesterone secretion are subject to regulation by pituitary gonadotropins, and, like androgens, they regulate their own production and release by feedback regulation of gonadotropin release. As the level of steroid hormone falls, more gonadotropin is released, increasing the level of steroid. As the level of steroid rises, the steroid curtails gonadotropin release.

Structures of human androgen, estrogen, and progesterone receptors are illustrated in Figure 3-28. All three receptors are encoded by genes consisting of eight exons and display structural similarities that suggest that all three receptors evolved from a common ancestral protein. The N-terminal domain of each receptor directly or indirectly influences the ability of the receptor to selectively bind to hormone-responsive genes. Exons B and C encode the DNA-binding domain of each receptor and

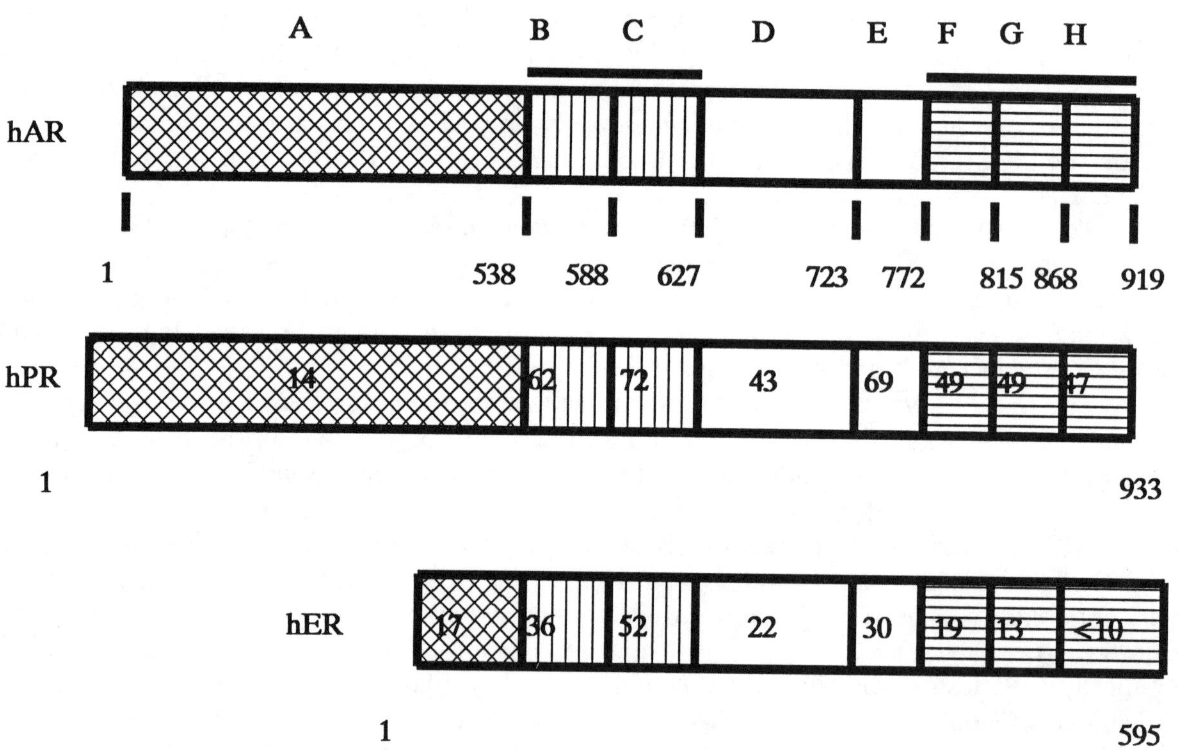

Fig. 3-28. Steroid receptors. hAR = human androgen receptor; hPR = human progesterone receptor; hER = human estrogen receptor. A-H = domains encoded by separate exons. A = N-terminal, B & C = DNA-binding, and F-H = hormone-binding domains, respectively. Numbers in hPR and hER domains indicate percent amino acid similarity to corresponding hAR domain.

recognizes the hormone response elements presented in Table 2-2 of Chapter 2. The hormone-binding the hormone-binding domain of each receptor is encoded by exons F-H. The human androgen and progesterone receptors appear to be more closely related to each other than to the estrogen receptor. If the hormone receptor genes evolved from a common ancestral receptor gene as has been proposed, the greater similarity of the androgen and progesterone receptors suggests a more recent divergence of these genes compared to the estrogen receptor gene.

Normal male and female fertility require the input of additional genes, some of which are autosomal and some sex-linked. Fertile women require two intact X chromosomes, and men must possess an intact X and Y to produce fully functional sperm. Furthermore, normal thyroid and pituitary function are also essential for normal fertility. For example, women who have thyroid deficiency are frequently infertile, and defective gonadotropin production or release disrupts the normal menstrual cycle. Deficiencies involving enzymes required for steroid biosynthesis by the adrenal cortex may lead to sex reversal. The enzyme, 21-hydroxylase, is essential for the production of cortisol. Deficient 21-hydroxylase activity results in cortisol deficiency. Lack of cortisol stimulates the release of adrenocorticotropic hormone (ACTH) by the pituitary. This increased ACTH causes the overproduction of other steroid hormones, including androgens, by the adrenal cortex. Increased androgen levels induce precocious puberty in young boys. Furthermore, since girls possess the androgen receptor, female infants may be masculinized by the effects of the increased androgen. 21-hydroxylase deficiency can be detected at birth, and its effects can be prevented by cortisol replacement therapy. Many states conduct a neonatal screening program for 21-hydroxylase deficiency to permit timely intervention and prevention of the abnormal effects of the deficiency. Other abnormalities of sexual differentiation will be described in later chapters.

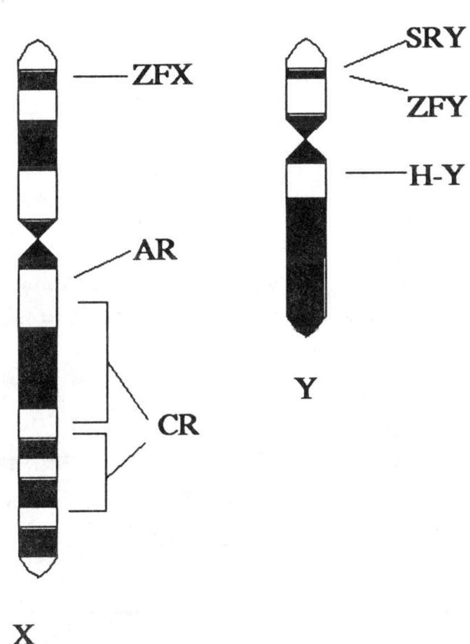

Fig. 3-29. X- and Y-linked loci participating in sexual differentiation. CR = critical regions. See text for explanation.

The X- and Y-linked loci participating in sexual differentiation are summarized in Figure 3-29. **SRY** is the current candidate for **TDF**, the gene encoding testis-determining factor. **ZFX** and **ZFY** are X- and Y-linked loci, respectively, which encode DNA-binding proteins containing zinc finger motifs. Although the functions of these proteins have not been established, they are believed to play a role at some later stage of sexual differentiation and/or fertility. **H-Y**, the Histocompatibility-Y locus, seems to be essential for spermatogenesis. **AR** is the androgen receptor locus, is subject to X-inactivation in females, and is expressed in both sexes. The brackets encompass two regions of the X that must be present in two copies for normal oogenesis and female fertility. Additional autosomal loci that play key roles in sexual differentiation include the 21-hydroxylase locus (**CYP21**: chromosome 6), the estrogen receptor locus (**ESR**: chromosome 6), and the progesterone receptor (**PR**), 5-α-reductase, and mullerian inhibitory substance (**MIS**) genes that have not yet been mapped to chromosomes.

Meiosis

The process of meiosis reduces the **diploid** number of chromosomes (two complete sets) of chromosomes (total = 46) to the **haploid** number (one complete set) (total = 23) (Fig. 3-30).

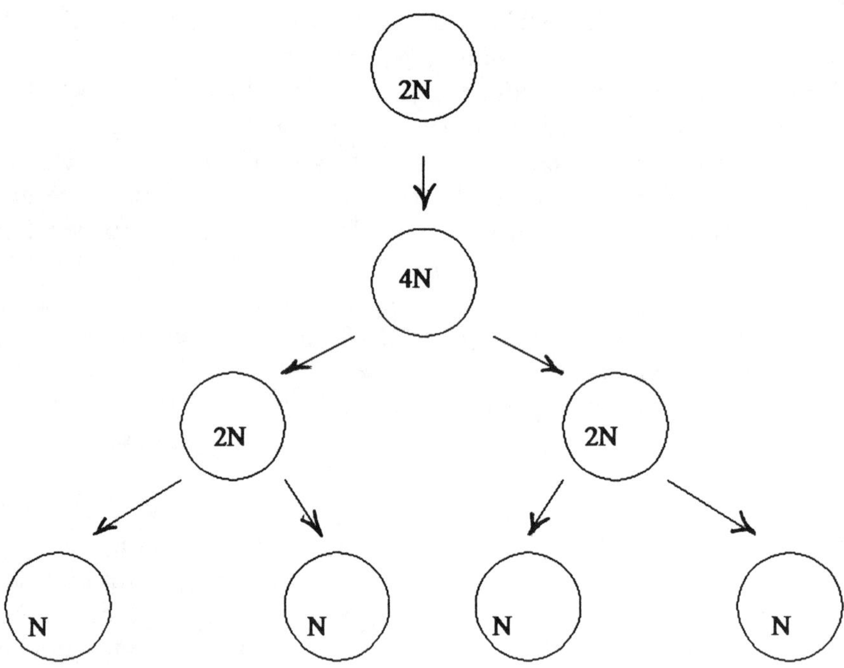

Fig. 3-30. **Effects of meiotic division on chromosome number.** Meiosis begins with chromosome replication and proceeds through two consecutive divisions generating four haploid (N) cells from the original parental cell (2N), where N = the haploid chromosome number (23 for humans).

Meiosis appears to have three effects upon the organism in which it occurs: 1) it maintains the correct chromosome number from one generation to the next; 2) meiosis maintains genetic variability in a population of organisms; and 3) it provides an opportunity for correction of errors that may have occurred in chromosomes during the replications that took place preceding each of the many mitotic divisions that formed the organism.

The human reproductive cycle is shown in Figure 3-31. The male **spermatozoa** (23 chromosomes) arise from progenitor cells known as **spermatogonia** (46 chromosomes) by meiotic division. Each spermatogonium gives rise to four spermatozoa or sperm. By contrast, each **oogonium** produces one mature egg or **ovum** and three **polar bodies** following meiosis. The polar bodies degenerate and do not participate in fertilization. Sperm and egg fuse to form the zygote (46 chromosomes) which subsequently divides mitotically to form the organism and the next generation of spermatogonia or oogonia.

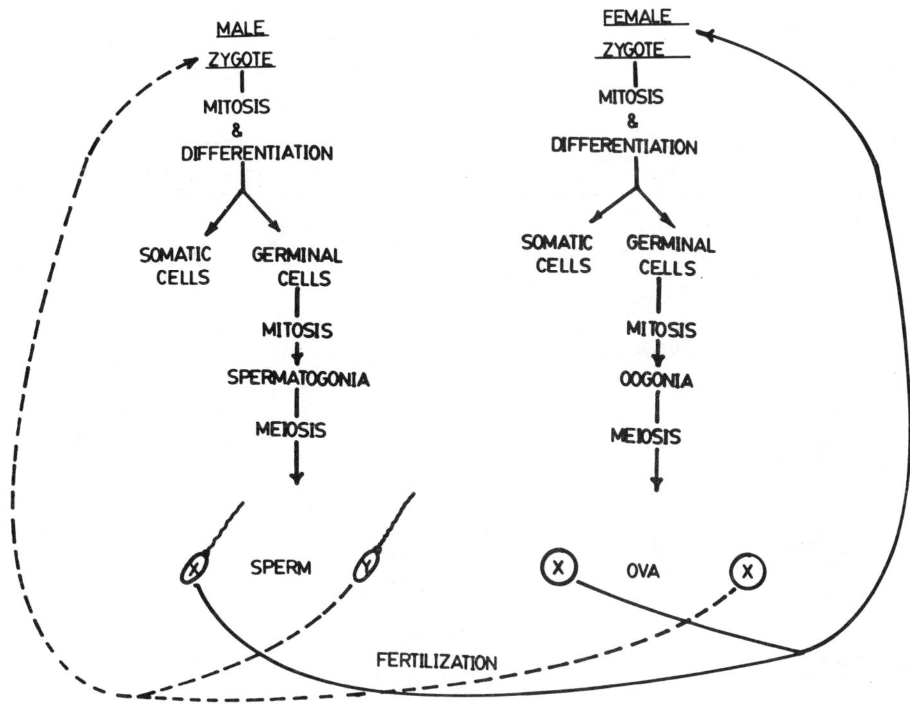

Fig. 3-31. Human reproductive cycle.

Spermatogonia and oogonia develop from more primitive germinal cells by a series of mitotic divisions. Female oogonia initiate meiosis before birth, and some may not complete meiotic division until 30-40 years later. Spermatogonia continuously enter meiosis from puberty until the seventies, and generally complete a meiotic cycle in forty days. Meiosis begins with replication of the chromosomes Fig. 3-32a. Two pairs of chromosomes, each with a single locus, are shown for clarity. Meiosis involves two divisions following chromosome replication, and the first division tends to be prolonged compared to the second division. Prophase 1 can be divided into five stages. **Leptotene** occurs first and is characterized by the appearance of extended thread-like chromosomes. Homologous chromosomes now associate by specific lengthwise pairing (**synapsis**). In males, the X and Y associate by pairing of their respective pseudoautosomal regions located near the ends of their short arms. The chromosomes continue to shorten during **pachytene** and appear as thick threads. The pachytene chromosomes are referred to as **bivalents**. Two bivalents are presented in the pachytene cell illustrated in Figure 3-32a; however, human pachytene cells contain 23 bivalents. During **diplotene** the synaptic pairing is relaxed as repulsive forces initiated at the centromere cause the chromatids to separate, and each bivalent is now observed to possess four chromatids, each with a single DNA double helix and its associated proteins. The bivalent is now called a **tetrad**. Each chromatid of the tetrad may be viewed as equivalent to the chromatid of a mitotic cell. The chromatids of the tetrad are held together by **chiasmata** which are interchromatid bridges connecting one maternally-derived chromatid and one paternally-derived chromatid. The chiasmata correspond to sites of exchange of genetic material between the maternal and paternal chromatids, and several may be found along the length of a chromosome. The actual exchange of

<u>MEIOSIS</u>

<u>PROPHASE   I</u>

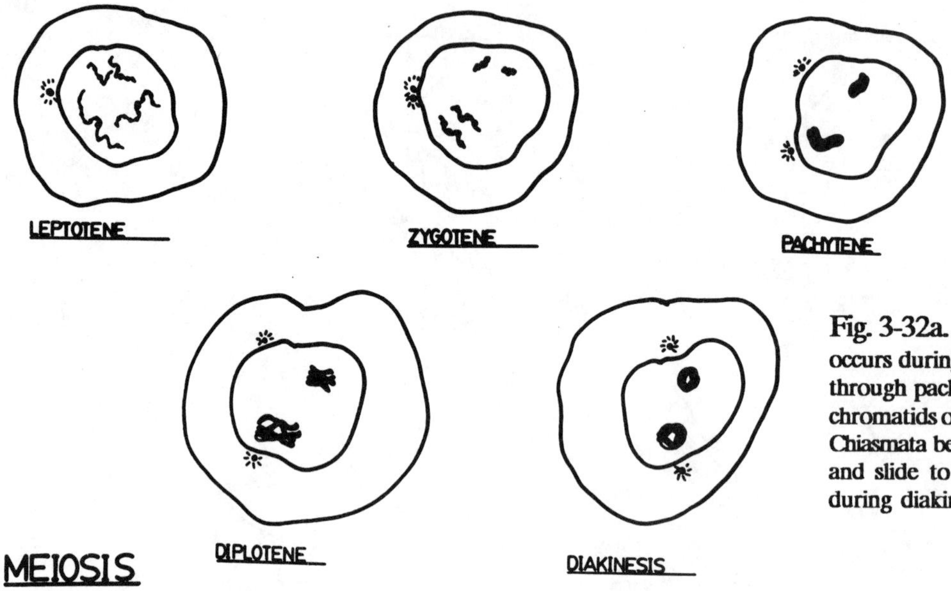

<u>LEPTOTENE</u>

<u>ZYGOTENE</u>

<u>PACHYTENE</u>

<u>DIPLOTENE</u>

<u>DIAKINESIS</u>

Fig. 3-32a. Meiosis I: Prophase. Synapsis occurs during late zygotene and continues through pachytene. During diplotene the chromatids of each tetrad repel each other. Chiasmata become visible during this period and slide to the ends of the chromatids during diakinesis.

<u>MEIOSIS</u>

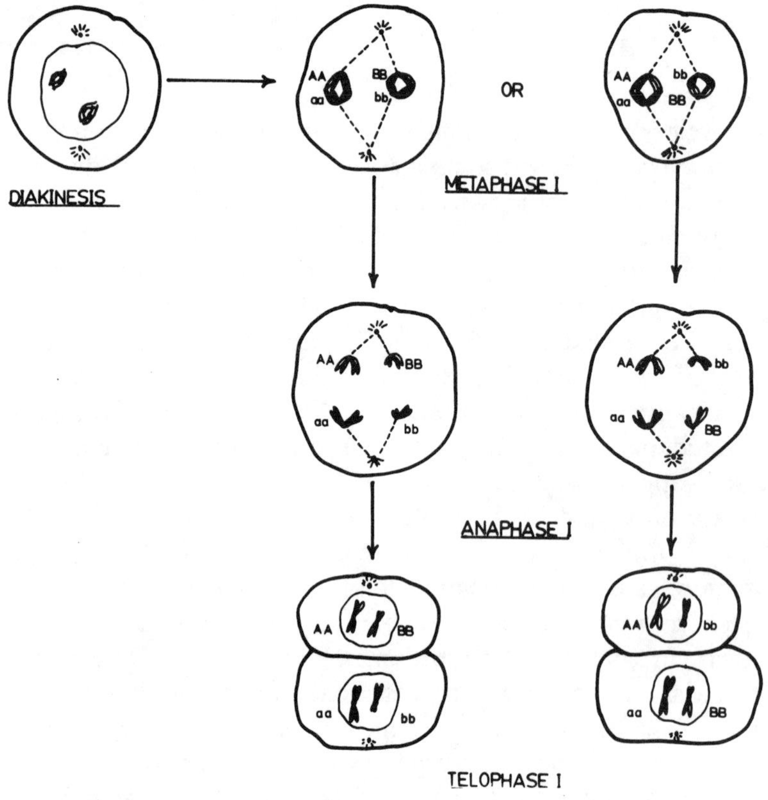

<u>DIAKINESIS</u>

AA
aa
BB
bb

OR

AA
aa
bb
BB

<u>METAPHASE I</u>

AA
aa
BB
bb

AA
aa
bb
BB

Fig. 3-32b. Meiosis I: Reduction division. The <u>maternal</u> (lower case letters) and <u>paternal</u> (upper case letters) alleles migrate to opposite poles of the dividing cell. Note that each daughter cell contains two copies of one of the two alleles at each locus. Since the parental cell had two different alleles at each locus, the genetic variability per cell has been reduced.

<u>ANAPHASE I</u>

AA
BB
aa
bb

AA
bb
aa
BB

TELOPHASE I

104

material occurs during synapsis of the chromosomes that is initiated during the zygotene stage. Presumably the stress imposed by the tight association and coiling of the chromatids leads to breaks in the respective strands and fusion of each broken fragment with the opposite chromatid (maternal or paternal, Fig. 3-33) resulting in an exchange of alleles. This genetic **recombination** produces chromosomes that contain arrays of genes that are different from those in other cells of the individual and different from those of either of his parents. This scrambling of genes as a consequence of recombination between homologous chromosomes is superimposed upon the usual independent assortment of genes that occurs during meiosis (Fig. 3-35).

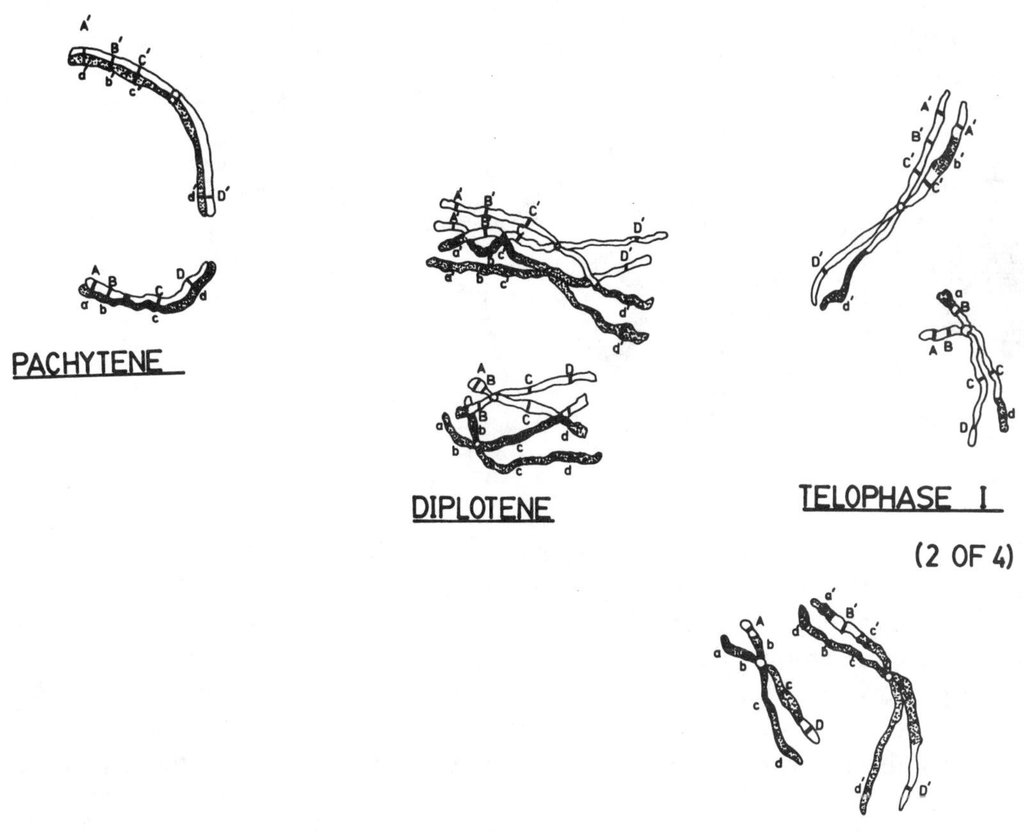

Fig. 3-33. Genetic effects of meiotic crossing over (recombination). Two of four possible chromosome combinations are shown for clarity. Notice the exchange of alleles between the maternal (stippled) and paternal (open) chromatids.

As prophase I of meiosis continues, the chiasmata appear to move toward the telomeres of the chromatids as the homologous chromosome pairs actually repel each other during **diakinesis**. Each tetrad aligns at the equatorial plane at metaphase I. Note that two alignments are possible in the simplified example using two chromosome pairs, and that the number of possible orientations is very large when all 23 tetrads (or 22 tetrads and the X-Y endwise pair in males) of the human meiotic cell are considered. **The maternal and paternal chromosome of each tetrad separate during anaphase** I and migrate to opposite poles of the dividing cell, and cell division is completed during telophase I (Fig. 3-32b, Fig. 3-34). This **reduction** division reduces the amount of allelic variability **per cell**. The parental cell had two sets of two different alleles each: **AaBb**. However, each daughter cell now has only one type of allele at each locus (**AAbb**; **AABB**; **aaBB**; or **aabb**). The amount of genetic material is halved: 4N -> 2N, where N is the haploid number of chromosomes (23).

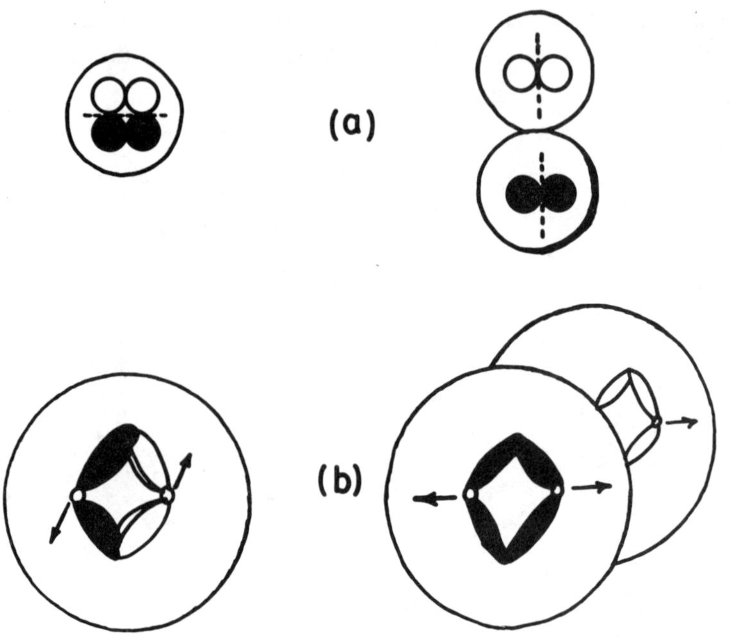

Fig. 3-34. Comparison of reductional (left)   and equational (right) divisions of meiosis. a) Top view; b) longitudinal view. Maternal chromosomes are shaded.

     The second meiotic division now commences (Fig. 3-35). During metaphase II, the spindle forms at right angles with respect to that of the meiotic spindle at metaphase I (Fig. 3-34). Chromatids of each maternal and paternal chromosome separate during anaphase II. Each daughter cell formed during telophase II contains one-half the number of chromosomes and one-half the amount of DNA present in a telophase I cell; however, the array of alleles is the same. Therefore, the second meiotic division is referred to as an **equational** division. Since each daughter cell has one-half the amount of DNA and one-half the number of chromosomes and genes present in somatic cells, it is said to be **haploid**.

     Meiotic division provides an opportunity for new genetic combinations to arise in the human population by independent assortment of genes located on different chromosomes. Suppose both husband and wife are **AaBb** as illustrated in Table 3-5 and Figures 3-32 and 3-35. Since the **A** and **B** loci are situated on different chromosomes, the **AB**, **Ab**, **aB**, and **ab** reproductive cells should be produced in approximately equal frequencies (Fig. 3-35). Combination of appropriate ova and sperm will produce the gene combinations presented in Table 3-5 in their children. Notice that 75% of the gene combinations (**genotypes**) are different from those of their parents. The genetic scrambling produced by recombination during prophase I is superimposed upon this variability.

     The third function of meiosis is to provide an opportunity for chromosome editing. As chromosomes pass through replication, differences in base sequences of intergenic regions may arise. Synapsis during prophase I aligns the maternal and paternal chromosomes point by point. Although chromosome editing has not been experimentally demonstrated in human cells, enzymes capable of comparing corresponding regions of the maternal and paternal homologous chromosomes could "correct" differences between them by using one chromosome as a reference for the other and "repairing" the exceptional sequences.

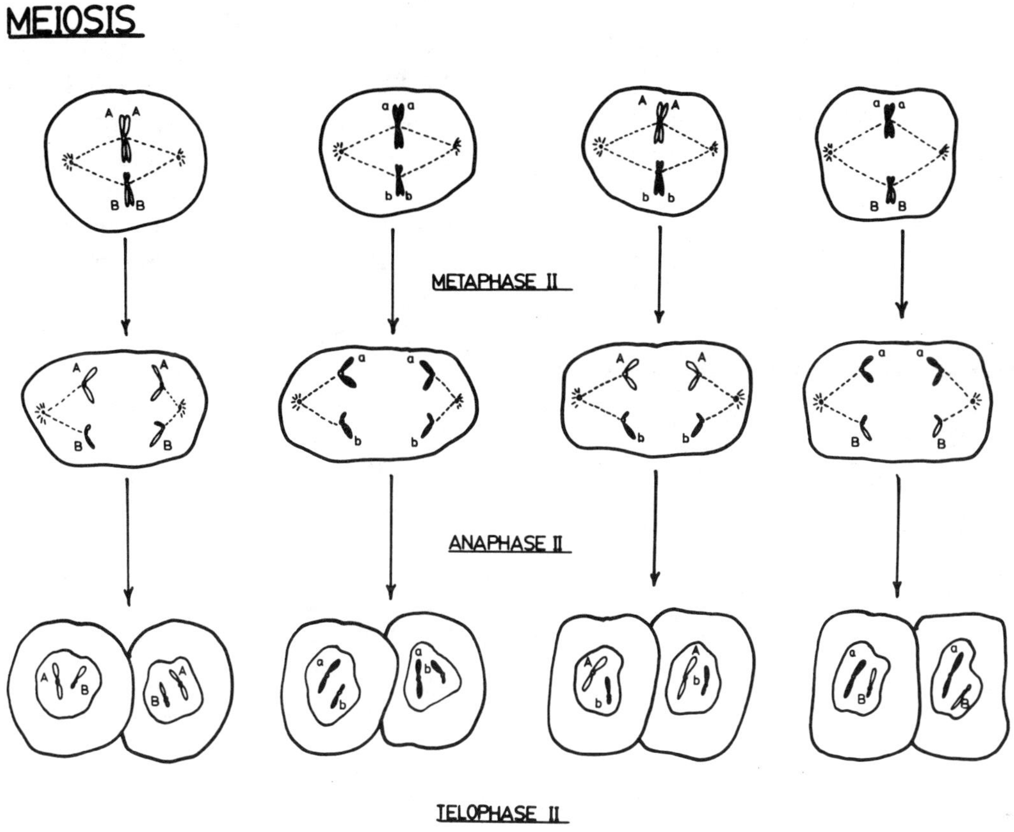

Fig. 3-35. Meiotic division (equational division). The second division occurs at right angles to the first. The equational division does not alter the genetic variability; however, the amount of genetic material is reduced from 2N to N.

Maturation of Gametic Cells

Male and female reproductive cells differentiate along different paths. The ovary contains hundreds of thousands of oogonia, each embedded in a follicle. These structures are formed prior to birth, and each oogonium enters meiosis during fetal development and arrests at diakinesis of prophase I (Fig. 3-36). These primary oocytes remain in prophase I until sexual maturity when one primary oocyte per month completes the first meiotic division and is ovulated as a secondary oocyte. This prolonged suspension of the oocyte in prophase I may be at least partially responsible for the increase in chromosomal anomalies among children of older couples. The meiotic spindle could be damaged by environmental agents, leading to nondisjunction (failure of separation) of homologous chromosomes and presence of too many or too few chromosomes in the ova produced by the older woman. Occasionally two primary oocytes may be ovulated during the same cycle. If both are fertilized, fraternal (dizygotic) twins may result. The second meiotic division is completed following fertilization. Only one of the four meiotic products of oogenesis participates in normal fertilization. The other three, the **polar bodies**,

Table 3-5. Potential Zygotes Produced by Two Parents Who Are Each <u>AaBb</u>.

	Spermatozoa			
Ova	AB	Ab	aB	ab
AB	AABB	AABb	AaBB	AaBb
Ab	AABb	AAbb	AaBb	Aabb
aB	AaBB	AaBb	aaBB	aaBb
ab	AaBb	Aabb	aaBb	aabb

degenerate. Notice that a major emphasis is placed upon conservation of cytoplasm during oogenesis. The cytoplasm contains the mRNAs, ribosomes, and nutrients required for direction of early embryogenesis.

Spermatogenesis is illustrated in Figure 3-37. Males do not initiate spermatogenesis until puberty, and approximately 40 days are required for completion of the development of spermatogonia into mature spermatozoa. The testis is composed of long convoluted seminiferous tubules and interstitial tissue. Spermatogenesis occurs within the seminiferous tubules, and the interstitial cells produce androgens. The spermatogonia line the seminiferous tubules, and successively more advanced cells are found as one moves from the

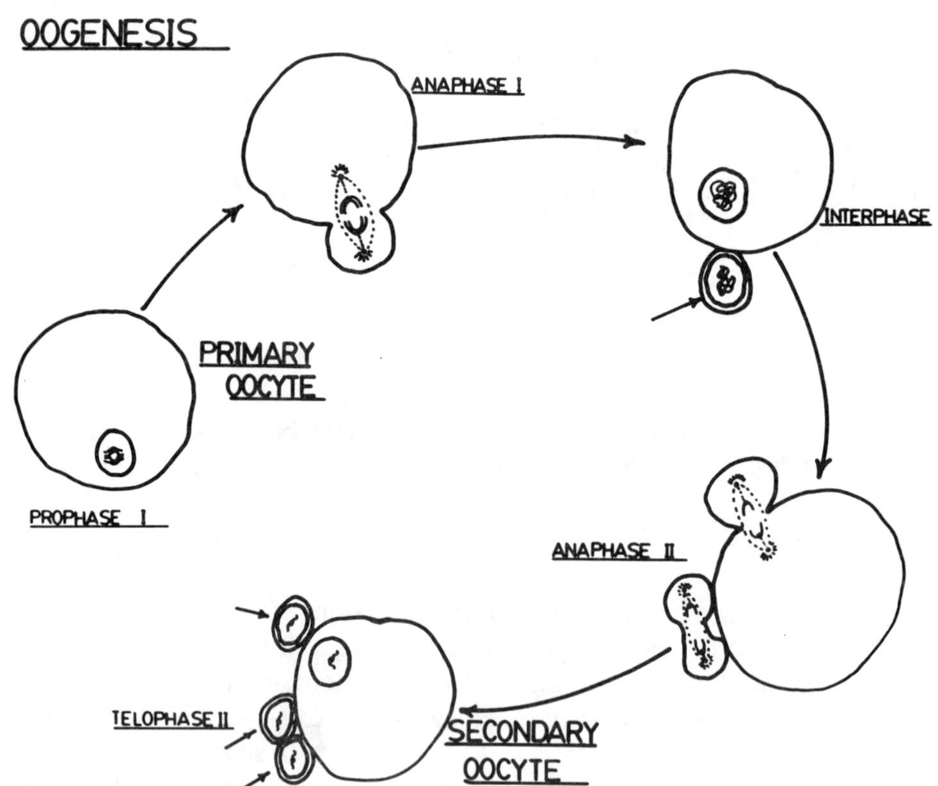

Fig. 3-36. Oogenesis. One pair of chromosomes is shown for simplicity. Small arrows indicate polar bodies.

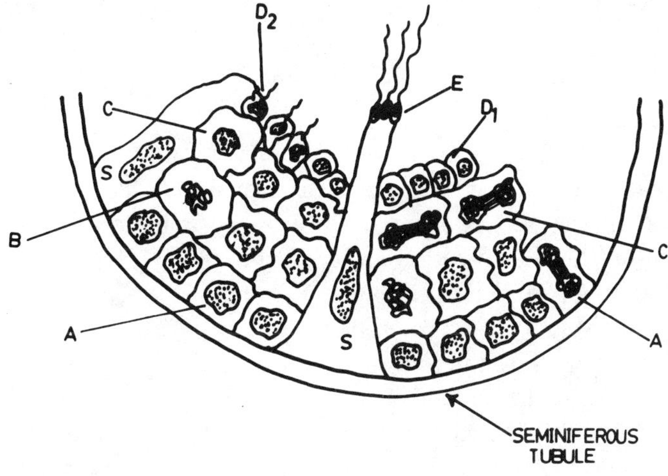

Fig. 3-37. Spermatogenesis. Spermatogonia (A) divide mitotically to form other spermatogonia or may develop into a primary spermatocyte (B) which enters meiosis and divides to form two secondary spermatocytes (C). Each secondary spermatocyte divides to produce two spermatids (D1) which subsequently interact with sertoli cells (S) and begin their differentiation into mature spermatozoa (E).

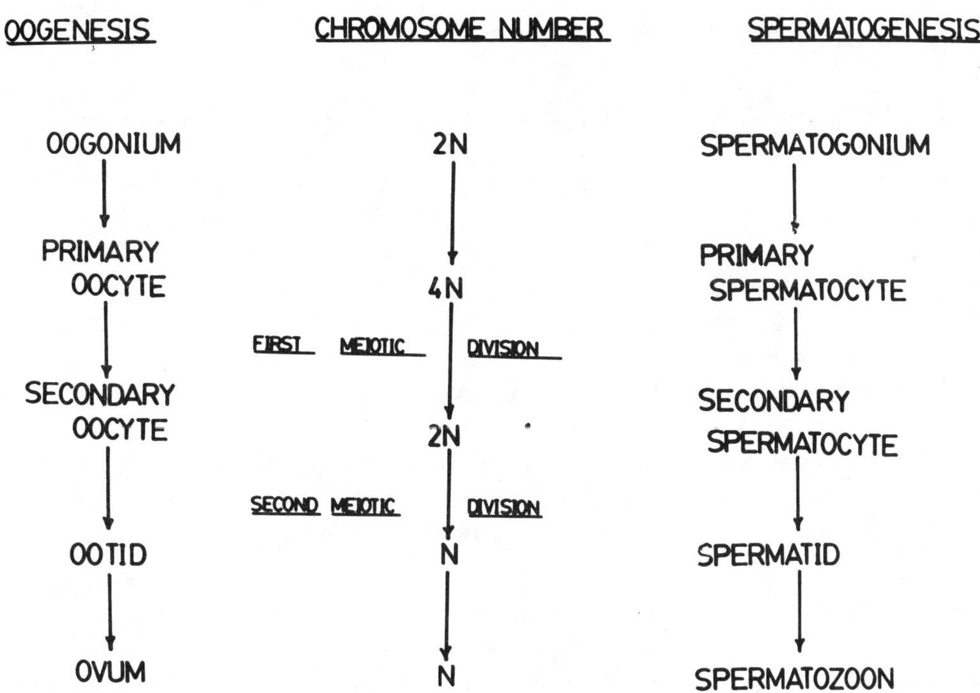

Fig. 3-38. Comparison of oogenesis, spermatogenesis, and their relationships to meiotic division. The second meiotic division occurs following ovulation in females and requires penetration of the ovum by a spermatozoon. Each oogonium differentiates into one ovum and three polar bodies. Each spermatogonium differentiates into four spermatozoa.

wall to the center of the tubule. The process of meiotic division is similar in male and female cells. However, two differences should be stressed: 1) male spermatogenesis is continuous from puberty through the seventies, while oogenesis is initiated prior to birth and not continued until ovulation; and 2) oogenesis produces one large ovum and three small polar bodies that degenerate, while spermatogenesis results in four sperm. Oogenesis, spermatogenesis, and their respective relationships are compared in Figure 3-38.

Fertilization

Ovulation releases the secondary oocyte, coated by a layer of follicle cells, into the Fallopian tube. **Fertilization** involves penetration of the secondary oocyte by a spermatozoon and usually occurs in the upper reaches of the Fallopian tube. Fertilization induces the second meiotic division of the the oocyte. Normally one sperm is succesful in penetrating the oocyte. Occasionally two sperm may fertilize the same oocyte, producing a triploid (3N) embryo. This condition is lethal. During fertilization, the entire sperm appears to penetrate the egg. The tail and male mitochondria degenerate, and the male pronucleus begins to swell and migrate toward the female pronucleus. The nuclear membranes of both pronuclei break down, and the male and female chromosomes replicate. The replicated male and female chromosomes line up in the same metaphase plane and complete the first mitotic division, forming the two-cell embryo. The degeneration of the male mitochondria shortly after fertilization results in the embryo inheriting all of its mitochondrial genes from its mother.

Autosomal and Sex-linked Inheritance

Autosomal genes are transmitted with equal frequency from the father to his sons and daughters (Fig. 3-39) and from the mother to her sons and daughters. When the father is heterozygous for the allele **A**, one-quarter of his sperm will carry **A** and the X chromosome and one-quarter of his sperm will possess **A** in combination with the Y chromosome. Notice how one-half of his sons and one half of his daughters inherit **A**. Similarly, when the mother is heterozygous for **A**, one-half of her sons and one-half of her daughters inherit **A**.

The middle panel displays the pattern characteristic for an X-linked gene. When the father is hemizygous for the X-linked allele, **A**, **all** of his daughters and **none** of his sons inherit **A**. By contrast, when the mother is heterozygous for the X-linked allele, **A**, one-half of her daughters and one-half of her sons inherit **A**.

Y chromosomes are usually not found in females. Therefore, genes located on the Y chromosome and outside of the pseudoautosomal region cannot occur in normal females. A father will transmit a Y-linked trait to all of his sons and none of his daughters.

Summary

Interphase chromosomes are long threads composed of protein, DNA, and a small amount of RNA. The DNA is present as a continuous double helix. When interphase chromosomes are extended by artificial means, they assume a "beads on a string" appearance. The core of each bead or nucleosome consists of histones, and the DNA helix is wound around the outside. Condensation of chromosomes to the rod-like structures observed during mitosis and meiosis involves supercoiling and folding. The initial stage of this condensation is mediated by histone H1, which connects nucleosomes. Phosphorylation of this histone promotes chromosome condensation.

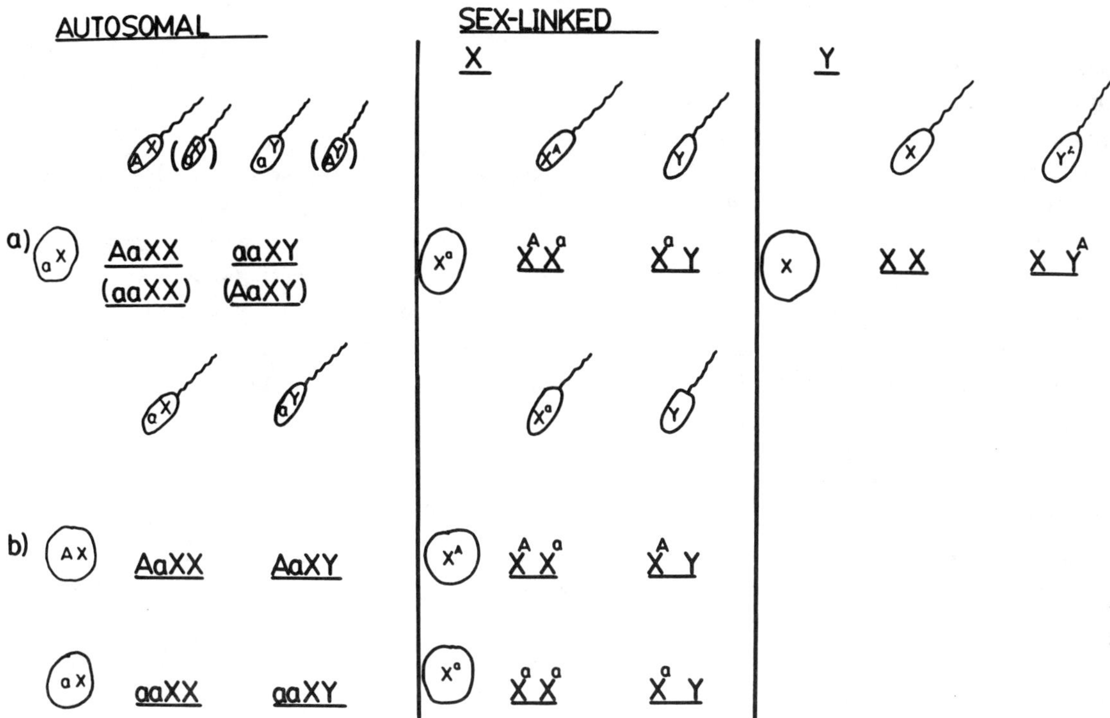

Fig. 3-39. Transmission of autosomal (left panel), X-linked (center panel), and Y-linked genes (right panel). The parentheses surrounding the genotypes in the upper left panel are formed by fertilization of the ovum by the sperm enclosed in parentheses. Notice how one-half of the sons and daughters inherit the A allele, regardless of whether the father (a) or mother (b) is heterozygous for A. If the allele A is located on the X-chromosome (X-linked), the father will not transmit that allele to his sons, but all of his daughters inherit A. A is transmitted from a heterozygous woman to one-half of her children, regardless of sex. Y-linked genes only occur in males. X = X chromosome; Y = Y chromosome.

Chromosomal DNA may occur in unique sequences (most protein-encoding genes), in tandem arrays of several copies (some protein-encoding genes and rRNA genes), as dispersed multiple copies of similar sequences (tRNA genes), and as moderately to highly redundant copies that do not appear to have a coding function. The moderate to highly repetitive DNA tends to be localized in telomeric, centromeric, and intercalary heterochromatic regions. Heterochromatin consists of highly condensed DNA that is genetically inactive and late replicating. Some heterochromatic regions are permanently inactive, while others may be active in certain cells or at certain times during development. X-inactivation is a a process that silences many genes on one of two X chromosomes in female cells. The inactive X is reactivated during oogenesis.

Histones are basic proteins that contribute to chromosome structure and participate in chromosome compaction. Several non-histone chromosomal proteins have been characterized. One of these, ubiquitin, is enriched in heterochromatic regions. Other non-histone proteins, including HMG-14 and HMG-17, are enriched in active gene regions and are believed to play an important role in nucleosome relaxation to permit transcription. Chromosomes are organized in domains which have several transcriptional units (genes), which are attached to the nuclear scaffold at their base, and which replicate as a unit. A second organizational structure of chromosomes has also been observed. Isochores are long stretches of DNA which share similar base sequences, may be GC- or AT-rich, and contain active genes embedded within them.

Centromeres (kinetochores) are specialized structures that serve as a point of attachment for the spindle fibers. Motor proteins are found at these sites and appear to be responsible for movement of chromosomes along the microtubules of the spindle. Human telomeres are located at the ends of chromosome arms, contain many tandem repeats of the sequence telomere-3'GGGATT-5'-, and prevent ends of chromosomes from sticking to one another. Telomeres tend to shorten during repeated chromosome replications. Telomerase extends telomere lengths. Aging appears to be associated with decreasing chromosome length, and chromosomes of cancer cells also possess unusually short telomeres.

The cell cycle can be divided into G_1 (phase of metabolic activity preceding S-phase), S-phase (period of DNA replication and histone synthesis), G_2 (pre-division stage), and M-phase (division of the cell into two daughter cells). Regulatory factors have been identified which control progression through the cell cycle. These include M-phase kinase which phosphorylates histone H1 and perhaps other nuclear proteins and S-phase promotional factor that induces DNA replication. M-phase kinase consists of two subunits, cyclin and p34. Inactivation of M-phase kinase occurs by the combined effects of phosphorylation of p34 and cyclin and degradation of cyclin.

Chromosome replication is preceded by relaxation of chromosome structure and unwinding of the double helix at the origin of replication which is located in the center of chromosomal domains. A short RNA primer is synthesized which serves as a focus for synthesis of DNA by DNA polymerase-α. When the DNA is elongating in the 5' to 3' direction, DNA polymerase-δ replaces the α polymerase and continues DNA synthesis. Multiple RNA priming and DNA synthesis by the α polymerase is required on the opposite side of the replication origin. The latter process produces multiple short fragments called Okazaki fragments. Removal of the RNA primers, gap-filling with DNA, ligation of the newly synthesized segments of DNA, and restoration of helical structure complete formation of the daughter double helices. Histones assort among the daughter DNA molecules in a conservative fashion. Nucleosomes contain all old or all new histones.

Mitosis produces two daughter cells which possess chromosome and gene complements identical to those of the parental cell. Stages of mitosis include prophase (period of chromosome shortening), metaphase (chromosomes align in the central plane), anaphase (chromosomes separate and migrate to opposite poles of the dividing cell), telophase (chromosomes arrive at poles; nuclear membranes form; and cytokinesis occurs, completing formation of the daughter cells). Human metaphase chromosomes are often used for diagnosis of systemic chromosome errors and those which are associated with certain leukemias and other cancers. Cells from appropriate tissues are cultured and processed for chromosome preparation. A variety of staining methods have been developed that selectively stain euchromatic (R-banding), heterochromatic (Q- and TG-banding), centromeric (C-banding), or nucleolar organizing regions (silver staining). Harlequin staining is employed to detect sister-chromatid exchanges. When small alterations in chromosome structure are suspected, cells will be harvested just prior to metaphase (pro-metaphase). Fragile X chromosomes are detected following growth of cells in a stressing medium containing FUdR (fluordeoxyuridine) and lacking folic acid. A convention has been developed for labelling karyotypes. The total chromosome number is placed first, followed by the sex chromosome complement and brief description of the abnormality, if any.

Human sex determination appears to depend upon both DNA-binding proteins and hormones. The testis-determining factor is a presumed DNA-binding protein encoded by a gene located in the sex-determining region of the Y chromosome. This protein most likely acts by activating genes essential for development of the testis. The fetal testis secretes two hormones, testosterone, which promotes further development of the mesonephric (Wolffian) ducts and other differentiation events in the testis and elsewhere, and mullerian inhibitory substance, which inhibits further development of the mullerian ducts. Females lack Y chromosomes and their **TDF** genes. Their indifferent fetal gonad differentiates into an ovary. The mullerian ducts form the Fallopian tubes, uterus, and part of the vagina. The Wolffian ducts degenerate in the absence of testosterone. Further sexual changes are mediated by hormones and their receptors. Testosterone is essential for spermatogenesis and collaborates with dihydrotestosterone in the formation of the secondary sexual structures of the male and other male-specific changes. These two steroids require a DNA-binding receptor, the androgen receptor, for their effects on gene expression. Progesterone and estrogens are synthesized and secreted by the ovary and are responsible for the secondary sex structures of females, other female-specific attributes, and for menstruation and pregnancy. Both of these female sex steroids act through different receptors that have sequence homology with the androgen receptor. All three steroid receptors possess domains which mediate specificity of DNA binding and hormone binding. Normal fertility requires presence of the appropriate normal sex chromosomes and expression of both sex-linked and autosomal genes that influence a variety of different sexual processes. Since women also express androgen receptor, sex reversal can occur when females are exposed to high androgen levels.

Meiotic division includes two successive cell divisions following DNA replication. The diploid chromosome number is reduced to the haploid value in ova and sperm. The first prophase is elongated and divisible into discrete stages. Synapsis of homologous chromosome pairs occurs during prophase I. Formation of chiasmata during this time is associated with exchange of alleles between chromatids of homologous chromosomes, generating new chromosomal arrays of genes that were not present in the parental cell. This genetic variation is superimposed upon the assortment of unlinked genes (genes on different chromosomes) that occurs among the four products formed by meiotic division. Female meiosis (oogenesis) is initiated prior to birth, and cells are arrested in prophase I for long periods of time. This delay is believed to be responsible for the greater tendancy of older mothers to produce ova with abnormal chromosome constitutions. Male meiosis (spermatogenesis) begins at puberty and continues through the seventies. There is no disruption of meiotic division during spermatogenesis. All four products of spermatogenesis develop into sperm. Oogenesis produces one ovum and three polar bodies. The latter three cells do not participate in normal fertilization.

Fertilization occurs in the upper region of the Fallopian tubes. The entire sperm appears to penetrate the egg. The male nucleus enlarges to form the male pronucleus, replicates its chromosomes, and migrates toward the female pronucleus. The female nucleus completes the second meiotic division following sperm penetration and enlarges while replicating its chromosomes. Both pronuclear membranes break down, and both sperm and egg chromosomes align in a common metaphase plane. The mitotic division is completed, and the two-cell embryo is formed. The sperm mitochondria degenerate. Therefore, the mitochondria of the embryo are of maternal origin.

Autosomal genes are transmitted with equal frequency to sons and daughters by either parent. X-linked genes are transmitted from father to daughter, and his Y-linked genes are inherited by his sons. Mothers transmit their X-linked genes with equal frequency to their sons and daughters. Females lack Y-linked genes.

Problems

1. If you labelled newly synthesized histones with radioactive lysine, how would you expect the label to be distributed over recently replicated, dispersed chromatin?

2. A patient has a total of 48 chromosomes, one Barr body, and two Y bodies. Which of the following karyotypes best fit this patient?

    a) 48,XXY,+21   d) 48,XYYY
    b) 48,XXYY   e) 48,XYY,+21

3. Transcriptionally-active genes often are nested in surrounding DNA consisting of tandemly repeated sequences. Certain diseases such as steroid sulfatase deficiency are frequently caused by deletions of DNA encompassing the enzyme locus. How might these deletions arise? Clues: meiosis, synapsis, recombination)

4. Suggest a mechanism whereby TG-banding produces its characteristic staining pattern.

5. DNA from which of the following cells would appear to be transcriptionally active if you were using DNase I and β-globin cDNA as a probe for gene activity? Note: DNase I will not digest condensed nucleosomes.

    a) hepatic cells   d) astrocytes (brain cells)
    b) renal cells   e) all of the preceding
    c) reticulocytes from bone marrow

6. A sperm carrying a minute deletion of the sex-determining region of the Y chromosome fertilizes an egg with a normal X chromosome and 22 normal autosomes. Speculate on the sexual development and fertility of the person developing from the conceptus.

7. Glucose-6-phosphate dehydrogenase (G6PD) is determined by an X-linked gene with several alleles. The iron transport protein, transferrin (TF), is encoded by an autosomal gene. Electrophoretic variants of the two proteins behave as codominants in heterozygotes (eg. a woman heterozygous for the **G6PD^A** and **G6PD^B** alleles expresses both G6PD-A and G6PD-B in her blood). Consider the phenotypes of the following couple and child:

	G6PD	TF
Woman	B	C
Man	A	D
Child(Boy)	A	CD

The boy has recently contracted a severe neurological disorder, and the parents are claiming that he is not their child and that a mixup occurred in the nursery shortly after his birth. How would you testify as the genetic expert in this case?

8. Which of the following techniques provide unambiguous identification of all 22 pairs of autosomes and the X and Y chromosomes?

    a)   TG-banding   c)   Harlequin staining
    b)   C-banding   d)   Q-banding

9. A female infant has trisomy (triplication of) 8. Express her karyotype in standard nomenclature.

a) 46,XX,8p+
b) 46,XX,8q+
c) 47,XX,+8
d) 47,XY,+8

10. A muscle biopsy was removed from a woman who is known to be heterozygous for an X-linked allele known to cause a form of Duchenne muscular dystrophy in which the muscle protein, dystrophin is absent. The muscle biopsy was stained with a fluorescent antibody directed against dystrophin. Fluorescent cells would be interpreted to be expressing dystrophin, while non-fluorescent cells lack the protein. The woman's muscle biopsy displayed patches of fluorescent muscle fibers interspersed with non-fluorescent fibers. How would you account for this result?

11. How would the effects of a failure of chromosome separation during anaphase I of meiosis differ from those of a non-disjunctional event at anaphase II?

12. An infertile man was karyotyped and found to be mosaic: 47,XXY/46,XY. His father is G6PD-A, and his mother is G6PD-B. The man is G6PD-AB. How would you explain the abnormal karyotype of this man?

Glossary of Terms

Acrocentric - chromosome having a centromere situated near one end.

Alleles - alternate forms of the same gene.

Amniocentesis - the procedure used to obtain amniotic fluid and amniocytes for genetic testing.

Anaphase - the period of cell division during which the chromatids migrate to opposite poles of the dividing cell.

Autosomal inheritance - inheritance of traits determined by genes located on autosomes.

Autosomes - the non-sex chromosomes (numbers 1-22).

Barr body - interphase nuclear mass corresponding to the inactive X chromosome.

Bivalents - associated pairs of meiotic chromosomes.

C-banding - technique for staining centromeric heterochromatin employing Giemsa stain after hot alkali digestion.

Cell cycle - the period of time and series of stages through which a cell passes in one generation.

Centrioles - the cellular elements that constitute the poles of the spindle apparatus.

Centromere (kinetochore) - constriction of the chromosome to which a spindle microtubule attaches.

Chiasmata - interchromatid bridges that are sites of gene exchange between chromatids of homologous chromosomes.

Chorionic villi - finger-like projections of the chorion which are sampled for prenatal diagnosis.

Chromosomal domain - a chromosome region containing several genes which replicates as a unit.

Chromosome - a term used to describe a chromosome element appearing during cell division. One chromatid consists of a complete double helix of DNA, protein, and a small amount of RNA. Each chromatid uravels during the subsequent telophase to become a chromosome in the daughter cell. Chromatids of a chromosome are attached in the region of the centromere.

CpG Island - a cluster of CpG residues located upstream or within the promoters of genes. These islands are unmethylated when the adjacent gene is actively transcribed and methylated when the adjacent gene has been transcriptionally silenced.

Crossing-over (recombination) - the process by which genes exchange positions relative to one another on homologous chromosomes.

Cytokinesis - cytoplasmic division.

Diploid - two complete sets for chromosomes (for human cells, 2N = 46).

Dynein - a motor protein found in the kinetochore which participates in chromatid movement during cell division.

Endoreduplication - one or more rounds of chromosome replication in the absence of nuclear division.

Euchromatin - "true-staining" chromatin with relaxed structure and containing transcriptionally active genes.

Fertilization - penetration of an egg by a sperm.

Genotype - the combination of genes and their alleles occurring within an individual or within a cell.

Haploid - one complete set of chromosomes (for humans, N = 23); one-half the number of chromosomes and genes present within a typical somatic cell nucleus.

Hemizygous - term referring to the occurrence of a single copy of a X-linked gene or Y-linked gene in a male; may also be used to refer to the presence of a single autosomal allele in a cell following deletion of its partner on the homologous chromosome.

Heterochromatin - "different-staining" chromatin containing transcriptionally silent genes and/or composed of highly condensed structure.
    Centromeric - heterochromatin near the centromere;
    Constitutive - heterochromatin possessing DNA that is not transcribed into RNA;
    Facultative - heterochromatin having genes that may be active at other times or in other cell

types (e.g. inactive X chromosome);
Intercalary - heterochromatic regions interspersed among euchromatic regions.

Heterologous chromosomes - chromosomes that are members of different pairs (eg. chromosome 1 and chromosome 13).

Heterozygote - individual carrying two different alleles at a locus.

Histones - family of basic chromosomes which contribute to chromosome structure and participate in chromosome condensation.

HMG proteins - chromosomal proteins of the non-histone fraction which possess rapid electrophoretic mobilities. These proteins appear to negatively or positively influence gene expression, depending upon type.

Homologous chromosomes - chromosomes of the same pair (chromosome 9 and chromosome 9).

Homozygote - Individual having two identical alleles at a given locus.

Hypersensitive site - chromosomal region that has relaxed structure and which is susceptible to nuclease digestion; hypersensitive sites are most commonly associated with actively transcribing genes and are believed to provide a "window" for entry of the transcriptional complex to the gene.

In situ hybridization - gene-specific annealing of a RNA or cDNA with the denatured DNA of an intact chromosome.

Interphase - period of the cell cycle occurring between cell divisions (G_o + G_1 + S + G_2).

Isochore - extended regions of DNA that have similar base compositions.

Karyotype - arrangements of mitotic or meiotic chromosomes according to size and appearance.

Locus - position of a gene on a chromosome.

Meiotic division - a form of cell division producing a reduction of gene and chromosome numbers to one-half those present in a typical somatic cell.

Metacentric - chromosome with a centrally placed centromere.

Metaphase - the stage of cell division associated with allignment of the chromosomes in the equatorial plane.

Mitosis - a form of cell division producing genetically identical diploid daughter cells.

Non-histones - family of acidic to neutral chromosomal proteins which appears to be involved with gene regulation.

Nuclear scaffold (nuclear matrix) - nuclear framework or superstructure to which the chromatin is attached.

Nucleosome - chromatin subunit consisting of a histone core surrounded by two turns of the DNA double helix.

Oogonia - diploid germ cells giving rise to ova.

Ova - mature female gametes.

Palindrome - segment of DNA in which the two strands have an axis of symmetry in the center such that the right half of one strand is the mirror image of the left half of the complementary strand and vice versa. Palindromes are usually several hundred to several thousand base pairs in length.

Phenotype - the biochemical, physiological, or morphological effect of a gene.

Polyploid - three or more complete chromosome sets.

Polyploidization - increase in chromosome number by duplication of complete chromosome sets as a consequence of abnormal chromosome segregation during cell division, cell and nuclear fusion, and/or multiple fertilization of an ovum.

Prometaphase-banding - preparation of extended chromosomes from cells at the prophase-metaphase boundary for study of small deletions and other structural alterations.

Prophase - the period of cell division during which chromosome condensation occurs.

Q-banding - use of quinacrine dyes under appropriate conditions to selectively stain heterochromatin.

R-banding - selective staining of GC-rich, euchromatic regions of chromosomes.

Replicon - unit of chromosome replication.

Satellite - chromosome segment attached by a stalk to one of the D- or G-group chromosomes. The stalk region possesses rRNA genes.

Satellite DNA - DNA which has a density different from that of the majority of DNA as a consequence of its base composition.

Sex chromosomes - chromosomes participating in sex determination (X and Y).

Sex determination - commitment of the indifferent fetal gonad to ovarian or testicular development. This stage is dependent upon a Y-linked regulatory gene, **TDF**.

Sex differentiation - hormone-dependent phase of sexual development that is responsible for maturation of internal and external male and female structures.

Sex-linked inheritance - inheritance of traits determined by genes located on the X or Y chromosomes.

Spermatogonia - diploid germ cells giving rise to spermatozoa.

Spermatozoa - mature male haploid gametes.

Spindle - apparatus consisting of spindle fibers, chromosomes, and centrioles that participates in the apportionment of chromosomes among the daughter cells.

Spindle fibers - microtubules that extend from centriole to centriole or from centriole to centromere and which assist chromosome assembly in the metaphase plane and chromosome migration to the poles of the dividing cell.

Submetacentric - chromosome having a centromere partially displaced toward one end.

Synapsis - sequence-specific pairing of homologous chromosomes during prophase I of meiosis.

Telomere - the end of a chromosome arm.

Telophase - the period during cell division when cytokinesis and reconstruction of cell structure occurs.

TG-banding - selective staining of heterochromatic chromosome regions with giemsa following treatment with trypsin.

Translocation - an exchange of chromosome segments between heterologous chromosomes.
    Balanced - translocation in which essential genes are retained in normal numbers.
    Unbalanced - translocation in which essential genes are present in excessive or deficient copies.

Trisomic - three copies of a specific chromosome.

X-inactivation - the mechanism responsible for prevention of expression of many X-linked genes situated on all but one of the X chromosomes of somatic cells.

Y-body - fluorescent interphase nuclear mass corresponding to the Y chromosome or a major portion of its long arm.

Zygote - a fertilized ovum.

Bibliography

1. Sutton, W.S. 1903. The chromosomes of heredity. Biol. Bull. 4:231-251.

2. Boveri, T. 1901. Ueber Mehrpolige Mitosen als Mittel zur Analyse des Zellkerns. Vererb. Phys. Med. Ges. N. F. 35:65-88.

3. Tjio, J.H. and A. Levan. 1956. The chromosome number of man. Hereditas 42:1-6.

4. Lejeune, J., M. Gauther and R. Turpin. 1959. Etude des chromosomes somatiques de neuf enfants mongoliens. C. R. Acad. Sci. Paris 248:1721-1722.

5. Weisbrod, S., M. Groudine and H. Weintraub. 1980. Interaction of HMG 14 and 17 with actively transcribed genes. Cell 19:289-301.

6. VandeBerg, J.L, E.S. Robinson, P.B. Samollow and P.G. Johnston. 1987. X-linked gene expression and X-chromosome inactivation: Marsupials, mouse, and man compared. Isozymes Curr. Top. Biol. Med. Res. 15:225-253.

7. Lock, L.F., N. Takagi and G.R. Martin. 1987. Methylation of the Hprt gene on the inactive X occurs after chromosome inactivation. Cell 48:39-46.

8. Driscoll, D.J. and B.R. Migeon. 1990. Sex difference in methylation of single-copy genes in human meiotic germ cells: Implications for X chromosome inactivation, parental imprinting, and origin of CpG mutations. Somatic Cell Molec. Genet. 16:267-282.

9. Brown, C.J. and H.F. Willard. 1990. Use of a polymerase chain reaction-based assay to address the transcriptional activity of genes on the active and inactive human X chromsome. Amer. J. Hum. Genet. 47(Suppl.):A109.

10. Bonne-Andrea, C., M.L. Wong and B.M. Alberts. 1990. In vitro replication through nucleosomes without histone displacement. Nature 343:719-726.

11. Burgers, P.M.J., R.A. Bambara, JL Campbell, L.M.S. Chang, et al. 1990. Revised nomenclature for eukaryotic DNA polymerases. Eur. J. Biochem. 191:617-618.

12. Tsurimoto, T., T. Melendy and B. Stillman. 1990. Sequential initiation of lagging and leading strand synthesis by two different polymerase complexes at the SV40 DNA replication origin. Nature 346: 534-539.

13. Vaughn, J.P., P.A. Dijkwel, L.H.F. Mullenders and J.L. Hamlin. 1990. Replication forks are associated with the nuclear matrix. Nucleic Acids Res. 18:1965-1969.

14. Tkachuk, D.C., C.A. Westbrook, M. Andreeff, A. Donlon, et al. 1990. Detection of bcr-abl fusion in chronic myelogenous leukemia by in situ hybridization. Science 250:559-562.

15. A proposed Standard System of Nomenclature of Human Mitotic Chromosomes, Denver, CO. 1960. Reprinted in: Chicago Conference: Standardization in Human Cytogenetics. National Foundation Birth Defects Original Article Series II(2):12-15, 1966.

16. Paris Conference (1971); Standardization in Human Cytogenetics. National Foundation Birth Defects Original Article Series VIII(7):1-46, 1972.

17. McLaren, A. 1990. What makes a man a man? Nature 346:216-217.

Additional Learning Resources

1. Alberts, B.M. 1990. Recipes for replication. Nature 346:514-515.

2. Bernardi, G. 1989. The isochore organization of the human genome. Annu. Rev. Genet. 23:637-661.

3. Lewin, B. 1990. Driving the cell cycle: M phase kinase, its partners, and substrates. Cell 61: 743-752.

4. Lyon, M.F. 1962. Sex chromatin and gene action in the mammalian X-chromosome. Amer. J. Hum. Genet. 14:135-148.

5. Miller, W.L. and Y. Morel. 1989. The molecular genetics of 21-hydroxylase deficiency. Annu. Rev. Genet. 23:371-393.

6. Voeller, B.R. The Chromosome Theory of Inheritance. New York: Appleton Century Crofts, 1968.

Chapter 4

Chromosomes and Disease

Approximately thirty years have passed since the chance discovery of methodological improvements that permitted enumeration and characterization of human chromosomes. Jerome Lejeune, appraised of the work of Tjio and Levan, was impressed with the striking similarities of children with Down's syndrome, then called mongolism, and suspected that these patients may be lacking a chromosome. Using a discarded microscope that was badly in need of repairs, he examined mitoses from these patients and discovered that they actually possessed an additional chromosome (1). Lejeune's discovery represented the first documentation of a human chromosome anomaly and initiated a period of scientific activity that was highlighted by rapid advances in understanding the chromosomal bases for a number of disorders, several of which affect a significant proportion of our population. A second burst of investigative activity was precipitated by the advent of banding techniques which greatly enhanced the sensitivity of cytogenetic analyses. Recent combination of molecular genetic techniques with more traditional cytogenetic procedures promises to bring about even more striking advances in our understanding of human chromosomes and the mechanisms by which chromosome anomalies affect human development.

Mechanisms Responsible for Numerical Chromosome Aberrations

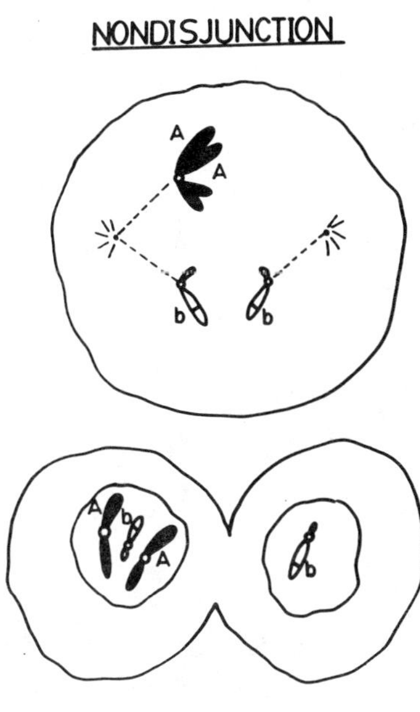

Nondisjunction refers to the failure of separation of chromosomes during anaphase of mitosis or meiosis. Mitotic nondisjunction (Fig. 4-1) results in **mosaic** individuals who possess two or more cell lines that differ with respect to their chromosome constitution. The phenotype of a mosaic person depends upon the relative frequency of the normal and **aneuploid** cell lines. Aneuploid cells have abnormal chromosome numbers. The example presented in Figure 4-1 illustrates a normal mitotic cell undergoing nondisjunction. One daughter cell inherits an excessive number of chromosomes and genes, while the other has a deficient number of chromosomes and genes. Nondisjunction usually affects the segregation of only one pair of chromosomes, such that the daughter cells in Figure 4-1 would contain 47 and 45 chromosomes, respectively. The seriousness of this nondisjunctional event depends upon the specific chromosome involved. If the chromosome is small or contains relatively few genes that are crucial for normal development, the cell

Fig. 4-1. Nondisjunction of mitotic chromosomes during anaphase. Nondisjunction of one pair (shaded) of chromosomes is shown.

123

lines may survive and contribute to a substantial proportion of the mature person. However, if the alteration of chromosome number effects an increase or decrease in dosage of a critical gene or genes, the cell line may be eliminated during the continued development of the individual. In general, trisomic cells tend to be less deleterious than monosomic cells.

Occurrence of nondisjunction during meiosis produces gametes that are **disomic** (two copies of the same chromosome) or **nullisomic** (lack a chromosome). Fusion of one of these cells with a gamete from the spouse will produce a **trisomic** (three chromosomes of one kind) or **monosomic** (only one chromosome of a kind) zygote. The fate of the zygote is dependent upon the nature of the chromosome imbalance. Only one monosomy is known to survive to birth (45,X). Even in this case, most 45,X concepti die before birth and often during the first three months of pregnancy. The odds for survival of trisomies depend upon the particular chromosome involved. Most autosomal trisomies severely handicap the patients in which they occur, cause fetal or early embryonic loss, and/or are fatal during infancy or childhood. Several of these problems will be discussed later in this chapter.

Nondisjunction of the X chromosomes of a woman is presented in Figure 4-2. Nondisjunction during Anaphase I results in migration of both the maternal and paternal X chromosome pairs to the same pole of the dividing cell. The second meiotic division occurs normally, and four types of ova may

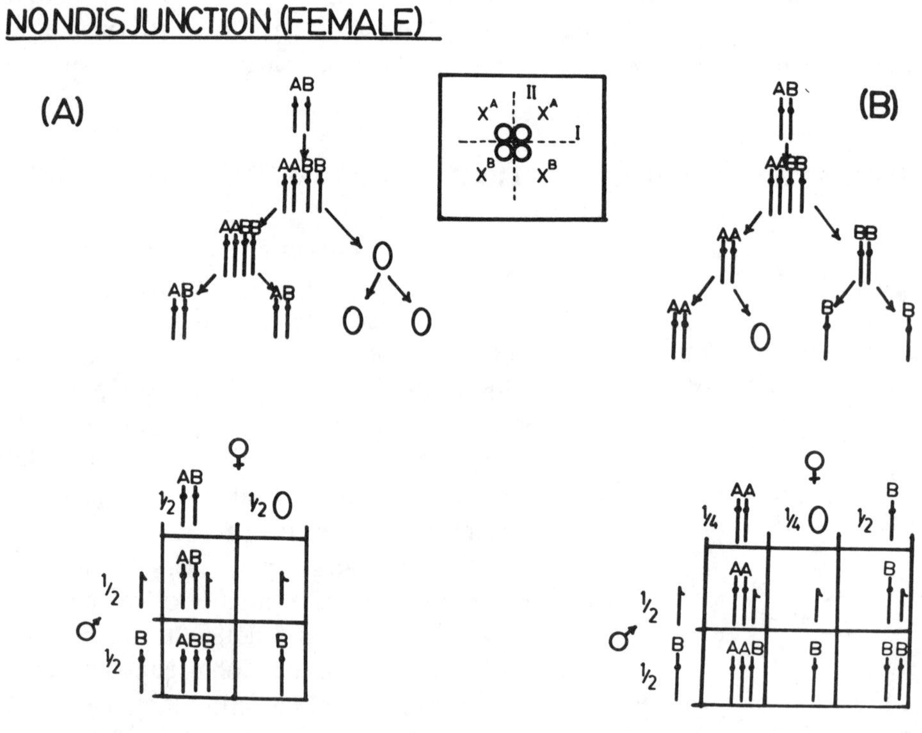

Fig. 4-2. Nondisjunction of sex chromosomes in the female during Anaphase I (A) and Anaphase II (B). The G6PD alleles are symbolized by A and B, respectively. The inset shows the division of the centromeres during Anaphase I and Anaphase II. Fractions refer to the chance that a given sex chromosome complement will occur in the ovum or spermatozoon. Segregation of chromosomes and fertilization products are displayed in the top and bottom, respectively, of each figure. The X and Y chromsomes are represented by a rod and half arrow, respectively.

be produced, two of which are disomic for the X chromosome and **G6PD^A/G6PD^B** (Note: **G6PD^A** codes for the rapid G6PD isozyme, and **G6PD^B** encodes the standard G6PD isozyme) and two of which lack an X chromosome and the associated G6PD allele. Three of the four meiotic products in each example would become polar bodies. Therefore, the probability that a particular chromosome and gene combination would enter a child depends upon the proportion of meiotic products in which it occurs and the chance that the cell differentiates into an ovum. Note that only the X chromosomes experienced nondisjunction in the two examples, and that the 22 pairs of autosomes segregated normally.

Suppose this woman marries a man who is **G6PD^B/Y**. Assuming meiosis occurred normally during spermatogenesis, this man would produce sperm carrying either the Y chromosome or the X chromosome bearing the **G6PD^B** allele. The lower portion of Fig. 4-2 illustrates the genotypes of the zygotes formed by fertilization of disomic and nullisomic ova by **G6PD^B**- or Y-bearing sperm. All four potential zygotes are aneuploid when nondisjunction occurs during Anaphase I. Zygotes lacking an X chromosome die during early embryonic development. More than 95% of 45,X females also die prior to birth. The 47,XXX and 47,XXY zygotes survive to term but display varying degrees of clinical abnormality. Therefore, all of the potential zygotes derived from an oogonium experiencing nondisjunction during Anaphase I would be clinically abnormal to some degree.

Nondisjunction during Anaphase II of meiosis in the female may result in the production of nullisomic, monosomic, or disomic ova depending upon which nucleus contributes to the mature ovum. Half of the possible zygotic combinations derived from fertilization of the respective ova by sperm from a **G6PD^B/Y** male are compatible with normal development (Fig. 4-2B). The trisomic zygotes (47,XXX and 47,XXY) will develop into individuals displaying varying degrees of clinical abnormality, while 45,X

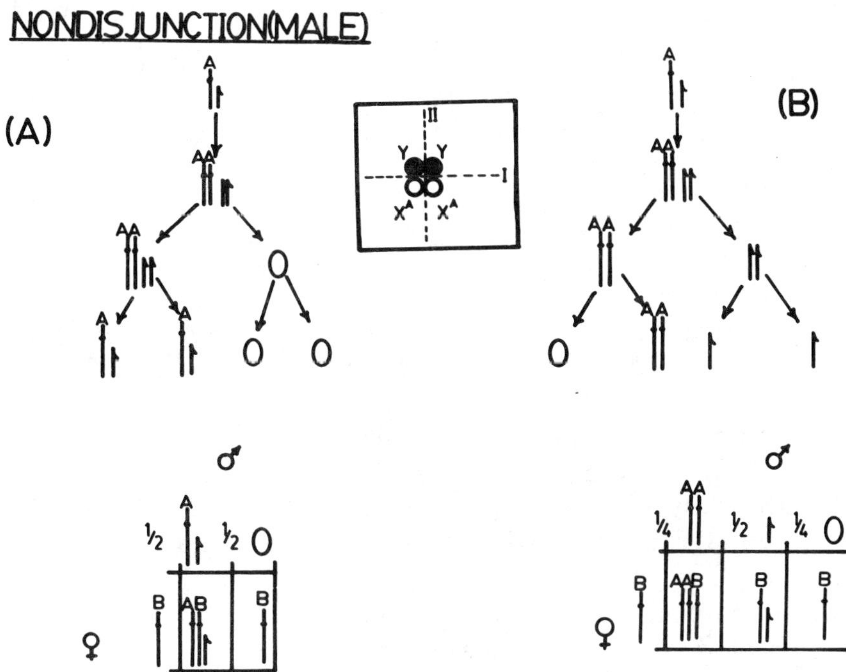

Fig. 4-3. Nondisjunction of the sex chromosomes during Anaphase I (A) and Anaphase II (B). Symbols are as described in the legend for Fig. 4-2.

and 45,Y are genetic lethals.

Consequences of meiotic nondisjunction in the male (Fig. 4-3) are similar to those of the female; however, all four meiotic products develop into spermatozoa. Meiotic nondisjunction during Anaphase I results in the production of two sperm carrying both the X and Y chromosomes and two sperm lacking sex chromosomes. All zygotes conceived by fertilization of normal ova by these sperm would be aneuploid. Half of the gametes produced following nondisjunction during Anaphase II contain normal chromosome complements, while the remainder are disomic or nullisomic. There is a 50% chance that zygotes conceived following fertilization of a normal egg by sperm produced following nondisjunction during anaphase II would contain aneuploid chromosome complements. Notice that a 47,XYY zygote may be produced following Anaphase II nondisjunction of the Y chromosomes.

Anaphase lag refers to a phenomenon in which one chromatid fails to migrate to the pole of a dividing cell (Fig. 4-4). The lagging chromosome may be excluded from the nuclei of the daughter cells and be degraded. The daughter cells would be disomic or monosomic for that chromosome. Alternatively, the lagging chromosome may be included in the wrong daughter nucleus (resulting in trisomic and monosomic daughter cells), or may be fortuitously included in the appropriate nucleus (resulting in two daughter cells with normal chromosome complements). Anaphase lag may occur during either mitosis or meiosis. Mosaicism is a common consequence of mitotic anaphase lag.

Structural Chromosomal Abnormalities

A second class of relatively common chromosomal abnormalities involve alterations of chromosome structure. These structural alterations are preceded by breakage of the chromosome strands. These breaks may be caused by certain chemicals, viruses, irradiation, or certain rare genes. A variety of structural alterations may arise from chromosome breaks (Fig. 4-5). **Inversions** require two breaks in the chromosome arms. The chromosome segment between the breaks rotates 180° to form a structurally modified chromosome. **Pericentric** inversions occur following breaks on either side of the centromere, while **paracentric** inversions do not involve the centromeric region. Inversions present problems during meiotic division, since pairing of the inverted chromosome with its normal homolog during synapsis is difficult. The inverted chromosome forms a loop to facilitate the pairing process (Fig. 4-6). If crossing over occurs within the inversion loop, chromatids with duplicated or deficient segments may be formed. This is particularly noticeable in paracentric inversions where one of the crossover products possesses two centromeres and is likely to break during a subsequent anaphase, and the other crossover product is acentric and will be lost during the subsequent cell division. Note that, in the case of both paracentric and pericentric inversion crossover products, the derivative chromosomes have duplications of certain regions and deletions of others. The synaptic pairing problems assoc-

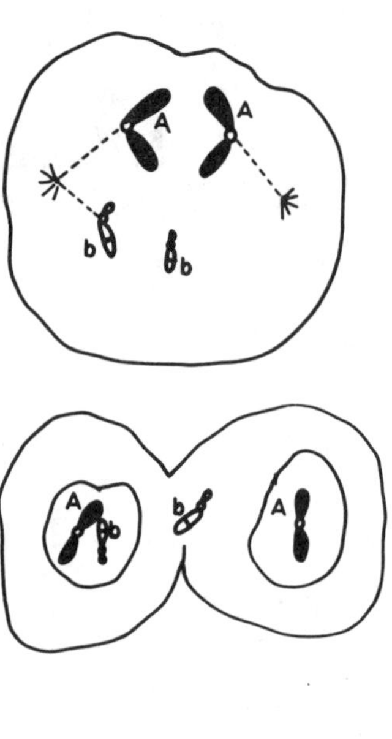

Fig. 4-4. Consequences of anaphase lag during mitotic division. The lagging chromosome has been excluded from the daughter nuclei and will be degraded.

iated with inversions tend to hinder recombination, and the abnormal products generated by crossing over within inversion loops are often lethal to the cells carrying them and reduce the likelihood of recovery of crossover products. As a result of these effects, recombination appears to be suppressed in inverted chromosome regions.

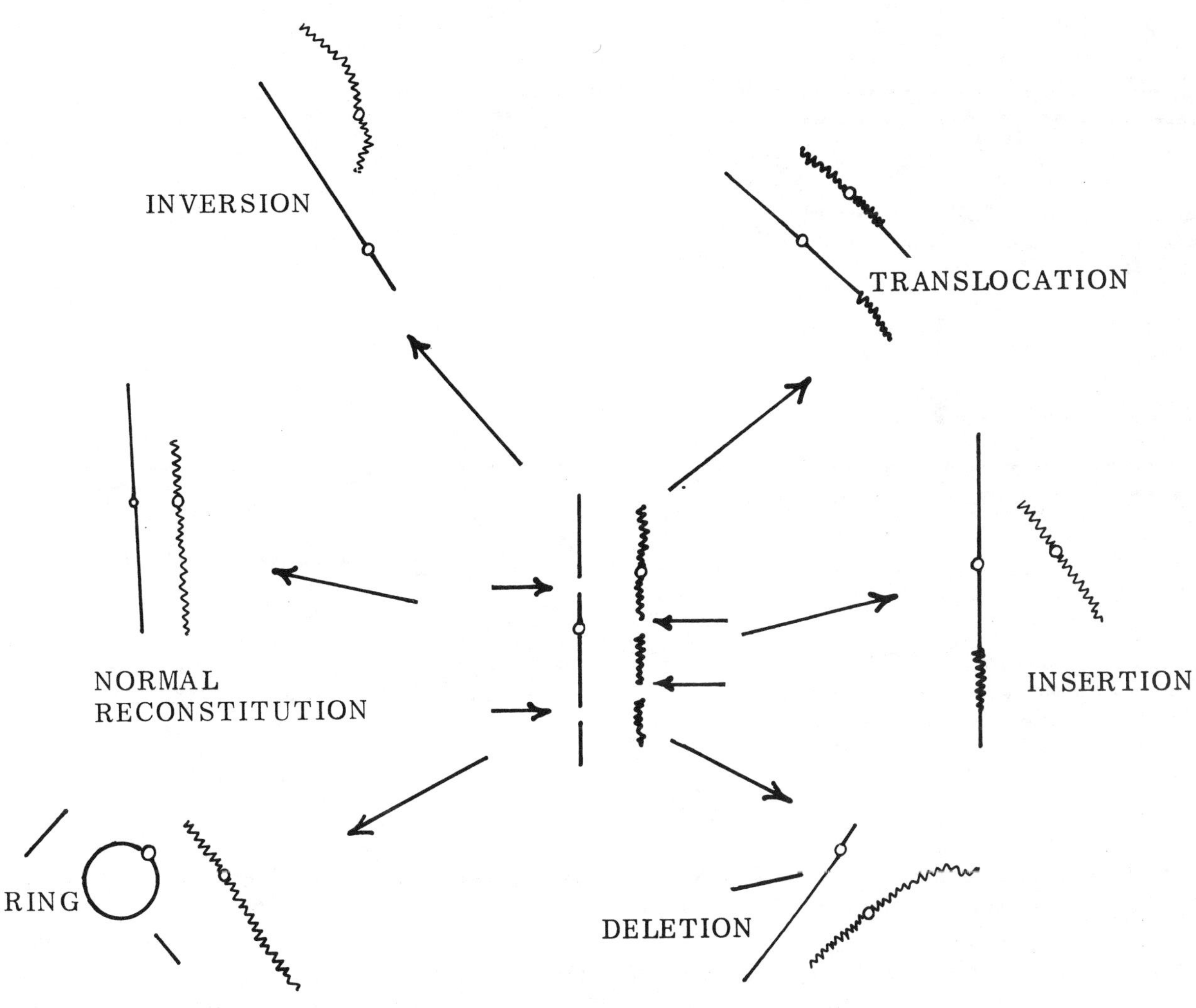

Fig. 4-5. Structural alterations arising from chromosome breakage. Central arrows indicate breakpoints in the chromosome arms. Six possible outcomes following reconstitution of the broken segments are displayed on the periphery of the figure.

Deletions occur when broken fragments of chromosomes fail to reconstitute. The acentric portion of the chromosome and the genes that it carries are lost during the subsequent anaphase. Ring

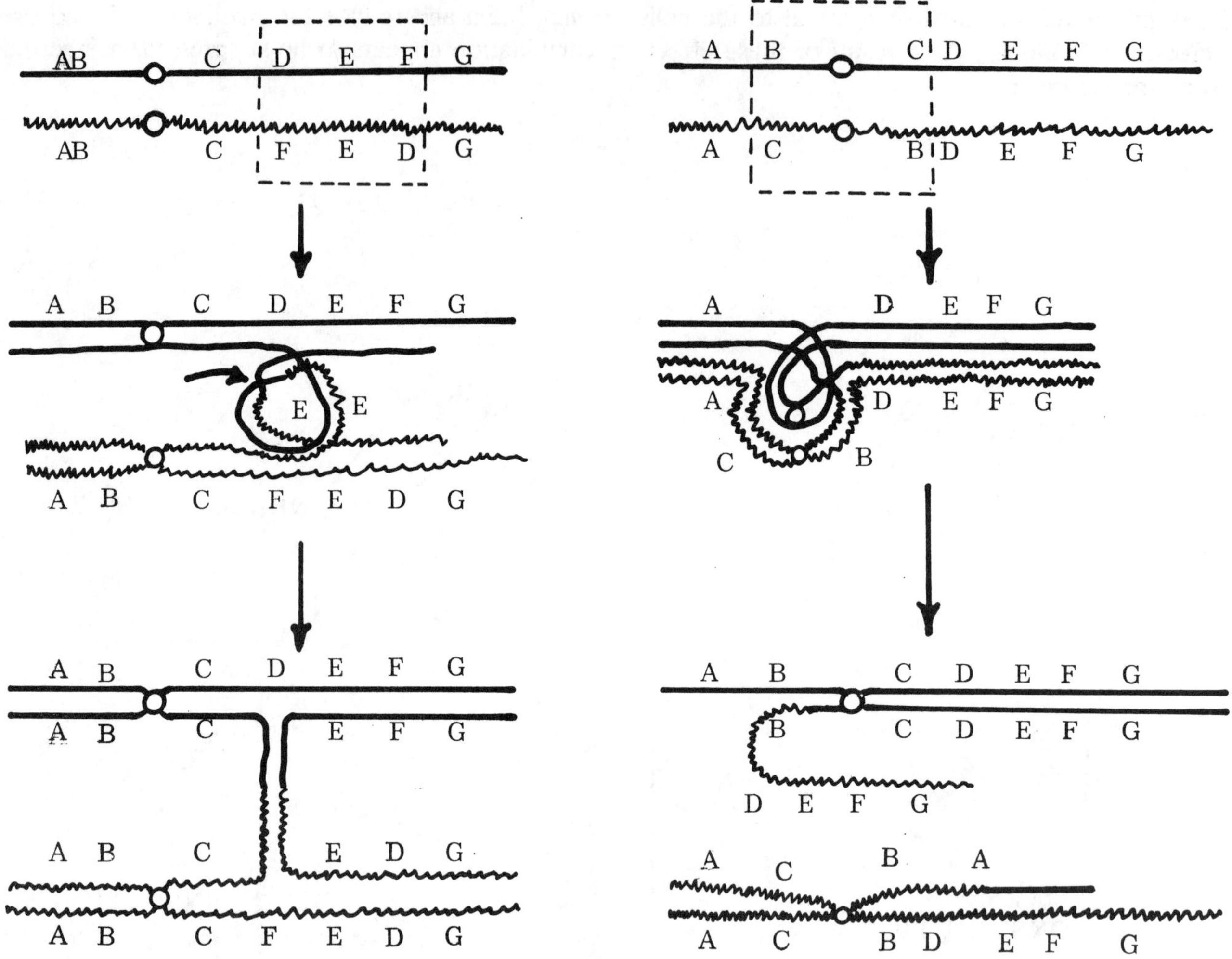

Fig. 4-6. Consequences of crossing over within inversion loops during meiosis. A paracentric inversion is displayed in the left panel and a pericentric inversion is presented in the right panel.

chromosomes are formed when two breaks occur, and the ends of the central piece fuse with one another. **Acentric** rings are lost during subsequent cell divisions. Although centric rings can be replicated and transmitted to daughter cells, the size and shape of the rings tend to vary from cell to cell. The fate of the terminal segments of the chromosome from which the ring was derived is determined by presence or absence of a centromere in the segment.

Exchange of chromosome segments between nonhomologous chromosomes is called a translocation (Fig. 4-5). Translocations are clinically significant, since pairing and segregation of the chromosomes during meiosis are unusual, and unbalanced chromosome complements can occur in the gametic products. Synapsis and segregation of balanced **reciprocal** translocation chromosomes is presented in Fig. 4-7. **Adjacent** segregation produces unbalanced chromosome complements in gametic products. For example, Daughter cell I derived from the first type of adjacent segregation contains two copies of certain chromosome segments and only one copy of others. Zygotes formed from this gamete and a normal gamete will be trisomic for certain genes, monosomic for some genes, and disomic for other genes. This genetic imbalance will cause clinical problems of variable severity, depending upon the specific genes involved. Therefore, although the carrier of this translocation (Fig. 4-7a: this person is referred to as a "balanced" carrier, since all genes are conserved although some are in an unnatural position) is usually free of clinical stigmata, concepti are subject to a higher risk for early embryonic loss, miscarriage,

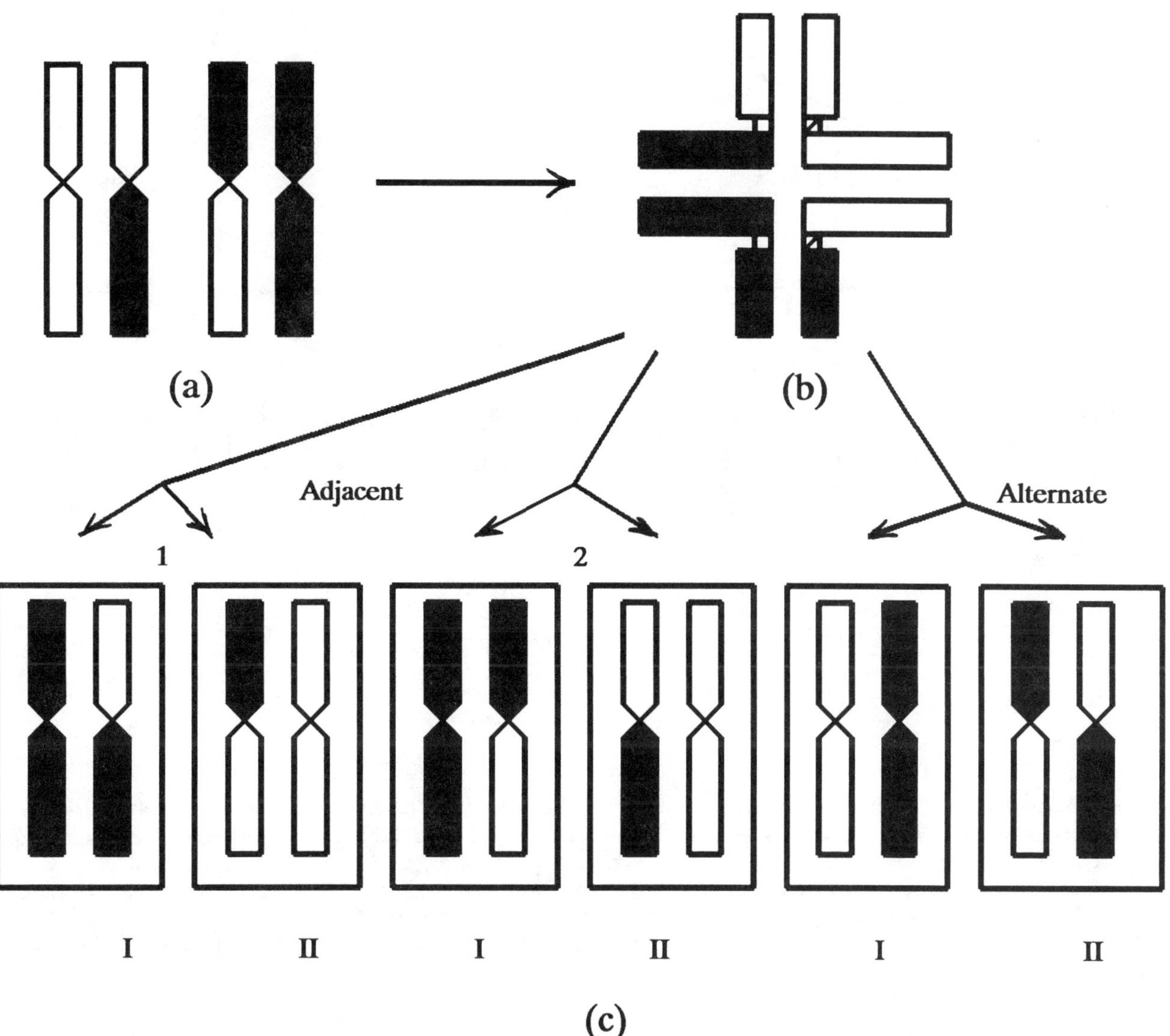

Fig. 4-7. Segregation of derivative chromosomes from a reciprocal translocation(a). Chromosomes form a quadriradial (b) during synapsis, and segregate according to three patterns during Anaphase I (c). Notice that only alternate segregation produces gametes with balanced chromosome complements. The remaining four gametes carry duplication-deficient chromosome complements. Chromosomes are drawn with closely opposed chromatids for clarity.

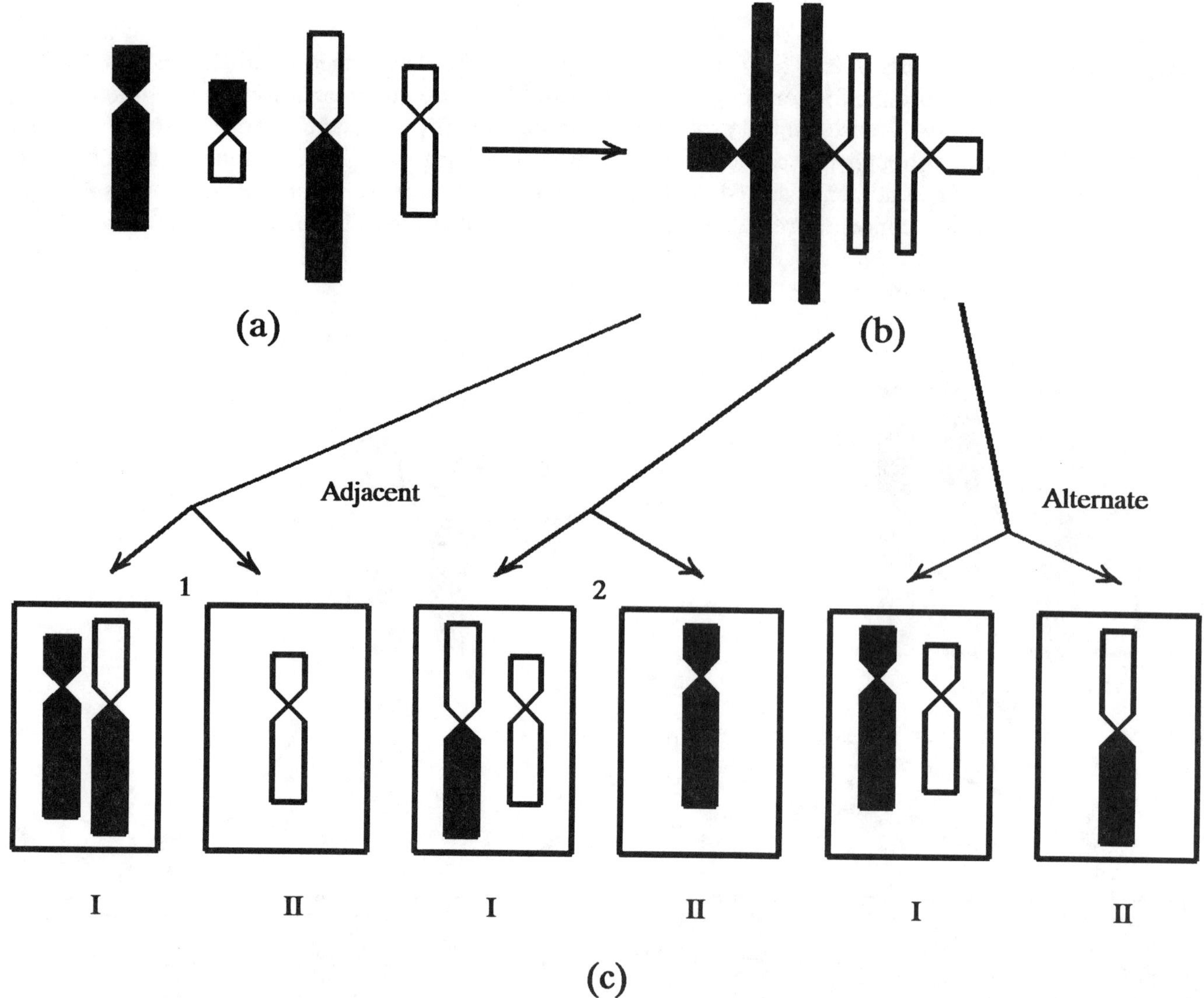

Fig. 4-8.   Robertsonian translocation involving acrocentric chromosomes 14 (shaded) and 21 (a).   Small centric fusion chromosomes (second chromosome from left) are lost during meiotic division. The three remaining chromosomes assemble into a chain of three at Metaphase I (b).   Note how the chromatids of each acrocentric long arm separate and pair with the corresponding arms of the translocation chromosome in the center. Six different gametes would be formed. The four gametes produced following adjacent segregation (c: left and center) contribute to unbalanced offspring. The gametes formed following alternate segregation (translocation chromosome moves to one pole, and chromosomes 14 and 21 move to the other pole) contribute to normal or balanced translocation carrier children.

or development of physical and/or mental handicaps.   Alternate segregation produces two types of gametes, one of which is normal and the other carrying the balanced reciprocal translocation.

     Robertsonian translocations involve two acrocentric chromosomes such as chromosomes 14 and 21 (Fig. 4-8). Breakage occurs near the centromere, and the long arms of the respective chromosomes fuse as illustrated. The small amount of short arm material that is lost during the formation of the translocation chromosome contains tandem repeats of the ribosomal RNA genes. Clusters of these genes are found on chromosomes 13-15, 21 and 22. Therefore, loss of two sets does not impair human development, and carriers of a "balanced" Robertsonian translocation (one normal 14, one normal 21, and the t(14q,21q) chromosome) appear clinically normal. However, these translocation carriers often experience a higher risk for pregnancy loss or birth of handicapped children. Children derived from gametes 1-I and 1-II and a normal gamete would produce concepti with partial trisomy of 14 and monosomy 14, respectively. Both conditions are lost during early development. Fetuses derived from gametes 2-I and 2-II and a normal gamete would have partial trisomy of 21 and monosomy 21, respectively. Trisomy 21 is called Down syndrome and will be described later in this chapter. Monosomy 21 is an early lethal. The last two gametes would give rise to a chromosomally normal person and a balanced translocation carrier, respectively.

Insertions require at least three breaks, two in one chromosome and one in a nonhomologous chromosome. The middle segment from one chromosome inserts between the fragments of the second chromosome (Fig. 4-5). If all genetic information is conserved following the exchange, the person in whom this mutation occurs will be clinically normal; however, he/she may produce gametes that carry unbalanced gene combinations. Chromosome banding methods have greatly increased the possibility for detection of insertions.

Isochromosomes arise following **transverse** division across the centromere (Fig. 4-9). The resulting chromatids contain either two identical copies of the short arm or two copies of the long arm. Segregation of the chromosome during meiotic division generates unbalanced gametes. A woman inheriting the isoXq chromosome and a normal X would most likely have impaired sexual development,

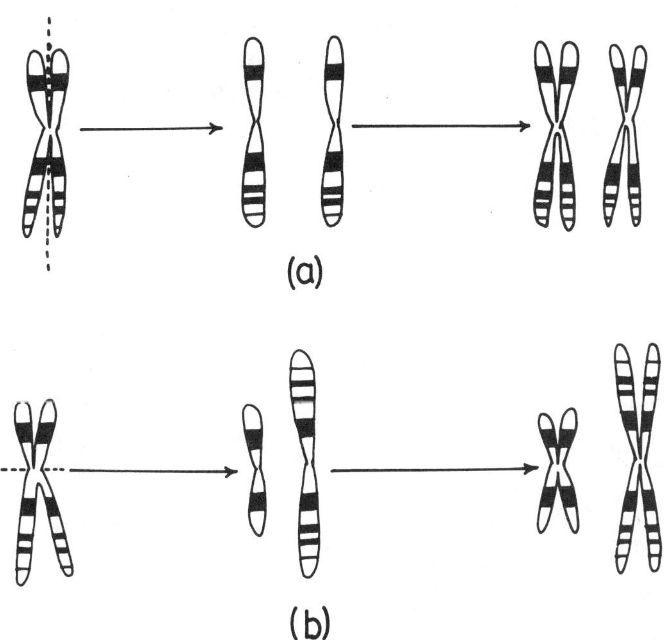

Fig. 4-9. Formation of an isoX chromosome. The X has been stained by a trypsin-giemsa procedure to show banding detail. a) Normal longitudinal segregation; b) transverse segregation generating an isoXp and isoXq chromosome.

be infertile, and runs a higher risk for mental retardation and behavioral problems compared to a woman with two normal X chromosomes.

Clinical Consequences of Chromosomal Imbalance

Since chromosomes carry a large number of genes, one would expect deviations from the normal chromosome number or gain or loss of chromosome segments to result in a variety of disturbances. These abnormalities present at the clinical level:

Mental Retardation	Sexual Ambiguity
Multiple Structural Anomalies	Growth Retardation
Infertility	Shortened Lifespan

These problems may vary from severe to quite subtle, depending upon the extent of chromosome gain or loss and whether genes located in the regions involved play major or minor roles in the determination of the respective traits. A collection of unusual clinical features is called a **syndrome**. Although each of the components of a syndrome may occur as isolated entities, each syndrome is unique by virtue of the constellation of anomalies it encompasses. Occurrence of a cluster of individually rare clinical features in a patient should be viewed as an indication for a possible chromosomal anomaly. For example, mentally handicapped persons with three or more physical malformations frequently possess a chromosomal abnormality. Careful examination of the face, hands, and feet often reveals clues that suggest presence of a chromosome error.

Many types of chromosome imbalance are incompatible with survival. Chromosome errors are a major cause of early embryonic and fetal loss, accounting for about 50% of spontaneous abortions that occur during the first three months of pregnancy. The most common anomalies appearing in spontaneous abortices include triploidy (often caused by double fertilization of an egg), monosomy X, and trisomy 16. Sex chromosomal trisomies (47,XXX; 47,XXY; 47,XYY) are rarely observed in abortion series. This trend most likely reflects both the protective effect of X-inactivation and the fact that the Y chromosome carries few genes that are essential for survival.

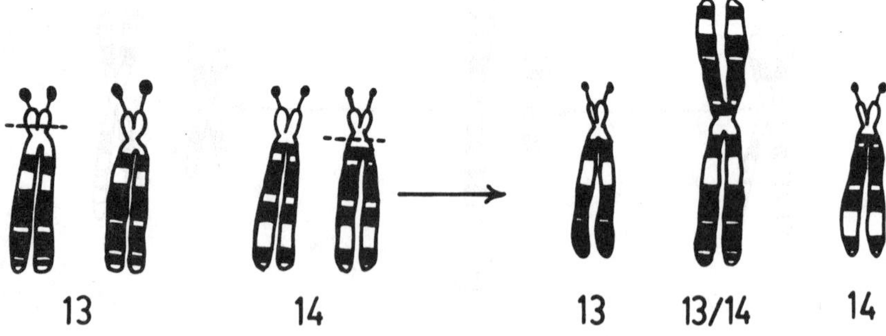

Fig. 4-10. Robertsonian translocation involving chromosomes 13 and 14. The dashed lines indicate putative breakpoints.

About 1/180 newborn possess major chromosome errors. Sex chromosome aneuploids are most frequent among the liveborn population (1/100 males, 1/700 females); however, monosomy X is relatively infrequent (1/2600 female infants). The Fragile-X chromosome is found in approximately 1/1000 males, contributing to mental retardation and characteristic physical problems (Martin-Bell syndrome). This chromsomal anomaly differs from the others with respect to its high risk of recurrence in families. The most common autosomal trisomy among liveborn infants is trisomy 21 (Down syndrome), appearing in about 1/800 newborn. Trisomy 13 (Patau syndrome) and trisomy 18 (Edward's syndrome) are much less frequent, reflecting their greater size and gene content.

Approximately 1/500 newborn carry balanced translocations. Robertsonian translocations are most common (Fig. 4-10). The figure illustrates a balanced Robertsonian translocation involving chromosomes 13 and 14. The chromosomes illustrated at the right of the figure would be expected to form a chain during Metaphase I that would be similar to that illustrated in Fig. 4-8b. The normal chromosome 13 and 14 would be at either end of the chain, and the translocation chromosome (t(13q,14q)) would be located in the center. Six possible gametes could be formed, four of which would be unbalanced. One gamete would have a normal chromosome complement, and one would carry the balanced translocation. One of the unbalanced gametes would have the normal 13 and the t(13q,14q) chromosomes. The conceptus derived from this gamete would be expected to have symptoms characteristic of Patau's syndrome (Trisomy 13), and a karyotype similar to that presented in Fig. 4-11 (left). Note that this patient has 46 chromosomes (46,XX,-14,+t(13q,14q)), including the translocation chromosome. Cells from the balanced translocation carrier contain 45 chromosomes (Fig. 4-11(right)).

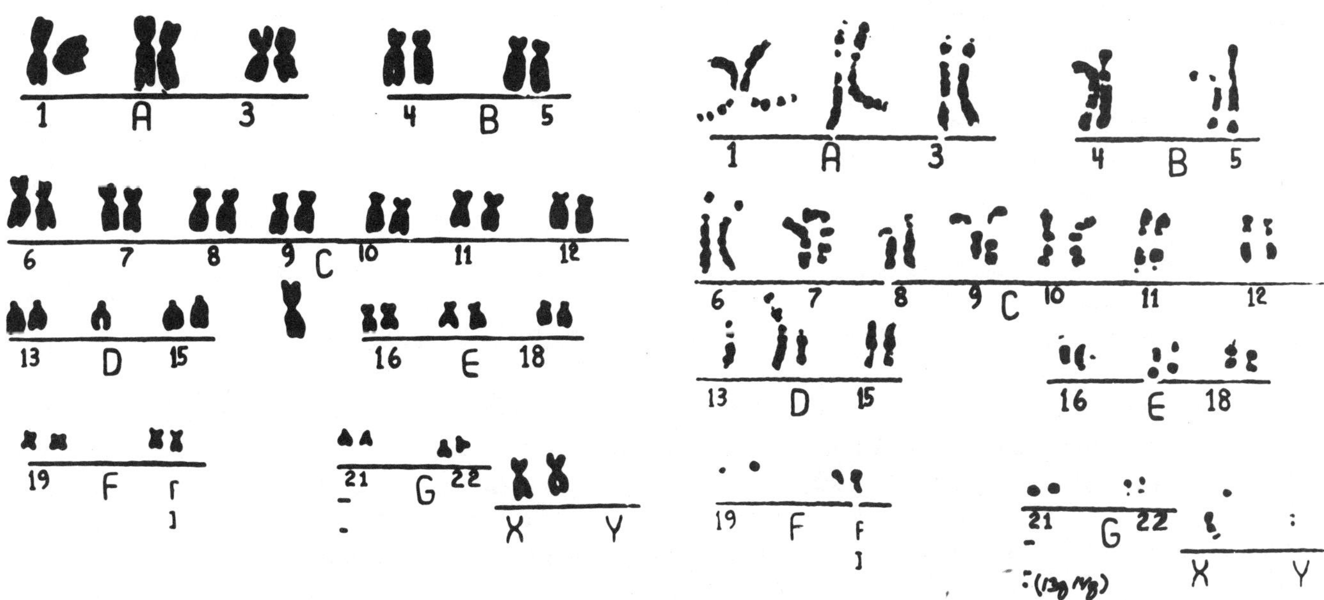

Fig. 4-11. Karyotypes of a patient with translocation Patau's syndrome (46,XX,-14,+t(13q,14q)) (left) and her balanced translocation carrier father (45,XY,-13,-14,+t(13q,14q)) (right). The short arms of the chromosomes fusing to form the translocation chromosome were lost during subsequent cell divisions.

Most individuals with Trisomy 13 die before birth. Those who survive have severe malformations of major organ systems that most commonly cause death before one month of age. The remaining three unbalanced gametes would result in the formation of zygotes with trisomy 14, monosomy 13, or monosomy 14. All of these combinations seem to be lost during early development and have not been observed among liveborn infants.

Balanced translocations are often transmitted for several generations (Fig. 4-12). The pedigree was encountered through the baby with Patau's syndrome (arrow). Half-shaded squares and circles represent balanced translocation males and females, respectively. Open symbols indicate persons with normal karyotypes. Symbols with dots represent persons who were not tested. Horizontal lines connect spouses, and vertical lines connect spouses with their descendants. This pedigree illustrates the need for testing members of a sibship (brothers and sisters) as well as other relatives of balanced translocation carriers, since they may be at higher risk for having children with multiple malformations. Note how

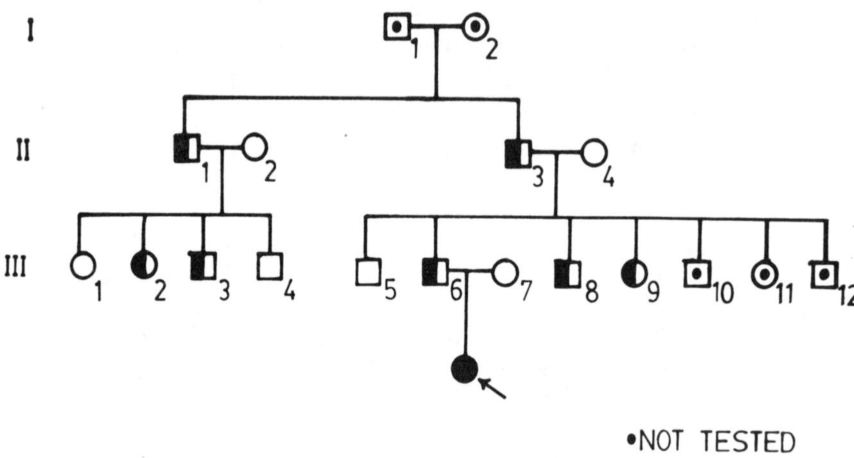

Fig. 4-12. Pedigree of a kindred segregating for a 13/14 translocation. The symbols are explained in the text.

approximately one-half of the clinically normal offspring of a balanced carrier parent inherit the balanced translocation.

The preceding example demonstrates one group in which an increased risk for chromosome disorders exists. In this case, the kindred was identified fortuitously through the birth of a child with a characteristic syndrome. In other instances, infertility, multiple abortions, or birth of two or more affected children may alert a clinician to the possible presence of a segregating translocation or inversion chromosome. Other groups that frequently possess chromosome abnormalities include the mentally handicapped (>15 years = 10%; <15 years > 15%), newborn with multiple malformations (= 10%), and individuals presenting with physical sexual anomalies, sexual ambiguity, or infertility (= 11% of infertile men; = 25% of women with primary amenorrhea).

EXAMPLES OF MAJOR CHROMOSOMAL ANOMALIES

Down Syndrome

More than ninety years elapsed between the initial description of Down syndrome by Langdon Down (1866) and Jerome Lejeune's discovery of 47 chromosomes in cells from these patients. Considerable variation is observed among children and adults with Down syndrome. About 30-40 percent are very severely affected and are miscarried before the end of the first trimester of pregnancy. Babies born with Down syndrome usually have poor muscle tone (hypotonia), simean creases (single transverse palmar line) on one or both hands, and may have a heart defect (usually a hole in the heart wall). As the infant develops, other features of the syndrome become more apparent. The head is usually broadened and flattend in the occipital region, and the ears are low set and malformed. Facial features are characteristic with unusual eyes bordered by medial epicanthal folds (Fig. 4-13), and speckling can be observed on the iris with suitable illumination. The nose is short and has a depressed bridge. The tongue is fissured and protruding, due to inadequate room in the floor of the mouth. Mental retardation of moderate severity is a later complication.

The hands of a Down syndrome child are short and broad and possess a single palmar crease (Fig. 4-14a). The little finger is unusually short (many Down's children lack the middle phalanx) and incurved (clinodactyly) (Fig. 4-14b). The finger and palm prints of these children are unusual. The first and second toes are separated by a wide gap (Fig. 4-15), and abnormal dermal patterns occur on the sole of the foot in the region of the large toe.

Down's children typically have short stature and hyperflexible and hyperextensible joints. Abdominal musculature is not well

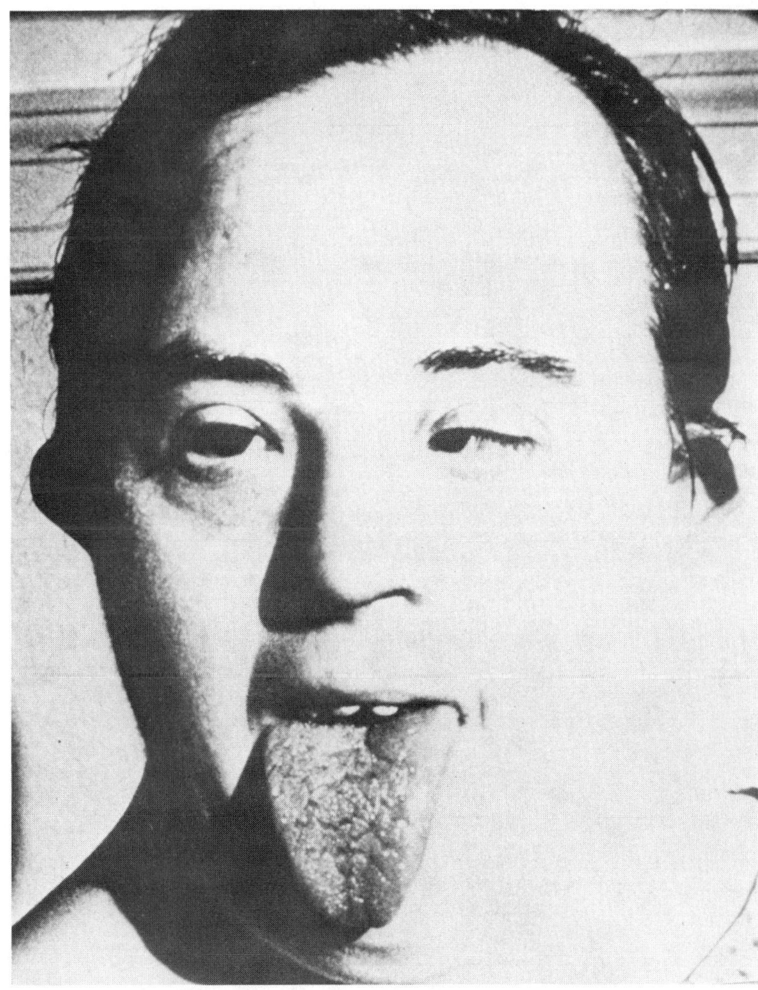

Fig. 4-13. Facial features of Down syndrome. Medial eyefolds are thick and overlapping (epicanthal folds), and strabismus is evident. The tongue is deeply fissured and appears too large. Note widely-spaced teeth.

developed, and hernias are not uncommon (Fig. 4-16).

Patients usually have abnormal external genitalia. Undescended testes, testicular degeneration, and small penis are frequently observed in males, and many are infertile. Females may have small or absent labia minora and rarely reproduce. Approximately one-half of the babies born to Down's women are also affected with Down symdrome.

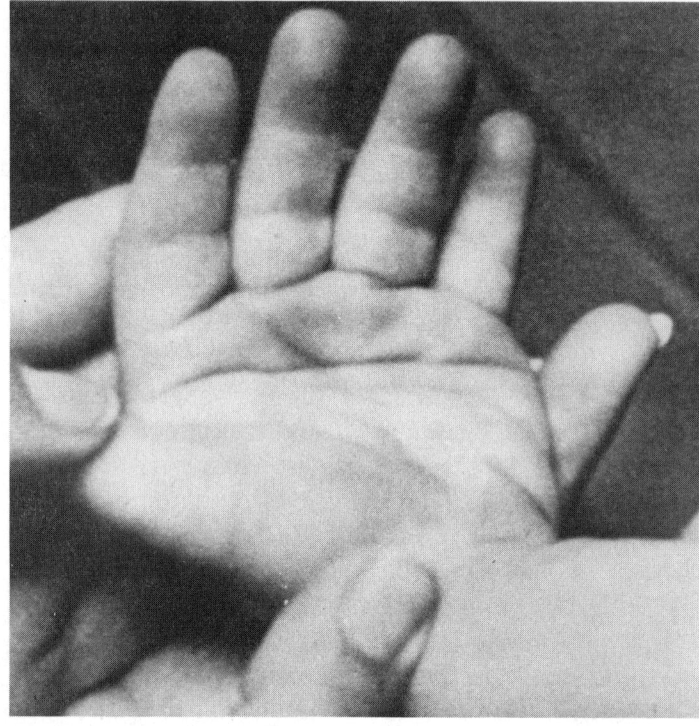

Fig. 4-14. Hands of Down syndrome patient. a) Palmar crease (right); b) Note the short, broad hand and presence of clinodactyly (below).

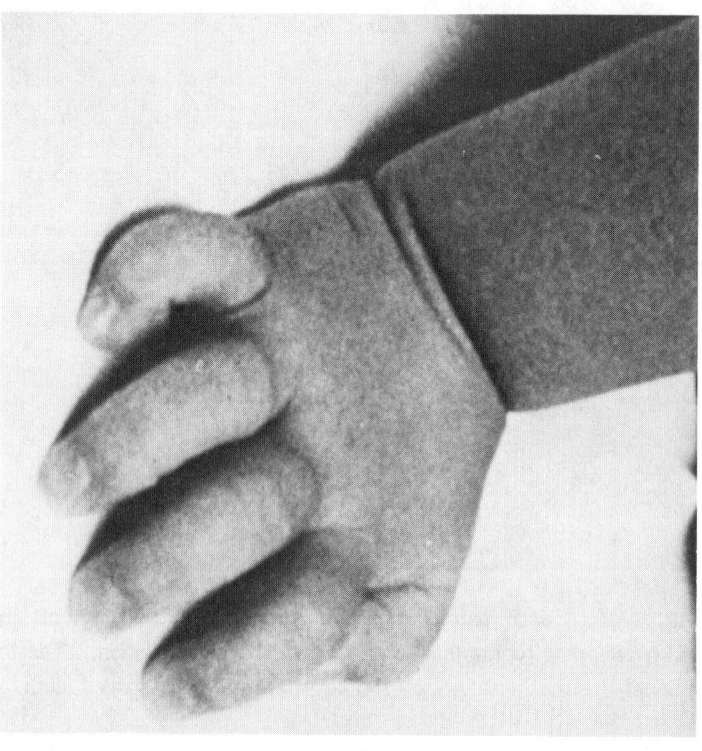

Children with Down syndrome do not necessarily exhibit all of the features described above. The syndrome is usually most characteristic during the juvenile period. Diagnosis of certain patients may be difficult in early infancy, particularly in mosaics, where the normal cell line may obscure many features of the syndrome.

Considerable progress has been accomplished with the treatment of Down syndrome persons. Infant surgery has proved successful in many cases, ameliorating many of the severe defects that contributed to infant mortality ony a few years ago. Aggressive management of respiratory infections which commonly afflict young Down's children has also succeeded in extending their life expectancy. Parents are taught how to exercise their children's musculature and to work on their coordination. Early educational enrichment has helped Down's children develop speech, language, and reading skills. Computers have proved quite useful in correcting speech defects associated with Down syndrome. Children tend not to hear or process short words such as articles and prepositions, often leaving them out of their speech. Computers emphasize these words and their pronunciation, improving their use and comprehension by Down's children. As they enter school, Down's kids receive both special education and the opportunity to mix with others their own age. The former practice optimizes development of the skills they have, while the latter allows them to develop social skills that are essential for normal rearing. Other children in the classroom appear to easily accept a Down's child as a playmate and equal.

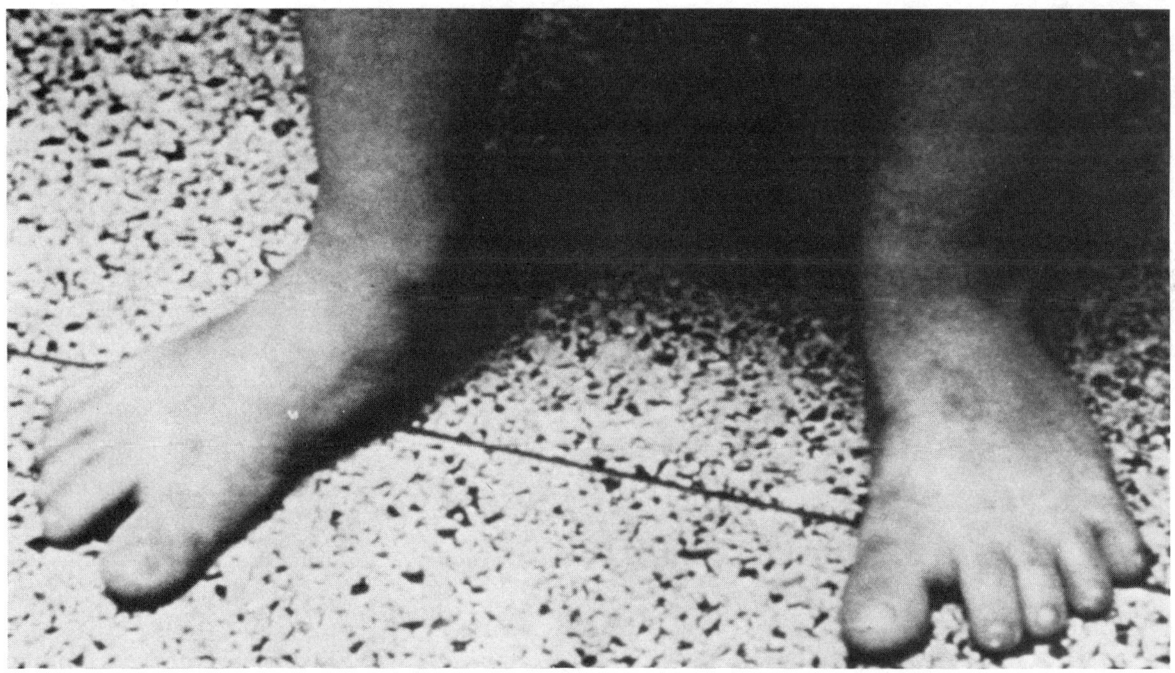

Fig. 4-15. Feet of Down syndrome patient. Notice widely-spaced first and second toes.

As they grow older, Downs' persons will continue to require some help, particularly with activities that require mathematic or arithmetic skills. Some continue to live with their families, while others may reside in group homes where daytime and nightime assistance is available from appropriately trained personnel. Many of the household tasks are performed by the residents. Sheltered workshops provide a source of

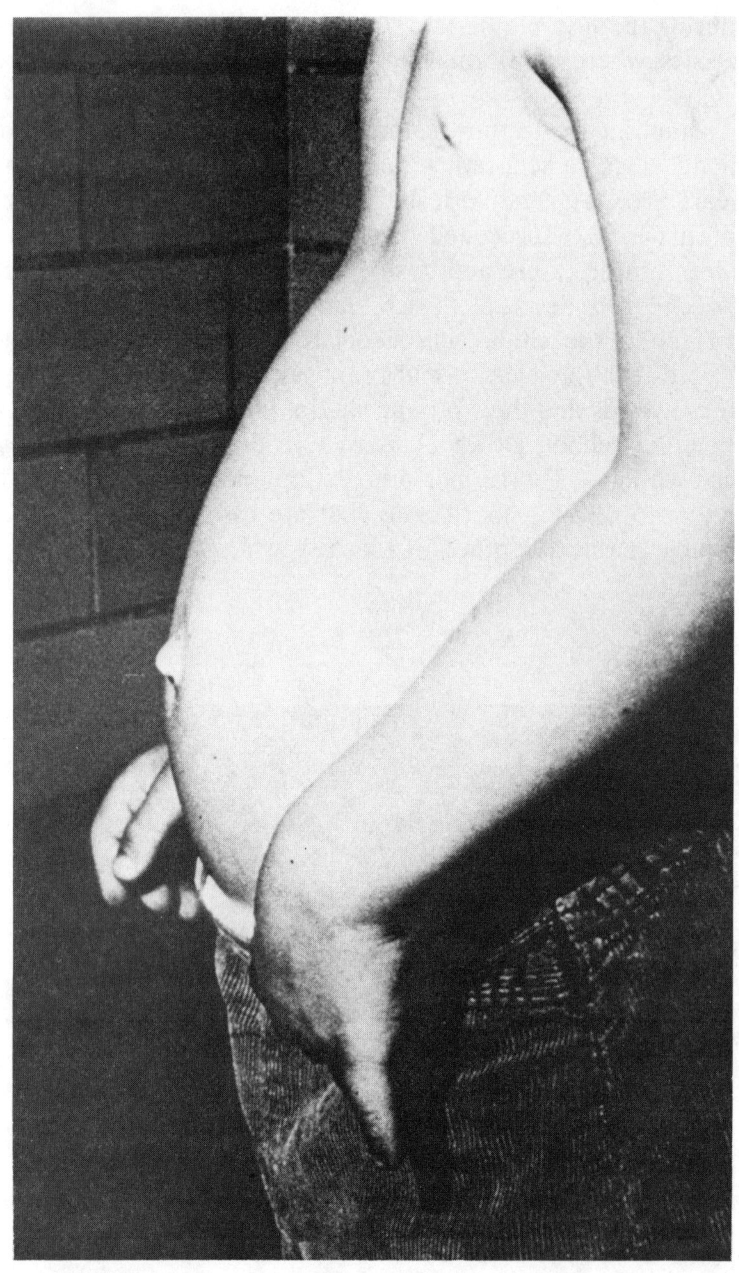

Fig. 4-16. Abdomen of Down syndrome child. Poor muscle development results in distension of the abdomen and tendency for umbilical hernia. Note recessed sternum (pectus excavatum).

income for individuals who are unable to find a place in the usual workforce; however, an increasing number are able to assume self-supporting positions in the workplace, and several have become successful in the performing arts. Recreational opportunities have improved for Down's and other handicapped persons. The Special Olympics and municipal park district programs have provided them with both a spirit of competition and the rewards of success.

As more Down syndrome patients are living to older ages, some of the later manifestations are becoming apparent. In addition to the reduced fertility mentioned above, older patients appear to have an increased risk for lymphatic cancer and leukemia and for syptoms resembling those of Alzheimer's disease. In the latter case, onset of Alzheimer's disease occurs at a much earlier age than is characteristic of the general population. Early diagnosis and treatment of the cancer and leukemia have met with some success.

Down syndrome results from a genetic imbalance imposed by the presence of three copies of chromosome 21. Recent evidence indicates that triplication of genes within a relatively small segment located on the distal half of the long arm is sufficient for manifestation of Down syndrome. More than 90 percent of Down syndrome patients have primary trisomy (47,XX or XY,+21). The three copies of chromosome 21 are separate (Fig. 4-17), and meiosis in these individuals would be anticipated to produce two types of gametes in equal proportion. One class would contain 23 chromosomes and a normal haploid karyotype, while the other would have 24 chromosomes including two copies of chromosome 21. A 50 percent risk for a Down syndrome child would be anticipated, and the few children who have been born to Down syndrome women appear to uphold this

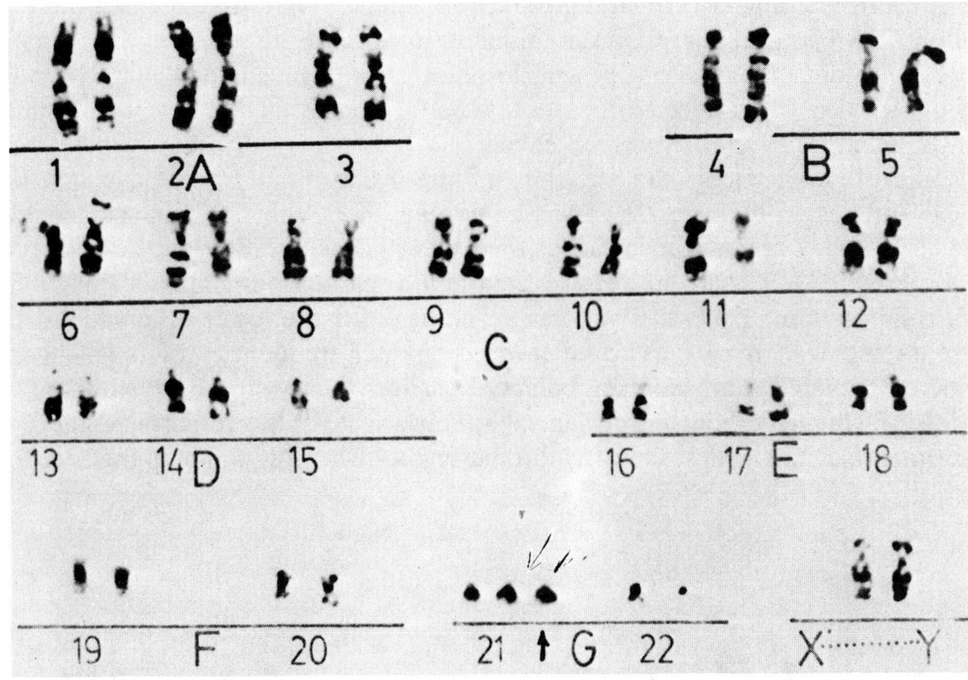

Fig. 4-17. Karyotype of a patient with primary trisomic Down syndrome (47,XX,+21). (Courtesy of S. Shafer)

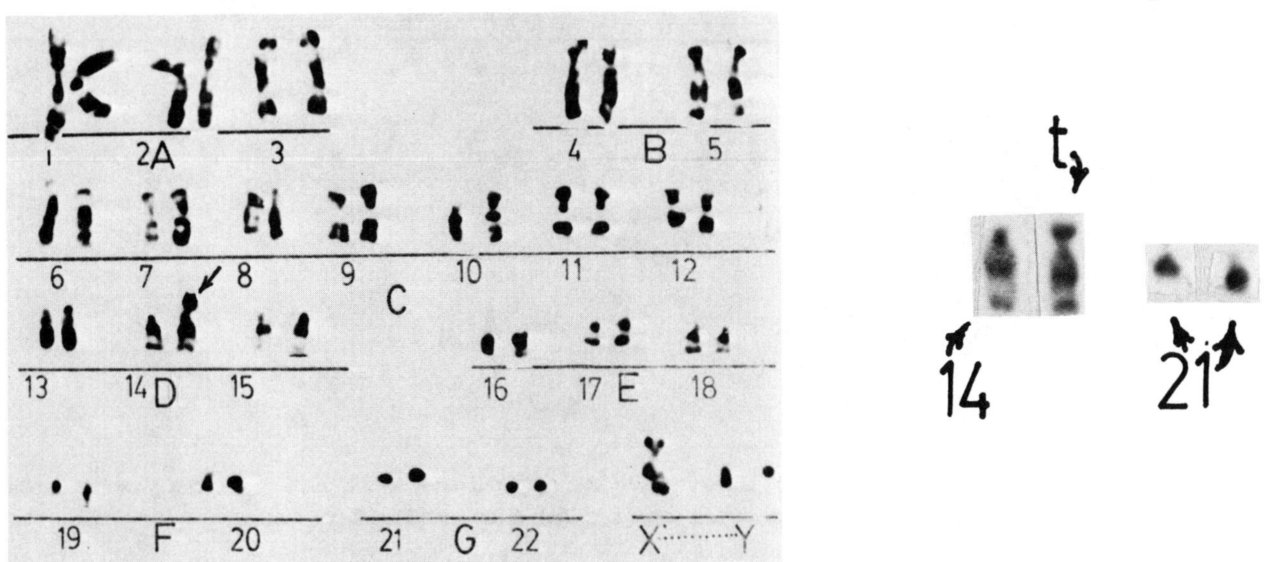

Fig. 4-18. Translocation Down's patient: 46,XY,-14,+t(14q,21q). (Courtesy of S. Shafer)

expectation.

Robertsonian translocations involving chromosome 21 and chromosomes 13, 14, 15, another chromosome 21, or chromosome 22 are observed in about 4% of Down syndrome children. The most common translocations typically involve chromosomes 14 and 21 (Fig. 4-18). Clinical features of patients with translocation Down syndrome are indistinguishable from those of patients with primary Trisomy 21. Approximately 40 percent of patients with translocation Down syndrome inherited their translocation chromosome from a balanced translocation parent. Therefore, it is advisable to test both parents when a translocation is found in a Down syndrome child.

Although somewhat less than one-third of the liveborn children of a balanced translocation parent would be anticipated to have Down syndrome (Fig. 4-8), the observed frequencies of affected children are much lower. When a woman carries the balanced translocation, the actual risk of occurrence or recurrence is about 10-15% percent. However, when a man carries the balanced translocation, less than 5% of his children develop Down syndrome. The basis for the lower observed risks is not known. Although alternate segregation appears to be favored, greater frequencies of unbalanced chromosome complements were recovered from sperm of balanced carriers than were observed among their liveborn children (see box). Therefore, either unbalanced sperm are less able to fertilize the egg, or there is stronger post-zygotic selection against Down syndrome zygotes when the father is the translocation carrier.

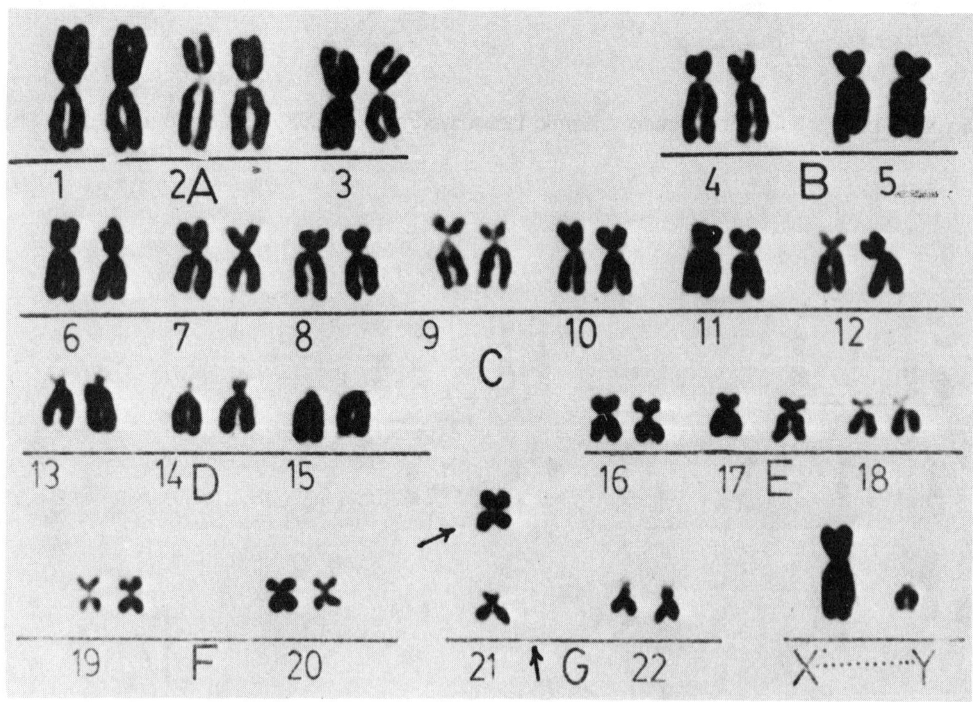

Fig. 4-19. G/G translocation Down syndrome. This karyotype was prepared prior to the discovery of chromosome banding; therefore, it is uncertain whether the translocation chromosome is a t(21q,21q) or t(21q,22q).

Recent studies of chromosomes containing two long arms from chromosome 21 suggest that most arise following transverse division across the centromere, producing an isochromosome. RFLP

Dr. R.H. Martin (1983) has developed a system for analyzing the chromosomes of human sperm. The sperm are capacitated _in vitro_ and used to fertilize hamster eggs. The fertilized eggs are briefly cultured to allow the human male pronucleus to swell, and the chromosomes are stained with a banding method. Several studies have been reported that have described the proportions of meiotic products produced by male balanced translocation carriers. Analyses of 24 sperm from a male balanced t(14q,21q) carrier revealed that 16 sperm had normal haploid chromosome complements, 4 had the balanced translocation, and 3 contained unbalanced translocation complements (3). Two studies have examined sperm from male t(13q,14q) balanced translocation carriers. Fifty percent of 78 sperm from one man contained normal chromosome complements, while 41 percent and 8 percent had balanced translocation and unbalanced karyotypes, respectively (4). The second study found 27% unbalanced, 38% balanced translocation, and 36% normal karyotypes among 121 sperm from a balanced t(13q,14q) carrier (5). Although the numbers of subjects and sperm analyzed were small, these studies suggest that alternate segregation is favored during male spermatogenesis. Similar studies of several reciprocal translocation carriers (t(4;17), t(5;13), t(6;7), and t(9;18)) demonstrated alternate segregation among 57 percent, 77 percent, 51 percent, and 35 percent of the sperm analyzed, respectively, indicating that proportionally more products from adjacent segregations occur than appears typical for Robertsonian translocations (6). These data suggest that there is a strong post-zygotic selection against unbalanced chromosome combinations, the strength of which varies according to the particular unbalanced combination inherited by the zygote. A recent comprehensive survey of 1237 fetuses at amniocentesis (7) revealed an incidence of Down syndrome of 10-15 percent and 1.5-5 percent among fetuses of balanced female (D,21) and male (D,21) translocation carriers, respectively. These frequencies indicate that the post-zygotic selection is occurring at an early stage, since the figures observed at amniocentesis are similar to those reported among liveborn children.

1-Hamster Egg/Human Sperm System for Study of Sperm Chromosome Complements.

analyses were used to study the structure and origin of the duplication 21q (dup(21q)) chromosomes from Down syndrome children of ten different couples(8). All ten cases represented new mutations. Eight of the ten dup(21q) chromosomes were found to have identical RFLP patterns on both arms of the chromosome, indicating they were isochromosomes. The remaining two patients had translocations in which the arms were derived from two different chromosome 21s. Four of the isochromosomes and both translocation chromosomes were maternal in origin, presumably representing germinal mutations.

Mosaicism for Trisomy 21 and a normal cell line occurs in approximately 1 percent of Down syndrome patients. The proportion of trisomic cells varies among individuals and among tissues within the same person. The clinical features of mosaic patients become less distinct as the proportion of abnormal cells declines, and diagnosis is difficult and often dependent upon chromosome analysis. Low grade mosaicism in a parent is associated with increased risk for recurrence of Down syndrome. The magnitude of the risk can be approximated by halving the average frequency of mosaicism estimated by sampling two or more cell types.

Down syndrome has been observed in all human populations and appears to occur in approximately 1/700 newborn in most populations studied. Maternal age at conception has a major effect upon the risk of occurrence of Down syndrome (Table 4-1), and the population frequency of Down syndrome may vary from one population to another and from one time to another within the same population. For example the incidence rises during times of economic depression or recession because

Table 4-1. Incidence of Down Syndrome as a Function of Maternal Age (9,10).

Maternal Age	Incidence
<20	1/2300
20-25	1/1600
26-30	1/1200
32-	1/800
37-	1/290
43-	1/100
47-	1/46

primary trisomic Down syndrome, the risk is relatively low and dependent upon maternal age at conception. The recurrence risks presented in Table 4-2 would appear to be appropriate estimates of the recurrence risk under these circumstances. As mentioned above, parental mosaicism of Trisomy 21 increases the risk of recurrence. Lymphocytes and skin fibroblasts may be cultured to obtain an average estimate of parental mosaicism. Suppose 30 percent of cultured lymphocytes and 10% of skin fibroblasts from the mother of a Down's child are trisomic. The estimate of trisomic oogonia would be 20 percent. One-half of the ova produced from these oogonia would be expected to be disomic for 21, and, when fertilized by a normal sperm, will produce a trisomic zygote. Therefore, the risk that the second child of this woman would develop Down syndrome would be about 10 percent. Since some Trisomy 21 fetuses die before birth, the risk may be somewhat lower. Presence of a balanced Robertsonian translocation in a parent is associated with increased risk of recurrence. For reasons discussed above, an estimate of 10-15 percent or 2-5 percent appears appropriate when the translocation is maternal or paternal, respectively.

couples often delay having children during these periods. Furthermore, populations which customarily delay marriage until the thirties will typically have higher incidences of Down syndrome than those where marriage at younger ages is more common. Studies using morphological chromosome markers and RFLPs have demonstrated that about two-thirds of Down's children occur as a consequence of nondisjunction at maternal Anaphase I which follows a state of prolonged meiotic arrest during Prophase I. This arrest is believed responsible for the maternal age effect. Other possible causes, such as advanced paternal age and inefficient selection against unbalanced zygotes by an ageing uterus, appear to be less important.

The risk that a second child will be born with Down syndrome varies with the type of cytogenetic anomaly present. When both parents have normal karyotypes, and their child has

Table 4-2. Influence of Birth of a Previous Down Syndrome Child and Maternal Age at Conception Upon Recurrence Risk (11).

Maternal Age	Recurrence Risk
<30	1/500
30-	1/450
32-	1/350
34-	1/200
36-	1/130
38-	1/65
40-	1/50
42-	1/35
44+	1/25

Trisomy 18 (Edward's syndrome) occurs among 1/7500 newborn. The actual incidence varies among populations depending upon their maternal age structure, since a maternal age effect similar to that observed in Down syndrome is also associated with Edward's syndrome. Chromosome 18 is larger than chromosome 21   (Fig. 4-20), presumably carries more genes, and would be expected to have

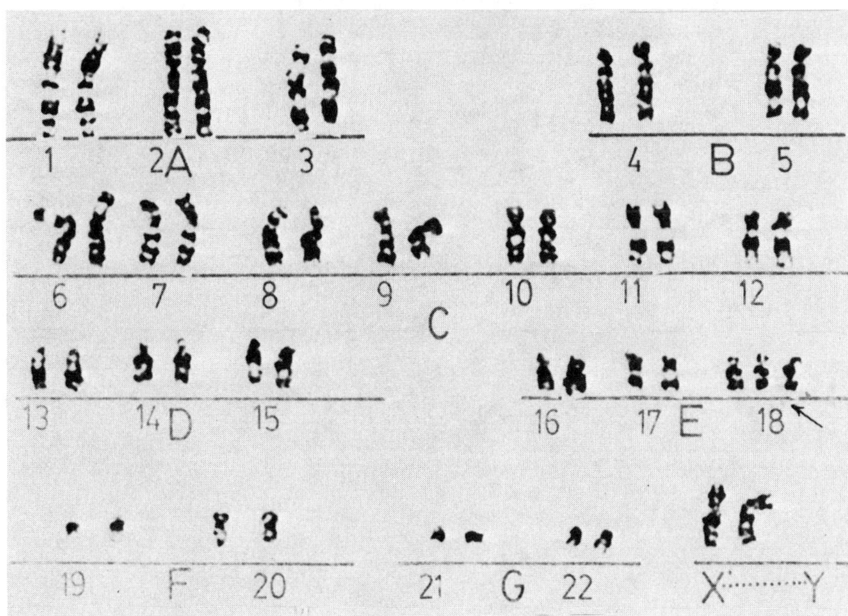

Fig. 4-20. Karyotype of a patient with Edward's syndrome. (Courtesy of S. Shafer)

relatively more severe clinical consequences. Prenatal lethality of the syndrome is greater than 98 percent. Babies who are born with the syndrome exhibit malformations of many organ systems. Edward's syndrome babies have elongated skulls, low-placed malformed ears, small chin, and various eye anomalies, including epicanthal folds (Fig. 4-21). Their muscles are very tense (hypertonic), and their hands are clenched with overlapping thumb and fingers (Fig. 4-22). Nails are underdeveloped, and dermal ridge patterns on their fingertips consist almost entirely of simple arches. Infants have a shortened sternum and heart defects that include patent ductus arteriosus and septal defects. Malformations involving the genitourinary system are often present and may present as double ureter, horseshoe kidneys, and/or undescended testes in males. The feet are unusual in many infants and are often described as "rocker-bottom" feet due to a dorsiflexed big toe and convex arch. Relatively few Trisomy 18 fetuses have been detected in reported abortice series, suggesting that prenatal lethality occurs at an early stage. Trisomy 18 is the most frequent chromosomal abnormality associated with perinatal death. Liveborn Edward's syndrome infants exhibit failure to thrive, and about 90% die within one year of birth. Improved medical care has prolonged the lives of an increasing number of Edward's patients. However, there is some debate concerning whether the quality of these persons lives has also improved. The economic, psychological, and emotional costs of rearing these severely handicapped children are great, placing a considerable strain upon the family unit. In spite of these problems, many parents and siblings of the infants are strongly attached to them, providing them with the love and care they need. Parent support

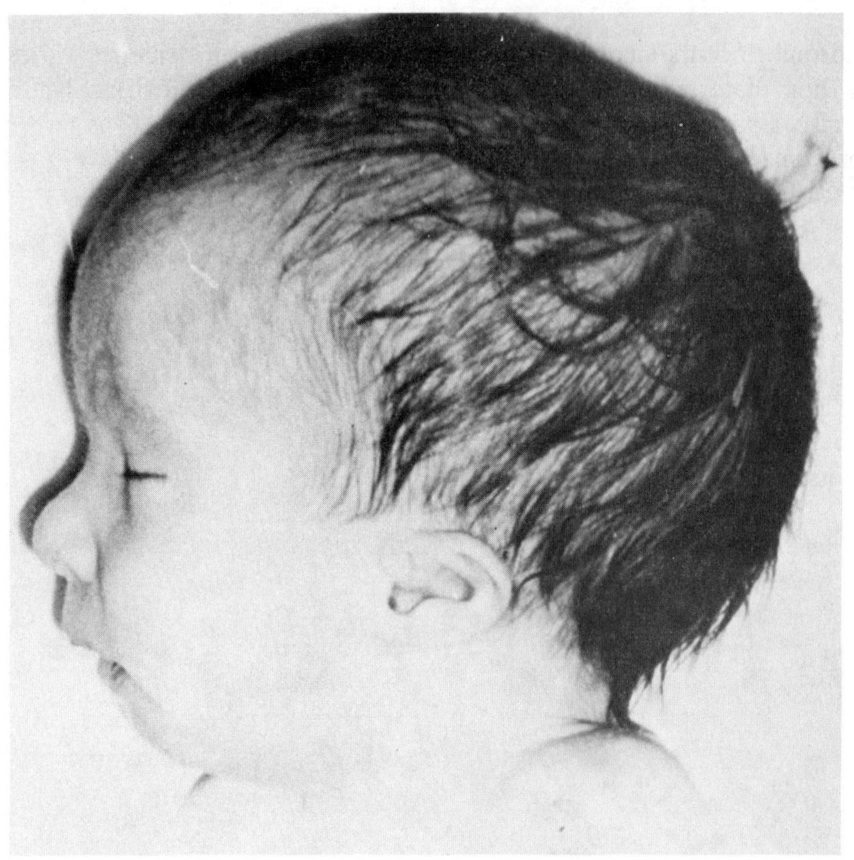

Fig. 4-21. Facial features and head of a patient with Edward's syndrome. Note depressed nasal bridge, receding chin, low-set malformed ears, and short neck. (Courtesy of Dr. S. Yasunaga)

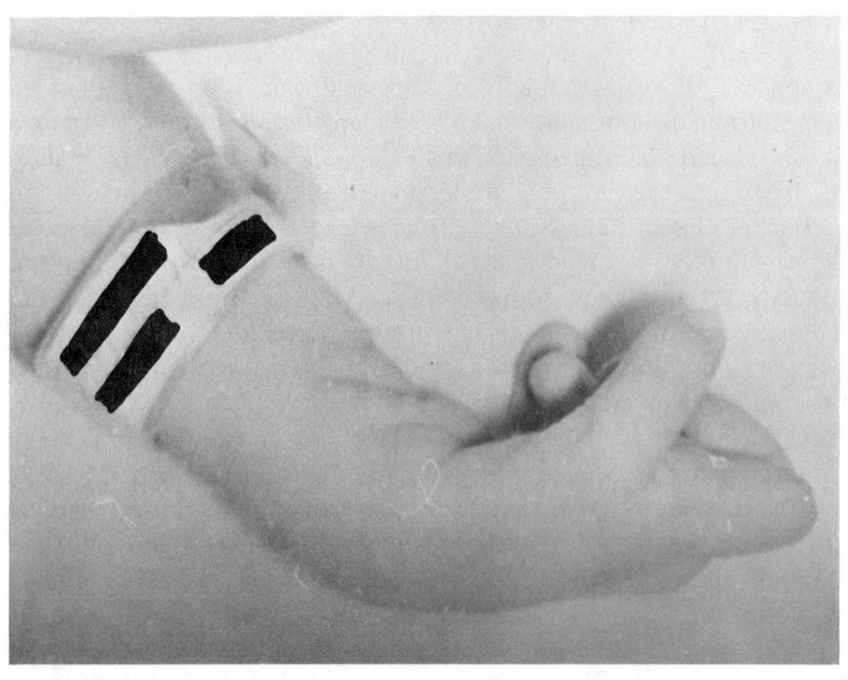

Fig. 4-22. Overlapping thumb and fingers of a patient with Edward's syndrome. (Courtesy of Dr. S. Yasunaga)

groups have recently appeared, greatly enabling parents to cope with the rearing of these children.

The cytogenetics of Edward's syndrome resembles that of Down syndrome. Most patients have primary trisomy, while a much lesser number are unbalanced segregants of reciprocal translocations or display mosaicism for Trisomy 18 and a normal cell line. Mosaic Edward's patients generally have milder symptoms and have longer lifespans. Males tend to be more severely handicapped by Trisomy 18 than females, and the syndrome occurs four-fold more frequently among newborn females. These trends suggest early developmental loss of male Edward's zygotes or less frequent conception of male Trisomy 18 zygotes.

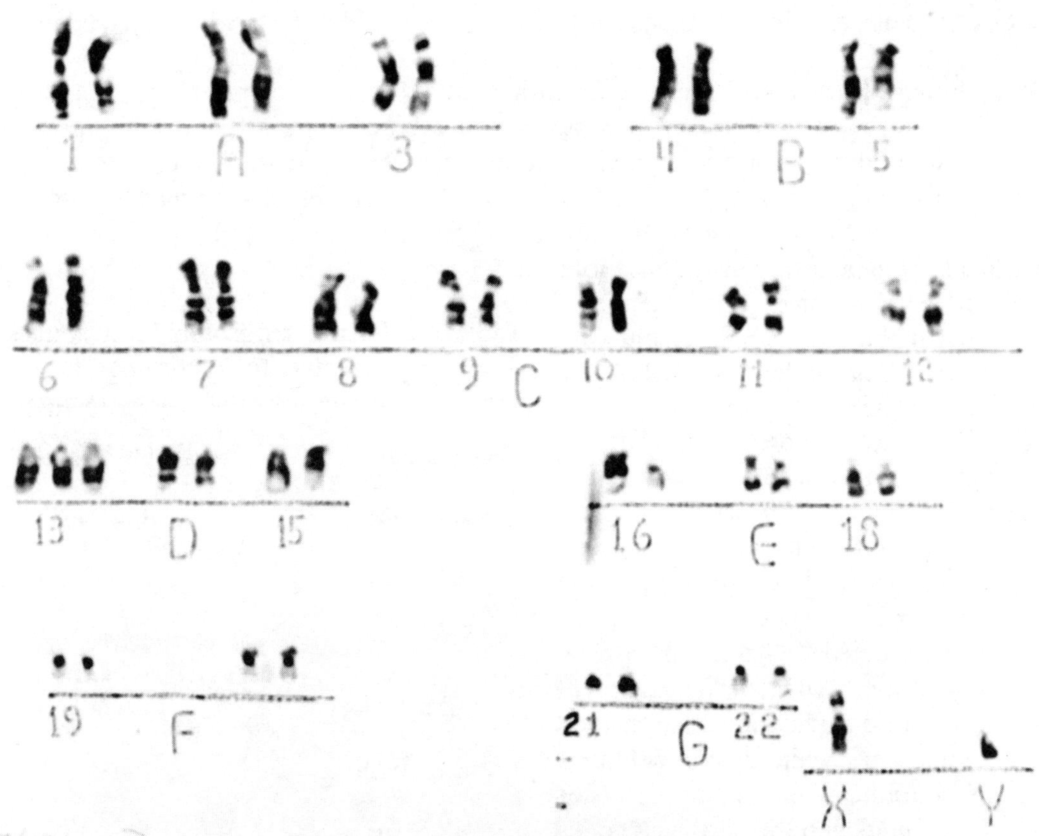

Fig. 4-23. Karyotype of a patient with primary trisomic Patau's syndrome. (Courtesy of S. Shafer)

Patau's Syndrome

Trisomy 13 (Patau's syndrome) occurs about half as frequently as Edward's syndrome. Unlike Edward's syndrome, clinical severity is similar in males and females, and the ratio of females to males with the syndrome is much lower than seen in Edward's syndrome. Although there is some overlap of clinical problems between Patau's and Edward's syndromes, certain characteristics distinguish them. Patau's

145

infants typically present with small or absent eyes, clefting of the lip and palate, cranial defects, and polydactyly (extra digits). Other malformations that occur more frequently in Patau's syndrome include absence of the olfactory bulbs, microcephaly, bicornuate uterus in females, and malrotation of the colon. Most patients die within the first three months, although some are living to older ages with improved medical care. Primary trisomic (Fig. 4-23), translocation, and mosaic forms of Patau's syndrome have been described. The latter group has milder handicaps and tends to live longer.

Unbalanced Duplications and Deficiencies of Autosomal Segments

Many syndromes have been described that occur as a consequence of partial trisomy or partial monosomy of various autosomes. The more common of these problems will be briefly described in this section. Cri-du-chat syndrome (Cat cry syndrome) is associated with a deletion of the short arm of chromosome 5 that encompasses band 15p15. These patients have a transient abnormality affecting the vocal chords and are often identified in the nursery as a consequence of their cat-like vocalizations. Other features of the disorder include mental retardation, "moon-like" face, widely-spaced eyes, and small skull.

A sampling of other deletions are listed in Table 4-3. A subtype of retinoblastoma, a child-hood cancer of the retina is observed in patients who are heterozygous for a deletion including band 13q14. The deletion unmasks a microdeletion or other recessive mutation in the **Rb** gene on the homologous chromosome, promoting entry of the cells into mitosis. Bone sarcomas also occur frequently in these patients. Alterations of the distal third of chromosome 13 (Fig. 4-24) illustrate the opposite phenotypic effects of deletions and duplications affecting the same chromosome region. Patients with trisomies that include this area present with polydactyly, while other individuals who are monosomic for the region lack thumbs.

The WAGR syndrome illustrates a phenomenon known as a **contiguous gene deletion syndrome**. These patients often have a deletion including the 11p13 band. Wilms tumor, aniridia, genitourinary anomalies, and mental retardation complete the syndrome. Each of these features are apparently determined by one of several closely situated genes in 11p13. The complete syndrome is often caused by deletions that include all of the contiguous genes, while smaller deletions present as either aniridia, Wilms tumor, or other combinations, depending upon the particular genes that are missing. Most patients are heterozygous for the respective deletions which render them **hemizygous** for mutant alleles on the homologous chromosome. Several other examples of contiguous gene deletion syndromes have been cited.

Table 4-3. Partial Autosomal Monosomies Producing Characteristic Syndromes.

Syndrome	Deleted Region
Angelman	15q11-q13
Cri-du-Chat	5p15
DiGeorge	22q11
Langer-Gideon	8q24
Prader-Willi	15q11-q13
Retinoblastoma	13q14
WAGR	11p13

Angelman syndrome and Prader-Willi syndrome may occur as a consequence of deletion of a small region of 15q. About half of the patients with Prader-Willi syndrome are heterozygous for a small deletion of this region that can be observed following prometaphase banding. This chromosome is

inherited from the father of the patient. When a similar deletion **is inherited from the mother**, the child presents with the features of Angelman syndrome. This phenomenon where the same genetic material causes different features in a patient, presumably as a consequence of maternal/paternal modifications of that material, is called **genomic imprinting**. Genomic imprinting has now been described for a number of different systems, some of which will be described in later sections of this text.

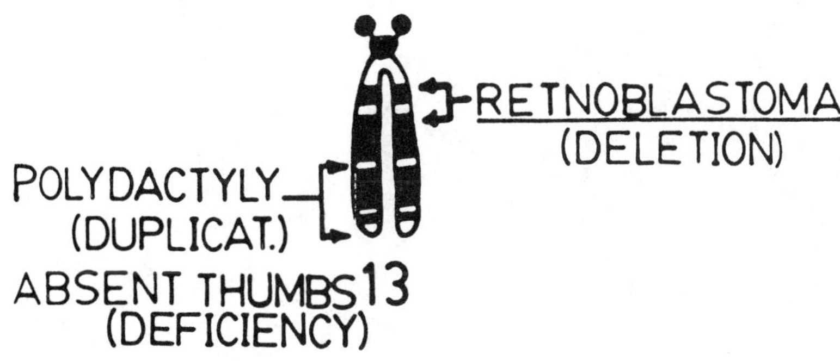

Fig. 4-24. Deletion and duplication of specific regions often produce opposite effects. Deletion of the indicated region of chromosome 13 results in deficient thumbs, while duplication of this region is associated with polydactyly.

Prader-Willi syndrome, characterized by mental retardation, marked obesity, and genital anomalies, can also be caused by the presence of two copies of the maternal chromosome 15. **Uniparental disomy** refers to the situation where an individual inherited both homologous chromosomes from the same parent and no homolog from the spouse. When a cell has two copies of the same homolog, the condition is called **uniparental isodisomy**. When the chromosomes are homologous but not identical, it is referred to as **uniparental heterodisomy**. About 20-30 percent of the cases of Prader-Willi syndrome appear to arise from maternal uniparental disomy, indicating that both maternal and paternal copies of the 15q11-q13 region are required for normal development. Another example of uniparental disomy has been reported in relation to cystic fibrosis, an autosomal recessive disorder. Approximately 1/500 of these patients appear to have inherited two copies of their mother's chromosome 7 carrying the cystic fibrosis allele. These patients exhibited signs of intrauterine growth retardation which is usually not a component of typical cystic fibrosis. One explanation for the occurrence of uniparental disomy is illustrated in Fig. 4-25. Individuals with uniparental disomy may have originated as trisomic zygotes, and the paternal chromosome was lost during an early mitotic division. Since trisomy 15 and trisomy 7 are lethal, the trisomic cell lines most likely disappeared during early development. Heterodisomy would be anticipated following a maternal Anaphase I nondisjunction, and isodisomy would occur following a maternal Anaphase II nondisjunction. A recent summary of genomic imprinting and uniparental disomy has been published (12).

The most common partial trisomy observed in humans is Trisomy 9p. The "pure" Trisomy 9p is observed following a break in a fragile site in the proximal long arm and nondisjunction of the piece containing the centromere or as the consequence of centromeric fusion of the short arm to a D- or G-group chromosome. These patients have flattening of the occipital region of the skull, a bulbous nose, and a "worried" expression on their faces. The palms of their hands are long and have single palmar creases. The second phalanges are short, and the second and third phalange may be fused. Nails tend to be claw-like. The patients tend to be moderately mentally retarded and may have moderate growth retardation. Their lifespans are essentially normal. Partial 9p trisomy may also occur as part of an unbalanced translocation in which other chromosome segments may be duplicated or deficient. These cases display a more complicated phenotype determined by the specific aneuploid segments involved.

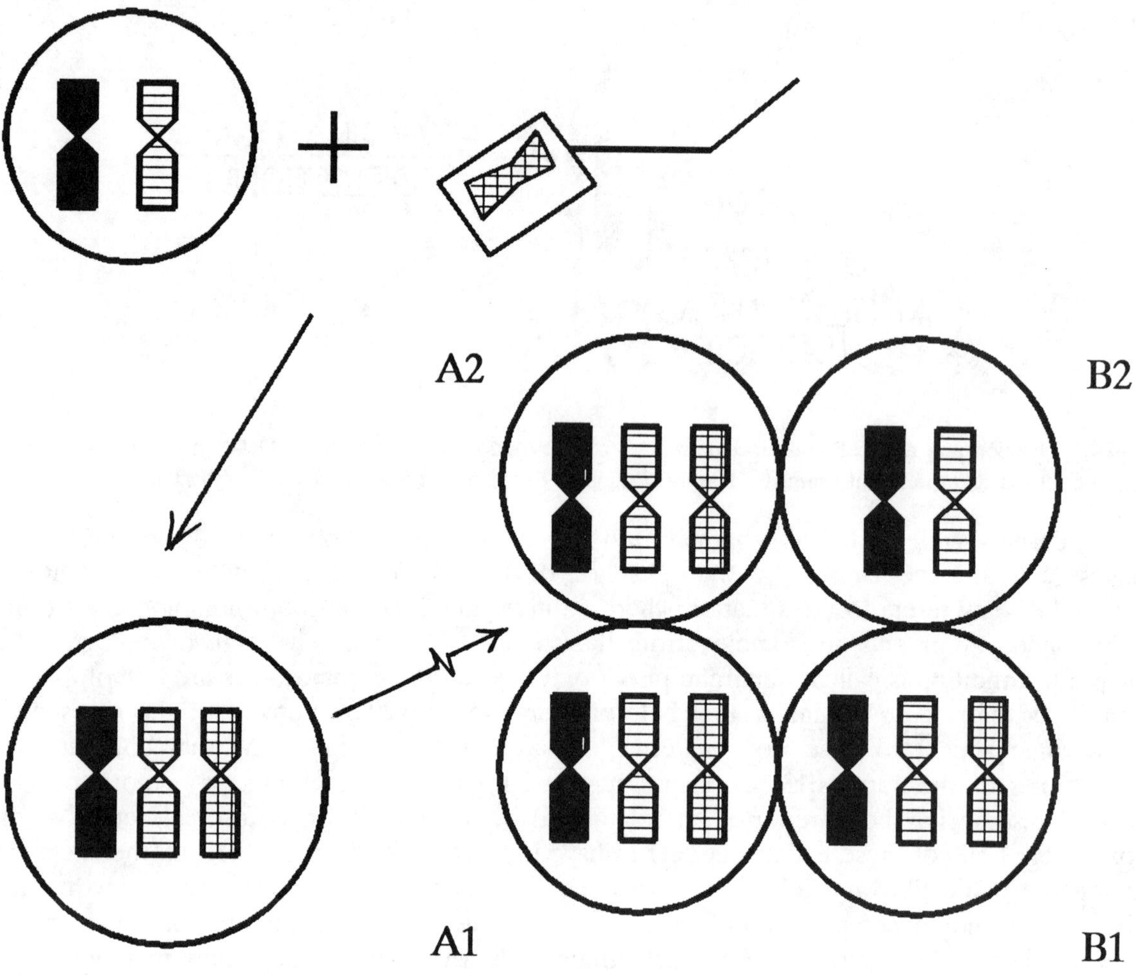

Fig. 4-25. Production of a uniparental heterodisomic individual. This person began as a trisomic zygote, inheriting two copies of the chromosome following nondisjunction during oogenesis at Anaphase I. Mitotic anaphase lag during the second division has led to the loss of the paternal chromosome in one of the four embryonic cells. Trisomic cells tend to divide more slowly than euploid cells. This tendency in combination with the probable lethal effects of the trisomy will result in eventual replacement of the trisomic cells by the uniparental heterodisomic cell line. Cells <u>A</u> and <u>B</u> were the products of the first mitotic division which occurred normally in this hypothetical example.

Many other autosomal partial trisomies are known. Although there may be clinical overlap among these conditions, each also includes features that are characteristic of the chromosome segments involved.

Sex Chromosome Abnormalities

More than 60 sex chromosome aberrations, including both numerical and structural anomalies of the X and Y chromosomes, have been described in the literature. The clinical features of these disorders are generally more benign than those of corresponding autosomal abnormalities involving chromosomes of similar size. This trend is largely attributable to inactivation of many X-linked genes on all but one of the X chromosomes in somatic cells and to the paucity of Y-linked genes. However, sex chromosome aberrations are a major cause of sexual ambiguity and infertility and affect about 24/10,000 liveborn infants (13).

Klinefelter's Syndrome

Klinefelter's syndrome is associated with a karyotype containing two or more X chromosomes and one or more Y chromosomes. The most common karyotype is displayed in Fig. 4-26. The severity

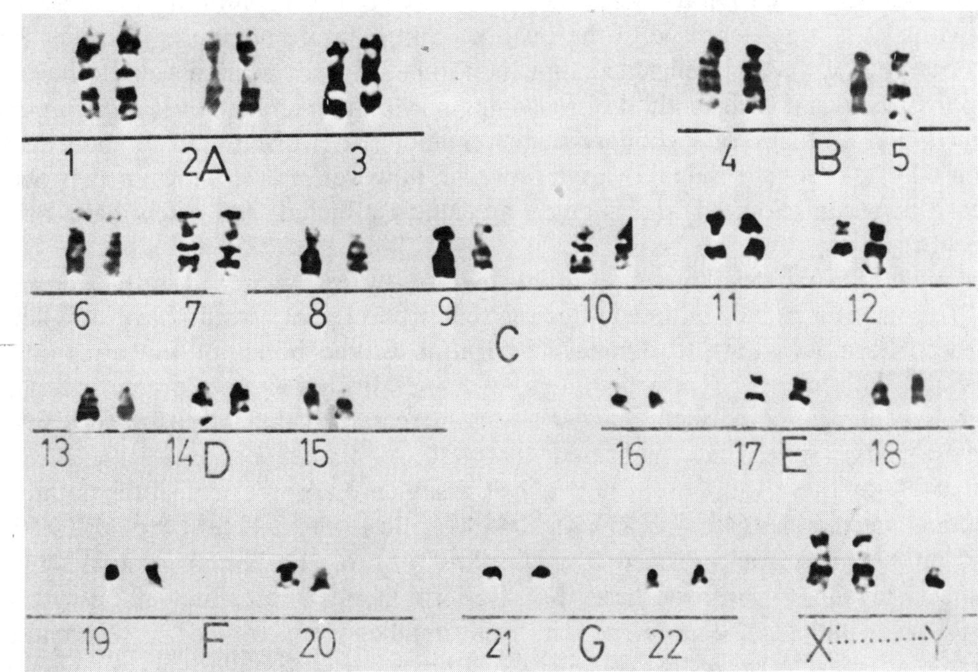

Fig. 4-26. Karyotype of a patient with Klinefelter's syndrome. (Courtesy of S. Shafer)

of the syndrome, particularly with respect to mental retardation, tends to increase with increasing numbers of X chromosomes. Patients with Klinefelter's syndrome rarely exhibit abnormal features prior to puberty, although retrospectively they may be found to have behavioral problems prior to this age (14). Following puberty, these individuals present with unusual sexual differentiation including small testes, aspermia, breast development, and feminine body profile. Urinary gonadotrophic hormones are often elevated. Clinical features are variable, with mildly affected patients able to complete spermatogenesis, and some may be fertile. Many of the mildly affected patients tend to be mosaic: 47,XXY/46,XY.

Severely involved individuals exhibit extreme degeneration of the seminiferous tubules following puberty, breast development, female hair pattern and voice pitch, broadened pelvis, and feminine body proportions. These men have small testes and are infertile. Although Klinefelter's men may have a diminished libido, they are capable of sustaining an erection and participating in intercourse. Many are initially detected at infertility clinics. Intermediacy of sexual development can lead to behavioral problems, and the frequency of Klinefelter's syndrome among criminals appears increased in comparison to the 46,XY male population (15).

The average age of mothers at conception of Klinefelter's patients is increased, suggesting that factors similar to those observed for autosomal trisomies are involved with the etiology of this syndrome. Average maternal age at conception is also elevated for mosaic (47,XXY/46,XY) individuals, indicating that these persons may have begun life as 47,XXY zygotes.

Turner Syndrome

Approximately 1/2000 newborn females and about 1/5 female abortices have Turner syndrome. Newborn Turner girls exhibit neck webbing, dorsal swelling of the hands and feet, shield-shaped chest, and widely-placed nipples. The swelling disappears shortly after birth. Turner patients do not display the usual pubertal changes and fail to menstruate. Urinary gonadotrophins are elevated, and connective tissue (streak gonads) is found in place of the ovaries. Short stature compared to other family members is commonly observed; however, intelligence appears normal. Turner women seem to have problems with visual depth perception, and their method of reasoning may remain childlike. If cases are missed at birth, delayed menstruation and lack of secondary sexual characteristics often lead to their detection. Some Turner women may have some residual ovarian function; however, menstruation may prove irregular and eventually cease to recur. Some Turner women are mildly afflicted, and a few have been reported to have children (16).

Secondary sexual development can be evoked by estrogen and progesterone replacement therapy. This treatment is usually initiated after the time when the girl would have completed her growth spurt, since these hormones tend to diminish elongation of the bones of the arms and legs. These treatments are not without risk. Turner women have a relatively low risk for uterine cancer; however, when estrogen is administered to them, cancer risk is increased to that faced by 46,XX women.

Several large series have indicated that half of Turner patients have a 45,X karyotype. Approximately 80% of these women did not inherit a sex chromosome from their father. About 45% of Turner patients are mosaics (45,X/46,XX; 45,X/46,XY; etc.). Five percent are 46,X,i(Xq), having one normal X and an isochromosome formed from the long arm of the X chromosome. Aborted Turner females are 45,X - leading to the hypothesis that liveborn Turner's cases may all be mosaics. Although many liveborn Turner patients are mosaic, a significant number seem to be 45,X. Normal cell lines were not demonstrable after extensive testing of several cell types. A typical Turner karyotype and a deletion variant are presented in Fig. 4-27.

Mosaics who have a 46,XY cell line or a cell line possessing a variant of the Y chromosome appear to run a 15-30 percent risk for tumors in the streak gonads. It is recommended that these gonads be removed from these patients to prevent later cancer development.

XXX Females

47,XXX females (Fig. 4-28) occur with a frequency of approximately 1/1000 newborn females. Oral epithelial cells from these women have two sex chromatin bodies, and their polymorphonuclear leukocytes exhibit two drumsticks. Both cytological markers correspond to the inactive X chromosomes

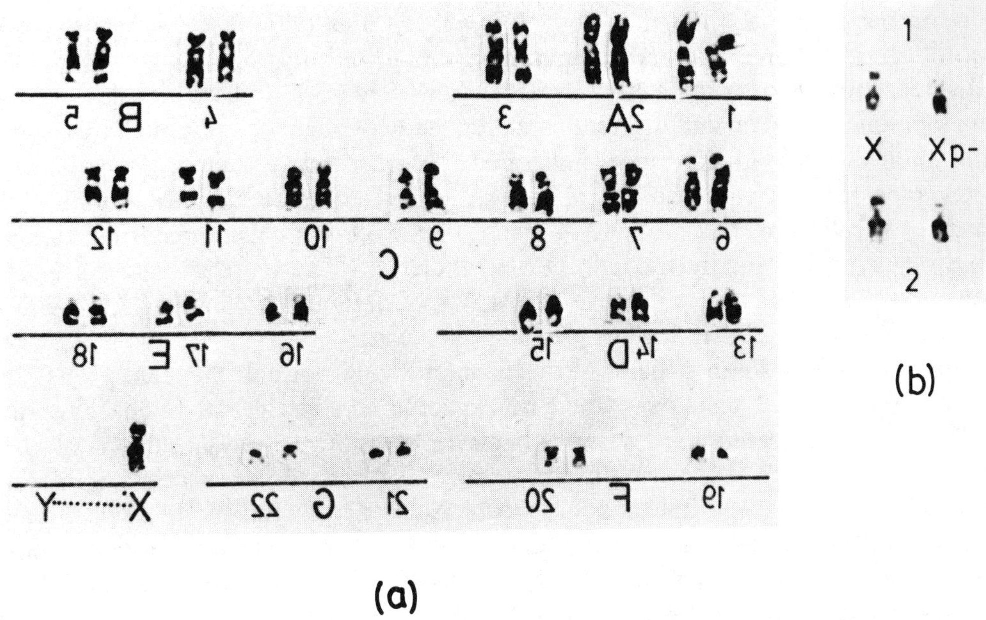

Fig. 4-27. Cytogenetics of Turner syndrome. a) Classical Turner syndrome karyotype (45,X); b) Partial karyotypes from two cells from a 46,XXp(-) variant. (Courtesy S. Shafer)

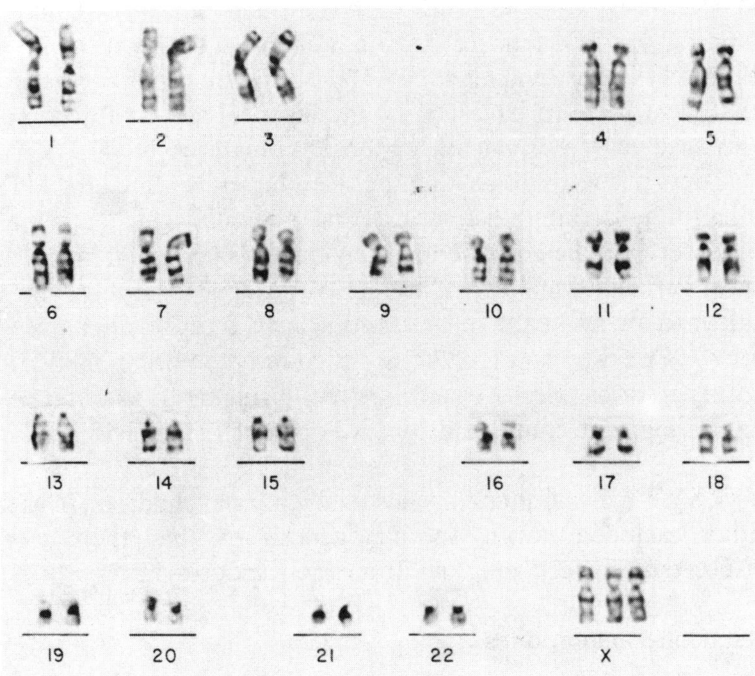

Fig. 4-28. Karyotype from a 47,XXX cell. (Courtesy of L. Goralski and R. Green)

occurring in interphase cells of these females. Clinical features of 47,XXX women vary considerably. Many are free of clinical stigmata, possess apparently normal fertility, and may be accidently ascertained during buccal smear surveys or routine karyotypic studies. Severely afflicted females exhibit disturbance of sexual development, social maladjustment, and reduced intelligence. Genitalia of such women are immature, and their breasts appear underdeveloped. Menstruation may be irregular or absent, and fertility is often impaired. Urinary gonadotrophin levels are frequently elevated. A strong maternal age effect is characteristically observed for XXX syndrome, with the mean maternal age at conception paralleling that observed in primary trisomic Down syndrome. The extreme variability among 47,XXX women presents many problems for predictive counseling of parents who have a 47,XXX infant or whose fetus was found to have 47,XXX following prenatal diagnosis.

Fertile 47,XXX women usually birth chromosomally normal children. This trend is rather surprising, since 50 percent of their ova would be expected to contain two X chromosomes. One-half of their sons would be expected to develop Klinefelter's syndrome, and one-half of their daughters would be anticipated to be 47,XXX like their mothers. The reason for this deviation from theoretical expectations is unknown; however, one possible mechanism may involve exclusion of the supernumerary X chromosomes from the ovum by partitioning them into the polar bodies formed during oogenesis. Why such a mechanism would be operant for the X chromosomes and not for chromosome 21 in Down syndrome women is not readily apparent.

47,XYY Males

47,XYY occurs in about 1/1000 newborn males. This syndrome has attracted some notoriety following discovery of an increased frequency of 47,XYY among certain incarcerated male populations. The 47,XYY male was conceptualized as a tall, criminally prone, mentally subnormal male with a severe case of acne. This clinical description has been tempered by the realization that most 47,XYY men do not commit crimes and may be free of the other stigmata. The clinical variability among 47,XYY men is similar to that observed for 47,XXX females. A comprehensive survey of 47,XYY males (17) revealed that they tend to be taller than corresponding 46,XY males; that their risk for incarceration is about twice that for 46,XY men; that most seem to have normal intelligence; and that acne may be somewhat more severe in 47,XYY males than in 46,XY males. Offenses committed by 47,XYY men tend to be non-violent and tend to be against property rather than people. Severely affected 47,XYY men often possess developmental anomalies of both internal and external genitalia which resemble those observed in Klinefelter's syndrome. Severely afflicted patients often exhibit signs of brain dysfunction and subnormal intelligence. This clinical variability makes counseling of parents regarding what to expect from their 47,XYY infants difficult, and many health professionals have faced a dilemma when confronted with parents of an infant with 47,XYY who was fortuitously encountered during Y-body surveys or chomosome studies performed for other reasons. Some counselors feel divulging the karyotype of their son may lead to parental bias during rearing that could lead to development of behaviors that would not otherwise occur.

Most fertile 47,XYY males father chromosomally normal children. This suggests that the extra Y chromosome is either excluded during spermatogenesis or that their sperm carrying two sex chromosomes are less successful in fertilizing ova from their spouses.

Hermaphrodites and Pseudohermaphrodites

The term **hermaphrodite** is derived from Hermes, a Greek god who served as a messenger and herald for other gods, and Aphrodite, the Greek goddess of love. Hermaphrodites are individuals who possess both testicular and ovarian tissue. The testicular and ovarian tissue may be confined to separate gonads, mixed in ovotestes, or a combination of both conditions may be found. The external habitus of hermaphrodites may be either female, male, or intermediate. Most hermaphrodites appear to have 46,XX karyotypes, presenting a problem for theories of male sex determination. Some cases have been found to be mosaics (46,XX/46,XY; etc.) with 46,XX cells in the ovarian tissue and 46,XY cells in the testicular tissues. Others have been found to have Y-chromosomal material in testicular tissue that is demonstrable by cytological or molecular techniques. Resolution of this sex-determination problem awaits identification of the sex-determining gene and development of sensitive probes that can reveal its presence in cells that appear to have normal 46,XX karyotypes.

Pseudohermaphrodites have external sexual features characteristic of one sex and gonadal tissue normally found in the opposite sex. One example of pseudohermaphroditism is provided by testicular feminization. Women who have a complete form of this syndrome have 46,XY karyotypes, external feminine features and testes. They have a mutation in the X-linked androgen receptor locus (AR) resulting in lack of androgen response. Therefore, their target tissues are insensitive to testosterone and dihydrotestosterone, as well as adrenal androgens. Consequently, their tissues develop along female patterns. Testicular feminized women usually lack axillary and pubic hair, because development of feminine sexual hair requires the actions of adrenal androgens. Testicular feminization and its variants will be discussed more fully in a later chapter.

The converse of testicular feminization is provided by congenital adrenal hyperplasia. 46,XX women possess androgen receptor levels comparable to those of men; however, their androgen levels are much lower. When exposure of female cells to high androgen levels occurs, masculinization may take place. About 1/14,000 newborn inherit a recessive deficiency of 21-hydroxylase activity. This enzyme is required for adrenal synthesis of both cortisol and aldosterone. The deficiency not only results in a lack of these hormones and consequent physiological problems, but also excessive production of adrenal androgens. The problem is exacerbated by disruption of normal feedback regulation of adrenal hormone production mediated by cortisol on ACTH release. Untreated males will undergo precocious pubertal changes, while females display varying degrees of external masculinization. Early identification and treatment by cortisol and aldosterone replacement have been partially but not completely successful in restoring normal sexual development and growth in the two sexes. New strategies are being developed (18) that promise to be more effective. Congenital adrenal hyperplasia will be discussed more extensively in a later chapter.

Pseudohermaphroditism or sexual ambiguity may also occur as a consequence of in utero exposure to exogenous androgen. If a pregnant woman develops an adrenal or ovarian tumor that produces excessive androgen or is administered exogenous androgen during pregnancy, the hormone crosses the placenta and causes varying degrees of masculinization of her female fetus.

Fragile X and the Martin-Bell Syndrome

The combination of mild to moderate mental retardation, characteristic speech defect, elongated face, large ears, and enlarged testes in males is called the Martin-Bell syndrome. This syndrome occurs in about 80% of males who possess a subtle chromosome anomaly. When cells from these patients are cultured under conditions of thymidine stress in the presence of low folic acid media, a proportion of their cells displays an unusual X chromosome containing a secondary constriction in the distal long arm at Xq27.3 (Fig. 4-29). This site is susceptible to breakage and has been designated a fragile site, and the Martin-Bell syndrome is also referred to as the Fragile-X syndrome. Mothers of Fragile-X patients are usually heterozygous for this fragile site, but have a lower proportion of cells that appear Fragile-X positive. Most of these heterozygous women have normal intelligence. Analyses of

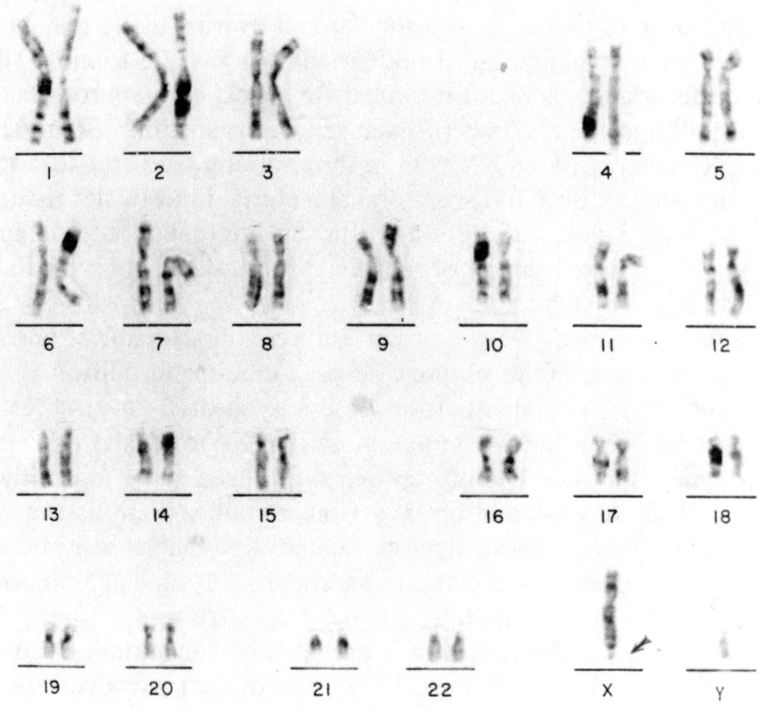

Fig. 4-29. The Fragile-X chromosome (arrow). (Courtesy L. Goralski and R. Green)

cultured skin fibroblasts from women who are known to be homozygous for the normal G6PD allele and who are carriers of the Fragile-X chromosome have revealed two types of cells, G6PD(+) and G6PD (-). The G6PD gene is located distal to the fragile site and is subject to X-inactivation. Therefore, the cells lacking G6PD activity have most likely lost the gene from the Fragile-X chromosome, and the allele on the normal chromosome has been silenced by X-inactivation. These observations suggest that one possible cause of the Martin-Bell syndrome in male patients is loss of genes distal to the fragile site in a majority of their cells.

The Martin-Bell syndrome is the second most frequent genetic cause of mental retardation. About 1/1250 males have the syndrome, and an additional 1/5000 carry the Fragile-X chromosome but lack symptoms (19). Therefore, the overall prevalence of the Fragile-X chromosome appears to be about 1/1000 males. One-third of all women who carry the Fragile-X chromosome appear to have mild to moderate mental retardation and appear to be present in the population at a frequency of approximately 1/2000 women. Therefore, the overall prevalence of Fragile-X carrier women is about 1/700. Persons with Martin-Bell syndrome do not reproduce. According to population theory which specifies that loss of the Fragile-X must be balanced by gain of new Fragile-X chromosomes, an unusually high mutation rate must exist to maintain this chromosome at such a high frequency in the population.

Laboratory diagnosis of the presence of the Fragile-X chromosome is difficult. Somewhat more than 40% of known Fragile-X carrier women (ie. those who have produced sons with the condition) display normal intelligence _and_ appear chromosomally normal - even when cytogenetic studies are very rigorous. This nonpenetrance of the Fragile-X chromosome in women presents major problems for genetic counselors who are advising female relatives of Fragile-X males regarding their risk for children with the problem. A second difficulty is experienced with the expression of a number of benign fragile sites and related morphological features located near Xq27.3 of the X chromosome. These normal X variants have required raising the theshold used to designate a male as Fragile-X positive. Certain individuals fall within a gray zone that ranges from 1-5 percent Fragile-X positive cells. These studies must be repeated one or more times to determine whether the person actually has the Fragile-X chromosome or a benign variant. These problems are especially acute when prenatal testing for Fragile-X is being performed. Use of closely linked RFLP markers may resolve some of the diagnostic problems, and work is progressing in this area.

A major question remains as to why some males with the Fragile-X chromosome escape its clinical consequences, and, conversely, why some female carriers develop mental retardation. Laird has proposed that **imprinting** of the Fragile-X chromosome in the individual's mother is responsible for penetrance of the Martin-Bell phenotype in some individuals and not others (20). One would expect a proportion of somatic cells in a female carrier of the Fragile-X chromosome to genetically inactivate the Fragile-X. According to Laird's theory, the Xq27.3 fragile site blocks **reactivation** of the Fragile-X chromosome that should occur prior to oogenesis. As a consequence, genes flanking the fragile site are not activated, and the lack of the products they determine contributes to the Martin-Bell syndrome. One way to test Laird's hypothesis is to compare the methylation status of genes flanking the fragile site in Fragile-X males with those of the corresponding genes in normal males. Methylation is used to consolidate X-inactivation, and the hypothesized transcriptional block might be caused by inappropriate methylation on the Fragile-X. Migeon's group studied the methylation patterns of HPRT, G6PD, P3, and GdX which are known to be near the fragile site and compared the methylation patterns of the Fragile-X chromosomes of eight males with those of normal males (21). No differences were observed, indicating that persistence of methylation near the fragile site is not responsible for the presence of Martin-Bell syndrome. This study does not eliminate the operation of other mechanisms to imprint the Fragile-X, particularly those that are not reversible as cells approach oogenesis.

Sex Chromosome Loss and Ageing

All men and women tend to lose their Y or one X chromosome, respectively, in a proportion of their cells as they age. This generates low grade mosaicism for 45,X which could potentially lead to misinterpretation at the clinical level. Particular problems are encountered in relation to leukemias and myeloproliferative disorders associated with loss of the Y chromosome. In older men, it is often uncertain whether the Y has been lost in relation to the cancer or as a consequence of normal ageing. Studies of normal cells from the same individual would partially resolve the uncertainty. Women who are over 30, should not be diagnosed as low grade Turner mosaics on the basis of a minor 45,X cell line when clinical features of the syndrome are absent.

Low grade mosaicism is also observed in somatic cells that may be of little clinical consequence if a similar proportion of germinal cells are not involved. The many cell divisions required during somatic differentiation are not error-free. Mitotic anaphase lag and mitotic nondisjunction will occasionally occure, producing minor cell lines with aneuploid complements. This type of mosaicism is encountered in everyone when sufficient numbers of cells are karyotyped.

Chromosomes and Human Pregnancy

Approximately one-seventh of all diagnosed human pregnancies terminate with spontaneous abortion. Many occur during the first trimester, and about one-half of these possess abnormal chromosome constitutions. The most common chromosome errors found in abortices are triploidy and 45,X (Table 4-4). Autosomal trisomies also occur in this group, and virtually all of the chromosomes have been involved. Trisomy 16 is the most common trisomy, while chromosomes 1, 19, and 20 are rarely involved, suggesting that these may be lost at very early stages.

Most of the chromosome aberrations occurring in spontaneous abortices are sporadic. However, certain types may be associated with a high risk of recurrence. Women experiencing repeated abortion during the first trimester of pregnancy should receive a thorough physical and endocrine evaluation. If these systems prove normal, **both** husband and wife should have their chromosomes examined. Tissue from the abortices should be cultured and karyotyped when available. Studies of families experiencing three or more abortions revealed that 12.7 percent carried a balanced translocation or inversion or were low grade mosaics for a lethal aneuploid cell line (23).

Two abnormalities are associated with chromosome complements that are entirely maternal or paternal in origin. Human ovarian teratomas are developmentally primitive tumors consisting of all varieties of embryonic tissue types. These tumor cells contain two complete sets of maternal chromosomes and no paternal chromosomes. By contrast, hydatiform moles are placental tumors formed by placental overgrowth and cystic degeneration of the chorionic villi. Tumor cells contain two complete sets of paternal chromosomes and no maternal chromosomes. Parental **imprinting** of chromosomes has been hypothesized to account for the differential effects of maternal vs. paternal chromosomes in these two anomalies. Studies of triploid pregnancies also support occurrence of parental imprinting of chromosomes. Most triploid fetuses appear to be formed by double fertilization of an ovum. These pregnancies which are characterized by the presence of two paternal sets of chromosomes have poorly developed embryonic tissues and well formed cystic placentas. By contrast, triploids containing two maternal sets of chromosomes are miscarried early in development and are associated with poor placental development. These four conditions indicate that the maternal chromosomes appear to be more involved with embryonic tissue development, while paternal chromosomes are of greater importance with respect to placental development. Both sets must be represented in normal dose for completion of a normal infant.

Table 4-4. Chromosomal Anomalies Occurring Among Spontaneous Abortices (22).

Type	Percent*
45,X	23.7
Other sex aneuploids	1.0
Autosomal Trisomies (Group)	49.8
Triploids	13.2
Tetraploids	4.2
Rearrangements:	
Balanced	0.3
Unbalanced	3.1
Other	4.5

* Percent of 287 chromosomally abnormal fetuses.

CHROMOSOMES AND EVOLUTION OF CANCER CELLS

Boveri (24) and Tyzzer (25) proposed that cancer was caused by somatic mutation. Several investigators subsequently recognized that cancer cells often contain abnormal chromosome complements, although the normal human chromosome number had not yet been established. Current evidence indicates that most cancer cells possess abnormal karyotypes at some time during their development (26), and that the virulence of the tumor is often related to the degree of departure from the normal chromosome number and configuration.

Many cancers evolve from a single **normal** cell and progress through a series of microscopic or submicroscopic mutations. This evolution of cancer cells is well illustrated by chronic myeloid leukemia (CML). CML is rarely inherited and arises from mutations in marrow cells that differentiate into granulocytes and related white blood cells (Fig. 4-30). The major mutation causing CML disrupts the

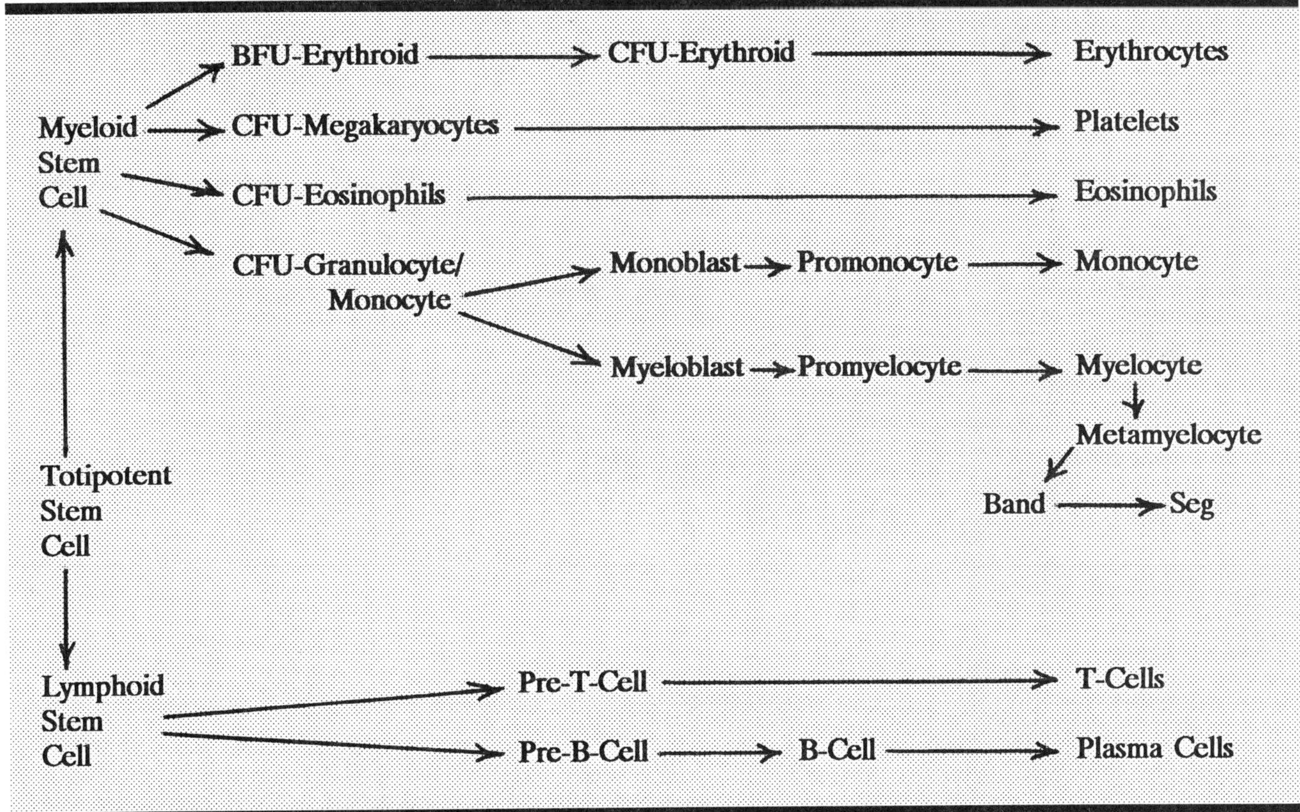

Fig. 4-30. Differentiation of hematopoietic cells. CFU = colony-forming unit; BFU = burst-forming unit.

normal differentiation pathway at the level of the conversion of the myelocyte to the metamyelocyte and results in the accumulation of immature cells in the marrow space and eventually in the circulation. This cancer may appear at any time ranging from less than two years to over seventy years of age. Periods of overt leukemia may alternate with remission for three or more years before the leukemia enters the acute stage and the terminal blastic crisis.

Marrow cells from more than 90 percent of CML patients were found to contain a small chromosome, called the Philadelphia or Ph[1] chromosome (27). The Ph[1] chromosome was found only in leukemic cells and was demonstrated by Rowley (28) to be formed by a balanced reciprocal translocation between chromosomes 9 and 22 (t(9;22)(q34;q11))(Fig. 4-31). The translocation interrupts two genes, the proto-oncogene **c-abl** on chromosome 9 and a 6 kilobase segment (**bcr** =breakpoint cluster region) near the lambda light chain region on chromosome 22 (Fig. 4-32). Both the **c-abl** and **bcr** genes encode protein kinases that appear to participate in regulation of cell division (Fig. 4-33). The translocation generates a fusion gene that encodes an abnormal kinase and places the fusion gene in a region of high transcriptional activity. A direct role for the fusion gene in CML was demonstrated by transfection studies in which the gene was introduced to mouse embryos (29). The transfected mice subsequently developed CML.

The Ph[1] may appear in marrow cells prior to clinical evidence for leukemia, and its abundance during subsequent stages is roughly proportional to the prevalence of Ph[1]-positive marrow cells. Evidence suggests that the leukemogenic change occurs in a single marrow stem cell, providing it with a growth advantage. The cell not only establishes a clone of cells in the marrow, but also is liable to accumulate further mutations (Fig. 4-34, Fig. 4-35) as the leukemia passes from the relatively benign to the acute

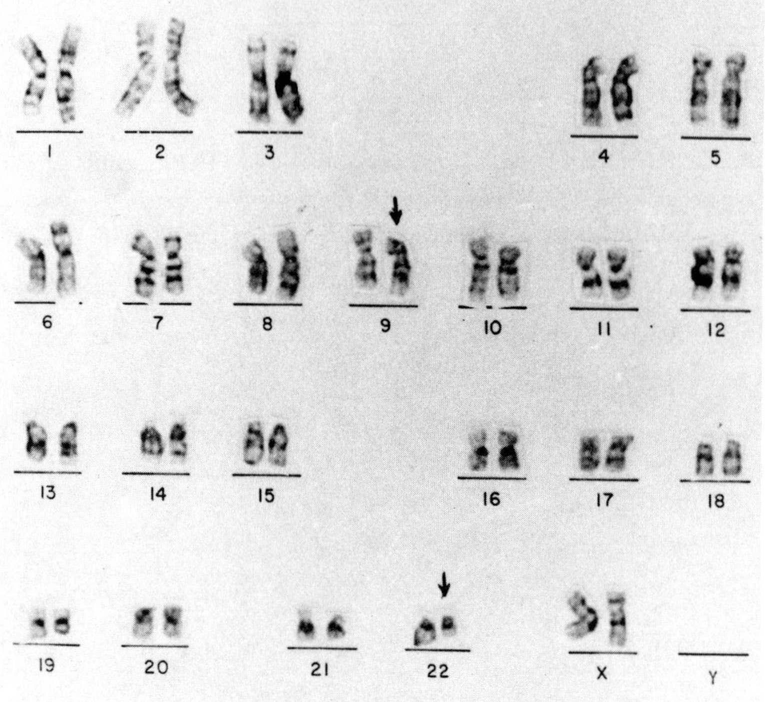

Fig. 4-31. Translocation associated with chronic myeloid leukemia. The arrows indicate the translocation chromosomes. (Courtesy L. Goralski and R. Green)

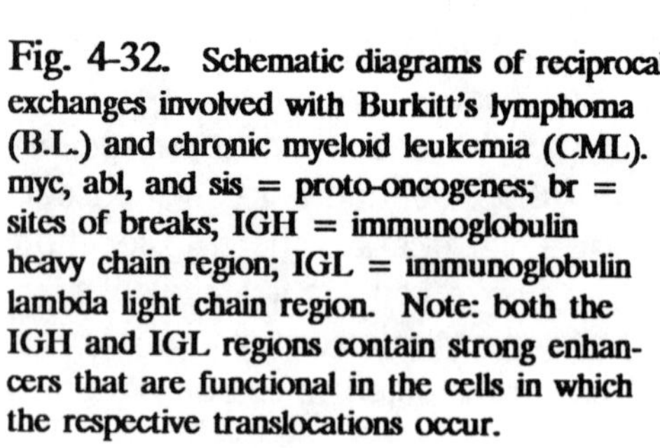

Fig. 4-32. Schematic diagrams of reciprocal exchanges involved with Burkitt's lymphoma (B.L.) and chronic myeloid leukemia (CML). myc, abl, and sis = proto-oncogenes; br = sites of breaks; IGH = immunoglobulin heavy chain region; IGL = immunoglobulin lambda light chain region. Note: both the IGH and IGL regions contain strong enhancers that are functional in the cells in which the respective translocations occur.

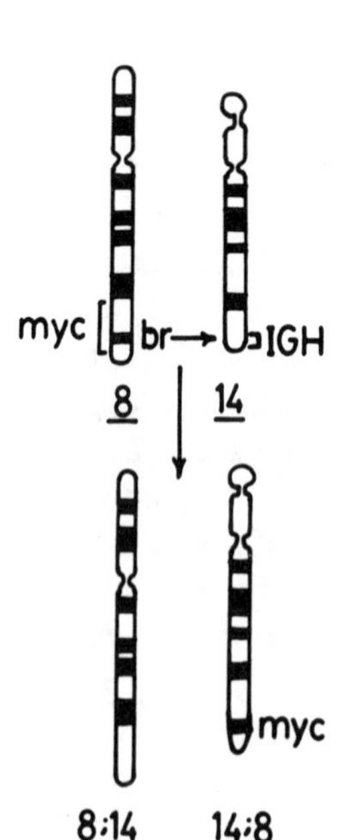

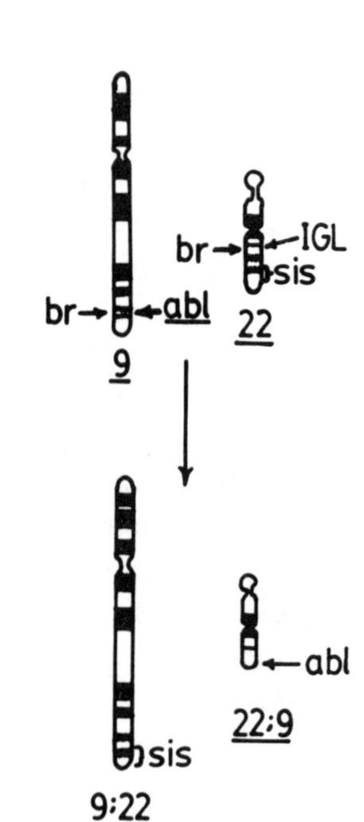

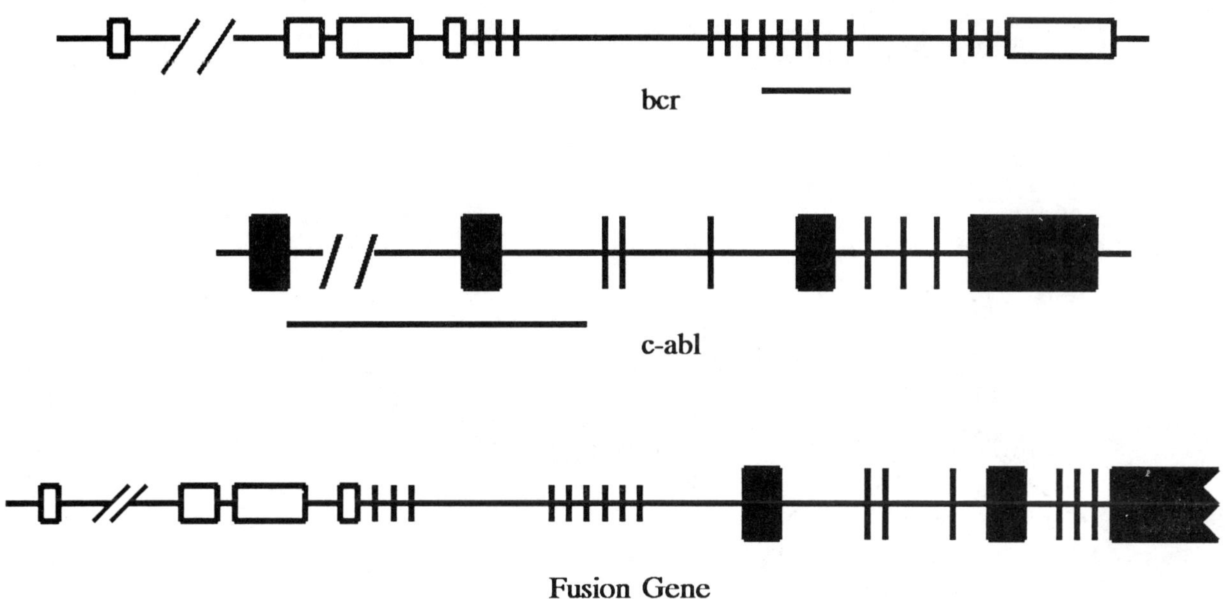

Fig. 4-33. Generation of a bcr-abl fusion gene by a t(9;22)(q34;q11) translocation. The horizontal bars below the bcr and c-abl proto-oncogenes specify the breakpoint regions in the respective genes. The c-abl gene is transcribed and processed into a 6.5kb mRNA which is translated into a 145kD tyrosine kinase. The bcr gene encodes a 190kD serine kinase. The CML fusion gene determines an 8.5kb mRNA and the p210(bcr-abl) kinase which has abnormal regulatory properties as described in the text. A related leukemia (acute lymphoblastic leukemia (ALL)) also involves a similar translocation; however, at least two fewer bcr exons occur in the fusion gene (i.e. break occurs at the left extreme of bar), and a smaller kinase is produced (p185(bcr-abl)). The loss of the N-terminal regions of the c-abl kinase is believed responsible for many of the abnormal properties of the kinase occurring in leukemic cells. p185(bcr-abl) has been observed to be a more potent carcinogen than p210(bcr-abl), presumably as a consequence of the more extensive deletion of bcr.

and blastic stages. As each clone gains an advantage, it overgrows prior clones, resulting in the replacement of those cells in the marrow. Toward the latter stage of the disease, leukemic cells enter the circulation and invade other tissues, producing secondary tumors (metastases).

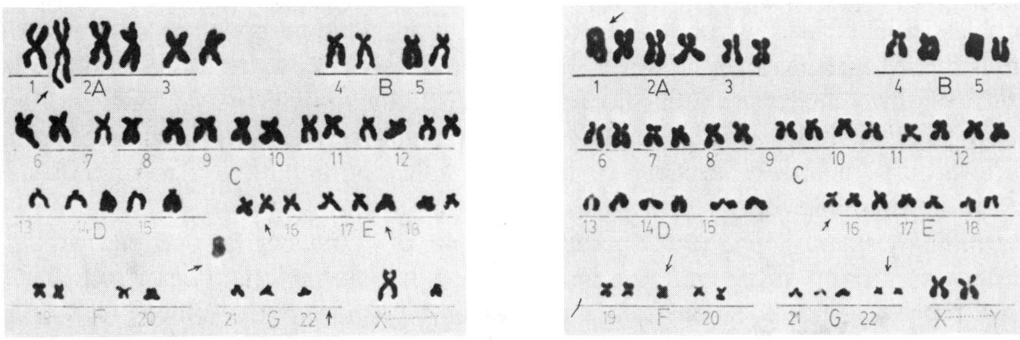

Fig. 4-34. Karyotypes of patients with more advanced CML. Notice that more extensive alterations have occurred which include both changes in number and structure of chromosomes. (Courtesy of S. Shafer)

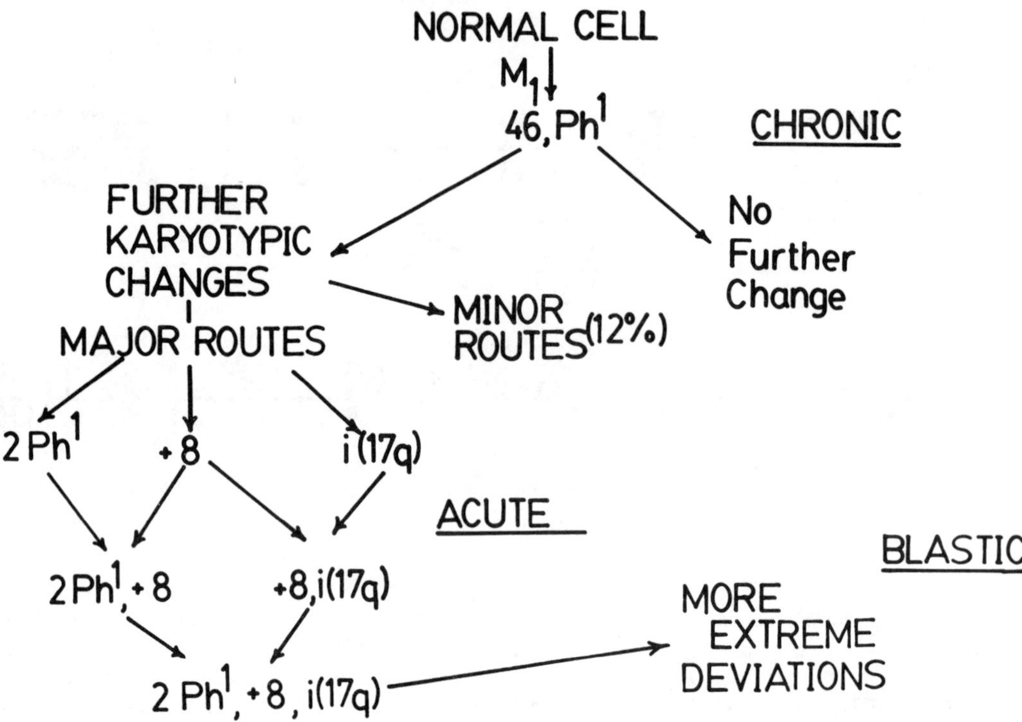

Fig. 4-35. **Chromosomal evolution of CML marrow cells.**

About 5-10 percent of CML patients lack the Ph¹ chromosome. In some of these, the **c-abl** and **bcr** genes seem to be rearranged by an as yet unidentified mechanism. No evidence for rearrangement has been found in some Ph¹-negative cases. Patients who possess the translocation chromosomes appear to respond better to therapy than those who lack them. Furthermore, the progress of the therapy can be monitored by following the frequency of Ph¹-positive cells in the marrow.

Structural modifications of chromosome 22 have appeared in a number of different cancers (Fig. 4-36). Translocations following breaks in the **bcr** gene have been identified in a number of different tumors. Most notable of these is the t(9;22)(q34;q11) translocation found in about 12% of patients with acute lymphoblastic leukemia (ALL). The reciprocal translocation appears identical to the CML translocation at the cytogenetic level; however, the fusion kinase lacks at least two additional exons from the **bcr** portion of the gene, contributing to its greater carcinogenic activity. Ewings Sarcoma and a constitutional translocation (apparently not associated with cancer) also occur following breaks in the **bcr** gene. Bilateral acoustic neurofibromas can be inherited following mutation in a gene (**BANF**) that has been mapped to chromosome 22.

Burkitt's lymphoma provides another example of a cancer associated with a reciprocal translocation moving a proto-oncogene to an immunoglobulin gene region. Tumor cells, but not normal cells, contain a reciprocal translocation transferring all or a portion of the **c-myc** proto-oncogene from chromosome 8 to the vicinity of the **IGH** (Immunoglobulin Heavy Chain Gene Cluster) enhancer on chromosome 14 (Fig. 4-32). Burkitt's lymphoma is the most frequent solid tumor occurring among African children and is associated with Epstein-Barr virus infection. This virus disrupts B-cell development, expands the target cell population involved with the lymphoma, and thereby increases their risk for the tumor. The **c-myc** gene contains a 400-500bp segment that is transcribed and processed into normal c-myc mRNA, but is not translated into the 49kD c-myc polypeptide. This polypeptide is apparently modified to a somewhat larger protein which dimerizes and exhibits DNA-binding activity. The c-myc protein is short-lived in normal cells, and levels of the protein appear to be under regulation by both pre-and post-transciptional factors. Post-transciptional regulation involves the 5' untranslated sequence of its mRNA. Analyses of the translocated **c-myc** gene have revealed alterations in this untranslated mRNA segment in a number of cases, and these alterations could lead to abnormal persistence of the myc RNA. This persistence, coupled with increased transcription of the gene promoted by the **IGH** enhancer, could produce unusually high levels of myc protein. Transfection experiments have indicated that the translocated myc gene is not self-sufficient for tumorigenesis; however, presence of both Epstein-Barr virus and the translocation induced B-cell tumors in mice. A second synergistic action of a virus and the translocated gene has been reported with respect to AIDS-associated B-cell lymphomas.

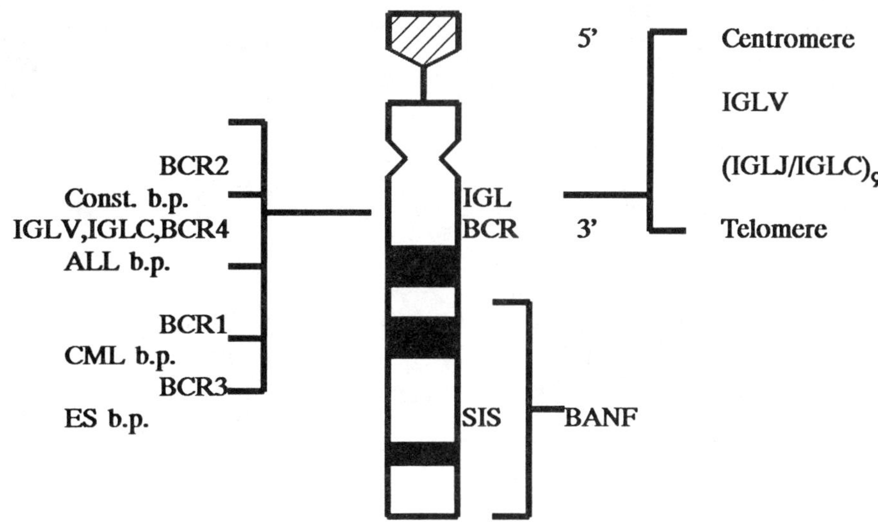

Fig. 4-36. Chromosome 22 and cancer. The BCR gene region has been expanded at the left, and the IGL gene cluster has been expanded at the right. The BANF gene has been localized to the zone enclosed by the bracket. Abbreviations: ALL = acute lymphoblastic leukemia; BANF = bilateral acoustic neurofibromas; BCR = breakpoint cluster region; b.p. = breakpoint; CML = chronic myeloid leukemia; Const. = constitutional translocation; ES = Ewings Sarcoma; IGL = immunoglobulin light chain gene cluster (V = variable sequences; J = joining sequences; C = constant sequences); SIS = simian sarcoma virus proto-oncogene (renamed: platelet-derived growth factor β-polypeptide gene (PDGFB)).

Numerous other translocations have now been found to be associated with leukemias and related cancers (Table 4-5), indicating that this is a common mechanism associated with tumorigenesis. The sites of the IGK, IGH, and IGL gene clusters on chromosomes 2, 14, and 22, respectively, are often involved in these rearrangements. The unusual recombination system involved with the differentiation of antibody genes is believed to be responsible for the errors leading to the translocations. This system is discussed in a later section of this text. Evidence has been gathered which suggests that leukemias and other cancers require several steps, and that different routes to the same end point may be followed (eg. acute lymphocytic leukemia and non-Hodgkin lymphomas have been seen to be associated with several

Table 4-5. An Abridged Series of Chromosomal Rearrangements Associated with Leukemias and Lymphomas.

Cancer	Rearrangement	Genes at Breakpoint	Patients with Anomaly
Acute Lymphoblastic	t(1;19)(q23;p13)	E2A(19p13)	8-30%
Leukemia (ALL)	t(4;11)(q21;q23)	ETS1(11q23)	5%
	t(8;14)(q24;q32)	MYC(8);IGH(14)	5%
	t(9;22)(q34;q11)	ABL(9);BCR(22)	12%
T-cell ALL	t(10;14)(q24;q11)	TCRA/D(14q11)	
Acute Myeloid	M2: t(8;21)(q22;q22)		
Leukemia (AML)	M3: t(1;17)(p36;q21)		
	t(15;17)(q22;q11-12)	RARA(17)	
	M4/5: t(10;11)(p14;q13-14)		
	t(11;17)(q23;q25)	ETS1(11q23)	
	M5: t(6;11)(q27;q23)	ETS1	
	t(8;16)(p11;p13)		
	t(10;11)(p11-15;q23)	ETS1	
Acute Promyelocytic			
Leukemia (APL)	t(15;17)(q22;q11.2-q12)	RARA(17q11.2-q12)	100%
Chronic Lymphocytic			
Leukemia (CLL)			
B-cell	t(2;14)(p32;q32)	IGH(14q32)	
	t(11;14)(q13;q32)	BCL1(11q13);IGH	
	t(14;17)(q32;q23)	IGH	
	t(14;18)(q32;q21.3)	IGH;BCL2(18q21.3)	
	t(14;19)(q32;q13)	IGH	
T-cell	inv(14)(q11q32)	TCR(q11);IGH	
Burkitt's Lymphoma	t(8;14)(q24;q32)	MYC(8q24);IGH	80%
(BL)	t(2;8)(p12;q24)	IGK(2p12);MYC	
	t(8;22)(q24;q11)	MYC;IGL(q11)	
Chronic Myeloid	t(9;22)(q34;q11)	ABL(9);BCR(22)	>90%
Leukemia (CML)			
Non-Hodgkin	del(6q)		20%
Lymphoma	t(11;14)(q13;q32)	BCL1(11);IGH(14)	10%
	t(14;18)(q32;q21)	IGH;BCL2(18q21)	20%*

*Note BCL2 is rearranged in 80% of follicular lymphomas

different rearrangements). A large number of different rearrangements have been observed in patients with acute myeloid leukemia (AML). AML has been subclassified into seven stages (M1-M7) according to severity. The large number of reported rearrangements suggest that evolution of this leukemia requires mutations in several critical loci. One hypothesis states that the order in which the mutations occur is unimportant, and that the particular temporal pattern of their occurrence varies from one patient to another. Although some evidence appears to support this hypothesis, the mutations observed in the various stages appear nonrandom. The nonrandom pattern suggests that certain mutations may lead to selection for others as the leukemia progresses from one stage to another. For example the translocation obsreved at stage M2 may provide an environment in which cells generating a second rearrangement involving chromosome 17 may have a selective advantage. These cells would overgrow the earlier clone, and have an environment favoring rearrangement of chromosome 11 and so on. This hypothesis is difficult to prove, since few studies have followed patients for sufficiently long intervals to monitor the appearance of the translocations as the leukemia evolves. Certain rearrangements appear during the development of more than one leukemia or lymphoma. For example, the translocation, t(8;14)(q24;q32), is involved in both Burkitt's lymphoma and ALL. This would seem to indicate that the mutated gene (**MYC**) regulates a developmental step shared by the cells from which the respective tumors arise.

Three terms have appeared in literature pertaining to the roles of oncogenes in cancer development. An **oncogene** is defined as a gene that contributes to the development of cancer. **Proto-oncogenes**, often abbreviated c-onc, are genes which have functions essential for normal development and function. Mutations in proto-oncogenes convert them to oncogenes, presumably by altering the timing and/or level of their expression. Retroviruses have been found which mediate mutations in proto-oncogenes in certain animal systems. These viruses infect the animal cells and insert their genomes into the host chromosomes. When the viruses subsequently extricate themselves from the host chromosomes, they may carry along portions of neighboring host genes. Subsequent infection of cells by the viral particles containing the hybrid genome (viral genome + incomplete host gene) can lead to transformation of the infected cells as a consequence of inappropriate expression of the modified host gene. This gene is referred to as a viral oncogene (v-onc).

As described above, chromosomal rearrangement has been demonstrated to alter the structure of certain genes at the sites of the breakpoints. Current research in this field is directed at cloning the breakpoints of translocations associated with leukemias, lymphomas, and other cancers to search for and characterize the genes that have been disrupted by the exchanges. Several examples of recent successes in this enterprise are presented in Table 4-5. For example, the gene **BCL2** has been found to encode an inner mitochondrial membrane protein (30). This protein somehow participates in changes that contribute to "programmed" death of B-cells, and the translocation and unusual expression of the translocated gene leads to prolonged B-cell survival and leukemia or lymphoma. Rearrangements involving 19p13 disrupt the **E2A** gene. This gene encodes two proteins, E12 and E47, which normally bind to the enhancer of the kappa light chain gene cluster on chromosome 2. Inordinate expression of the translocated **E2A** gene may lead to abnormal regulation of a neighboring gene or other genes which contribute to leukemia development (31). The α-retinoic acid receptor gene (**RARA**) is located in bands 17q11.2-q12. The t(15;17)(q22;q11.2-q12) translocation has been observed in 100 percent of patients presenting with acute promyelocytic leukemia. This translocation interrupts the first exon of the **RARA** gene, altering its product (32). The altered receptor presumably has abnormal properties, interfering with normal regulation of the developmental path leading to segmented granulocytes or neutrophils (Seg, Fig. 4-30). Another interesting rearrangement, inv(14)(q11q32) places components of the immunoglobulin heavy chain gene cluster (**IGH**) into juxtaposition with elements of the T-cell receptor gene cluster (**TCR**) on chromosome 14. This rearrangement occurs in leukemic cells of some patients presenting with T-cell chronic lymphocytic leukemia. A different rearrangement, t(10-14)(q24;q11), disrupts the V_α -

$D_\delta 2$ region of the **TCR** gene cluster in band 14q11. This translocation occurs in leukemic cells from persons with T-cell ALL.

Origin of Chromosomal Anomalies

Evidence implicating the chromosomal alterations described above in leukemia or lymphoma development is steadily accumulating. A number of agents have been identified which cause chromosome breaks, and at least some of these have been demonstrated to be potent cancer-causing agents (**carcinogens**). Radiation has long been known to damage chromosomes, and such breakage tends to be a problem among patients receiving radiation therapy to cure their cancers. Tumor cells divide much more rapidly than normal cells; therefore, radiation and chemotherapeutic agents have a more severe effect upon cancer cells than upon normal tissues. However, radiation may occasionally induce mutations in normal cells surrounding the irradiated area, causing secondary tumors. Chemotherapeutic agents have a more generalized effect, since the treatment area cannot be confined; and many more cells are exposed to the potentially damaging chemicals. Lymphoid tissues retain their ability to divide in adults, and leukemias or lymphomas may occur in persons receiving chemotherapy for other cancers. Individuals accidentally exposed to high radiation doses often develop leukemias later in life as a consequence of chromosomal rearrangements or other mutations induced by the radiation.

Unusual intrinsic nuclear properties can also contribute to translocations and inversions. Antibody genes and T-cell receptor genes employ an intricate recombination system to form mature functional genes. The recombination mechanism builds mature genes by altering the relative positions of their components (V-, D-, J-, and C-region elements). Rearrangements normally involve elements on the same chromosome and occur only in T- and/or B-cells and their developmental precursors. When this recombination process occasionally involves not only the usual chromosome but also affects other chromosomes in their vicinity, translocations can occur, contributing to the cancers arising from these cell types.

Viruses have been viewed as cancer-causing agents for a long time. Certain viruses, especially retroviruses, have been demonstrated to cause cancer in certain experimental animals. However, viral involvement with human cancers has been difficult to detect. Exceptions include the EB virus (Epstein-Barr) and AIDS viruses described above. Even in these cases, the respective viruses are not capable of inducing the cancers by themselves, but must interact with one or more mutations in the genomes of the cells giving rise to the cancer.

Chemical carcinogens, particularly those which mimic radiation effects, are capable of causing chromosome damage, predisposing to rearrangements. As mentioned above, some of these materials are used for cancer chemotherapy. Others may be natural components of our environment or arise during normal cellular metabolism. Still others are added to our environment in the form of industrial wastes, certain insecticides or herbicides, or may be produced by cooking or other food processing.

Regardless of the type of environmental insult, human cells possess surveillance and repair systems which destroy abnormal cells or repair damaged DNA. These systems are usually very effective in reducing our cancer risk. However, certain genetic diseases such as xeroderma pigmentosa, Bloom's syndrome, Fanconi's anemia, and ataxia telangiectasia, which seem to be associated with abnormal repair, and immune deficiency diseases, which compromise the surveillance system, increase cancer risk.

Cancer is more frequent among older than among younger persons. This trend probably has several causes, including longer exposure to carcinogens, gradual demise of our immune systems, and age-related compromise of our repair processes. These factors interact to promote the accumulation of mutations in our cells. When the appropriate combination of mutations is achieved, cancer results. The tendency for a particular aggregate of mutations to be localized to the tumor cells and the initial slow development of many tumors increase the possibility of early detection and treatment, enhancing the probability of recovery.

SUMMARY

Numerical chromosomal abnormalities occur when the chromosomes fail to separate or migrate to the appropriate poles of the dividing somatic or germinal cell. Nondisjunction produces daughter cells with either an excessive or deficient number of chromosomes. Depending upon the fate of the lagging chromosome, anaphase lag and nondisjunction may result in a similar distribution of chromosomes among daughter nuclei, or anaphase lag may produce daughter nuclei with normal and deficient complements, respectively. Occurrence of anaphase lag and nondisjunction during meiotic division is responsible for chromosomal errors that uniformly involve all of the cells of the conceptus derived from the abnormal sperm or egg. Fortunately, anaphase lag or nondisjunction affect relatively few gametic cells, leading to the sporadic occurrence of numerical chromosome errors in most families. Mosaicism is a consequence of nondisjunction or anaphase lag during mitotic division. Selective involvement of some tissues and not others and differential survival of the cell lines produced by the abnormal chromosome segregation often make detection of mosaicism difficult. Since a large number of cell divisions occur during development, and since a certain proportion of these divisions are susceptible to nondisjunction or anaphase lag, all people are mosaics. Most usually have a great preponderance of normal cells, and abnormal cell lines account for less than one percent of their cells. A few people will have greater numbers of aneuploid cells, presenting with variable features of syndromes characteristic of the particular chromosome involved.

Structural rearrangements of chromosomes frequently occur following breakage of chromosomes. Breaks may be caused by radiation, viruses, or certain chemicals. These rearrangements may occur in balanced state, and persons carrying them appear clinically normal. However, balanced carriers of the inversion or translocation are at higher risk for experiencing reproductive problems and/or having handicapped children. These concepti inherit unbalanced chromosome combinations containing duplications and deficiencies of chromosome regions participating in the rearrangements.

Down syndrome, Patau's syndrome, and Edward's syndrome are associated with trisomy 21, trisomy 13, and trisomy 18, respectively. These syndromes typically involve multiple systems and include moderate to profound mental retardation. Each syndrome occurs in primary trisomic, mosaic, and translocation forms. The latter type is associated with an increased recurrence risk when a balanced translocation occurs in one of the parents of the affected child. In the case of Down syndrome, sex of the balanced carrier parent influences recurrence risk, with a higher risk occurring when the mother is a carrier. Clinically normal parents who are mosaic for a minor trisomic cell line also have an increased risk of recurrence of the trisomy that approximates one-half the average prevalence of the abnormal cell line.

Sex chromosome aberrations tend to have less severe clinical effects than their autosomal counterparts. This trend reflects the protective effect of X-inactivation and the paucity of major Y-linked genes. Problems associated with sex chromosome aneuploidies include abnormal sexual development, infertility, mental retardation, and abnormal behavior. The array of syptoms and their severity differ among patients with the same or very similar chromosome complements, making predictive counseling difficult.

Deletions of chromosomal material generally have more severe consequences than do duplications of these segments. Duplications and deficiencies of the same chromosome segment generally have opposite phenotypic effects. Although several partial monosomies involving autosomes (eg. Cri-du-Chat, 5p-) have been reported, but they are generally rare. Partial trisomies, including 9p+, are somewhat more common. Complete autosomal monosomies are not observed among the liveborn, unless they are present with a normal cell line. Most fetuses with 45,X die before birth, and mosaics are more frequent among liveborn than miscarried Turner syndrome individuals.

Chromosomal abnormalities constitute a common cause of spontaneous miscarriages. The most frequent anomalies include 45,X, triploidy, and autosomal trisomies. Every autosome is represented among the latter group. However, the larger chromosomes and F-group chromosomes are infrequent, suggesting loss at earlier stages. Sex chromosome trisomies are infrequent in abortice series, indicating that most survive to term. Although trisomies and monosomies resulting in early pregnancy termination are usually non-recurrent, about 13% of parents experiencing three or more miscarriages carry balanced chromosome rearrangements or are mosaic for a minor trisomic or monosomic cell line.

Cancer cells generally evolve from normal cells. Chromosome rearrangement is a major cause of leukemias and lymphomas and occasionally of other tumors. Translocations and inversions contributing to cancer development disrupt critical regulatory genes, move these genes to regions of high transcriptional activity, or both. Two or more mutations are required for cancer development, and genetic changes occurring within tumor cells may interact with each other or with environmental factors such as viruses to bring about transformation of the normal cell to a cancer cell. The same cancer or very similar cancers often have different rearrangements during their evolution, suggesting that tumorigenesis may be accomplished by several mutational routes. Some evidence suggests that occurrence of a particular chromosomal rearrangement may favor the occurrence of a second rearrangement, leading to a specific sequence of mutational events during cancer evolution. As mutations continue to accumulate in cancer cells, the virulence of the cancer increases.

Chromsomal anomalies often arise spontaneously. Agents believed to contribute to the occurrence of chromosomal abnormalities include radiation, viruses, certain chemicals, and the aging process. The body possesses surveillance systems that detect and eliminate abnormal cells. Repair enzymes effectively correct DNA damage. Persons who have genetic diseases affecting their immune system or DNA repair mechanisms have an increased risk for cancer.

PROBLEMS

1. Two patients with Down syndrome have the following karyotypes:

   Patient 1:   47,XX,+21   Patient 2:   47,XX,+21/46,XX

   Which of the two patients would be anticipated to have the most severe clinical manifestations? Why?

2. Why do sex chromosome aneuploidies tend to be less severe than autosomal aneuploidies involving chromosomes of similar size?

3. The father of a Down syndrome child is a balanced carrier for a t(14q,21q) translocation. What is the chance that he and his wife will have a second child with Down syndrome?

   a.   100%   d.   10-16%
   b.   40-60%   e.   2-8%
   c.   30-39%

4. A woman is a balanced carrier for a Robertsonian translocation involving chromosomes 13 and 14. Diagram the chromosome orientation that would be observed at metaphase I of meiosis.

5. List the gametic chromosome complements produced by this woman.

6. Comment on the following statement. Young couples with two Down syndrome children are victims of fate and need not worry about recurrence until after age 40.

7. A 25-year-old woman has short stature, webbing of the neck, and lacks feminine breast development. She had irregular menstruation until age 22, but has not experienced her period since that time. Test results are:

Thyroid Hormones: normal   Karyotype: 45,X
17-ketosteroids: normal   Laparoscopy: streak gonads
Sex Chromatin: negative

What is your diagnosis?

(8-15) Match the following syndromes with their appropriate karyotypes.

8.	Edward's syndrome	a.	47,XY,+13
9.	Cri-du-Chat	b.	47,XX,+21
10.	Klinefelter's syndrome	c.	46,XY,t(9;22)(q34;q11)
11.	Chronic myeloid leukemia	d.	45,X
12.	Patau's syndrome	e.	46,XX,del(5)(p15)
13.	Down syndrome	f.	47,XY,+18
14.	Turner syndrome	g.	47,XXY
15.	Martin-Bell syndrome	h.	47,XYY
		i.	47,XXX
		j.	46,X,fra(X)(q27)

16. 46,XX males have been reported in the literature. How may this lack of correspondence between genotype and phenotype have occurred?

17. A man has mosaicism for trisomy 18:

skin fibroblasts: 70% 46,XY; 30% 47,XY,+18
lymphocytes: 90% 46,XY; 10% 47,XY,+18

Estimate the theoretical probability that this man may father an infant with Edward's syndrome. How would the observed risk compare with your theoretical risk? If they are different, explain why.

GLOSSARY OF TERMS

Acentric - without a centromere

Anapahase lag - failure of a chromatid to migrate to pole of dividing cell during anaphase

Aneuploid - abnormal dosage of chromosome(s) or chromosome segment(s)

Contiguous gene syndrome - components of a syndrome are determined by abnormal dosage of two or
more closely linked genes. The term is applied to small deletions/duplications of chromosomal
material.

Deletion - loss of chromosome material

Disomic - two copies of a chromosome

Euploid - normal gene or chromosome dosage

Fragile site - a secondary constriction on a chromosome demonstrable by special methods which is
susceptible to breakage

Hermaphrodite - a person with male and female gonads or gonadal tissues

Imprinting - parental alteration of chromosome structure and/or function

Insertion - addition of chromosomal material between chromosome segments

Inversion - double break in a chromosome followed by 180° rotation of the central piece and rejoining
paracentric - centromere excluded from the inversion
pericentric - centromere included in the inversion

Isochromosome - chromosome consisting of two identical arms

Marker - benign structural chromosome variant

Maternal age effect - tendency for trisomies to increase with maternal age at conception

Monosomic - one copy of a chromosome present

Mosaic - mixture of two or more cell lines

Nondisjunction - failure of separation of chromatids during cell division

Nullisomic - lack chromosome

Oncogene - a gene capable of causing cancer

Proto-oncogene - a normal cellular gene structurally similar to an oncogene, but which plays a role
in normal regulation of cell division or metabolism

Pseudohermaphrodite - an individual whose external sexual appearance is opposite to the gonadal sex

Ring chromosome - circular chromosome formed by two breaks and fusion of the ends of the middle
segment

Translocation - breaks in non-homologous chromsomes and exchange of segments distal to the break
      reciprocal - all genetic material involved in the translocation is conserved
      Robertsonian - translocation in which the long arm of one chromosome is fused to the centromere and long arm of a second chromosome; generally involves acrocentric chromosomes

Triploid - presence of three complete chromosome sets

Trisomic - three copies of a chromosome are present

Uniparental disomy - both homologous chromosomes are inherited from the same parent
      heterodisomy - different homologs were inherited
      isodisomy - two copies of the same chromosome were inherited

BIBLIOGRAPHY

1. Lejeune, J. 1959. Le mongolisme. Premier example d'aberration autosomique humaine. Ann. Genet. Semaine Hop. 1:41-49.

2. Martin, RH. 1983. A detailed method for obtaining preparations of human sperm chromosomes. Cytogenet. Cell Genet. 35:253-256.

3. Balkan, W and RH Martin. 1983. Segregation of chromosomes into the spermatozoa of a man heterozygous for a 14;21 Robertsonian translocation. Am. J. Med. Genet. 16:169-172.

4. Pellestor, F, B Sele, and H Jalbert. 1987. Chromosome analyses of spermatozoa from a male heterozygous for a 13;14 Robertsonian translocation. Hum. Genet. 76:116-120.

5. Martin, RH. 1988. Cytogenetic analysis of sperm from a male heterozygous for a 13;14 Robertsonian translocation. Hum. Genet. 80:357-361.

6. Pellestor, F, B Sele, H Jalbert, and P Jalbert. 1989. Direct segregation analysis of reciprocal translocations: A study of 283 sperm karyotypes from four carriers. Am. J. Hum. Genet. 44:464-473.

7. Daniel, A, EB Hook, and G Wolf. 1989. Risks of unbalanced progeny at amniocentesis to carriers of chromosome rearrangements: Data from United States and Canadian laboratories. Am. J. Med. Genet. 31:14-53.

8. Antonarakis, SE, PA Adelsberger, MB Petersen, et al. 1990. Analysis of DNA polymorphisms suggests that most de novo dup(21q) chromosomes in patients with Down syndrome are isochromosomes and not translocations. Am. J. Hum. Genet. 47:968-972.

9. Penrose, LS and GF Smith. Down's Anomaly. London: Churchill, 1966.

10. Collman, RD and A Stoller. 1962. A survey of mongoloid births in Victoria, Australia, 1942-1957. Am. J. Public Health 52:813-829.

11. Stevenson, AC, BCC Davison, and MW Oakes. <u>Genetic Counseling</u>. Philadelphia: JB Lippincott Co., 1970.

12. Hall, JG. 1990. Nontraditional inheritance. Growth, Genetics, & Hormones 6(4):1-4.

13. Evans, HJ. 1977. Chromosome anomalies among livebirths. J. Med. Genet. 14:309-312.

14. Nielsen, J, S Bjarnason, U Friedrich, A Froland, VH Hansen, and A Sorensen. 1970. Klinefelter's syndrome in children. J. Child. Psychol. Psychiat. 11:109-119.

15. Casey, MD, LJ Segall, DRK Street, and CE Blank. 1966. Sex chromosome abnormalities in two state hospitals for patients requiring special security. Nature 209:641-642.

16. Bahner, F, G Schwarz, DG Harnden, PA Jacobs, <u>et al.</u> 1960. A fertile female with XO sex chromosome constitution. Lancet ii:100-101.

17. Witkin, HA, SA Mednick, F Schulsinger, E. Bakkestrom <u>et al.</u> 1976. Criminality in XYY and XXY men. Science 193:547-555.

18. Cutler, GB and L Laue. 1990. Congenital adrenal hyperplasia due to 21-hydroxylase deficiency. N. Eng. J. Med. 323:1806-1813.

19. Brown, WT. 1990. The fragile X: Progress toward solving the puzzle. Am. J. Hum. Genet. 47:175-180.

20. Laird, CD. 1987. Proposed mechanism of inheritance and expression of the human fragile X-syndrome of mental retardation. Genetics 117:587-599.

21. Khalifa, MM, AL Reiss, and BRM Migeon. 1990. Methylation status of genes flanking the fragile site in males with the Fragile-X syndrome: A test of the imprinting hypothesis. Am. J. Hum. Genet. 46:744-753.

22. Alberman, ED and MR Greasy. 1977. Frequency of chromosomal abnormalities in miscarriages and perinatal deaths. J. Med. Genet. 14:313-315.

23. Byrd, JR, DE Askew, and PG McDonough. 1977. Cytogenetic findings in fifty-five couples with recurrent fetal wastage. Fertil. Steril. 28:246-250.

24. Boveri, T. 1914. <u>Zur Frage der Entstehung Maligner Tumeren</u>. Jena: Gustav Fischer, p. 64.

25. Tyzzer, EE. 1916. Tumor immunity. J. Cancer Res. 1:125.

26. Mittelman, F and G Levan. 1976. Clustering of aberrations to specific chromosomes in human neoplasms. II. A survey of 287 neoplasms. Hereditas 82:167-174.

27. Nowell, PC and DA Hungerford. 1960. A minute chromosome in human chronic granulocytic leukemia. Science 132:1197.

28. Rowley, JD. 1973. A new consistent chromsomal abnormality in chronic myelogenous leukemia identified by quinacrine fluorescence and giemsa staining. Nature 243:290-293.

29. Daley, GQ, RA VanEtten, and D Baltimore. 1990. Induction of chronic myelogenous leukemia in mice by the P210$^{bcr/abl}$ gene of the Philadelphia chromosome. Science 247:824-830.

30. Bentley, GA, G Boulot, MM Riottot, and RJ Poljak. 1990. Bcl-2 is an inner mitochondrial membrane protein that blocks programmed cell death. Nature 348:334-336.

31. Mellentin, JD, C Murre, TA Donlon, PS McCaw, et al. 1989. The gene for enhancer binding proteins E12/E47 lies at the t(1;19) breakpoint in acute leukemias. Science 246:379-382.

32. Borrow, J, AD Goddard, D Sheer, and E Solomon. 1990. Molecular analysis of acute promyelocytic leukemia breakpoint cluster region on chromosome 17. Science 249:1577-1580.

ADDITIONAL LEARNING RESOURCES

1. de Grouchy, J and C Turleau. <u>Clinical Atlas of Human Chromosomes</u>. 2nd ed., New York: John Wiley and Sons, 1984.

2. Fryna, JP. 1984. The fragile X syndrome. Clin. Genet. 26:497-528.

3. Emery, AEH and Rimoin DL. <u>Principles and Practice of Medical Genetics</u>. 2nd ed.,New York: Churchill Livingstone Inc, Chap. 17-21, 1990.

Chapter 5

Genes and Families

The origins of human genetics extend backward many centuries to the time of the ancient Hebrews. The Talmud exempted boys from circumcision who were born to mothers who had produced two sons with bleeding disease (hemophilia) and likewise exempted sons of the boys' maternal aunts (1). This indicates that the Jews of the time recognized that this bleeding tendency was transmitted to sons through normal females.

The first account of an autosomal dominant trait, polydactyly, was reported by Maupertius in 1742 (2), and transmission of X-linked recessive traits was clearly described by Nasse in 1820. Although Joseph Adams distinguished between dominant and recessive inheritance in his book, <u>A Treatise on the Supposed Hereditary Properties of Disease</u>, published in 1814 and appeared to recognize many features of human genetics and their relevance to medicine, most medical personnel and biologists of that period did not fully appreciate the significance of heredity (3).

Charles Darwin (4) recognized heredity as a major factor in the origin of variability among species and as a major factor in evolution; however, he apparently had little comprehension of the mechanisms of inheritance. Francis Galton (1822-1911), a distinguished British physician and scientist and a cousin of Charles Darwin, attempted to apply scientific laws to the inheritance of complex traits such as intelligence and height. He is credited with founding eugenics and biometrics. He proposed comparing expression of traits in identical and non-identical twins as a method for assessing the relative contributions of heredity and environment to variation of human traits. However, Galton underestimated the importance of environmental influence on human traits.

Initial comprehension of the inheritance of discontinuous traits, those which occur as discrete entities, is credited to Gregor Mendel, an Austrian monk. Mendel studied the inheritance of seven pairs of traits in the garden pea and reported his results in 1865. His experiments demonstrated the laws of **segregation, independent assortment,** and **dominance** which will be discussed later in this chapter. Mendel's work remained obscure until it was rediscovered in the first decade of the twentieth century. The laws of inheritance established by Mendel were subsequently recognized to apply to a large number of human traits. Johannsen is credited with coining the terms **gene** and **genotype** which now enjoy widespread use in genetics. These investigators and those who followed them in the initial part of this century clarified the patterns of autosomal and X-linked inheritance, elucidated their genetic mechanisms, and determined the ratios of offspring of specific genotypes derived from various matings. This information is used today to determine genetic risks.

This chapter deals with the inheritance of discontinuous traits, where there is a clear difference between alternative phenotypes. Approximately 2000 of these traits have been described in humans, and many of these are associated with some degree of pathology. These traits are often said to be determined by "single genes." However, with the exception of X-linked genes in the male and of Y-linked genes, all nuclear genes occur in pairs, one located on each of the two homologous chromosomes. Therefore, most discontinuous traits are determined by the sum of the effects of two alleles at a single locus. If this fact is remembered, comprehension of the various forms of inheritance discussed below will come more easily.

AUTOSOMAL INHERITANCE

Human cells contain 22 pairs of autosomes and, therefore, two copies of each autosomal gene. Different structural forms of the same gene are called **alleles**, and the position of a gene on a chromosome is termed the **locus** (plural: **loci**) of that gene. As discussed in Chapter 2, the coding sequences of a gene (the exons) express themselves through a mRNA intermediate that is translated into the amino acid sequence of a polypeptide. This polypeptide is subsequently modified to form the mature protein. Therefore, most genes effect their expression by determining the structures of proteins which contribute in some way to the structure and function of our cells. Alleles of a gene differ in terms of their nucleotide sequence. Many of these differences are ultimately reflected in structurally modified proteins that produce effects in the cell which are distinct from those of the standard (most common) allele in the population. This relationship between DNA sequence, amino acid sequence, and physiological effect provides the link between the gene and the **phenotype**, the discernible effect produced by a pair of alleles.

Autosomal Codominant Inheritance

The gene **GYPA** (Glycophorin A blood group gene) encodes the amino acid sequence of a protein appearing on the surfaces of our red blood cells. This protein is referred to as a blood group antigen. The **GYPA** gene occurs as two alleles, **GYPA^M** and **GYPA^N**, that have been mapped to chromosome 4 (Fig. 5-1). Three possible combinations of alleles (**genotypes**) of the **GYPA** alleles can occur: **GYPA^{M}GYPA^M**, **GYPA^{M}GYPA^N**, and **GYPA^{N}GYPA^N**. Persons who possess two identical alleles are called **homozygotes**, while people who have two different alleles are known as **heterozygotes**. The relationships between gene, allele, genotype, and phenotype for the **GYPA** system are demonstrated in Fig. 5-2.

Meiosis apportions only one chromosome of each homologous pair and only one allele of each pair to a specific gamete. Therefore, under normal conditions, a parent can contribute only one of two alleles to a given child. **GYPA^{M}GYPA^M** parents can produce only **GYPA^M** gametes, and **GYPA^{N}GYPA^N** parents produce only **GYPA^N** gametes. Therefore, the following marriages would produce children with the genotypes:

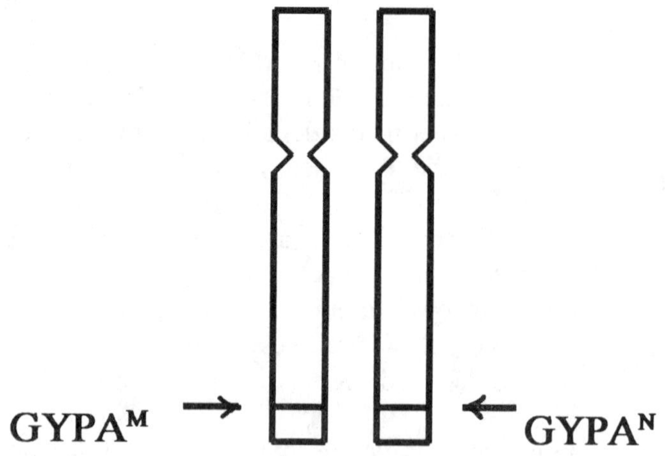

Fig. 5.1. Chromosomes 4 from an individual heterozygous at the GYPA locus.

Marriage	Children
1) GYPA^{M}GYPA^M x GYPA^{M}GYPA^M	All GYPA^{M}GYPA^M
2) GYPA^{N}GYPA^N x GYPA^{N}GYPA^N	All GYPA^{N}GYPA^N
3) GYPA^{M}GYPA^M x GYPA^{N}GYPA^N	All GYPA^{M}GYPA^N

GENE   ALLELE   GENOTYPE   PHENOTYPE

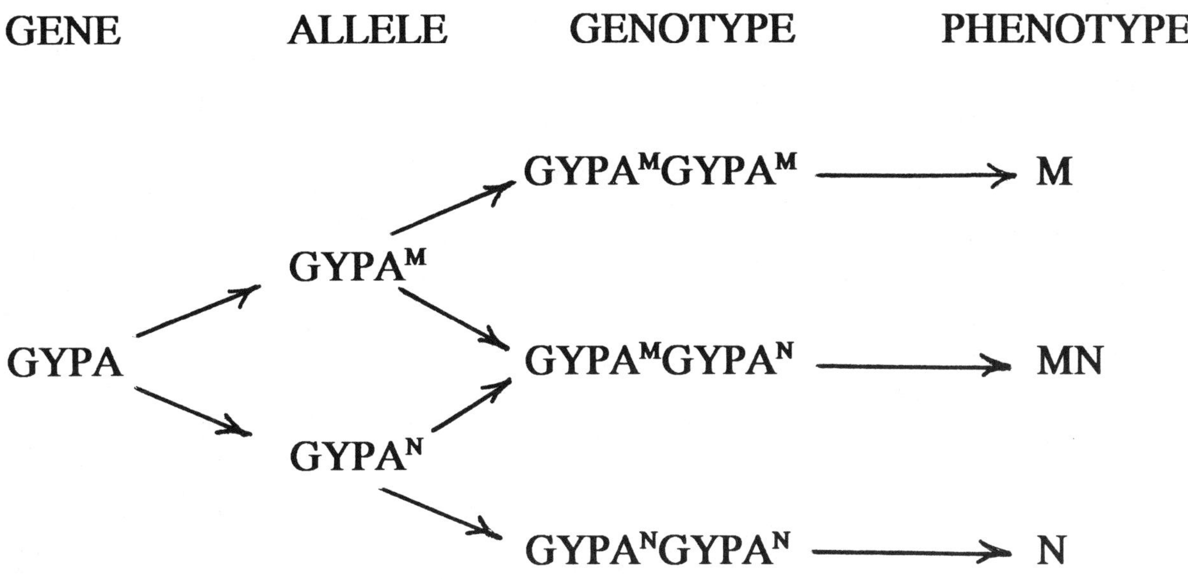

Fig. 5-2. Relationships among the terms gene, allele, genotype, and phenotype.

Marriages (1) and (2) illustrate an important characteristic of homozygous single-gene traits. These traits breed true, that is, parents homozygous for the same allele only produce children with a phenotype identical to their own. Furthermore, marriage (3) demonstrates that parents homozygous for alleles producing different traits have children who are all heterozygous for the respective parental alleles. When each of these alleles determines a distinct phenotype (eg. M and N antigens) in heterozygotes, the trait is said to be **codominant.**

Marriage of a heterozygous parent to a homozygous parent (Fig. 5-3) will result in two genotypic classes of children. In the example presented, one-half of the children will be **GYPA^{M}GYPA^M** (M phenotype) and one-half will be **GYPA^{M}GYPA^N** (MN phenotype). The corresponding marriage of a N parent to a MN parent will produce 1/2 N and 1/2 MN children.

One feature that distinguishes autosomal traits from their sex-linked counterparts is the tendency for a parent to transmit alleles determining an autosomal trait to his or her sons and daughters with approximately equal frequency. For example, the chance that a child will inherit **GYPA^N** from the man in Fig. 5-3 is 1/2, regardless of whether the child is male or female.

Marriages of two heterozygotes (Fig. 5-4) generate three kinds of offspring. One-half will be homozygous for one or the other allele, and one-half will be heterozygous like their parents. The reappearance of the M and N phenotypes was interpreted by Mendel as an indication that the alleles responsible for the traits were separating during the formation of the gametes:

> "It is now clear that the hybrids form seeds having one or other of the two differen-
> tiating characters..." (5)

This separation of the alleles during gametogenesis is termed the **Law of Segregation.** This segregation occurs at all loci, producing a large number of different gametic types.

175

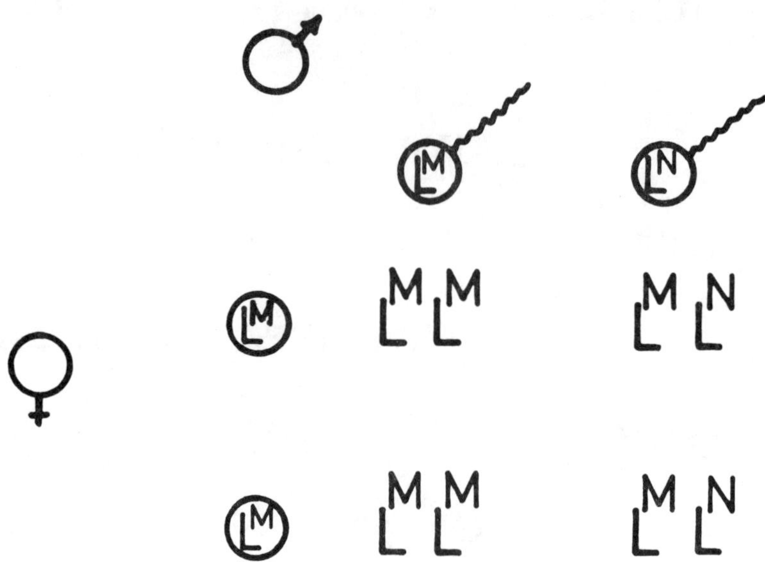

Fig. 5-3. Segregation of alleles during formation of gametes and reassociation in the zygotes conceived by a homozygous woman married to a heterozygous man. The expected ratio of homozygous to heterozygous children is 1:1. (Note: L = GYPA)

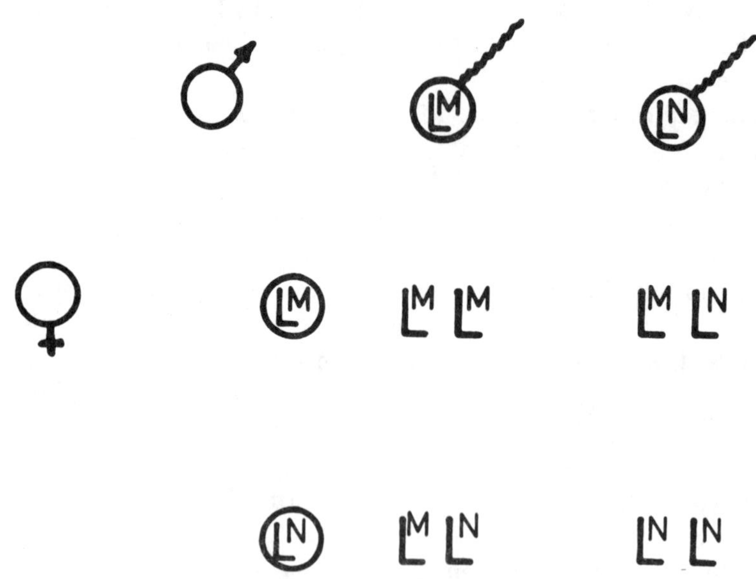

Fig. 5-4. Segregation of traits among children of two heterozygous parents. The proportion of M:MN:N is 1:2:1. (Note: L = GYPA)

Suppose we now consider two loci, the **GYPA** locus on chromosome 4 and the **GOT1** locus on chromosome 10. GOT1 encodes cytosolic glutamic-oxalacetic transaminase and occurs as two common alleles **GOT1^a** and **GOT1^b** which determine fast and slow isozymes, respectively. Since these genes are linked to **nonhomologous** chromosomes, a doubly heterozygous parent will produce four different gametes in approximately equal frequency:

Genotype		**Gametes**		
M/M;a/b	**M;a**	**M;b**	**N;a**	**N;b**

Only the superscripts of the alleles are presented for brevity. Genes occurring on nonhomologous chromosomes are separated by a semicolon, and alleles are separated by a slash. Assuming equal participation of each of these gametes in fertilization, two doubly heterozygous parents would produce the zygotes illustrated in Table 5-1. Notice that 4/16 of the children will be **MM**, 8/16 **MN**, and 4/16 **NN**, which reduces to the 1:2:1 ratio described above. Similarly, the ratio of **aa:ab:bb** is also 1:2:1. This

Table 5-1. Independent Assortment of Alleles at the GYPA (Chromosome 4) and GOT1 (Chromosome 10) loci. (Alleles are represented by their respective superscripts.)

Ova	Sperm			
	M;a	**M;b**	**N;a**	**N;b**
M;a	M/M;a/a	M/M;a/b	M/N;a/a	M/N;a/b
M;b	M/M;a/b	M/M;b/b	M/N;a/b	M/N;b/b
N;a	M/N;a/a	M/N;a/b	N/N;a/a	N/N;a/b
N;b	M/N;a/b	M/N;b/b	N/N;a/b	N/N;b/b

example illustrates Mendel's second law, the **Law of Independent Assortment:**

"...the relation of each pair of different characters in hybrid union is independent
of the other differences in the two parental stocks." (5)

In other words, alleles at two unlinked loci assort independently from one another during gametogenesis. Since both MN blood type and GOT electrophoretic mobility are inherited as autosomal codominant traits, the genotypic and phenotypic ratios are identical (note: f = fast GOT; s = slow GOT):

Ratio	Genotype	Phenotype	Ratio	Genotype	Phenotype	Ratio	Genotype	Phenotype
1/16	**M/M;a/a**	M;f	2/16	**M/N;a/a**	MN;f	1/16	**N/N;a/a**	N;f
2/16	**M/M;a/b**	M;f-s	4/16	**M/N;a/b**	MN;f-s	2/16	**N/N;a/b**	N;f-s
1/16	**M/M;b/b**	M;s	2/16	**M/N;b/b**	MN;s	1/16	**N/N;b/b**	N;s

Human families are characteristically small, and all possible genotypes will not occur in the family of a given set of parents. Therefore, the type of inheritance must be inferred from phenotypes of individuals occurring within sufficiently large groups of related persons (**kindreds**) and from several different kindreds. A shorthand method for presenting family data, the **pedigree**, is often used for such purposes. Symbols that are commonly used for pedigree construction are listed in Fig. 5-5. While

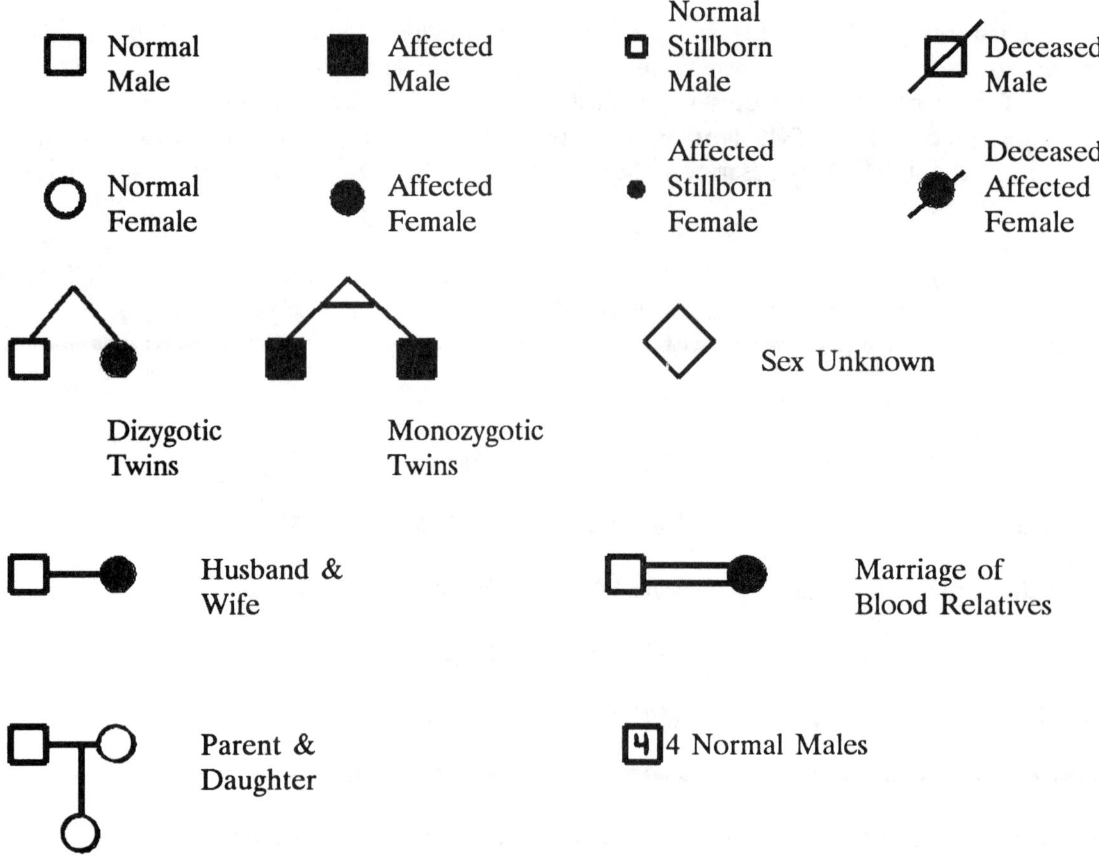

Fig. 5-5. Symbols commonly employed for pedigree construction. Dizygotic and monozygotic twins are derived from two fertilized eggs and one fertilized egg, respectively.

kindreds span several generations, **nuclear families** designate parents and their children. Brothers and sisters are called **sibs** (derived from siblings), and a combination of sibs is a **sibship**. The individual who alerts an investigator to a family segregating for a trait is designated the **proband**. Several rules are followed when constructing pedigrees: 1) where possible, the male symbol of a married couple is placed to the left of the female symbol; 2) sibs are arranged in order of birth; 3) all members of a generation are numbered from left to right using arabic numerals; 4) generations are indicated by Roman numerals; and 5) if the birth order within a sibshib is unknown, members are connected by a horizontal wavy line:

A sample pedigree illustrating the inheritance of the MN blood groups is presented in Fig. 5-6. Several important points are illustrated by this pedigree. 1) II-4 tranmitted the M phenotype to his sons: III-1, III-3, and III-5. **This male-to-male transmission excludes X-linked inheritance, since males give their X chromosomes to their daughters.** 2) Transmission of the M phenotype from II-4 to his daughters, III-4 and III-7 excludes Y-linkage, **since women do not have Y chromosomes.** These two observations indicate that the M phenotype is determined by an autosomal gene. 3) The appearance of traits M and N in successive generations, and 4) the expression of both traits in the children of II-4 and II-5 are most compatible with the expectations of autosomal codominant inheritance.

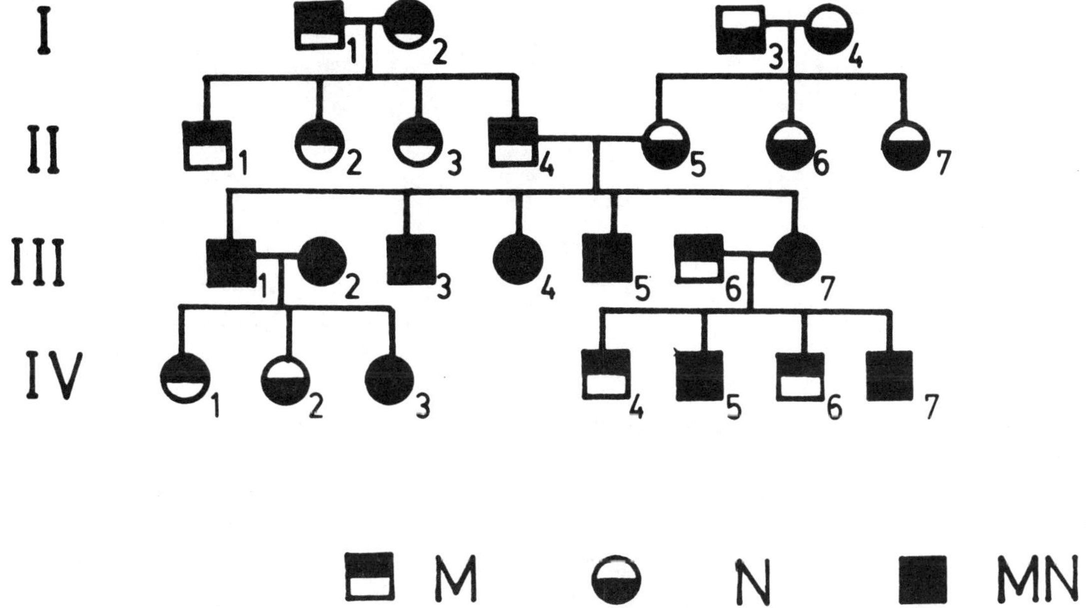

Fig. 5-6. Pedigree segregating for the MN blood group.

Autosomal Dominant Inheritance

The two alleles, **GYPA^M** and **GYPA^N**, presented in the previous example both exerted a discernible effect in the heterozygote whose phenotype included expression of antigens characteristic of both homozygotes. **Autosomal dominant** inheritance differs from codominance with respect to the phenotype of the heterozygote. The phenotypic effects of one allele tend to obscure the effects of the other, such that the phenotype of the heterozygote more closely approximates that of one homozygote than that of the other. Huntington's disease is inherited as an autosomal dominant trait and afflicts about

1/10,000 persons. Because of its low frequency, most marriages producing individuals at risk for Huntington's disease involve one normal homozygote and an individual heterozygous for the Huntington's allele (Fig. 5-7). The risk for Huntington's disease approximates 50 percent, regardless of the sex of the affected parent or the sex of the child. This high risk places a considerable psychological burden upon members of kindreds segregating for this trait. The disease usually does not appear until the mid-thirties and presents with neurological deterioration that does not respond readily to treatment. Death usually occurs about fifteen years after onset. Although age of onset may be fairly uniform within kindreds, it may vary from under 12 to more than 60 years of age among unrelated individuals. Many parents have completed their families prior to appearance of the illness in themselves and not only must bear the burden of their own risk, but also that associated with knowing they may have passed the gene on to their children. Alcoholism and suicide occur more frequently in Huntington's disease families than in the general population and reflect the psychological stress experienced by these families.

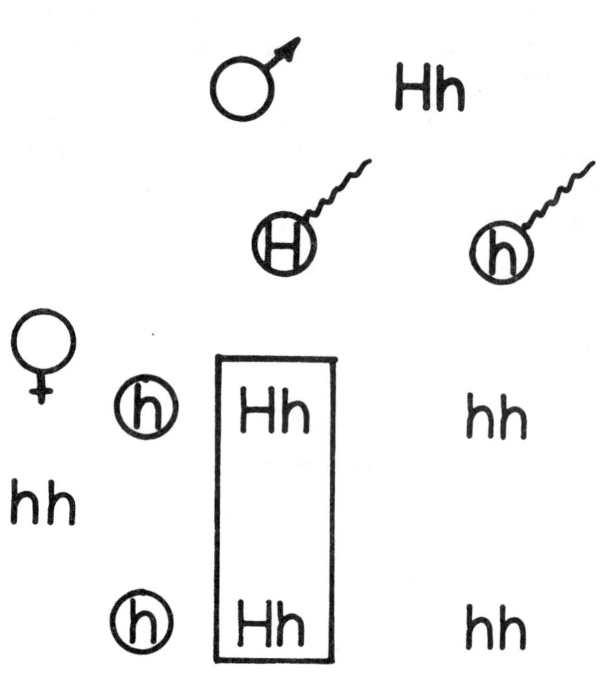

Fig. 5-7. Inheritance of Huntington's disease (HD). Like most autosomal dominant traits, HD is relatively infrequent (1/10,000); and affected individuals are usually heterozygous. H = HD allele; h = normal allele. (Rectangle encloses persons at risk for HD.)

A typical pedigree for Huntington's disease is illustrated in Fig. 5-8. III-5 is under the age of onset, so his genotype is uncertain as are those of his three children. His risk for Huntington's is 1/2. If he develops the disease, the risk for each of his children is likewise 1/2. Therefore, the risk that both he and IV-1 will develop Huntington's disease is $(1/2)(1/2) = 1/4$.

The pattern of inheritance of Huntington's disease can be inferred from this pedigree. There are two occasions of father to son transmission, and the trait was present in two women. Notice the occurrence of the trait in multiple generations, and that approximately one-half of the children of an affected parent contracted the illness. Occasionally parents or siblings of a patient erroneously interpret genetic risk estimates and their associated probabilities. For example, III-5 may incorrectly assume that, since three of his siblings have already developed Huntington's disease, his risk for the illness is minimal because his family has "used up" its 50 percent liability. It is important to stress that the risk of 1/2 applies to each sibling, regardless of the phenotypes of older brothers and sisters.

Two persons in a large Venezuelan kindred have been inferred to be homozygous for the Huntington's allele. Their neurological symptoms did not appear to be more severe than those exhibited by their heterozygous relatives. This pattern, where the phenotype of the dominant homozygote is comparable to that of the heterozygote, is commonly observed for traits that have been studied in experimental systems. However, the generality of this trend in humans is not known. Many dominant traits are rare, and homozygotes have not been described. Several exceptions to the equality of

patients include moderate to profound mental retardation, growth retardation, seizures, eczema, and an unusually light complexion compared to other family members. Since PKU only occurs in homozygotes, it is said to be **recessive.** Almost all parents of PKU individuals are heterozygotes or carriers of a **PAH°** allele. The genotypes, phenotypes, and proportions of children conceived by two heterozygotes are presented in Fig. 5-9. Three of the four genotypes have at least one standard allele and normal phenotypes, and one-fourth of the offspring will be homozygous for a **PAH°** allele and at risk for PKU. The risk of 1/4 for PKU remains the same for each conception, provided the parents are known to be carriers. The risk for heterozygosity for the **PAH°** allele is 1/2 for each conception. **However, if the child has a normal phenotype, risk for carrier status is 2/3.** The latter fraction represents the proportion of normal siblings of a PKU patient who are heterozygous and should be used in risk calculations when the phenotype of the sibling is known.

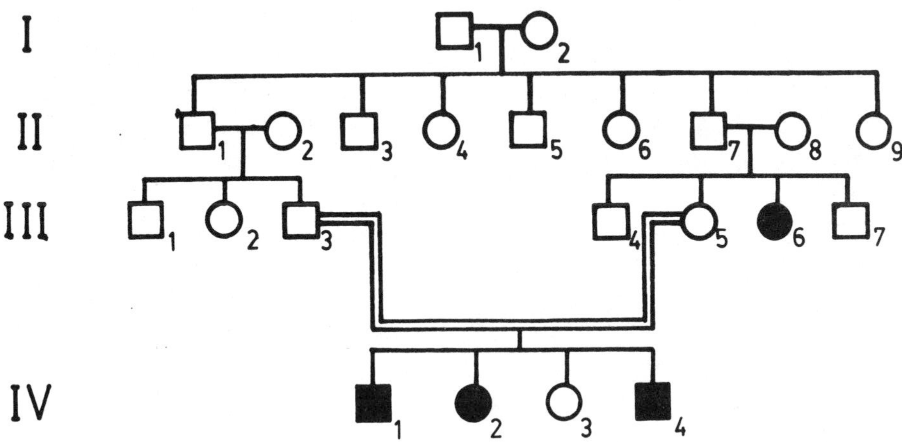

Fig. 5-10. Pedigree of a kindred segregating for PKU. III-3 and III-5 are first cousins.

A sample pedigree of a kindred segregating for PKU is presented in Fig. 5-10. This pedigree illustrates another feature of recessive inheritance. Most recessive diseases are quite infrequent. The incidence of PKU among newborn Caucasians in the United States is about 1/14,000. Other recessive diseases may occur as rarely as 1/1,000,000 births. Since the respective recessive diseases are infrequent to rare, carriers are also infrequent, although much more common than the diseases themselves. For example the carrier frequency for PKU approximates 1/60 persons. III-3 and III-5 are first cousins, and III-5 has a PKU sister. Suppose we go back in time to a point where III-3 and III-5 are newly married and are contemplating starting a family. What is their risk for a PKU child? This risk is calculated:

Probability (P) for a PKU child = (P III-3 is heterozygous)(P III-5 is heterozygous)(1/4)

Since III-5 has a normal phenotype and has a PKU sister, her chance for heterozygosity is 2/3. Both II-7 and II-8 are heterozygous for a PKU allele. Therefore, either I-1 or I-2 must have carried the gene. The probability that either I-1 or I-2 would transmit the PKU allele to II-1, and that II-1 would subsequently pass the allele to II-3 is (1/2) x (1/2) = 1/4. Therefore, the chance that III-3 and III-5 will have a PKU child is:

$$P = (1/4) \times (2/3) \times (1/4) = 1/24.$$

This risk is about 600-fold higher than the risk experienced by two unrelated persons. In general, **blood relatives have a much higher chance for having offspring with a recessive disease than experienced by unrelated spouses.** As many as one-half of children afflicted with a rare recessive illness are products of consanguineous (blood-related) marriages. **Presence of consanguinity in a pedigree should alert the investigator to the possibility that the illness occurring in the kindred may be recessive.**

In this particular example, both III-3 and III-5 proved to be carriers and had three PKU children. Although risk for a recessive child to two carrier parents is 1/4, **chance has no memory;** and the actual proportion of recessive children in such families may vary from 0 to all. When counseling parents relevant to recessive or other types of inherited problems, it is important to communicate the interpretation of genetic risk figures effectively.

We have approached the phenotypic effects of the alleles at the **PAH** locus from two different vantage points. When the phenotype is defined as enzyme activity, an intermediate or additive inheritance pattern of inheritance was observed. However, when the clinical state of the patient (PKU) was considered as the phenotype, PKU was recessive to the normal phenotype (or the normal phenotype was dominant to the PKU phenotype). These examples illustrate the importance of limiting terms like "dominant", "codominant", "intermediate", and "recessive" to the phenotype and not extrapolate them to a particular gene or allele.

Independent Assortment of Autosomal Dominant Traits

ABO blood type is determined by three common alleles at the **ABO** locus on chromosome 9. The Type A and Type B phenotypes are both dominant to Type O and codominant with respect to each other. **ABO^A**, **ABO^B**, and **ABO^O** determine the respective antigen types. The Rh blood group antigens are encoded by the **RH** locus on chromosome 1. Although this locus is complex and possesses many different alleles, the system is often simplified for medical purposes. Rh(+) is dominant to Rh(-). Genotypes of children from a group of marriages in which both parents were doubly heterozygous at the **ABO** and **RH** loci are displayed in Table 5-2. In each case, the genotypes of both parents were **ABO^{A}ABO^O;RH$^+$RH^O**. A single-locus two-allele system is assumed for Rh for simplicity. Notice how the children fall into four phenotypic classes which exhibit a **9:3:3:1** ratio:

Phenotypic Class	Proportion
A, Rh(+)	9/16
A, Rh(-)	3/16
O, Rh(+)	3/16
O, Rh(-)	1/16

This ratio applies to all similar situations, provided the loci are unlinked, there is no interaction between loci, and all genotypes can be recovered (i.e. no prenatal lethality). If two loci are linked (on the same chromosome), the chromosomal arrays of alleles which occur in the parents will predominate in their

Table 5-2. Independent Assortment of Alleles Determining Two Autosomal Dominant Traits.

Ova	Sperm			
	A;R	A;r	O;R	O;r
A;R	A/A;R/R	A/A;R/r	A/O;R/R	A/O;R/r
A;r	A/A;R/r	A/A;r/r	A/O;R/r	A/O;r/r
O;R	A/O;R/R	A/O;R/r	O/O;R/R	O/O;R/r
O;r	A/O;R/r	A/O;r/r	O/O;R/r	O/O;r/r

Note: A = ABO^A; O = ABO^O; R = RH$^+$; r = RH^O.

children, leading to a distortion of the 9:3:3:1 ratio (see Chapter 9).

Interaction between two unlinked loci can also distort the 9:3:3:1 ratio. Examples of interacting gene systems are presented in Chapter 10. One of these interactions involves the H locus on chromosome 19 and the **ABO** locus. The A and B antigens are formed by a series of metabolic steps. The **H** locus encodes an enzyme which catalyzes the production of H antigen. This antigen is then converted to either A antigen or B antigen by the enzymes determined by **ABO**^{A} or **ABO**^{B}, respectively. **hh** persons cannot make A nor B antigen because they lack the substrate required by the **ABO**^{A} and

Table 5-3. Assortment of Alleles at the ABO and H Loci.

Ova	Sperm			
	A;H	A;h	O;H	O;h
A;H	A/A;H/H	A/A;H/h	A/O;H/H	A/O;H/h
A;h	A/A;H/h	<u>A/A;h/h</u>	A/O;H/h	<u>A/O;h/h</u>
O;H	A/O;H/H	A/O;H/h	O/O;H/H	O/O;H/h
O;h	A/O;H/h	<u>A/O;h/h</u>	O/O;H/h	O/O;h/h

ABO^B gene products. Consider the possible genotypes of children of **ABO**^{A}**ABO**^{O};Hh parents (Table 5-3). The three genotypes that are underlined include at least one **ABO**^{A} allele; however, they are also homozygous **hh**. Therefore, their phenotype will superficially appear to be Type O. This type of interaction transforms the 9:3:3:1 ratio to a 9:3:4 ratio. In other words, the phenotypic class A(+), H(-) is missing.

If certain genotypes die before they can be observed, the phenotypic ratio will also be altered.

Achondroplasia is an autosomal dominant trait involving dwarfism consequent to defective growth of the long bones of the limbs. Persons who are homozygous **AA** die <u>in utero</u> or at birth. Suppose children derived from marriages of two Rh(+) achondroplastic dwarfs (A/a;R/r x A/a;R/r) were examined:

Phenotype	Proportion	
	Expected	Observed
Achondroplasia, Rh(+)	9/16	6/12
Achondroplasia, Rh(-)	3/16	2/12
Normal Stature, Rh(+)	3/16	3/12
Normal Stature, Rh(-)	1/16	1/12

If all genotypes were equally viable, a 9:3:3:1 ratio would be anticipated. However, **AA** is lethal, removing all genotypes carrying this combination from the observed children. Notice how this lethality reduces the phenotypic ratio of Achondroplasia:Normal Stature to **2:1**.

Multiple Alleles and Interpretation of Allelism by Pedigree Analysis

Genetic studies of populations using a variety of methods have revealed that almost all loci are occupied by families of alleles. Most of these alleles are quite rare; however, a few may become relatively frequent as a consequence of selective pressures favoring heterozygotes or as a result of founder effect and random drift (see Chapter 13).

The most abundant adult hemoglobin, Hemoglobin A_1, (HbA), provides an excellent example of a multiple allelic system. Hemoglobin is a red blood cell protein that functions in the transport of O_2 from the lungs to the tissues. Hemoglobin is a tetrameric protein consisting of two α and two β polypeptide chains (molecular formula: $\alpha_2\beta_2$). The α globin locus is situated within a cluster of several related genes on chromosome 16, while the β globin gene is one of several β-like genes on chromosome 11. Both loci possess multiple alleles. A number of β globin alleles are frequent among members of populations who reside in areas currently infested with malaria or which were infested in the past. These alleles include β^S in central Africa, the Saudi Arabian peninsula, and west India, β^C in west central Africa, β^D in India, and β^E in southeast Asia. Each of these alleles provided persons who were heterozygous for one of these alleles and the standard allele, β^A, with protection against malaria, allowing them to live longer and have larger families than non-heterozygotes residing in the same region. This **heterozygous advantage** led to the spread of the mutant alleles throughout these populations, and they continue to be relatively common among their descendants who have emigrated to non-malarial areas including the United States.

Figure 5-11 illustrates a kindred segregating for Hemoglobin S ($\alpha^A_2\beta^S_2$)(=HbS), HbC ($\alpha^A_2\beta^C_2$), and HbA ($\alpha^A_2\beta^A_2$). This pattern of inheritance would be anticipated if the genes determining HbS, HbC, and HbA were allelic. This can be demonstrated by examining the nuclear family:

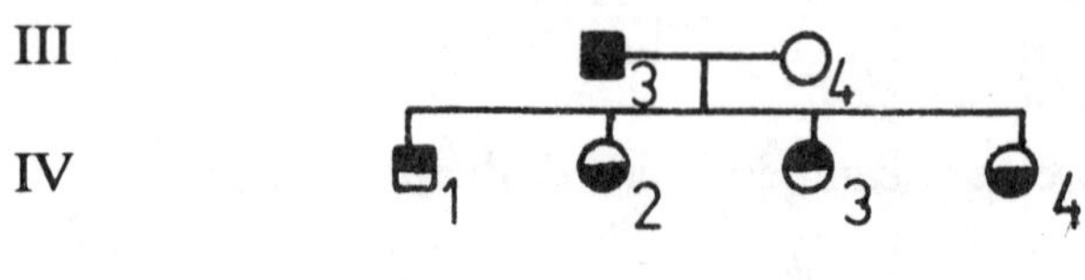

Genotypes of III-3 and III-4 are $\beta^S\beta^C$ and $\beta^A\beta^A$, respectively. III-3 can transmit **only one of his two alleles** to any given son or daughter, whereas III-4 would pass β^A to all of her children. Note how the genotypes of each of their children reflect this transmission pattern. The genotypes of IV-1, 2, 3, and 4 are $\beta^S\beta^A$, $\beta^C\beta^A$, $\beta^S\beta^A$, and $\beta^C\beta^A$, respectively. No children should occur who are $\beta^A\beta^A$ or $\beta^S\beta^C$. **If both or neither of two dominant or codominant traits present in an individual are transmitted to the same child, the traits are most likely determined by non-allelic genes.**

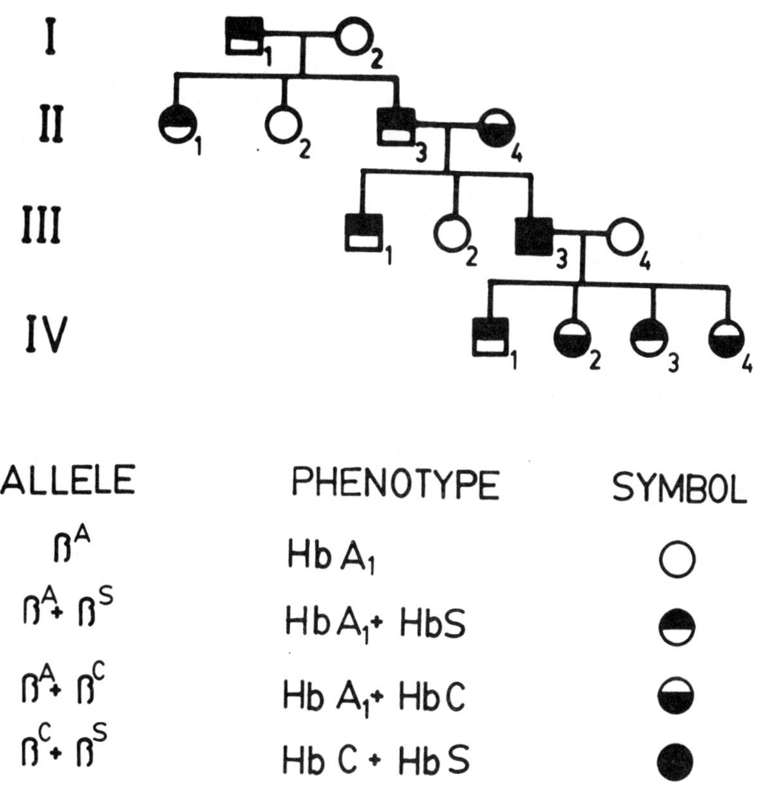

ALLELE	PHENOTYPE	SYMBOL
β^A	Hb A$_1$	◯
$\beta^A + \beta^S$	Hb A$_1$ + HbS	◑
$\beta^A + \beta^C$	Hb A$_1$ + HbC	◒
$\beta^C + \beta^S$	Hb C + Hb S	●

Fig. 5-11. Demonstration of allelism using a multi-allelic system: the β globin locus.

SEX-LINKED INHERITANCE

Traits determined by genes located on the sex chromosomes often exhibit an unequal sex distribution and an asymmetrical transmission pattern from parent to child. The dissimilar sex chromosome complements of the two sexes lead to gene dosage differences between males and females, and influence the frequencies of affected males and females among children of affected male x normal female versus normal male x affected female marriages. Females possess two X chromosomes, and males have one X and one Y chromosome per cell. Therefore, for X-linked genes, three genotypes can potentially occur in females and only two in males: Female = **AA, Aa,** or **aa**; Male = **A** or **a.** Males are said to be **hemizygous**, since they can have only one copy of each X-linked gene. Two other rules regarding sex-linked inheritance should be remembered:

1. Males cannot normally transmit an X-linked trait nor its gene to their sons (fathers pass their Y chromosomes to their sons and their X chromosomes to their daughters).

2. Females cannot express Y-linked traits nor transmit genes responsible for these traits to their offspring (females lack Y chromosomes).

X-linked Dominant Inheritance

Figure 5-12 illustrates transmission of an X-linked dominant trait to children from reciprocal marriages. Affected females are assumed to be heterozygous, since most X-linked dominant traits are rare. Approximately one-half of the daughters and one-half of the sons born to an affected mother will inherit the trait. An affected father transmits the trait to **all** of his daughters and **none** of his sons.

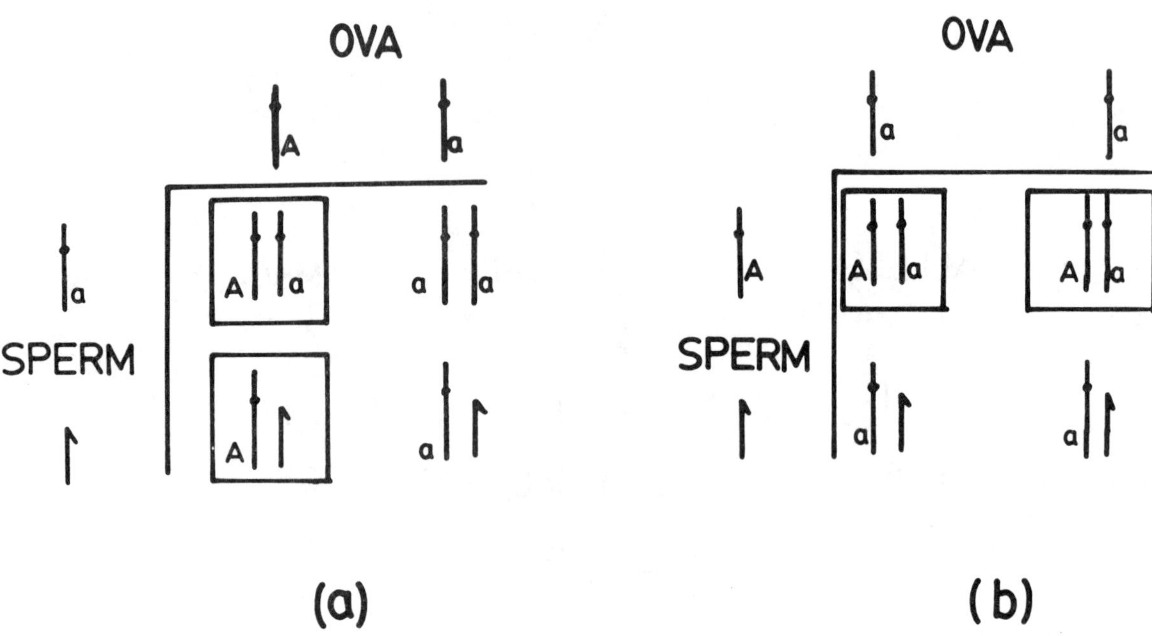

Fig. 5-12. Transmission of an X-linked dominant trait to children of affected parents. a) Affected mother x normal father; b) normal mother x affected father. The X chromosome is represented by a vertical bar, and the Y chromosome is symbolized with a half arrow. The squares enclose affected children.

A typical pedigree demonstrating X-linked dominant inheritance is presented in Fig. 5-13. Notice that the trait was not transmitted from father to son (six opportunities), and that all daughters of an affected father inherited the trait. By contrast, one-half of the daughters and one-half of the sons of an affected mother developed the trait. Although autosomal dominant inheritance is not rigorously excluded by this pedigree, it is much less likely. The probability that an autosomal dominant trait would not be transmitted to any of six sons of an affected male is 1/64 $(=(1/2)^6)$.

X-inactivation can influence expression of X-linked dominant traits in females. For example,

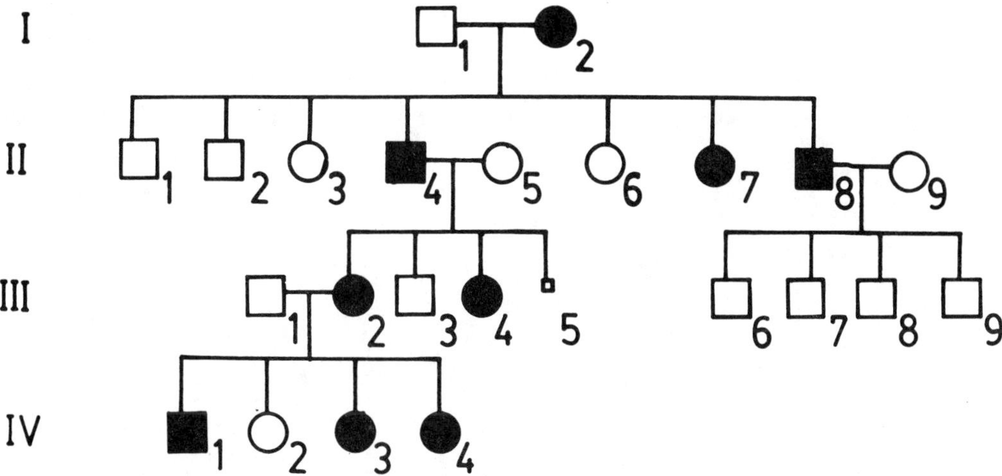

Fig. 5-13. **Pedigree illustrating X-linked dominant inheritance. Individual III-5 was a stillborn male lacking the trait in question.**

a form of vitamin D-resistant rickets (hypophosphatemia) is inherited as an X-linked dominant trait. The severity of skeletal changes associated with this disease is quite uniform among males; however, heterozygous women are usually less extensively involved and variation from essentially normal to very severely affected females has been observed. This more moderate and variable phenotype can be explained by the relative proportion of cells possessing the active versus the inactive hypophosphatemia allele. Amelogenesis imperfecta, a disorder affecting tooth enamel, provides a second example of the effect of X-inactivation on alleles determining X-linked dominant traits in heterozygous females. Two types of the disease occur: one in which the enamel is unusually soft (hypomaturation form) and a second in which the enamel is unusually thin (hypoplastic form). Affected males exhibit uniformly involved teeth, while affected females possess teeth in which patchy areas of abnormal enamel are interspersed among regions of normal enamel.

Since all daughters and no sons of an affected male inherit an X-linked dominant trait, significantly more affected women would be expected in a population where the trait is relatively frequent. This argument has been used in support of an X-linked dominant component in the etiology of bipolar affective disorder (manic-depressive psychosis)(6).

X-linked Recessive Inheritance

More males than females are affected with X-linked recessive traits because males cannot counteract the effects of the mutant genes through heterozygosity. Females are rarely afflicted with X-linked recessive traits, since these traits are usually rare and females must be homozygous, structurally hemizygous (the normal allele is deleted from one X-chromosome), or functionally hemizygous (the normal allele in heterozygotes is inactivated in a majority of cells) for the allele causing the recessive disease.

Inheritance of an X-linked recessive trait, ocular albinism, is diagrammed in Fig. 5-14. Fifty percent of the sons of a heterozygous female (Fig. 5-14a) will develop ocular albinism, and half of her

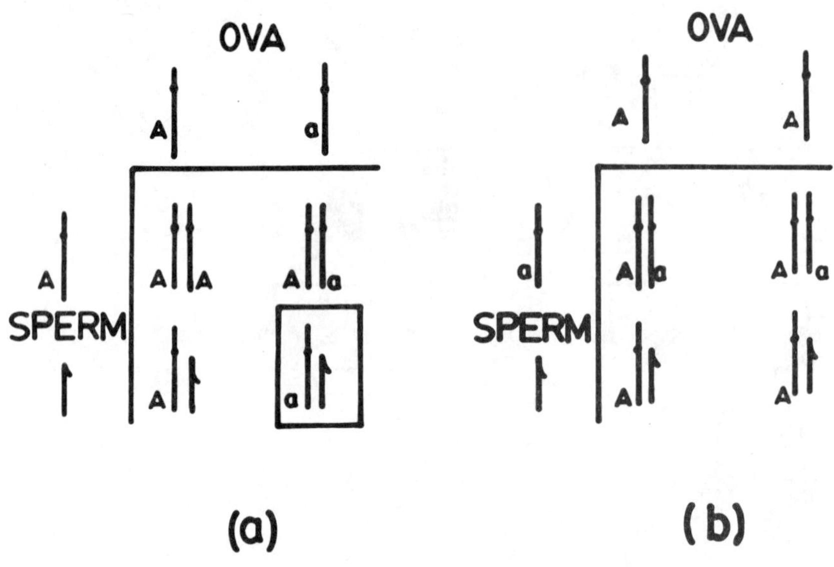

Fig. 5-14. Inheritance of ocular albinism, an X-linked recessive trait.
a) Marriage of a carrier female to a normal male; b) marriage of an affected
male to a normal female. Symbols are identical to those in Fig.5-12.

daughters will be carriers of the allele. Affected fathers do not transmit the trait to their sons; however, all of their daughters are heterozygotes and at risk for having affected sons. This inheritance pattern is evident in Fig. 5-15. Notice how affected males are connected by normal-appearing but heterozygous females.

Several X-linked recessive disorders, such as Duchenne muscular dystrophy, prevent reproduction of males. In these cases, the female side of the family will be positive for the disease, while the male side is usually negative. A pedigree depicting inheritance of Duchenne muscular dystrophy is presented in Fig. 5-16. Notice how the Duchenne allele was transmitted through clinically

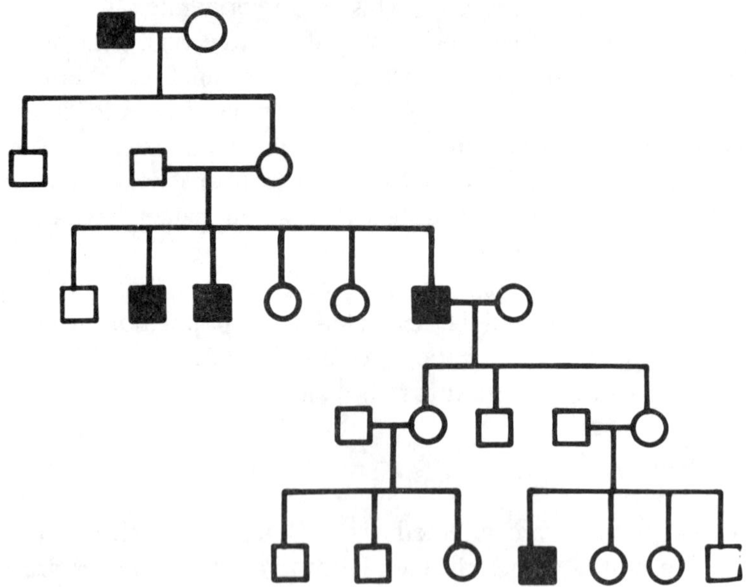

Fig. 5-15. X-linked recessive inheritance of ocular albinism.

normal females in generations II and III. Occurrence of the disease in II-7, II-8, and III-2 and in IV-3 indicate that II-6 and III-5 are obligate heterozygotes. IV-4 may also be heterozygous, and the normal

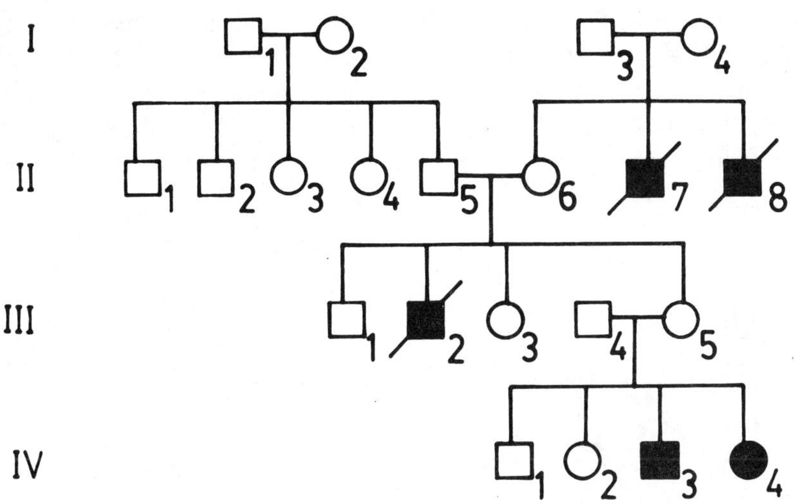

allele was inactivated in the majority of her muscle cells. Alternatively, one of her X-chromosomes may have suffered a deletion of the normal allele present in her father during spermatogenesis or during her early development.

Y-linked Inheritance

Expression of Y-linked genes should be limited exclusively to males. Molecular analysis of Y-chromosomal DNA has revealed three types of sequences: Y-specific, sequences occurring on both the X and Y chromosomes, and sequences that are found on all chromosomes (see Chapter 9). One of the Y-specific

Fig. 5-16. Inheritance of Duchenne muscular dystrophy. This pedigree is compatible with X-linked recessive inheritance. The possible genetic status of IV-4 is discussed in the text.

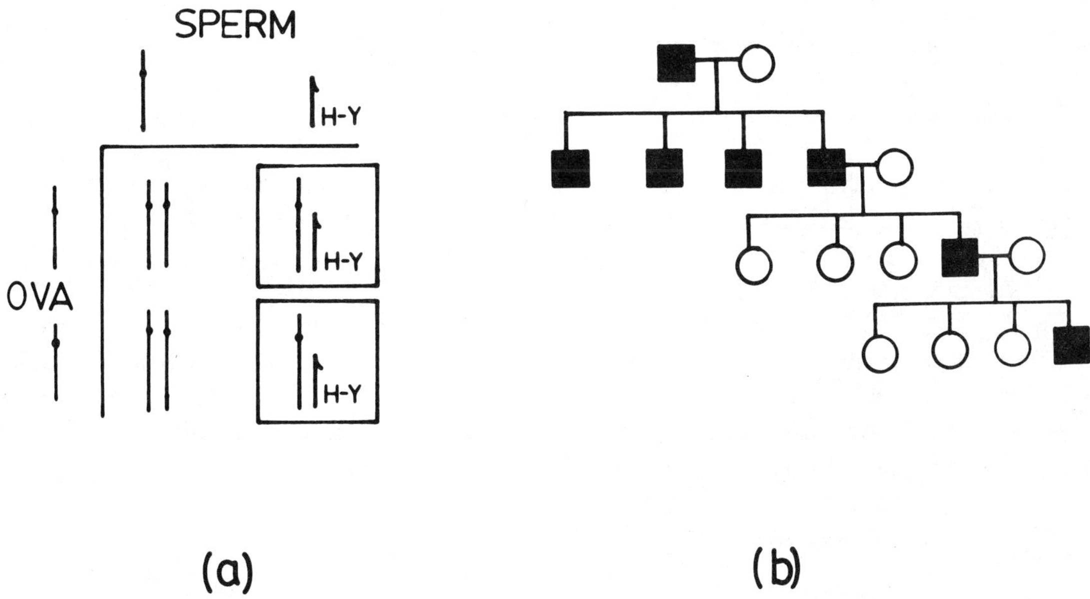

Fig. 5-17. Inheritance of a Y-linked trait, H-Y antigen. a) Gene transmission; b) pedigree.

sequences located on the proximal long arm of the Y chromosome appears to encode a male-specific antigen, H-Y. Inheritance of H-Y is illustrated in Fig. 5-17. Note how all sons and no daughters inherit the trait from their father. Several other Y-specific sequences have been identified in the **TDF** region of the short arm. Another trait, hypertrichosis of the ears, also appears to be determined by a Y-linked locus. The hypertrichosis allele promotes excessive growth of hair on the pinnae of the ears. All male and no female descendants of an affected male develop the "hairy ears". Segregation of this trait is illustrated in Fig. 5-18. Note how none of the daughters of an affected male transmitted the hypertrichosis allele to their sons.

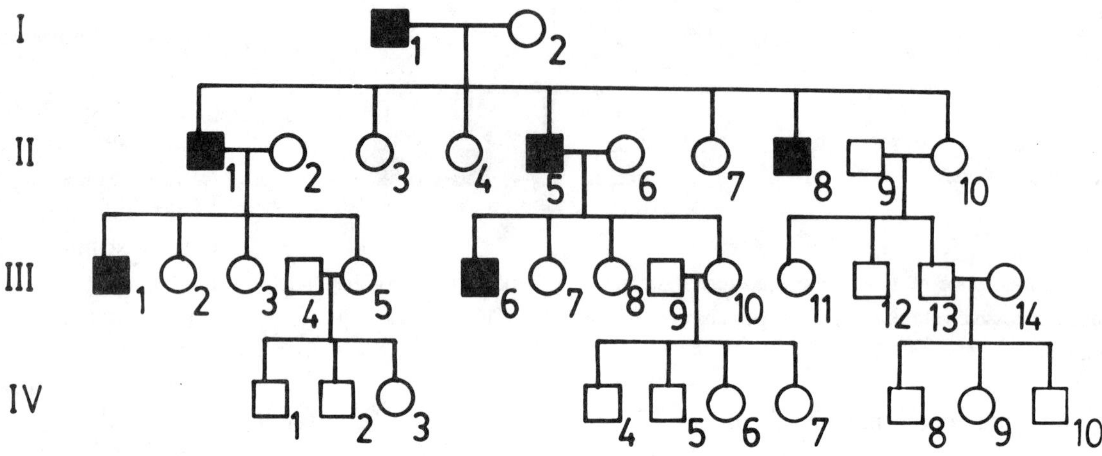

Fig. 5-18. Transmission of a presumed Y-linked trait, hypertrichosis of the ears.

Effect of Nondisjunction upon Transmission of X-linked Traits

The previous discussion of X- and Y-linked traits assumed normal segregation of these chromosomes during meiotic division. Nondisjunction of these chromosomes during the first or second meiotic division will lead to violations of the usual rules of inheritance of X-linked traits (Fig. 5-19). Comigration of the X and Y chromosomes to the same pole during anaphase I results in the production of sperm which possess both an X and Y chromosome and ultimate transmission of paternal X-linked genes from father to son. A son conceived following fertilization of an ovum possessing the slow $G6PD^B$ allele by a sperm carrying both the X and Y chromosome of his father (Fig. 5-19b) would inherit the X-linked $G6PD^A$ allele from his father. By contrast, a daughter derived from the sperm lacking an X chromosome would fail to inherit the $G6PD^A$ allele from her father.

Preferential Expression of Autosomal Traits in One Sex

Two types of inheritance are characterized by preferential expression of a trait determined by an autosomal locus in one sex or the other. **Sex-limited inheritance** refers to the inheritance of traits that are **exclusively** expressed in one sex and not the other. Prolapsed (forward-tilting) uterus and accumulation of fluid (hydrometrocolpos) is often inherited as a sex-limited autosomal trait. A hypothetical pedigree is presented in Fig. 5-20. The consanguinity occurring in this pedigree supports recessive inheritance. The transmission of the allele responsible for the trait through at least two

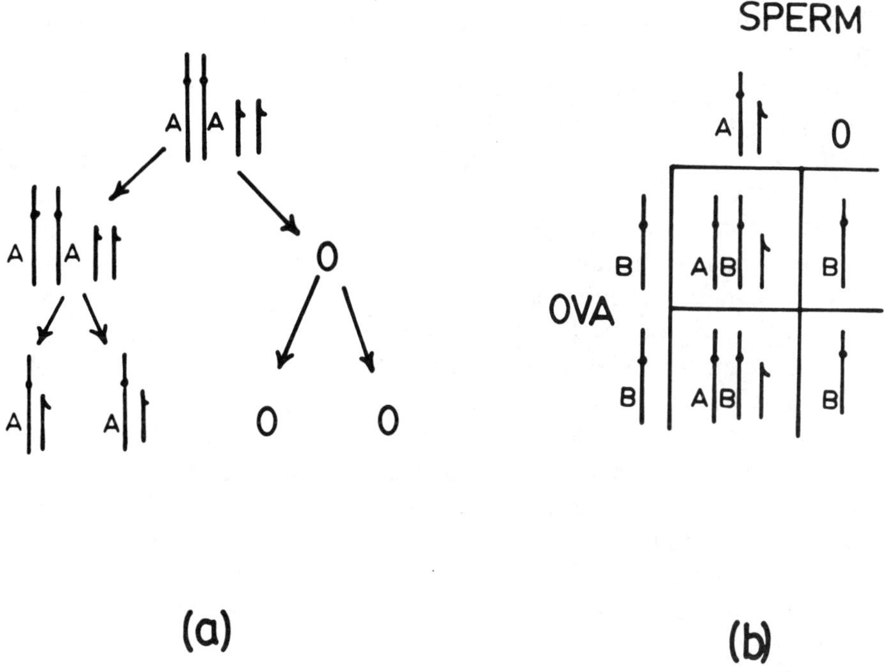

Fig. 5-19. Effect of nondisjunction upon transmission of an X-linked trait. a) Nondisjunction during anaphase I; b: possible products of conception and their G6PD phenotypes. A = fast isozyme; B = standard isozyme.

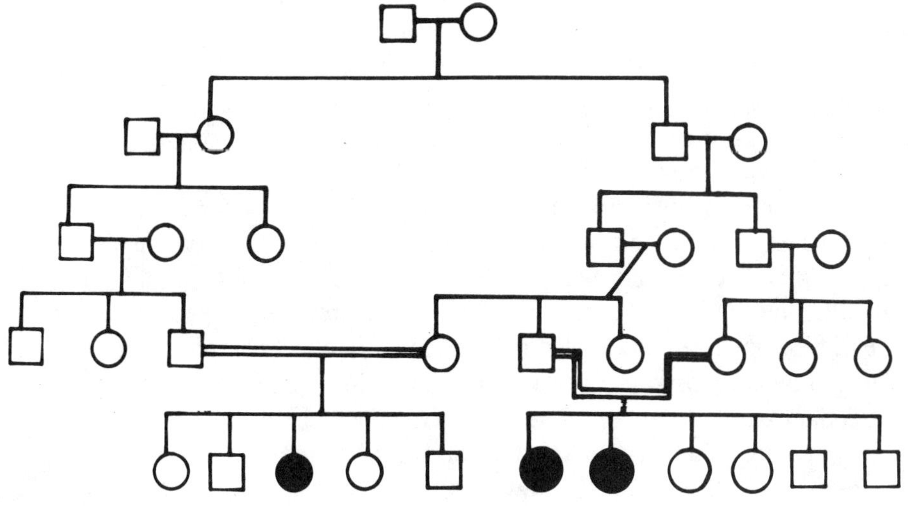

Fig. 5-20. Inheritance of a sex-limited autosomal recessive trait.

successive males in each line of descent excludes X-linkage.

Testicular feminization provides a second example of a sex-limited trait; however, this trait is caused by a recessive mutation in the androgen receptor gene (**AR**), which is X-linked. Males who are hemizygous for an **AR°** allele lack functional receptor and have an external female appearance due to androgen insensitivity. Females who are functionally or structurally hemizygous for the gene go undetected as a consequence of very minor or no symptoms of the receptor deficiency.

Sex-influenced traits occur more often in one sex than the other. Many of these traits are caused by one or more autosomal loci. Pattern baldness is a frequently cited example of a sex-influenced trait. Two or more loci are involved in the inheritance of a predisposition to this trait, and its expression is largely dependent upon male androgen levels and a normal androgen response. Numerous occasions of father to son transmission have been documented in a large number of pedigrees, and the genetic tendency for the trait may be transmitted through a non-bald female. Although occasional bald women are encountered, most women in these pedigrees do not display baldness. Those who do are believed to carry an unusually large number of predisposing genes. Bald women tend to have sons who are all bald; however, their daughters may or may not develop baldness, depending upon the hair pattern of their father and the relative number of bald-prone genes they inherit from their mother and their father.

An example of a less complicated sex-influenced trait is provided by Alport's syndrome, an X-linked dominant trait. A preponderance of females are affected by Alport's syndrome which includes deafness and nephritis. All daughters of affected males are affected. Females tend to be more mildly affected than males, and the excess of Alport females seems to be caused by early loss of males who have severe ear and kidney defects.

MOLECULAR BASES OF DOMINANCE AND RECESSIVITY

The terms dominant, codominant, intermediate (additive), and recessive have been applied to inheritance of phenotypes characteristic of certain genotypes. The heterozygote has been used as a point of reference. If a trait occurs in the heterozygote, it is said to be dominant. If a trait only occurs in homozygotes, the trait is recessive. Intermediate inheritance refers to those situations where the phenotype of the heterozygote is approximately midway between those of the homozygotes and is usually restricted to quantitatively measurable effects of a gene. Codominance applies to those cases where each allele in a heterozygote produces a discrete effect (e.g.: blood group antigens or different electrophoretic forms of hemoglobin (HbA and HbS)). It is clear from these examples that the expression or nonexpression of a trait in a heterozygote is dependent upon the relative impact of the mutant allele and its interaction with the environment in which it functions.

Sickle cell anemia is an autosomal recessive trait. The occurrence of the sickling of the red cells requires interactions among hemoglobin tetramers. The "oily" patches on the surface of the hemoglobin molecule that are caused by the β^S allele associate with similar regions of neighboring molecules to form long hemoglobin stacks that deform the red cell membrane. Heterozygotes, $\beta^S\beta^A$, produce three types of hemoglobin molecules (1/4 $\alpha_2\beta^S{}_2$, 1/2 $\alpha_2\beta^S\beta^A$, and 1/4 $\alpha_2\beta^A{}_2$). Three-quarters of the hemoglobin formed by heterozygotes cannot participate in formation of the intermolecular stacks, and sickling does not occur, except under extremely low oxygen tensions.

Most enzymes are produced in substantial excess of metabolic needs. Although persons who are heterozygous for a PKU allele (**PAH°**) have half the normal amount of phenylalanine hydroxylase activity, this activity is more than adequate to carry on normal metabolism. Therefore, PKU only occurs in untreated homozygotes, rendering the disease recessive. Most other inherited metabolic diseases also display autosomal recessive inheritance. However, a number of metabolic diseases are inherited as autosomal dominant traits. Some of these conditions are caused by deficiencies of enzymes that catalyze regulatory steps in metabolic pathways. Many of these steps are located at branch points in metabolic pathways and control the flow of metabolites into alternative routes. Regulatory enzymes are synthesized in barely adequate amounts, such that presence of one deficiency allele in heterozygotes will sharply reduce the level of activity below the minimum required for normal metabolism and cause a metabolic disease in the heterozygote. Acute intermittent porphyria is inherited as a dominant trait. Heterozygotes present with chronic attacks of abdominal pain and neurological signs including mental confusion and paralysis. The disease is quite variable, and attacks are often triggered by drugs such as sulfonamides, barbiturates, and certain steroids. Some patients have episodes without exposure to known triggering agents. The disease is caused by deficiency of the rate-limiting enzyme, uroporphyrinogen synthetase, which catalyzes the conversion of porphobilinogen to uroporphyrinogen. Both of these compounds are intermediates in the pathway leading to the formation of heme, a component of hemoglobin that interacts with O_2. The enzyme deficiency is detectable in certain relatives of porphyric patients who are also apparent heterozygotes but have not yet experienced porphyric episodes.

Mutations in genes encoding subunits of multimeric proteins may lead to dominantly inherited diseases. Substitution of nonpolar amino acids within the hydrophobic core of hemoglobin by polar amino acids or of small amino acids by much larger amino acids in the same region alter the three-dimensional structure of the globin subunit involved. Because of the cooperativity of the hemoglobin molecule, where the function of the entire protein is the sum of the interaction of its parts, a change in the conformation of one subunit impacts on the function of the others. Alternatively, the hemoglobin may denature and precipitate. An individual who is heterozygous for one of these mutations affecting the hemoglobin core will produce two types of chains, normal and mutant, and a majority of his hemoglobin molecules will contain at least one mutant chain. Examples of dominantly inherited hemoglobin disorders include HbZurich, HbM, and HbRanier. HbZurich is associated with a drug-induced anemia. The hemoglobin is unstable, readily denatured, and precipitates, forming inclusions called Heinz bodies. Red cells containing HbZurich and Heinz bodies are cleared from the circulation, causing anemia. HbM is produced by a group of hemoglobin mutations which affect either the α or β chain in a region interacting with the heme ring. The resulting amino acid substitutions cause the heme iron to be stabilized in the +3 state. HbM fails to pick up O_2 in the lungs. By contrast HbRanier is produced as a consequence of an amino acid substitution near the C-terminal end of the β-globin molecule. This region interacts with neighboring globins, and the substitution spoils cooperativity among the globin subunits. HbRanier binds O_2 but will not release O_2 efficiently at the tissue level. Each of these disorders will be discussed more extensively in a later chapter.

Many connective tissue disorders are dominantly inherited. Connective tissue contains proteins, the collagens, that occur as trimers. Extensive interactions occur among the collagen chains in these trimeric molecules to produce the protein networks observed in the ground substance of connective tissue. Heterozygous persons produce both normal and abnormal collagen chains, and seven-eighths of their collagen trimers will contain at least one abnormal chain and are functionally abnormal.

SAMPLING BIAS AND ITS EFFECTS UPON INHERITANCE OF HUMAN TRAITS

Human families typically have about two children per couple. This small size reduces the chance that all possible combinations of alleles present at a given locus in the parents will occur in their children. More than half of the two-child families of two heterozygous parents (3/4 x 3/4 = 9/16) will lack a recessive child. Therefore, if heterozygote x heterozygote marriages are identified by birth of recessive children, all couples with exclusively phenotypically normal children will not be ascertained.

The proportion (**q**) of recessive children born to two heterozygous parents is given by:

$$\mathbf{q} = \frac{\text{number of recessive children born to these parents}}{\text{total number of children born to these parents}}$$

and would approximate 1/4 if all families of two heterozygous parents, including those with no recessive children, were represented in the sample. Omission of families with only normal children from the sample results in an inflation of the estimate of **q**. The inflation may be extreme enough to approach 1/2, such that the disorder, which is actually recessive, may appear to be caused by dominant mutations in the germ line of one of the parents.

A second type of bias is introduced by the referral behavior of medical personnel. Most recessive diseases are infrequent. Specialists located in medical centers are often referred patients by family physicians, pediatricians, or obstetricians who are relatively unaware of the nature of these recessive disorders. Referrals will more likely be made when two or more affected children are born into the same family or when clinical features of the disease in question is unusually severe. These referral patterns often lead to a biased view of the inheritance of a trait or may create an impression that persons of a particular genotype are severely afflicted. Persons of that same genotype who lack symptoms or are mildly involved will not be encountered by these genetic specialists.

A third type of bias occurs when a stratified selection system is used to ascertain persons with a genetic problem. For example, mentally handicapped children with a particular genetic syndrome may be identified by screening a school population. A family with three affected children, each in a different grade, is more likely to be ascertained by such a sampling approach than is a family with one affected child.

Statistical procedures for dealing with sampling bias have been developed for use in appropriate situations. These methods may be found in texts dealing with quantitative genetics (7,8).

DETERMINATION OF RECURRENCE RISKS FOR SINGLE GENE TRAITS

Two types of risks apply to couples who are concerned about a genetic disease. One of these is the **risk of occurrence** of the disease in one of their future children. This type of risk pertains to couples who are considering beginning a family or who have one or more normal children. The **risk of recurrence** refers to the chance that a child subsequent to an affected sibling will develop the same genetic problem. The determination of risk is dependent upon a correct diagnosis and an understanding of the type of inheritance responsible for transmission of the trait.

Suppose a young couple is planning a family and is concerned about the risk for Rh hemolytic disease in a future child. The husband (II-1) is Rh(+), and his wife (II-2) is Rh(-) (Fig. 5-21). The Rh antigen may be assumed to be determined by an autosomal locus, and the Rh(+) phenotype is dominant to Rh(-). The man (II-1) has an Rh(-) mother; therefore, the man must be heterozygous (Rr). His wife, who is Rh(-), is **rr**. Hence, the risk of occurrence of Rh(+) in their first child is 1/2. It would be advisable to recommend that this woman take Rhogam (a commercial anti-Rh preparation) to protect her against being sensitized to form her own antibodies by the potential Rh(+) fetus.

The genotypes of II-1 and II-2 were able to be deciphered from their family history. However, genotypes of parents cannot always be determined in this manner. Therefore, the risk of occurrence must be calculated using probability theory. A pedigree illustrating inheritance of Huntington's disease was presented in Fig. 5-8. The probability that III-5 inherited the **H** allele from II-3 is 1/2. If III-5 has

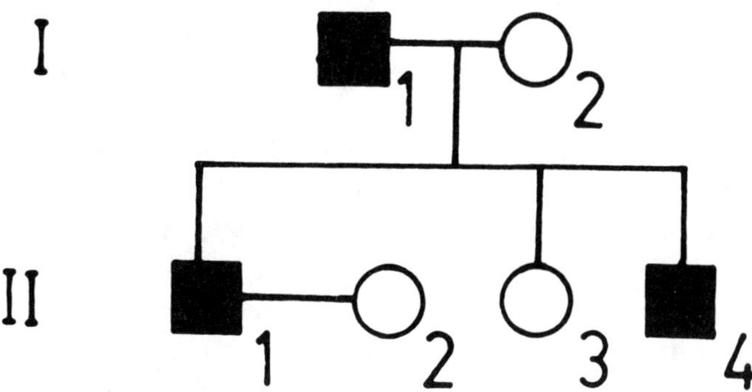

Fig. 5-21. **Pedigree of a family segregating for Rh antigen.** Individuals represented by the filled symbols are Rh(+), and those indicated by the open symbols are RH(-).

inherited **H**, the chance that he will transmit that gene to IV-1 is also 1/2. The occurrence of Huntington's disease in IV-I requires the presence of **H** in **both** III-5 **and** IV-1. The probability (P) of occurrence of **simultaneous** events is equal to the **product** of their separate probabilities. Therefore, the risk of occurrence of Huntington's disease (HD) in IV-1 can be calculated:

Risk (HD in IV-1) = P(III-5 is **Hh**) x P(III-5 transmitted H to IV-1) = 1/2 x 1/2 = 1/4.

This risk was calculated as a **prior probability.** That is, it was determined **before** information (medical tests, symptoms, etc.) was available that might provide a clue regarding the genotypes of either III-5 or IV-1.

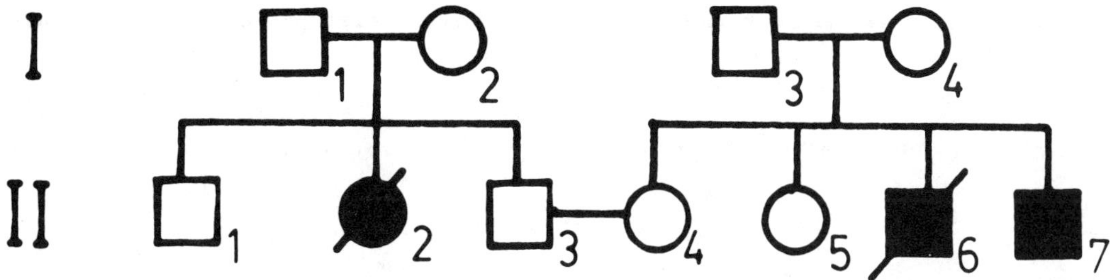

Fig. 5-22. Inheritance of cystic fibrosis (CF). Filled symbols indicate persons with CF.

The pedigree presented in Fig. 5-22 portrays the inheritance of CF, an autosomal recessive disease. What is the risk of occurrence of CF in children of II-3 and II-4? All four individuals in generation I are assumed to be heterozygous for the CF allele. **Both II-3 and II-4 are phenotypically normal but have CF siblings.** Therefore, the probability that both are heterozygous for the CF allele is 2/3 x 2/3 or 4/9. Remember that 3 out 4 of the children produced by heterozygous parents will have a normal phenotype, and two out of three of these normal-appearing children will be heterozygous (Fig. 5-9)! The risk of occurrence of CF in a child of II-3 and II-4 is calculated:

$$P \text{ (CF child)} = P \text{ (II-3 a carrier)} \times P \text{ (II-4 a carrier)} \times 1/4 = 2/3 \times 2/3 \times 1/4 = 1/9$$

Suppose the first child of this couple develops CF. This establishes both II-3 and II-4 as heterozygotes; therefore, the **risk of recurrence** of CF in a second child is 1/4.

Both branches of the kindred presented in Fig. 5-22 contained CF children. More commonly, one of the prospective parents has a positive family history for CF, and the other does not. Suppose II-4 had married a man who had no CF relatives. The man's chance for being heterozygous is no greater than that of any other individual in the population. CF occurs with a frequency of about 1/1600 newborn Caucasians in this country. It can be shown that the carrier frequency approximates 1/20 people. The

---

Hardy and Weinberg developed an algebraic proof that alleles with recessive effects would not be replaced by alleles with dominant effects in a population of intermarrying indivuals, but would attain an equilibrium in which the genotypes AA, Aa, and aa occurred with frequencies of p^2, 2pq, and q^2, respectively, where p is the frequency of A, and q is the frequency of a. For cystic fibrosis, the frequency of the disease closely approximates the frequency of the recessive homozygote and is equal to q^2. Therefore, the carrier frequency (2pq) can be obtained:

$$q^2 = 1/1600$$

$$q = 1/40$$

$$2pq = 2(39/40)(1/40) = 1/20.$$

This approximation fits most recessive diseases and their carrier frequencies and is commonly used in genetic counseling situations. Since p is very close to one for most of these disorders, the square root of the population frequency of the disease is doubled to estimate the carrier frequency.

Population Frequency of Heterozygotes

risk of occurrence of CF in a child of this couple is:

$$2/3 \times 1/20 \times 1/4 = 1/120.$$

Notice how the risk has markedly decreased. Couples usually are not deterred from having their own children when risks drop below 1/20; however, the risk calculated for the couple described in the preceding paragraph, 1/9, may often motivate prospective parents to pursue adoption or artificial insemination by donor as alternatives to having their own children.

X-linked recessive disorders such as Duchenne muscular dystrophy tend to involve males, while females are capable of transmitting the disease but are rarely afflicted with it. Fig. 5-23 illustrates a kindred segregating for this form of muscular dystrophy. The occurrence of muscular dystrophy in both II-5 and III-1 implicates II-6 as an obligate carrier of a Duchenne allele. The prior probability that III-4 is a carrier is 1/2. What is the probability that she will produce a dystrophic son? The risk of

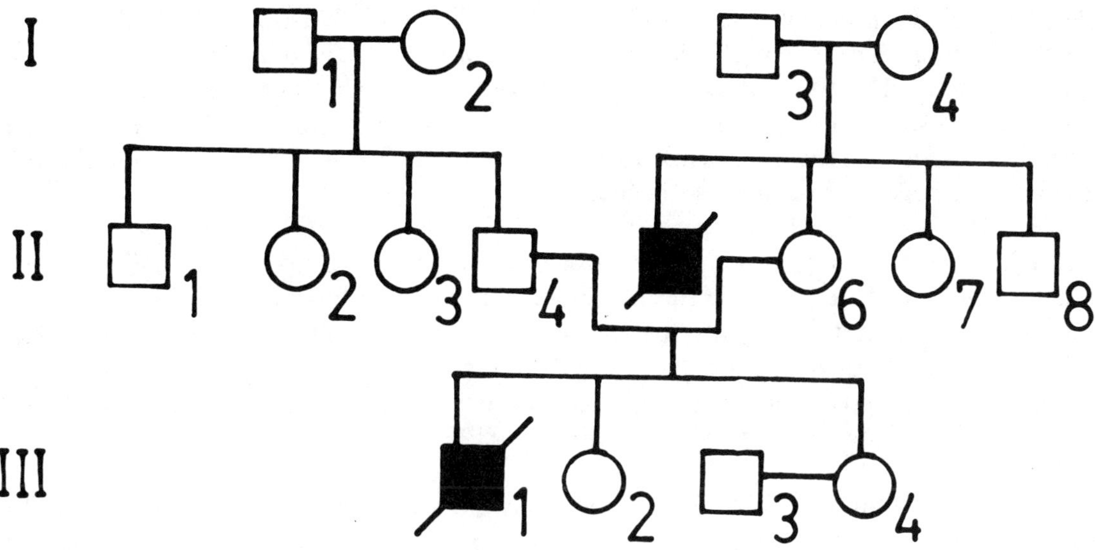

Fig. 5-23. **Inheritance of Duchenne muscular dystrophy. Dystrophic males are represented by filled symbols.**

occurrence may be calculated:

$$P \text{ (dystrophic son)} = P \text{ (III-4 carrier)} \times P \text{ (son)} \times P \text{ (transmission of dystrophic allele)}$$
$$= 1/2 \times 1/2 \times 1/2 = 1/8.$$

Use of Posterior Probabilities to Estimate Genetic Risk

Physicians and other health professionals who participate in genetic counseling often have information available which can influence the probabilities calculated above. When clinical data are available that make it more or less likely that a person is a carrier of a dominant or X-linked recessive allele, **posterior probabilities** should be used to assess risk of occurrence of a genetic problem. Cases where posterior probabilities should be employed include:

For X-linked recessive traits:

1. Normal test result for a woman who may be a carrier
2. Birth of normal sons to a suspected carrier woman

For an autosomal dominant:

1. Cases where an autosomal dominant trait exhibits reduced penetrance (defined below)

For example, III-4 in Fig. 5-23 may have been tested for serum creatine phosphokinase (CPK) activity. Serum activity of this enzyme (CPK is largely restricted to muscle) is usually low in normal persons; however, CPK is very high in dystrophic males during the early stages of their disease as a consequence of leakage from the degenerating muscle fibers. Effects of X-inactivation of the normal dystrophin allele leads to an elevation of CPK in about 2/3 of heterozygous women. Suppose III-4 has normal levels of serum CPK. What is the probability that III-3 and III-4 will have a dystrophic son? Bayesian methods may be employed to calculate the probability that III-4 is a carrier, given she has a normal test result. Bayesian probability considers the possible interpretations of available pedigree data and weights the probability of each possible genotype for the person at risk according to the likelihood that the respective genotypes would present with the phenotypic effect (e.g. normal CPK test) that has been observed. This type of problem can be solved in stepwise fashion:

1. What is the prior probability that III-4 is a carrier (based upon her family history)?
2. If III-4 is a carrier, what is the likelihood that she will have normal CPK?
3. What is the joint probability that she is both a carrier and will have cormal CPK?
4. Repeat (1) through (3) for the alternative genotype.
5. Calculate the posterior probability that III-4 is a carrier given her positive family history for Duchenne muscular dystrophy and normal CPK.

This approach is demonstrated in Table 5-4. Notice that the posterior probabilty (calculated **after** some informative aspect of the phenotype (normal CPK) is known) is lower than the prior probability determined above and is more suitable for this particular couple, since it considers the results of the CPK

Table 5-4. Bayesian Estimates of Genetic Risk: Posterior Probability that III-4 is a Carrier of the Duchenne Allele (Fig. 5-23) Given she has a Normal CPK level.

Genotype		DMD/DMD^o	DMD/DMD
1)	Prior Probability	$P(A) = 1/2$	$P(B) = 1/2$
2)	Likelihood Normal Serum CPK	$a = 1/3$	$b = 1$
3)	Joint Probability (1) x (2)	$P(A)a = 1/6$	$P(B)b = 1/2$
4)	Posterior Probability	$\dfrac{PA(a)}{P(A)a + P(B)b} = \dfrac{1/6}{1/6 + 1/2} = \dfrac{1}{4}$	

The risk of occurrence of muscular dystrophy in their first child becomes:

$$P(\text{muscular dystrophy}) = P(\text{III-4 carrier}) \times P(\text{son}) \times P(\text{transmission}) = 1/4 \times 1/2 \times 1/2 = 1/16.$$

Information concerning the genotype of a potentially heterozygous woman can also be derived from her sons. A woman at risk for being a carrier of X-linked recessive Charcot-Marie-Tooth disease and who has three normal sons is presented in Fig. 5-24. The posterior probability that II-8 is a carrier is 1/9 (Table 5-5), which is considerably lower than her prior probability of 1/2 and reflects the fact that a carrier is less likely to produce normal sons than is a normal homozygote.

Both of these examples demonstrate that posterior probabilities are often lower than corresponding prior probabilities. However, in neither case was the possibility of carrier status eliminated.

If one of the women birthed an affected child, her carrier status is immediately established; and her risk of recurrence would be 1/4 for each subsequent pregnancy.

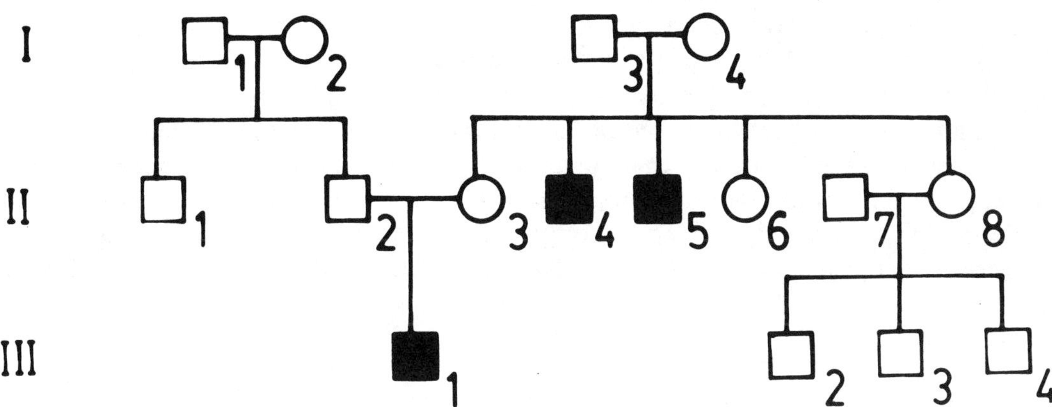

Fig. 5-24. Kindred segregating for X-linked recessive Charcot-Marie-Tooth disease.

Table 5-5. Effect of Normal Sons Upon the Probability that a Woman (II-8) is a Carrier of an X-linked Allele (Fig. 5-24).

	Carrier	Non-carrier
Prior Probability	1/2	1/2
Likelihood 3 Normal Sons	$(1/2)^3 = 1/8$	$(1)^3 = 1$
Joint Probability	1/16	1/2
Posterior Probability	$\dfrac{1/16}{1/16 + 1/2} = \dfrac{1}{9}$	

Bayesian methods are also appropriate in cases where dominant diseases are not fully penetrant. **Penetrance** refers to the frequency with which the phenotype characteristic of a specific genotype occurs in persons who have inherited that genotype. In these situations, a **predisposing** genotype increases the risk that its bearer will contract a trait; however, actual appearance of the trait depends upon both the environment and genetic "background" in which the genotype occurs. **Genetic background** includes other genes present in the individual which influence the expression of the gene primarily responsible for the predisposition. For example, Hb Zurich is associated with a drug-induced anemia. The anemia usually does not occur unless the Hb Zurich person is exposed to the appropriate drug (environmental agent). The anemia is said to be non-penetrant in the Hb Zurich person who is never exposed to precipitating drugs. It is theoretically possible, that occasional Hb Zurich individuals may not develop anemia when exposed to a drug that causes anemia in most other Hb Zurich persons. This subset of non-penetrant people could possess other genes that promote rapid clearance of the drug from their systems before it could have a detrimental effect.

Inheritance of an autosomal dominant trait with reduced penetrance is presented in Fig. 5-25. The index case (arrow) is concerned about the possibility that he and his wife may have affected children. What is the probability that the index case is a carrier of the allele responsible for the trait? Phenotypically normal persons in this pedigree may have a normal genotype or **may be non-penetrant carriers of the mutant gene.** The presence of the abnormal allele may be revealed by either the presence of the trait or by **transmission of the trait by an asymptomatic carrier to his or her child.** Notice that the latter occurred on two occasions. A total of 16 people either had the trait or transmitted its gene to their children. Therefore, an estimate of non-penetrance would be 1/8. The posterior probability that IV-16 is a carrier of the allele responsible for the trait is 1/9 (Table 5-6). The risk of occurrence in

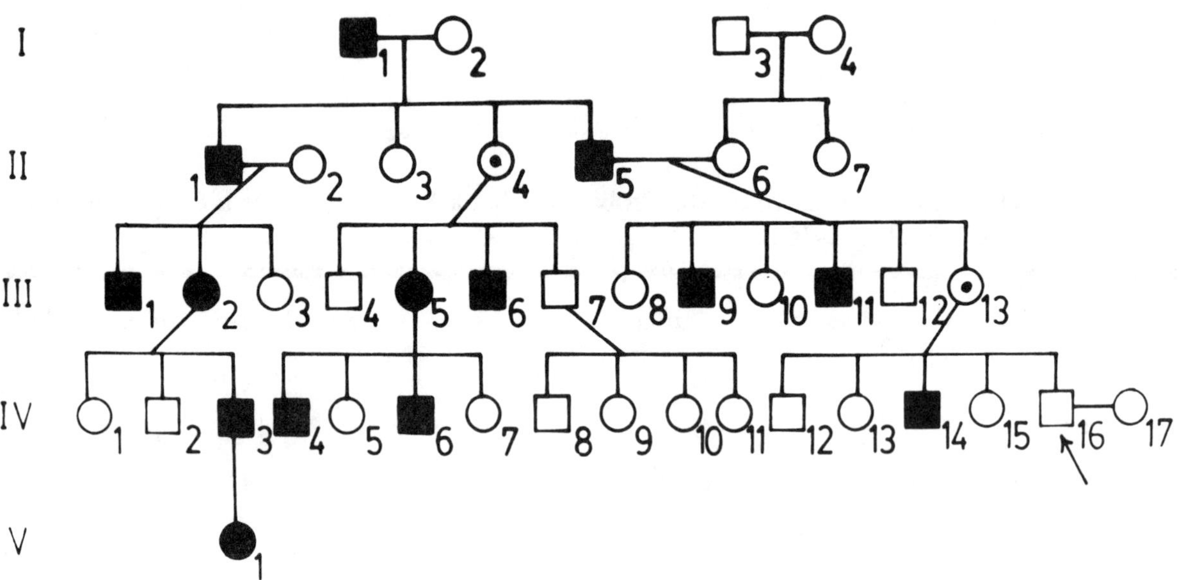

Fig. 5-25. Inheritance of an autosomal dominant trait with reduced penetrance. Note that the trait was not expressed in two individuals (II-4 and III-13) who are known to have carried the allele by virtue of their affected descendants.

his first child is approximated by:

P(affected child) = P(carrier) x P(transmission) x P(penetrance) = 1/9 x 1/2 x 7/8 = 7/144.

This risk may be interpreted in two ways, depending upon the predilection of the index case. If he is alarmed by the high frequency of the trait among his relatives, he may be reassured by the relatively low magnitude of the risk estimate. On the other hand, if he was assuming that his children would be normal because he did not have the trait, he may consider the risk of approximately 1/20 too high to justify fathering his own children.

Table 5-6.   Posterior Probability that an Individual (IV-16, Fig. 5-25) is a Carrier of an Allele Determining an Autosomal Dominant Trait Exhibiting Reduced Penetrance.

	Carrier	Non-carrier
1) Prior Probability	1/2	1/2
2) Likelihood of Normal Phenotype	1/8	1
3) Joint Probability	1/16	1/2
4) Posterior Probability	$\dfrac{1/16}{1/16\ +\ 1/2}\ =\ \dfrac{1}{9}$	

SUMMARY

Inheritance patterns of human traits are influenced by whether they are determined by autosomal or sex-linked genes and by the relative effect of the standard and mutant alleles on the phenotype of heterozygotes. Genes determining autosomal traits are transmitted from fathers to sons and daughters with equal frequency. However, X-linked genes are transmitted by men to all of their daughters and none of their sons. Y-linked traits and genes determining them occur exclusively in males.

When the effects of an allele penetrate in the heterozygote, the trait determined by that allele is said to be dominant. Traits which occur only in homozygotes are recessive. Codominant traits are determined by alleles which each have a discernible effect in the heterozygote. Intermediate inheritance patterns are usually displayed by quantitative aspects of gene expression such as the level of enzyme activity or the quantity of a protein. In either case, the heterozygote possesses a phenotype that approximates one-half that of the normal homozygote.

Three questions should be raised when deciphering inheritance patterns from pedigrees:

1) Are there examples of transmission of the trait or its allele from father to son? If there are no such occurrences, the trait is most likely X-linked.
2) Do females develop the trait or transmit the allele responsible for the trait? If yes, the trait is not Y-linked.
3) Are two or more individuals in a direct line of descent affected by the trait? If so, then the trait is most likely dominant.
4) Do consanguineous parents who are themselves normal have affected children? Consanguinity often implicates autosomal recessive inheritance.

Combinations of answers to these questions can be used to infer inheritance. For example, when the answers to questions (1) through (3) are all yes, the trait is an autosomal dominant.

Risks of occurrence of human traits are dependent upon the genotypes of the respective parents and the penetrance of the trait. When the genotypes of the parents are known, risks of occurrence and recurrence are identical and are derived from straightforward Mendelian ratios. In the case of autosomal dominant traits, the affected individuals are assumed to be heterozygous unless there is evidence to the contrary. Prior or posterior probabilites are used to estimate risk. Prior probabilities are employed in all cases excepting the following: 1) reduced penetrance of an autosomal dominant trait, 2) normal test result for a female suspected of carrying an X-linked allele; 3) or when a suspected carrier of an X-linked allele has normal sons and no sons with the trait. Posterior probabilities calculated by Bayesian methods should be used in each of these cases. When dealing with autosomal recessive traits, it is important to remember that 2/3 of the normal siblings of an affected person are heterozygous. The chance for a person with a negative family history for a recessive trait being heterozygous for the allele determining that trait approximates twice the square root of the population frequency of the recessive trait.

Inheritance patterns can be distorted by nondisjunction. For example, nondisjunction of the X and Y chromosomes during Anaphase I can lead to father-to-son transmission of X-linked genes. Sampling bias can also influence genetic ratios. If a sample of families segregating for an autosomal recessive trait is ascertained through the occurrence of recessive children, the ratio of affected:total offspring can be inflated to levels approaching those characteristic of dominant mutations.

PROBLEMS

1. A rare metabolic storage disease has been observed in seven families. Three of the families had two affected children, while the remainder had one affected child. Six of the ten children with the disease were boys. None of the parents had signs of the problem, although four couples were blood relatives. What type of inheritance is most likely responsible for this defect?

2. Consider the following pedigree:

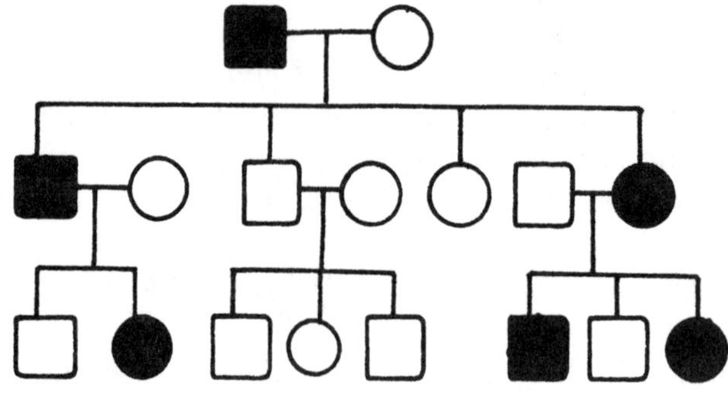

What type of inheritance appears to be responsible for transmission of this trait?

3. A neurological disease is known to be inherited as an autosomal dominant with 60% penetrance. What is the probability that a normal-appearing woman with an affected father and a normal-appearing man with no family history of the disorder will have a child who will develop the neurological disease? Assume the husband and wife are both beyond the age when symptoms of the disease first appear.

4. Tay-Sachs disease afflicts about 1/3600 newborn of couples in which both spouses are of Ashkenazi Jewish descent. Suppose a man whose sister died from Tay-Sachs disease marries an Ashkenazi Jewish woman who has no evidence of the disease among her relatives. What is the probability that they will have a Tay-Sachs infant?

5. A family was observed to be segregating for two hemoglobin variants:

Are the variant hemoglobins determined by allelic genes?

6. What type of inheritance is most likely responsible for transmission of the trait segregating in the following pedigree?

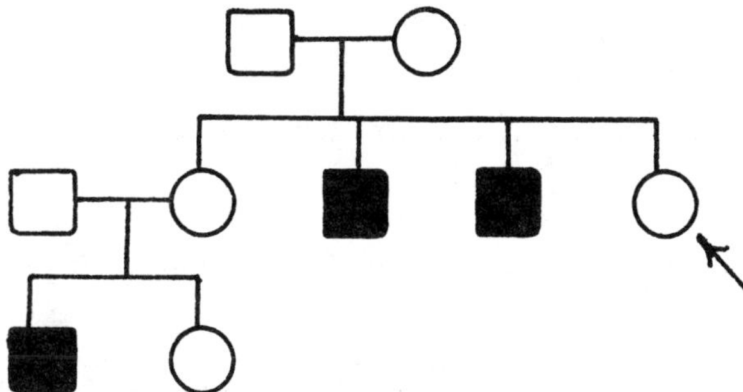

7. What is the probability that the index case (problem 6, arrow) will have an affected son?

8. Suppose the woman in problem 7 and her husband have a normal son. What is the probability that their next child will inherit this trait?

9. List the characteristics of X-linked dominant inheritance.

10. How would X-inactivation affect expression of an X-linked dominant trait?

11. Fragile-X syndrome (Martin-Bell syndrome) appears to be transmitted as an incomplete X-linked recessive. About 90 percent of males with the fragile-X chromosome have Martin-Bell syndrome, while about 30 percent of women known to be heterozygous for the fragile-X chromosome exhibit some degree of mental retardation. You have been commissioned by the federal government to study the relationship between the fragile-X chromosome and its associated syndrome. One way to collect a sample for study is to screen adult workshops for persons with the fragile-X chromosome. How might this means of ascertainment affect the outcome of your study?

12. A large study was performed to determine the type of inheritance responsible for a certain trait. Several large pedigrees from the United States were included in the study. Consider the following data:

| Marriage | | Phenotypes of Children | | | |
Father	Mother	Normal Sons	Affected Sons	Normal Daughters	Affected Daughters
Normal	Normal	25	0	30	0
Affected	Normal	22	0	1	26
Normal	Affected	10	12	13	11
Affected	Affected	3	2	0	6

What type of inheritance seems to be responsible for this trait?

GLOSSARY OF TERMS

Allele - one of two or more forms of a gene.

Carrier - a person who is heterozygous for an allele with recessive effects or who is heterozygous for a dominant allele exhibiting reduced penetrance.

Codominant - inheritance in which the phenotype characteristic of each of two alleles appears in the heterozygote.

Dominant - inheritance pattern in which the phenotype characteristic of an allele is expressed in the heterozygote, concealing the effects of the second (recessive) allele.

Genotype - the combination of alleles occurring at one or more loci of an individual.

Hemizygote - male carrying a single X-linked allele.

Heterozygote - a person who has two different alleles at a locus.

Homozygote - a person who has two identical alleles at a locus.

Intermediate inheritance - inheritance in which the phenotype of the heterozygote is quantitatively midway between those of the respective homozygotes.

Kindred - a group of blood-related individuals extending over several generations.

Linked - loci that are located on the same chromosome.

Locus - the position of a gene on a chromosome.

Nuclear family - parents and their children.

Occurrence risk - the chance that a hereditary trait will occur for the first time in a sibship.

Pedigree - representation of a kindred by appropriate symbols indicating phenotype, sex, and relationship.

Penetrance - the occurrence of a phenotype characteristic of a specific genotype in an individual having that genotype.

Phenotype - the morphological, physiological, or chemical trait specified by a genotype.

Recessive - inheritance in which the trait determined by an allele appears only in an individual homozygous or hemizygous for that allele.

Recurrence risk - the chance that a trait will appear in a later born sibling of an affected person.

Sex-influenced trait - the more frequent occurrence of a trait in one sex.

Sex-limited trait - the occurrence of a trait exclusively in one sex.

Siblings - brothers and sisters.

BIBLIOGRAPHY

1. Stern, C. 1973. <u>Principles of Human Genetics</u>. 3rd ed. San Francisco: WH Freeman, p. 316.

2. Glass, B. 1959. Maupertius, pioneer of genetics and evolution. In: <u>Forerunners of Darwin: 1745-1859</u>. Eds. B. Glass, O. Temkin, and W. Straus, Jr., Baltimore: Johns Hopkins Univ. Press, pp. 51-83.

3. Motulsky, AG. 1959. Joseph Adams (1756-1818). A forgotten founder of medical genetics. Arch. Int. Med. 104:490-496.

4. Darwin, C. 1859. <u>The Origin of Species</u>. London: John Murray, 6th ed. (reprinted many times).

5. Mendel, G. 1865. Experiments in plant hybridization. Reprinted in: <u>Classic Papers in Genetics</u>. 1959. Ed. JA Peters. Englewood Cliffs, NJ: Prentice-Hall, Inc. pp. 1-20.

6. Cadoret, RJ and G Winokur. 1975. X-linkage in manic-depressive illness. Ann. Rev. Med. 26:21-25.

7. Cavalli-Sforza, LL and WF Bodmer. <u>The Genetics of Human Populations</u>. San Francisco: WH Freeman and Co., 1971, pp. 851-860.

8. Vogel, F and AG Motulsky. <u>Human Genetics</u>, 2nd ed., New York: Springer-Verlag, 1986, pp. 648-658.

Chapter 6

Mutation and Its Consequences

The genetic message stored in our nuclei is copied with great accuracy during each replication to insure uniformity of hereditary information among all of our cells and continuity of human characteristics from generation to generation. However, our genes consist of DNA which is subject to gradual change following exposure to certain chemical and physical agents. A heritable change in the structure or quantity of DNA is called a **mutation,** and the agents causing mutations are known as **mutagens.**

Mutagenic Agents

Mutagens may be intrinsic or extrinsic. Intrinsic mutagens include highly active chemicals that are produced as intermediates or by-products of normal metabolism. These chemical mutagens are usually produced in trace quantities and are short-lived; however, they are believed to account for a significant proportion of the **spontaneous mutations** that occur in humans and other organisms. Spontaneous mutations are those hereditary changes in DNA structure or dosage that occur in the absence of perceptible mutagenic agents. Extrinsic mutagens include diagnostic, therapeutic, and occupational irradiation, solar ultraviolet and other extraterrestrial irradiation, and a variety of chemicals arising from industrial, agricultural, medical, and dietary sources. A large proportion of extrinsic mutagens are introduced to our environment by human technology. These extrinsic mutagens are responsible for **induced mutations.**

Irradiation may be ionizing or nonionizing. α-rays, X-rays, gamma rays, and cosmic rays are forms of ionizing radiations. The energies and abilities to penetrate living tissue vary widely among these four types of radiation. Penetration of a cell leads to the formation of positively and negatively charged molecules (ion pairs) which can subsequently damage DNA. Although DNA may occasionally suffer direct hits by these radiations, most damage is caused by secondary effects of the ion pairs formed by impacts with other chemicals in the cell. Ultraviolet irradiation is nonionizing and exerts its effects through the creation of thymidine dimers in DNA. This structural alteration subsequently promotes pairing errors during replication. Sunlight provides an important source of ultraviolet radiation, and the incidences of skin cancer in areas of intense sun exposure is increased compared to those in geographic regions at more polar latitudes. The relationships among radiation, mutation, and cancer will be discussed later in this chapter.

Radioactive isotopes are high energy forms of elements that dissipate their energy by emitting α-, β-, and/or gamma radiation. Chromosome breakage and rearrangement are common sequelae to such radiation exposure. Radiation is frequently employed as an anti-cancer therapy, successfully killing cancer cells because they are actively dividing in an otherwise non-dividing tissue. However, repeated exposure to therapeutic radiation can occasionally damage normal cells in areas adjacent to the tumor, leading to cancerous growth. Patients receiving radiation therapy should be monitored for subsequent appearance of these radiation-induced tumors.

A variety of chemicals such as benzpyrene, ethylene oxide, nitrogen mustards, proflavine, benzene, 5-bromouracil, and nitrous acid are known to induce mutations. These chemicals act through a number of different mechanisms to modify the bases in DNA, deform the DNA double helix, or cause breakage of chromosomes. Representative examples of these effects will be presented later in this chapter. Chemical mutagens are inadvertently introduced to our environment as the result of improper waste disposal, components of certain herbicides and insecticides, as constituents of motor vehicle exhaust, and as the result of spills and other accidents. Although the initial levels of these pollutants may be quite low, food chains can concentrate them to high levels. Humans are often the last link in the food chains and are vulnerable to the effects of these mutagens. Fetuses, infants, and young children are especially sensitive to mutagens because they have a large number of dividing cells.

Certain drugs or drug combinations are utilized for cancer therapy. These drugs kill cancer cells by damaging their DNA, generating fatal mutations. Cancer cells are highly susceptible to these chemical mutagens as a consequence of their high division rate; however, normal cells are not immune to their effects. Unlike radiation damage which is localized to regions surrounding the irradiated tumor, the anti-cancer chemicals move throughout the body and are capable of damaging cells at sites far removed from the tumor being treated. Indications for using these anti-neoplastic agents are obvious; however, patients who are of reproductive age should be advised of their side effects, and all patients exposed to these mutagens should be monitored for secondary tumors. Exposure of pregnant women to such agents is especially conducive to induction of mutations in the fetuses they are carrying. Counseling of these women relevant to potential fetal damage is strongly advised to assist them in making decisions regarding the continuation of the pregnancy.

A number of chemicals may be harmless in themselves, but these **promutagens** can be converted to mutagens by enzymes found in liver and certain other tissues. Promutagens include dibenzanthracene, methyl cholanthrene, and benzpyrene. These chemicals are normally inactivated by a liver microsomal detoxification system. Highly active epoxides are generated as intermediates, and these epoxides can cause mutations. The list of promutagens has grown markedly during recent years.

The pronounced rise in human exposure to potentially harmful agents in our environment that has occurred during the last few decades has led to vigorous action by governmental and private agencies to curtail such exposure. Industry and hospitals have been encouraged to reduce occupational exposure of their personnel. Air and water pollution have been gradually reduced by cooperative governmental and industrial efforts. The alterations required for properly treating water- and airborne wastes are expensive and have increased consumer costs. However, most consumers are willing to bear this burden in favor of a better environment for themselves and for future generations. Recognition of side effects of some drugs has spurred more responsible prescription practices by physicians and implementation of mutation screening programs by pharmaceutical firms and the federal government for detection of potentially harful effects of new products. Although the release of potentially beneficial drugs is often delayed by these programs and has been subject to public controversy, the overall value of the screening programs seems to warrant their continuation.

Detection of Mutation and Screening Potential Mutagens and Promutagens

Screening for mutagenic effects of chemicals utilizes a number of test systems. Viruses, bacteria, and the fuitfly, <u>Drosophila</u> <u>melanogaster</u>, have been used for some time to demonstrate the mutagenicity of radiation and chemical agents. For example, bacteria that possess a metabolic lesion in a certain pathway are irradiated and plated on growth medium lacking the metabolite they cannot synthesize. Bacterial cells that have experienced a mutation that corrects the metabolic lesion will grow and form colonies. Cells which have not sustained a mutation or which have mutations affecting other pathways will not grow in this medium. The mutation rate is approximated by the total number of colonies observed divided by the number of cells exposed to the irradiation. Mutagenicity of products colonies observed divided by the number of cells exposed to the irradiation. Mutagenicity of products of promutagens can be screened by adding the promutagen to a rat liver homogenate prior to exposure of the bacteria. The liver enzymes metabolize the promutagen to mutagenic intermediates which subsequently induce mutations in the bacteria. Results from these tests should be interpreted with caution, since a substance that proves mutagenic in bacteria may not be mutagenic in humans.

Human cells grown in culture can also be employed to screen for mutagens. In one type of test, mutant cell lines which lack an enzyme required for growth under defined conditions are exposed to the suspected mutagen and cultured in the defined medium. Cells lacking mutations that correct the original defect fail to grow, while those which have experienced a corrective mutation grow in the medium. A second test utilizes cell lines that have normal genetic constitutions. Mutations that have been incurred following exposure to a suspected mutagen may alter the drug resistance, nutritional requirements, or temperature sensitivity of these cells.

The **host-mediated assay** employs an experimental animal that has been injected with bacteria that possess a known mutant phenotype. The animal is then treated with the agent under study, and, after a period of time, the bacteria are recovered from the animal and screened for mutations that correct the original mutant phenotype. Comparison of mutation rates in the recovered bacteria with those in similar bacteria that were directly exposed to the potential mutagen permits an assessment of whether the animal can detoxify the mutagen before mutation occurs. Alternatively, this method can be used to determine whether the animal can convert a promutagen to a mutagen.

Yeast (<u>Saccharomyces</u> <u>cerevisiae</u>) and fungi (<u>Neurospora</u>) can also be employed for screening chemicals for mutagenicity. These organisms are used in a manner similar to that described for bacteria and can be included as an indicator organism in the host-mediated assay.

Cytogenetic studies can be used to screen for potential mutagens. Chromosome breakage and rearrangement in cultured lymphocytes and other human cells serve as indicators of mutation. The frequency of chromosomal aberrations in mutagenized cells is compared with that in untreated control cells. A variation of this system involves use of live animals. Blood is taken from the animal prior to exposure to serve as a control. After exposure, a second blood specimen is collected and evaluated for cytogenetic damage caused by the suspected mutagen.

The **dominant lethal test** provides a means for estimating the effects of radiation or chemical mutagens upon loci that are critical for early development. Male mice are treated with the agent and then mated to untreated virgin female mice at 1-3, 4-5, and 6-8 weeks after treatment. Female mice are sacrificed 12-13 days after conception and examined for corpora lutea (provides an estimate of early fetal death) and total implants (include late fetal deaths and living fetuses). The three mating intervals permit screening for mutations occurring during post-meiotic and premeiotic stages of spermatogenesis, respectively. A similar protocol employing rats has also been used for dominant lethal tests.

MOLECULAR MECHANISMS OF MUTATION

Base Substitutions

One common type of mutation involves substitution of one base for another. Substitutions of one purine for another purine or one pyrimidine for another pyrimidine are called **transitions**. Purine-pyrimidine interconversions are referred to as **transversions**. Some mutagens, known as **base analogs**, may actually replace natural bases in the DNA (Fig. 6-1). Both natural bases and base analogs can alternate between different structural forms (tautomers)(Fig. 6-2). The amino tautomer of adenine is favored under normal circumstances, and this tautomer preferentially pairs with thymine. By contrast, 5-bromodeoxyuridine (BrdU) spends most of its time in the enol form. This tautomer preferentially pairs with

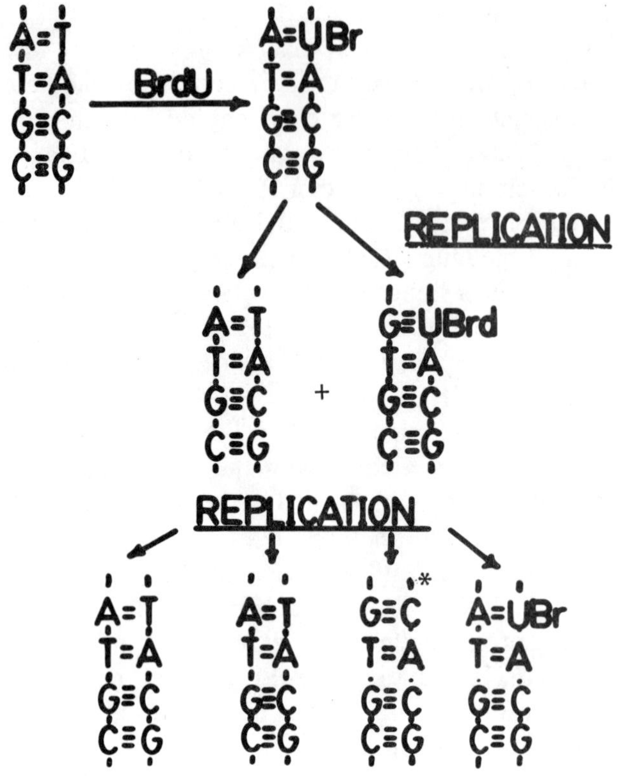

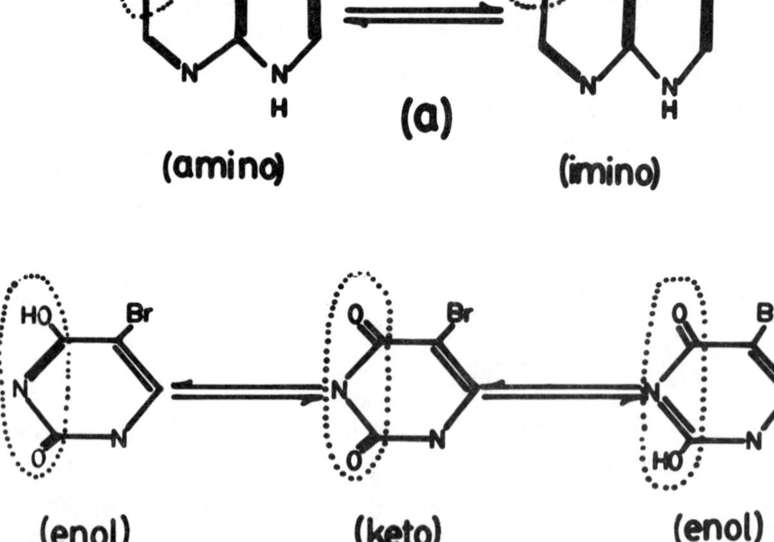

Fig. 6-1. Transition induced by 5-bromo-deoxyuridine (BrdU). UBr = keto form; UBrd = enol form; * = mutant double helix.

Fig. 6-2. Tautomeric forms of adenine (a) and 5-bromouracil (b). The amino form of adenine preferentially pairs with thymine; however, the imino form may pair with cytosine. The keto form of 5-bromouracil pairs with adenine, while the enol form pairs with guanine.

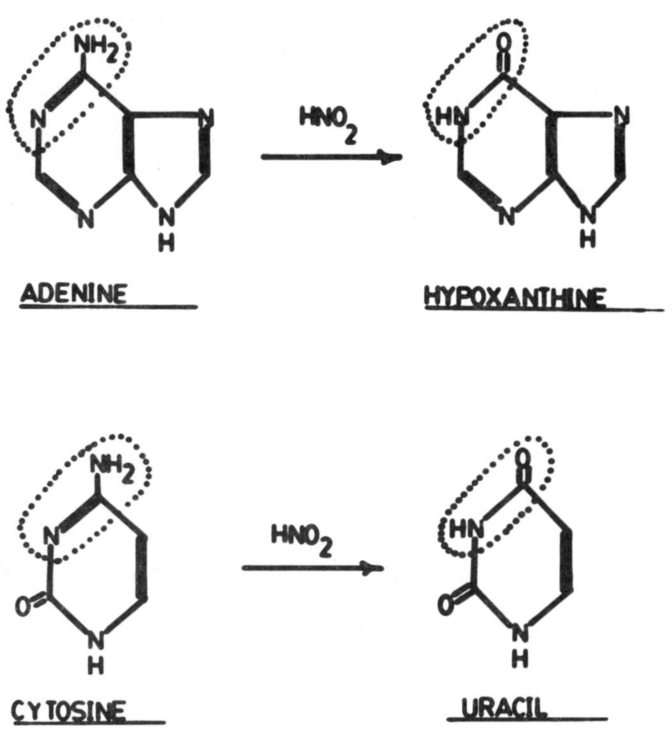

Fig. 6-3. Effect of nitrous acid (HNO_2) upon purines and pyrimidines. Hypoxanthine pairs with cytosine.

guanine. BrdU replaces some of the thymidine residues in DNA. Wherever such a replacement occurs, the original A-T pair is replaced by a G-C pair after two rounds of replication. This effect of BrdU will be "spotty", since the base can tautomerize back to the keto form which pairs with adenine. The overall effect is the production of a sectored cell population in which most of the cells will contain the normal gene and a small number will have the mutant allele. If the mutation is sustained by a gametic stem cell, there is a possibility that the mutation will be transmitted to the next generation. The actual probability that the mutation will be transmitted depends upon the size of the mutant cell sector. This sector is usually small, and many mutations are not transmitted to the next generation. The probability that two children of a parent sustaining a germinal mutation will inherit the mutant allele is very small; however, occasional families are encountered in which this rare event has apparently occurred.

Nitrous acid causes transitions by removing amino groups from purines and pyrimidines and replacing them with keto groups (Fig. 6-3). The end effect of these changes is the substitution of an A-T base pair with a G-C pair and a C-G base pair with an A-T pair (Fig. 6-4). Note that hypoxanthine pairs like guanine, and that replication is required for the base substitution to take place. Guanine is also modified by nitrous acid, being converted to xanthine. The latter base pairs like guanine, but creates a weak point in the double helix, since it forms two bonds with cytosine rather than the usual three. Note that the effects of nitrous acid are irregularly distributed, and that some of the susceptible bases remain unaffected.

Intercalating Agents and Frameshift Mutations

Proflavine is a three-ringed planar compound that inserts between the stacked bases of the double helix. The intercalation of the proflavine molecules results in the deletion or addition of bases during the following replication. The mechanism through which the intercalation occurs is unknown, but it apparently does not involve breakage of the sugar-phosphate backbone of the DNA. Removal or addition of bases in combinations other than multiples of three will shift the translational reading frame of the mRNA derived from the mutant allele, changing the sense of its coded message (Fig. 6-5). These **frameshift** mutants determine polypeptides which have a normal amino acid sequence up to the amino acid corresponding to the site of alteration. That amino acid and most of the subsequent amino acid residues will differ from those of the normal polypeptide. Occasionally, a frameshift mutation will generate a premature stop signal, resulting in a short version of the gene product. Alternatively, many frameshift mutations convert the normal translational stop to a triplet encoding an amino acid, and translation will continue until another stop is encountered in the mRNA. Many of these unusually large proteins are associated with abnormal function and/or stability.

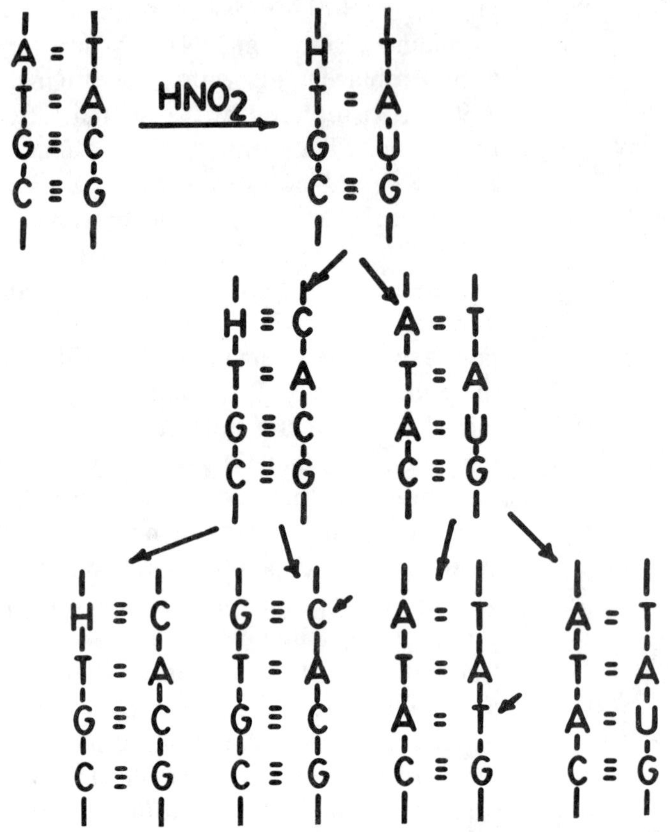

Fig. 6-4. Base pair substitutions induced by nitrous acid. H = hypoxanthine. Note that all bases susceptible to nitrous acid are not necessarily modified.

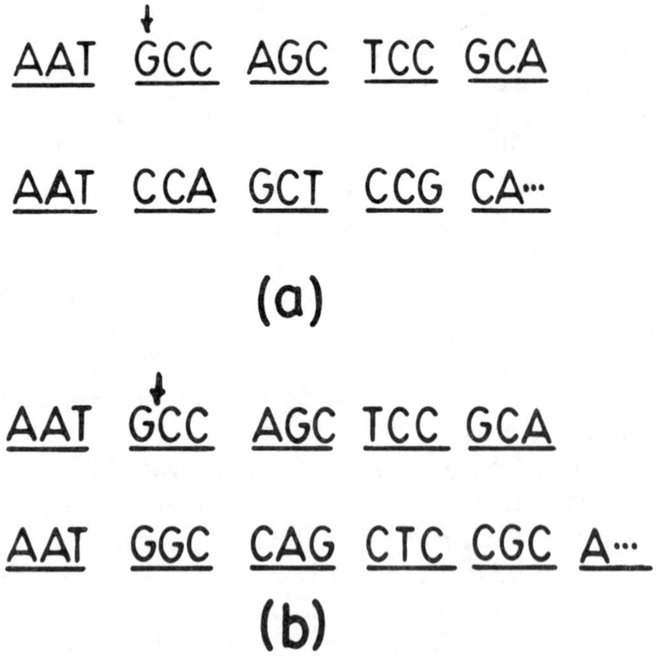

Fig. 6-5. Frameshift mutations. Deletion (a) or addition (b) of bases caused by certain mutagens such as proflavine distort the translational reading frame of the mRNA. Triplets are underlined, and arrows indicate the site of the added or deleted base.

214

Mutations Caused by Non-Ionizing Radiation

Ultraviolet irradiation induces an unusual base-pairing between adjacent pyrimidines (Fig. 6-6). For example, adjacent thymines will form a dimer connected by a four carbon (cyclobutane) ring.

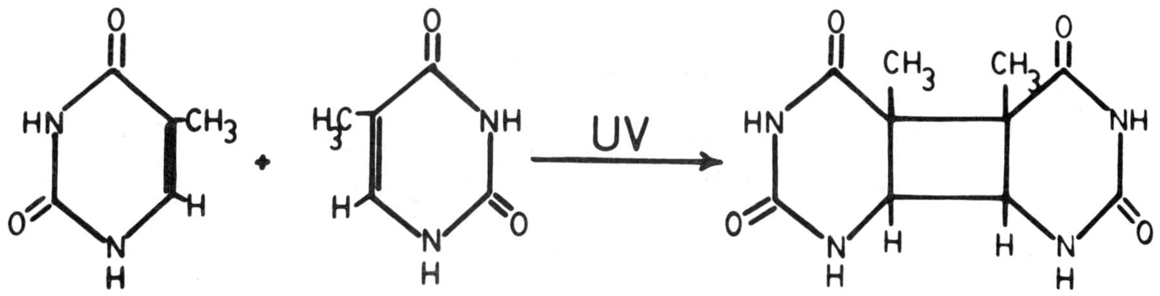

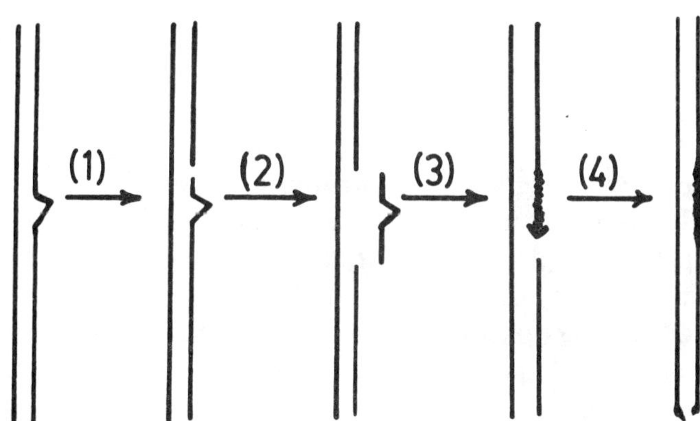

Fig. 6-6. Formation of cyclobutane ring between adjacent thymine residues following ultraviolet radiation.

This structure produces a kink in the DNA double helix, interfering with normal replication and transcription. Our cells are usually protected from harmful effects of ultraviolet irradiation by a repair system that recognizes and corrects the molecular lesions produced by UV. This type of repair is illustrated in Fig. 6-7. The UV-damaged DNA is cut on either side by an enzyme known as an endonuclease, and the damaged segment is removed from the chromosome by an exonuclease. New

Fig. 6-7. Excision repair of ultraviolet-damaged DNA. The damaged DNA (thymine dimers) is indicated by a kink in the chain. 1) Endonuclease cleavage of strand; 2) excision of dimers; 3) replication of DNA by polymerase; 4) ligation of new to old DNA.

DNA is synthesized by a DNA polymerase using the intact strand as a template, and the new DNA is linked to the ends of the old strand by a ligase (joining enzyme).

This repair process eliminates a majority of the ultraviolet damage from sun-exposed skin cells of normal individuals; however, continued exposure of skin that is not adequately protected by sun screens to intense sunlight can lead to the gradual accumulation of UV-induced mutations and cancer later in life. Rarely, persons are encountered who have inherited lesions in their excision repair system (Table 6-1). Xeroderma pigmentosa (XP) patients cannot effectively repair UV-damaged DNA, as a

Table 6-1. Xeroderma Pigmentosa Variants (1).

Complementation Group	Clinical Phenotype	Residual Repair (%)
A	Neurological	2-5
B	Neurological	3-7
C	Classical	10-20
D	Neurological	25-50
E	Classical	40-50
F	Classical	18
G	Neurological	<2
H	Neurological	30
I	Neurological	15-40

consequence of mutations affecting excision repair. Two groups of patients are observed: one in which there is an exquisite sensitivity to sunlight leading to skin and eye degeneration and skin cancer, and a second in which these problems are present in combination with neurological deterioration. Each group is heterogeneous, with each subtype possessing a different mutation affecting excision repair. The fact that these mutations involve different genes is supported by nuclear complementation studies (Fig. 6-8). Cells from two patients are fused to form dikaryons (cells with two nuclei). When the patients from which the different nuclei are derived have mutations affecting **non-allelic genes**, such as those encoding the enzymes catalyzing steps (1) and (2) in excision repair, the nuclei **complement** one another. The nucleus from patient (1) is homozygous for a XP mutation preventing reaction (1), but has an intact gene encoding enzyme (2). Conversely, the nucleus derived from patient (2) has an intact gene encoding the enzyme for reaction (1), but is homozygous for a XP mutation preventing reaction (2). The normal mRNAs from the two nuclei are translated in the dikaryon to form the respective enzymes, and repair occurs normally. By contrast, two nuclei derived from patients having XP mutations in the **same** gene **would not complement** in this system. XP patients must avoid strong sunlight and other sources of UV irradiation to prevent skin cancer. Some XP mutations have been traced to New England colonists. They wore protective robes and avoided daylight. Some evidence suggests these people may have been

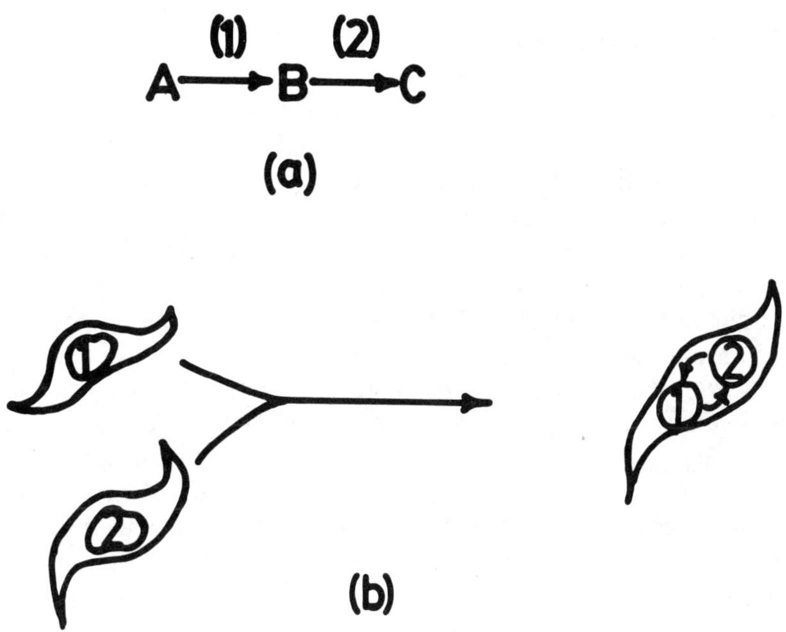

Fig. 6-8. Complementation test demonstrating that cells from two patients with XP are homozygous for mutations in non-allelic genes encoding enzymes catalyzing different steps in excision repair (a). When cells from patient (1) (lack enxyme 1) are fused with cells from patient (2) (lack enzyme 2), the nuclei complement one another. Nuclei from patient (1) produce normal mRNA for enzyme (2) and vice versa, leading to normal excision repair in the dikaryon (b).

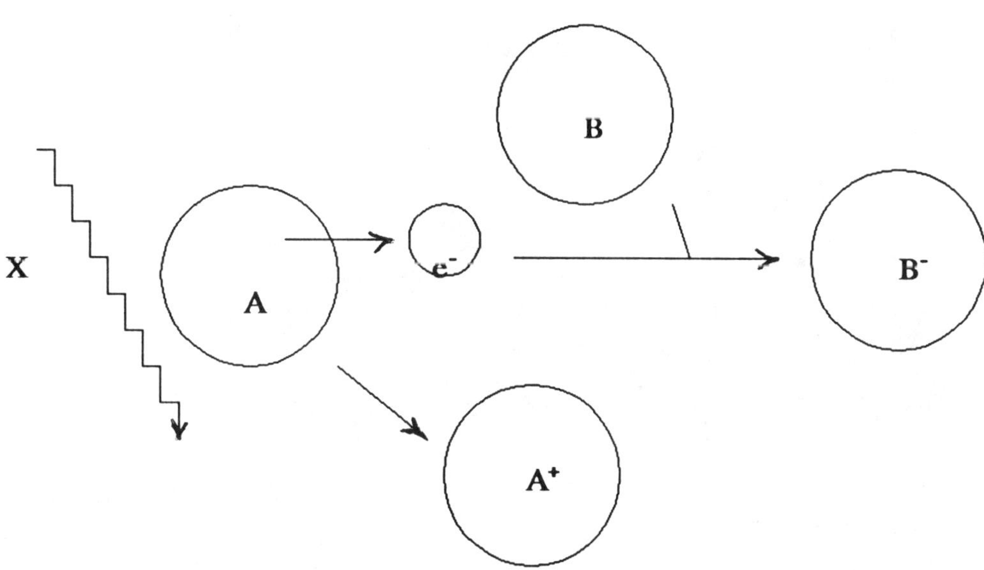

Fig. 6-9. Molecular effects of ionizing radiation. X-ray (X) strikes or passes near molecule A causing ejection of an electron and production of a positive ion. The electron may cause ejection of a second electron from another molecule or may be captured by molecule B producing a negative ion.

mistaken for witches because of their nocturnal behavior

Mechanism for Induction of Mutations by Ionizing Radiation

Ionizing radiations such as X-rays and gamma rays exert their effects by either forming ion pairs by ejecting electrons from molecules (Fig. 6-9) or by exciting electrons and promoting instability of chemical bonds. Excitation of electrons participating in chemical bonds makes the bonds vulnerable to breakage and increases the likelihood that molecules containing them will interact with neighboring molecules. Ionizing radiations have different energies, markedly affecting their abilities to penetrate tissues. Alpha particles have a mass approximating helium nuclei, have relatively low energies, and rarely penetrate far beyond the surface of tissues. However, the relatively large size of these particles leads to a high density of ionizations along a short path. Most alpha particles can be stopped by a sheet of paper, and materials emitting this form of radiation are not much of a problem unless they are inhaled, ingested, or accidentally spilled on the skin. β-rays, which include X-rays, possess a wide range of energies and correspondingly exhibit considerable variation in their penetration properties. Ionizations tend to cluster near the ends of their paths of tissue entry. Gamma rays have high energies and are particularly troublesome with respect to the damage they cause to internal organs and to the fetus of a pregnant woman. Like β-rays, most of the ionizations caused by gamma irradiation clusters near the end of the path of entry. Cosmic rays have very high energies and rarely cause mutations.

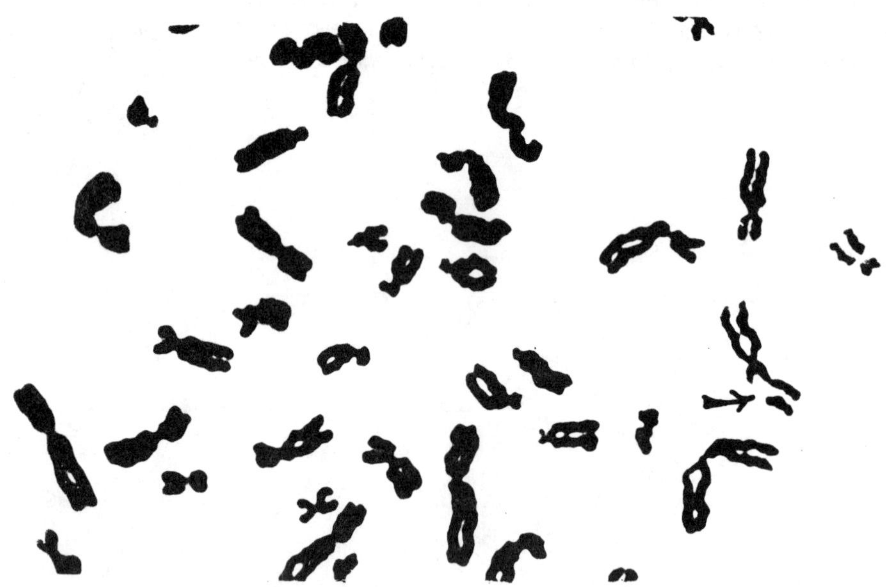

Fig. 6-10. Chromosome break (arrow).

One of the most common visible effects of ionizing radiation is chromosome breakage (Fig. 6-10). Chromosome breaks may rejoin in their original configuration. Alternatively, the segment distal to the break may be lost, and the genes contained by the segment will be absent from all of the cells descended from the cell in which the break occurred. Multiple chromosome breaks may lead to inversions, ring chromosomes, or translocations. These chromosomal rearrangements were discussed in

Chapter 4.

Gene mutation does not require direct hits by the ionizing radiation upon DNA. The radiation may perturb other molecules which secondarily act upon DNA to cause mutation. Water is the most abundant molecule in the cell and can be converted to hydrogen peroxide by high energy particles. These peroxides are highly reactive, participate in chemical modification of DNA, and cause strand breakage.

The amount of radiation exposure and the type of cell involved determine the extent of genetic damage. Radiation is measured in several ways. The **roentgen** (r) measures radiation exposure. The **rad** measures absorbed dose of radiation. One rad corresponds to 100 ergs of energy per gram of tissue, and is capable of producing 1.8×10^{12} ion pairs per cubic centimeter of tissue. The actual number of ion pairs produced is dependent upon the type of radiation and the characteristics of the tissue receiving the irradiation. Extensive experimentation has demonstrated the following:

1. **The amount of genetic damage is proportional to the dose of radiation received.**

2. **The production of chromosomal rearrangements is proportional to the power of the dose (e.g. production of rearrangements requiring two breaks (translocations, inversions, rings) increases with the square of the dose).**

3. **There is no safe threshold of radiation exposure below which mutation will not occur.**

Mammalian cells can repair radiation damage if given adequate time. For example, a dose rate of 90r per minute produced more mutations than a comparable cumulative exposure administered at a dose rate of 90r per week (2). Tissues comprised of actively dividing cells are more susceptible than tissues composed of non-dividing cells to mutations induced by ionizing radiation and other mutagens. For example, radiation exposure of embryonic or fetal tissues can have drastic consequences because most of these cells are dividing rapidly, and the mutation is consolidated by replication before the damaged DNA can be repaired. By contrast, many adult tissues are more resistant to mutagens because they rarely divide, and the repair systems have ample opportunity to replace the damaged DNA. Exceptions include adult lymphoid cells and cells lining the gastrointestinal tract. These cells retain high division rates, and are highly suceptible to radiation-induced mutation as well as mutations caused by chemical mutagens.

MOLECULAR CONSEQUENCES OF MUTATION

The effects of mutation upon gene products are schematically represented in Table 6-2. Base substitutions produce comparatively minor changes in the gene which may or may not result in an appreciable phenotypic effect. The degeneracy of the genetic code is protective in the sense that an amino acid is often encoded by more than one codon. Base substitutions involving the third base of a triplet often generate synonymous codons instructing incorporation of the same amino acid in the polypeptide. Furthermore, base substitutions may result in the substitution of "equivalent" amino acids in the polypeptide encoded by the gene sustaining the mutation. For example, substitution of an alanine for a glycine may not have much of an effect upon the function of the protein because the two amino acids are similar in size and have similar properties. By contrast, substitution of a polar for a nonpolar amino acid located in the hydrophobic core of a protein may seriously alter its conformation and promote instability, resulting in protein deficiency. Other types of drastic changes include the replacement of small amino acids by bulky amino acids, replacement of hydrophilic amino acids on the protein surface by hydrophobic amino acids, and substitution of cysteine for other amino acids. Cysteine forms disulfide bonds with other cysteine residues, producing an unnatural bend in the structure of the protein. Examples of these changes and their phenotypic effects are presented in Chapter 7.

Base substitution may have more profound effects upon polypeptides. The codons for several amino acids

Table 6-2. Schematic Representation of Effects of Various Mutations Upon a Gene and Its Product.

Type of Change	Sequence	Example
None	The Big Old Dog Was His	Hb A$_1$
Base Substitution	The B**ag** Old Dog Was His	Hb S
Triplet Deletion	The Old Dog Was His	Hb Freiburg
Triplet Insertion	The Big **Bad** Old Dog Was His	Hb Grady
Duplication	The Big Old Dog Was His The Big Old Dog Was His	Haptoglobin-2
Inversion	The Dlo Gib Dog Was His	
Frameshift	The Bi*o Ldd Ogw Ash Is-	Hb Wayne

Hb = Hemoglobin; * = site of nucleotide deletion.

differ from chain terminators by one nucleotide. Substitutions of a key nucleotide in one of these codons may lead to the formation of a premature terminator and the synthesis of a truncated protein or no protein at all. Alternatively, substitution of certain nucleotides in the normal termination codon of a gene may result in its replacement by a codon specifying an amino acid. Translation of the mutant mRNA will continue until another terminator is encountered, producing an abnormally long polypeptide. In some cases these elongated proteins do not appear to cause significant problems; however, there are other examples where a clinical disease is associated with the presence of the elongated polypeptide. Hemoglobin Constant Spring contains an α-globin chain that has 172 amino acids rather than the normal 142. Glutamine occupies position 142 in Hemoglobin Constant Spring. The codons for glutamine are **CAA** and **CAG**. Both can be derived from the chain terminators **UAA** and **UAG** by a single base change. This abnormal α-globin is associated with deficiency of α-globin chains and anemia.

Insertion or deletion of triplets or multiples of three bases will produce polypeptides having the corresponding gain or loss of amino acids. Hemoglobin Grady contains a duplication of three amino acids in the α-globin chain (Glu-Phe-Thr at positions 116 through 118). Hemoglobin Freiburg lacks the valine residue normally found at position 23 of the β-globin chain. Amino acid sequences on the N-terminal and C-terminal sides of these alterations are normal.

By contrast, frameshift mutations arise when additions or deletions of one base or combinations of bases other than three or multiples of three occur. The consequences of this type of mutation were illustrated in Fig. 6-5. Frameshift mutations produce sequences which match those of the standard alleles up to the point of alteration. However, the base sequences of the remainder of the mutant alleles are out of phase with the remainder of the standard alleles. Several outcomes may be anticipated. 1) The amino acid sequence of the polypeptide encoded by the altered region will bear little resemblance to the normal polypeptide sequence. 2) A premature terminator may be generated, resulting in a truncated protein. 3) The DNA sequence of one or more splice junctions may be affected, or a "cryptic" splice junction may be uncovered leading to abnormal splicing of the mRNA. 4) The normal chain termination codon may be changed to an amino acid-encoding codon. Examples of frameshift mutants include Hemoglobins Wayne (172 amino acids), Cranston (157 amino acids), and Tak (157 amino acids). These three mutant hemoglobins differ from the previously described chain termination mutants (such as Hemoglobin Constant Spring) by virtue of their abnormal sequences prior to the position of the original chain termination codon.

α-like globin genes are clustered on chromosome 16, and β-like globin genes are clustered on chromosome 11. The organization of exons and the base sequences of genes within each cluster are very similar. Consequently, the globin genes within clusters on homologous chromosomes may mispair during meiosis. Recombination within mispaired genes can generate "fusion genes" which possess portions of the sequences of each of the genes from which they were derived (Fig. 6-11). Hemoglobin containing the non-alpha globin consisting of the N-terminal portion of the δ-globin gene and C-terminal segment of the β-globin gene is known as Hemoglobin Lepore, and several different types (including Hb Lepore[Boston], Hb Lepore[Hollandia], and Hb Lepore[Baltimore]) have been reported. Note that the

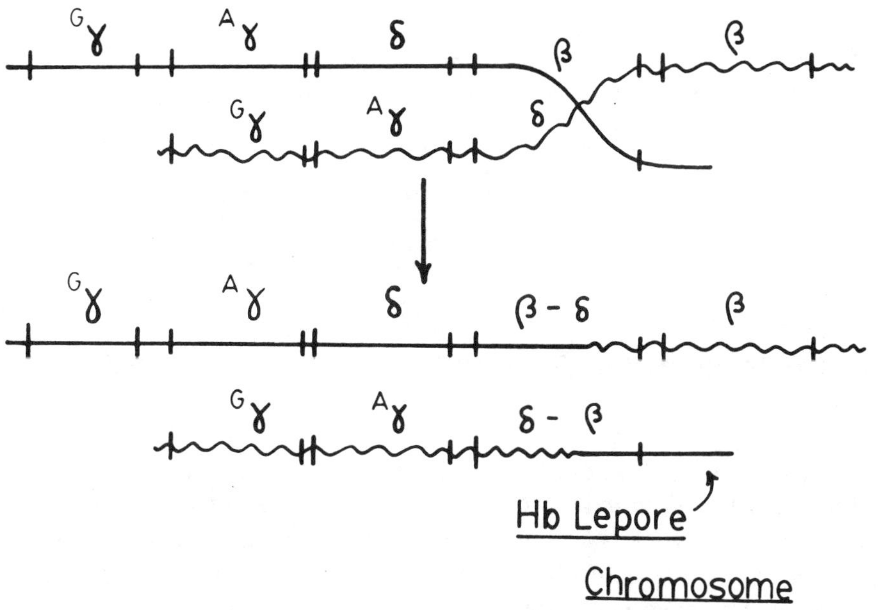

Fig. 6-11. Mispairing of the δ- and β-globin genes and intragenic recombination to form a fusion gene. The crossover produces two types of fusion genes. The $\delta\beta$-fusion gene encodes a globin that has an N-terminal segment corresponding to δ-globin , and a C-terminal sequence corresponding to β-globin. Hemoglobins encoded by the $\delta\beta$-fusion gene are known as Lepore hemoglobins.

chromosome carrying the δ-β fusion gene lacks normal δ- and β-globin genes. Persons who are heterozygous for this abnormal chromosome region develop symptoms characteristic of $\delta\beta$-thalassemia, in which both types of chains are deficient and are replaced by the abnormal Lepore globin. Homozygotes for the abbreviated chromosome are very rare. The reciprocal chromosome carrying intact δ- and β-globin genes and the $\beta\delta$-fusion gene (examples include Hb O^{Congo} and Hb Miyada) has also been observed in heterozygous state. Persons carrying this chromosome rarely have clinical problems.

Fusion gene products have also been found which contain portions of the $^{A}\gamma$- and β-globin genes. Hb Kenya possesses a non-alpha chain consisting of the N-terminal $^{A}\gamma$ sequence and C-terminal β-globin sequence. The Hb Kenya chromosome lacks normal $^{A}\gamma$-, δ-, and β-globin genes. Patients heterozygous for this chromosome will exhibit symptoms of $\gamma\delta\beta$-thalassemia, an anemia characterized by a reduction of all three chains.

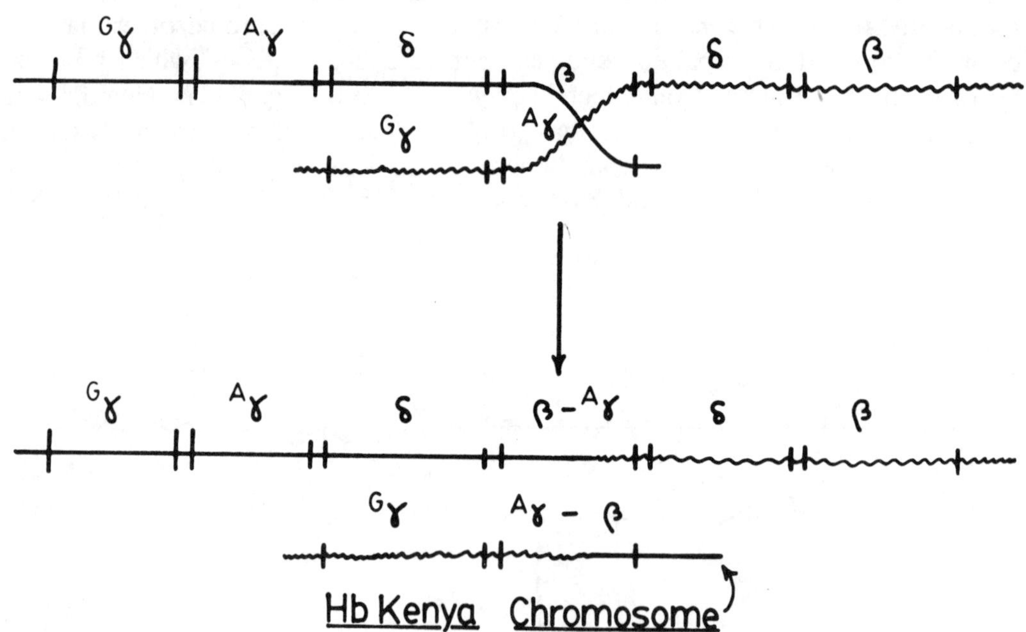

Fig. 6-12. Formation of the Hemoglobin Kenya fusion gene. The chromosome carrying the $^{A}\gamma$-β fusion gene lacks normal $^{A}\gamma$-, β- and δ-genes.

The α-like globin gene cluster on chromosome 16 includes two α-globin loci that determine identical polypeptides. The DNA sequences of these genes are almost identical, and misalignment may occur. Recombination **between** adjacent α-globin genes can generate chromosomes with one or three α-globin loci (Fig. 6-13). Both of these chromosome types have been observed, and the chromosome possessing the single α-globin locus has been found in certain α-thalassemics (anemia associated with deficient α-globin chains). This form of unequal, **intergenic** recombination may occur more than once, with one of the products lacking the α-globin locus. As discussed in Chapter 7, gene deletion secondary to unequal, intergenic recombination is believed responsible for a majority of cases of α-thalassemia in many areas of the world. The severity of α-thalassemia increases as the number of residual α-globin genes decreases.

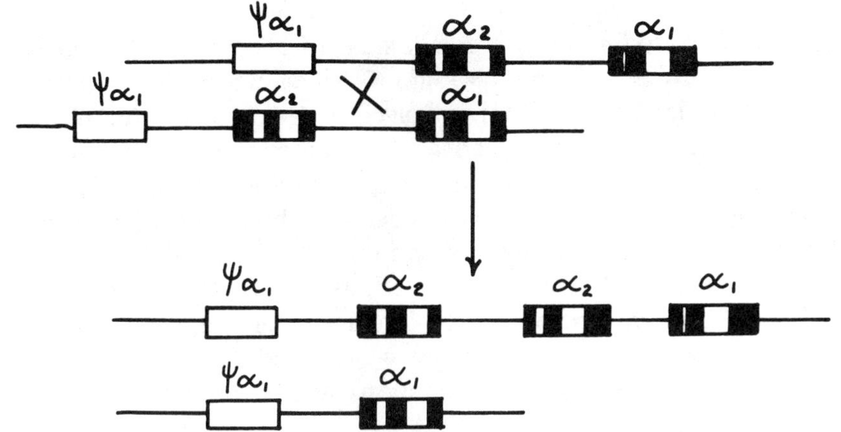

Fig. 6-13. Recombination between misaligned α-globin loci.

Mutational Hotspots

Mutations are generally viewed as random events. That is all sites within a gene are about equally likely to experience a change in nucleotide sequence. Although most data support this view, some regions of a gene may appear more likely to suffer mutation than others. Mutational "hotspots" in genes may occur for a number of different reasons. A major cause of a deviation of the pattern of **observed** mutations from the expected random pattern is that the functional consequences of a mutation can differ markedly from one intragenic site to another. For example, exons of a gene that encode surface amino acids that are not critically related to the function of an enzyme will appear to display a higher mutation rate than sites within an exon that determine the structure of the active site of an enzyme. Although mutations involving nucleotides within the exon encoding the active site may occur as often as those in the exon encoding non-essential surface amino acids, the active site mutations lead to the death of the cells in which they occur, and are not recovered. That is, mutation is generally random, but selective pressures against mutations that have lethal effects may lead to different observed rates of mutation in various regions of the same gene.

There are cases where the nucleotide sequence of a gene can render certain sites more mutable than others. CpG islands provide one example of a mutational hotspot. These islands are subject to methylation as described in Chapter 2. Deamination of a methylated cytosine leads to the formation of thymine, such that a CG pair is eventually replaced by a TA pair in the gene. Many reports have described mutations in CpG islands of genes that have lead to human diseases. Genes may be surrounded by runs of short repetitive sequences. These sequences can cause misalignment of alleles of that gene during meiosis and deletion of segments or all of the gene from one of the homologous chromosomes. The X-linked gene encoding the enzyme, steroid sulfatase, is surrounded by a number of these short repeats. More than 80% of patients with steroid sulfatase deficiency lack all or major portions of the **STS** gene.

PHYSIOLOGICAL CONSEQUENCES OF GENE MUTATION

The physiological consequences of gene mutations are dependent upon the site in the polypeptide affected by the change and by the relative importance of the role played by the gene product in metabolism carried on by the cell, cell structure, or other function related to cell survival. Genes are generally divided into three or more exons, each encoding a different domain of a protein. Although all amino acids of a protein contribute to its structure, some play more critical roles than others. Mutations that alter sites encoding amino acids which are intimately involved with any of the following may seriously compromise the function of the protein involved: 1) active site of an enzyme, 2) coenzyme binding site of an enzyme, 3) sites that interact with other proteins and/or lipids in a membrane, 3) sites participating in subunit interactions of a multimeric protein, 5) sites that serve as points of attachment for carbohydrates or other ligands, 6) targeting sequences (e.g. signal peptide), and 7) substitutions that change the overall structure of protein domains. The latter class of mutations include substitutions of hydrophobic amino acids in the core of a protein by hydrophilic amino acids, the replacement of hydrophilic surface amino acids by hydrophobic amino acids, and changes which distort the structure of helical regions of the protein. Specific examples of these types of mutations will be provided in later chapters.

Mutant proteins may be synthesized at a slower rate than the respective normal proteins. For example, $\beta^S\beta^A$ heterozygotes produce β^A- and β^S-globin chains in a ratio approximating 60:40. The disparity is also observable at the mRNA level; however, it is not known why fewer translatable copies of β^S mRNA are present. Evidence derived from studies of Hb Lepore suggest that control elements near the mutant genes are not responsible for the slower rate of mutant globin synthesis. Normal normoblasts synthesize β- and δ-globin chains in a ratio of about 97:3. The β^{Lepore} allele possesses the promoter of the δ-globin gene. Therefore, Hb Lepore would be expected to account for about 1.5% (1/2 x 3%) of the hemoglobin synthesized by heterozygotes. Surprisingly, Hb Lepore accounts for ten-fold more of the hemoglobin produced by heterozygotes than anticipated. The $\alpha^{Constant\ Spring}$ allele and its mRNA differ from the α^A allele and its mRNA by a single nucleotide. Hb Constant Spring is synthesized at a relatively slow rate compared to Hb A_1. Either the $\alpha^{Constant\ Spring}$ mRNA is translated more slowly, or the $\alpha^{Constant\ Spring}$-globin is more unstable than the normal globin chain.

ESTIMATION OF MUTATION RATES

Mutation rates are usually expressed as the number of mutations occurring in a gene per gamete per generation. Mutation rates are most readily estimated for autosomal dominant traits that have high penetrance, are readily diagnosed, are not lethal before the reproductive period, and are not often mimicked by environmentally caused conditions. Direct estimates of mutation rates can be obtained by screening a population for affected persons who are born to normal-appearing parents. Suppose 50 out of a total of 5,000,000 children born to normal parents were subsequently found to develop an autosomal dominant trait. The 5,000,000 children were derived from the fusion of 10,000,000 gametes. Therefore, the mutation rate may be calculated:

$$\text{Mutation rate} = \mu = (\text{Number of Mutations}) \div (\text{Total Number of Gametes})$$

$$= (50) \div (2(5,000,000))$$

$$= 5 \times 10^{-6} \text{ mutations/gamete/generation}$$

If the dominant trait is not fully penetrant, the calculation can be corrected accordingly. Suppose the penetrance in the preceding example was 80% rather than the 100% previously assumed. The missed cases of the trait can be included in the estimate of the mutation rate:

$$\mu = (\text{number of mutations} \div \text{penetrance}) \div \text{total number of gametes}$$

$$= (50 \div 0.8) \div (10,000,000) = 6.25 \times 10^{-6}$$

A representative series of mutation rates is presented in Table 6-3. Most of the estimates fall between 1 and 100 per locus per million gametes per generation. Considerable variation is evident among the numbers of mutations occurring at different loci. Furthermore, many of the estimates are considerably higher than the average rate (1×10^{-6}) for loci encoding structural proteins and enzymes derived from diverse species. Several factors are most likely responsible for the unusually high human

Table 6-3. Representative Mutation Rates for Autsomal Dominant and X-linked Recessive Traits (3).

Trait	Population	Mutation Rate
Autosomal Dominant		
Achondroplasia	Denmark	1×10^{-5}
	Germany	$6\text{-}9 \times 10^{-6}$
Retinoblastoma	England	
	Michigan	
	Switzerland	
	Germany	$6\text{-}7 \times 10^{-6}$
	Japan	8×10^{-6}
Neurofibromatosis	Michigan	1×10^{-4}
	Moscow	$4.4\text{-}4.9 \times 10^{-5}$
Intestinal Polyposis	Michigan	1.3×10^{-5}
Polycystic Kidney Disease	Denmark	$6.5\text{-}12 \times 10^{-5}$
X-linked Recessive		
Hemophilia A	Germany	5.7×10^{-5}
	Finland	3.2×10^{-5}
Duchenne Muscular Dystrophy	Utah	9.5×10^{-5}
	Wisconsin	9.2×10^{-5}
	Great Britain	$4.3\text{-}10.5 \times 10^{-5}$

mutation rates. Loci which mutate at a high rate are more likely to be included in a study of mutation rates than are loci which mutate at low rates. This sampling bias will inflate the overall mutation rate. Inclusion of **phenocopies**, environmentally caused cases of a trait that is usually inherited, will also inflate the mutation rate. Phenocopies may resemble the genetic disorder very closely, preventing their exclusion from the sample, particularly if the environmental causative agent is not recognized. Heterogeneity of a trait may also contribute to unusually high estimates of mutations rates. For example, two different genes are now known to be involved in the production of neurofibromatosis. One form can be distinguished from the other by the occurrence of acoustic neuromas in affected persons, while the second gene mutation is not usually associated with acoustic neuromas. Mutations at both loci may have been included in the studies entered in Table 6-3. This type of error is especially likely when the molecular basis of the trait is unknown, as is the case for three of the conditions listed in the table. The high mutation rate for Duchenne muscular dystrophy can be partially attributed to the large "target size" provided by the gene. The dystrophin gene encompasses more than two megabase pairs, and the probability that one or more of these bases may be damaged by a mutagen is much higher than that for an average-sized gene. It is also noteworthy that Becker muscular dystrophy is now known to be caused by mutations in the dystrophin gene. Becker muscular dystrophy was not included with Duchenne muscular dystrophy when the studies listed in Table 6-3 were conducted. This exclusion would be expected to lead to an underestimate of the mutations occurring at the dystrophin locus. The relatively long generation time of humans compared to Drosophila and mouse can also contribute to relatively high human mutation rates compared to those observed for these two species. The prolonged exposure of human cells to mutagens would be anticipated to result in relatively more mutations per locus.

Mutation rates can also be estimated by indirect methods. These estimates are subject to the same biases as the direct estimates; however, they are more appropriate for determining mutation rates for autosomal recessive and X-linked recessive traits. Application of indirect methods will be discussed in a later chapter.

MUTATION AND CANCER

Many **carcinogens** (cancer-causing agents) are known to be mutagens, suggesting a strong link between mutation and cancer. Cancer has diverse origins (Fig. 6-14). Certain cancers show little familial aggregation and are believed to be largely environmental. The majority of cancers display moderately increased frequencies among relatives; however, the patterns of inheritance are not supportive for Mendelian transmission. Breast cancer provides an example of this type of environmental-genetic cancer, which accounts for the majority of human cancers. Environmental agents are implicated in the etiology of these cancers by demographic studies. Breast cancer rates are low among Asian women compared to Caucasian women. However, the difference in rates disappears within two generations following migration of Asian women to the United States, suggesting that one or more carcinogens that are less common in Asia are more prevalent in the United States. Several other types of environmental-genetic cancers have been associated with lifestyles of people. For example, gastric cancer is more common among Japanese than among U.S. Caucasians. The difference has been traced to the manner in which Japanese prepare certain foods. Use of tobacco products has also been linked to cancers. Oral and lung cancers are markedly increased among people who smoke, use snuff, or chew tobacco. Some smokers are at an especially high risk for lung and other cancers as a consequence of their heritage. Aryl-hydrocarbon hydroxylase (AHH) is a microsomal enzyme that participates in the detoxification of pro-carcinogens present in cigar and cigarette smoke. Persons who inherit the capability to synthesize large amounts of this enzyme or who produce isozymes with unusually high activities will produce high levels of carcinogenic intermediates , increasing their cancer risk relative to people who have normal AHH activities. Genetic determination of AHH levels is complex, partially as a consequence of several isozymes possessing this activity and partially as a consequence of different levels of regulation.

Genetic cancers account for about 10 percent of the total, generally have early onset, and tend to be uniformly distributed among different populations. Well-characterized examples include retinoblastoma, xeroderma pigmentosa, and Wilms tumor. In each case, individuals have inherited a germline mutation that markedly increases their cancer risk. Mutations are required to transform normal cells into cancer cells (4). Therefore, cancer development is dependent upon repeated exposure to

	GENETIC	CARCINOGEN
GENETIC	+	–
ENVIRONMENTAL		
(1)	+	+
(2)	–	+
SPORADIC	–	–

Fig. 6-14. **Origin of Cancer Development.**

carcinogens that induce mutations. Furthermore, liability to mutation plays a significant role in cancer development. For example, human cells possess systems that repair damaged DNA, protecting the cell against mutation. Inherited defects in repair systems, such as that described for xeroderma pigmentosa (XP), increase liability for mutation and cancer development. Persons homozygous for a XP mutation are said to be **genetically predisposed.** Exposure to ultraviolet light, a carcinogen, at high doses and for long periods will eventually promote the occurrence of skin cancer on exposed surfaces of most people. However, XP individuals lack systems for the repair of UV damage and are exquisitely sensitive to even mild to moderate UV irradiation. Other genetic predispositions for cancer which are related to defective repair of damaged DNA include Fanconi's anemia, Bloom's syndrome, and ataxia telangiectasia. Interactions of carcinogens with genetic predisposition can be used as a paradigm to explain the spectrum of cancer etiology. The diverse environmental agents known to contribute to cancer development share a common link: most are capable of causing mutation. In the case of environmental cancers, the environmental agents are relatively common, and genetic predisposition is essentially nonexistent. Environmental-genetic cancers result from interaction of common environmental agents with a relatively common and perhaps complex genetic predisposition. Genetic cancers represent interaction of a relatively rare, usually Mendelian, genetic predisposition with an environmental carcinogen.

Clonal Origin of Cancer

Many cancers appear to arise from single cells. Typing of certain tumor cells arising in women who were heterozygous for the A and B isozymes of glucose-6-phosphate dehydrogenase (G6PD) revealed the exclusive occurrence of G6PD-A or G6PD-B in the respective tumors. Both isozymes were never observed in cells from the same tumor. The G6PD locus is X-linked and subject to X-inactivation. Therefore, the tumor cells must have arisen from a single cell in which X-inactivation had committed the cell to synthesizing either the A or the B isozyme. A schematic diagram of the process believed to transform a normal cell to a malignant cell is presented in Figs. 6-15 and 6-16. The damage produced by a carcinogen may be repaired or may persist as a chromosome break, gene deletion, base substitution, or other mutation. Repair of damage is favored in slowly dividing cells. However, damage may be consolidated as a mutation in rapidly dividing cells before repair can be completed. Once the mutation has occurred, it will be inherited by all progeny cells, and these cells are now predisposed toward tumor development. At this stage, the predisposed cells appear normal both morphologically and biochemically. If one of the predisposed cells is exposed to the same or another carcinogen, the damage incurred may produce a second mutation which subsequently transforms the cell into a tumor cell. Benign tumor cells are morphologically and biochemically distinguishable from thier normal cell counterparts and are typically slow growing and non-invasive. However, occurrence of one or more additional mutations will convert

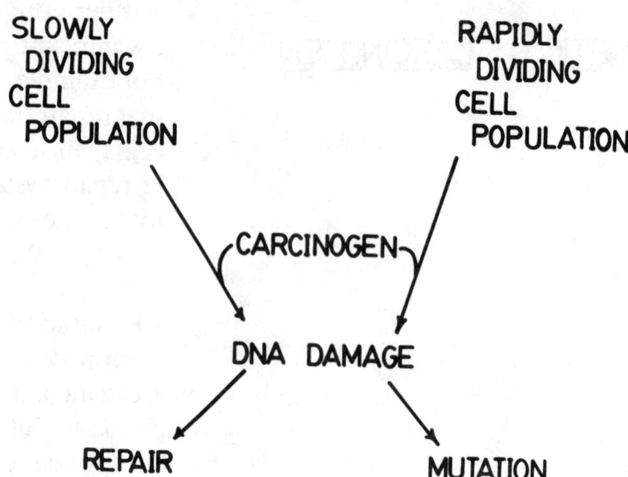

Fig. 6-15. Fates of cells exposed to carcinogens.

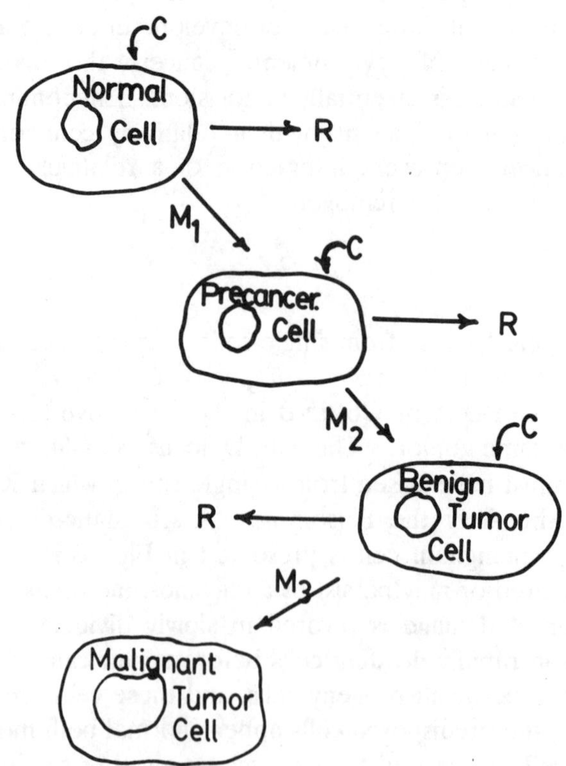

Fig. 6-16. Transformation of a normal cell to a malignant tumor cell. C=carcinogen; M=mutation; R=repair.

228

the benign tumor cell to a highly invasive and lethal malignant tumor.

This pathway of cancer evolution has developed from studies of childhood cancers: retinoblastoma, neuroblastoma, and Wilms tumor. Retinoblastoma occurs in both unilateral and bilateral forms. The unilateral tumors usually are sporadic but occasionally may be associated with a family history of the tumor. Bilateral cases of retinoblastoma, accounting for about 40% of all cases, are frequently inherited as an autosomal dominant trait with about 80 percent penetrance. This cancer pattern has been explained by proposing that families afflicted with the hereditary cancer are segregating for a germline mutation predisposing toward retinoblastoma (first mutation in Fig. 6-16). Some of these patients appeared to have a small deletion of chromosome 13, while others were suspected of having a point mutation within the same region of chromosome 13. Actual transformation of predisposed retinal cells to retinoblastoma cells required a second mutation, presumably in the same region. By contrast, sporadic cases of retinoblastoma were hypothesized to occur in persons who lacked the germline mutation. Therefore, their retinal cells had to undergo two mutations within the same region to become transformed to retinoblastoma cells. Occurrence of two mutations is far less likely than occurrence of one, leading to a considerably enhanced tumor risk in predisposed individuals (Table 6-4). Hereditary

Table 6-4. Relative Risk for Retinoblastoma in Genetically Predisposed and Non-predisposed Persons(4).

Subject	Risk
Normal Child	3×10^{-5}
Predisposed Child	3-4 tumors/child
Relative Risk	$3 \div (3 \times 10^{-5}) = 10^{5}$

and sporadic retinoblastoma are compared in Fig. 6-17.

Considerable progress has been made since Knudson's theory (4) relating the occurrence of two or more mutations with the evolution of cancer cells from normal cells. Studies of translocations involving chromosome 13 narrowed the region responsible for predisposition to retinoblastoma to band 13q14.2, and the gene has been named **RB1**. Other work has demonstrated that **RB1** was deleted from both chromosomes in a majority of retinoblastomas, but at least one intact **RB1** allele was present in normal cells from the same individual. Persons who were members of retinoblastoma kindreds often had one of their **RB1** alleles deleted in their normal cells, while persons with retinoblastoma but no family history of the tumor had two intact **RB1** genes. This pattern provided support for Knudson's hypothesis, and implicated the **RB1** gene as a candidate gene for an **anti-cancer** gene that was compromised by mutation in retinoblastoma. Proteins encoded by anti-cancer genes play regulatory roles in normal cells which prevent those cells from transforming into cancer cells. Mutations (deletions, base substitutions, etc.) which incapacitate **both** alleles of an anti-cancer gene promote conversion of the normal cell to a cancer cell. In the case of retinal cells, **RB1** plays a pivotal role in preventing retinal cells from entering mitosis. Therefore, only two "hits", one in each **RB1** allele, are required for development of retinoblastoma.

The **RB1** gene has been isolated and characterized. It is transcribed into a 4.7 kb mRNA that is translated into a protein known as p105-RB. This protein is expressed in all normal tissues, and is absent from retinoblastoma cells. p105-RB appears to prevent non-dividing cells from entering mitosis.

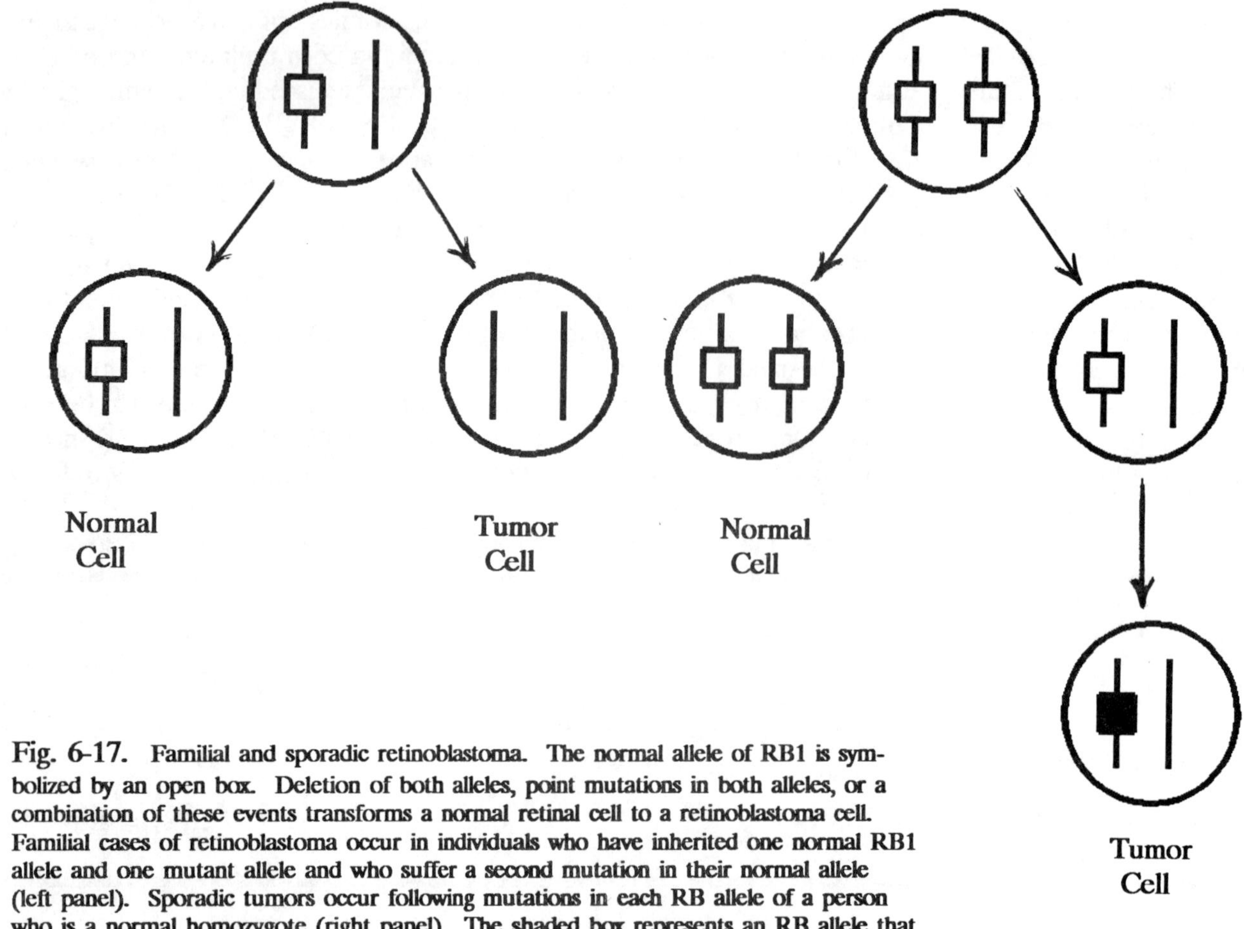

Normal
Cell

Tumor
Cell

Normal
Cell

Tumor
Cell

Fig. 6-17. Familial and sporadic retinoblastoma. The normal allele of RB1 is symbolized by an open box. Deletion of both alleles, point mutations in both alleles, or a combination of these events transforms a normal retinal cell to a retinoblastoma cell. Familial cases of retinoblastoma occur in individuals who have inherited one normal RB1 allele and one mutant allele and who suffer a second mutation in their normal allele (left panel). Sporadic tumors occur following mutations in each RB allele of a person who is a normal homozygote (right panel). The shaded box represents an RB allele that sustained a point mutation, such as a base substitution, preventing its expression.

Several viruses are capable of transforming certain cells to cancer cells. Virus-encoded proteins, including E1A protein (adenovirus type 5), Large T-antigen (SV40), and E7 protein (human papillomavirus type 16), participate in the transformation process. Studies have shown that these proteins form complexes with p105-RB, inducing DNA replication and cell division of the infected cells. Furthermore, p105-RB appears to be unphosphorylated in G_0 and G_1 cells, and is phosphorylated just prior to entry of the cells into S-phase (5,6). At the end of G_2, a phosphatase desposphorylates p105-RB. This cyclic phosphorylation and dephosphorylation of p105-RB and its incapacitation by the viral proteins are illustrated in Fig. 6-18.

p105-RB occurs in essentially all cells, and mutations in **RB1** are also observed in other tumors, including osteosarcoma (which occurs more frequently in patients who have had retinoblastoma) and breast cancer. However, mutations affecting other anti-cancer genes are found in these tumors as well, suggesting that p105-RB may be one of several proteins that are essential for prevention of DNA replication in these cells. At least one other protein has been identified that also complexes with E1A in certain cultured cells (7). E1A must bind to both p105-RB and p300 in these cells to induce DNA synthesis. This result indicates that regulation of the entry of cells into cell division may be more complicated in some cells than in others, and that several anti-cancer genes may have to be incapacitated

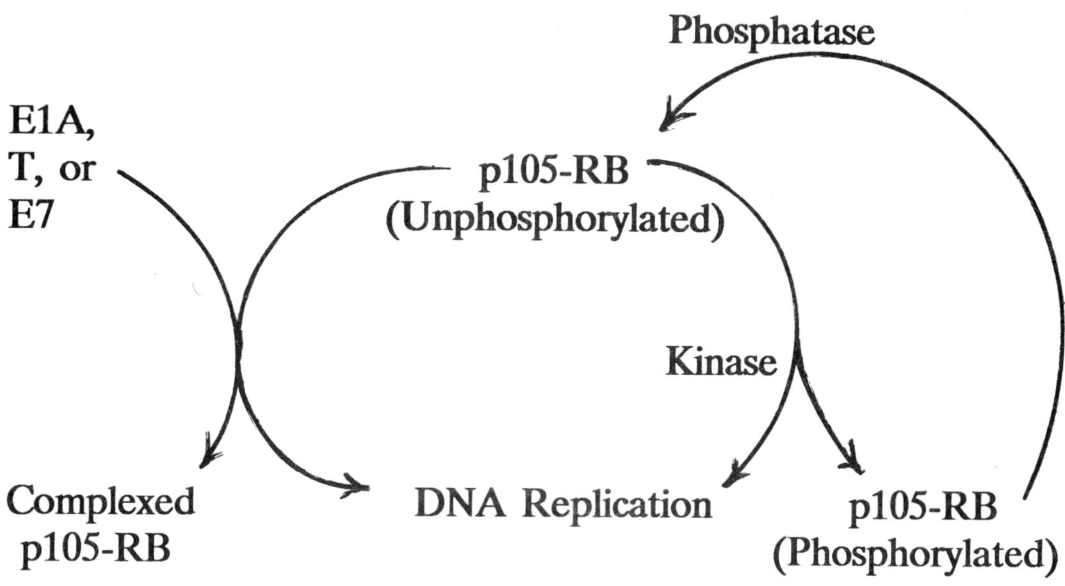

Fig. 6-18. Participation of p105-RB in prevention of DNA synthesis and entry into mitosis. p105-RB undergoes a cyclic phosphorylation and dephosphorylation. The phosphorylation of p105-RB near the end of G_1 is followed by initiation of DNA synthesis. Dephosphorylation of p105-RB following G_2 results in suppression of DNA synthesis in G_0 and G_1 cells. Virus-encoded proteins (E1A, T, and E7) can promote DNA synthesis by complexing with p105-RB, preventing its action.

by mutations before tumors will be induced in these tissues.

Several other anti-cancer genes have been identified, and the proteins they encode are currently being characterized. Table 6-5 summarizes what is known about some of these loci. Wilms' Tumor, a cancer of the kidney, resembles retinoblastoma with respect to the occurrence of both familial and sporadic forms. Three different genes may be involved in tumor development. The best studied example, **WT1**, appears to encode a zinc-finger protein of unknown function. This gene does not appear to be mutated in familial cases. **WT1** is expressed during both renal and gonadal development, and the failure of its apparent involvement in familial Wilms' Tumor may be related to its gonadal expression. Persons experiencing a **WT1** mutation might be rendered infertile. A second locus, currently unnamed, has been mapped to 11p15. The 11p15 locus is also intact in familial cases of Wilms' Tumor, implying the existence of the third locus which actually predisposes toward the tumor. The **NF-1** locus encodes a large protein exceeding 2400 amino acids in length whose function is not known. The structure of the NF-1 protein suggests that it may act as an activator for an enzyme, GTPase. This form of neurofibromatosis affects about 1/3500 people. Some carriers of the mutated **NF1** gene do not develop neurofibromas, but have five or more coffee-colored spots on their skin (cafe-au-lait spots). The **P53** gene is expressed in many cell types and has been found to be involved in several tumors, some of which are listed in Table 6-5. A rare syndrome, the Li-Fraumeni syndrome, is characterized by a dominant predisposition to these tumors, and affected members of these kindreds possess a mutated **P53** gene. The second mutation results in development of the cancers. **P53** encodes a protein, p53, which has some properties that suggest it may be a transcriptional activator. Perhaps p53 prevents entry of cells into mitosis by activating other genes that suppress mitosis. Phosphorylation of p53, complexing of p53 with viral proteins, or deletion or other mutational inactivation of p53 may release this suppression, resulting in tumor development. The **FAP** gene is associated with a dominantly inherited condition, familial

Table 6-5. Anti-cancer genes.

Gene	Site	Product	Inheritance	Tumors
RB1	13q14.2	p105-RB	Recessive*	Retinoblastoma;Osteosarcoma;Carcinoma of Breast, Bladder, Lung
P53	17q12-13.3	p53	Recessive*	Astrocytoma;Osteosarcoma;Carcinoma of Breast, Lung, Colon
WT1	11p13			Wilms' Tumor (Kidney)**
NF1	17q11.2		Dominant	Neurofibromatosis Type I
DCC	18q21		Recessive	Colon Carcinoma
FAP	5q21-22		Recessive*	Colon Carcinoma
MEN-I	11q13		Dominant	Multiple Endocrine Tumors

*Predisposition to tumor may be inherited as a dominant; however, tumor development requires mutations in both alleles.
**At least three genes may be involved in Wilms' Tumor.

adenomatous polyps of the colon. These polyps evolve into carcinomas later in life, apparently as a result of a mutation in the other **FAP** allele. Early diagnosis and surgical removal of the polyps prevents the later carcinomas.

Common adult tumors appear to require several mutations during the transformation from normal to tumor cells. For example, many breast tumors have been found to have dominant mutations resulting in amplification of the proto-oncogenes **NEU** and **MYC** and recessive mutations that occur in **RB1** and in unidentified genes located within the chromosome regions 11p and 3p. In some cases, certain mutations observed in adult cancers appear to precede others. However, the particular order of mutations most likely varies during the evolution of many other tumors, suggesting that the final cluster of mutations may be more important than the route followed during their accumulation.

SUMMARY

Mutation generates new alleles, alters the number of gene copies, or causes new groupings of existing genes. These effects increase the genetic diversity of the human population. Mutations may be spontaneous or induced. Spontaneous mutations reflect a natural change in our genetic material that occurs as the result of exposure of our DNA to transient chemicals arising as intermediates during cellular metabolism, to the sun's rays, and to other mutagens that are constituents of our natural environment. Induced mutations occur following exposure to mutagens introduced to our environment from human sources. Certain industrial, agricultural, dietary and therapeutic agents may possess mutagenic activity or may be converted to mutagens by cellular enzyme systems.

Ionizing radiation breaks chemical bonds by forcing the ejection of electrons from atoms and forming ion pairs. These ions are chemically active and interact with other molecules. Direct hits on

DNA are not necessary for mutations, since ionized molecules from other sources can secondarily effect changes in DNA. Nonionizing radiation such as unltraviolet light causes the formation of thymidine dimers, promoting mispairing and replication errors. Chemical agents may mimic radiation, causing breakage of DNA strands and subsequent rearrangements or may produce base substitutions. Frameshift mutations are caused by addition or deletion of bases. The sense of the gene is changed from the point of alteration onward. Frameshifts generate alleles which encode proteins that have runs of amino acids that are very different from those of the normal protein. An addition followed by a base deletion (or vice versa) may result in restoration of the correct reading frame downstream from the compensatory change. The mutant polypeptide would have normal amino acid sequences flanking the sites of the two alterations, but an unusual sequence between them. Frequent outcomes of frameshift mutations are the generation of termination codons at unusual places in the RNA transcribed from the gene and the production of unnatural splice junctions which result in abnormal mRNAs.

Physiological consequences of mutations are dependent upon the sites within the polypeptides affected by the mutations. Proteins may be rendered nonfunctional or unstable by mutations. Alternatively, substitution of residues by similar amino acids may not affect the three-dimensional structure or function of the protein. Rarely, mutation may produce an allele that encodes a protein which has funtional properties that are superior to those of the standard protein.

Mutation rates may be calculated by direct or indirect methods and generally fall within the range of 1 to 100 per locus per million gametes per generation. Many observed mutation rates appear high compared to those of other species and may reflect bias toward more mutable loci, genetic heterogeneity, inclusion of environmental phenocopies, large target size, and/or the relatively long human generation time.

There is an intimate relationship between mutation and cancer. Many childhood cancers appear to follow a pattern in which two or three mutations convert a normal cell to a malignant tumor cell. Retinoblastoma serves as a paradigm for illustrating the role of mutation in childhood cancer. About 15% of all retinoblastomas exhibit autosomal dominant inheritance with reduced penetrance. Familial cases differ from sporadic cases by virtue of the inheritance of one of the two mutations required for retinoblastoma through the germline. These "predisposed" individuals require only one somatic mutation to develop the tumor. Sporadic cases require two somatic mutations for tumor development. **RB1**, an anti-cancer gene, encodes a protein that retards entry of cells into mitosis. Mutations in both **RB1** alleles lead to loss of this protective protein and results in rapid cell division.

Adult cancer appears to require more mutational "hits", and several anti-cancer genes and proto-oncogenes appear to be involved. The mutational path leading from a normal cell to an adult cancer cell often differs among individuals.

PROBLEMS

1. You have successfully sequenced a mutant protein. The amino acid sequence is normal from residue 1 through 34 and from residue 52 to the end of the chain. Residues 35 through 50 do not match sequences from any known protein variant and do not resemble that of the standard chain. How did this mutation arise?

2. The Lepore hemoglobins contain a non-α chain consisting of an amino terminal portion derived from the δ chain and a carboxy terminal portion derived from the β chain. Speculate on the chromosomal relationship of these loci.

3. Why is leukemia a common late effect of radiation exposure?

4. Numerous attempts to produce superior organisms by inducing mutations have failed. Why?

5. Why are mutations medically important?

6. Russell demonstrated that oocytes and spermatids are most sensitive to radiation. Spermatogonia were radioresistant and possessed effective repair mechanisms capable of correcting 2/3 of the radiation-induced damage. A somewhat larger fraction of damage incurred by oocytes was also repaired. Suppose a man and woman both receive identical doses of radiation (measured at the gonadal level). Given that one spermatogenic cycle takes about 10 days and that sperm can remain viable for approximately 42 days after their maturation, who would be more likely to transmit a mutation to their children if conception occurred six months following radiation exposure? Why?

Match the following mutagens with their nuclear effects:

7. nitrous acid	a. protein denaturation
8. X-rays	b. alkylation
9. base analogs	c. ionization
10. ultraviolet light	d. base substitution
11. nitrogen mustards	e. excitation

12. In spite of the efficient repair systems possessed by spermatogonia and oocytes, some damage escapes repair. Given that female meiosis remains arrested at prophase I for long periods of time and that male meiosis occurs continuously for more than 60 years, predict the relative effects of male and female gonadal exposure to mutagens upon the mutation rate for the human species.

GLOSSARY OF TERMS

Anti-cancer gene - gene that retards the development of cancer. Many of these seem to act by preventing cells entry into mitosis.

Base analog - an unnatural nucleotide that replaces A, T, C, or G and which promotes transitions or transversions.

Carcinogen - a cancer-causing agent.

Complementation - the cooperation of two mutant nuclei within the same cell or two mutant cells in the same culture to produce a normal phenotype. Complementation is interpreted to mean that the mutations involve nonallelic genes.

Dominant Lethal Test - mutagenized males are mated with non-mutagenized females and progeny are recovered at intervals post-conception. Loss of progeny at these stages is interpreted to reflect the effects of dominant mutations caused by the mutagen.

Frameshift - deletion or insertion of a base or bases changes the reading frame of the gene such that the coding sense of the base sequence following the site of alteration is changed.

Genetic predisposition - genotype that makes it more likely for a person having that genotype to develop a trait or disease than someone who has a different genotype.

Host-mediated assay - a test organism is deliberately introduced to a rodent, and a suspected promutagen is given to the rodent. If the chemical is a promutagen, the rodent will metabolize it to a mutagen, causing mutations in the test organism.

Mutagen - agent capable of causing mutation.

Mutation - heritable change in the structure of DNA. Note that cell division is required for consolidation of a mutation.
Induced - mutations caused by agents added to the environment by humans.
Spontaneous - mutations caused by naturally occurring intrinsic or extrinsic mutagens.

Phenocopy - environmentally caused condition resembling a hereditary trait.

Promutagen - a nonmutagenic agent that is converted to a mutagen by metabolic processes.

Transition - substitution of one pyrimidine for another pyrimidine or one purine for another purine.

Transversion - substitution of a purine for a pyrimidine or vice versa.

BIBLIOGRAPHY

1. Cleaver, JM and KH Kraemer. 1989. Xeroderma pigmentosum. in <u>The Metabolic Basis of Inherited Disease</u>. 6th ed., Scriver, CR, AL Beaudet, WS Sly and D Valle eds., New York: McGraw-Hill, p. 2961.

2. Russell, WL, LB Russell and EM Kelly. 1958. Radiation dose rate and mutation frequency. Science 128:1546-1550.

3. Vogel, F and R Rathenberg. 1975. Spontaneous mutation in man. Adv. Hum. Genet. 5:223-318.

4. Knudson, AG. 1977. Genetics and etiology of human cancer. Adv. Hum. Genet. 8:1-66.

5. Buchkovich, K, LA Duffy and E Harlow. 1989. The retinoblastoma protein is phosphorylated during specific phases of the cell cycle. Cell 58:1097-1105.

6. Ludlow, JW, J Shon, JM Pipas, DM Livingston and JA DeCaprio. 1990. The retinoblastoma susceptibility gene product undergoes cell cycle-dependent dephosphorylation and binding to and release from SV40 large T. Cell 60:387-396.

7. Howe, JA, JS Mynryk, C Egan <u>et al.</u> 1990. Retinoblastoma growth suppressor and a 300-kDa protein appear to regulate cellular DNA synthesis. Proc. Natl. Acad. Sci. USA 87:5883-5887.

ADDITIONAL LEARNING RESOURCES

1. Bishop, JM. 1991. Molecular themes in oncogenesis. Cell 64:235-248.

2. Cohen, MM and HP Levy. 1989. Chromosome instability syndromes. Adv. Hum. Genet. 18:43-149.

3. Crow, JF and C Denniston. 1985. Mutation in human populations. Adv. Hum. Genet. 14:59-123.

4. Knudson, AG. 1977. Genetics and etiology of human cancer. Adv. Hum. Genet. 8:1-66.

5. Marshall, CJ. 1991. Tumor suppressor genes. Cell 64:313-326.

6. Neel, JV and SE Lewis. 1990. The comparative radiation genetics of humans and mice. Ann. Rev. Genet. 24:327-362.

7. Sancar, A and GB Sancar. 1988. DNA repair enzymes. Ann Rev. Biochem. 57:29-67.

8. Sawyers, CL, CT Denny and ON Witte. 1991. Leukemia and the disruption of normal hematopoiesis. Cell 64:337-350.

9. Stanbridge, EJ. 1990. Human tumor suppressor genes. Ann. Rev. Genet. 24:615-657.

Chapter 7

Genes, Hemoglobins and Anemias

Hemoglobins have played an invaluable role in human genetics with respect to understanding the molecular basis of genetic disease. These interesting proteins continue to be exploited by molecular geneticists for experiments designed to study regulation of human genes during development and as a model system to explore the feasibility of gene manipulation for the correction of human diseases. Hemoglobins display a high degree of polymorphism, and many of the common globin alleles are associated with malarial resistance. Heterozygotes for these mutations enjoy a selective advantage in areas where falciparum malaria continues to be a problem. Globin genes have also been employed as "molecular clocks" to study human evolution. This chapter illustrates the central importance these proteins have with respect to human health and their contributions to our understanding of gene action and the effects of mutations upon protein function.

STRUCTURE AND FUNCTION OF NORMAL HEMOGLOBIN

Hemoglobin, a major constituent of erythrocytes, is a heterotetramer consisting of two α-like chains and two β-like chains (Fig. 7-1). The major adult hemoglobin, HbA_1, has the molecular formula $\alpha_2\beta_2$. Each α chain possesses 141 amino acids, and each β chain has 146 amino acids. Hemoglobin

Fig. 7-1. Structure of the hemoglobin molecule. The molecule is displayed in three dimensions. The monomers represented by solid lines project in front of the page, while the monomers indicated by dashed lines are located behind the plane of the page. The rectangles symbolize the planar heme rings that are located in a pocket of each chain. The circles represent iron molecules which paticipate in oxygen exchange.

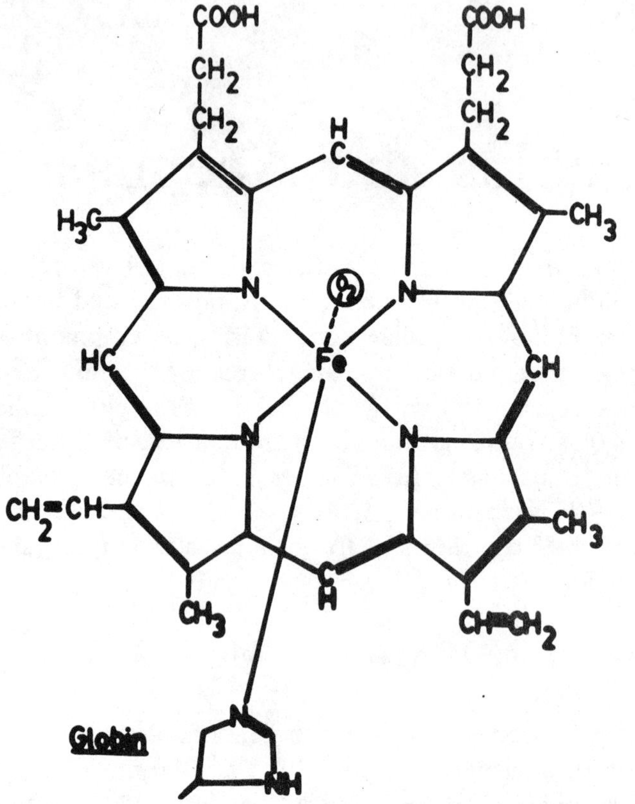

Fig. 7-2. Planar structure of heme, the oxygen-binding moiety of hemoglobin. A critical histidine of one of the globin chains interacts with the heme iron, stabilizing it in the ferrous (+2) state. This charge is required for proper oxygen binding to the iron.

plays a vital role in the transport of oxygen from the pulmonary alveoli to other tissues. Each α and β monomer encloses a heme ring in a cleft of the chain. Oxygen binds to the iron that is centered within this heme ring (Fig. 7-2). The charge of the iron must be +2 for the binding and release of oxygen to occur normally. The iron of more than 99% of the hemoglobin molecules is maintained in the +2 state by an enzyme, methemoglobin reductase (note: methemoglobin contains ferric (+3) iron).

Muscle contains a protein, myoglobin, which is structurally related to the α and β chains (Fig. 7-3). This protein serves as an oxygen store in muscle cells.

The α and β subunits of hemoglobin interact with one another to facilitate both the binding and release of oxygen. When one globin chain binds oxygen, a conformational change occurs in this globin molecule which affects the adjacent globin subunits, enhancing their oxygen-binding capabilities such that the second, third and fourth molecules of oxygen are bound more easily than the first. This effect is illustrated in Fig. 7-4. The partial pressure of oxygen is high in the lungs, promoting a rapid binding of oxygen by hemoglobin. Oxygen pressure in tissues is low, promoting oxygen release by hemoglobin. When oxygen is released by a globin chain, it reverts to its original conformation. This structural change impacts on the neighboring globin subunits, facilitating their oxygen release. This cooperative behavior of the subunits of hemoglobin is especially well suited for its carrier role.

The normal function of hemoglobin is dependent upon the amino acid sequences of its respective globin subunits. The amino acids on the surface of the hemoglobin molecule are polar and hydrophilic, enhancing the attraction of water molecules and making the hemoglobin soluble in the erythrocyte cytosol. Substitutions of nonpolar amino acids for polar amino acids may decrease the solubility of hemoglobin, promote abnormal interaction with other hemoglobin molecules, or a

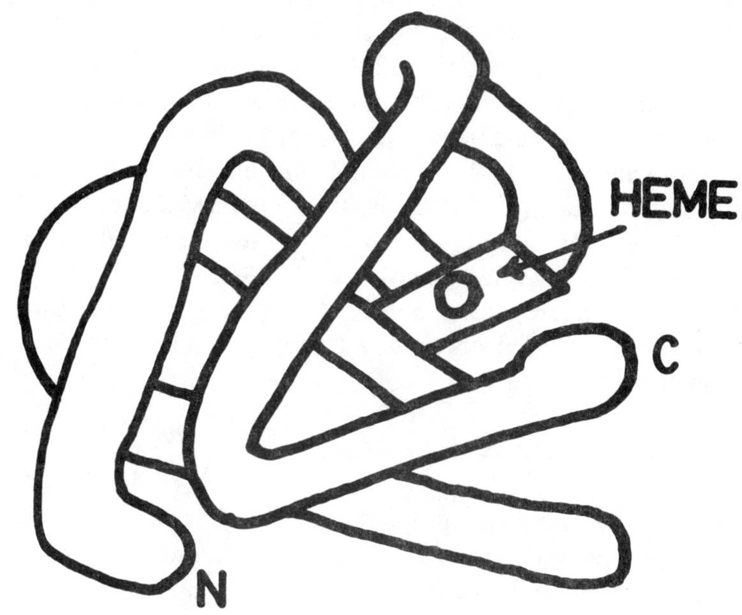

Fig. 7-3. Structure of myoglobin, an oxygen-binding protein in muscle. Although the amino acid sequence of myoglobin is somewhat different from those of α- and β-globin, the three dimensional structure of the three molecules is quite similar.

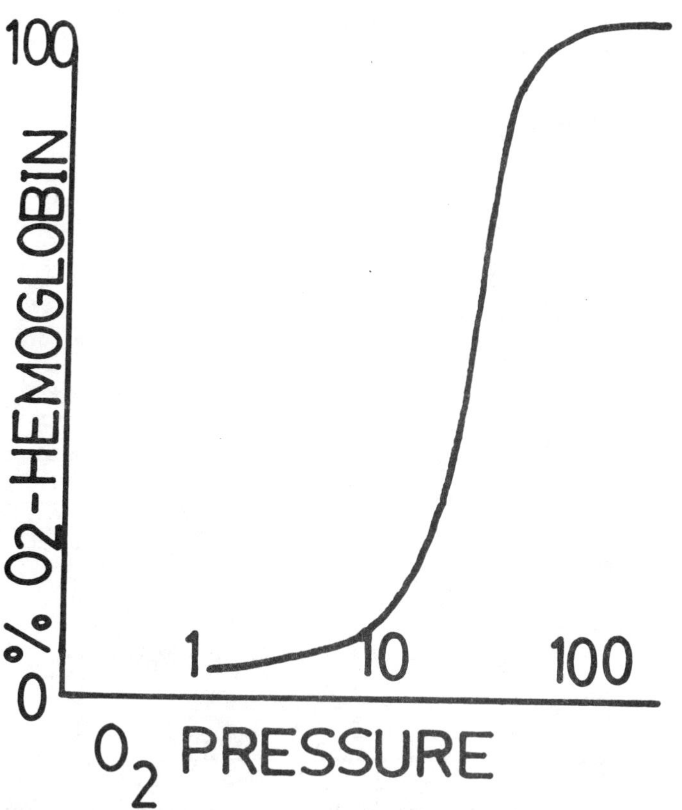

Fig. 7-4. Interaction of hemoglobin with oxygen. At low oxygen pressures (characteristic of most tissues) hemoglobin rapidly releases oxygen (left portion of curve). At high oxygen pressures (such as those in the pulmonary alveoli) hemoglobin effectively binds oxygen (right portion of curve).

combination of these events. Substitutions of charged surface amino acids by neutral residues or by amino acids of opposite charge will alter the surface charge and electrophoretic mobility of hemoglobin, permitting diagnosis of the presence of these unusual hemoglobins. Many amino acids on the surfaces of the globins contact other globin molecules of the tetramer. Changes in these contact amino acids can modify the "watch spring" cooperativity among the globin chains and disrupt normal oxygen binding and release. Amino acids located in the heme pocket tend to be nonpolar and hydrophobic. Amino acid substitutions replacing these hydrophobic residues with hydrophilic amino acids may change the shape of the pocket, interfering with iron-oxygen linkage or causing instability of the hemoglobin tetramer. Examples of these mutations will be presented in a later section of this chapter.

The intermolecular attractions among the α and β subunits involve both electrostatic and hydrophobic interactions. These bonds can be dissociated under usual physiological conditions to form half molecules and the respective monomers:

$$\alpha_2\beta_2 \; \rightleftharpoons \; 2\,\alpha\beta \; \rightleftharpoons \; 2\alpha + 2\beta$$

The tetramer, dimers, and monomers are in equilibrium with one another, and movement from one state to another occurs quite rapidly. The equilibrium can be distorted by mercaptoethanol, urea, and certain other agents which reduce hydrogen bonds, favoring the accumulation of the free monomers. Clinical laboratories will often incorporate these agents into their electrophoretic systems, permitting them to

Table 7-1. Interrelationships among the multiple globin loci and the polypeptide chains they determine.

Developmental Stage	Site of Synthesis	Active Loci	Hemoglobin	Formula
Embryonic	Yolk Sac	Zeta, Gamma*, Epsilon, Alpha	Portland	$\zeta_2\gamma_2$
			Gower I	$\zeta_2\epsilon_2$
			Gower II	$\alpha_2\epsilon_2$
Fetal	Liver, Spleen, Bone Marrow	Gamma 136 Gly, Alpha, Gamma 136 Ala	F^{Gly}	$\alpha_2\gamma_2^{136\ Gly}$
			F^{Ala}	$\alpha_2\gamma_2^{136\ Ala}$
Adult	Bone Marrow	Beta, Alpha, Delta	A_1	$\alpha_2\beta_2$
			A_2	$\alpha_2\delta_2$

*Subtypes of Hb Portland and Hb Gower I may be formed in a manner similar to HbF.

identify the mutant globin chain responsible for a patient's clinical problem.

Multiple Types of Hemoglobin are Synthesized During Development

The sites of erythropoiesis shift during development (Table 7-1;Fig. 7-5). Cells in the yolk sac, a structure outside the embryo proper, is the initial location for hemoglobin synthesis (1). The zeta (an α-like globin gene) and the ϵ-globin (a β-like globin gene) genes function during the earliest embryonic stages, and the gamma- and α-globin genes are activated near the end of the embryonic period. Therefore, Hb Gower I is most likely the earliest hemoglobin made, while Hb Portland and Hb Gower II represent transition hemoglobins.

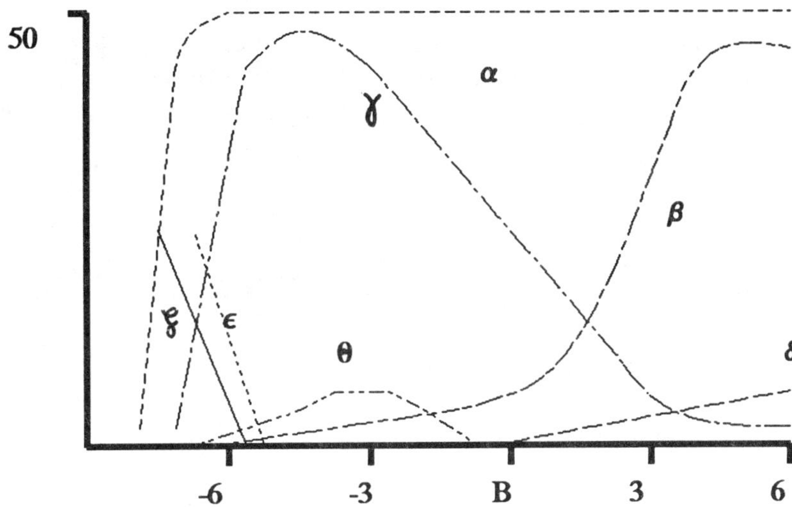

Fig. 7-5. Developmental variation of the expression of α-like and β-like globin loci (modified from Huehns et al. (2)). The vertical axis represents the percent of the total globin chains represented by each chain type at a given developmental age. The horizontal axis lists the age of development in months from the time of birth (B). The θ-globin locus has recently been found downstream from the α1 gene and is expressed during fetal development and at a low level.

A shift in globin gene expression occurs following implantation. The site of erythropoiesis and hemoglobin synthesis are transferred to the fetal liver, spleen and bone marrow. Coincident with this change in the site of erythropoiesis, the epsilon- and zeta-globin loci are down regulated; and the α- and gamma-globin loci are switched on. These changes result in the replacement of the embryonic globins by fetal hemoglobins, HbF. The hemoglobin shift is not effected by the replacement of embryonic erythrocytes containing solely embryonic globins by cells containing exclusively HbF. Transition cells can be found which contain not only embryonic and fetal hemoglobins, but also the transition hemoglobins, Hb Portland and Hb Gower II. This observation indicates that the up regulation and down regulation of globin loci occur gradually, and that a considerable period of overlap of gene expression exists.

Studies have indicated that there are two gamma-globin loci (3), each encoding a protein that differs by a single amino acid at the C-terminal end. One of these globins possesses glycine at that position, while the other encodes an alanine residue at position 136. The two globins appear to have equivalent function. Two α-globin loci per haploid chromosome set have also been identified (4). Their globin chains have identical amino acid sequences.

Within about six months after birth the hemoglobin composition of an infant's blood attains a pattern typical for normal adults. Erythropoiesis is restricted to the bone marrow, and the α-, β- and δ-globin loci remain active, while expression of the gamma-globin locus falls to about one percent of the total β-like globin chains synthesized. β-globin expression far exceeds δ-globin gene expression such that the two adult hemoglobins, HbA$_1$ ($\alpha_2\beta_2$) and HbA$_2$ ($\alpha_2\delta_2$), usually occur in a ratio of about 97:3. This low expression of the δ-globin locus has been interpreted to suggest that this gene may be on its way to evolutionary extinction and relegation to pseudogene status.

Regulation of the Hemoglobin Developmental Program

The orchestration of globin gene expression during development appears to be accomplished at the transcriptional level. The α-like globin genes are clustered on chromosome 16 in their order of developmental appearance (Fig. 7-6). The β-like genes display a similar gene organization on

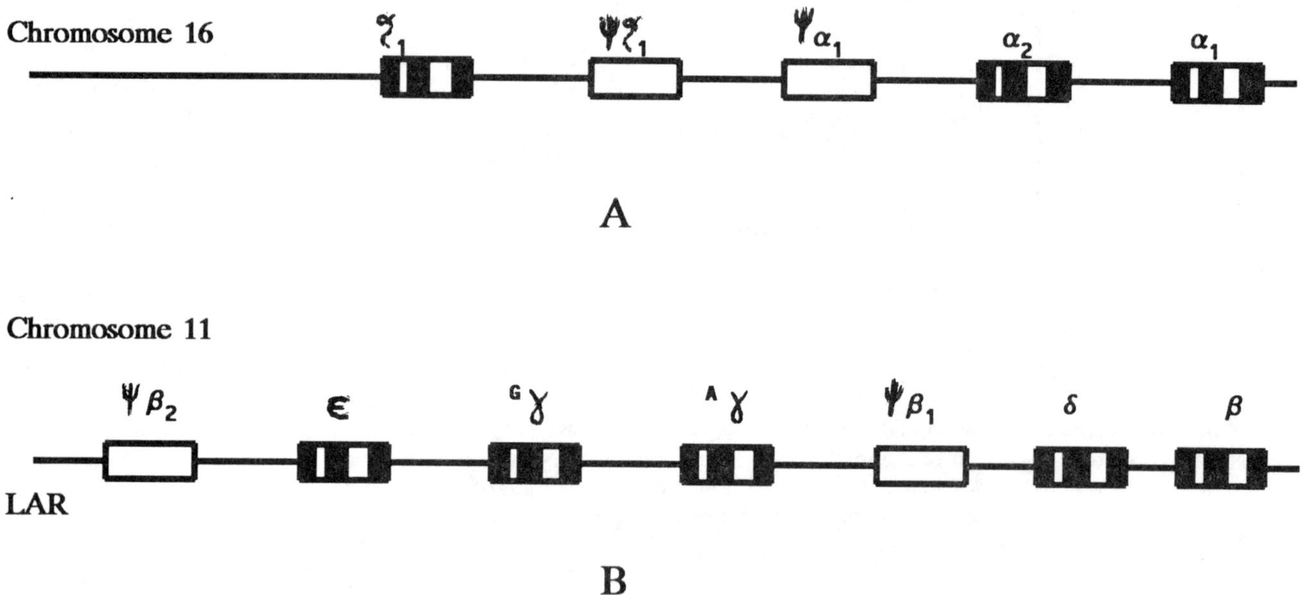

Fig. 7-6. Linkage groups of α-like and β-like globin genes. The genes are situated in order of their developmental expression. Four pseudogenes have been found, two in each linkage group. LAR = locus activating region.

chromosome 11. Studies using β-like globin genes that were introduced to mouse embryos demonstrated that the genes were expressed in a tissue- and temporal-specific manner provided promoter regions of the respective human genes were included in the gene constructs (5). This particular study, which is one of several similar reports from different laboratories, used gene constructs that contained the entire $^A\gamma$-

globin gene (including 1.3 kb and 370 bases of its upstream and downstream sequences, respectively), the entire β-globin gene (including 0.8 kb and 2.5 kb of its upstream and downstream sequences, respectively), and a $\gamma\beta$-globin fusion gene (including the 1.3 kb $^A\gamma$-globin upstream sequence and the 2.5 kb β-globin downstream sequence). The human $^A\gamma$-globin gene construct was expressed in erythroid cells during the murine embryonic period, and the human β-globin gene construct was expressed in erythroid cells in adult mice. In both cases, the human globins were expressed at a much lower level than their stage-specific murine counterparts. The human fusion gene was expressed in erythroid cells throughout murine development and at a lower level than the corresponding murine genes. These experiments demonstrated that sufficient information was present in flanking regions of the respective human genes to permit their response to tissue- and developmental-specific transcription factors produced by the murine cells. Furthermore, there must be sequences in the β-globin downstream segment that can activate transcription from the $^A\gamma$-globin gene promoter at times (adult cells) when that locus is largely silent. The low level of human globin gene expression indicated that other regions, perhaps in the globin gene cluster are required for the normal expression of these genes.

Deletions of the upstream end of the β-like globin cluster prevent expression of any of the downstream genes (see discussion of thalassemias in a later section of this chapter), suggesting that elements in this area are essential for activation of transcription. Furthermore, gene constructs containing sequences upstream from the ϵ-globin locus spliced to promoters of downstream globin loci resulted in their high level expression in transgenic animals and in cells transfected in culture. The region involved contains an enhancer element and perhaps other sites that are critical for normal in vivo expression of the β-like gene cluster. Digestion of the β-like globin region with DNase I has demonstrated the presence of a hypersensitive site in this region that appears developmentally stable, i.e. relaxed structure is present throughout development. A similar developmentally stable hypersensitive region is located downstream from the β-globin gene. Although there are also hypersensitive sites closer to each β-like globin gene within the cluster, the latter sites are not present at all stages of development. The hypersensitive region that appears essential for high level expression of β-like globin genes has been designated the **LAR** (locus activating region; Fig. 7-6). When the LAR was deleted from the chromosome, sequences flanking the site of the deletion, including the entire β-like region, were DNase I-resistant, and displayed late replication in an erythroid environment (6). By contrast, all of these sequences were early replicating when the LAR was intact. These experiments suggest that sequences, including the enhancer, in the LAR up regulate β-like globin gene expression by acting over long distances. The mechanism by which the LAR brings about its effects is unknown. A similar LAR may be present on chromosome 16 which modulates α-like globin gene expression; however, these studies are less complete.

HEMOGLOBINOPATHIES

Sickle cell anemia is a hemolytic disorder that is associated with abnormal red cells which change from the normal biconcave disk shape to a "sickle" shape in an oxygen-poor environment. This sickling property was first reported by Herrick in 1910 (7) and again by Hahn and Gillespie in 1927 (8). The sickling phenomenon appeared to occur in two genotypes (9,10). Most persons whose cells exhibit this property under appropriate conditions are heterozygous for the β^S allele, usually do not develop anemia, and are said to have **sickle cell trait**. Heterozygotes are frequent in malarial areas, where up to 20% of the population may be heterozygous, and still relatively common among descendants of these populations who have emigrated to regions where malaria is no longer a problem. It is estimated that about 9% of Black Americans are heterozygous for this gene. Sickling also occurs in individuals who are homozygous for the β^S allele. Homozygotes have anemia, may have severe hemolytic crises, and develop other complications as a consequence of the chronic effects of the disease. As many as two percent of the population in central Africa die from **sickle cell anemia** during childhood or early adolescence. The frequency of sickle cell anemia among Black Americans approximates 1/500. It is important for the reader to distinguish among three terms that appear in the literature. **Sickle cell anemia** occurs in persons who are $\beta^S\beta^S$, while **sickle cell trait** is found in heterozygotes, $\beta^S\beta^A$. These heterozygotes do not develop anemia nor other complications unless placed under unusually low oxygen partial pressures. **Sickle cell disease** is a more general term that refers to a clinical problem associated with sickling of erythrocytes and occurs in persons of a number of different genotypes, including those in which β^S is present in combination with another abnormal β-globin allele such as β^C.

Red cells of persons heterozygous or homozygous for the β^S allele were found to contain an unusual hemoglobin. Linus Pauling and his associates demonstrated that this hemoglobin differed in charge from normal hemoglobin by electrophoretically separating the two hemoglobins from heterozygotes (11). The abnormal hemoglobin, HbS, migrated more slowly than the normal hemoglobin, HbA_1 at alkaline pH. The slower electrophoretic migration of HbS was subsequently found to be caused by an amino acid substitution (valine --> glutamic acid) at position 6 near the N-terminus of the β chain (12,13). The β-globin was digested to form peptides which were separated by electrophoresis and then by chromatography (use of solvents to separate materials) at right angles to the direction of electrophoretic migration. The resulting peptide "fingerprints" from the β^S- and β^A-globin chains were stained

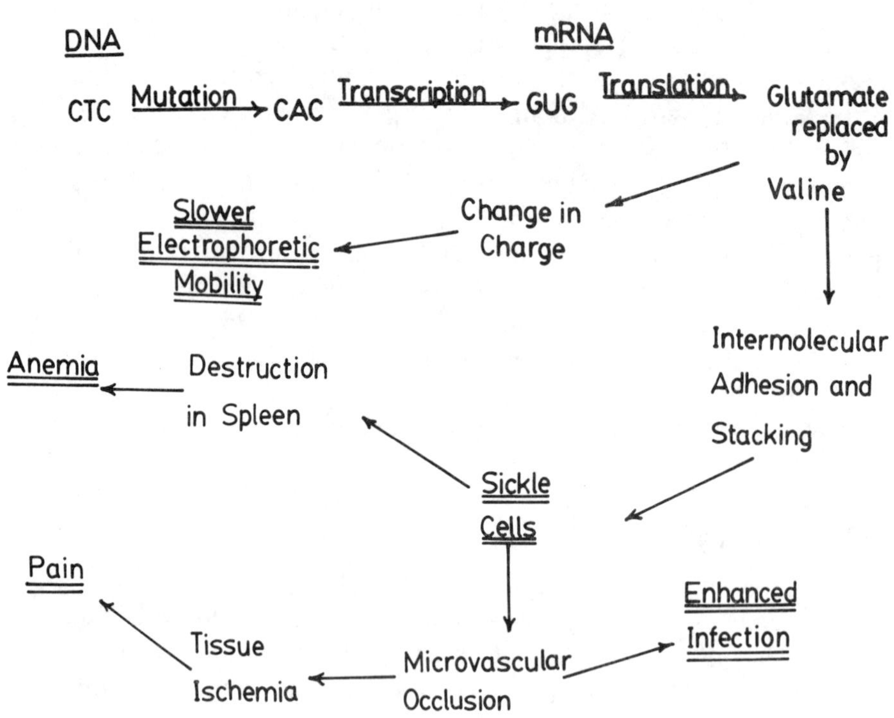

Fig. 7-7. Relationship between the base substitution responsible for the β^S allele and the clinical disease, sickle cell anemia.

and found to have a single peptide difference. The spots that differed between the two globins were cut out and analyzed, revealing the aminoacid substitution.

More recent work has demonstrated that a single base change is responsible for production of sickle cell anemia, **CTC --> CAC** (Fig. 7-7). The mRNA from this region would have the codons **GAG** and **GUG** which encode glutamic acid or valine, respectively. This amino acid position is located on the surface of the β-globin chain. The placement of a valyl residue at this site results in an oily patch in an otherwise hydrophilic environment. When oxygen is released by the hemoglobin molecule, the change in conformation places the valine in a position where it can link to a corresponding valine on a neighboring hemoglobin tetramer. These interactions promote stacking of hemoglobin into filaments which deform the red cell membrane, producing the sickle cell. These abnormal cells are subject to accelerated destruction by the spleen resulting in anemia. Furthermore, blockage of capillaries by sickle cells increases risk for infection in the extremities, and interferes with delivery of oxygen and nutrients to tissues, especially in the abdominal region where it is associated with pain. This chain of events is called a **sickle cell crisis.** The frequency and severity of crises vary among individuals. Crises may be brought on by factors which reduce the blood oxygen (fever associated with infection, strenuous exercise, etc.). In some cases, crises may occur in the absence of apparent triggering events. Red cells of β^S homozygotes exhibit reversible sickling as a course of their normal circulation. The cells sickle on the venous side of the capillary bed, and circularize on the arterial side following oxygenation of the blood. Complications arise when the arterial oxygen pressure falls, preventing recircularization.

Early detection of sickle cell anemia and aggressive management of infection have markedly improved the lives of these patients. Average lifespan has increased to more than 30 years, and the quality of their lives is much better than was the case a few years ago. Most β^S homozygotes are able to lead normal, productive lives. The genotypes associated with sickle cell anemia and sickle cell trait can now be detected before birth (Fig. 7-8). Restriction sites for the enzyme MstII occur upstream from the β-globin gene, within the first exon (corresponding to the sixth amino acid in β-globin) and downstream from the β-globin gene. The same mutation that produces the β^S allele destroys the MstII site in the first exon, such that the β^S allele will be present on a larger restriction fragment than the β^A allele (Fig. 7-8a). Cells are obtained by chorionic villus sampling (ca. 8 weeks) or amniocentesis (ca. 16 weeks), and DNA is extracted for analysis. The DNA is digested with MstII, and the fragments are separated by electrophoresis producing a smear (Fig. 7-8b). The gel is subjected to Southern blotting using a labelled β-globin cDNA probe. All three genotypes can be differentiated using this diagnostic procedure. β^A forms 1150bp and 200bp fragments, while β^S is found on a 1350bp fragment. The 200bp fragment is not shown on the gel, since it migrates very rapidly and exits the gel at the (+) end.

Inspection of Fig. 7-8a reveals that the HbC mutation (β^C allele) cannot be detected with this method, because the mutation affects the N position of the MstII restriction site where any base can be tolerated without loss of cleavage. Both the β^S and β^C alleles can be detected by hybridization with allele-specific oligonucleotide probes using conditions of high stringency. The sequence in the region of the mutation is used to prepare a labelled complementary oligonucleotide. DNA is extracted from the fetal cells and digested with a restriction enzyme to make fragments of a manageable size range. The fragments are applied to several different lanes of a gel, separated electrophoretically and transferred from the gel to a membrane for Southern analysis. The membrane is cut into several sections parallel to the direction of separation. One section of the membrane is hybridized with an oligonucleotide probe derived from the β^A allele, and the others are hybridized with one of the probes derived from the mutant alleles, β^S or β^C. Hybridization conditions assure that the probe will anneal only with the specific allele from which it is derived. If a band is observed in the section that has been probed with the β^S-derived oligonucleotide and in the section hybridized with the β^C-derived oligonucleotide, but not the section probed with the β^A-derived oligonucleotide, the genotype would be inferred to be $\beta^S\beta^C$. In a similar way, all possible genotypes can be determined from the particular combination of hybridization signals

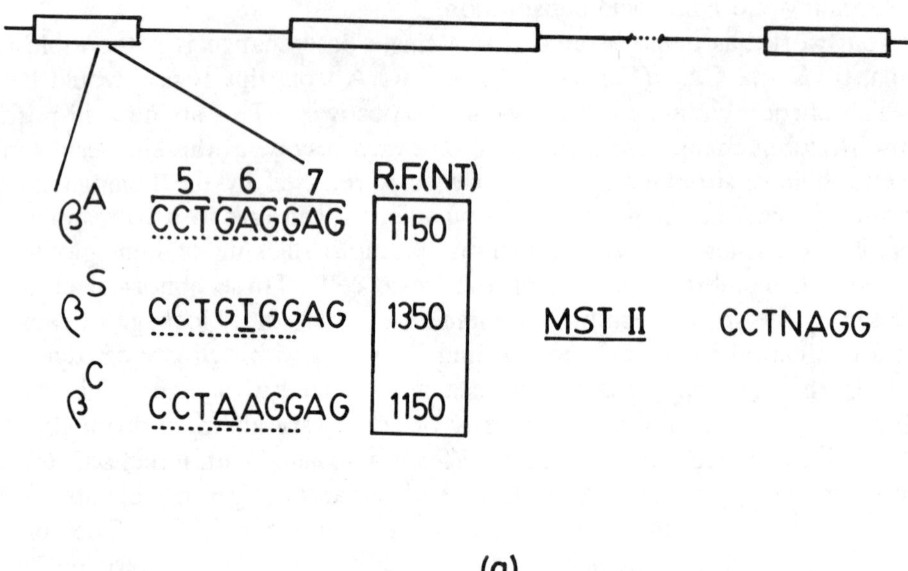

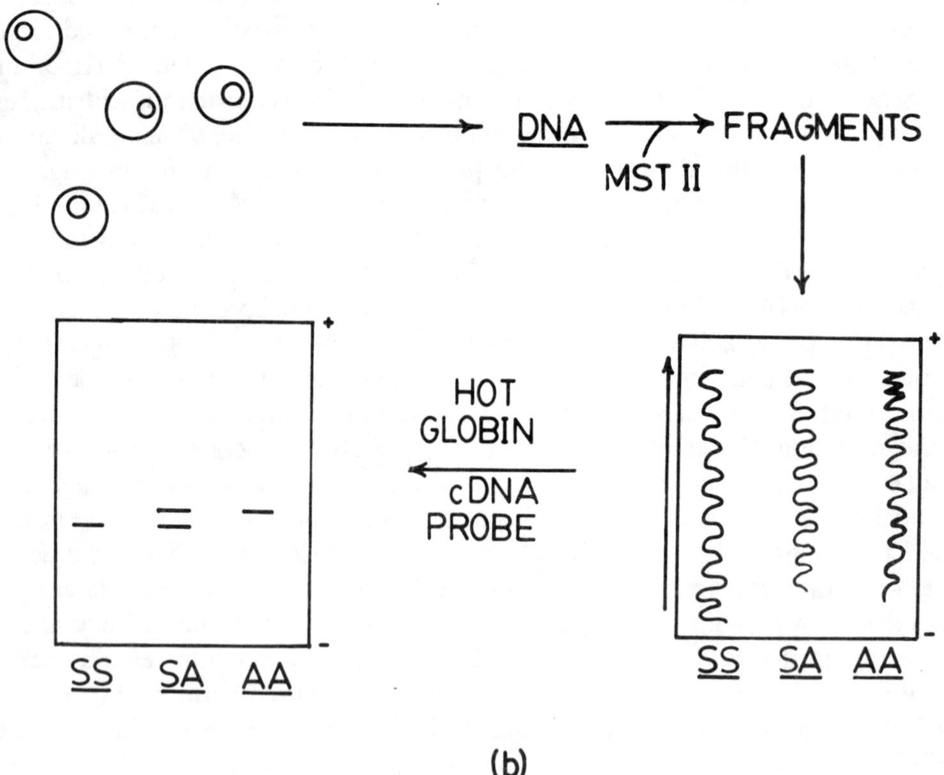

Fig. 7-8. Prenatal diagnosis of sickle cell anemia. a) MstII cleavage site and corresponding sequences within the first exon of three β-globin alleles; b) Procedure for determining genotypes from amniotic fluid cells (14).

obtained (15).

The relationship between the molecular alterations and clinical disease illustrated by sickle cell anemia provides a conceptual framework for understanding the relationship between mutant genes and inherited metabolic diseases in general. Since the discovery of HbS, more than 130 hemoglobin variants have been described. Most of these proteins differ from HbA_1 by single amino acid replacements. Many other systems have now been reported in which the majority of the proteins that have been sequenced display single amino acid substitutions. The hemoglobinopathies will be used as models to illustrate the relationships among genes, the proteins they encode, and the effects of representative mutations upon protein function.

Physiological Consequences of Amino Acid Substitutions

Sickle cell anemia is caused by an amino acid substitution on the surface of the β-globin chain (Fig. 7-9). Sickle cell anemia is inherited as an **autosomal recessive** because the clinical complications require an **interaction among different hemoglobin tetramers.** The HbS tetramers occur in sufficient quantity to form the aggregates responsible for deformation of the erythrocytes in homozygotes, but usually not in heterozygotes (sickle cell trait). In order for the aggregates to form, each Hb molecule

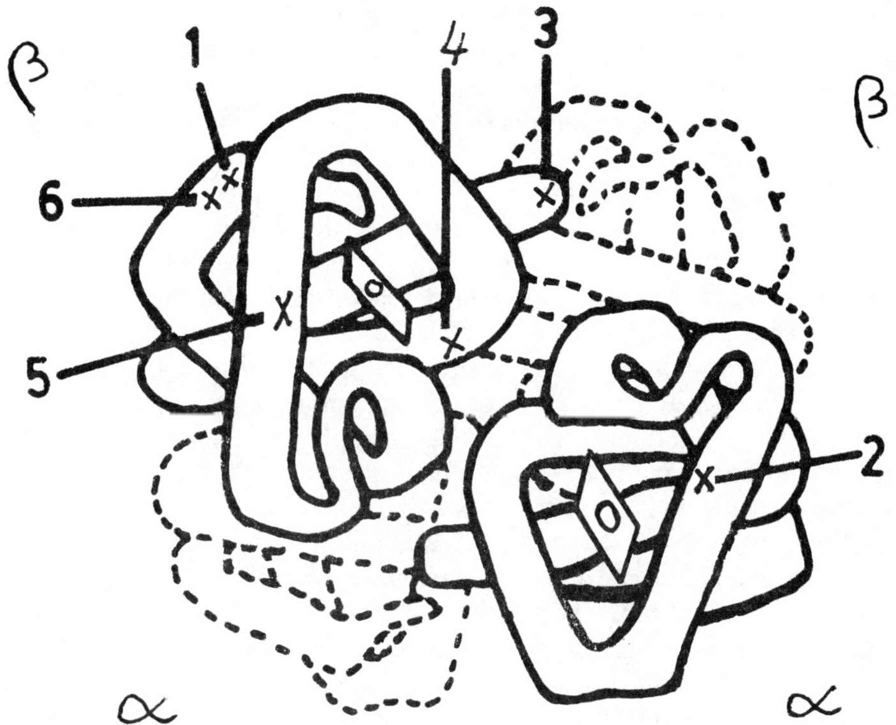

Fig. 7-9. Sites of representative amino acid substitutions in hemoglobin giving rise to different physiological effects. 1) Sickle cell anemia and HbC disease; 2) HbM Boston; 3) Hb Hiroshima; 4) Hb Kansas; 5) Hb Zurich; 6) Hb Porto Alegre.

must possess two β^S-globin chains, enabling it to form a hydrophobic interaction with the preceding and following hemoglobin molecule in the stack. $\beta^S\beta^A$ heterozygotes produce three types of molecules: $\alpha_2\beta^A_2$, $\alpha_2\beta^S\beta^A$ and $\alpha_2\beta^S_2$ in the approximate ratio of 1:2:1. Neither the HbA$_1$ nor the hemoglobin containing both β^A and β^S chains can participate in aggregate formation; therefore, sickle cell anemia does not occur in heterozygotes.

Hemoglobin C (HbC), $\alpha_2\beta^C_2$, possesses a lysine at position 6 of the β-globin chain. This basic amino acid imparts properties to the hemoglobin that are distinct from those of both HbA$_1$ and HbS. The molecule migrates more slowly than HbS in an electrical field, and HbC of $\beta^C\beta^C$ homozygotes displays a reversible crystalization in oxygen-rich blood. The crystals are redissolved on the venous side of the capillary bed. This trait, HbC disease, is also inherited as an autosomal recessive, since the crystalization requires an interaction among HbC molecules. Clinical symptoms of HbC disease are mild, and $\beta^S\beta^C$ **genetic compounds** (persons who are heterozygous for two different mutant alleles) suffer an anemia that is milder than sickle cell anemia.

Several dominantly inherited hemoglobinopathies have been described. Two histidine residues in the heme pocket of each α- and β-globin chain participate in the linkage of the heme-iron complex to the respective globin molecules. A group of mutant hemoglobins possess other amino acids in place of one of the histidine residues. For example, HbM Boston has one of the histidine residues of the α chain (histidine 58 located at position #2 in Fig. 7-9) replaced by tyrosine. This substitution has relatively little effect when the heme iron is in the ferrous state ($+2$); however, when the iron is oxidized ($+3$), it forms a bond with the tyrosyl oxygen forming a stable complex that prevents reduction of the iron. Since effective binding of molecular oxygen by globin requires ferrous iron, two of the subunits of HbM Boston are incapable of transporting O$_2$. Furthermore, secondary conformational changes imposed by the tyrosine substitution in the α-globin chains interfere with oxygen binding by the heme irons of the neighboring β-globin chains. Failure of oxygen transport leads to tissue oxygen deprivation. These patients tend to have a blue coloration (cyanosis) due to the large fraction of deoxygenated hemoglobin in their circulation. HbM Boston is one of several mutant hemoglobins that contain predominantly ferric iron (**methemoglobin**). These mutations invariably affect amino acids, such as histidine, that interact with the heme iron. Methemoglobinemia can also be caused by deficiency of an enzyme, methemoglobin reductase, which converts the heme iron from the ferric to the ferrous form. The cyanosis and other clinical complications associated with HbM Boston and related HbM disorders are frequently inherited as autosomal dominant traits. Heterozygotes produce both mutant (α^M or β^M) globin chains and normal globin chains. Three out of every four hemoglobin tetramers will possess at least one mutant chain, and, as a consequence of the cooperative behavior of the globin subunits, these molecules will display abnormal oxygen transport. Therefore, the clinical complications will appear in heterozygotes. Homozygotes for these mutations have not been observed, possibly as a consequence of the rarity of the respective mutations and possible lethality of the homozygous state.

A second class of dominant hemoglobin mutations involve amino acid substitutions at sites involved with α-α, α-β, or β-β contacts within the hemoglobin tetramer. Hb Hiroshima has an aspartic acid residue in place of the usual histidine at position 146 of the β chain (site 3, Fig. 7-9). This change increases the affinity of the hemoglobin for oxygen, impairing its release at the tissue level. The low tissue oxygen stimulates the release of a hormone, erythropoietin, which promotes release of more red cells from the bone marrow. However, these cells have the same abnormal hemoglobin, so the problem is not corrected. As the cycle continues, the number of red cells in the circulation rises, and polycythemia results (Fig. 7-10). Excessive red cells (polycythemia) increase blood viscosity and contribute to high blood pressure and associated complications. A second type of mutation, Hb Kansas replaces the asparagine at position 102 of the β-globin chain with threonine (site 4, Fig. 7-9). This alteration

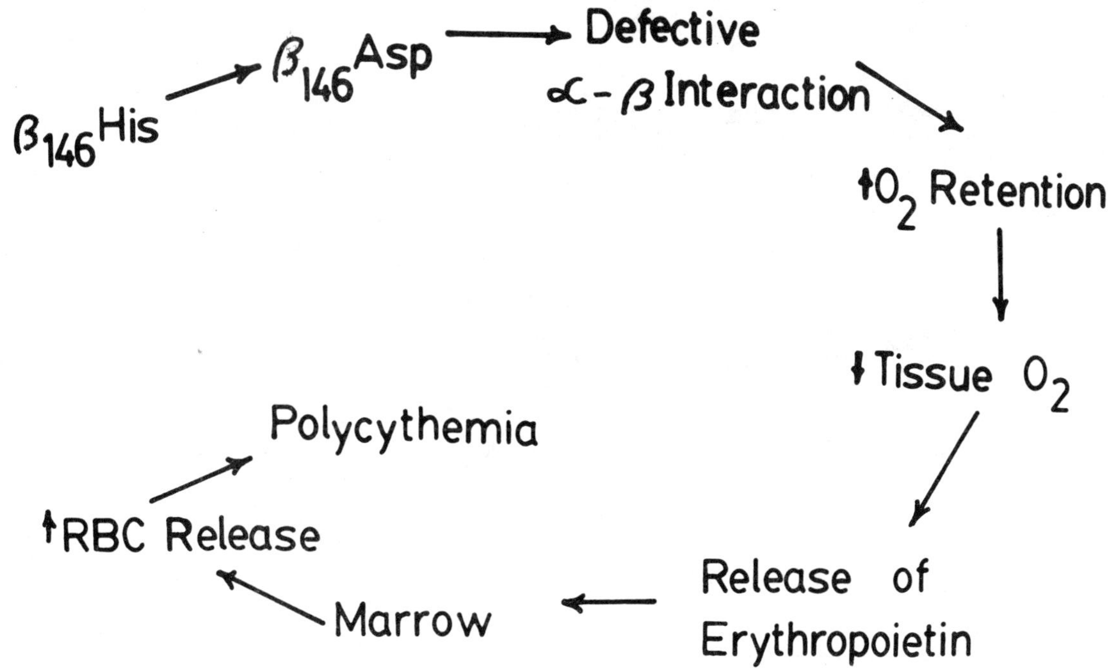

Fig. 7-10. Hb Hiroshima exhibits an increased affinity for oxygen leading to a lack of oxygen at the tissue level. Erythropoietin is released and stimulates red cell production and release by the bone marrow. The excessive number of red cells (polycythemia) in the circulation can lead to other circulatory complications.

reduces the oxygen affinity of the mutant hemoglobin, resulting in a decreased red cell count and anemia. Both hemoglobinopathies, Hb Hiroshima and Hb Kansas, are inherited as dominant traits as a consequence of the presence of at least one mutant globin chain in a majority of hemoglobin molecules and the resulting disruption of cooperativity among subunits of the involved tetramers.

A third type of dominant hemoglobinopathy is caused by molecular instability. Amino acid substitutions which replace hydrophobic amino acids of the globin core with hydrophilic amino acids or which involve interchange of bulky amino acids for compact amino acids (or vice versa) in this region cause a distortion of the three dimensional structure of the hemoglobin molecule, rendering it unstable. The unstable hemoglobin precipitates to form granular inclusion bodies called Heinz bodies. Cells containing these structures are especially susceptible to destruction by the spleen, and patients present with chronic anemia. Hb Zurich (site 5, Fig. 7-9) is an interesting example of these instability mutations, since the anemia associated with its presence is often drug-induced. Another type of unstable hemoglobin, Hb Porto Alegre (site 6, Fig. 7-9), has a serine at position 9 of the β-globin chain replaced by a cysteine. Cysteine, a sulfur-containing amino acid, forms disulfide (-S-S-) bridges with cysteine residues of neighboring hemoglobin tetramers to form insoluble octameric aggregates. These octamers precipitate and ultimately result in the demise of the cell and anemia.

Heterozygotes and Hybrid Hemoglobin Molecules

As described in the preceding section, heterozygotes who carry mutant alleles affecting the α- or β-globin chains produce both the mutant and standard globin chain (Fig. 7-11). Hybrid proteins

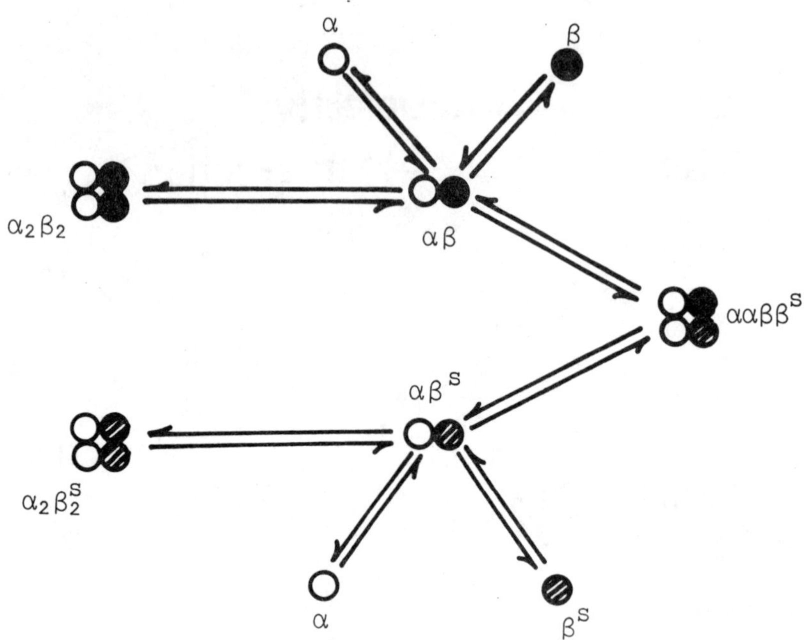

Fig. 7-11. Formation of hybrid molecules in heterozygotes.

containing both mutant and standard globin chains are formed within red cells in addition to hemoglobins containing exclusively mutant or standard globins. These hybrid hemoglobins are usually not observed following electrophoresis. When the rate of dissociation of tetramers and dimers occurs rapidly, the monomers are separated according to their charge, forming pools of mutant or standard globin chains. These chains reassociate to form either the standard hemoglobin (e.g. HbA_1) or mutant hemoglobin (HbS). The hybrid is prevented from forming because the necessary β^A and β^S have been sequestered in different regions of the support used for electrophoresis. Occasionally an amino acid substitution may slow down the dissociation rate. In this case, a three-banded pattern that includes the hybrid hemoglobin molecules would be observed.

Influence of Structural (Exon) Mutations upon Rates of Protein Synthesis

The quantity of HbA_1 exceeds that of HbS and of HbC in persons with sickle cell trait or HbC trait, respectively. In the case of sickle cell trait, HbA_1 accounts for about 60% of the total hemoglobin present in erythrocytes. Recent data indicate that reticulocytes and other red cell progenitors appear to either have more β^A-mRNA than β^S-mRNA, or that β^A-mRNA is more efficiently translated than β^S-mRNA. Therefore, the single base change not only seems to affect the structure of the protein, but also influences either the transcription of the β^S allele, the processing of its nascent RNA, or the translation of its mRNA.

Thalassemias

The preceding section addressed the effects of mutations within exons of globin genes. These mutations primarily affect the function of globin, resulting in the production of near-normal quantities of a defective protein. Thalassemia, derived from the Greek word for sea (thalassa), is an anemia that is caused by a deficiency of globin chains. The disease was first described in the Mediterranean region, hence its name. The various thalassemias are named according to the deficient chain (Table 7-2). The globin deficiency may be caused by a variety of different mutations, most of which involve bases that are located outside of the exons of the genes involved. The effects of these mutations have a cis effect, i.e. they affect the expression of the globin gene on the same chromosome and generally not the allele on the homologous chromosome. In some cases the gene encoding the globin chain may be deleted from the chromosome.

Table 7-2. Thalassemias.

Thalassemia	Deficient Globin
α-thalassemia	α
β-thalassemia	β
γ-thalassemia	γ
$\delta\beta$-thalassemia	δ,β

Three terms have appeared in the older literature that are not as frequently used in more recent publications. Thalassemia major refers to a severe anemic state in which the cells are small and have low quantities of hemoglobin, the spleen is enlarged, and there are characteristic changes in the bone marrow. The enlarged spleen reflects compensatory changes associated with increased red cell destruction, and the enlarged marrow space found in patients occurs in an abortive attempt to restore the circulating red cells to more normal numbers. Thalassemia minor is a less severe clinical disorder and is characterized by relatively mild features that are often specific for the type of globin chain involved. Thalassemia intermedia refers to a clinical phenotype intermediate between those of thalassemia major and minor.

α-Thalassemias

α-Thalassemias are clinical disorders that are associated with varying degrees of α-globin deficiency. As mentioned previously, the α-globin loci are duplicated in most human populations with the standard chromosome carrying two tightly linked α-globin loci. The most common forms of α-thalassemia are caused by deletions of these loci (16). mRNA from normal human reticulocytes (erythrocyte precursors) was purified and used to prepare cDNA. The cDNA was used as a probe to search for α-globin sequences in DNA extracted from spleen cell nuclei of severely affected α-thalassemic patients. The α-thalassemic DNA was denatured and hybridized with radiolabelled normal cDNA. Since α-globin mRNA accounts for about half of the total globin mRNA present in normal reticulocytes, α-globin gene deletions should be detectable by this technique as a decreased amount of radioactivity in the double-stranded DNA fraction. The experiments demonstrated a reduced quantity of label in the double-stranded fraction which was interpreted to indicate a complete or partial loss of α-globin genes from the α-thalassemic cells. Subsequent experimentation has shown that most patients with α-thalassemia lack two or more α-globin genes. The relationship between the numbers of α-globin genes lost and the severity of the thalassemia is presented in Table 7-3. Deletion of all four α-globin genes renders the patient incapable of synthesizing HbA_1, HbF, and Hb GowerII, since all of these hemoglobins contain α chains. Little is known about embryonic development of these infants; however, the other two embryonic hemoglobins most likely protect these α-thalassemic embryos from experiencing serious problems. Two unusual hemoglobins occur in these α-thalassemic fetuses and newborn: Hb Barts, a gamma chain homotetramer, and HbH (β_4). Both of these abnormal hemoglobins function poorly in oxygen exchange, and tissues suffer oxygen deprivation. Affected infants die either before birth, or are

Table 7-3. Relationship between clinical severity of α-thalassemia and number of deleted α-globin genes.

Number of Deleted Genes	Hemoglobins Synthesized	Clinical Features
4	Hb Barts ($\pm$ HbH)	Hydrops Fetalis Early Death
3	Largely HbA$_1$ 2-40% HbH $\pm$ Hb Barts	Enlarged Spleen Abnormal Blood Smear Moderate Anemia(Variable)
2	HbA$_1$, Normal or Decreased HbA$_2$, 2-8% HbH	Abnormal Blood Smear
1	HbA$_1$, HbA$_2$, $\pm$ HbH	Normal

born with gross edema and a variety of other abnormalities and die during the neonatal period. Thalassemia in which there are no α-globin chains present is referred to as α^0-thalassemia.

α^0-thalassemics retain low levels of embryonic hemoglobin synthesis. Hb Portland, containing zeta and gamma chains, accounts for about 1-5% of the total hemoglobin present. The unusual persistence of this embryonic hemoglobin may represent an abortive attempt to compensate for the lack of α-chains by continuing expression of the embryonic locus. Apparently compensation is not complete because elements promoting the normal down regulation of the zeta locus are still present on the thalassemic chromosomes.

When one functional α-globin gene is retained in the patient's cells, moderate anemia is often found, although considerable variation has been observed, ranging from mild to severe anemic phenotypes. These observations have been interpreted to suggest that both genetic background (other nuclear genes) and environmental factors play a role in development of the particular thalassemic phenotype in a specific patient. Larger amounts of Hb Barts are found, and an enlarged spleen is present. The blood smear appears abnormal due to the presence of cells containing granules formed from precipitated Hb Barts. Loss of one or two α-globin genes is easily tolerated, and no clinical complications are seen. Patients with two α-globin genes have a small amount of Hb Barts at birth, and occasionally will have a small number of cells with Hb Bart granules. Thalassemias in which some residual α-chain synthesis occurs are called α^+-thalassemias.

More extensive deletions of the α-like globin region on chromosome 16 have been observed in certain Mediterranean populations. Some of these patients are heterozygous for a chromosome lacking the zeta-globin gene, the zeta-globin pseudogene, the alpha-globin pseudogene, the α^2-locus, and the anterior third of the α^1-locus. It is unlikely that homozygotes for this extensive deletion will be encountered, since they would be incapable of forming normal embryonic, fetal and adult hemoglobins and would most likely die during embryonic development.

Although the duplicate α-globin genes encode identical proteins and have identical coding sequences, their 3' untranslated regions exhibit moderate sequence divergence. This sequence dissimilarity permits estimation of their relative expression at the mRNA level. α^2-mRNA is two- to three-fold more prevalent than α^1-mRNA, suggesting that deletion of the α^2-globin locus would have a more severe impact upon the physiology of an individual than deletion of the α^1-globin locus. Studies of α-

thalassemics indicate that α^2-globin deletions are more prevalent than α^1-globin gene deletions, supporting this expectation.

The origin of α-like globin gene deletions is related to their similar base sequences. Meiotic misalignment of these loci followed by recombination between the mispaired loci generates chromosomes with too many and too few genes (see Chapter 6). Inheritance of two of the latter chromosomes may lead to α-thalassemia.

α-thalassemia may also be caused by base substitutions, small deletions or small insertions within and around the α-globin loci. These thalassemias include mutations within promoter sequences, at splice junctions, and at critical sites within exons. One of the latter mutations leads to the formation of a highly unstable α-chain. Another mutation eliminated five nucleotides near the 5' splice junction of the first intron. Deficient α-globin synthesis was caused by abnormal splicing and resulting mRNA deficiency. Similar mutations will be discussed more fully in relation to β-thalassemia.

The preceding discussion demonstrates that α-thalassemia is a heterogeneous disease. α-thalassemic patients may have various combinations of chromosomes containing gene deletions, nonfunctional α-globin genes, or a combination of these alternatives (Fig. 7-12). The relative severity

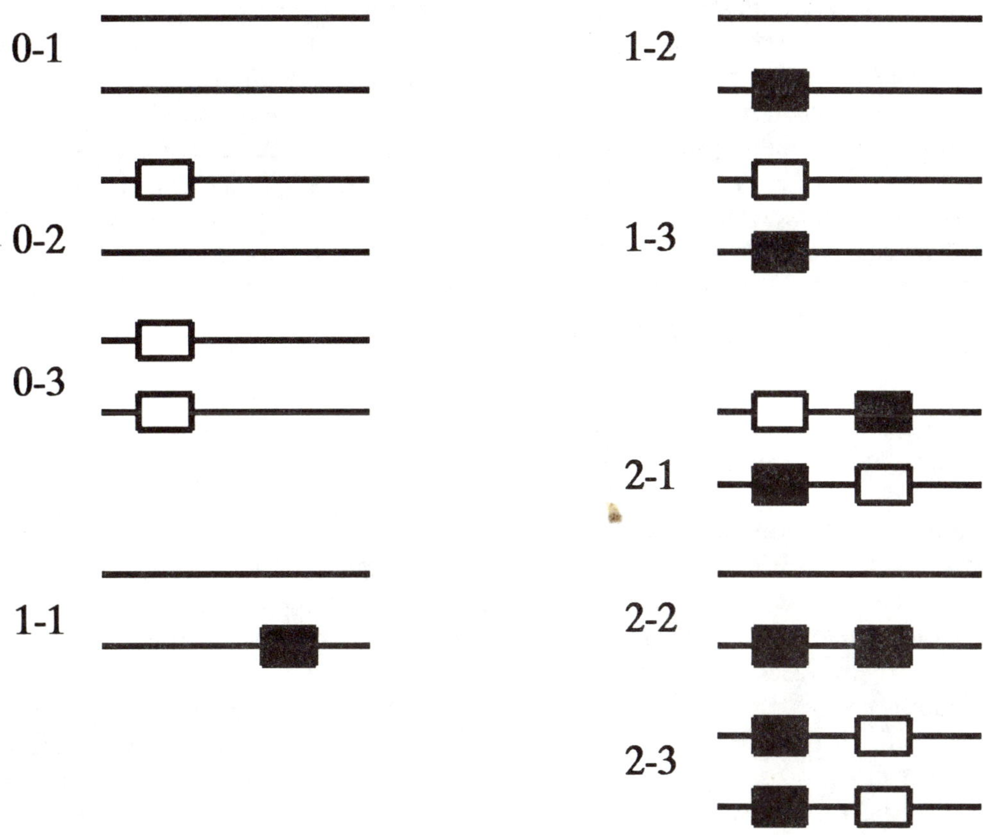

Fig. 7-12. Combinations of gene deletions and other mutant α-globin genes that may be found in α-thalassemics. Chromosomes on which no α-globin loci are present are indicated by straight lines. Non-functional genes are represented by open boxes, and functional genes have been shaded. The α^2-globin locus is placed at the left and the α^1-globin locus at the right of each chromosome, respectively. The first number in each pair represents the number of functional α-globin loci present in the nucleus of each patient.

of the clinical phenotype is not only determined by the numbers of residual functional loci, but also by whether the intact alleles occur at the α^1-globin locus or α^2-globin locus. For example, patients 1-1 and 1-2 both retain a single functional α-globin allele. However, patient 1-1 would be expected to have a more severe disease than patient 1-2 due to the greater contribution of the α^2-globin locus to the mRNA pool in normal cells. Similarly, patient 2-3 would be anticipated to have near normal levels of α-globin, since both α^2-globin loci are present; whereas, significant reductions of α-globin chains would most likely occur in patients 2-1 and 2-2. Predictions regarding the severity of a particular chromosome combination should be made with caution. Most common deletions spanning the α^2-globin locus and its upstream flanking region, but leaving the α^1-globin gene and its upstream region largely intact (bottom chromosome, individual 1-1), tend to be associated with a partial compensatory increase in the expression of the remaining locus. However, there is no increased expression of the α^1-globin locus on a chromosome in which the α^2-globin gene has been inactivated by a point mutation. These observations suggest that the deletion removed one or more sequences that usually participate in the down regulation of the α^1-globin gene.

Many thalassemia mutations occur quite frequently in areas infested with falciparum malaria, such as southeast Asia, central Africa, and tropical regions of South America. Furthermore, these mutations continue to be frequent among Mediterranean populations in regions where this malaria was once a problem. In each of these regions, persons heterozygous for various thalassemia chromosomes had a survival and reproductive advantage and spread their alleles among their offspring. The distribution of chromosome types varies among these populations, reflecting the chance occurrence of a particular α-globin locus mutation in a region and its subsequent increase in frequency due to **heterozygous advantage**. Chromosomes that have lost both α-globin loci are frequent among southeast Asian and Mediterranean populations, but relatively rare in Africa. The greater frequency of the "double deletion" chromosome is partially responsible for the higher frequency of the hydrops fetalis phenotype in these populations and its relative rarity among Africans. The most common α-thalassemic chromosome in all α-thalassemic groups lacks the α^2-globin gene and retains the α^1-globin gene.

β-Thalassemias

β-Thalassemias are associated with absent (β^0) or reduced quantities (β^+) of β-globin. The reduced levels of β-globin result in: 1) a relative excess of α-globin chains, 2) decreased quantities of HbA_1, and (3) somewhat increased amounts of HbA_2 and HbF. The quantities of HbA_2 and HbF vary from cell to cell and are insufficient to compensate for the deficiency of HbA_1. The excess α-globin chains aggregate and form insoluble complexes or inclusion bodies in nucleated red cell precursors Fig. 7-13). These cells are usually sequestered and destroyed by the spleen with accompanying enlargement of this organ. Cells remaining in the circulation have very low hemoglobin contents. The bone marrow expands the compartment responsible for production of erythrocytes. Impact of these changes upon the skeleton is striking, with enlargement of the bones of the face, a "hair-on-end" X-ray image of the skull bones and a weakening of the long bones, making these patients more susceptible to fractures. These symptoms are most characteristic of persons homozygous for β^0-thalassemia mutations. Symptoms in individuals homozygous for β^+-mutations or those who are genetic compounds for β^0- and β^+-thalassemic mutations tend to be milder. Heterozygotes for a normal allele and a thalassemia mutation are often asymptomatic. Clinical diagnosis of heterozygosity can often be made on the bases of reduced mean cell hemoglobin and mean cell volume and increased HbA_2 levels. Treatment of severe β-thalassemia usually involves combined use of transfusions and administration of iron chelating agents. The latter are necessary to prevent iron deposition in the heart, liver, kidney and other tissues. When chelating compounds are not employed, iron deposition in the heart usually results in death before age 30.

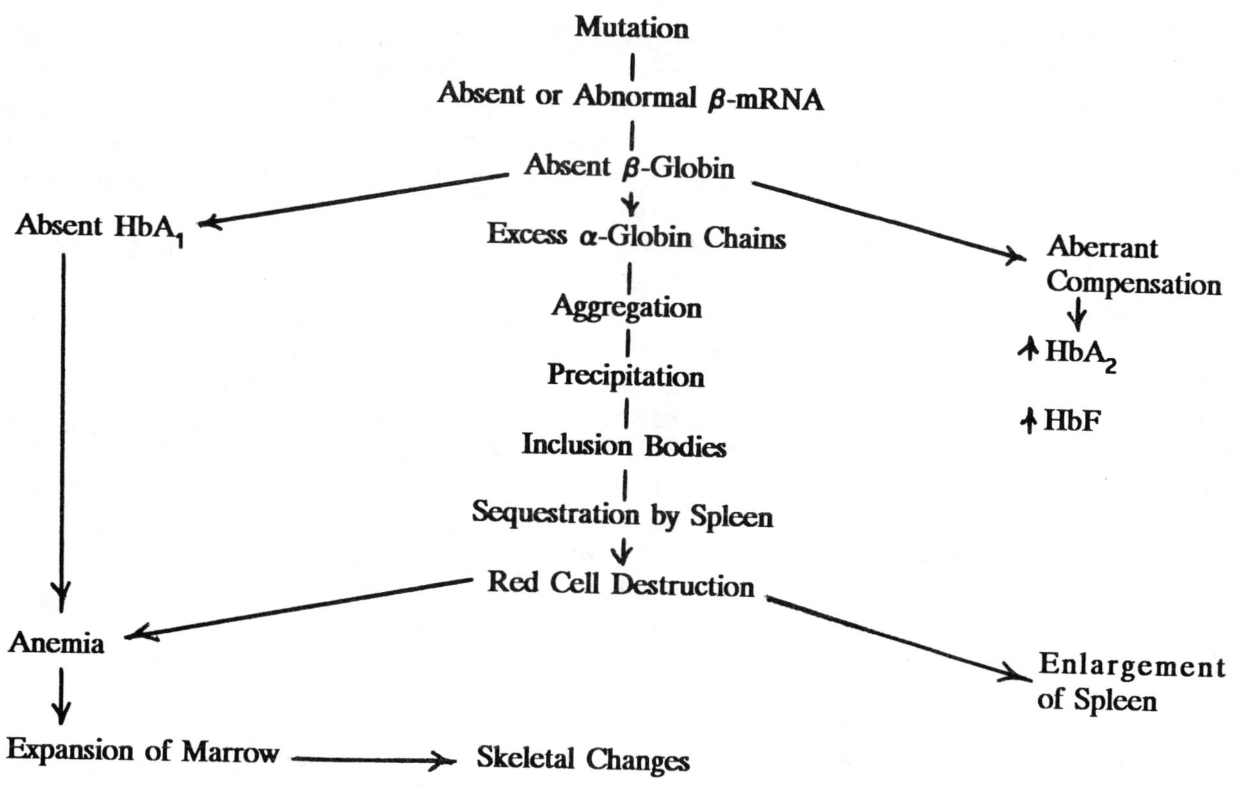

Fig. 7-13. Relationship between the molecular changes in β^0-thalassemia and the clinical phenotype.

The molecular basis of β-thalassemia has been the subject of intensive investigation during recent years. Unlike α-thalassemias, β-thalassemias are rarely caused by gene deletion. β-thalassemias are heterogeneous at the molecular level. They may be caused by mutations in the β-globin promoter, within exons, at intron-exon junctions, or in the AATAAA box. Occasionally intragenic recombination between mispaired β- and δ-globin genes will result in a disorder resembling β-thalassemia. Several of these mutations are summarized in Fig. 7-14. Mutations in region 1 (promoter) affect transcription of the β-globin allele affected. A C --> T transition at position -88 of the promoter results in a sharp reduction of transcription and generates a β^0 allele. This mutation has been observed in both African

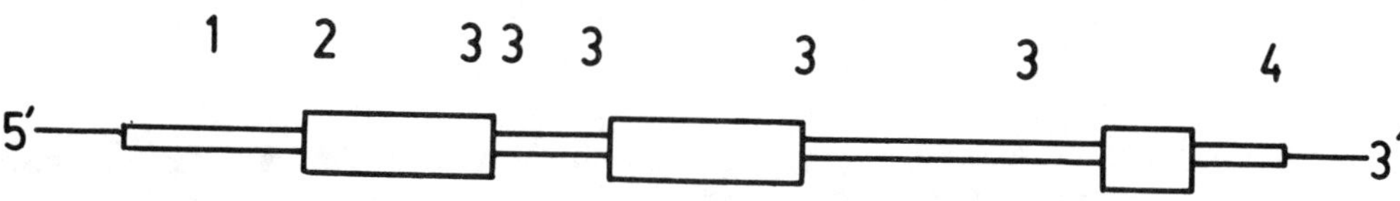

Fig. 7-14. Representative mutations generating β-thalassemia alleles, 1) Mutations in the promoter; 2) mutations generating premature translational stop signals resulting in truncated globin molecules; 3) mutations affecting RNA splicing; and 4) mutations within the transcriptional stop (AATAAA) signal.

and Asian Indian populations. A second mutation (**A-->G**) at position -29 (ATA box) decreases transcription of the allele and produces a common β^+-thalassemia in Africans. Two mutations that generate premature translational stops (region 2) occur frequently. One of these converts a lysine codon (**AAG**) at position 17 to **TAG**. This mutation is common among southeast Asians and produces a β^0-thalassemia. Another mutation affects position 39, converting an amino acid encoding triplet to a translational stop. This β^0 allele is very common among Mediterranian populations.

Several mutations have been described which alter the splicing pattern of the β-globin pre-mRNA (3 in Fig. 7-14). These mutations may abolish donor or acceptor sites, create new donor or acceptor sites, and/or uncover "cryptic" splice sites that are not used unless the usual site is modified. If a base substitution alters either the 5' **GT** or the 3' **AG** of the intron, splicing is abolished, and β^0-thalassemia results. Nucleotides near the **GT** and **AG** doublets also are conserved and constitute splice site consensus sequences. Mutations in these regions generally produce β^+-thalassemias. For example, a base substitution near the 5' end of the second intron may abolish or markedly diminish the ability of this donor site to participate in RNA splicing. The efficiency of splicing drops, resulting in a deficiency of β-globin mRNA. Furthermore, the spliceosome may "scan" the intron looking for a downstream cryptic site and form an mRNA that has additional bases located between the bases encompassed by the second and third exons (Fig. 7-15). If the addition is "in-frame" a globin may be translated from this mRNA that has an unusual sequence. Out-of-frame splicing may lead to a drastically altered protein or may generate a premature translational stop, producing an unstable β-globin.

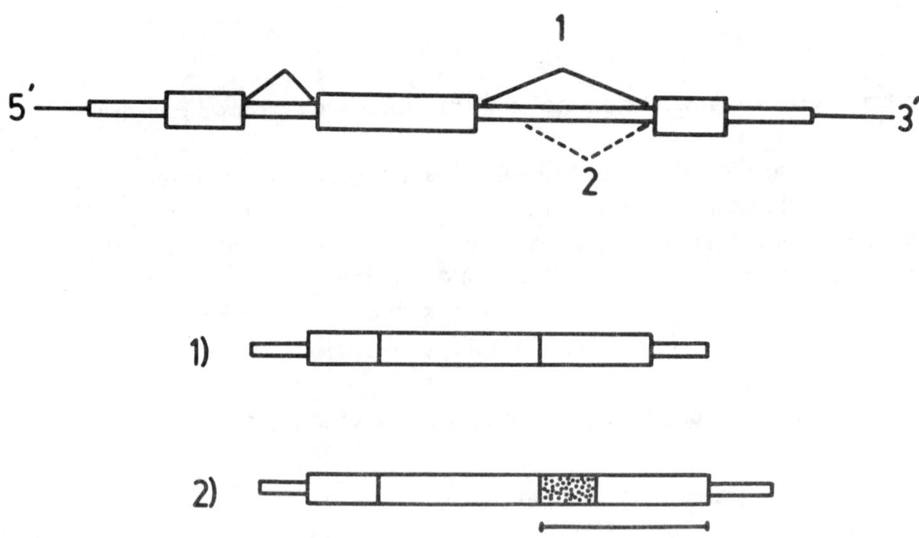

Fig. 7-15. Effects of a mutation affecting a donor splice site consensus sequence. The normal splicing process is illustrated by (1), producing the mRNA in the center of the figure. Mutation within the consensus sequence forces the spliceosome to scan the intron until it finds a cryptic donor site, and splicing will follow pattern (2) producing the mRNA at the bottom of the figure. The stippled box represents the additional intron sequence that has been spliced into the mRNA.

The final type of thalassemia mutation illustrated in Fig. 7-14 (region 4) occurs in the transcriptional stop/polyadenylation signal. An African mutation involving a **T-->C** transition is responsible for a β^+-thalassemia. This mutation drastically reduces the quantity of β-globin mRNA produced, and small quantities of an elongated RNA have been found in reticulocytes that are most likely not polyadenylated and are unstable.

Several deletions of varying size have been observed in β-thalassemics. One of these that removes 619 bases from the 3'end of the β-globin gene is common among SIND populations in Pakistan and India. Most of the other deletions are rare and encompass from 1.3 to about 10kb of DNA.

Intragenic recombination within mispaired β- and δ-globin genes was described in Chapter 6. Hb Lepore, the $\delta\beta$-fusion gene product, has abnormal properties and is expressed at low levels, since it is under control of the δ-globin promoter region. These patients present with a β^+-thalassemia. Several other fusion proteins have been described which arise by a similar mechanism and which are associated with β^+-thalassemias.

β-thalassemia mutations tend to have **cis** effects, i.e. they affect only the β-globin gene in or around which they occur. Fig. 7-16 presents a kindred that is segregating for both β^S and β^0 alleles. The person indicated by the arrow is the son of a man with sickle cell trait ($\beta^S\beta^A$) and a woman who is heterozygous for β^0-thalassemia ($\beta^0\beta^A$). The boy has the genotype, $\beta^S\beta^0$. The thalassemia mutation blocks expression of the β-globin allele on that chromosome, resulting in the occurrence of sickle cell anemia in the patient. Many other examples of patients who are genetic compounds are known. One

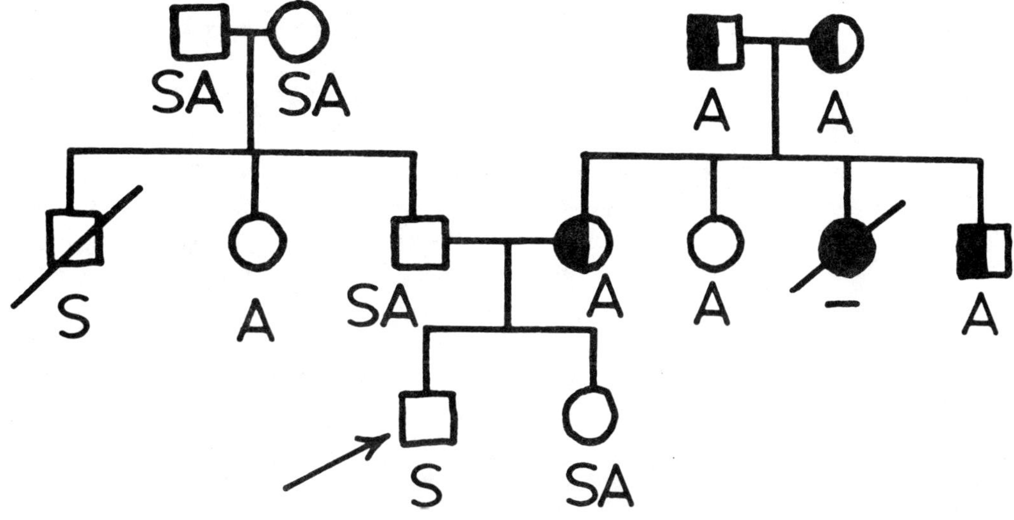

Fig. 7-16. Pedigree of a kindred segregating for sickle cell anemia and β-thalassemia. The hemoglobin phenotypes are presented below the pedigree symbols. A shaded symbol indicates homozygous β^0-thalassemia, and half-shaded symbols denote individuals with elevated HbA$_2$ and who are $\beta^0\beta^A$.

of these is common among southeast Asians. HbE ($\alpha_2\beta^E_2$) is the most common hemoglobin variant occurring in this region of the world. Persons who are homozygous for the β^E allele develop a mild β-thalassemia. The base substitution in codon 26 not only results in a replacement of glutamic acid by lysine, but also generates a cryptic splice site. The efficiency of splicing is reduced, resulting in a mild reduction of β-globin mRNA. However, when the β^E allele occurs in combination with a β^0 allele, moderate to severe β-thalassemia is found.

Hereditary Persistence of Fetal Hemoglobin

A large number of deletions including two or more β-like globin genes and their surrounding sequences have been reported. Several of these are illustrated in Fig. 7-17. Deletions which are quite small and include only the β- and δ-globin genes produce $(\delta\beta)^0$-thalassemia. Clinical symptoms of this disease are quite similar to those of β-thalassemia, since HbA_2 constitutes such a small fraction of adult hemoglobin. One exceptional deletion occurring in Spanish populations is quite large, extending more than 50kb downstream from the β-globin locus. Although there is some elevation of HbF in $\delta\beta$-thalassemics, compensation is incomplete, and the level of HbF varies among erythrocytes.

By contrast, deletions that extend further upstream from the δ-globin locus but exclude the $^A\gamma$-locus are expressed as **hereditary persistence of fetal hemoglobin (HPFH)**. Individuals heterozygous for the deleted chromosome display deficiencies of β- and δ-globin chains coupled with increased γ-globin synthesis. About 10-35% of the hemoglobin occuring in these heterozygotes is HbF. Furthermore, there appears to be a uniform increase in HbF in all erythrocytes. Homozygotes possess normal hemoglobin levels, and all of their hemoglobin is HbF. This interesting group of deletions results in derepression of the gamma-globin genes and total compensation for the missing β- and δ-globin chains, such that HPFH persons are free of clinical complications. There does not appear to be a simple explanation for the different consequences of the $(\delta\beta)^0$-deletions and the HPFH deletions. One possibility may be related to the continued presence of sequences in the region downstream from the β-globin gene in the $\delta\beta$-globin deletions and not in many of the HPFH deletions. These sequences could down regulate expression of the gamma-globin genes. However, the Spanish $(\delta\beta)^0$-deletion mentioned above also deletes these sequences; and there are at least two HPFH deletions that terminate near the β-globin gene. Therefore, this argument does not appear to hold. Another possibility might involve sequences located in the region ranging from upstream from the β_1-pseudogene to the vicinity of the δ-globin promoter. These sequences may also exert negative control of expression of gamma-globin genes. This possibility also appears incorrect, since a Japanese variant of $(\delta\beta)^0$-thalassemia also deletes the region speculated to play a regulatory role, and an African HPFH deletion leaves a majority of this region intact. In fact, the latter deletion is included within the region spanned by the Japanese deletion!

A number of non-deletion forms of HPFH have also been found. One of these occured following a base substitution at -202 of the $^G\gamma$-globin gene. This alteration resulted in continued high-level expression of this gene in adults. A second group of HPFH point mutations affects the $^A\gamma$-globin gene promoter. Base substitutions at -117 and -196 result in continued expression of this gene in adults at unusually high levels. In some of these mutations, up regulation of the gamma-globin locus does not occur unless there is a downstream β^S allele. The mechanism by which these changes cause unusual expression of these genes is unknown. One possibility may be that the elements involved are response elements for regulatory proteins encoded by other regions of the β-like gene cluster which normally down regulate expression of the fetal genes in adult normoblasts and other red cell precursors. The mutations would then render the site unresponsive to these proteins. The interaction with the β^S allele is difficult to explain.

More extensive deletions have also been described. Several of these result in loss of not only the δ- and β-globin genes, but also one or more of the gamma-globin genes, and, in some cases, the ϵ-globin gene. Homozygosity for these large deletions has not been observed, possibly as a consequence of their lethality and relative rarity. The English deletion affecting expression of all β-like globin genes (lower deletion, Fig. 7-17) is of particular interest, since the $^A\gamma$-, δ- and β-globin loci are intact but not expressed. This observation suggested that there may be a regulatory region governing expression of all globin loci upstream from the $^A\gamma$-locus. This **locus activating region**, situated approximately 60kb upstream from the β-globin locus, has been demonstrated to be essential for normal expression of the

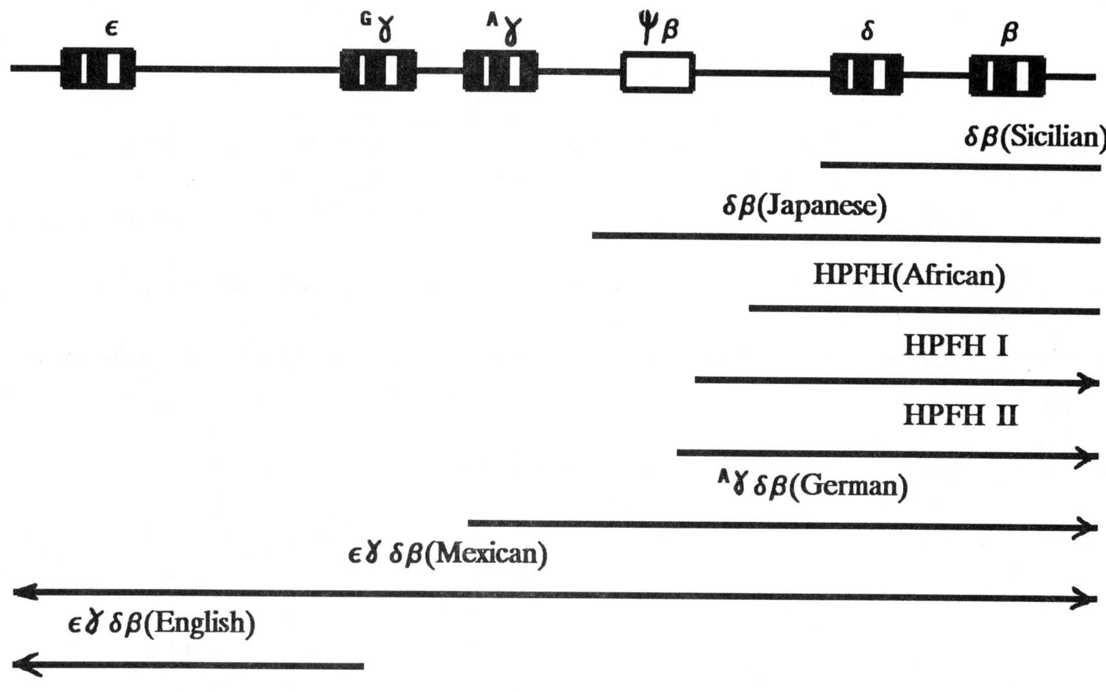

Fig. 7-17. **Complex thalassemias caused by deletions spanning two or more globin loci.** The extent of the deletions is indicated by horizontal bars below the cluster. Arrowheads indicate that the deletions extend further in the direction indicated. The absent globin chains are listed above the bars. HPFH = hereditary persistence of fetal hemoglobin.

downstream genes (see Forrester et al., (6)). There also appears to be a **LAR** for the α-like globin gene cluster, since deletions of the zeta-globin upstream flanking region (most notably from a -28kb to -67kb (17)) block normal expression of the intact downstream α-globin loci.

The occurrence of the **LAR** and other globin gene controlling elements in all nuclei and their expression only in red cell progenitors imply that these sites must be accessible in expressing tissues and inaccessible to transcribing enzymes in nonexpressing tissues. One or more trans-acting proteins might be involved with the tissue and temporal specificity of globin gene expression. Evidence for the existence of one such factor has been obtained from studies of a family segregating for a non-deletion form of HPFH(18). β^S and β^0 alleles were segregating in the kindred in addition to HPFH. Two persons were β^S/β^A heterozygotes, exhibited normal levels of HbS, but had 5.3 and 5.2% HbF levels which are about five or more times as high as levels anticipated in persons with sickle cell trait. A β^A/β^0 heterozygote (β-thalassemia trait) had unusually high levels of HbF (6.8%). A β^S/β^0 genetic compound had HbS, HbA$_2$, and HbF in the ratio of 80:4:16. Segregation of the mutation responsible for the HPFH phenotype appeared to be segregating independently from the β-like globin gene cluster, and this suspicion was confirmed by DNA analysis of the region within and around the cluster to rule out other alternatives. Studies of cultured erythroid cells have demontrated that appropriate gamma-globin gene expression requires the interaction of one or more proteins with the gamma-globin promoter(19). The protein(s) are absent when gamma-globin genes are expressed and present when these genes are silent. The mutation in this family may have been incurred by a locus encoding such a protein. Its absence results in the continued expression of the gamma-globins in adult cells, irregardless of their β-globin locus genotype.

Haplotypes and Globin Mutations

An array of closely linked genetic markers is called a **haplotype.** Since the markers are tightly linked, allelic variation at these sites tends to be maintained together, permitting their use for diagnosis of hemoglobin disorders and for tracing the evolutionary origins of the respective globin mutations that

Table 7-4. A series of common haplotypes composed by restriction site polymorphisms in the β-like globin gene cluster.

Location:	--ϵ---------------$^G\gamma$----$^A\gamma$------ β-----------δ-----β---------------------------------						
Sites:	Hc	Hd	Hd	Hc	Hc	A	B
Haplotype:							
I	+	--	--	--	--	+	+
II	--	+	+	--	+	+	+
III	--	+	--	+	+	+	--
IV	--	+	--	+	+	--	+
V	+	--	--	--	--	+	--
VI	--	+	+	--	--	--	+
VII	+	--	--	--	--	--	+
VIII	--	+	--	+	--	+	--
IX	--	+	--	+	+	+	+

Restriction Sites: Hc = Hinc II; Hd = Hind III; A = Ava II; B = Bam HI.

are nearby. A series of relatively common haplotypes found in human populations that can be demonstrated by the use of four restriction enzymes is presented in Table 7-4. Each haplotype differs with respect to the presence or absence of a number of these sites. Each site is polymorphic, and most populations contain several different haplotypes for the β-like globin region. Interpretation is usually dependent upon the assumption that the "haplotype framework" was present in the chromosome first, and that the respective globin mutations occurred later. Derivation of haplotype-globin allele combinations from studies using the four restriction enzymes must be done with care. A given allele may be associated with haplotype I in one population and with haplotype IV in another population. A country such as the United States has often been settled by peoples from diverse origins, and haplotype-globin allele associations will also be heterogeneous. Furthermore, occasional recombinations can also alter these associations.

Some thalassemia mutations are associated exclusively with a specific haplotype. A β^0 mutation caused by a **GT-->AT** mutation in the donor site of intron 1 and which is found in Mediterranean populations is found on haplotype V (Table 7-4). By contrast, the common mutation in Mediterranean populations described above that introduces a translational stop at position 39 is found in association with haplotypes I, II, VII and IX. Either this mutation occurred several times, or its relationship with the original haplotype on which it occurred has been altered by "gene conversion" events (see below). An example of the same mutation occurring more than once in different populations is provided by the β^S mutation. A polymorphic HpaI restriction site has been identified which is located between the AvaII and BamHI sites of chromosome 11 (Table 7-4). Digestions of this DNA with HpaI result in the production of 13kb or 7.6kb (or occasional 7.0kb) restriction fragments containing the β-globin gene. One origin of the β^S mutation can be traced to West Africa where it apparently arose on the chromosome lacking the HpaI site and which generates the large fragment. This mutation conferred a selective advantage to heterozygotes, and the 13kb-β^S association became common while the original 13kb-β^A combination became infrequent. Africans from this region migrated to Sicily, North Africa, and the southern tip of the Saudi Arabian peninsula. By contrast, the β^S allele in East Africans and of emigrants from this region is associated with the 7.6kb (7.0kb) fragment, implicating a separate mutational event. RFLP studies using several restriction enzymes of Kuwaiti and Indian groups indicate that they share a third haplotype that is distinct from both African mutations.

Broader screening of West African and Jamaican populations using a larger number of restriction enzymes to probe the structure of the β-globin gene region has indicated that the β^S allele occurs in association with several different haplotypes. This result raises the possibility that the original β^S-haplotype array may have been altered by **gene conversion**. Gene conversion refers to the nonreciprocal exchange of chromosome segments. Reciprocal recombination produces two genetic combinations that are identical to the original arrays and two genetic combinations that have new arrays of genetic material (see Chapter 3). Gene conversion results in three products that resemble the original configuration and one that has a new array of genes or markers. The mechanism by which this nonreciprocal exchange is achieved is unknown; however, a possible mechanism is illustrated in Fig. 7-18. The example shows the transfer of the HpaI restriction site to a chromosome carrying the β^S allele. As a consequence of the nonreciprocal exchange, one of the β^S alleles will now be located on the small HpaI restriction fragment. However, structural features flanking the region of conversion will have the original characteristics of the original β^S chromosome. Gene conversions appear to occur frequently in both the α-like globin gene region and the β-like globin gene region.

Prenatal Diagnosis of Thalassemias

Deletion forms of α-thalassemias may be diagnosed using DNA "titration" studies (Fig. 7-19). DNA from fetal cells is digested with a restriction enzyme such as EcoRI to manageable sizes, and the digested DNA is applied to nitrocellulose in varying amounts. At least two series are applied per cell extract. One lane is hybridized with radiolabelled or biotinylated β-globin cDNA and the other with an α-globin cDNA probe. The lane treated with the β-globin probe serves as a control to make sure the DNA has not been degraded during isolation and treatment. Four possible outcomes may be obtained using the α-globin cDNA probe. If no α-globin genes are present in the nuclei of the cells from

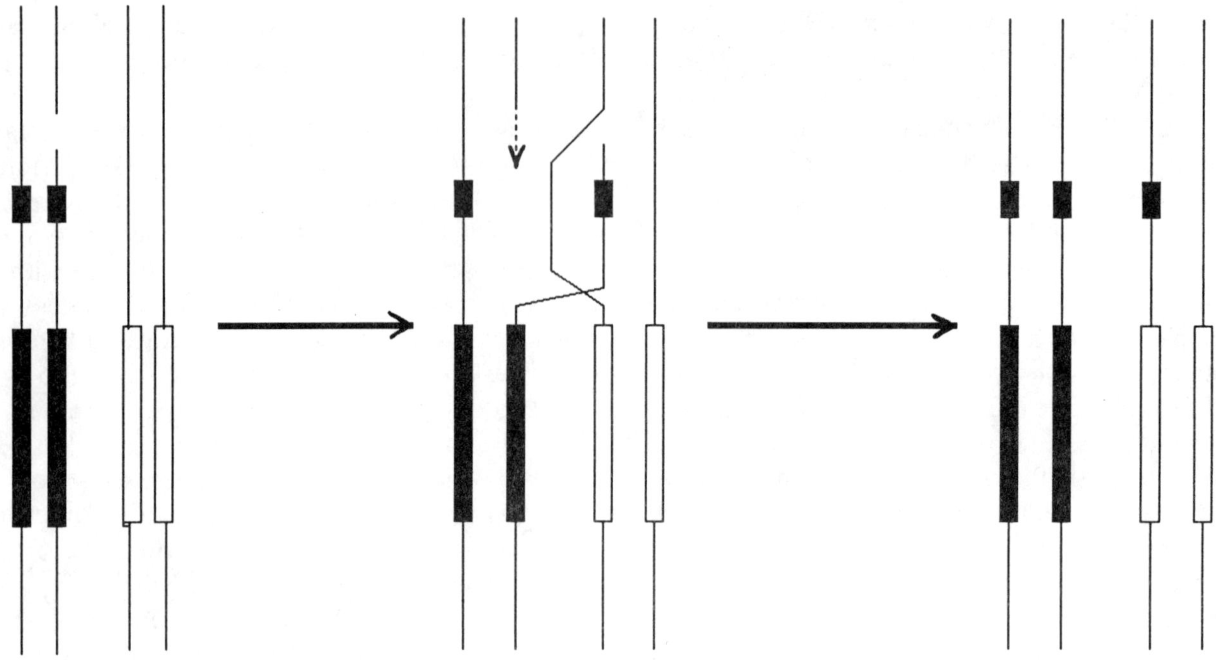

Fig. 7-18. Gene conversion event transferring a HpaI restriction site to a chromosome carrying the β^S allele. Left: a nick in one strand of a double helix appears, and a portion of the broken strand invades the neighboring double helix, displacing one of its chains (center). The connection of the invading HpaI site with its original β^A allele is broken, and the gap is filled by DNA replication using the intact strand as a template. The displaced flanking region of the β^S allele is excised from the strand, and the HpaI site is ligated to the β^S allele. The products of the gene conversion are displayed in the right panel. Note how three of the four DNA chains are present in their original structural forms, and that the rearranged chromosome will have a small segment of the original β^A chain inserted into the β^S DNA chain. The β^A allele is symbolized by a shaded long rectangle, and the β^S gene is represented by an open rectangle. The HpaI site is indicated by a short shaded rectangle. Each double helix is displayed in its uncoiled state.

which the DNA was obtained, no hybridization signal will appear in any of the application bands (right track). As the number of residual genes increases, hybridization signals will appear in the application zones containing 15, 10, and/or $5\mu g$ of digested DNA.

Nondeletion forms of thalassemia may be diagnosed by RFLP analysis, provided the parents have "informative" combinations of chromosomes. This procedure exploits the presence of a restriction site near one allele and the absence of that site from the chromosome carrying the other allele. When the DNA is extracted, digested with the appropriate restriction enzyme(s), and the fragments are separated by electrophoresis, one allele will migrate with a large fragment; and the other allele will occur on a small fragment. Fig. 7-20 illustrates the application of a "double digest" approach for the diagnosis of a β-thalassemia mutation. Two haplotypes are segregating in this kindred, and the two β-thalassemic individuals possess only haplotype VII which lacks a BamHI site. This indicates that the β^+ allele is linked to haplotype VII. The couple in generation II both have β-thalassemia trait and run a 25% risk for having a child with β^+-thalassemia. If the couple requests prenatal diagnosis, chorionic villus cells or amniocytes may be used as a source of fetal DNA. The DNA would be digested with both AvaII and

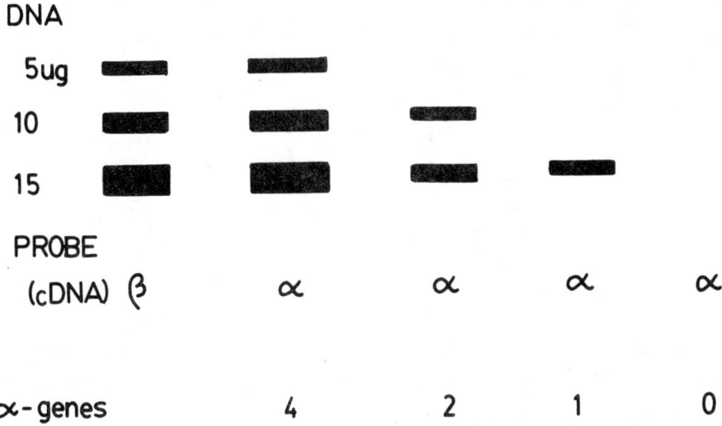

Fig. 4-19. Estimation of the number of α-globin genes by nitrocellulose blotting. DNA is extracted and digested with EcoRI, and varying quantities of the digest are applied to nitrocellulose as indicated. One series of applications for each sample (left lane) are probed with a β-globin cDNA to insure the DNA has not degraded during isolation and treatment. The remaining four lanes show four possible outcomes when the remaining series of applications are hybridized with α-globin cDNA. No hybridization signal would be observed in any of the application zones if all four α-globin genes have been deleted. By contrast, the hybridization profile for DNAs possessing three or four α-globin genes would closely resemble the profile for the β-globin control.

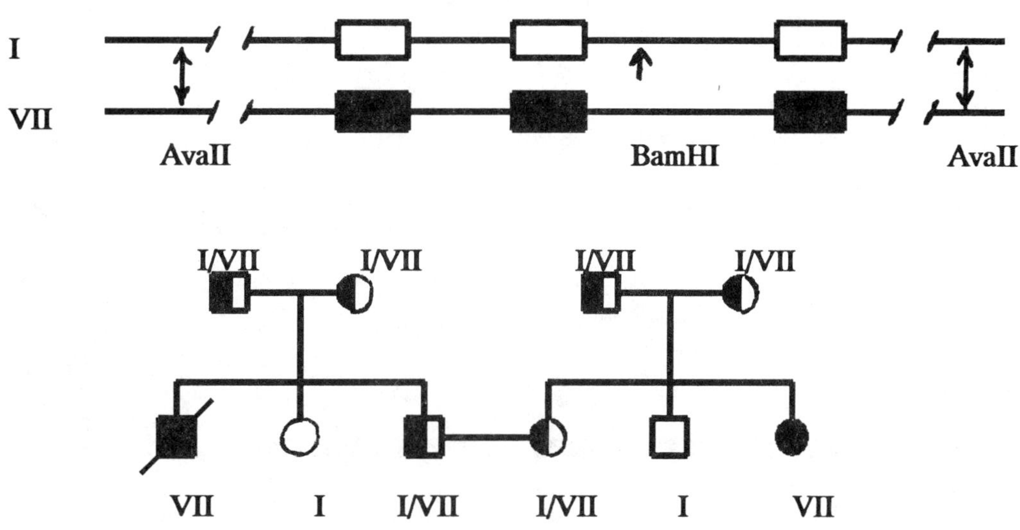

Fig. 7-20. Diagnosis of β-thalassemia by using RFLPs. Chromosomes are shown at top. The β-thalassemia allele (shaded) lacks a BamHI site. Below: Pedigree segregating for β-thalassemia. Patients are shaded, and carriers are represented by half-shaded symbols. Haplotypes are listed above or below pedigree symbols.

BamHI, and the fragments would be separated by electrophoresis. After Southern blotting using a β-globin cDNA, a diagnosis of all three genotypes is possible. If the fetal cells produce a small β-globin fragment or a small and large fragment, the genotype would be $\beta^A\beta^A$ or $\beta^A\beta^+$, respectively. However, the presence of only the large β-globin fragment would indicate homozygosity for the β^+ allele. Since the BamHI restriction site is within the second exon of the β-globin gene, it is unlikely that an error would be made as a consequence of a recombination event that would transfer the site from haplotype I to haplotype VII.

The term "informative" was used in the preceding paragraph to refer to the occurrence of unambiguous allele-haplotype combinations in key members of the pedigree. Suppose the thalassemic girl was found to have both haplotypes I and VII? This would indicate that each of her parents was carrying the β^+ allele on a different haplotype, and it would be difficult to determine which allele-haplotype combination was present in the woman at risk for having a thalassemic child. This combination of relatives and haplotypes would be an example of an uninformative pedigree.

GLUCOSE-6-PHOSPHATE DEHYDROGENASE DEFICIENCY

There are many other causes of anemia, both hereditary and acquired. Another hereditary form of anemia that is found in high frequency in falciparum malarial areas and among descendants of these populations who have migrated elsewhere is glucose-6-phosphate dehydrogenase (G6PD) deficiency. This enzyme is encoded by a X-linked gene that is subject to X-inactivation in females and which appears to be expressed in many different tissues. However, the enzyme is especially important to erythrocytes, since these cells lack mitochondria and must rely upon the hexose monophosphate shunt (HMS) for reducing substances required for stability of the red cell membrane. G6PD catalyzes the first reaction that commits glucose-6-phosphate to the HMS, and G6PD deficiency results in a lack of reducing

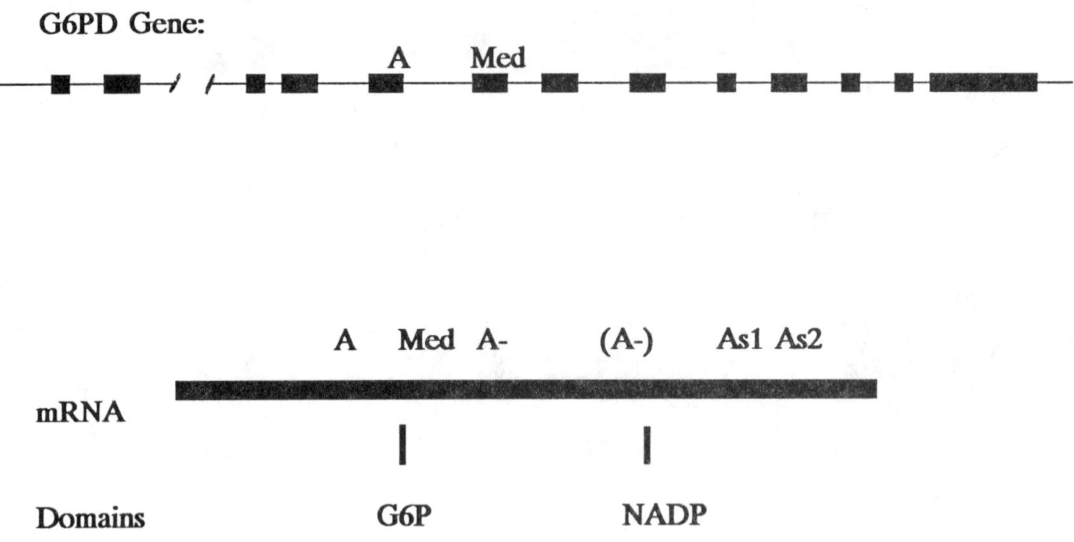

Fig. 7-21. Glucose-6-phosphate dehydrogenase gene and its mRNA. Approximate sites of common mutations are entered above the gene and mRNA: A = G6PD-A variant; A- = most common African G6PD deficiency allele; (A-) = less common African G6PD deficiency allele; As1 and As2 = common Asian deficiency alleles; Med = Mediterranean deficiency allele. Domains: G6P = active site region; NADP = coenzyme binding site.

substances and fragility of the erythrocyte membrane, resulting in anemia. The biochemistry of this important system was presented in Chapter 3 in relation to mosaicism caused by X-inactivation.

G6PD is a homodimeric enzyme that is encoded by the gene presented in Fig. 7-21. The gene consists of 13 exons that are transcribed into a mRNA of about 1600 nucleotides in length. The standard form of the enzyme in all populations is G6PD B. The "Mediterranean" allele was formed by a C-->T transition resulting in the replacement of a hydrophilic amino acid (serine) by a large hydrophobic amino acid (phenylalanine) near the active site of the enzyme. The enzyme encoded by the mutant allele has very low activity but displays favorable kinetics, such that the lifespan of red cells in which the enzyme occurs is about normal. Anemia usually occurs under stresses imposed by infections and other factors. As the name implies, the Mediterranean variant (a G6PD B- isozyme) was initially found in Mediterranean populations, with the allele frequency approaching 30% in some of these populations. The high frequency of the allele has been attributed to heterozygous advantage in females in these areas which once had infestations of falciparum malaria. The Mediterranean variant has been "rediscovered" in three other regions (Dallas, Birmingham and Sassari). All four proteins were initially thought to be determined by different alleles; however, allele sequencing revealed all were the result of the same type of mutation. The G6PD Mediterranian mutation is believed to have occurred twice: once in the Mediterranian and once in India. This opinion is supported by the occurrence of an ancient "silent" mutation involving a second nucleotide in the Mediterranean G6PD deficiency allele, but not in the Indian (Sassari) G6PD deficiency allele. G6PD Mediterranean interacts with a condition known as favism. The fava bean is a common dietary component in the Mediterranean region, and some people develop an anemia after eating the beans. This bean-induced anemia is called favism, and all persons with favism also have G6PD deficiency. However, not all people with G6PD deficiency develop favism when exposed to the beans. This suggests that another hereditary or environmental factor in addition to G6PD deficiency is required for the development of favism.

G6PD deficiency is also common in Africa and among African Americans where it occurs in about 11% of African American males. African populations possess four G6PD alleles. The most common is **G6PD^B**. **G6PD^A** is found in 20-40%, displays a rapid electrophoretic mobility as a consequence of an asparagine-->aspartic acid substitution at position 376, and **is not** associated with anemia. The two G6PD deficiency alleles arose following a second mutation in the **G6PD^A** allele. The most common has a valine-->methionine substitution at position 68, while the less common has a leucine --> proline substitution at position 323. The more common **G6PD^{A-}** allele has been found in seven different geographical areas; however, all seven alleles can be traced to a common origin in Africa. Anemia usually is not present unless induced by stress. Several drugs, including acetanelid, methylene blue, nitrofurantoin, and primaquine will induce anemia in people hemizygous for one of these **G6PD^{A-}** alleles. Several other drugs have been said to cause drug-induced anemia in G6PD A- patients; however, many of these, including aspirin, isoniazid, phenytoin, and quinine, do not and can be safely prescribed to G6PD A- individuals. The G6PD A- enzyme has a higher residual activity than G6PD Mediterranean.

G6PD deficiency also occurs frequently among southeast Asians who reside in malaria-infested regions. Two mutations account for a substantial proportion of G6PD deficiency in southeast Asian populations. Both involve replacement of arginine residues that are located near the C-terminus of the enzyme (arginine #459-->leucine and arginine #463-->histidine). The high frequency of G6PD deficiency alleles in southeast Asian populations can be attributed to heterozygous advantage in females.

Although G6PD is a homodimeric enzyme, females who are heterozygous for alleles specifying polypeptides with different charges (ie. **G6PD^A/G6PD^B**) do not form a hybrid band consisting of an A- and a B-type subunit. The G6PD gene is subject to X-inactivation, so only one allele is expressed per cell. The A-type and B-type subunits do not have an opportunity to combine to form the heterodimer. The two homodimeric bands appearing in hemolysates from female heterozygotes reflect the heterogeneous origin of the circulating erythrocytes. Some are descendants of cells in which the **G6PD**[A] allele was inactivated, while others differentiated from cells in which the **G6PD**[B] allele was inactivated.

GENES AND ISOZYMES

Approximately one-fourth of the mutations producing amino acid substitutions in a protein result in a change of its net molecular charge that is demonstrable by electrophoresis. Despite this limitation, electrophoresis continues to be an important means for identification of mutations affecting the structure of enzymes and other proteins. Since enzymes are typically present in small quantities, sensitive histochemical or fluorescent dyes must be used to locate the enzyme bands in electrophoretic gels. Three general procedures are used to locate zones of enzyme activity following their electrophoretic separation (Fig. 7-22). A synthetic substrate may be prepared that resembles the structure

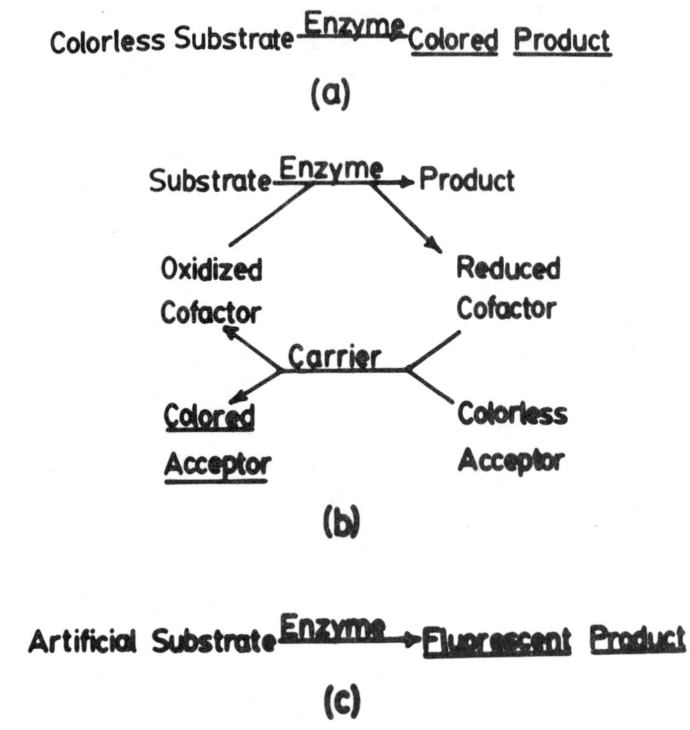

Fig. 7-22. Methods utilized for locating enzymes in electrophoretic gels: a) Direct production of a colored product; b) coupled reaction; c) production of a fluorescent product (visualized under ultraviolet light).

of the natural substrate. The enzyme catalyzes the conversion of the colorless substrate to a colored product, permitting location of the enzyme band. A second procedure uses a coupled reaction. Enzymes such as dehydrogenases reduce a coenzyme (NAD, NADP) during the course of the reaction. The coenzyme can be reoxidized by adding a colorless acceptor molecule, which is changed to a colored acceptor form during the reoxidation process. This system can be used indirectly to locate enzymes that are not dehydrogenases. For example, the enzyme phosphoglucomutase (PGM) can metabolize glucose-1-PO$_4$ to glucose-6-PO$_4$. It is difficult to locate PGM with a direct stain; however, the PGM-catalyzed reaction can be coupled with that catalyzed by G6PD. Glucose-1-PO$_4$, G6PD and its coenzyme (NADP), and the colorless acceptor are used to locate PGM. PGM converts glucose-1-PO$_4$ to glucose-6-PO$_4$, and G6PD oxidizes the glucose-6-PO$_4$, setting the chain of reactions in Fig. 7-22(b) in motion and locating the position of PGM in the gel. The third technique employs a synthetic nonfluorescent substrate that is converted to a fluorescent product by the enzyme. When the gel is viewed under ultraviolet light, the enzyme bands can be seen.

Interpretation of Isozyme Patterns

Many enzymes occur as monomers (contain a single polypeptide). The properties of these proteins do not favor association with identical polypeptides or with polypeptides encoded by other loci. Fig. 7-23 illustrates typical patterns (**zymograms**) of monomeric isozymes occurring in two homozygotes and a heterozygote. Notice that the heterozygote possesses a two-banded pattern with one band corresponding to the band occurring in each of the homozygotes, indicating codominant expression of

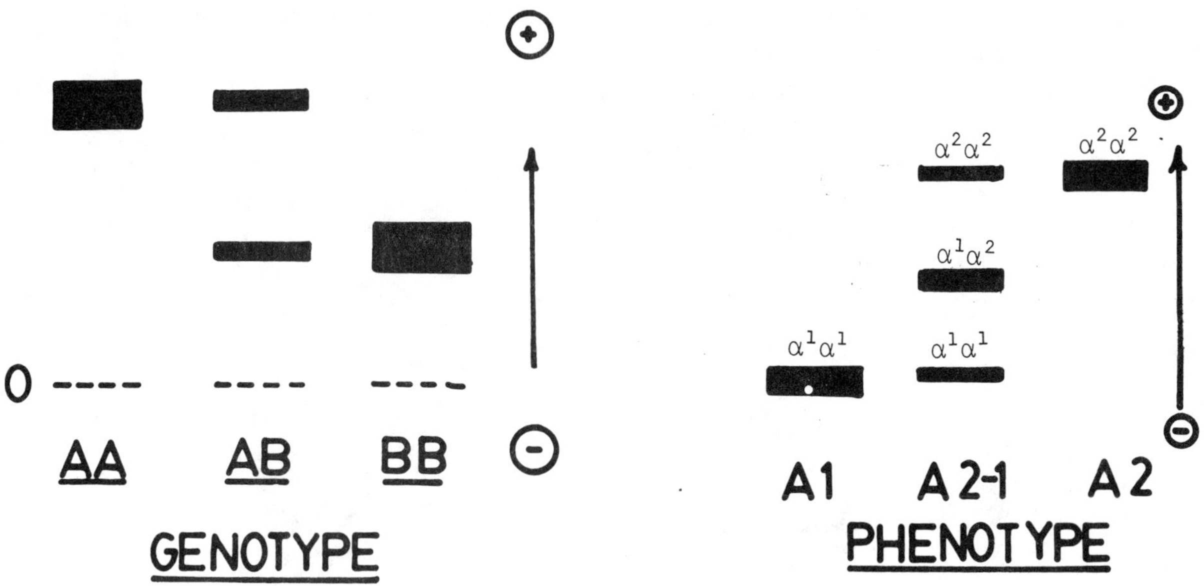

Fig. 7-23. Isozyme patterns of individuals homozygous for the standard allele (A) or mutant allele (B) and a heterozygote (AB). O = origin where samples were applied prior to electrophoresis. + and - denote the positive and negative ends of the gel, respectively.

Fig. 7-24. Zymogram of peptidase A isozymes and their subunit compositions.

the trait. Examples of monomeric enzymes include carbonic anhydrase, PGM, and "red cell" acid phosphatase.

A number of dimeric enzymes have been described which have been observed to exhibit electrophoretic variability in human populations. These include glycerol-3-phosphate dehydrogenase, glutamate-oxalacetate transaminase (GOT), alcohol dehydrogenase, G6PD, and peptidase A. Isozyme patterns for peptidase A are presented in Fig. 7-24. An equilibrium between the monomers and dimers occurs in the cell that is similar to that described for hemoglobin. However, in the peptidase example, the rate of dissociation of the dimers into monomers is relatively slow, permitting the observation of the hybrid band in the gel. The heterozygote produces both homodimers ($\alpha^1\alpha^1$ and $\alpha^2\alpha^2$) and the heterodimer ($\alpha^1\alpha^2$). If both the α^1 and α^2 monomers are equally prevalent and have comparable activities, the heterozygous isozyme profile will be symmetrical. The dimers can be separated into their subunits by treatment with urea and mercaptoethanol. When these reagents are subsequently removed, the subunits will reassociate into dimers. If peptidase A from a person with the A1 phenotype and from a person with an A2 phenotype are mixed, dissociated into their respective monomers by adding urea and mercaptoethanol, and then the urea and mercaptoethanol are removed, then the monomers will reassociate into dimers resembling the heterozygous pattern after electrophoresis. Although the presence of a three-banded pattern following electrophoresis of an enzyme from a heterozygote implies dimeric structure, a two-banded heterozygous pattern does not exclude the possibility of dimeric structure. For example, women heterozygous for a mutation altering the charge of G6PD display only two bands in their hemolysates. The lack of a hybrid band is caused by X-inactivation of one allele in each cell, preventing the opportunity for the different subunits to come together to form the hybrid band. A second cause of a two-banded pattern for a dimeric enzyme extracted from heterozygous tissue is a rapid dissociation of the dimer into monomers. The different-charged monomers are separated by the electrical field before they can reassociate into a hybrid molecule.

Formation of hybrid enzymes may prove to be beneficial in certain cases. For example, certain patients may have a dimeric enzyme that is catalytically ineffective. Introduction of normal enzyme to such individuals may promote some restoration of activity following dissociation of the normal and abnormal enzymes and reassociation of the monomers into a partially functional heterodimer (Fig. 7-25).

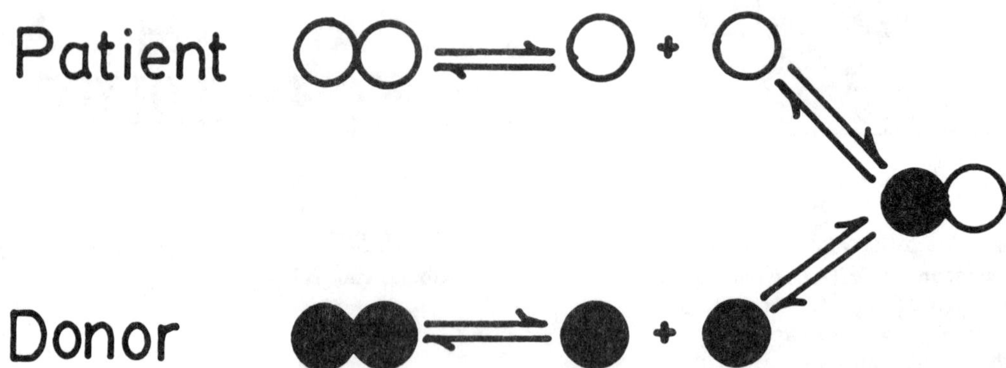

Fig. 7-25. Correction of a hereditary metabolic disease by subunit exchange between the patient's enzyme and infused normal enzyme.

Evidence that such subunit exchange actually occurs _in vivo_ has been derived from studies of patients with Gaucher's disease. This disorder is associated with formation of an abnormal enzyme, glucocerebrosidase, a β-glucosidase participating in the degradation of complex lipids known as cerbrosides. Infusion of purified glucocerebrosidase into Gaucher's patients raised glucocerebrosidase activity to levels that exceeded the sum of the patient's original activity and that of the infused enzyme. One way in which this could happen would be through subunit exchange between the patient's defective enzyme molecules and the infused normal enzyme, resulting in restoration of at least partial activity to the patient's enzyme population.

Fig. 7-26. *Zymograms of lactic dehydrogenase. Left to right: Human plasma samples (lanes 1-3), mouse kidney, and deer liver. The positive pole is at the top .*

A variety of other multimeric enzymes are known to occur in humans. Lactic dehydrogenase (LDH) catalyzes interconversion of pyruvic acid and lactic acid and plays a major role in muscle metabolism. The enzyme is a tetramer with subunits encoded by the loci **LDHA** and **LDHB**. The relative expression of these two loci varies among tissues. **LDHA** is more active in anaerobic tissues such as liver and skeletal muscle, while **LDHB** is more active in aerobic tissues such as cardiac muscle and kidney. This leads to a skewing of the isozyme patterns according to which locus is most active (Fig. 7-26). The human plasma enzyme (left three lanes) is derived from liver, and the relative intensities of the bands more closely resemble those from deer liver than the isozyme profile of mouse kidney LDH in which the isozymes containing primarily B subunits are enriched. By convention, the isozymes are numbered from top to bottom: LDH-1 through LDH-5 (Fig. 7-27). LDH-1 is a homotetramer consisting of B subunits, while LDH-5 contains only A subunits. When equal amounts of LDH-1 and LDH-5 isozymes are mixed together, treated with mercaptoethanol and urea, and then the mercapto-ethanol and urea are removed, the subunits reassociate in a random fashion forming all five isozymes and generating a symmetrical pattern. This symmetrical pattern is observed in early human fetal tissues; however, as differentiation continues, either the **LDHA** or **LDHB** locus predominates, producing the skewed patterns observed in adult tissues.

Mutation at either the **LDHA** or **LDHB** locus resulting in production of a subunit differing in charge from the standard subunit produces a complex zymogram (Fig. 7-28). All isozymes possessing the subunit involved in the mutation will radiate into multiple bands which differ with respect to their relative proportions of mutant and standard subunits. The effects of such a mutation would be expected to vary among tissues in relation to the relative expression of the LDH locus experiencing the mutation. For example, a mutation affecting **LDHB** would be anticipated to have a more profound effect upon kidney and heart than upon liver and skeletal muscle.

The previous examples have demonstrated that multiple isozymes may be produced by heterozygosity at a single locus (peptidase A), by contributions of two loci to multimeric enzymes (LDH), or by a combination of both (LDH). Furthermore, LDH isozyme patterns vary among tissues and within tissues at different developmental stages. Multiple isozymes may also be produced following post-translational modification. Four loci encode PGM isozymes (**PGM1**, **PGM2**, **PGM3** and **PGM4**), and each locus determines the structure of a monomeric enzyme. **PGM4** is expressed only in lactating breast tissue. The other three loci appear to be expressed more widely. Comparison of PGM patterns of erythrocyte hemolysates with those of placental, leukocyte or fibroblast extracts indicates that several of the PGM isozymes arise as a consequence of partial degradation of isozymes as the cells age (Fig. 7-29). Nucleated cells replenish their enzymes before the aging effects become predominant. However,

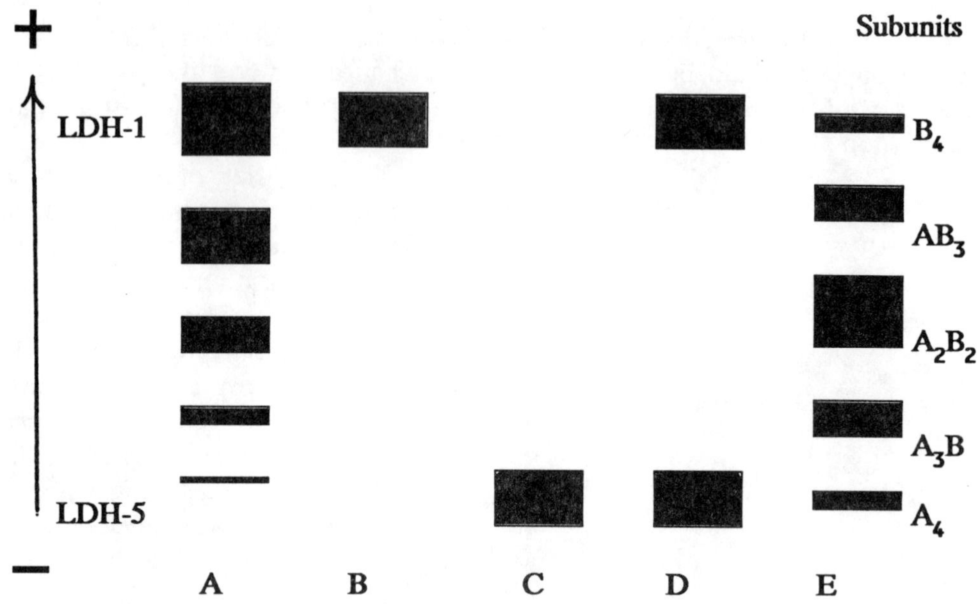

Fig. 7-27. Subunit composition of LDH isozymes. A) Kidney zymogram; B) Isolated LDH-1; C) Isolated LDH-5; D) Mixture of LDH-1 and LDH-5; E) Mixture of LDH-1 and LDH-5 that has been dissociated in the presence of mercaptoethanol and urea and permitted to reassociate into tetramers following removal of these agents.

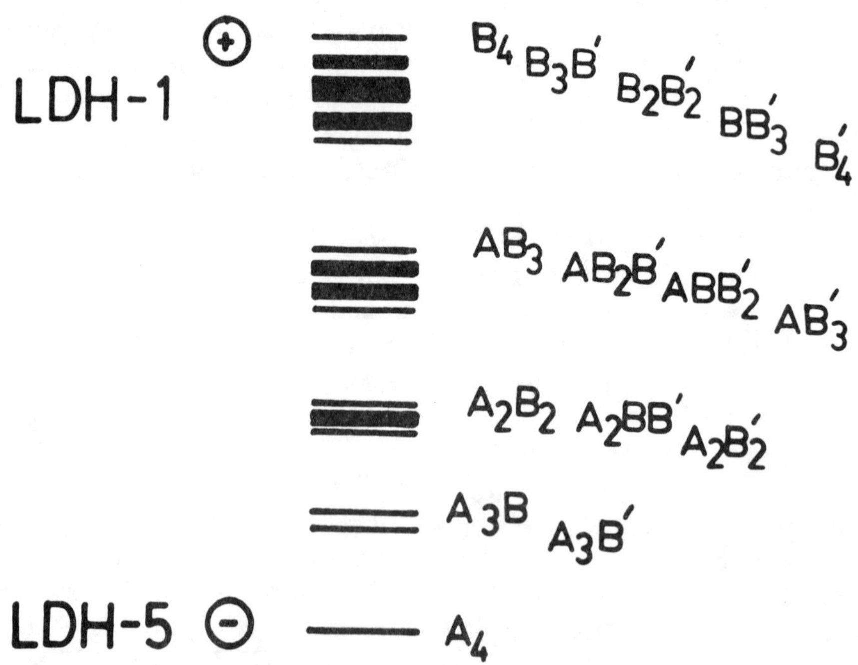

Fig. 7-28. Heterozygosity for a mutation affecting the molecular charge of the B subunit. The altered patterns for kidney LDH are shown. The subunit composition is presented at the right. B = standard subunit; B' = mutant subunit.

the anucleate red cells must utilize their existing PGM for the duration of their lifespan, and the aged enzyme molecules accumulate.

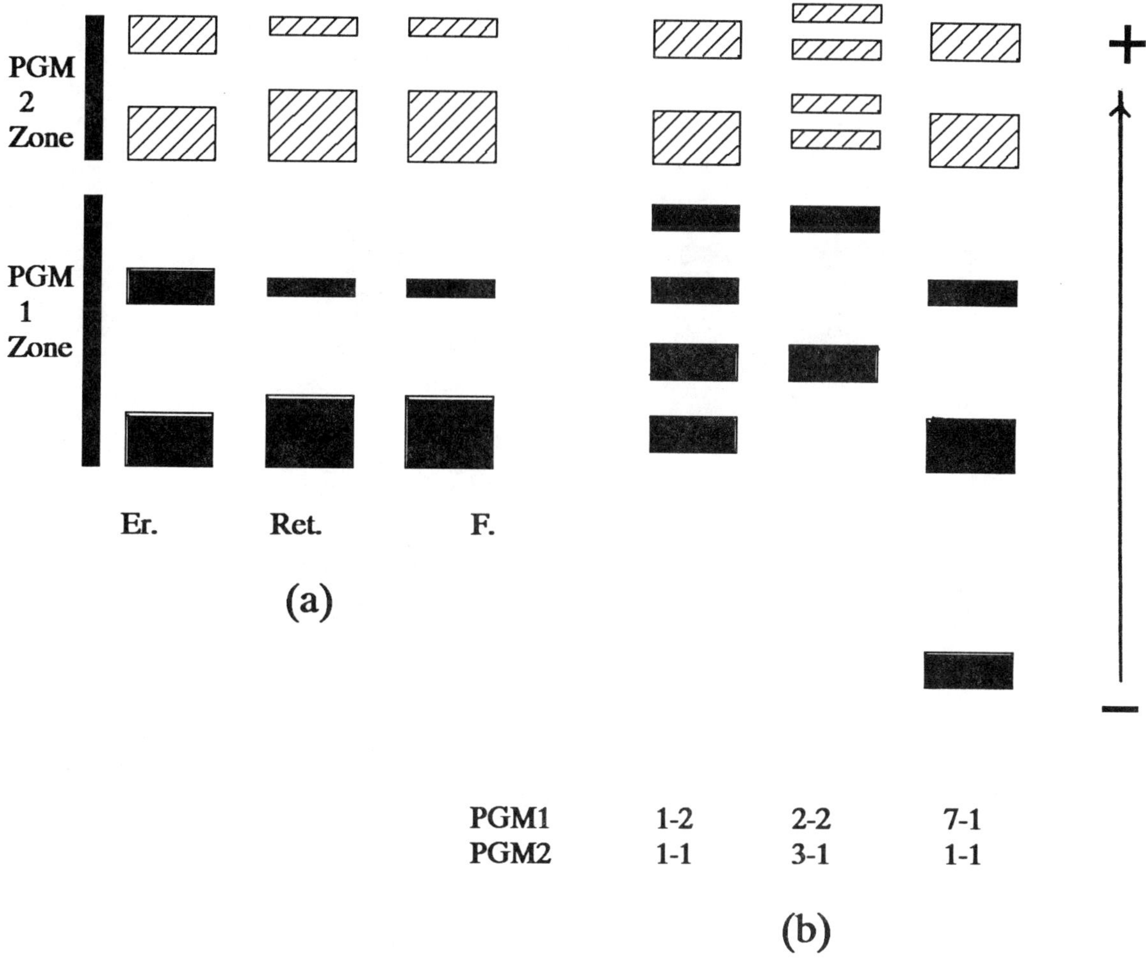

Fig. 7-29. Phosphoglucomutase isozymes. a) Comparison of the PGM1 1-1 and PGM2 1-1 isozymes in erythrocytes (Er.), reticulocytes (Ret.) and fibroblasts (F.). Notice how the lower band is enriched in the reticulocytes (enzyme still being synthesized) and fibroblasts (nucleated and enzymes still being synthesized). b) Comparison of three different PGM1 and PGM2 phenotypes in erythrocyte hemolysates. PGM1 isozymes are indicated by solid bars and PGM2 isozymes by hatched bars in both panels.

Nucleoside phosphorylase, a trimeric enzyme, provides a second example of an enzyme subject to post-translational modification. Erythrocyte isozyme patterns are considerably more complex than those of fibroblasts (Fig. 7-30). The multiple bands exhibiting more rapid mobilities than the corresponding fibroblast isozymes may arise following deamidation of one or more glutamine or asparagine residues. A similar mechanism has been proposed for the secondary isozymes of PGM.

Electrofocusing, a procedure that separates proteins on the basis of their isoelectric points, can resolve enzymes such as arylsulfatase A into multiple charge isomers. These isomers appear to differ with respect to their oligosaccharide chains. The more acidic bands contain more neuraminic acid than the basic isoforms. Treatment of the enzyme with neuraminidase prior to electrofocussing condenses the pattern to one or a few basic isoforms.

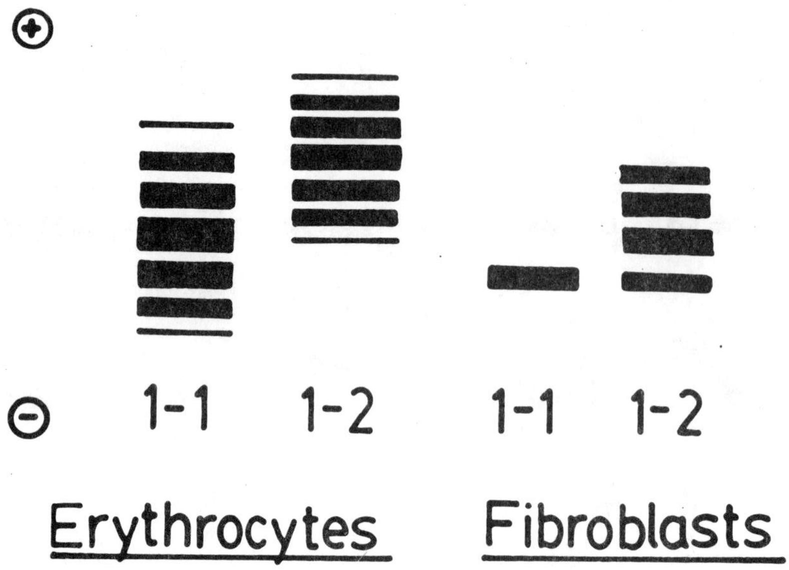

Fig. 7-30. Post-translational modification of nucleoside phosphorylase isozymes. Aging enzyme in the erythrocytes is converted to more rapidly migrating isozymes than are found in fibroblasts which have a younger enzyme population. The four isozymes in heterozygous fibroblasts (1-2 phenotype) consist of trimers that differ with respect to their relative numbers of NP-1 and NP-2 subunits.

SUMMARY

The relationship between mutation and human disease is illustrated by the hemoglobinopathies. Sickle cell anemia is caused by a single base substitution which results in the replacement of a polar amino acid by a hydrophobic amino acid on the exterior surface of the molecule. Under low oxygen pressures corresponding sites of different HbS molecules interact, causing a stacking of hemoglobin tetramers. These long chains deform the red cell into a sickle shape. The spleen destroys the sickled cells, contributing to the anemia experienced by these patients. Sickle cells also obstruct capillaries, leading to increased risk of infection, oxygen deprivation in tissues and pain. The severity of sickle cell anemia varies among patients, suggesting that other factors influence the disease. The mutation that produces the amino acid substitution also alters the restriction site for the enzyme MstII. This alteration can be used for prenatal diagnosis of both sickle cell anemia and sickle cell trait. Many other mutations have been described which produce amino acid substitutions. Their effects are determined by the type of susbstitution and the site at which the substitution occurs and include increased or decreased oxygen affinity, instability of the molecule, or abnormal interactions with other hemoglobin tetramers.

Humans form several different hemoglobins during their development that differ with respect to their α-like chains, β-like subunits, or both. Three functional α-like loci are closely linked on chromosome 16. The zeta-globin locus is expressed during the early embryonic period, while the α^1- and α^2-loci are expressed from the latter part of the embryonic period throughout life. Orchestration of α-like globin gene expression is accomplished by interactions among the LAR located upstream from the zeta-globin locus, proximal promoter elements, sequences downstream from the α^1-globin gene, and by a number of trans-acting proteins encoded by loci unlinked to the α-like globin gene cluster. The β-like globin genes are clustered on chromosome 11 in the order of their developmental expression. Regulation of this cluster is accomplished by mechanisms similar to those that regulate the α-like globin gene cluster. The normal developmental sequence of hemoglobin expression is: **Embryonic** (yolk sac) - Hb Gower1, Hb Gower2, Hb Portland; **Fetal** (liver, spleen, bone marrow) - HbF; **Adult** (bone marrow) - HbA_1 and HbA_2. The sites of erythropoiesis during each stage are given in parentheses.

Thalassemias are anemias that are caused by deficient globin chain production. α-thalassemias are usually caused by deletions of α-globin loci. The severity of the anemia is inversely proportional to the number of α-globin genes remaining in the genome of the patients. By contrast, β-thalassemias are most commonly caused by point mutations, most commonly base substitutions, within the β-globin gene. Failure of β-thalassemia gene expression may occur at the transcriptional, RNA processing, or translational levels, depending upon the particular site of the base alteration. Large deletions encompassing two or more β-like genes have been observed. $\delta\beta$-thalassemia is caused by variable deletions of these two genes and, in some cases, their downstream regions. Compensation by increasing the level of fetal hemoglobin is unsuccessful, with low quantities of HbF and intercellular HbF variation being seen. Hereditary persistence of fetal hemoglobin (HPFH), a clinically benign condition, is characterized by continued synthesis of relatively large quantities of HbF during childhood, adolescence and adulthood.

Certain hemoglobinopathies (HbS, HbC, HbD, and HbE) are unusually common in certain populations. The increased frequencies of these mutations have been traced to a selective advantage they provide heterozygotes in malarial areas. Some thalassemia alleles also display this phenomenon.

Studies of both globin clusters have revealed polymorphisms of restriction sites within and around the various globin genes. A particular array of restriction sites on a chromosome is referred to as a haplotype. Haplotypes may be used for tracing the evolution of globin gene mutations and for prenatal diagnosis of certain anemias caused by globin mutations.

Glucose-6-phosphate dehydrogenase catalyzes the step commiting glucose metabolites to the hexose monophosphate shunt, an important source of reduced $NADP^+$. The HMS is the only source of $NADPH + H^+$ available to erythrocytes, and G6PD deficiency is an important cause for susceptibility to hemolytic anemia. Anemias may occasionally occur spontaneously, but most are triggered by infection or exposure to certain drugs. Several G6PD deficiency alleles are especially common in certain populations. This increased frequency arose as a consequence of heterozygous advantage in females in malarial areas.

Enzymes usually occur as monomers, dimers, trimers, or tetramers. Comparison of isozyme patterns of homozygotes and heterozygotes can often be used to infer the subunit structure of an enzyme. Variation in isozyme patterns may be contributed by multiple allelism at a single locus, alleles at two or more loci, or by various post-translational modifications of the enzyme and its subunits. Isozyme patterns may differ among tissues and during development of a specific tissue.

PROBLEMS

1. Enzyme B from most persons migrates as a single band (Phenotype I) following electrophoresis and histochemical staining for enzyme activity. About 1/2500 persons have a form of Enzyme B that migrates more rapidly than the usual form (Phenotype II). Marriages of persons with Phenotype I to individuals with Phenotype II produce children who all have an Enzyme B that migrates in a three-banded isozyme pattern. Two bands correspond to the isozymes present in their parents. The third band is most prevalent and exhibits a mobility intermediate to those of the parental isozymes. Propose a subunit structure for enzyme B and explain the phenotype present in the children.

2. One of your patients has severe sickle cell anemia. Her mother has sickle cell trait, and her father has a mild chronic anemia characterized by somewhat reduced levels of HbA_1 and increased quantities of HbA_2 and HbF. The amounts of HbF never exceed 3% of the total and vary among erythrocytes. A paternal uncle died from a severe hemolytic anemia. The hematology report for this uncle mentioned the presence of target cells, absence of HbA_1, and elevations of HbA_2 and HbF. What is the most likely explanation for the presence of sickle cell anemia in your patient?

3. Predict the effects of an amino acid substitution at a site associated with the heme pocket.

4. Why do mutations affecting β- and δ-globin loci express their major effects during mid to late infancy and not earlier?

5. A young boy has a severe G6PD deficiency. Reticulocyte G6PD activities are about 10-fold higher than G6PD activities of most of his circulating red cells, but are still less than 20% of normal. Molecular studies have failed to identify any major deletions or insertions in the gene, and the quantity or G6PD mRNA in reticulocytes appears normal. Given that reticulocytes are anucleate precursors of erythrocytes, and that reticulocytes can still synthesize their proteins, provide an explanation for the cause of the G6PD deficiency in this boy.

6. How can one tissue express a different array of isozymes from that occurring in another tissue?

7. A man and woman were referred for prenatal diagnosis. Both have a normal phenotype, but each had a sibling die from severe β-thalassemia. Consider the following pedigree. β-like globin cluster haplotypes are presented below the symbols. Shaded symbols represent β-thalassemia.

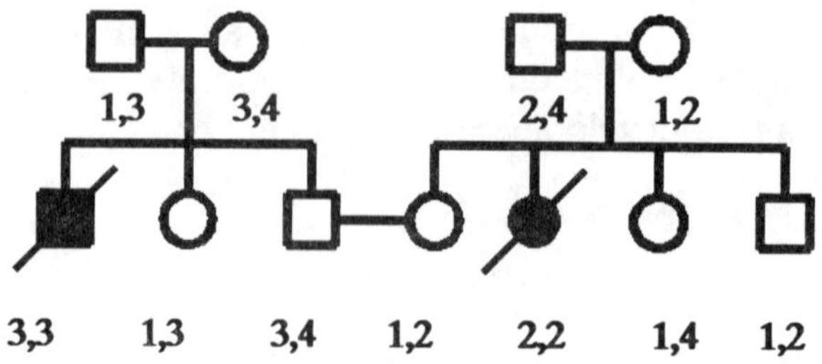

Suppose the fetus of the couple in the second generation is found to have haplotypes 2 and 3. How would you advise this couple regarding their risk for a thalassemic child?

GLOSSARY OF TERMS

Chromatography the separation of a mixture of molecules by a solvent system (and support medium) that exploits differences among those molecules.

Cis effect effect of a regulatory element upon a gene on the same chromosome.

Electrofocusing separation of molecules on the basis of their isoelectric points (pH at which the net molecular charge is 0).

Electrophoresis separation of charged molecules on the basis of their surface charge.

Gene conversion a nonreciprocal form of genetic recombination.

Genetic compound person heterozygous for two different mutant alleles.

Haplotype closely linked array of genes or restriction sites that are rarely separated by recombination.

Heterozygous advantage persons heterozygous for a mutant allele transmit more of their genes to the next generation than do either of the homozygotes.

Isozymes structurally different forms of an enzyme.

Locus activating region regions upstream from the zeta- and epsilon-globin loci which participate in the regulation of all downstream globin loci on the same chromosome.

Methemoglobin hemoglobin containing iron that is in the oxidized (+3) state.

Trans effect a regulatory factor that influences the expression of a gene on the same chromosome and/or on the homologous or other nonhomologous chromosome.

Zymogram pattern of isozymes; usually applied to isozymes separated by electrophoresis.

BIBLIOGRAPHY

1. Weatherall, DJ and JB Clegg. 1976. Molecular genetics of human hemoglobin. Ann Rev Genet 10:157-178.

2. Huehns, ER, N Dance, GH Beavan, <u>et al.</u> 1964. Human embryonic hemoglobins. Cold Spring Harbor Symp Quant Biol 29:327-331.

3. Schroeder, WA and THJ Huisman. 1974. Multiple cistrons for fetal hemoglobin in man. Ann NY Acad Sci 241:70-79.

4. Nute, PE. 1974. Multiple hemoglobin α-chain loci in monkeys, apes, and man. Ann NY Acad Sci 241:39-60.

5. Kollias, G, N Wrighton, J Hurst, and F Grosveld. 1986. Regulated expression of human $^Aγ$-, β- and hybrid δβ-globin genes in transgenic mice: Manipulation of the developmental expression patterns. Cell 46:89-94.

6. Forrester, WC, E Epner, MC Driscoll, T Enver, et al. 1990. A deletion of the human β-globin locus activation region causes a major alteration in chromatin structure and replication across the entire β-globin locus. Genes & Development 4:1637-1649.

7. Herrick, JB. 1910. Peculiar elongated and sickle shaped red corpuscles in a case of severe anemia. Arch Intern Med 6:517-521.

8. Hahn, EV and EB Gillespie. 1927. Sickle-cell anemia: report of a case greatly improved by splenectomy; experimental study of sickle cell formation. Arch Int Med 39:233.

9. Neel, JV. 1949. The inheritance of sickle cell anemia. Science 110:64-66.

10. Beet, EA. 1949. The genetics of the sickle cell trait in a Bantu tribe. Ann Eugen 14:279-284.

11. Pauling, L, HA Itano, SJ Singer and JC Wells. 1949. Sickle cell anemia, a molecular disease. Science 110:543-548.

12. Ingram, VM. 1957. Gene mutations in hemoglobin: the chemical difference between normal and sickle cell hemoglobin. Nature 180:326-328.

13. Ingram, VM. 1959. Abnormal human hemoglobins. III. The chemical difference between normal and sickle cell hemoglobins. Biochim Biophys Acta 36:402-411.

14. Chang, JC and YW Kan. 1982. A sensitive new prenatal test for sickle cell anemia. N Eng J Med 307:30-32.

15. Studencki, AB, BJ Conner, CC Impraim, et al. 1985. Discrimination among the human $β^A$, $β^S$ and $β^C$-globin genes using allele-specific oligonucleotide hybridization probes. Am J Hum Genet 37:42-51.

16. Ottolenghi, S, WG Lanyon, J Paul, et al. 1974. Gene deletion as the cause of α thalassemia. Nature 251:389-392.

17. Liebhaber, SA, E-U Griese, I Weiss et al. 1990. Inactivation of human α-globin gene expression by a de novo deletion located upstream of the α-globin gene cluster. Proc Natl Acad Sci USA 87: 9431-9435.

18. Martinez, G, A Novelletto, H. Di Rienzo et al. 1989. A case of hereditary persistence of fetal hemoglobin caused by a gene not linked to the β-globin cluster. Hum Genet 82:335-337.

19. Mantovani, R, N Malgaretti, B Giglioni <u>et al.</u> 1987. A protein factor binding to an octamer motif in the γ-globin promoter disappears upon induction of differentiation and hemoglobin synthesis in K562 cells. Nucleic Acids Res 15:9349-9364.

ADDITIONAL LEARNING RESOURCES

1. Beutler, E, Glucose-6-phosphate dehydrogenase deficiency. N Eng J Med 324:169-174.

2. Weatherall, DJ, JB Clegg, DR Higgs, and WG Wood. 1989. The Hemoglobinopathies. in <u>The Metabolic Basis of Inherited Disease</u>, CR Scriver, AL Beaudet, WS Sly, and D Valle eds., 6th ed., New York: McGraw-Hill, Chapter 93.

Chapter 8

Inborn Errors of Metabolism

Sir Archibald Garrod coined the term, **inborn errors of metabolism,** in 1908 (1). Garrod observed that patients with a rare disorder, alcaptonuria, excreted a compound in the urine which turned black when the urine was allowed to stand for a period of time. The chemical responsible for the color reaction was identified as homogentisic acid which was oxidized to the black color upon contact with air. When homogentisic acid was fed to persons exhibiting the unusual urine response, it was quantitatively excreted in the urine. However, when homogentisic acid was fed to normal persons, it was metabolized and was not excreted in the urine. Garrod proposed that alcaptonurics lacked an enzyme responsible for catabolizing homogentisic acid. As a result of this block, the acid accumulated in the tissues, spilled into the bloodstream and was quantitatively excreted into the urine. Normal individuals possessed the necessary enzyme, homogentisic acid oxidase, and did not accumulate the acid. The actual enzymatic lesion was demonstrated 50 years subsequent to publication of Garrod's inborn error theory. Garrod also showed that homogentisic acid could be induced in alcaptonurics by feeding them protein and particularly by feeding them the amino acids phenylalanine and tyrosine. Garrod correctly inferred that homogentisic acid was a degradation product of these amino acids (Fig. 8-1). Alcaptonurics have surprisingly few

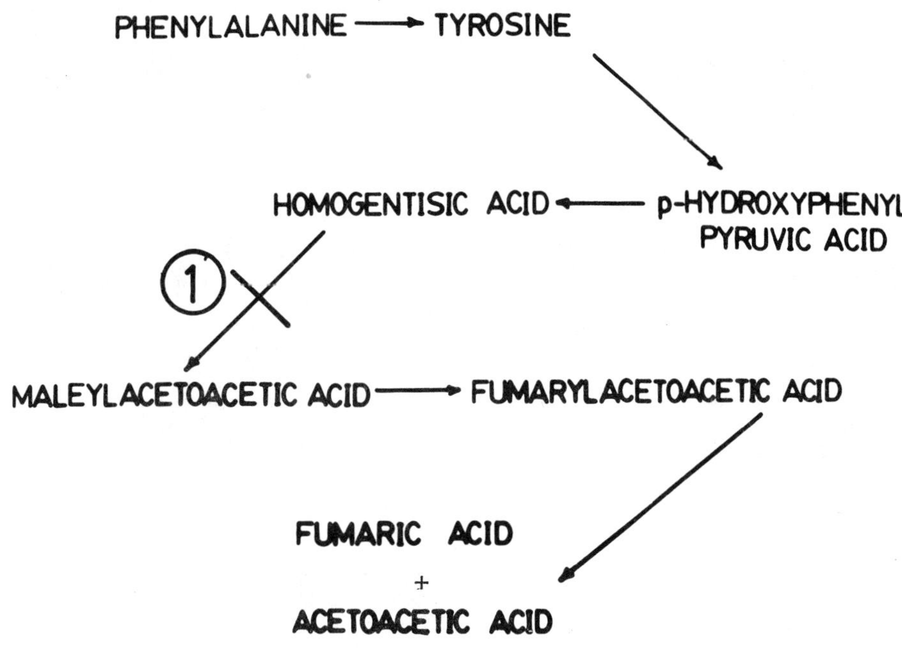

Fig. 8-1. Degradation of phenylalanine. 1) Site of the metabolic block responsible for alcaptonuria.

clinical complications. Patients have increased risk for arthritis which appears to be caused by accumulation of a byproduct of homogentisic acid in cartilage and other connective tissues. Garrod observed that alcaptonuria frequently occurred in siblings and that parents of these affected children were often blood relatives. After consulting with Bateson, a prominent geneticist of that period, Garrod interpreted the cause of alcaptonuria to be homozygosity for a Mendelian factor. We now recognize alcaptonuria as an autosomal recessive disorder. The concept of an **inborn error of metabolism** can now be described as follows. Patients inherit a genotype which produces a deficiency of functional enzyme. The enzyme deficiency prevents conversion of a metabolite from one form to another. The metabolite accumulates and either directly or indirectly produces the clinical signs of the disease.

PHENYLKETONURIA AND THE MOLECULAR BASIS OF PLEIOTROPISM

Folling (2) described a clinical disorder, phenylketonuria (PKU), which includes mental retardation, microcephaly, light pigmentation, seizures, skin rash and a variety of other problems. The disease is generally quite rare, but appears more frequently among certain European populations, French Canadians, and certain Chinese groups. About 1/8000 Caucasian newborns in the United States are at risk for PKU, which is inherited as an autosomal recessive trait. Patients characteristically have high levels of blood phenylalanine and excrete byproducts such as phenylpyruvic acid in their urine. The latter compound can be detected by using an acid ferric chloride reagent which produces a green color in the presence of phenylpyruvate.

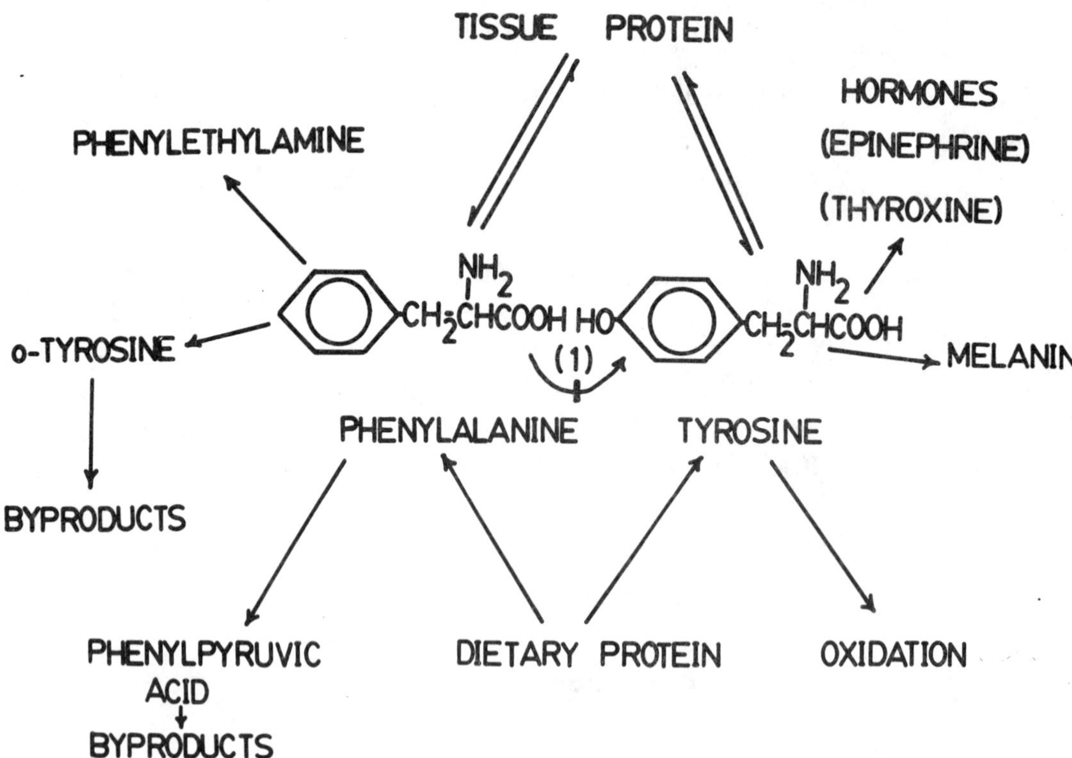

Fig. 8-2. Fate of phenylalanine. 1) Block occurring in phenylketonuria.

The metabolic block in PKU patients occurs at the step where phenylalanine is hydroxylated to form tyrosine (Fig. 8-1). Phenylketonurics lack phenylalanine hydroxylase activity, a liver-specific enzyme. Phenylalanine is ingested as a component of dietary proteins, cannot be metabolized to tyrosine by PKU individuals, and accumulates in hepatocytes. Excess phenylalanine enters the bloodstream and is transported to other tissues. The amino acid or its byproducts (Fig. 8-2) (phenylethylamine, phenyl-pyruvic acid, etc.) accumulate in tissues and cause damage to the central nervous system, producing the seizures and mental retardation observed among untreated patients. Excess phenylalanine also inhibits certain enzymes responsible for the production of neurotransmitters from tryptophan, and unusual tryptohan metabolites appear in the urine of untreated phenylketonuric children. Phenylpyruvate and other acidic derivatives of phenylalanine are skin irritants and are most likely responsible for the skin rash that appears in these children. Excessive levels of phenylalanine inhibit tyrosinase, an enzyme complex that participates in the formation of a body pigment, melanin. Consequently, PKU children usually display lighter pigmentation than other family members. Kidney cells metabolize phenylalanine to phenylpyruvate which is quantitatively excreted in the urine. Tyrosine is also a component of dietary protein. Therefore, the inability of the PKU to synthesize this amino acid does not present major problems. Phenylketonuria provides an excellent example of the phenomenon known as **pleiotropism**: the production of several phenotypic effects by a mutation in a single gene. Phenylalanine hydroxylase occupies a central point in metabolism. Accumulations of its substrate impact upon many other systems, producing the many symptoms observed in PKU patients.

Diagnosis of Phenylketonuria

Overflow of phenylalanine into the blood produces hyperphenylalaninemia, and is used to diagnose the condition. Blood from newborn infants is spotted onto filter paper and sent to a central testing laboratory that monitors newborn infants for the presence of several preventable or treatable metabolic disorders. A disk is punched from the blood spot and placed on an agar medium that contains bacteria requiring tyrosine for growth, a carbon source, and a competitive inhibitor of bacterial phenylalanine hydroxylase (β-2-thienylalanine). The bacteria convert the carbon source to phenylalanine; however, the β-thienylalanine occupies the active site of bacterial phenylalanine hydroxylase, preventing the conversion of phenylalanine to tyrosine. Consequently, the bacteria fail to grow. PKU blood contains high levels of phenylalanine. This amino acid diffuses from the blood spot to the surrounding agar, dislodges the β-thienylalanine from the active site, and permits tyrosine synthesis. Bacterial growth around the blood disk constitutes a positive test for PKU. Normal blood contains much lower quantities of phenylalanine, and bacterial growth rings are much smaller or nonexistent. This screening method is called the **Guthrie test.**

A positive Guthrie test is the first step in the diagnosis of PKU. The test is repeated to eliminate false positives due to normal variation of the late-developing phenylalanine hydroxylase system. Infants who are slow-developers will test positive on the first occasion, but will have a normal test result when it is repeated. **Benign hyperphenylalaninemia** is a condition that occurs almost as frequently as PKU. Infants with this condition have higher than normal blood phenylalanine concentrations; however, these concentrations are considerably below those observed in PKU infants and are not associated with clinical disabilities. True PKU infants have phenylalanine hydroxylase activites that approximate zero, while phenylalanine hydroxylase activites of infants with the benign variant generally exceed 5% of normal. These two conditions can be distinguished by tests of liver biopsy specimens for enzyme activity. Babies with abnormalities affecting tyrosine metabolism will also test positive with the Guthrie method. These disorders can be distinguished from true PKU by appropriate tests. These examples illustrate the need for caution when interpreting results of "screening" tests. The metabolite detected by the test may be elevated for a variety of reasons, only a few of which are of serious consequence to the patient. Furthermore, drugs or other metabolites may exist that will provide a false positive test response. In all cases of positive test results, confirmatory testing is mandatory.

Treatment of Phenylketonuria

Early detection of phenylketonuria is desirable, since the disease is preventable by dietary restriction of phenylalanine. When blood phenylalanine is kept near the normal range by eliminating proteins containing substantial amounts of phenylalanine, mental and physical development occur normally. The skin problems, seizures, and other complications of PKU are also prevented. It has been customary for physicians to remove children from the restrictive diet at about age seven if such practice does not result in regression of the child's development. However, two discoveries have motivated some researchers to recommend that the diet be maintained after this time. When patients are removed from the diet, their blood phenylalanine returns to high values. Although gross mental deterioration does not seem to occur after age seven, evidence suggests that patients display subtle neurological dysfunction if their blood phenylalanine levels are not maintained below 1mM. Furthermore, women who were treated in the past for PKU and not maintained upon a restrictive diet during pregnancy run a high risk for abnormal pregnancy outcome. These women have high blood phenylalanine concentrations, and the placenta compounds the situation by further concentrating phenylalanine on the fetal side. Many such pregnancies terminate in abortion, and liveborn children have an increased risk for mental retardation and birth defects. Studies have indicated that maintenance of maternal blood phenylalanine levels between 0.2 and 1mM by special diets and supplementing the diet with multivitamins may prevent these pregnancy complications.

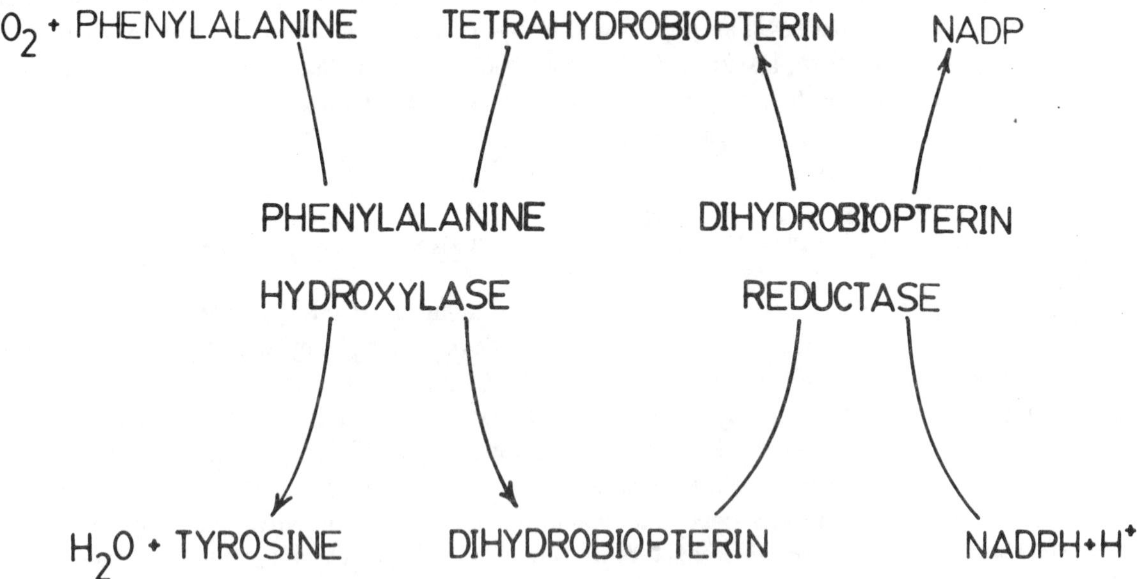

Fig. 8-3. Phenylalanine hydroxylase system.

Biochemical Variants of Phenylketonuria

Although about 95% of phenylketonuria is caused by mutations in the gene (**PAH**) encoding phenylalanine hydroxylase, mutations in enzymes associated with this system can produce similar symptoms. The phenylalanine hydroxylase system involves more than one enzyme (Fig. 8-3). Normal phenylalanine hydroxylase activity requires adequate quantities of the coenzyme, reduced dihydropteridine (tetrahydrobiopterin = BH_4). The enzyme dihydropteridine reductase (DHPR) converts the inactive coenzyme to its active form. Occasional PKU patients are encountered who lack adequate activities of this enzyme. Since BH_4 is required by two other central nervous system enzymes, tyrosine hydroxylase and tryptophan hydroxylase, effects of DHPR deficiency are more widespread than those of "classical" PKU. Lack of BH_4 results in deficiencies of neurotransmitters such as serotonin and dihydroxyphenylalanine (DOPA). Dietary restriction of phenylalanine is insufficient to correct the problem experienced by patients with DHPR deficiency. Experimental BH_4 replacement combined with supplementation of the deficient neurotransmitters is being tried; however, it is uncertain whether this approach will be successful.

BH_2 is synthesized from sepiapterin and other precursor metabolites. Deficiencies of any of the enzymes participating in the production of sepiapterin or its conversion to BH_2 would be expected to result in PKU and other complications. Deficiencies of two enzymes have been reported, and treatment parallels that for DHPR deficiency.

Molecular Genetics of Phenylketonuria

The human phenylalanine hydroxylase gene, **PAH**, spans about 90kb of DNA and has been mapped to chromosome 12q22-q24.1. Most PKU patients found in populations where the disease is rare are genetic compounds. However, several alleles have been identified which occur more frequently (Fig. 8-3). The Caucasian missense mutation in exon 12 and the splice junction mutation in intron 12 appear to have arisen among Celtic peoples residing in central Europe. As these people migrated to northern and western Europe, they carried the mutant alleles with them. **Founder effect**, a phenomenon in which one or a few founders of new populations carry a mutation that subsequently becomes unusually common

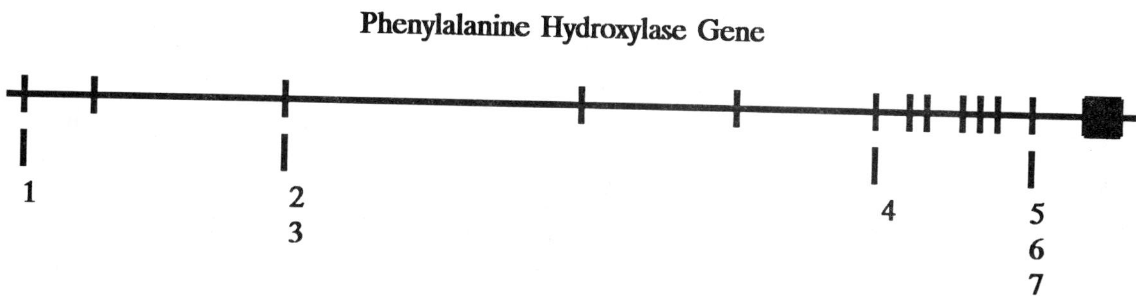

Fig. 8-3. The phenylalanine hydroxylase gene and representative mutations. 1) French-Canadian mutation (aa1: Met–>Val); 2) exon 3 is deleted in all Yemenite Jewish patients; 3) Chinese (aa111: Arg–>Ter); 4) Chinese (aa204: Tyr–>Cys); 5) Chinese (aa413: Arg–>Pro); 6) European and U.S. Caucasians (G–>A at donor splice site, intron 12); 7) European and U.S. Caucasian and French-Canadian (aa408: Arg–>Trp). All of these mutations generate PAH[0] alleles.

following closed breeding of the group for several generations, is believed to have led to the present-day frequencies of these two alleles among descendants of these peoples. A similar founder effect is believed to be responsible for the three relatively common PAH^0 alleles among French-Canadians. The missense mutation in exon 12 that is observed in both the United States and among French-Canadians is associated with different haplotypes, suggesting that gene conversion or recombination may have occurred after the mutation (3). Alternatively, the same mutation may have occurred more than once during the origin of these different populations. The most striking example of founder effect is provided by PKU among Yemenite Jews. The disease is due to the same mutation in all cases examined.

All of the mutations listed in Fig. 8-3 produce PAH^0 alleles. In the case of the Chinese mutation in exon 1, methionine is replaced by valine. This mutation would be expected to prevent normal initiation of translation. The mutation in the donor splice site of intron 12, the most common PAH^0 allele in the United States, frequently causes a skipping of the 12th exon, producing a truncated mRNA. Other abnormal splicing products are often observed in association with this mutation.

PKU mutations represent one extreme of a continuum of mutations in the **PAH** gene. Many mutations at this locus generate alleles that encode proteins that have lost much, but not all, of their activity. One of these mutations has a French origin and involves a $G{-}{>}A$ transition in exon 7 which results in the replacement of glutamate by lysine at position 280. The mutant enzyme retains about 2-3% of normal activity. Persons homozygous for this mutation (often products of first cousin marriages) have a higher tolerance for phenylalanine and either lack signs of PKU or have more moderate expression of this disease. Persons who are genetic compounds for this allele and a PAH^0 allele generally develop signs of classical PKU if untreated.

PKU illustrates one of the hallmarks of many genetic diseases. Investigations at the molecular level have revealed that PKU is a heterogeneous disease. What was once considered to be a simple Mendelian recessive disorder has proved to be much more complex. Many different alleles exist at the **PAH** locus. A patient may develop PKU as the result of homozygosity for any one of these alleles; however, it is much more likely that a PKU may be a genetic compound. Superimposed upon this multi-allelic system is heterogeneity due to genetic lesions involving associated enzymes such as DHPR. Diseases that display genetic heterogeneity involving both multiple allelism and mutations at multiple loci are the rule rather than the exception in medical genetics.

LYSOSOMAL STORAGE DISEASES

Lysosomes provide a means for concentrating individually infrequent cellular proteins. These intracellular organelles are derived from the Golgi apparatus and contain numerous enzymes that are involved with the catabolism of foreign substances entering the cell by either receptor-mediated endocytosis or phagocytosis. Lysosomes fuse with organelles such as endosomes that are produced by either of these processes and digest the materials they contain. Numerous complex molecules, including constituents of cell membranes, are subject to degradation as part of the dynamic processes that constantly remodel membranes and organelles. Enzymes that degrade these materials are encapsulated in lysosomes. The substrates enter the lysosomes where they are degraded to smaller metabolites which are recyled through other metabolic pathways. The following group of metabolic diseases occur as a consequence of deficiencies of lysosomal enzymes. The substrates enter the lysosomes; however, since the enzymes that normally degrade these materials are absent, the substrates accumulate in these organelles. Cells become packed with lysosomes that are filled with these metabolites, and eventually die, producing the clinical symptoms characteristic of lysosomal storage diseases.

GM$_2$-Gangliosidoses

Tay-Sachs disease is one of several disorders that occur as a consequence of the accumulation of GM$_2$-ganglioside. Gangliosides are polar lipids that are important constituents of cell membranes in the central nervous system (CNS). Gangliosides contain a lipid component, sphingosine, coupled to a fat and a carbohydrate chain (Fig. 8-4). These compounds are degraded stepwise by a series of lyso-

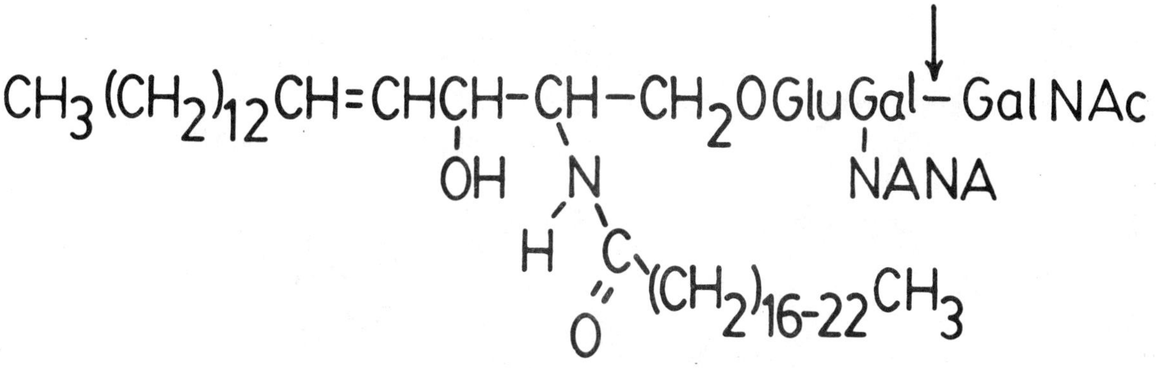

$$CH_3\,(CH_2)_{12}CH=CHCH-CH-CH_2OGluGal\overset{\downarrow}{-}GalNAc$$

with OH on the first substituted carbon, an N substituent leading to H–N and a carbonyl $\overset{O}{C}(CH_2)_{16-22}CH_3$, and NANA appended below the Gal.

Fig. 8-4. Structure of GM$_2$-ganglioside. The arrow indicates the bond cleaved by hexosaminidase A, the enzyme deficient in GM$_2$-gangliosidoses. Glu = glucose; Gal = galactose; NANA = N-acetylneuraminic acid; GalNAc = N-acetylgalactosamine.

somal hydrolases which sequentially remove the sugars appended to sphingosine (Fig. 8-5). The ganglioside substrates accumulate within lysosomes when their respective hydrolases are deficient. GM$_2$-gangliosidoses occur when the lysosomal enzyme, hexosaminidase A, is deficient. These storage diseases occur in three general forms: infantile, juvenile, and adult. The infantile forms totally lack hexoasminidase A activity, while patients with the juvenile- and adult-onset types retain some residual enzyme activity. Tay-Sachs disease (TSD) is one of the better-known GM$_2$-gangliosidoses. This storage disease occurs more commonly among Ashkenazi Jewish populations and certain French-Canadian groups, where its frequency may approach 1/3600. The incidence of TSD among members of other groups is about 1/300,000. The relatively high incidence in the two populations mentioned above is believed to be due to founder effect, although heterozygous advantage has not been rigorously excluded as a possibility in some cases. Carriers of TSD alleles number about 1/30 among Ashkenazi Jewish citizens of the United States.

Infants destined to develop TSD appear normal at birth. Signs of neurological deterioration begin to appear at about 3-6 months as an exaggerated startle response to noise. Awareness of their environment progressively diminishes, and neuromuscular control deteriorates. By one year of age signs of blindness, deafness and severe mental retardation are apparent. Death ensues prior to five years of age. Patients with TSD exhibit localized retinal degeneration producing a "cherry red spot" in the macular region. The red spot consists of capillaries that are visible on a gray background composed of dead cells engorged with GM$_2$-ganglioside. Deposition of the ganglioside in lysosomes produces a series of concentric spherical or ovoid structures that are very characteristic when examined by electron microscopy. Since GM$_2$-ganglioside is most prevalent in the CNS, the major features of TSD are limited to this system.

As stated previously, the enzymatic lesion in TSD involves hexosaminidase A. This enzyme hydrolyzes GM$_2$-ganglioside to GM$_3$-ganglioside. Enzymes exist which catalyze the removal of N-acetyl-

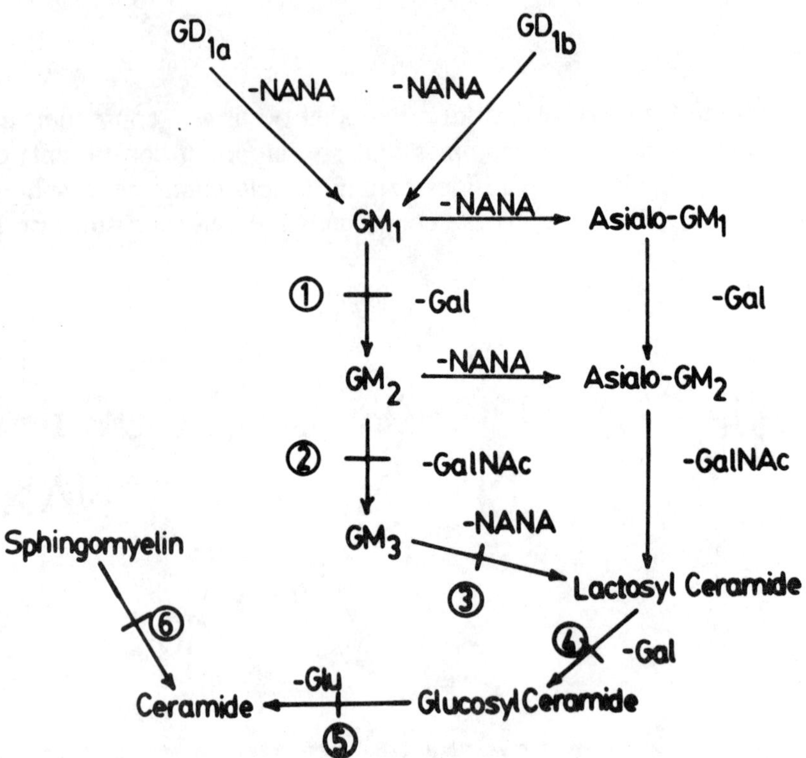

Fig. 8-5. Catabolic pathway for disposal of gangliosides. Numbers indicate genetic blocks (2 = TSD). GD = disialoganglioside; GM = monsialoganglioside; ceramide = sphingosine + fatty acid; other symbols as in Fig. 8-4.

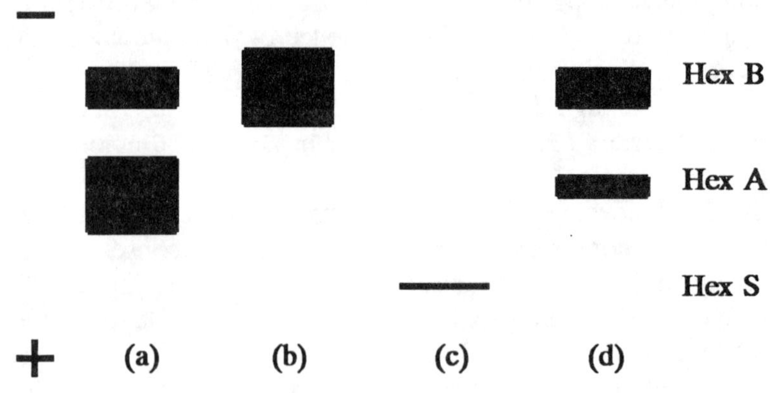

Fig. 8-6. Isozymes of hexosaminidase occurring in normal (a), TSD (b), Sandhoff's disease (c), and juvenile GM_2-gangliosidosis (d) cell extracts.

neuraminic acid from both GM$_1$- and GM$_2$-gangliosides. Although this may appear to indicate that the metabolic block occurring in TSD may be bypassed, the capacity of the alternate pathway is very limited and cannot efficiently remove the accumulating GM$_2$-ganglioside.

Several isozymes of hexosaminidase have been described. Two of these, hexosaminidases A and B, contribute the majority of enzyme activity (Fig. 8-6). Lanes (b) and (c) contain hexosaminidases from a TSD patient and from a patient with a second type of infantile GM$_2$-gangliosidosis known as Sandhoff's disease (SD). SD does not show the ethnic pattern displayed by TSD, but is found at a low frequency in all populations. Patients with TSD lack hexosaminidase A activity and have normal or elevated hexosaminidase B activity. By contrast, SD patients lack both hexosaminidase isozymes and have a small amount of an isozyme known as hexosaminidase S. Patients with juvenile GM$_2$-gangliosidosis

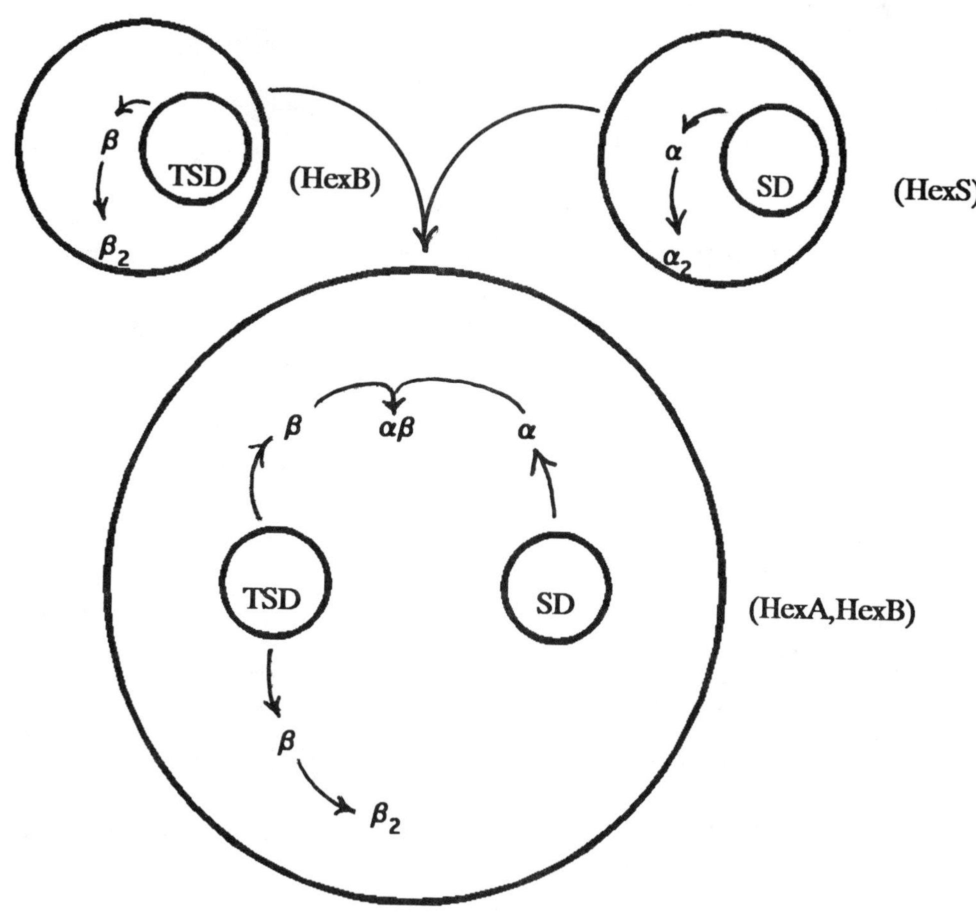

Fig. 8-7. Internuclear complementation demonstrating that the mutations producing TSD and SD are nonallelic. TSD cells which produce a normal β subunit and HexB are fused with SD cells which make a normal α subunit and HexS. The heterokaryon combines the β subunit determined by the TSD nucleus with the α subunit encoded by the SD nucleus to make HexA.

usually have reduced hexosaminidase A activity and normal hexosaminidase B activity (lane d). SD patients are clinically indistinguishable from TSD patients; however, the juvenile onset disease has a slower rate of progression.

Beutler and Kuhl (4) were able to generate all three hexosaminidase isozymes by repeated freezing and thawing of purified hexosaminidase A, demonstrating that hexosaminidase A shared one or more subunits with the other two isozymes. Further biochemical characterization identified two subunits, α and β, in hexosaminidase preparations; and the three isozymes were assigned the following subunit formulae: hexosaminidase A ($\alpha\beta$), hexosaminidase B (β_2), and hexosaminidase S (α_2). Nuclei from TSD and SD patients exhibited complementation in dikaryons, indicating that the two diseases were caused by mutations in different genes (Fig. 8-7). Both genes, **HEXA** and **HEXB**, contain 14 exons which are similarly placed, suggesting the two genes evolved following duplication of an ancestral gene (5). The **HEXA** locus on chromosome 15 specifies the structure of the α chain, while the **HEXB** gene on chromosome 5 encodes the β chain. Post-translational processing of both subunits is extensive. Both chains are glycosylated as they traverse the ER-Golgi system and undergo limited proteolysis following entrapment in lysosomes. Both the α and β subunits occur as three segments joined by disulfide bonds in the mature enzymes (6). A more complete description of hexosaminidase A processing was presented in Chapter 2.

The mutations for the various types of GM_2-gangliosidoses may now be deciphered (Fig. 8-8). SD patients have mutations in the **HEXB** gene. Since the β subunit occurs in both hexosaminidases

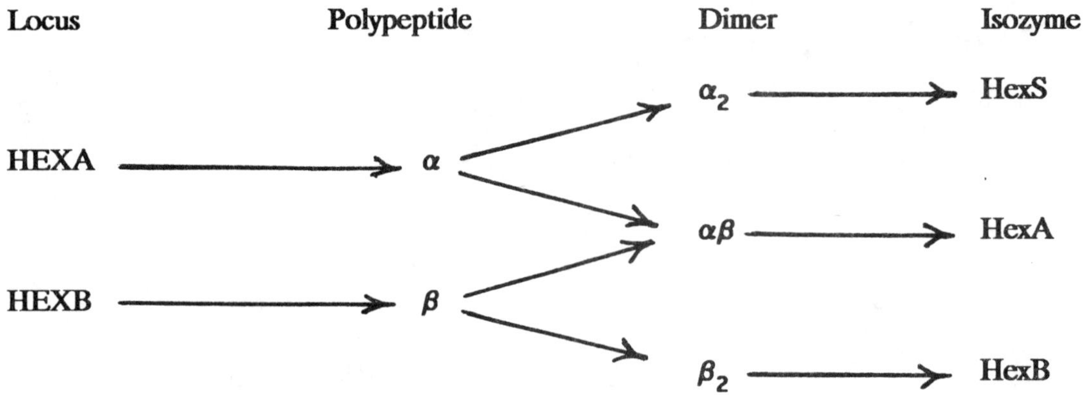

Fig. 8-8. Genetic determination of the hexosaminidase isozymes. TSD results from mutations in the HEXA gene. Since the α polypeptide encoded by this gene occurs in HexA and not HexB, TSD is associated with HexA deficiency. SD occurs as a consequence of mutations in the HEXB gene. Since the β subunit encoded by this gene occurs in both HexA and HexB, both activities are deficient in SD patients. The excess α subunits of SD patients dimerize to form HexS.

A and B, both isozymes are deficient. TSD occurs following mutations in **HEXA**. Since the α-subunit does not occur in HexB, TSD patients have an exclusive deficiency of HexA. HexS is formed by dimerization of the normal α subunits formed by SD patients.

Hexosaminidase A is more sensitive to heat than hexosaminidase B. This difference has been exploited to screen for TSD heterozygotes and homozygotes. Normal persons have about 70% hexosaminidase A in their blood serum. Heating of the serum at 50°C will destroy the hexosaminidase A while leaving hexosaminidase B intact. Therefore, about 30% of normal serum hexosaminidase activity will remain after heating. By contrast, all of the hexosaminidase activity of TSD serum is contributed by hexosaminidase B. Consequently, almost all of TSD hexosaminidase activity should remain after heating.   Heterozygotes will fall midway between these two values with about 65% heat-resistant hexosaminidase.

Two limitations to the interpretation of these tests have been recognized. Pregnant women produce a hexosaminidase isozyme, hexosaminidase P, which is synthesized in liver and enters the blood stream. Hexosaminidase P is heat-stable like hexosaminidase B, and the greater proportion of stable isozyme will cause a pregnant normal woman to fall into the heterozygous range. When a suspected carrier is known to be pregnant or is taking oral contraceptives, leukocytes are used as the source of the hexosaminidase isozymes. Hexosaminidase P does not occur in white blood cells and will not interfere with the test. A second limitation occasionally occurs when a patient has the B1 variant of TSD. The mutant enzyme has lost its ability to hydrolyze GM_2-ganglioside and other acidic substrates, but

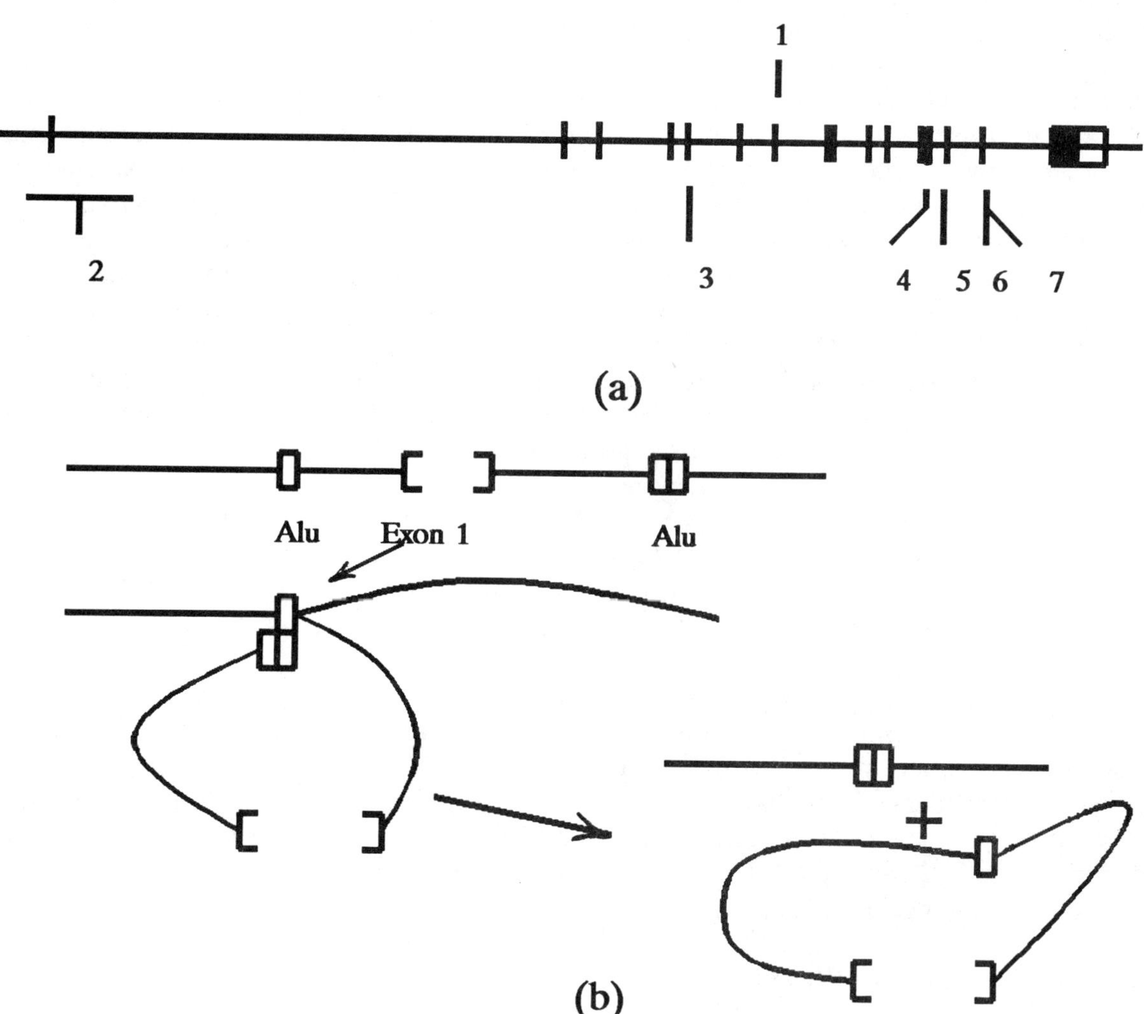

Fig. 8-9.   Molecular genetics of GM_2-gangliosidoses. a) Representative mutations: 1) Juvenile GM_2-gangliosidosis; 2,4,5)TSD mutations in which both α-mRNA and α chains are absent; 3,6,7) TSD mutations associated with a defective α chain. b) Alu recombination leading to deletion of exon 1. See text for discussion.

retains its ability to hydrolyze the synthetic substrates used to test activity and other neutral substrates. Persons with this mutation will appear normal in the usual test systems employed for detection of TSD. When the B1 variant is suspected to be present in a family, the natural GM_2-ganglioside must be used to assess enzyme activity. The converse mutant, where the enzyme has lost its ability to hydrolyze the synthetic test substrate but retained its ability to cleave GM_2-ganglioside, has also been observed. When this type of defect is being considered, a two step protocol is followed. The synthetic substrate is used first. If a result suggesting TSD is obtained, the result is confirmed using GM_2-ganglioside.

Molecular Genetics of Tay-Sachs Disease

The **HEXA** gene contains 14 exons, and the PCR technique has led to the identification of a number of mutations producing the various forms of infantile, juvenile, and adult GM_2-gangliosidoses that are associated with exclusive loss of hexosaminidase A activity. A representative sampling of these mutations is displayed in Fig. 8-9(a). Mutation 1 is a rare base substitution that results in a replacement of glycine at position 269 with serine. Both are neutral amino acids, and the effect of the amino acid substitution is the formation of an enzyme that has reduced activity. This mutation presents as a juvenile- or adult-onset problem with relatively slow progression and considerable inter-patient variability of manifestations of the disorder. The remaining mutations cause TSD. Two of these mutations are very common among Ashkenazi Jewish TSD patients. Mutation 4 involves a 4bp insertion in exon 11. The insertion causes a frameshift and generates a premature chain terminator. These patients lack both α-mRNA and α chains. The reason why no mRNA is present is unknown. The second common mutation (#5) replaces the **G** with a **C** at the donor splice site of intron 12. Consequently, splicing of the pre-mRNA occurs abnormally. Intron 12 may be included in the RNA, or exon 12 may be deleted from the RNA. Occasionally other upstream exons may be deleted. Regardless of the outcome of splicing, no mature mRNA can be detected; and there are no α chains present.

Tay-Sachs disease is also relatively frequent among certain French-Canadians. One of these mutations appears to have been caused by meiotic mispairing of three Alu elements that flank the first exon. Recombination within the mispaired region (Fig. 8-9(b)) led to deletion of the first exon and substantial portions of the upstream region and the first intron (a total of 7.6kb of DNA). These TSD patients lack both α-mRNA and α chains. This mutation may be one of several occurring in the French-Canadian population.

The B1 variant is caused by a base substitution in exon 5 (mutation 3). A base substitution converts arginine[178] to histidine. Two additional mutations are rare. Mutation 6 involves a base substitution converting glutamate[482] to lysine. The resulting α chain is highly unstable and never exits the endoplasmic reticulum. The other mutation (#7) deleted one base from exon 13. The resulting frameshift removes the last 23 amino acids from the α chain. The truncated α chain is degraded before it leaves the endoplasmic reticulum. Both of the rare mutations were observed in Italian children who were products of consanguineous marriages.

Molecular Genetics of Sandhoff's Disease

Similar studies of **HEXB** mutations producing SD are in progress, and several mutations have been characterized. These mutations display a range of structural changes and phenotypic effects that are similar to those described for TSD. One deletion of the **HEXB** gene is particularly remarkable in that 50kb of DNA are lost, including the promoter and the first five exons of the gene (7). This patient lacked both β-mRNA and β chains, presenting with an infantile-onset disease.

SD may also occur in juvenile- and adult-onset forms. One mutation has been observed in at least three unrelated individuals (8). Two of these people had juvenile-onset disease, and one had a normal clinical phenotype at the time of testing. A **G-->A** transition in intron 12 of the **HEXB** gene produced a new splice site, resulting in the inclusion of eight additional amino acids in the β subunit. This subunit appeared to have an abnormal conformation and was not processed to the mature β forms. The juvenile-onset cases had a small amount of hexosaminidase A activity (3-6% of normal) and no hexosaminidase B activity. Apparently some of the β-pre-mRNA was spliced normally, accounting for the synthesis and processing of a small amount of normal β chain. This normal β chain was paired with α chains that were present in excess. The asymptomatic patient had higher residual hexosaminidase A activities (9-10% of normal) which may also have arisen from low level normal β-pre-mRNA splicing. This person may develop SD later in life. Another currently normal individual was found to have a duplication that spanned the acceptor splice site of intron 13. An alternate splice site was produced by this mutation leading to an in-frame addition of 18 nucleotides to the mRNA. Apparently, the normal splice sites were used infrequently to generate normal mRNA, accounting for sufficient hexosaminidase A activity to produce a normal clinical phenotype in the girl possessing the mutation in combination with a CRM(-) allele inherited from the other parent. No hexosaminidase B was found. This girl may develop SD symptoms at a later time.

Activator Proteins, Hydrophilic Enzymes and Hydrophobic Substrates

The glycolipid substrates catabolized by many lysosomal hydrolases are strongly hydrophobic. The enzymes, including hexosaminidase A, are hydrophilic. Consequently, assistance is needed to bring the enzyme and its natural substrate into intimate contact, enabling the enzyme to cleave the necessary

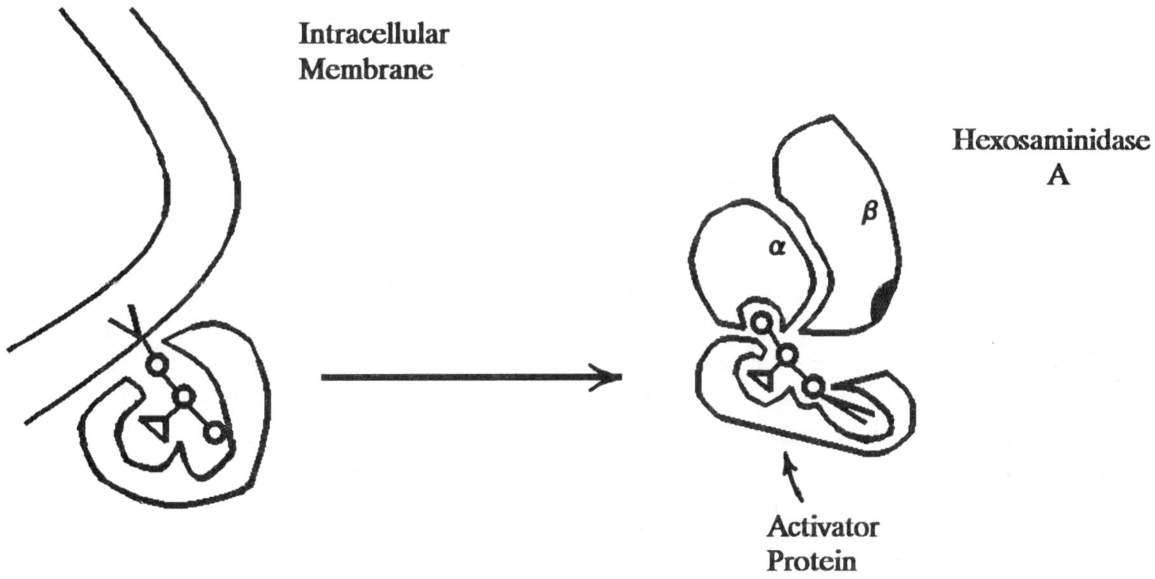

Fig. 8-10. GM$_2$-activator protein. The activator protein transports the substrate from the membrane (left) to the lysosome (right) where it adjusts the fit of the hydrophobic substrate to the active site of the hydrophilic enzyme. The dark patch on the lower right surface of the β subunit of hexosaminidase A represents the site for hydrolysis of synthetic substrates.

bonds. The assistance is often provided by accessory proteins called **activator proteins.** Hexosaminidase A utilizes GM_2-activator protein to facilitate hydrolysis of GM_2-ganglioside (Fig. 8-10). GM_2-activator protein transports the ganglioside from the membrane to the lysosome, where the activator protein adjusts the fit of the ganglioside to the active site of hexosaminidase A. Patients are known who have normal hexosaminidase A, but who lack functional GM_2-activator protein. Hexosaminidase A hydrolyzes the test substrates in a normal manner, but fails to cleave GM_2-ganglioside unless a detergent is added to the assay system. The detergent substitutes for the activator protein, making the carbohydrate of the ganglioside more accessible to the active site of the enzyme. These patients exhibit signs of GM_2-gangliosidosis and are said to have an "AB-variant" form of the disease.

Other activator proteins have been identified. Two of these, sphingolipid activator protein-1 (SAP-1) and sphingolipid activator protein-2 (SAP-2), assist several different lysosomal hydrolases. SAP-1 is required for normal activity of GM_1-ganglioside β-galactosidase and arylsulfatase A toward their natural substrates, while SAP-2 assists certain other lysosomal hydrolases. Both SAP-1 and SAP-2 are encoded by the same gene, transcribed into the same mRNA, and translated into a pre-activator protein. The protein is subsequently processed to form the respective activator proteins.

Protective Proteins and Lysosomal Enzyme Complexes

The enzyme, GM_1-β-Galactosidase, catalyzes the degradation of GM_1-ganglioside to GM_2-ganglioside (Fig. 8-5). This enzyme is known to occur in normal lysosomes in two forms: aggregated in a large complex with a sialidase which catalyzes removal of sialic acid (also known as neuraminic acid) and as a monomer. About 90% of β-galactosidase activity is associated with the multimeric aggregate in normal lysosomes (Fig. 8-11). Enzyme components of the aggregate are interconnected by a **protective protein** that is synthesized as a large precursor and processed into two smaller polypeptides. Patients with a storage disease, galactosialidosis, lack **both** GM_1-β-galactosidase and sialidase activities. The mutation responsible for this rare condition has been traced to the autosomal gene encoding the protective protein. Patients lacking the protective protein cannot assemble the complex. The monomeric forms of the enzymes are unstable and are rapidly degraded, leading to the dual enzyme deficiency (9). This β-galactosidase not only hydrolyzes GM_1-ganglioside, but also cleaves galactosyl residues from a number of other substrates, including glycosaminoglycans that are found in connective tissue. The respective substrates usually metabolized by β-galactosidase and sialidase accumulate in lysosomes, causing the demise of the patients.

Golgi Processing Enzymes and Multiple Enzyme Deficiencies

Lysosomal enzymes are targeted to lysosomes by a mannose-6-phosphate receptor system (Chapter 2). Lysosomal enzymes enter the lumen of the endoplasmic reticulum where N-linked oligosaccharides of the "high mannose" type are added. As the enzymes traverse the Golgi stack, the oligosaccharides are modified by other enzymes. Two of these are especially relevant to normal operation of the receptor system. A phosphotransferase transfers a phosphorylated sugar to certain mannose residues on the immature enzymes. The sugar is then cleaved from the molecule, exposing the mannose-6-phosphate group. Enzymes possessing the mannose-6-phosphate marker are targeted to lysosomes by a specific receptor that binds to the group and conveys the enzymes to a prelysosomal compartment. The enzymes are released into the compartment, and the receptors are recycled back to the Golgi. In some cells, lysosomal enzymes are also secreted into the extracellular space. Mannose-6-phosphate receptors on the surfaces of neighboring cells internalize the secreted hydrolases containing the mannose-6-phosphate marker by receptor-mediated endocytosis.

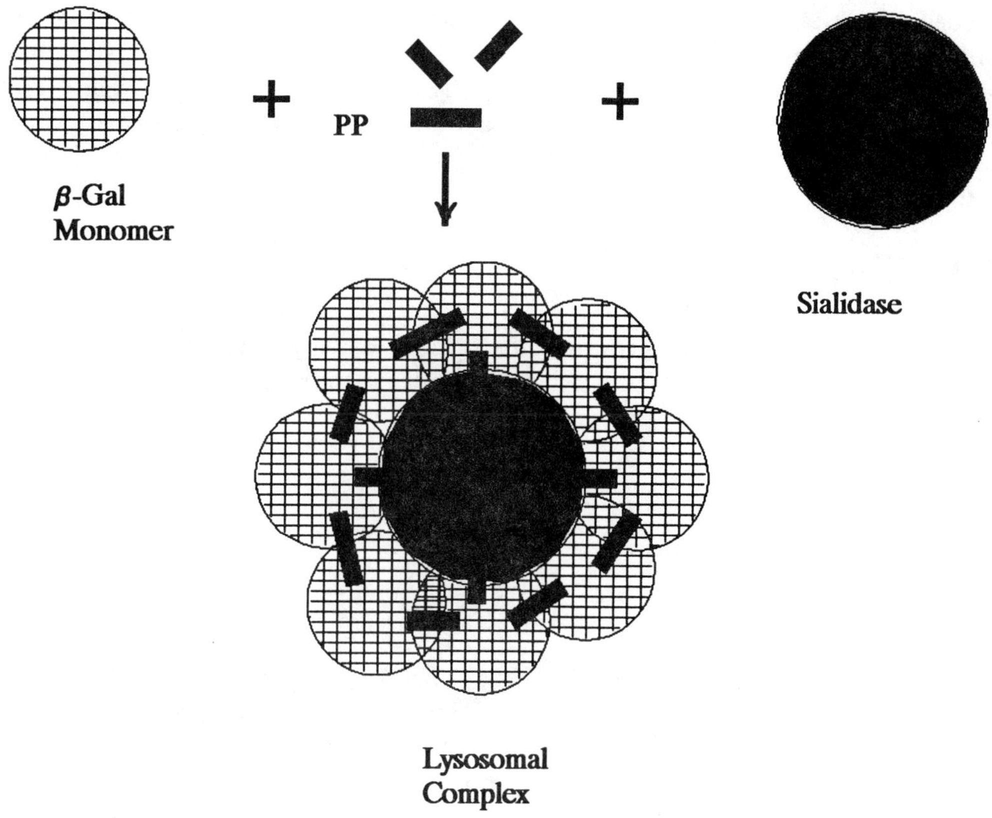

Fig. 8-11. Schematic diagram of β-galactosidase/sialidase complex. Several β-galactosidase (β-Gal) monomers are coupled to one or more sialidase units by protective proteins (PP).

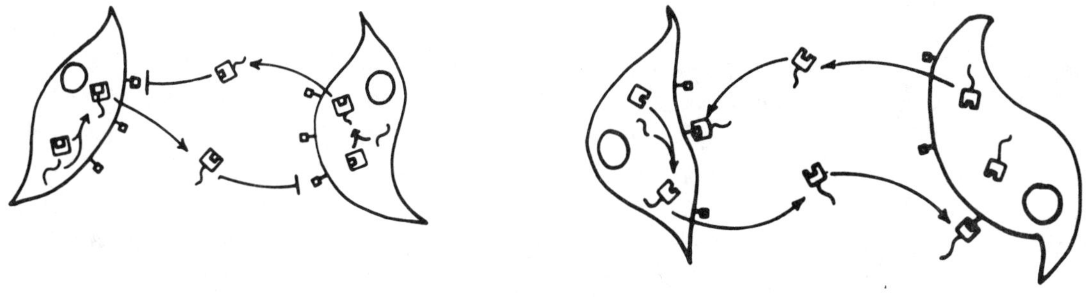

Fig. 8-12. Lysosomal enzyme processing in cells of I-cell patients (left) and those of normal persons (right).

A rare storage disease, Mucolipidosis II (I-Cell disease), occurs in patients who lack the phosphotransferase. Many lysosomal enzymes do not enter the lysosomes and tend to accumulate in the extracellular space of certain tissues (Fig. 8-12). Patients accumulate a variety of complex substrates of these enzymes and usually die during the first decade of life. Activities of some enzymes such as glucocerbrosidase and acid phosphatase tend to be normal, presumably because these enzymes do not share the mannose-6-phosphate system. In some tissues of I-Cell patients, lysosomal hydrolase activities are essentially normal. The reason for this observation is unclear, but may suggest the existence of alternate targeting pathways in these tissues.

MUCOPOLYSACCHARIDOSES

Proteoglycans are important constituents of connective tissue, where they contribute to the structure of the extracellular matrix. A typical proteoglycan is presented in Fig. 8-13. Proteoglycans

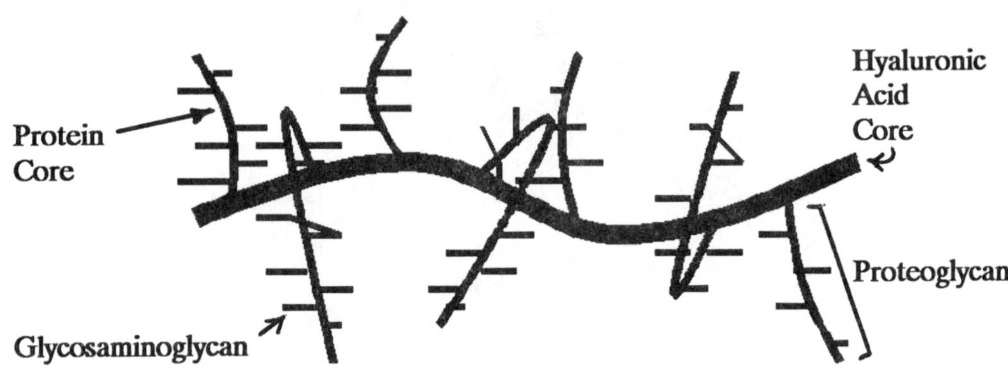

Fig. 8-13. Segment of a proteoglycan aggregate. The proteoglycans are joined to a core composed of hyaluronic acid by two linker proteins (not shown). Each proteoglycan resembles a bristle brush, having a protein core from which the glycosaminoglycans radiate.

are degraded by a variety of lysosomal hydrolases which participate in the stepwise removal of the sugars forming their glycosaminoglycan chains and by proteases that attack their protein cores. **Mucopolysaccharidoses** are a group of rare storage diseases that are caused by deficiencies of the lysosomal hydrolases which degrade glycosaminoglycans. Although most of these hydrolases are targeted directly to lysosomes, about 10% are secreted and endocytosed by the mannose-6-phosphate receptor system.

Several mucopolysaccharidoses have been clinically and enzymatically defined (Table 8-1). These diseases have been clinically arranged on the basis of the relative involvement of the eyes, skeletal system, the type of glycosaminoglycan accumulating in lysosomes, excreted glycosaminoglycan derivatives, and severity of mental retardation. For example, patients with Morquio syndrome have severe skeletal malformations but usually have normal intelligence. Subtypes of this syndrome are now recognized that differ according to their biochemical cause. Hurler's patients display severe physical and mental defects, while Scheie patients seem to have milder manifestations. An intermediate Hurler-Scheie disease has been observed. These three clinical disorders are most likely due to allelic mutations at the α-L-

Table 8-1. Classification of Mucopolysaccharidoses.

Syndrome	Enzyme Deficiency	Inheritance
Hurler	α-L-Iduronidase	AR
Scheie	Same	AR
Hurler-Scheie	Same	AR
Hunter	Iduronate sulfatase	XLR
Sanfilippo A	Heparan N-sulfatase	AR
Sanfilippo B	α-N-acetylglucosaminidase	AR
Sanfilippo C	Acetyl-CoA:α-glucosaminide acetyltransferase	AR
Sanfilippo D	N-acetyl glucosamine-6-sulfatase	AR
Morquio A	Galactose-6-sulfatase	AR
Morquio B	β-galactosidase	AR
Maroteaux-Lamy	Arylsulfatase B	AR
Sly	β-Glucuronidase	AR

iduronidase locus, and all three varieties represent homozygosity for different alleles, although genetic compounds most likely occur in some cases.

Diagnosis of mucopolysaccharidoses is ultimately based upon the pattern of glycosaminoglycan derivatives excreted in the urine, the clinical appearance of the patient, and the particular enzyme deficiency that is present. When urine from one of these patients is spotted on filter paper and treated with toluidine blue, a purple spot appears on a blue background. When a positive test is observed, the identities and concentrations of the compounds are determined with more sophisticated techniques, and the deficient enzyme is identified following tissue culture of skin fibroblasts. One diagnostic procedure utilizes a **complementation test** (Fig. 8-14 and 8-15). A proportion of the lysosomal hydrolases that degrade glycosaminoglycans are secreted by skin fibroblasts and taken up by their neighbors in tissue culture. Cells from the patient are co-cultured with cells from cell lines derived from patients with known mucopolysaccharidoses. If the cells of the patient and those from the cell line have a deficiency in the same enzyme, they will not complement (Fig. 8-14). Lack of complementation would establish the diagnosis as being the same as or allelic with the disease present in the person from which the cell line was derived. Complementation of the cells (as demonstrated by normal catabolism of the test substrate) would indicate that the patient has a different disease caused by a non-allelic mutation.

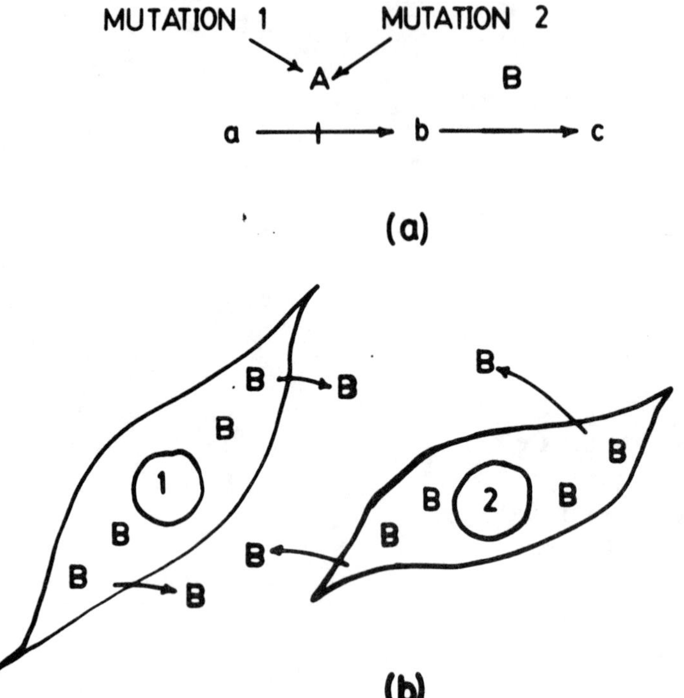

Fig. 8-14. Complementation test demonstrating that two mutations affect the same metabolic step and presumably the same locus. a) Metabolic pathway; b) Lack of complementation when both types of mutant cell are cultured together. A and B are enzymes that are secreted and taken up by neighboring cells by receptor-mediated endocytosis. Neither cell produces functional enzyme A which is required for complementation to occur.

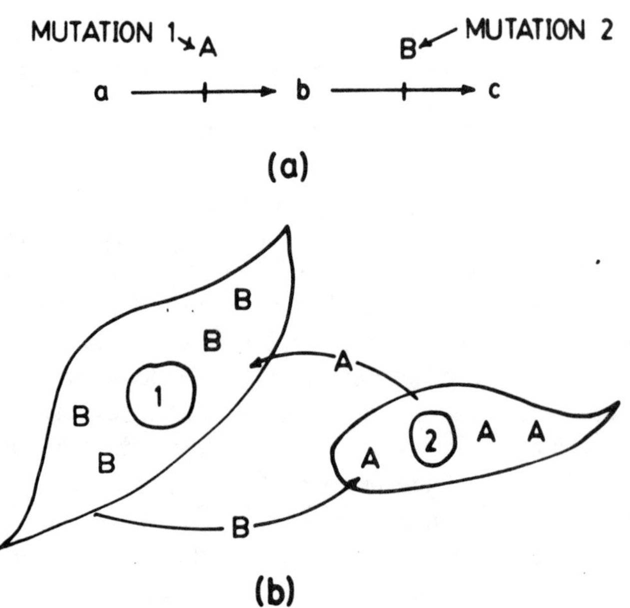

Fig. 8-15. Complementation of cells possessing blocks at alternate sites in a metabolic pathway as a consequence of nonallelic mutations. a) Metabolic pathway; b) Complementation in co-cultured cells. Enzyme A secreted by cell 2 corrects the defect in cell 1 and vice versa.

Genetic heterogeneity refers to the cause of the same or very similar phenotype by different mutations. Genetic heterogeneity due to multiple allelism is illustrated by Hurler, Scheie, and Hurler-Scheie syndromes. All three of these clinically different phenotypes are caused by different mutations in the same gene. Patients who are products of consanguineous marriages are most likely homozygous for one of these mutations. However, patients who have unrelated parents are probably genetic compounds for two rare mutant alleles. Genetic heterogeneity due to nonallelic mutations is exemplified by Sanfilippo syndrome. Mutations at any one of four different loci can produce the phenotype characteristic of this disorder.

VITAMIN DEPENDENCIES

Vitamins are organic compounds that cannot be synthesized by human cells and which are precursors of coenzymes. Coenzymes facilitate enzyme action. Vitamin requirements vary among people; however, daily dietary requirements of most persons fall within a relatively narrow range (Fig. 8-16).

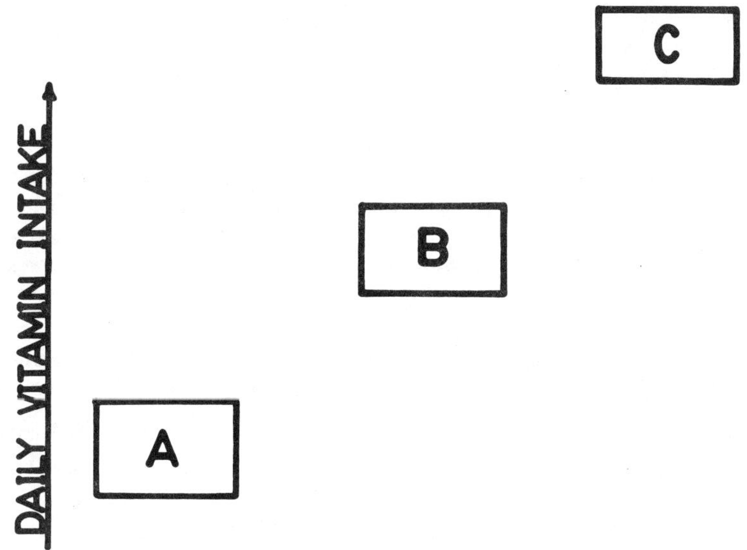

Fig. 8-16. Daily vitamin intake. A) Vitamin deficiency; B) Recommended daily allowance for normal persons; C) Vitamin dependency.

Vitamin deficiencies occur when environmental sources of the vitamin are inadequate to meet daily needs. Examples include scurvy which was a common disease among peoples who did not consume adequate quantities of foods enriched in Vitamin C, and rickets, a bone disease, which affected children who lived in industrial cities in temperate climates. These diseases are largely preventable by adhering to a balanced diet or a diet with Vitamin D supplement in areas of reduced sunlight. Occasional patients present with symptoms of vitamin deficiency when they are ingesting sufficient vitamins to meet normal daily requirements. These individuals have an unusually high vitamin requirement as the result of abnormal vitamin absorption, transport, processing, or coenzyme-enzyme binding (Fig. 8-17). Excessive metabolic demand for a vitamin is referred to as a **vitamin dependency**.

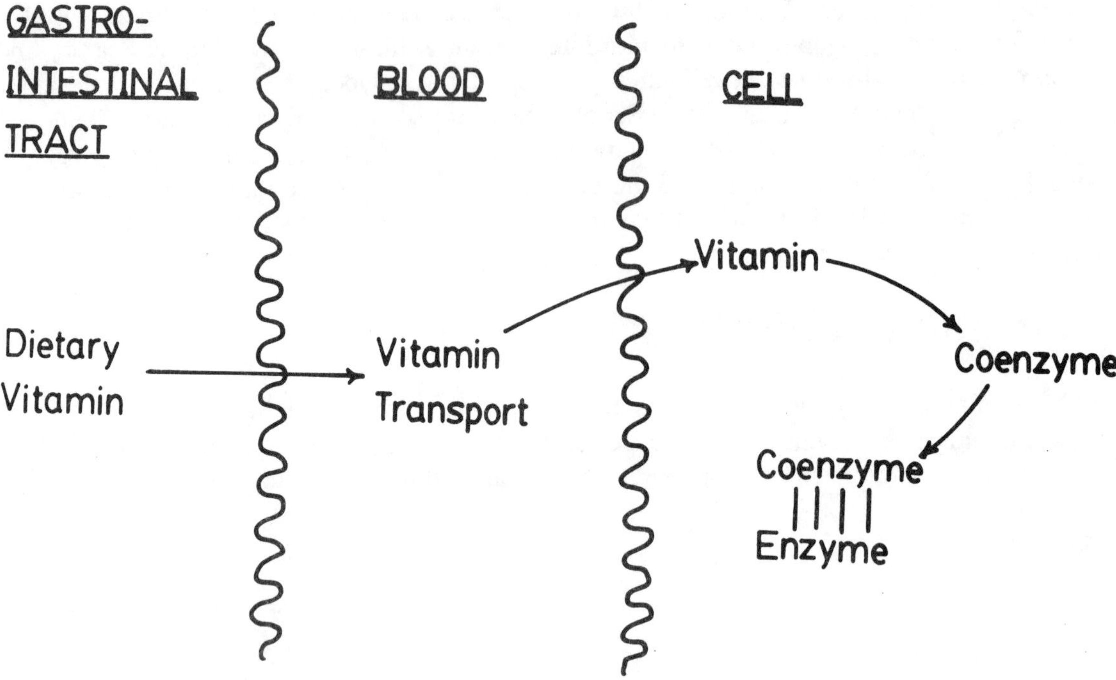

Fig. 8–17. Fate of dietary vitamin. Gastrointestinal absorption: Vitamin B_{12} (cobalamin) is synthesized by bacteria in the gut, and its absorption is facilitated by intrinsic factor, a glycoprotein that is synthesized by certain cells in the lining of the stomach. Excessive use of antibiotics can kill the gut bacteria and lead to pernicious anemia which can be produced by Vitamin B_{12} deficiency. Vitamin transport: Vitamin D is carried by Gc protein (Vitamin-D binding protein). Cellular metabolism: Vitamins are transformed to coenzymes by a series of metabolic reactions that are enzyme-catalyzed.

Homocystinuria is a rare autosomal recessive disorder in which urinary homocystine excretion is increased. Homocystinurics have an enhanced risk for thromboembolic episodes and present with features resembling dominantly inherited Marfan syndrome. The extremities are elongated, and their fingers and toes are spider-like, a condition known as arachnodactyly. Patients also have dislocated lenses which also occur in Marfan syndrome. Homocystinuria and Marfan's syndrome are **genocopies**, similar phenotypes caused by mutations at two different loci. Homocysteine is an intermediate in sulfur metabolism (Fig. 8-18). Accumulation of homocysteine can occur as a consequence of any one of several metabolic blocks. Classical homocystinurics have a deficiency of the enzyme cystathionine synthase (#1, Fig. 8-18). Homocysteine accumulates, is nonenzymatically oxidized to homocystine, and is excreted in the urine. Methionine, a precursor of homocysteine also tends to be increased in the blood of these patients. The biochemical cause of the life-threatening thromboembolic problems experienced by homocystinurics is unknown.

Classical homocystinuria occurs in Vitamin B_6-responsive and nonresponsive forms. Pyridoxal phosphate, a coenzyme for cystathionine synthase, is synthesized from Vitamin B_6. Some homocystinurics have a mutation that affects the coenzyme-binding site of cystathionine synthase. These patients can be treated with high doses of Vitamin B_6. This therapy increases the amount of coenzyme and overcomes the binding defect sufficiently to restore a normal phenotype. Cystathionine synthase, like many other enzymes, is overproduced. Therefore, it is not necessary to replace all of the missing activity. Studies indicate that restoration of 5% of normal activity by vitamin supplementation is adequate to prevent the adverse homocystinuric phenotype. The nonresponsive form of cystathionine synthase deficiency presumably occurs as a consequence of mutations that alter the active site of the enzyme, affect its production, or otherwise interfere with its normal catalytic function.

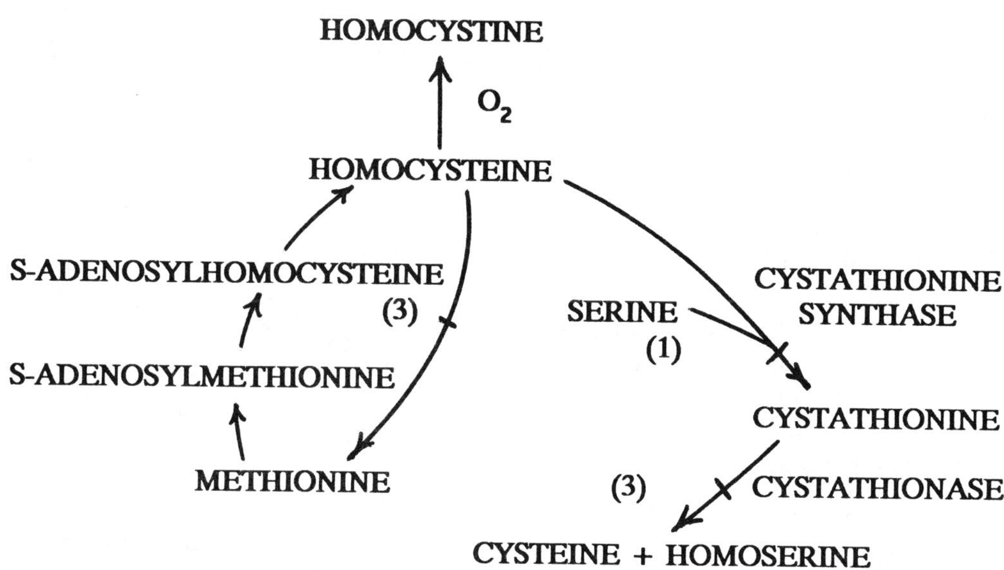

Fig. 8-18. Synthesis of cysteine. The metabolic block responsible for homocystinuria may occur at (1) or (3). Cystathioninuria results from deficient cystathionase activity (3).

Homocystinuria may also occur as a consequence of a block in the conversion of homocysteine to methionine. One type of patient with this deficiency appears to have a modified apoenzyme having deficient catalytic activity that is not vitamin-responsive. However, two other forms of the disease are vitamin responsive. Folic acid corrects one form by overcoming a partial deficiency of an enzyme participating in the synthesis of methyltetrahydrofolate from folic acid. Methyltetrahydrofolate is the coenzyme for the methyltransferase that catalyzes reaction 3. The methyltransferase also requires a second coenzyme, methylcobalamin, which is derived from Vitamin B_{12}. The second group of vitamin-responsive patients requires increased Vitamin B_{12} to overcome a partial deficiency of an enzyme that transforms cobalamin to methylcobalamin. These examples further stress the problems genetic heterogeneity creates for medicine and the importance of thorough biochemical evaluation for precise diagnosis of the specific biochemical lesion, since three of the five lesions respond to vitamin therapy. Two other defects, cystathioninuria and methylmalonic aciduria, also occur in vitamin-resistant and vitamin-responsive forms. Some patients with the former disease respond to increased Vitamin B_6, while a portion of cases of the latter disease are effectively treated with Vitamin B_{12}.

INBORN ERRORS OF DRUG METABOLISM

Glucose-6-phosphate dehydrogenase deficiency affects the intracellular milieu of the red cell, predisposing patients with this deficiency to hemolytic anemia when exposed to certain drugs. G6PD deficiency was discussed in Chapter 7. Drug sensitivity may also occur as the result of a deficiency for a drug-metabolizing enzyme. Isoniazid is a drug that is frequently used for treatment of tuberculosis. Isoniazid is inactivated by acetylation (Fig. 8-19). Homozygosity for a frequent allele is associated with

$$\text{ISONIAZID + ACETYL CoA} \xrightarrow{\text{ACETYL TRANSFERASE}} \text{ACETYLISONIAZID + CoA}$$

Fig. 8-19. Metabolic inactivation of isoniazid.

slow acetylation of the drug. These individuals have low acetyl transferase activity and, if they are exposed to isoniazid, may develop toxic side effects to the unacetylated form which persists longer in their systems.

Serum Cholinesterase Deficiency

Succinylcholine is used as a muscle relaxant during surgery. Detoxification of this drug is accomplished via hydrolysis by an enzyme, serum cholinesterase (pseudocholinesterase). Serum cholinesterase activity is encoded by the **CHE1** and **CHE2** loci which have been mapped to chromosomes 3 and 2, respectively. Two additional genes were detected by in situ hybridization with **CHE1** cDNA probes. These loci have been named **CHEL1** (cholinesterase-like) and **CHEL3** and have been assigned to chromosomes 3 and 16, respectively. **CHE1** determines the majority of serum cholinesterase activity. The exact function of the **CHE2** locus is unknown. About 8% of most populations contain an unusual serum isozyme, referred to as the C5 variant. One hypothesis regarding **CHE2** function views the **CHE2** gene as inactive in most individuals, but encoding a polypeptide in some persons which interacts with the **CHE1** catalytic subunit to produce the variant pattern. The functions of the cholinesterase-like genes is unknown.

Several mutant alleles at the **CHE1** locus have been described. **CHE1^u** is the standard allele in all populations. About 1/2000 Caucasians are homozygous for an atypical allele, **CHE1^a**, that encodes an enzyme with a reduced affinity for succinylcholine. As a result, the drug remains in the circulation for a longer period of time and can cause extended muscular paralysis and interfere with normal respiration (apnea). The atypical enzyme also has a reduced affinity for dibucaine, an inhibitor of cholinesterase, and dibucaine is used to screen for the atypical form of cholinesterase deficiency. Enzyme activity is determined in the presence and absence of dibucaine, and percent inhibition of activity by dibucaine is recorded. **CHE1^{a}CHE1^a** homozygotes exhibit a dibucaine number (% inhibition) of about 20, while heterozygotes and normal homozygotes have dibucaine numbers approximating 60 and 80, respectively. **CHE1^{a}CHE1^a** persons do not have clinical difficulties unless they are exposed to the relaxant. This common allele has an A-->G transition resulting in a replacement of aspartate[70] by glycine (10). A second mutant allele **CHE^f** encodes a fluoride-resistant form of serum cholinesterase. Individuals homozygous for this allele have increased succinylcholine sensitivity; however, their clinical symptoms appear to be less severe than those experienced by persons homozygous for the atypical allele. A third form of serum cholinesterase deficiency is associated with unusual enzyme segregation patterns. This type of mutation is referred to as the "silent" form. Although infrequent among Western Europeans and populations derived therefrom, silent alleles occur frequently in Eskimo and in certain Asian Indian groups. Two or more silent alleles are believed to exist. Some of these mutations appear to be null alleles (no enzyme), while others encode a protein with less than 10% of normal activity. One of these **CHE1^s** alleles was sequenced following PCR amplification (11). The region glycine^{115}-->glycine^{117} is encoded by three successive **GGT** triplets. Such repetitive regions have been observed to be "hotspots" for frameshift mutation in other organisms and in other genes in humans. Seven individuals from two unrelated families possessed an apparent double mutation. A T-->A transversion occurred in the third triplet, and an additional **G** was inserted after the **A.** The net result of the two mutations was the generation of a translational stop downstream from the mutated sites.

Combinations of these four alleles, as well as others that have been reported, generate a continuous distribution of serum cholinesterase activities in the population. Most of these genotypes will not be associated with adverse succinylcholine reactions; however, genotypes that reduce the total cholinesterase activity below 30-40% will display some degree of succinylcholine sensitivity. The marked variation in the distribution of certain of these alleles among different populations serves as a warning for persons considering international marketing of drugs. If other mutations affecting drug detoxification systems display a similar heterogeneity of distribution, drugs found to be safe for one population may prove harmful to certain members of a genetically different population.

Not all mutations reduce enzyme activity. A rare variant at the **CHE1** locus, the Cynthiana variant, actually produces three times the activity encoded by the usual allele. The increased activity appears to reflect enhanced synthesis of the enzyme, rather than altered catalytic activity. Such patients display accelerated clearance of succinylcholine from their system and must be given higher doses to bring about muscle relaxation.

INBORN ERRORS OF CHOLESTEROL METABOLISM

Cholesterol has been recognized as a major factor in the etiology of heart disease. Cholesterol is enriched in meat, milk and products derived from these sources. Vegetables, fish, and certain other food have low cholesterol contents. In addition to dietary cholesterol, cells can manufacture their own cholesterol from acetyl-CoA. Hydroxymethylglutaryl-coenzyme A reductase (HMG CoA reductase) catalyzes the step that commits metabolites to the cholesterol pathway and is an important regulatory enzyme in cholesterol biosynthesis. This enzyme is sensitive to intracellular cholesterol levels. As the level of intracellular cholesterol rises, activity of HMG-CoA reductase falls, diminishing cholesterol biosynthesis. The reverse trend occurs when intracellular cholesterol falls. In this way, the cell adapts its synthesis of cholesterol to its cholesterol requirements.

Dietary cholesterol impacts upon cholesterol biosynthesis through an elaborate absorption-transport-internalization system (Fig. 8-20). Dietary cholesterol enters the enterocytes (intestinal cells) where it is esterified and incorporated along with triglycerides and a small lipoprotein (ApoB48) into chylomicrons. The chylomicrons enter the enterohepatic circulation, where they are modified into chylomicron remnants by the enzyme lipoprotein lipase. The chylomicron remnants enter hepatocytes, and their triglycerides and cholesterol esters are transferred to a lipoprotein particle called VLDL (very low density lipoprotein). The VLDL particle exits the liver, and some VLDL particles are further modified by the enzyme LCAT (lecithin-cholesterol acyltransferase) to LDL (low density lipoprotein particle) which is enriched in cholesterol esters. A schematic representation of an LDL particle is presented in Fig. 8-21. The interior of the particle is occupied by cholesterol esters, while much of the surface consists of a phospholipid-cholesterol membrane. The lipoprotein, B100, is situated at one end of the particle and participates in receptor binding as described in the next section. B100 is also a constituent of VLDL particles. Most of the remaining VLDL is primarily involved with triglyceride

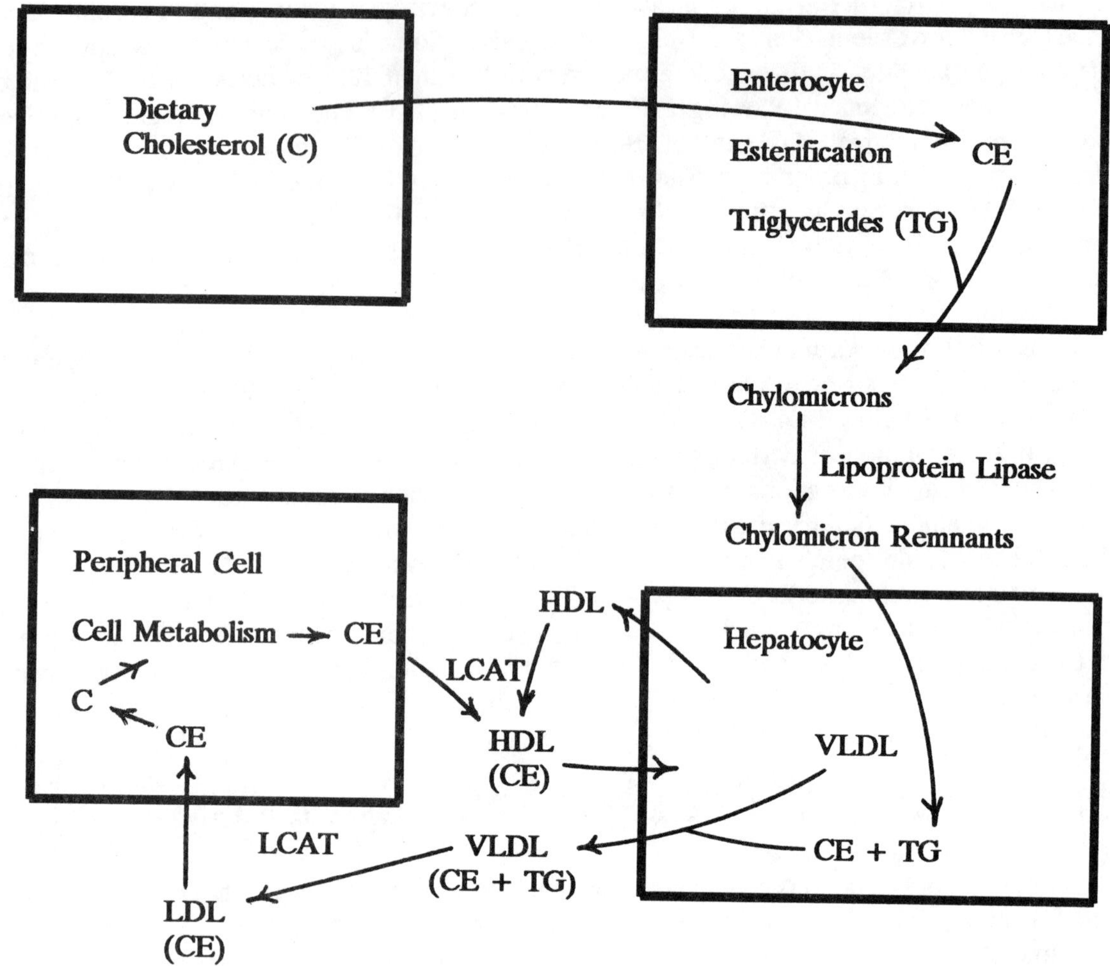

Fig. 8-20. Fate of dietary cholesterol and cholesterol exchange among tissues. CE = cholesterol esters; VLDL = very low density lipoprotein; HDL = high density lipoprotein; LDL = low density lipoprotein; LCAT = lecithin-cholesterol acyltransferase.

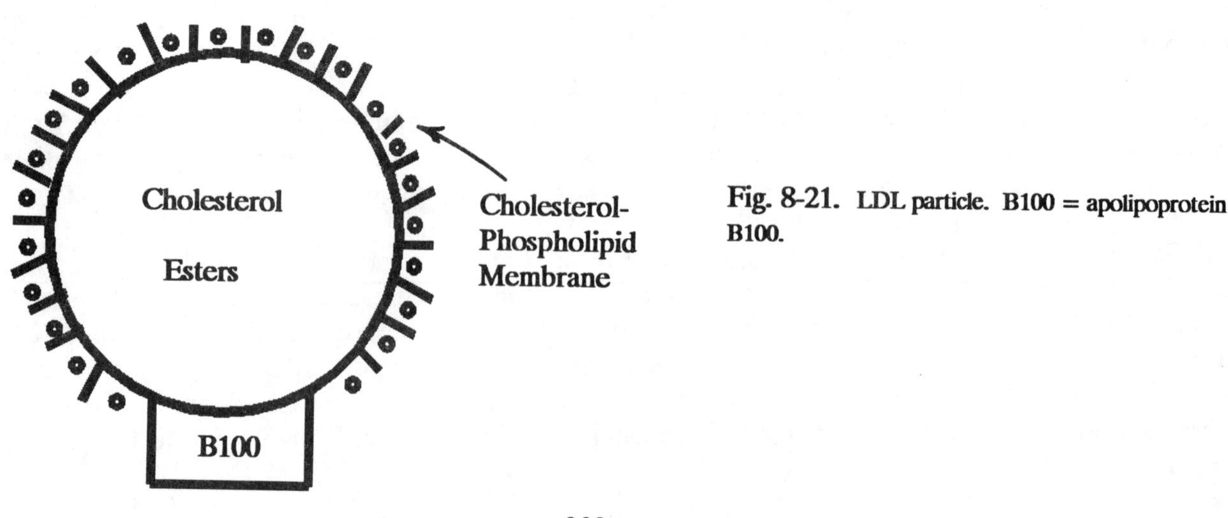

Fig. 8-21. LDL particle. B100 = apolipoprotein B100.

transport.

Intracellular and extracellular cholesterol levels are linked through the LDL-receptor system (Fig. 8-22). The B100 component of the LDL particle binds in a specific manner with the LDL-receptor

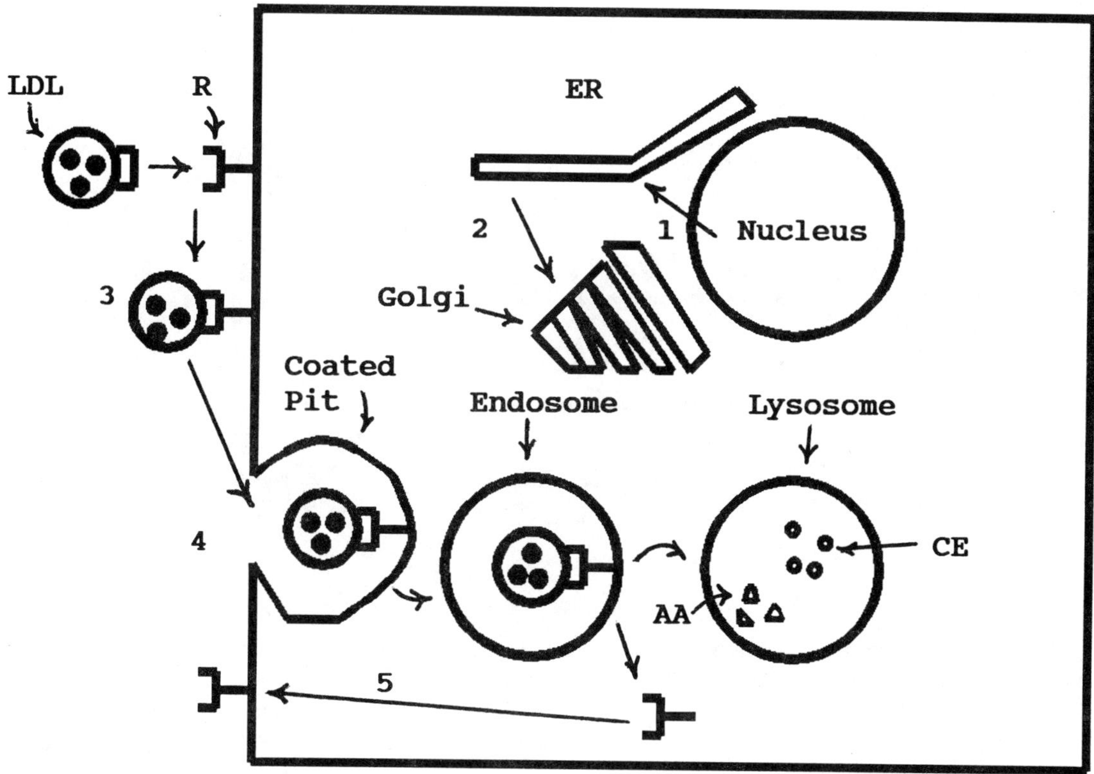

Fig. 8-22. Internalization and processing of cholesterol by peripheral cells. The LDL particle binds to the LDL receptor, and the LDL-receptor complexes cluster in coated pits. The complex is internalized into endosomes which undergo acidifcation and fusion with lysosomes. The receptor is recycled to the cell surface, and the LDL particle is degraded, releasing its cholesterol esters. Deesterified cholesterol inhibits HMG-CoA reductase, limiting cholesterol biosynthesis. 1) Synthesis of the LDL receptor; 2) transport of the immature receptor for processing in the Golgi; 3) binding of the LDL particle to the receptor; 4) clustering of the receptor-LDL complexes in coated pits; 5) recycling of the LDL-receptor. AA = amino acids; CE = cholesterol ester; R = receptor.

which migrates with other LDL-receptor complexes to coated pits. The complexes are internalized, localizing in the endosome. The endosome is acidified and ultimately fuses with a lysosome where the LDL is degraded into its constituent amino acids, and the cholesterol esters are hydrolyzed. Free cholesterol enters cell metabolism and inhibits HMG-CoA reductase. The LDL-receptors recycle to the cell surface. Excess peripheral cholesterol is reesterified by LCAT and transported by HDL (high density lipoprotein) back to the liver where it can be excreted.

Mutations Affecting Cholesterol Transport

Several mutations have been identified in the **APOB** gene which has been mapped to 2p24-p23. This gene encodes the ApoB100 protein containing 4536 amino acids that is found in VLDL and LDL particles (Fig. 8-23). Two general types of mutant alleles appear to occur at the **APOB** locus. Null mutations (cross-reacting material negative (CRM⁻); note: CRM usually corresponds to an inactive form of the protein that can be identified with an antibody against the normal protein) have been

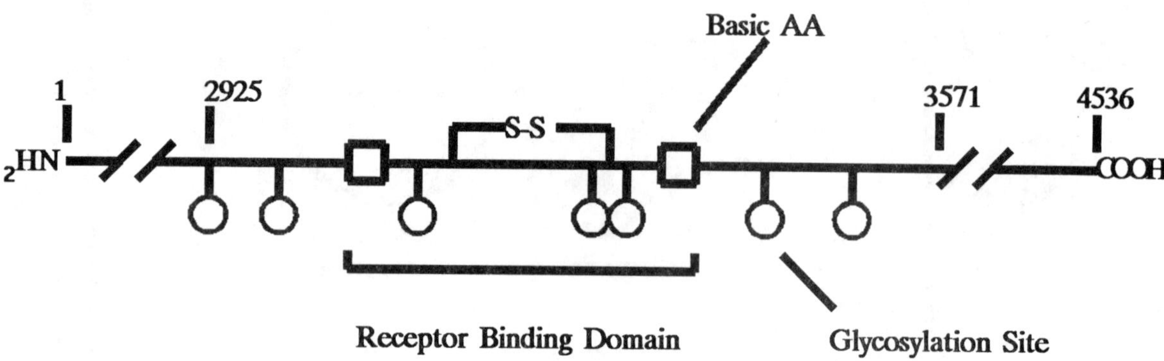

Fig. 8-23. ApoB100. This protein is a constituent of VLDL and LDL. The portion of the molecule that binds with the LDL receptor is flanked by two basic amino acid (AA) motifs.

identified which, when homozygous, result in absence of ApoB100 and an inability to transport cholesterol and triglycerides from the intestine to the liver and from the liver to peripheral cells. Patients who lack ApoB100 do not have VLDL nor LDL. Fat is not absorbed properly from the intestinal lumen, and triglycerides accumulate in intestinal cells. Patients develop retinal and neurological degeneration and have abnormal red cell morphology (acanthocytosis: spiked projections from red cell surface). This syndrome is known as abetalipoproteinemia. Treatment of these patients involves maintenance on low fat diets and the administration of tocopherol which retards worsening of the retinal and neurological degeneration. Some patients appear to produce ApoB100 mRNA and have some ApoB100 protein. These patients may have a mutation that interferes with normal processing and secretion of the ApoB100 protein. A second group of mutations is associated with the presence of low ApoB100 levels. This group is apparently quite heterogeneous, with some of these alleles producing symptoms in homozygotes that resemble those of homozygosity for null alleles, while others are associated with milder disease.

ApoB48, a protein consisting of the amino-terminal 2152 amino acids of ApoB100, is a major component of chylomicrons. The protein is synthesized in enterocytes and is also encoded by the **APOB100** gene. The structure of the **APOB100** gene is identical in hepatocytes and enterocytes; however, the enterocyte mRNA transcribed from the **APOB100** gene contains a U in place of the C in codon 2153. The base substitution must take place either at the time of transcription, or during pre-mRNA processing. The result is the replacement of codon 2153 with a translational stop signal that leads to the synthesis of the truncated protein. This single base modification which occurs during or after transcription is called **RNA editing.** RNA editing has also been observed in a number of other eukaryotes. ApoB48 deficiency which results from mutations in the amino-terminal half of the **APOB100** gene result in a lack of chylomicrons, interfering with fat absorption and transport to the liver. ApoB48 deficiency is inherited as an autosomal recessive trait.

HDL participates in the transport of cholesterol from the periphery back to the liver, where the cholesterol is subsequently excreted. The HDL particle contains apolipoprotein A-1 (ApoA1) which is encoded by the **APOAI** gene that is part of a gene cluster of lipoprotein-encoding genes on the long arm of chromosome 11. Circulating HDL levels appear to be subject to both environmental and genetic factors. For example, regular and sustained exercise has been demonstrated to increase plasma HDL and is associated with a reduced risk for coronary disease. Genetic control of HDL levels appears to be accomplished by additive effects of several genes. The cluster of lipoprotein genes on chromosome 11 seems to be intimately involved with this regulation. Rare mutations in the **APOAI** locus are known which act in an apparent dominant fashion to increase HDL levels, providing protection against heart disease. By contrast, ApoA1 deficiency and resulting low HDL concentrations are inherited as an austosomal recessive trait and are associated with increased risk for heart attacks.

One of the most common lipoprotein abnormalities, called **severe combined hyperlipidemia,** occurs in about 1/50 to 1/200 persons. This disorder is inherited as an autosomal dominant and is associated with elevations of LDL, VLDL or both particles and increases in cholesterol and triglycerides. Patients have an increased risk for heart disease, accounting for about 10% of cases of early onset myocardial infarctions. The cause of this disorder has not been fully determined; however, recent studies have mapped the gene responsible for the problem to the **APOAI-APOCIII-APOAIV** gene cluster on 11q23-q24 (12).

Familial Hypercholesterolemia and the LDL-receptor

Familial Type IIa hypercholesterolemia is characterized by elevated LDL-cholesterol, tendon xanthomas (swellings occurring on the elbows, knees and certain other locations), and increased risk for heart attacks. The disease is inherited as an autosomal dominant, with heterozygotes experiencing their first myocardial infarction before age 45-50. About 1/500 people are heterozygous for this trait and account for about 5-10% of early onset heart attacks. Rarely, individuals may inherit an abnormal allele from each of their heterozygous parents. LDL-cholesterol levels of these persons tend to be 4-6 times normal, and they may suffer their first heart attack during late childhood or early adolescence. Most die before the end of their third decade of life.

This form of hypercholesterolemia is caused by a large group of mutations that affect the LDL-receptor (Fig. 8-24). Most patients with the very severe form of Type IIa hypercholesterolemia are genetic compounds for two different mutant alleles. The mutations may be apportioned among 5 groups. About half of the reported alleles result in a failure of synthesis of the LDL-receptor (step 1, Fig. 8-22). This class includes promoter mutations, premature stop codons, splicing mutations, and deletions of various exons of the gene. One rare mutation was caused by the insertion of an Alu element in exon 15. The second class of mutation (#2; Fig. 8-22) interferes with the transport of immature receptors from the endoplasmic reticulum to the Golgi apparatus. The receptors are degraded before they enter the Golgi complex. Most of these mutations result from base substitutions or small deletions that promote abnormal folding of the receptor molecule. Another class affects binding of the receptor with ApoB100 of the LDL particle (#3; Fig. 8-22). These mutations affect the ligand-binding domain of the protein that is encoded by five exons (LB; Fig. 8-24). Mutations in exon 18, encoding the cytoplasmic domain of the receptor, prevent normal clustering of receptors in coated pits. The final category of mutation interferes with normal recycling of receptors (#5; Fig. 8-22) and involve changes in exons 7-14 which share sequence homology with the epidermal growth factor precursor gene.

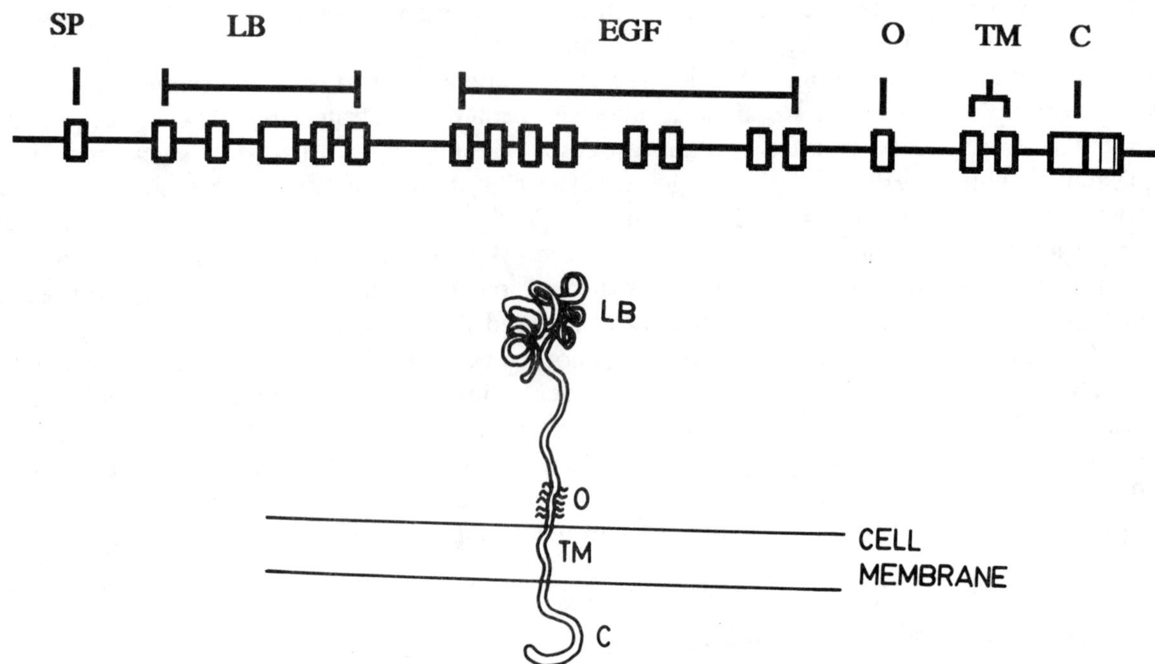

Fig. 8-24.   The LDL-receptor and its structural gene.   The gene (top) contains 18 exons that are clustered according to the receptor domains they encode.   SP = signal peptide; LB = ligand-binding domain (binds to LDL particle); EGF = domain having homology with the epidermal growth factor precursor protein; O = domain linked to oligosaccharide chains; TM = membrane-spanning region; and C = cytoplasmic domain.   A model of the receptor is provided in the lower portion of the figure.

Hypercholesterolemia and Myocardial Infarction

Hypercholesterolemia is a multifactorial disorder of cholesterol metabolism.   Single-gene components of hypercholesterolemia include mutations producing elevated circulating LDL-cholesterol (LDL-receptor defects and severe combined hyperlipidemia) and decreased circulating HDL-cholesterol (ApoA1 deficiency).   Elevated blood cholesterol is believed to trigger a series of events that culminates in the formation of arterial plaques that occlude the coronary arteries (Fig. 8-25).   For example, an inherited LDL-receptor defect retards normal cholesterol uptake by many types of peripheral cells, leading to unusually high blood cholesterol levels.   Smooth muscle cells in the arterial walls can accumulate this cholesterol by a mechanism that does not require LDL-receptors.   The cholesterol accumulates, killing the cells and building arterial plaques.   This process is known as **atherosclerosis**. Migrating white blood cells known as macrophages have special receptors that are distinct from the LDL-receptor.   These cells also accumulate cholesterol and are transformed into foam cells that further contribute to the development of the arterial plaques.   The plaque continues to grow, eventually blocking the artery. The block in the circulation of the heart muscle impairs delivery of vital oxygen and nutrients and the removal of waste materials. Consequently, the heart muscle deteriorates, producing the heart attack and its associated pain.

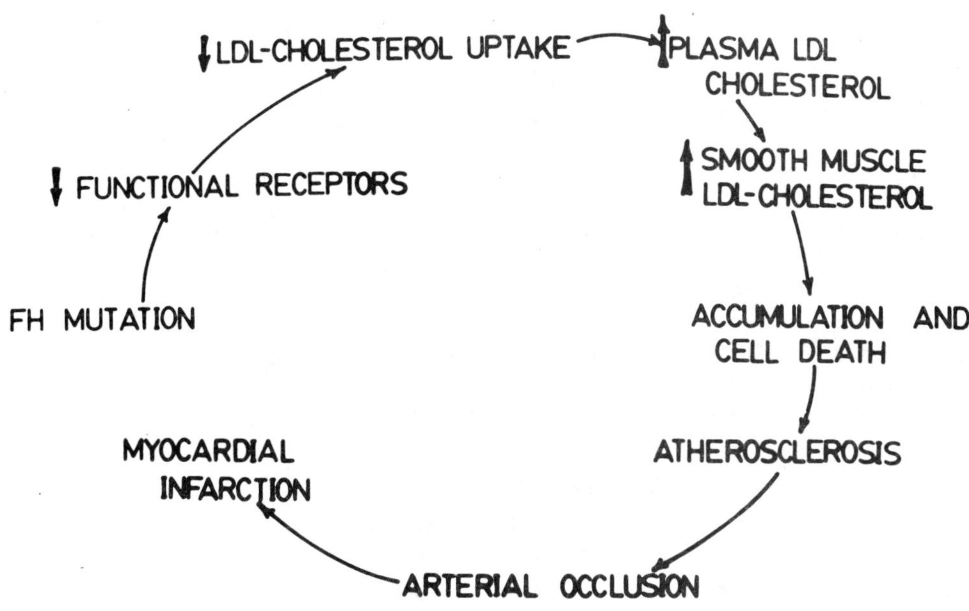

Fig. 8-25. Relationship between familial hypercholesterolemia and pathogenesis of myocardial infarction.

SHARED AND PRIVATE TRANSPORT SYSTEMS

Receptor molecules have been implicated as mediators for translocation of a variety of substances across cell membranes. In addition to LDL-cholesterol transport, receptors participate in immunological recognition and stimulation of immune responses, in transport of carbohydrates such as glucose, and in transport of amino acids across membranes of a variety of cells.

Cystinuria is an autosomal recessive disease in which excessive quantities of the amino acid cystine are excreted in the urine. This amino acid has a low affinity for water and precipitates, forming cystine stones. These stones occlude the kidney ducts, damaging the organ and causing considerable pain. Although renal stones were considered to be the only complaint of cystinuric patients, cystinuria tends to be about 10-fold more frequent among the mentally handicapped, occurring in one per thousand. These retarded individuals may have a subtype of cystinuria that is associated with mental retardation, while many cases of the disease display normal intelligence (13).

Patients with cystinuria exhibit a deficiency of absorption of cystine and dibasic amino acids (ornithine, lysine and arginine) by intestinal and/or kidney proximal tubular cells. The disease appears heterogeneous, since patients can be classified into groups based upon their patterns of excretion of urinary and intestinal amino acids (Table 8-2). These data suggest that cystinurics share a defective transport system responsible for absorption of cystine, ornithine, arginine and lysine by kidney and intestinal cells. Physiological studies support participation of a **group transport system** in the absorption of these four amino acids. Family studies indicate that many cystinurics are homozygous for various mutations that affect this common carrier. Heterogeneity has been observed among the effects of the

Table 8-2. Heterogeneity of Cystinuria.

	Amino Acids Elevated	
Type	Kidney	Intestine
I	Cys, Lys, Arg, Orn	Cys, Lys, Arg, Orn
II	Cys, Lys, Arg. Orn	Cys*, Lys, Arg, Orn
III	Cys, Lys, Arg, Orn	Partial Defect or Normal
Relative Severity	II > III > I	I > II > III

* Partial defect; Cys = cystine; Lys = lysine; Arg = arginine; Orn = ornithine.

alleles that are involved. For example, comparison of the intestinal transport of these amino acids has permitted the identification of at least three alleles that are responsible for cystinuria. Furthermore, some heterozygotes excrete quantities of the four amino acids in their urine that are intermediate between those of the homozygote and normal values. Heterozygotes for other alleles display normal urinary amino acid excretion profiles. Genetic compounds for different cystinuria alleles have also been reported who display excretion patterns that are intermediate with respect to those of the corresponding cystinuric homozygotes. The molecular basis for the cystinurias is unknown; however, the locus may encode the protein carrier normally utilized by the four amino acids.

Hyperdibasicaminoaciduria occurs in four varieties: 1) associated with cystinuria (described above), 2) Type 1 hyperdibasicaminoaciduria (French-Canadian), 3) Type 2 hyperdibasicaminoaciduria (lysinuric protein intolerance), and 4) lysine malabsorption syndrome. The type I transport error is inherited as an autosomal dominant and has been observed in only one French-Canadian kindred. Urinary arginine, lysine and ornithine are elevated; however, urinary cystine excretion is normal. Most persons with the Type I defect do not experience clinical complications. Lysine malabsorption syndrome has been reported in only one patient. This patient exhibited excessive lysine excretion; however, cystine, arginine and ornithine excretion were normal. She had seizures, mental retardation and severe growth failure. Lysine protein intolerance (LPI) is a rare disorder, with a major proportion of patients having Finnish ancestry. All three dibasic amino acids are elevated in urine, depleted in blood, and poorly absorbed from the gut. Lysine is most prominently affected. Cystine excretion is normal. LPI patients are usually normal until they are removed from a milk diet. When placed on a high protein diet, they grow poorly, are at increased risk for mental retardation, and display a variety of other complications. Favorable response to a low protein diet and administration of citrulline has been observed, and this may be the treatment of choice. The exact cause of LPI is unknown.

Cystinosis is a rare disorder caused by defective cystine transport across the lysosomal membrane. The cystine is retained within this organelle and forms cystine crystals. Cystine accumulation occurs in a variety of different tissues including kidney, cornea of the eye, and liver. Patients require kidney dialysis by 6-12 years of age. The rates of accumulation of cystine vary among tissues, leading to variable involvement of the respective organ systems. The molecular basis for this autosomal recessive disease is unknown, and genetic heterogeneity may be present.

These examples indicate that multiple carrier proteins are most likely involved with amino acid transport across various cell membranes. Some carry groups of chemically related amino acids, while others are specific for individual amino acids such as cystine and lysine. The latter carriers are often referred to as "private" carriers, and these transporters generally have a wide distribution and low capacity. Group carriers most commonly have high capacities and display restricted distributions (e.g. renal and gastrointestinal epithelial cells).

CHRONIC OBSTRUCTIVE PULMONARY DISEASE (COPD)

COPD or emphysema is a frequent adult malady and has both environmental and genetic causative components. The roles of atmospheric pollution and cigarette smoking in respiratory disease have been well documented. The genetic predisposition to COPD is less well understood and most likely involves the interaction of a number of different genes. This section will address two relatively frequent problems that include respiratory disease as one component and which are determined by mutations at major loci.

α-1-Antitrypsin Deficiency

α-1-Antitrypsin (AT) deficiency is determined by an autosomal locus, and clinical complications of the deficiency are inherited as an autosomal recessive trait. Serum contains a protease inhibitor which appears to be involved in protection of lung alveolar membranes. AT is synthesized in the liver, coupled with oligosaccharides and secreted into the blood which transports AT to lung tissues. During inflammation, neutrophils, macrophages and other white blood cells release proteases onto alveolar surfaces to clear up irritating debris and microorganisms. Homozygotes and genetic compounds for certain AT deficiency alleles cannot effectively inhibit these proteases and sustain progressive damage to their alveoli and loss of respiratory capacity. Patients with AT-deficiency have low serum AT activity, and this activity is associated with a series of protein bands that migrate more slowly than the standard AT pattern (Fig. 8-26). Only patterns of homozygotes are shown for clarity. Individuals with serum levels of AT that fall below 60% of normal appear to have an increased risk for COPD (Table 8-3), especially those with the ZZ phenotype. COPD occurs at a much earlier age and is more rapidly progressive in the genetically predisposed persons than in MM people. The MZ phenotype is associated with borderline-low levels of AT; however, the relevance of this phenotype to risk for COPD

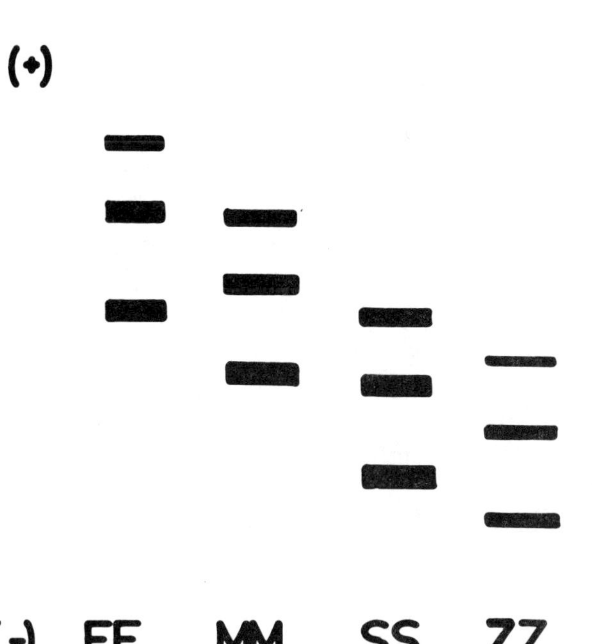

Fig. 8-26. α-1-Antitrypsin electrophoretic phenotypes. F = fast; M = usual; S = slow; Z = ultraslow deficiency variant.

Table 8-3. **Amount of Immunoreactive* α-1-Antitrypsin Occurring in Various Antitrypsin Phenotypes (14,15).**

	Phenotype					
	MM	MS	MZ	SS	SZ	ZZ
Immunoreactive Antitrypsin*	100	79	61	53	40	15
Frequency Among**:						
Healthy Adults	898	41	29	1	1	<1
COPD Adults	768	55	110	12	6	43

* Amount of protein reacting with antibody against purified MM α-1-antitrypsin (expressed as % of normal).
**Expressed as cases per 1,000 people (numbers do not add up to 1000, since rare phenotypes have been excluded).

is controversial, and the consensus of current opinion does not favor an enhanced risk for COPD in individuals with the MZ phenotype (16).

The pathogenesis of COPD in ZZ patients is believed to occur by the following route. Emphysematous lesions have been induced in lungs of hamsters by introduction of a white cell enzyme, elastase. Elastase is capable of attacking alveolar structure and producing the typical changes associated with emphysema. AT inhibits elastase activity. Patients with the ZZ and SZ phenotypes have little inhibitor, and leukocyte elastase and perhaps other proteases will persist for longer periods of time in lungs of these patients. Chronic release of proteases from leukocytes and macrophages during the course of respiratory infections and in response to air pollutants produce progressive alveolar damage in AT-deficient persons. The effect of cigarette smoke would lead to a prediction that smokers who have the ZZ or SZ phenotype run an especially high risk for COPD. Smokers with the ZZ phenotype have been found to develop problems 10 to 15 years earlier than nonsmokers with the same phenotype.

AT-deficiency also appears to be causally related to certain types of liver cirrhosis, especially in newborn infants. Granules of a substance that has been demonstrated to be a precursor form of AT accumulate in hepatocytes and contribute in some way to malfunction and cell death. Some form of liver disease occurs in about 20% of ZZ infants. In many cases, jaundice (yellowish skin color) is apparent by four months of age and disappears at about six months; however, complications of cirrhosis appear during late childhood or early adolescence. Most ZZ persons displaying liver malfunction during infancy die in late adolescence. The reason why most ZZ individuals escape liver involvement is unknown.

The gene for AT (**PI**) has been mapped to chromosome 14q32.1 where it is closely linked to **AACT** which encodes a closely related protein, α-1-antichymotrypsin. The **PI**^{Z} and **PI**^{S} alleles contain base substitutions that result in replacement of glutamate^{342} by lysine and glutamate^{264} by valine, respectively. The substitution in the Z variant involves an amino acid located at a critical bend in the protein that participates in the formation of a salt bridge that is essential for normal molecular stability. Site-directed mutagenesis that restored the salt bridge resulted in normal secretion of the Z variant in an artificial system. The results of the experiments suggest that the substitution of lysine for glutamic acid interferes with normal folding of the molecule and promotes AT aggregation, producing the granular liver deposits.

Cystic Fibrosis

Cystic fibrosis (CF) is a relatively common disease among populations derived from northern and western Europe. Approximately 1/20 Caucasians are heterozygous for a CF allele, and about 1/1600 to 1/2000 newborn Caucasian infants develop CF, an autosomal recessive disorder. The disease primarily involves exocrine glands and is associated with abnormal chloride efflux from the apical surface of cells composing these glands. Patients have high chloride levels in their sweat, secrete a thick mucus in their lungs which serves as a rich growth medium for bacteria, and display impaired secretion of pancreatic enzymes. Severely affected patients experience digestive failure, intestinal obstruction, have difficulty breathing, and suffer repeated respiratory infections. Unless aggressive action is taken to treat these symptoms, death often occurs at an early age. Considerable phenotypic variation is observed among patients, with some having both respiratory and digestive components of the disease, while others have predominantly respiratory or digestive disease. Some CF persons are very mildly involved and may not be encountered until later in life when they show signs resembling emphysema. Fertility is impaired in both CF men and women. With modern advances in treatment, the average life expectancy for a CF individual is about 25-30 years.

The gene involved with CF has been recently mapped to chromosome 7, cloned, and sequenced. The gene spans about 250kb of DNA and has 27 exons. The gene encodes a protein, cystic fribrosis transmembrane regulator (CFTR), which is a membrane protein that apparently interacts with other proteins of the chloride channel on the apical surface. A model of the structure of CFTR is presented in Fig. 8-27. The protein crosses the membrane 12 times and has two domains that

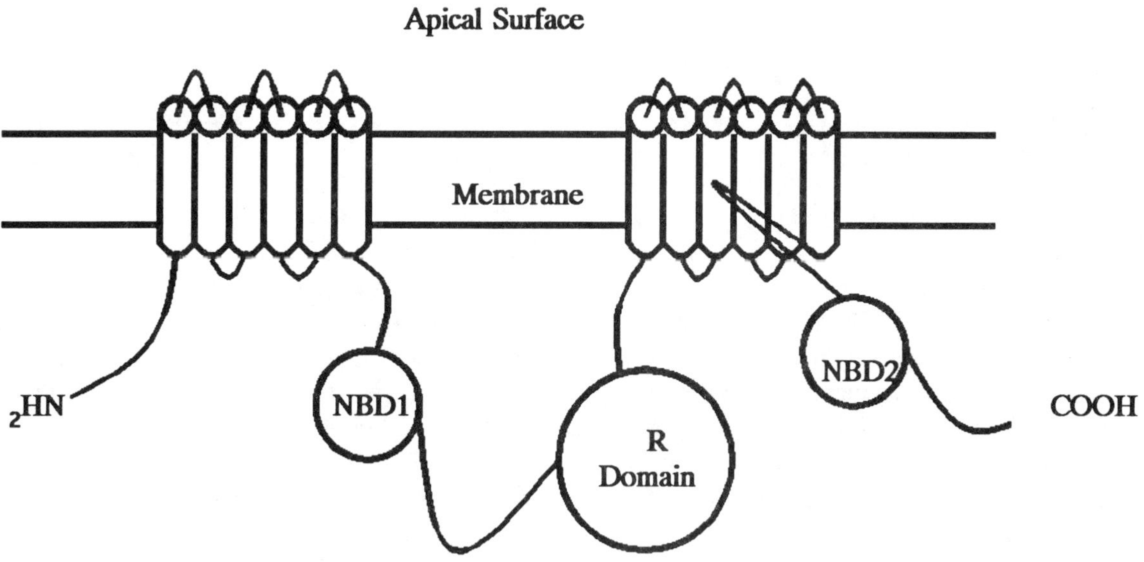

Fig. 8-27. The Cystic Fibrosis Transmembrane Regulator protein. The most common mutation, accounting for about 70% of the CF alleles in the United States encodes a protein lacking a phenylalanine from the first nucleotide-binding region. NBD = nucleotide-binding domain; R Domain = regulatory domain.

have sequences resembling those of proteins known to bind nucleotides. A third cytoplasmic domain, the R domain, may have a regulatory function. The way in which the CFTR protein affects chloride efflux is unknown. It is believed to interact with other proteins in the presence of cyclic AMP to facilitate movement of chloride through the ion channel. Cultured CF airway cells lack functional CFTR, and ion movement through the chloride channel is blocked (Fig. 8-28). When these cells were transfected with a normal CFTR gene, ion efflux was restored in the presence of cyclic AMP. However, when the cells were transfected with a Δ F508 allele, the most common CF allele which lacks a phenylalanine in the first nucleotide-binding domain, ion efflux was not observed (17). These

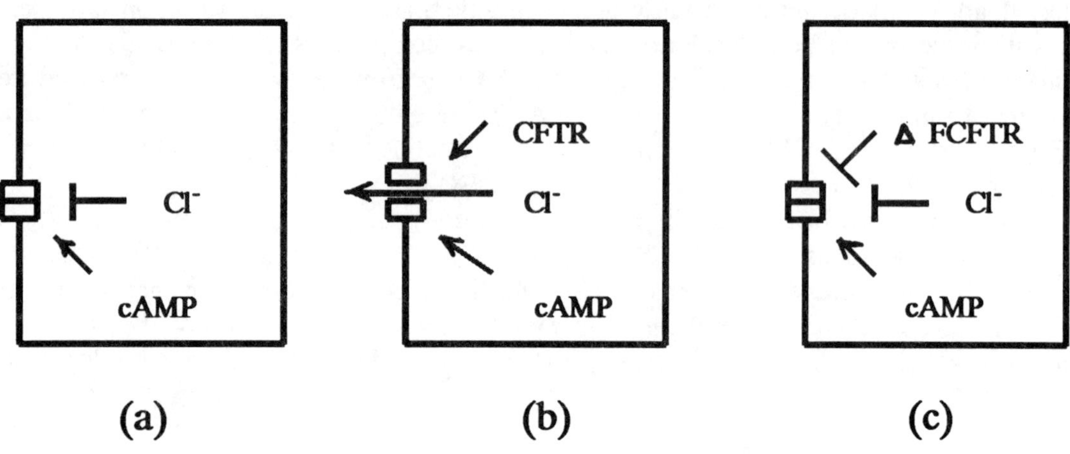

Fig. 8-28. Chloride efflux from the apical surface of cultured CF airway cells (17). a) Cultured CF airway cells do not exhibit Cl⁻ efflux; b) cultured cells transfected with the normal CFTR gene display chloride efflux; c) cultured cells transfected with the Δ F508 CFTR allele do not exhibit chloride efflux. cAMP = cyclic AMP; Δ FCFTR = Δ F508 CFTR allele.

experiments demonstrated that a normal phenotype could be restored to CF airway cells by an intact CFTR gene, and that the common CF allele of this gene was incapable of facilitating chloride efflux.

Although 70% of CF alleles are accounted for by the Δ F508 mutation, more than 60 other alleles have been identified, with the most common accounting for less than 1% of the total. This heterogeneity presents major difficulties for population screening for CF heterozygotes, since only 70% will be detected by probes for the Δ F508 allele. However, family studies can be done using Δ F508 CFTR probes. Most patients who have the common CF allele display relatively severe pancreatic dysfunction, while other mutations are often associated with milder pancreatic disease. Furthermore, mutations affecting the first nucleotide-binding domain appear to have more serious phenotypic complications than those involving the second nucleotide-binding domain.

DYSTROPHIN AND THE X-LINKED MUSCULAR DYSTROPHIES

Duchenne muscular dystrophy (DMD) affects about 1/3500 newborn males. Muscle degeneration begins soon after birth and symptoms are apparent during early childhood. Muscle enzymes such as creatine phosphokinase and aldolase are elevated in serum of Duchenne males. Signs of muscle weakness include a difficulty in rising from the floor. Patients tend to "climb up their legs" as they try to stand. Although muscles such as the calf muscles appear strong and hard, these muscles are actually deteriorating. Cross sections of dystrophic muscle reveal centrally-placed nuclei and marked variation of fiber diameter. Normal skeletal muscle has peripheral nuclei, and its fibers tend to have comparable diameters. Infiltration of DMD muscle by connective tissue is also apparent. Patients are usually confined to wheelchairs by age 12 and die during late adolescence or early adulthood as a consequence of spread of the dystrophy to respiratory and heart muscles. A second form of X-linked muscular dystrophy, Becker muscular dystrophy (BMD), occurs less frequently, has a later onset, and a milder course. Patients usually are not confined to wheelchairs until their forties or fifties. Although DMD patients do not reproduce, fertility of BMD patients is about 70% of normal.

DMD and BMD are determined by allelic mutations in a gene that has been mapped to Xp21.3-Xp21.1 by deletion analysis of unbalanced translocations affecting this region. This approach compares the deleted X-chromosomal material in patients who have muscular dystrophy as a consequence of loss of the **DMD** gene. The area of minimal overlap of deleted material in the DMD patients is inferred to contain the **DMD** gene. The gene was cloned and characterized and found to span about 2.3 megabases of DNA containing about 70 exons. The protein determined by the **DMD** gene, dystrophin, has been localized to the cytoplasmic surface of the sarcolemma. The function of dystrophin is unknown; however, it is believed to interact with other muscle membrane proteins to maintain the integrity of the sarcolemma.

The majority of DMD and BMD patients have deletions or insertions in the **DMD** gene (18). These alterations appear to occur in two clusters, involving exons that encode the N-terminal regions of dystrophin or exons near the middle of the gene. The reason why some mutations result in DMD and some cause BMD is not clear; however, BMD usually results from in-frame deletions and insertions, while DMD mutations frequently generate frameshifts. The **DMD** gene is also expressed in brain. When the N-termini of the brain and skeletal muscle dystrophins were compared, they were found to contain different amino acid sequences. This result and further sequencing studies support the use of different promoters in brain and muscle, with resulting inclusion of different N-terminal exons in the respective mRNAs. About 30% of DMD patients are also mentally handicapped. It is possible that the mutations responsible for this subtype of DMD may involve brain-specific exons.

INBORN ERRORS OF COLLAGEN METABOLISM

Collagen is an important protein constituent of the extracellular matrix. Multiple collagens are known and more than twenty collagen-encoding genes have been identified in the human genome. The following discussion will be limited to certain mutations affecting Type I and Type IIII collagen. Type I collagen is formed by extensive processing of Type I procollagen (Fig. 8-29). Type I procollagen consists of two globular domains that are separated by an extended triple-helical region. Each procollagen molecule contains two proα1I chains (encoded by the **COL1A1** locus) and one proα2I chain (encoded by the **COL1A2** locus). The triple-helical domain requires the sequence: -Gly-Pro-Hyp-Gly-Pro-HyLys-Gly-X-Y-)$_n$ for its normal formation (note: **HyP** = hydroxyproline; **HyLys** = hydroxylysine; and **X** and **Y** refer to any amino acid). The globular N-terminal and C-terminal propeptides are cleaved from the procollagen, and the triple-helical portion contributes to the collagen molecule. Extensive molecular packing produces the collagen fibril.

Type I collagen is a predominant collagen in skin, bone, tendons and arteries. A number of mutations have been described which affect this collagen and generally produce a heterogeneous disease known as **osteogenesis imperfecta**. A number of point mutations in **COL1A1** that result in defective or deficient proα1I chains have been described. These lesions tend to have relatively mild effects and display autosomal dominant inheritance patterns. Heterozygotes have normal stature, and, although they are subject to bone fractures, there is minimal disfigurement. Other features of this form (Type I) of osteogenesis imperfecta include blue coloration of the sclera of the eyes and otosclerosis (a form of deafness). Not all patients exhibit this triad of symptoms. For example, about half have hearing

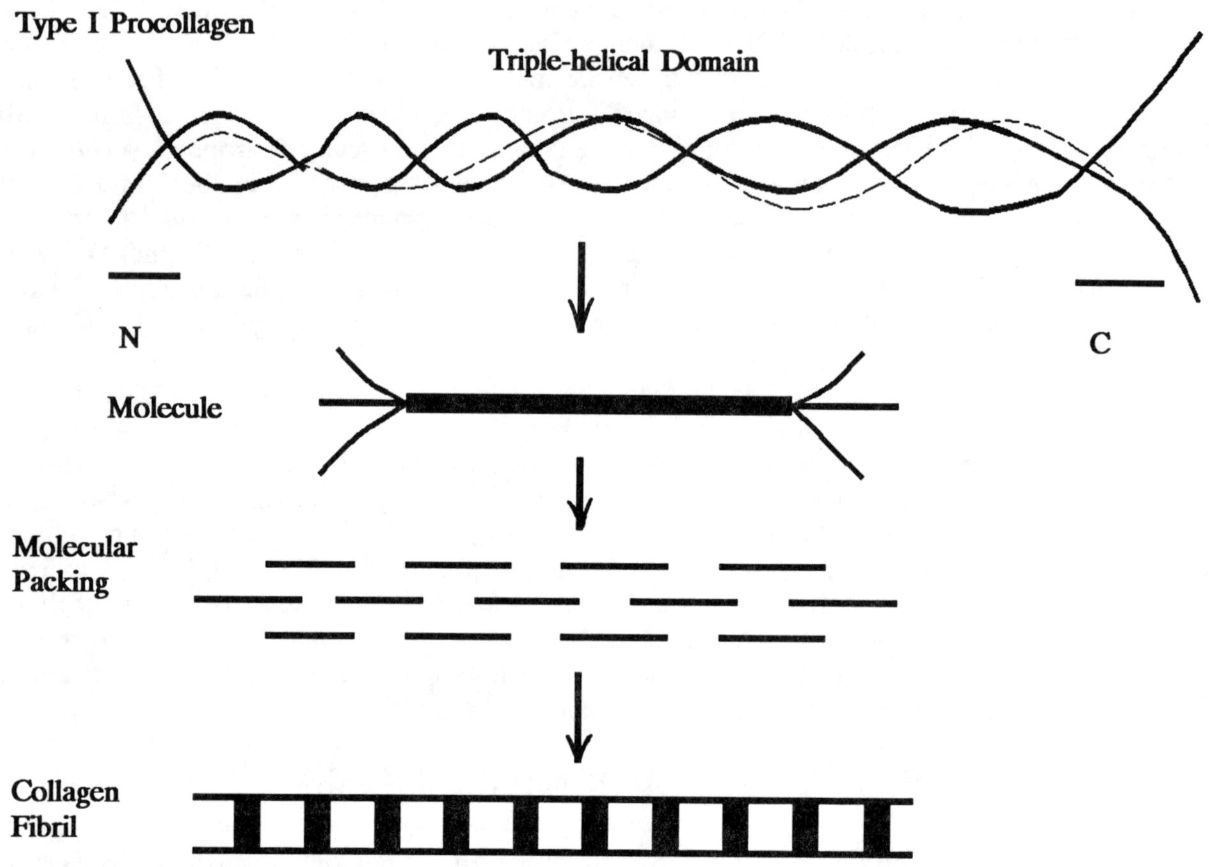

Fig. 8-29. Assembly of Type I collagen. Two proα1I (solid line) chains and one proα2I chain form a trimer that has N-terminal (N) and C-terminal (C) propeptides separated by a triple-helical domain. The propeptides are cleaved to form the collagen molecule, and these molecules are tightly packed to compose the collagen fibril. Hydroxylation of prolines and lysines within the triple-helical domain by prolyl hydroxylase and lysyl hydroxylase, respectively, occurs during this process and stabilizes the collagen fibril.

impairment. Type I osteogenesis imperfecta afflicts about 1/15,000 persons.

Type II osteogenesis imperfecta is less common and results from more severe mutations that often involve multiple exon insertions in or deletions from either the **COL1A1** or **COL1A2** genes. A second cause of Type II osteogenesis imperfecta is the replacement of glycine residues in the triple-helical domain of the molecule. These mutations are either inherited as autosomal recessives, or

314

represent new dominant mutations. Type II disease is especially severe, disfiguring, and most commonly lethal before or shortly after birth. Several other varieties of osteogenesis imperfecta are known.

Ehlers-Danlos syndrome encompasses several distinct diseases that are caused by abnormal collagen. Patients with Type I Ehlers-Danlos syndrome display skin hyperextensibility, "cigarette paper" scars on their skin, and hypermobility of their large joints. The specific collagen defect responsible for this syndrome is unknown. Ehlers-Danlos Type I is inherited as an autosomal dominant. Type IV Ehlers-Danlos syndrome is characterized by thin skin, normal skin extensibility and joint mobility, and susceptibility to arterial, uterine and bowel rupture. This disease is caused by abnormal synthesis or secretion of Type III collagen following mutations in the **COL3A1** gene. Type III collagen is enriched in skin, arterial walls and the uterus. Autosomal recessive lysyl hydroxylase deficiency is associated with soft velvet-like and hyperextensible skin, hypermobile joints, scoliosis, and ocular rupture. This form of Ehlers-Danlos syndrome is classified as Type VI.

MITOCHONDRIAL DISEASES

Cells may contain from one to more than 1000 mitochondria, depending upon their metabolic activity. Therefore, mutations in mitochondrial genes will tend to take time to have a detectable effect upon the phenotype of most cells and tissues. Most persons with mitochondrial gene mutations will be mosaic for the mutation and will exhibit variable severity of the associated phenotype. Furthermore, mitochondria are passed through the maternal gametes to the next generation, and mutations affecting mitochondrial genes will display maternal inheritance. Susceptibility of tissues to mitochondrial mutations will depend upon the energy demands of particular cell types. The central nervous system, skeletal muscle, heart, kidney and liver have particularly large energy demands and would be anticipated to be more strongly affected by these mutations than other cell types.

Leber's Hereditary Optic Atrophy

Persons with Leber's hereditary optic atrophy experience a relatively sudden loss of vision following death of the optic nerve. This relatively rare disease is maternally inherited. One common type of mutation found in kindreds segregating for this form of blindness involves a **G-->A** transition in a mitochondrial gene encoding a subunit of mitochondrial NADH dehydrogenase. This mutation changes an essential arginine to histidine. Deficient NADH dehydrogenase activity interferes with mitochondrial respiration, reduces cellular ATP, and contributes to the death of optic nerve cells.

Myoclonic Epilepsy with Ragged Red Fibers (MERRF)

This rare epilepsy displays extensive variability among patients and is maternally inherited. Energy metabolism in muscles of these patients is abnormal, and the variability indicates that their muscle cells most likely contain mixed populations of normal and mutant mitochondria. The alteration may have such severe effects that it may only be able to be maintained in cells containing these mixed mitochondrial populations. The specific mutation responsible for this disease has not been identified.

Diseases Caused by Extensive Deletions of Mitochondrial DNA

Two syndromes, Kearns-Sayres syndrome and chronic progressive external ophthalmoplegia, are caused by large deletions of mitochondrial DNA. The mitochondrial chromosome is transcribed into large RNAs containing sequences of several genes. Therefore, deletions anywhere in the DNA molecule would be anticipated to disrupt expression of a number of different genes. The deletions occurring in these two diseases have severe effects and can be tolerated only when normal mitochondria are also present. In mixed populations of mitochondria, DNA molecules containing deletions replicate more rapidly than normal mitochondrial DNAs. Therefore, over time, the mutant molecules will become more abundant. Furthermore, the oocyte amplifies the number of its mitochondria by rapid replication of mitochondrial DNA. Many mitochondrial mutations occur during oogenesis. Deletions that arise during this time could recruit a significant number of mitochondria and cause problems for individuals who are derived from these cells.

Mitochondrial mutations may be influenced by nuclear genes. For example mutation in a nuclear gene encoding a protein required for mitochondrial DNA replication would be anticipated to influence the accuracy of this process. One type of eye defect known to occur as a consequence of mitochondrial DNA deletions is transmitted as an autosomal dominant (19). A second nuclear gene-mitochondrial DNA interaction may be represented by the discovery of a mitochondrial DNA depletion syndrome (20). Four infants in three unrelated families developed a fatal mitochondrial disease. In two unrelated infants, mitochondrial DNA depletion was observed in kidney and skeletal muscle, while in the third family, one infant exhibited this depletion in skeletal muscle and the other in liver. Examination of the mitochondrial DNAs extracted from these tissues revealed normal structure of the mitochondrial DNA replication origins. Presumably some nuclear-encoded factor was interfering with replication of the mitochondrial DNA. Leber's hereditary optic neuropathy exhibits considerable variation in severity, and recent evidence suggests that a gene located on the proximal short arm of the X-chromosome may influence the clinical severity of this disease (21). Neither the identity of the gene nor its mode of action are known.

Mitochondrial DNA has a much higher mutation rate than nuclear DNA. Accumulation of these mutations would lead to a slowing of mitochondrial respiration that may contribute to the aging process.

SUMMARY

Inborn errors of metabolism involve blocks of synthetic or degradative pathways at critical points. These blocks are usually caused by homozygosity for an allele which determines a defective enzyme or which fails to produce the enzyme. The former possibility accounts for the majority of metabolic errors. When normal enzyme is purified and used to raise antibodies against the enzyme protein, treatment of cell extracts from patients with metabolic diseases reveals two groups. Many of these patients are CRM(+), possessing the catalytically inactive enzyme. Others are CRM(-), lacking enzyme protein. CRM(+) patients may have a better response to enzyme replacement therapy, regardless of whether the enzyme is infused into the patient or whether somatic gene replacement is conducted. In either case, the patient will not mount an immune rejection of the replacement enzyme. By contrast, the immune system of CRM(-) patients has never experienced the enzyme, and antibodies may be produced against the replacement enzyme.

Other types of phenotypic modification that have been used to treat metabolic diseases include substrate limitation (PKU), product replacement (congenital adrenal hyperplasia and diabetes), vitamin supplementation (vitamin-responsive metabolic diseases), drug therapy (treatment of Wilson's disease patients with copper chelating compounds), and organ transplantation (cystinuria). Most nonregulatory enzymes are produced in considerable excess of metabolic needs; therefore, it is not necessary to restore enzyme activity to normal levels.

The metabolic step that is blocked in a particular disease may be responsible for the synthesis of a product required by a number of cellular processes. The central importance of this step provides the molecular basis for pleiotropism, since all of these processes will be affected by lack of the metabolite, and the deficiency will have a variety of phenotypic effects.

Relative severity of inborn errors of metabolism may vary among patients. For example, a small fraction of untreated PKU patients have normal intelligence. The precise mechanisms responsible for this phenotypic variation among persons sharing the same metabolic block are unknown. Part of the variation may be due to differences in "genetic backgrounds." Genes at other loci may intensify or reduce the severity of the enzyme deficiency caused by homozygosity for the PKU allele. The overall effect of these other loci is dependent upon the particular array of alleles at each locus. Allelic heterogeneity may be responsible for a major portion of phenotypic variation of metabolic errors. Large families of alleles exist at most loci, and most of these alleles are rare. Studies of familial hypercholesterolemia and other metabolic errors have revealed that many patients are genetic compounds, having two different mutant alleles. The severity of the trait is determined by the combination of alleles at the major locus responsible for the trait. Additional heterogeneity arises from the occurrence of similar phenotypes in patients who are homozygous or heterozygous for mutations at different major loci. For example, Sanfilippo syndrome may occur as a consequence of a deficiency of any one of four different enzymes required for catabolism of glycosaminoglycans. Cellular or nuclear complementation can be used to discriminate between mutations at the same or different loci.

Membrane transport usually requires recognition by specific surface receptors that have protein components. Type IIa familial hypercholesterolemia (LDL-receptor deficiency) and cystinuria provide instructive examples of genetic defects influencing transport processes. Heterogeneity of LDL-receptor defects suggests that patients may respond differently to dietary and drug regimens. Studies of cystinuria and related disorders have revealed that molecules may be transported by high capacity group transporters and by low capacity private transporters.

Chronic obstructive pulmonary disease (COPD) has a number of genetic and environmental causes. α-1-antitrypsin deficiency appears to contribute to COPD as the result of decreased neutralization of protease activity in inflamed lung regions. Chronic exposure of lung alveoli to the excessive protease activity leads to their gradual destruction and decreases the surface area available for respiratory exchange. Cystic fibrosis is associated with obstruction of lung passages by viscous mucus. The mucus interferes with exchange of gases and provides a growth medium for bacteria. Defective chloride efflux appears responsible for the features of cystic fibrosis. CFTR facilitates movement of chloride ion through the ion channel in some undefined way. CF patients lack functional CFTR.

Both Duchenne and Becker muscular dystrophies are caused by allelic mutations in the X-linked **DMD** gene which encodes the muscle protein, dystrophin. This protein is not only found on the cytosolic side of the sarcolemma, but is also transcribed in the central nervous system. Many DMD and BMD patients have deletions or insertions involving the **DMD** locus. Most of the alterations are in-frame in the Becker patients, while DMD patients commonly have frameshifts.

Collagen disorders display both autosomal dominant and autosomal recessive inheritance. Autosomal dominant types generally involve small alterations. Since the collagen molecule consists of trimeric components, most molecules of a heterozygote will contain at least one mutant subunit. Recessive collagen disorders involve either more drastic modifications of collagen genes or deficiencies of enzymes involved with processing of collagen.

Mitochondrial disorders have a pronounced effect upon tissues that have high energy demands. Considerable variation of the disorders is observed and is derived from the presence of mixed mitochondrial populations containing varying numbers organelles with mutant or normal mitochondrial DNA molecules. Furthermore, the number of mitochondrial DNA molecules per mitochondrion also varies, and the proportion of mutant and normal DNAs within an organelle can differ among mitochondria. Mitochondrial mutations appear to be subject to unspecified nuclear genes which may encode factors that influence replication of mitochondrial DNA.

PROBLEMS

1. A patient is excreting large amounts of all amino acids and has an enlarged liver and spleen. Signs of severe mental deterioration, poor motor development, and failure to thrive are present. Liver and brain biopsies reveal dense-staining cellular inclusions in neuronal elements. Liver and skin fibroblast enzyme A activities are less than 1% of control values. Skin fibroblasts from the patient's mother and father and from a normal sister possessed enzyme A activities that were intermediate between those of the patient and those of control fibroblasts. Both parents have a normal phenotype. An older brother of the patient died two years ago from problems similar to those of the patient. Discuss the pathogenesis of this disorder.

2. Discuss how you would determine the site of the molecular lesion responsible for the above patient's disease.

3. Define genetic predisposition and provide an example.

Match the following diseases with the appropriate enzyme deficiency.

4. Hurler's syndrome
5. Isoniazid sensitivity
6. Phenylketonuria
7. Tay-Sachs disease

a. Hexosaminidase A
b. Phenylalanine hydroxylase
c. Iduronidase
d. An acetyl transferase

8. How would you distinguish between an internal cellular metabolic error (e.g., an enzyme deficiency) and a transport error affecting the kidney proximal tubule (e.g., cystinuria)?

9. From a therapeutic viewpoint, would you prefer to have a patient with an enzyme deficiency be CRM(+) or CRM(-)? Why?

10. In an attempt to diagnose a metabolic disease, heterokaryons were constructed using cells from the patient and cells from cell lines established from patients with known diseases. The heterokaryons were tested for their ability to metabolize a substrate known to accumulate in the respective diseases and in the patient's cells. What is your diagnosis (+ = substrate metabolized by heterokaryons; - = substrate not metabolized.)

Source of cell line nucleus	A	B	C	D	E
Heterokaryon	+	-	-	-	+

GLOSSARY OF TERMS

Complementation test - cells from two patients are cocultured. If the patients possess mutations at different loci, the cells will grow and metabolize normally. If the mutations affect the same locus, the cells will exhibit an abnormal phenotype. The complementation test may also be performed using heterokaryons containing nuclei from each patient.

Genocopy - a) an inherited form of a disease that usually has an environmental cause; b) two diseases that share the same or similar phenotypes but are caused by mutations at different loci.

Genetic background - the combination or milieu of genes in which major loci function.

Genetic compound - individual possessing two mutant alleles at the same locus.

Inborn error of metabolism - disease which is caused by an inherited deficiency of an enzyme , coenzyme, receptor, or transporter.

Pleiotropism - a single gene produces several phenotypic effects by influencing a process or reaction that is essential for normal development or operation of the systems affected.

Vitamin dependency - an individual requiring more than the recommended daily allowance (normal dietary requirement) of a vitamin for normal metabolism.

BIBLIOGRAPHY

1. Garrod, AE. 1908. Inborn errors of metabolism. Lancet ii:1-7, 73-79, 142-148, 214-220.

2. Folling, A. 1934. Uber Ausscheidung von Phenylbrenstraubbensaure in den Harn als Stoffwechsel-anomalie in Verbindung mit Imbezillitat. Hoppe-Seylers Z Physiol Chem 227:169-176.

3. John, SWM, R Rozen, CR Scriver, et al. 1990. Recurrent mutation, gene conversion, or recombination at the human phenylalanine hydroxylase locus: evidence in French-Canadians and a catalog of mutations. Am J Hum Genet 46:970-974.

4. Beutler, E and W Kuhl. 1975. Subunit structure of human hexosaminidase verified: interconvertibility of hexosaminidase isozymes. Nature 258:262-264.

5. Proia, RL. 1988. Gene encoding the human β-hexosaminidase β chain: Extensive homology of intron placement of the α- and β-chain genes. Proc Natl Acad Sci USA 85:1883-1887.

6. Hubbes, M, J Callahan, R Gravel and D Mahuran. 1989. The amino-terminal sequences in the pro-α and -β polypeptides of human lysosomal β-hexosaminidase A and B are retained in the mature isozymes. FEBS Lett 249:316-320.

7. Bikker, H, FM van den Berg, RA Wolterman, et al. 1990. Distribution and characterization of a Sandhoff disease-associated 50-kb deletion in the gene encoding the human β-hexosaminidase β-chain. Hum Genet 85:327-329.

8. Dlott, B, A d'Azzo, DVK Quon, and EF Neufeld. 1990. Two mutations produce intron insertion in mRNA and elongated β-subunit of human β-hexosaminidase. J Biol Chem 265:17921-17927.

9. Hoogeveen, AT, FW Verheijen, and H Galjaard. 1983. The relation between human lysosomal β-galactosidase and its protective protein. J Biol Chem 258:12143-12146.

10. McGuire, MC, CP Nogueira, CF Bartels, et al. 1989. Identification of the structural mutation responsible for the dibucaine-resistant (atypical) variant form of human serum cholinesterase. Proc Natl Acad Sci USA 86:953-957.

11. Nogueira, CP, MC McGuire, C Graeser, et al. 1990. Identification of a frameshift mutation responsible for the silent phenotype of human serum cholinesterase, Gly 117(GGT-->GGAG). Am J Hum Genet 46:934-942.

12. Wojciechowski, AP, M Farrall, P Cullen, et al. 1991. Familial combined hyperlipidemia linked to the apolipoprotein AI-CIII-AIV gene cluster on chromosome 11q23-q24. Nature 349:161-164.

13. Gold, RJM, MJ Dobrinski and DP Gold. 1977. Cystinuria and mental deficiency. Clin Genet 12:329-332.

14. Pierce, JA, B Eradio and TA Dew. 1975. Antitrypsin phenotypes in St. Louis. J Am Med Assoc 231:609-612.

15. Mittman, C and J Lieberman. 1973. Screening for α_1-antitrypsin deficiency. Israel J Med Sci 9: 1311-1318.

16. Morse, JO. 1978. Alpha-1-antitrypsin deficiency, II. N Eng J Med 299: 1099-1105.

17. Rich, DP, MP Anderson, RJ Gregory, et al. 1990. Expression of cystic fibrosis transmembrane conductance regulator corrects defective chloride channel regulation in cystic fibrosis airway epithelial cells. Nature 347:358-363.

18. Den Dunnen, JT, PM Grootscholten, E Bakker, et al. 1989. Topography of the Duchenne muscular dystrophy (DMD) gene: FIGE and cDNA analysis of 194 cases reveals 115 deletions and 13 duplications. Am J Hum Genet 45:835-847.

19. Zeviani, M, S Servidei, C Gellera, et al. 1989. An autosomal dominant disorder with multiple deletions of mitochondrial DNA starting at the D-loop region. Nature 339:309-311.

20. Moraes, CT, S Shanske, H-J Tritschler, et al. 1991. mtDNA depletion with variable tissue expression: a novel genetic abnormality in mitochondrial diseases. Am J Hum Genet 48:492-501.

21. Vilkki, J, J Ott, M-L Savontaus, et al. 1991. Optic atrophy in Leber hereditary optic neuro-retinopathy is probably determined by an X-chromosome gene closely linked to DXS7. Am J Hum Genet 48:486-491.

ADDITIONAL LEARNING RESOURCES

1. Scriver, CR, AL Beaudet, WS Sly and D Valle. 1989. The Metabolic Basis of Inherited Disease. 6th ed., New York: McGraw-Hill, Vol I & II.

Chapter 9

Linkage and the Organization of the Human Genome

Gregor Mendel conducted a series of experiments designed to determine whether inheritance of genes for one trait can influence the inheritance of genes for a different trait. He crossed pea plants which were heterozygous for two different traits and observed the transmission of these traits to their progeny. His observations led to the Principle of Independent Assortment: the genes for one trait (e.g., **A, a**) will be apportioned to gametes during meiosis uninfluenced by the genes for another trait (e.g., **B, b**). This principle implies that an organism with a particular combination of genes for one trait will not necessarily have a specific combination of genes for a second trait (e.g., presence of brown hair does not preclude presence of blue eyes in the same individual). Mendel recorded occasional exceptions to this principle; however, he failed to provide an adequate explanation for the cosegregation of genes determining certain pairs of traits.

Sutton and Boveri's postulation of the Chromosome Theory of Inheritance (1,2) provided an explanation for the exceptions reported by Mendel fifty years earlier. Sutton and Boveri correlated the meiotic behavior of genes (and their alleles) with that of chromosomes during meiosis (Table 9-1).

Table 9-1. Behaviors of Genes and Chromosomes During Meiosis.

Genes	Chromosomes
1. Occur in pairs (alleles)	1. Occur in pairs (homologues)
2. Alleles segregate from one another during meiosis.	2. Homologous chromosomes segregate from one another during meiosis
3. Genes for different traits independently assort from one another during meiosis.	3. Non-homologous chromosomes independently assort from one another during meiosis.

Subsequent experiments demonstrated that a chromosome was a composite of many genes, each gene governing a specific trait. Furthermore, members of any one pair of homologous chromosomes were always composed of the same series of genes physically attached (**linked**) to one another. Only the alleles of the respective genes varied among individuals. Chromosomes provided the physical mechanism whereby the meiotic behavior of genes could be explained. Humans have 22 pairs of autosomes and one pair of sex chromosomes. Each chromosome pair possesses its own unique array of genes. The array of genes assigned to a chromosome is called a **linkage group**.

Consider the pairs of homologous chromosomes presented in Fig. 9-1. **A** and **B** are linked to the same chromosome, while **a** and **b** are linked to its homologue. This arrangement may be written:

AB/ab, where the slash separates genes on one chromosome from their alleles on the homologous chromosome. The particular linkage arrangement pictured in Fig. 9-1 is designated linkage **in coupling** or **cis** linkage. The alternate arrangement, **Ab/aB** is called linkage **in repulsion** or **trans** linkage.

Linkage is a violation of the Principle of Independent Assortment, since genes occupying the same chromosome will enter the same gamete. For example, when the linkage array in a parent is **AB/ab**, the most frequent gametes produced by this parent will be **AB** and **ab**. The **Ab** and **aB** combinations will be less frequent and arise following **crossing-over** and subsequent rearrangement of linked genes. This gene rearrangement is called **recombination**, and the rearranged chromosomes (which now carry the gene arrays **Ab** and **aB**) are called recombinant chromosomes. Gametes (or individuals) receiving recombinant chromosomes are called **recombinants**, while gametes (or individuals) possessing the original parental gene arrays are termed non-recombinants.

Recombination occurs during synapsis, or tight association, of homologous chromosomes during Prophase I of meiosis. The chromatids of homologous chromosomes become intertwined during this period, and the stresses generated by this twisting action may promote breakage of the chromosome arms. These breaks are usually healed correctly by repair enzymes; however, if two breaks occur at corresponding points in homologous chromosomes (Fig. 9-2), exchange of genes between homologues can occur. All four chromosomes have an equal probability of entering the gamete(s) derived from this meiotic cell.

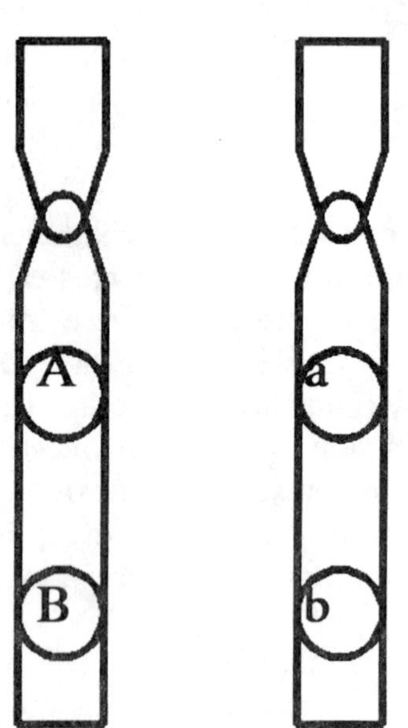

Fig. 9-1. Linked genes in cis array.

Not all cells entering meiosis will experience the recombination event producing the four gametes illustrated in Fig. 9-2. The frequency of recombination between two genes is directly related to the physical distance between them. The larger the distance between the two loci, the more frequently they will recombine. For any two loci, the recombination frequency will not exceed 50%, since, in most cases, only two of the four chromatids participate in recombination.

The frequency of recombination is usually determined by crossing a double heterozygote with a recessive individual (Table 9-2). If the respective traits are determined by unlinked genes, all four types of offspring should be observed in **about equal** frequency (independent assortment). On the other hand if the two traits are determined by allelic genes, the doubly affected parent can transmit **only one or the other allele** to an individual descendant. **If the traits are linked, the two most frequent genotypes among the offspring will be those of the parents, and an occasional offspring will have the recombinant array.**

The physical distance between two loci is measured in map units or **centimorgans (cM)**. One centimorgan represents 1% recombination. This distance is calculated from the observed recombination frequency (Table 9-3). The order of many different loci can be determined by performing a series of crosses. The percent recombination between respective pairs of loci is used to establish their linear order or **linkage map**.

A. H. Sturtevant (3) is credited with the first demonstration of linear order of genes along the chromosome. Sturtevant's method can be illustrated using a hypothetical three-locus system. Two stocks

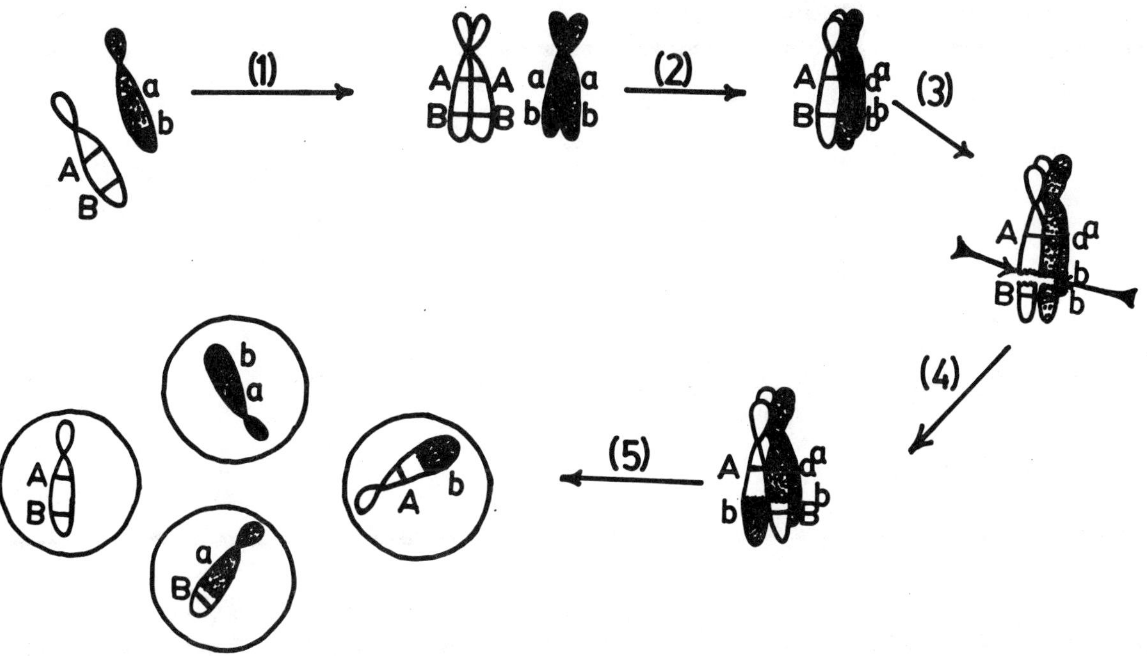

Fig. 9-2. Crossing-over and recombination. Stress generated during synapsis can produce breaks in chromatids of homologous chromosomes. If breaks occur at corresponding sites (arrows) in non-sister chromatids, reciprocal exchange of genes between these chromatids may occur. Gametes derived from the cell experiencing this recombination will be of four types: AB, Ab, aB and ab. For simplicity, the homologus chromosomes are shown as paired but not physically intertwined. 1) Chromosome replication; 2) synapsis; 3) breakage; 4) exchange and joining of segments; 5) gametic products of meiotic division.

Table 9-2. Independence, Linkage, and Allelism.

	Independence		Linkage (AB/ab)		Allelism	
Marriage:	AaBb x aabb		AB/ab x ab/ab		A^AA^B x A^0A^0	
Gametes:	¼ AB all ab		many AB all ab		½ A^A all A^0	
	¼ Ab		many ab		½ A^B	
	¼ aB		few Ab			
	¼ ab		few aB			
Offspring:	¼ AaBb, ¼ Aabb		many AB/ab, few Ab/ab		½ A^AA^0	
	¼ aaBb, ¼ aabb		many ab/ab, few aB/ab		½ A^BA^0	

of fruit flies differing with respect to three X-linked traits are crossed (Table 9-4). Stock 1 is homozygous for genes **A**, **B**, and **C** which determine X-linked dominant traits, while stock 2 is homozygous for their respective recessive alleles (**a**, **b**, **c**). Males possess only one X chromosome; therefore, recombination is limited to Prophase I of female meiosis. Heterozygous females are constructed by crossing **ABC/ABC** females with **abc/Y** males. These heterozygous females are subsequently mated to to the F_1 (**abc/Y**) males to obtain an F_2 generation. The complications of

Table 9-3. Determination of Map Distance.

Parental Cross: AaBb x aabb

Offspring:

Genotype	Number	Type
AaBb	250	Parental
Aabb	10	Recombinant
aaBb	15	Recombinant
aabb	225	Parental

$$\% \text{ Recombination} = \frac{\text{total recombinations}}{\text{total offspring}} \times 100 = \frac{25}{500} \times 10 = 5\%$$

Map Distance = 5 cM

---

dominance are eliminated by recording the phenotypes and numbers of hemizygous F_2 males who inherited their X chromosomes from the F_1 females. If recombination occurred in these F_1 females, then the F_2 males will directly reflect the exchanges in their phenotypes. Consider the hypothetical data presented in Table 9-4. The parental or nonrecombinant phenotypes are represented by the two most

Table 9-4. Linkage Analysis of Three X-linked Traits.

Parental Stocks: 1 = ABC   2 = abc

Parental Cross: ABC/ABC Female x abc/Y Male

F_1 Generation: Females = ABC/abc

Males = ABC/Y

F_1 Cross: ABC/abc Female x ABC/Y Male

F_2 Generation:

Phenotype	#Males	Phenotype	#Males
ABC(parental)	120	Abc	36
ABc	3	abC	1
aBC	40	aBc	8
AbC	12	abc(parental)	180
		Total:	400

frequent F_2 classes (ABC, abc). The two rarest classes (ABc, abC) represent **double crossovers** and were formed following exchanges between the central locus and each of the flanking loci. The remaining classes are all single crossovers. The three loci can now be ordered. Exchanges between **A** and **B** produced the AbC, aBC, Abc, and aBc offspring for a total of 96. Exchanges between **A** and **C** generated the ABc, aBC, abC, and Abc offspring, a total of 80 recombinants. Finally, exchanges between **B** and **C** resulted in 24 recombinant descendants (ABc, AbC, abC, and aBc). The relative sizes of the totals place the C locus in the center. The correct order, **ACB** or **BCA**, cannot be inferred without examining linkage relations with a fourth locus near this cluster.

The relative distances between loci can be calculated using Sturtevant's formula. First calculate the distance between **A** and **C**:

$$\frac{(3 + 40 + 36 + 1)}{400} \quad x \quad 100 \quad = \quad 20cM$$

The relative distance between **B** and **C** is:

$$\frac{(3 + 12 + 1 + 8)}{400} \quad x \quad 100 \quad = \quad 6cM$$

The distance between **A** and **B** is equal to the sum (26cM) of the **A-C** and **B-C** intervals or can be calculated directly:

$$\frac{(2(3) + 12 + 40 + 36 + 2(1) + 8)}{400} \quad x \quad 100 \quad = \quad 26cM$$

Notice that the numbers of **ABc** and **abC** flies were doubled, since they each resulted from two crossover events. The order of the three loci and their relative interlocal distances are presented in Fig. 9-3.

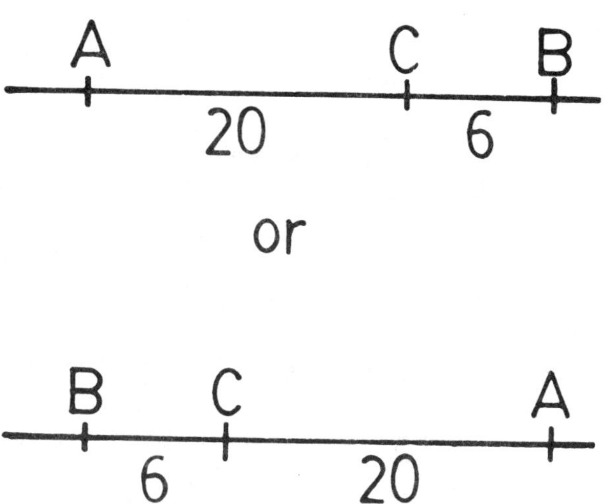

Fig. 9-3. Spatial relationships of A, B, and C. Either arrangement is possible given the data in Table 9-4. Outside loci are required to determine whether A is to the left or right of C.

In summary, linkage problems for experimental organisms can be solved by the Sturtevant method using appropriate genetic crosses. The noncrossovers, single crossovers, and double crossovers are identified on the basis of their relative frequencies. The central locus is identified by comparing interlocal distances between pairs of loci. The relative distances between loci are then expressed as percent of crossovers between the respective pairs. It can be demonstrated that the maximum crossover distance measurable by this procedure is 50%. At this value, the number of crossovers is equivalent to the number of noncrossovers, resembling the outcome of the segregation of unlinked genes. Two genes that are on the same chromosome, but too far apart to demonstrate linkage can be shown to be linked by locating a gene that is between them. If each of the widely separated loci are linked to a third locus, then they are linked to each other.

A phenomenon known as **interference** may also complicate mapping of loci. The occurrence of one crossover is believed to physically limit the occurrence of another crossover nearby. This **positive interference** reduces the observed frequency of multiple crossovers. If crossovers are independent events, then the probability of two crossovers should approximate the product of their individual probabilities. For example, the expected frequency of double crossovers between **A** and **B** (Fig. 9-3) is (0.20)(0.06) or 0.012. The observed frequency was 0.01. The percent of the theoretically possible double crossovers that are actually observed is called the **coincidence** ((0.01 ÷ 0.012) × 100 = 83%). Multiple crossovers and interference contribute to a deviation of predicted map distances from those obtained by summation of distances between sequential pairs of loci determined by genetic experiments.

Inversions have been observed to suppress the frequency of observed recombination in the chromosome region involved. One factor contributing to the lower observed recombination frequency may be the unbalanced products that frequently are produced. When the inversion is present in heterozygous form, the normal and the inverted chromosome regions pair during meiosis by forming a loop (see Chapter 4). Recombination within the loop leads to the formation of recombinant chromosomes that possess gene duplications or deficiencies. Many of the offspring who inherit these duplications or deficiencies die before detection, decreasing the observed frequency of recombination.

In many species, recombination frequency varies between sexes. This is true for both fruit flies and humans, where the linkage map for each female chromosome appears longer than that for the corresponding male chromosomes. The mechanism responsible for this observation is unknown.

MOLECULAR BASIS OF RECOMBINATION

Inspection of appropriate meiotic preparations reveals bridges or chiasmata between homologous chromatids of a specific tetrad. These chiasmata appear to represent sites of exchange of homologous segments between the two chromatids. The actual mechanism by which chromatids are exchanged in meiotic cells is unknown; however, several models for recombination have been proposed. Studies of bacterial, viral, and fungal systems suggest that recombination occurs by a breakage and rejoining mechanism that may or may not be accompanied by DNA synthesis, depending upon the system investigated. Numerous models for recombination have been proposed, and two of these will be briefly discussed. The Holliday model for genetic recombination (Fig. 9-4) involves the production of a special intermediate in which single-stranded DNA links the maternal and paternal DNA double helices. This intermediate has been observed in electron micrographs. Transverse cleavage of the intermediate will lead to a restoration of the parental array of genes on either side of the recombination site. However, vertical cleavage results in a recombination of the flanking genes. Since transverse and vertical cleavage seem to occur equally often, the Holliday model predicts that this type of physical exchange will produce a rearrangement of flanking genes about 50% of the time. When recombination occurs, the Holliday model predicts a reciprocal exchange of genes flanking the site of DNA modification. That is, the

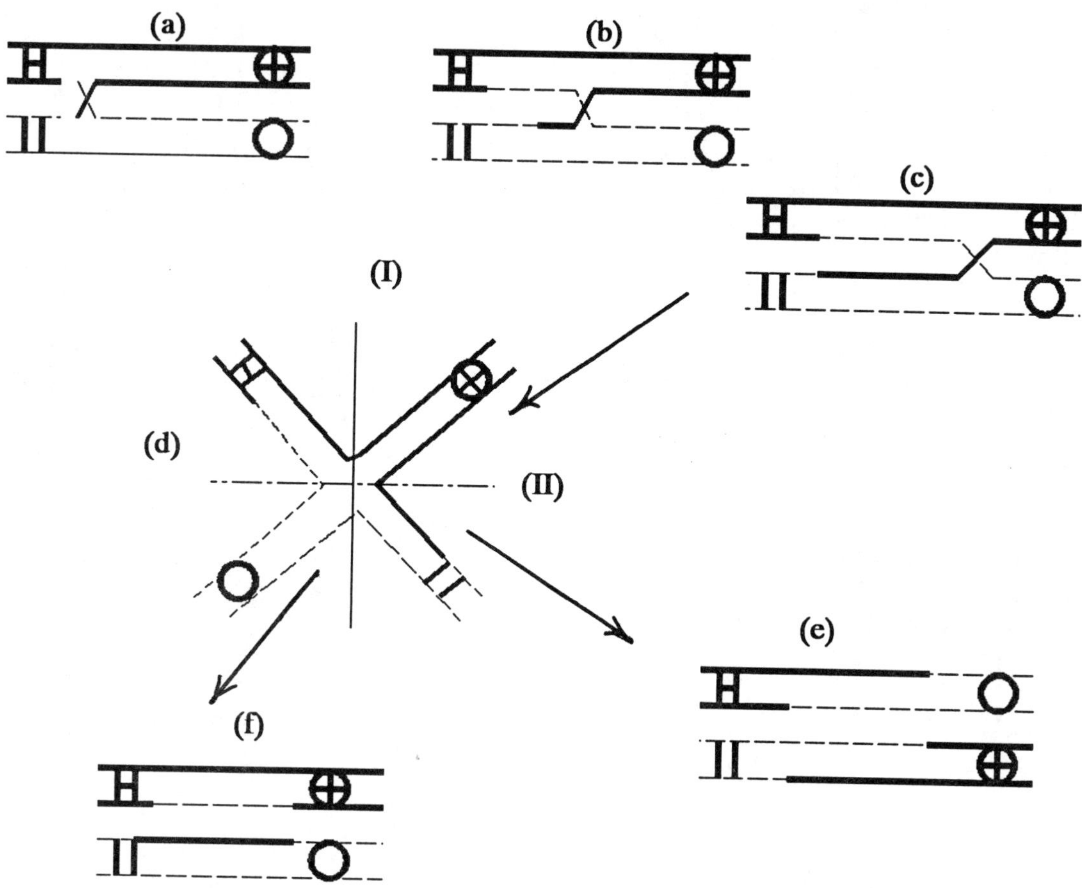

Fig. 9-4. The Holliday model for recombination. Only two of the four chromatids (those that actually are involved in the recombination event are shown). a) A nick occurs in one strand of each homologus DNA double helix, and the ends invade the other chromatid. b) The invading ends are connected to the respective homologues by a DNA polymerase and ligase to form a branch. c) The branch migrates. d) As the ends of the chromatids separate, the chromatids rotate around the branch point to form the Holliday intermediate in which the different arms of the respective double helices are held together by single-stranded DNA. e) If the intermediate is cleaved vertically (I), the flanking genes exhibit recombinant arrays (e). However, if transverse cleavage of the intermediate takes place (II), the parental arrays of the flanking genes are maintained (f).

parental arrays and the recombinant arrays should occur in a 2:2 ratio.

Exceptional ratios have been observed in certain cases where all of the gametic products of meiosis can be recovered. For example, scoring of the spores of certain fungi have revealed 3:1 and 1:3 ratios of certain alleles within the region of recombinational exchange. This nonreciprocal type of recombination is often called **gene conversion**. One model for gene conversion leading to a 3:1 ratio is displayed in Fig. 9-5. Only two of the four chromatids, those that are participating in the conversion event, are shown for clarity. A nick occurs in one DNA strand of the paternal double helix (left), and one end invades the maternal double helix (right), displacing one of the maternal strands. The displaced strand (and the allele it carries) is excised (right panel), and the paternal invading piece is cleaved from the rest of the paternal strand and ligated to the maternal strand. The gap left in the paternal double helix is filled by DNA replication using the partner strand as the template. Notice how another round of replication will be required to generate three paternal alleles and one maternal allele at the central

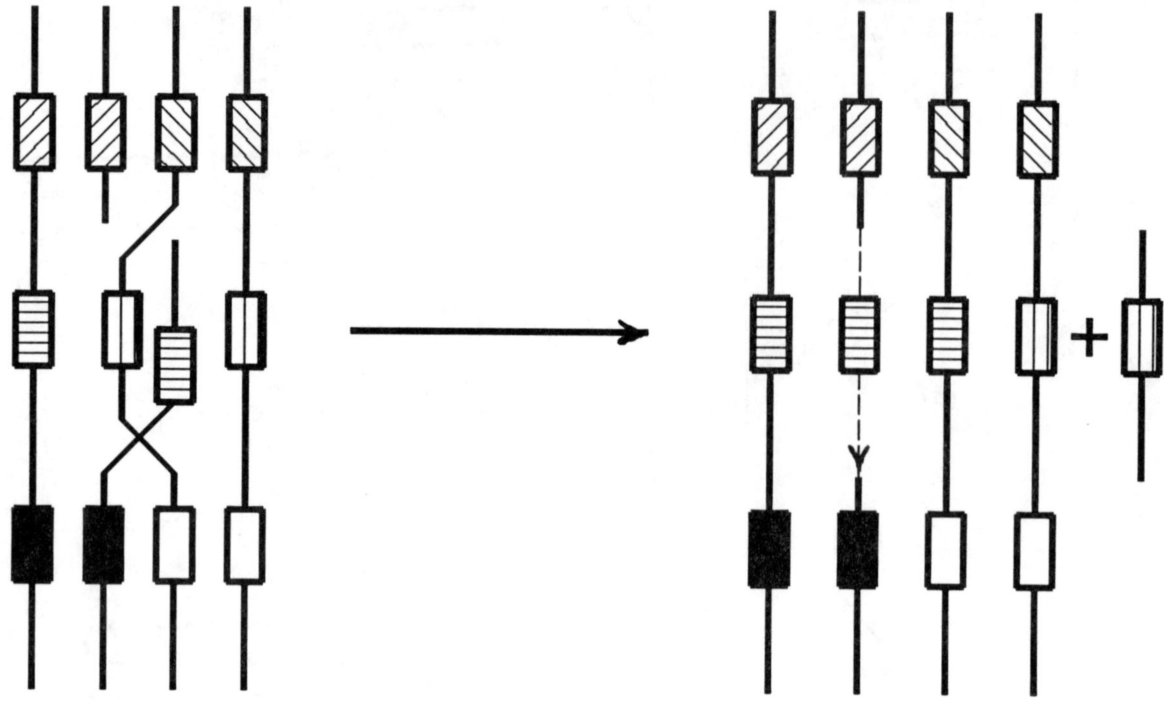

Fig. 9-5. Gene Conversion. The two central chromatids of a tetrad are shown. In both the left and right panels, the paternal double helix is represented by the left-hand two strands and the maternal double helix by the two right-hand strands. Each parental chromatid contains three genes represented by rectangular boxes, and alternate shading patterns represent different alleles of each gene. A nick occurs in one of the two paternal DNA strands (left panel), and one broken end invades the maternal double helix, displacing one maternal strand. The displaced maternal strand is excised and degraded (right panel), and the invading paternal segment is excised from the remainder of the paternal DNA and ligated to the maternal DNA double helix. The gap in the paternal double helix is filled by DNA replication using the intact strand as a template. Subsequent replication of the chomosomes will generate a 3:1 ratio of paternal to maternal alleles at the central locus.

locus. Several other models for gene conversion have also been proposed.

Somatic recombination or mitotic recombination has also been observed. Since homologous chromosomes do not pair during mitosis, somatic recombination is a much rarer event than meiotic recombination.

DETECTION OF LINKAGE IN HUMAN SYSTEMS

The previous discussion of linkage largely has been limited to methods employed for experimental organisms where parental genotypes can be controlled and adequate numbers of progeny can be scored for their recombinant or parental genotypes. Since these approaches cannot be used for human systems, several alternative approaches have been developed for human gene mapping.

Pedigree Analysis and the LOD Method

The determination of the linkage of two or more loci is dependent upon those loci being polymorphic; i.e. having two or more common alleles. Furthermore, at least one of the alleles must specify a dominant phenotype, or the phenotypes must exhibit codominance.

A pedigree analysis involving polymorphic loci requires informative marriages that typically resemble a testcross or backcross performed in experimental genetics: **AaBa x aabb**. This type of marriage has the greatest power to discriminate among independent, linked and allelic alternatives. Although other marital combinations may provide linkage data, they are not as desirable as the marriage of a parent with both traits to one who lacks both traits. Haldane and Smith (4) were the first to propose a mechanism whereby the likelihood of seeing a particular pedigree based upon a varying **recombination frequency (theta)** could be calculated. This method was elaborated by C.A.B. Smith (5) and especially by N.E. Morton (6-10). Morton was a major developer of the method to be described in this section: the **lod score method.** Linkage between loci is suggested by nonrandom assortment of pairs of traits among children of informative marriages (review Table 9-2). Fig. 9-6 illustrates a family segregating for traits A and B. Trait A (symbolized by A̲) is dominant to the absence of trait A (a̲), and trait B (B̲) is dominant to the absence of trait B (b̲). Inspection of the pedigree reveals two patterns:

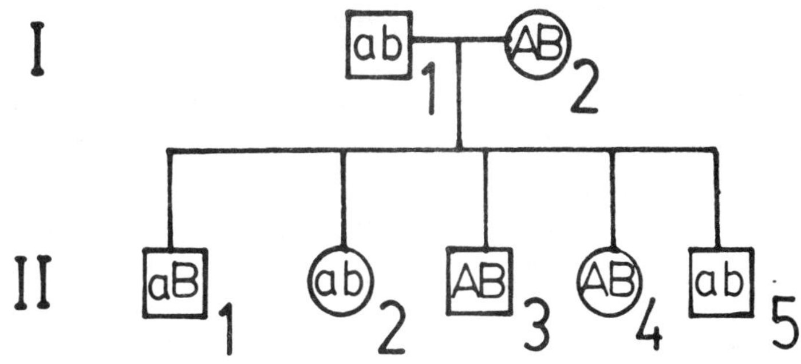

Fig. 9-6. Pedigree of a family segregating for the phenotypes A̲, B̲, a̲ and b̲. A̲ is dominant to a̲, and B̲ is dominant to b̲. The phenotypes are enclosed in the pedigree symbols and are explained more fully in the text.

1) I-2 transmitted both A̲ and B̲ to two of her children, and failed to transmit either trait to two of her children; and 2) only one child (II-1) has a phenotype different from those of his parents. The first pattern eliminates the possibility that A̲ and B̲ are determined by allelic genes. Pattern 2 suggests that the two dominant traits are most likely determined by two linked genes, since assortment of independent genes would have produced four different phenotypic combinations in about equal frequency.

The phenotypes of the children can now be used to infer the genotypes of the parents. The genotype of 1-1 is **ab/ab**. The genotypes of II-2 and II-5 are also **ab/ab**. Since one **a** and one **b** allele must have been inherited from I-2, she is doubly heterozygous **AaBb**. II-3 and II-4 are also doubly heterozygous: **AB/ab**. II-1 is **aB/ab**. Note that the chromosomal array (cis or trans) of 1-2 cannot be conclusively inferred from two-generation data. Since the chromosomal array cannot be inferred, both possibilities must be entertained when calculating the strength of linkage between two loci. The **lod score**

method compares the probability of observing a sequence of births given linkage of the genes determining those traits with that of observing the same combination of births if the genes determining the traits are unlinked:

$$P \text{ (linkage} = \theta) = \frac{p_1\theta \times p_2\theta \times p_3\theta \; *** \times p_n\theta}{p_1(\frac{1}{2}) \times p_2(\frac{1}{2}) \times p_3(\frac{1}{2}) \times \; *** \times p_n(\frac{1}{2})}$$

where P = the relative probability, θ = the frequency of recombination, and $p_1\theta$ = the probability that a given genotype would be produced given linkage of θ. In practice, θ is allowed to vary from 0 (allelic) to 0.5 (independent).

If **A** and **B** are cis in I-2, then the marriage in Fig. 9-6 may be represented: **ab/ab x AB/ab**. A cis array in I-2 implies that II-1 is a recombinant, and that the remaining children are nonrecombinants. Let θ equal the recombination frequency, and $(1-\theta)$ be the probability that there is no recombination. The probability of observing the family given cis linkage of **A** and **B** in I-1 is:

$$P(F_1)_{cis} = \theta(1-\theta)^4.$$

If **A** and **B** are trans in I-2, then the marriage may be represented: **ab/ab x Ab/aB**. II-1 is now a nonrecombinant, and the remaining chidren are recombinants. The probability of observing this family given trans linkage of **A** and **B** is calculated:

$$P(F_1)_{trans} = \theta^4(1-\theta).$$

Since the linkage array in I-2 is unknown, the average of these two probabilities is taken for the probability of linkage at strength θ:

$$P_\theta = \frac{1}{2}[\theta(1-\theta)^4 + \theta^4(1-\theta)].$$

Similarly, for two unlinked genes:

$$P_{\frac{1}{2}} = \frac{1}{2}[\frac{1}{2}(\frac{1}{2})^4 + (\frac{1}{2})^4(\frac{1}{2})]$$

and Z, the **log of the odds**, is given by

$$Z = \log_{10}\{[\frac{1}{2}(\theta(1-\theta)^4 + \theta^4(1-\theta))] \div 1/32\}$$

$$= \log_{10}\{16[\theta(1-\theta)^4 + \theta^4(1-\theta)]\}.$$

This formula is now used to determine Z for each value of θ listed in Table 9-5. For example, for $\theta = 0.1$:

$$Z = \log_{10}\{16[(0.1)(0.9)^4 + (0.1)^4(0.9)]\} = \log_{10}\{1.051\} = 0.050.$$

Notice how the Z score for this family achieves a maximum at $\theta = 0.2$, suggesting that the two loci are about 20cM apart.

A more accurate estimate of θ can be obtained by combining Z scores from several generations of a large kindred, from several different nuclear families, or a combination of these approaches. Since

Table 9-5. Illustration of the Lods Method for Determining Linkage of the Traits Segregating in the Pedigree in Fig. 9-6.

θ	θ^4	$(1-\theta)^4$	$16[\theta(1-\theta)^4 + \theta^4(1-\theta)]$	Z
0	0	1	0	-
0.1	0.0001	0.6561	1.051	0.050
0.2	0.0016	0.4096	1.331	0.124
0.3	0.0081	0.2401	1.243	0.094
0.4	0.0256	0.1296	1.075	0.031
0.5	0.0625	0.0625	1.000	0.000

Z is a logarithm, the Z scores for different family units can be summed to provide an aggregate Z score for each recombination fraction (Table 9-6). An equation is developed for each family in a manner similar to that illustrated on the preceding page, and Z is calculated for each family. The Z scores are then totalled for each value of θ. A $Z \geq 3$ is interpreted to indicate linkage at that particular recombination frequency. For example, in Table 9-6, $Z_T = 3.79$ or the odds are about 6200:1 in favor of linkage at a distance of about 20cM.

Table 9-6. Calculation of Cumulative Z Scores.

	Families				
	1	2	3	n	
θ	Z_1	Z_2	Z_3 •••	Z_n	Z_T
0	----	0.00	0.01	0.02	0.55
0.1	0.05	0.11	0.06	0.10	1.45
0.2	0.12	0.21	0.11	0.21	3.79
0.3	0.09	0.15	0.11	0.20	2.23
0.4	0.03	0.08	0.05	0.10	1.02
0.5	0.00	0.02	0.01	0.00	0.45

Application of the lods method occasionally reveals two groups of families: those that support linkage of two loci and those that appear to exhibit independent assortment of the two genes involved. This dichotomous distribution indicates that one of the traits being studied is determined by at least two loci. For example, certain families segregating for Rh blood group antigens and a red cell trait, elliptocytosis, indicated that the **RH** and **EL** genes were closely linked ($\theta < 0.1$), while the other group of families exhibited a Z value that reached a maximum at $\theta = 0.5$. These data have been interpreted to indicate that two elliptocytosis loci exist, **EL1** and **EL2**, with **EL1** closely linked to **RH**. Both **EL1** and **RH** are now known to reside on chromosome 1.

The lod score method is extensively used today to map loci and their flanking RFLPs to human chromosomes. Caution must be used when interpreting these linkage data. Z scores can become quite impressive; however, when an exceptional individual is found, the values plummet to 0. It is best to correlate results of several different linkage techniques, where possible, before reaching a firm conclusion regarding the relationship between two loci and their relative positions on chromosomes.

Deletion Mapping and Linkage Studies

Regional assignment of genes to chromosomes can be accomplished through deletion mapping.

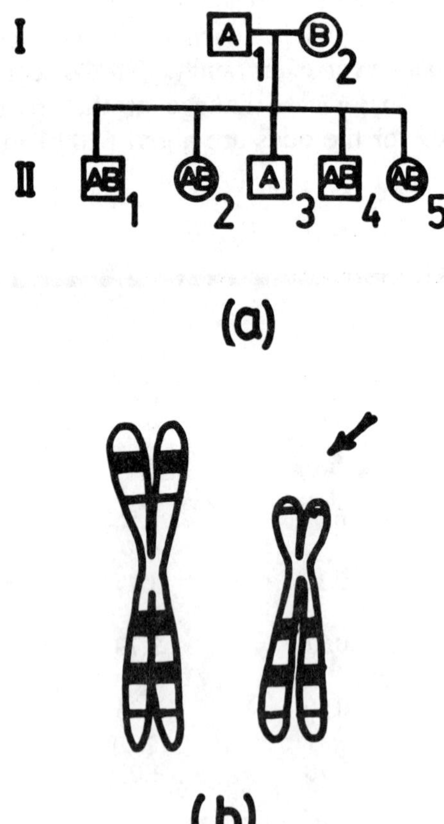

(a)

(b)

Fig. 9-7. Deletion mapping of a human gene. a) Pedigree demonstrating that an individual is hemizygous at an autosomal locus. Traits A and B are determined by alleles with codominant effects. b) Partial karyotype illustrating a deletion (arrow) of the short arm of a chromosome. The missing allele in II-3 was apparently located on the absent segment, and provisional assignment of the locus to that region can be made.

Individual II-3 (Fig. 9-7a) is apparently hemizygous for the allele determining trait A which is known to be inherited as an autosomal codominant. I-1 and I-2 must have the genotype $A^A A^A$ and $A^B A^B$, respectively; therefore, II-3 should have inherited an allele of each type. II-3 also carries an unusual chromosome (Fig. 9-7b) that lacks a major portion of the short arm. By deduction, A^B must have been located on the missing segment. Provisional assignment of the **A** locus may be made to this segment of the chromosome. The assignment must be confirmed by locating other families with deletions in this area and by other methods such as somatic cell hybrids (see next section) constructed from cell lines carrying balanced translocations involving this chromosome. Deletion mapping has been successfully used to locate many genes and to find "candidate" genes for human traits. For example the region containing the dystrophin gene on the X chromosome was located in this manner (11).

Further localization of loci to chromosome bands and sub-bands can also be accomplished by deletion mapping (Fig. 9-8). Seven patients are known to have translocations producing hemizygosity for sections of the chromosome illustrated. Each of these patients has symptoms of a particular syndrome known to be caused by an autosomal recessive mutation. These patients apparently are hemizygous for one recessive allele and possess a deletion of the locus from the homologous chromosome. By comparing the extent of the deletions in the seven patients, the most likely region of the locus can be inferred. Various molecular methods can be used to move from either end of the minimal region to locate a candidate gene responsible for the syndrome.

Somatic Cell Hybridization

Somatic cell hybridization may be defined as the fusion of somatic cells from the same (intraspecific) or from different (interspecific) species. Somatic cell fusion results in the formation of **heterokaryons** that contain one or more nuclei from each parental cell type (Fig. 9-9). The nuclei fuse to form a **syncaryon** which has complete sets of chromosomes from each parental cell. The parental cell lines are chosen such that one cell line has been cultured for a long period of time, and the other has been recently cultured. Single cell clones are established from the synkaryons. Cells of each clone will have a complete chromosome set derived from the established cell line and varying combinations of chromosomes derived from the freshly cultured parental cell line. For human gene mapping, an established mouse or hamster cell line is used along with a fresh human cell line. Persistence of a human chromosome and persistence of a human gene product imply that a human gene required for expression of that human product is present on the human chromosome remaining. Human cell lines which have been used for somatic cell hybridization include skin fibroblasts, tumor cells and lymphocytes.

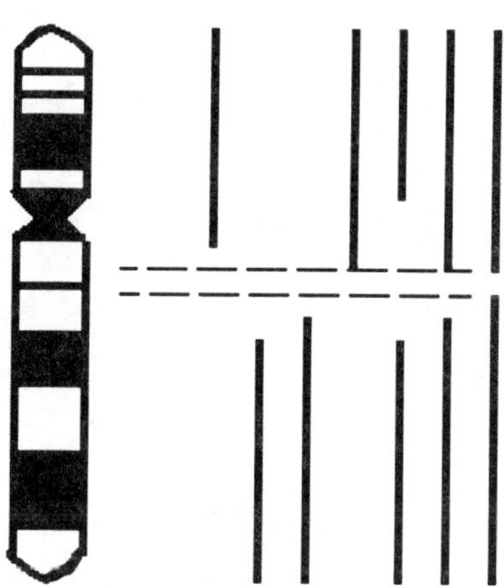

Fig. 9-8. Localization of a gene to a subregion of a chromosome arm. Seven patients with a particular syndrome known to be inherited as an autosomal recessive are encountered who have translocations involving the chromosome indicated at the left. The vertical bars indicate the segments of the chromosome remaining in cells that are heterozygous for the deleted translocation chromosome derivative. The region of minimal deletion overlap (dashed lines) indicate the approximate position of the recessive allele on the normal homologue.

Interspecific fusion occurs as a rare event <u>in</u> <u>vitro</u>. Therefore various techniques have been developed to enhance or induce cell fusion. These include the addition of heat-inactivated Sendai virus or chemical agents (polyethyleneglycol; PEG) that induce membrane fusion in the cell culture. The virus

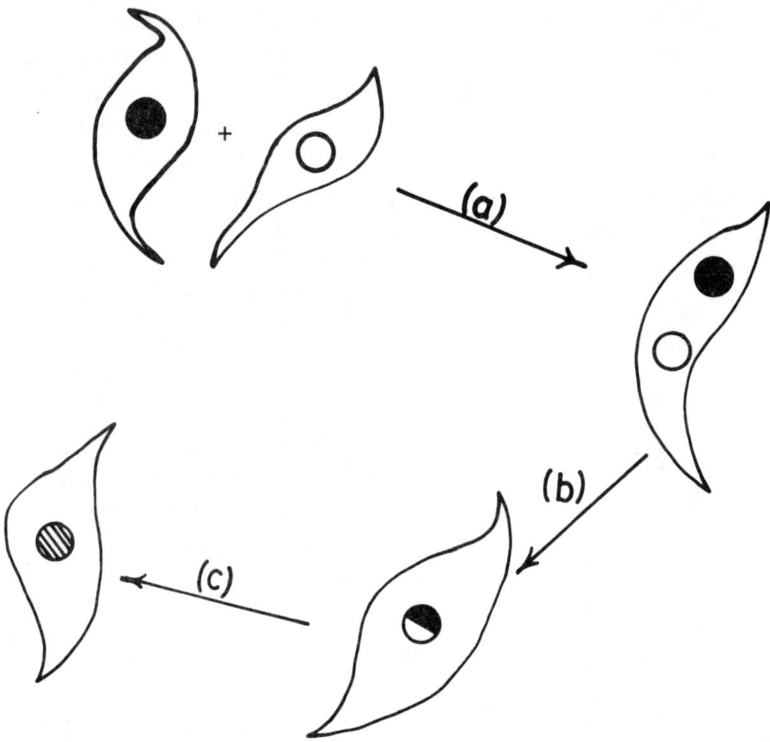

Fig. 9-9. Cell fusion. a) Fusion to form heterokaryon; b) Fusion of nuclei to produce a synkaryon; c) Proliferation and segregation of chromosomes to form hybrid cells.

appears to increase cell contact and promote merging of cell membranes. PEG seems to cause agglutination (clumping) of cells and subsequent membrane fusion. Additional agents that have been used for cell fusion include lecithin and other lipid derivatives.

Cell fusion is enhanced by these agents; however, hybrids still occur at low frequency, requiring use of special growth media that favor hybrid growth and discourage survival and proliferation of the parental cells from which the hybrids were derived. The parental cells used for fusion experiments often differ with respect to their nutritional requirements, temperature optima, adhesive ability, response to intercellular contact, aging rates, and ability to grow in suspension. These differences are exploited in the development of growth conditions that impede growth of the parental types and promote growth of the hybrid cells. The following example illustrates the use of intergenic complementation to select for hybrid cells.

The drug, azaguanine (AG), is metabolized by normal cells that possess the enzyme HPRT (hypoxanthine phosphoribosyltransferase). Products of this metabolism interfere with nucleic acid synthesis, resulting in death of the cell. Therefore, cells that are deficient for this enzyme (HPRT⁻) can be isolated by growing a mixture of cells in a medium containing AG.

Bromodeoxyuridine (BrdU) produces its effects by incorporation into DNA at sites previously occupied by thymidine, promoting base substitutions during subsequent rounds of replication. BrdU is phosphorylated prior to incorporation by the enzyme, thymidine kinase (TK). Cells possessing TK will die when grown in BrdU-supplemented media. Cells that are TK$^-$ survive.

Using AG and BrdU, human cell lines that are HPRT$^-$TK$^+$ and mouse cell lines that are HPRT$^+$TK$^-$ are obtained. Fusion of one of the human and one of the mouse cell lines eventually produces synkaryons that are HPRT$^+$TK$^+$ (Fig. 9-10). The synkaryons are grown in a special selection

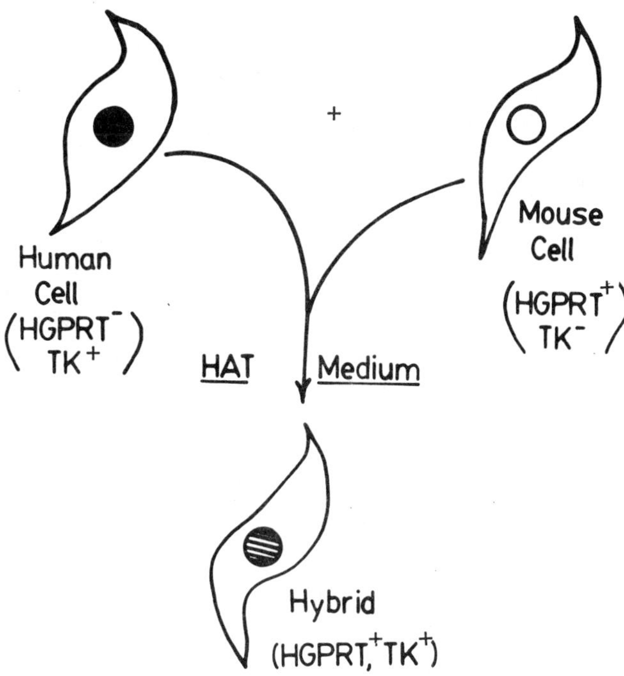

Fig. 9-10. Use of HAT medium (hypoxanthine-aminopterin-thymidine) to select for hybrid cells. HGPRT = HPRT.

medium containing an inhibitor of folic acid reductase known as aminopterin. Tetrahydrofolic acid is essential for _de novo_ synthesis of nucleic acids; therefore, when this inhibitor is included in the medium, alternative sources of purines (e.g. hypoxanthine) and pyrimidines (thymidine) must be used. Neither of the parental cell types can proliferate because they lack one of the enzymes required for utilization of these alternate purine and pyrimidine sources; however, hybrid cells contain both enzymes and proliferate in the selection medium.

The hybrid cells derived from human and mouse cells usually lose human chromosomes as they continue to proliferate. Different hybrid cell clones retain different human chromosomes (as few as one or two) and a full complement of mouse chromosomes. A number of human and mouse enzymes possess different charges and can be resolved by electrophoresis. Furthermore, human and mouse chromosomes can be distinguished by virtue of their unique staining characteristics. Cells from a given hybridization experiment are cloned to obtain an array of colonies that contain a large number of different combinations of human biochemical markers and human chromosomes. Each clone is characterized with respect to its enzyme phenotype and its karyotype. This permits detection of associations between the presence of a chromosome and the enzyme and the absence of that chromosome and the same enzyme (Fig. 9-11). The associations can be verified by subcloning of the original clones to see if the associations persist. Pairs of human enzymes or other biochemical markers can also be studied using a similar comparison, permitting the demonstration that genes essential for expression of these markers are **syntenic** (located on the same chromosome).

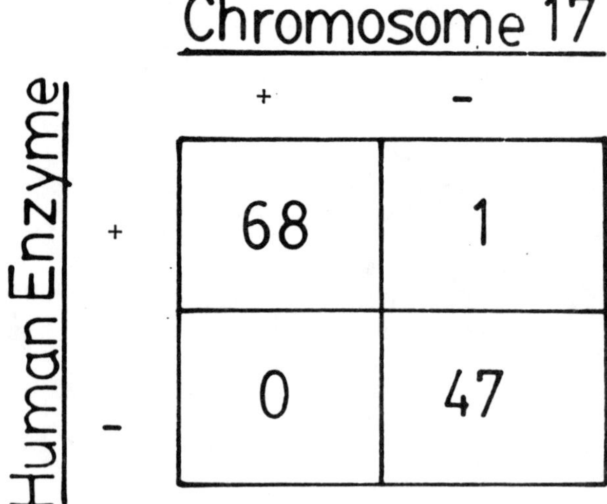

Fig. 9-11. Scoring of cloned hybrid cells for presence of a human enzyme and human chromosome 17. Numbers within each box represent the number of clones expressing the phenotype indicated. + = present; - = absent.

It is useful to accumulate data on the chromosome composition of many different clones and use this information in conjunction with marker analysis (note: a marker is an easily scored biochemical phenotype that is determined by a polymorphic gene). Consider the data in Table 9-7. If a marker is present in clones A, B and C, but not in D, then a gene controlling expression of that marker may be tentatively assigned to chromosome 5.

Use of human cell lines possessing structurally altered human chromosomes has permitted regional mapping of human genes by somatic cell hybridization. For example, Ricciuti and Ruddle (12) assigned the loci for the enzymes phosphoglycerate kinase (PGK1), G6PD and HPRT to the long arm of the X chromosome, and the gene for nucleoside phosphorylase (NP) to the proximal long arm of chromosome 14 using this method. A human cell line heterozygous for a balanced translocation (Fig. 9-12) involving the X chromosome and chromosome 14 was fused to a mouse cell line, and the hybrid clones were karyotyped and scored for these four human enzymes. All clones containing the intact X and/or chromosome t2 expressed PGK1, HPRT and G6PD. All clones possessing the intact chromosome 14 and/or chromosome t2 expressed NP. Clones having chromosome t1 lacked all four enzymes. This pattern is consistent with the above

Table 9-7. Chromosome Content of a Panel of Hybrid Clones.

	Human Chromosome Number				
Clone	1	2	3	4	5
A	+	+	-	-	+
B	+	+	+	+	+
C	+	-	-	-	+
D	+	-	-	-	-

assignments.

Use of somatic cell hybrids to assign linkage relationships has rapidly expanded the human chromosome map. The procedure has been modified to permit transfer of single chromosomes of chromosome segments to hybrid cells, permitting a considerable degree of resolution.

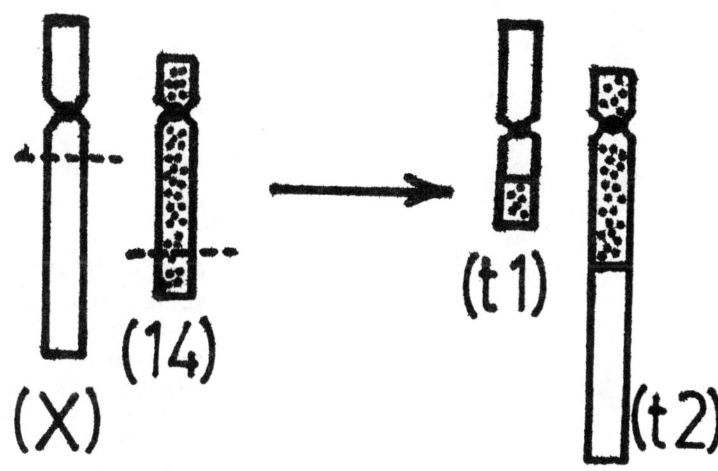

Fig. 9-12. Translocation used by Ricciuti and Ruddle (12) to regionally map PGK1, G6PD, HPRT and NP. A human cell line heterozygous for this balanced translocation was fused to mouse cells to construct somatic cell hybrids.

In Situ Hybridization

In situ hybridization is being used to locate the position of genes on metaphase chromosomes. This technique utilizes labelled cDNA probes to search for genes complementary to those sequences. Metaphase chromosome spreads are denatured, and the slide is flooded with the labelled probe. Stringent hybridization conditions are employed to insure that the probe hybridizes only with the appropriate gene sequence (Fig. 9-13), and unhybridized probe is washed from the slide. If a radiolabelled probe is employed for hybridization, the slide is overlayed with an appropriate photographic film, and the radioactive decay from the probe will expose the film at the site of hybridization. The slide is subsequently stained with a banding procedure and photographed. The autoradiogram is superimposed upon the photograph of the stained chromosomes to locate the gene. This method is somewhat limited by the decay characteristics of the radioactive isotope used to label the probe. Low radioactivities make it difficult to distinguish a hybridization signal from general "background" exposure of the film. Autoradiography is most efficient for locating tandemly repeated genes, such as rRNA genes, histone genes, and globin genes.

A recent modification of this procedure has employed fluorescent cDNA probes. This method is highly sensitive and has been used to localize a number of single-copy genes (those that occur in one copy per haploid chromosome set) on metaphase chromosomes. The procedure is similar with regard to the hybridization steps; however, the locations of the hybridized probes can be viewed with a fluorescent microscope. The hybridization signals appear as a bright pair of dots on a dull fluorescent background when an appropriate fluorescent counterstain is used to define the chromosome arms.

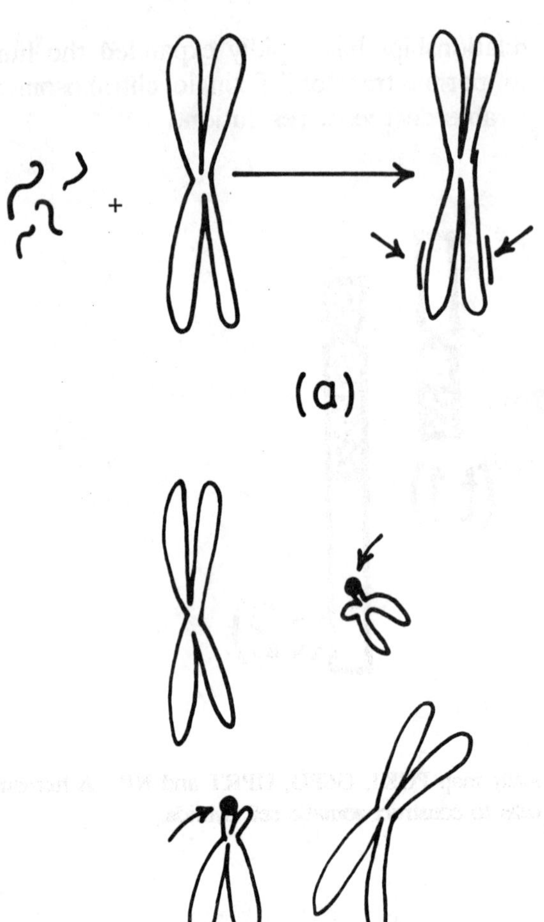

Fig. 9-13. Use of <u>in</u> <u>situ</u> hybridization to map loci to specific chromosomes. a) Incubation of a radiolabelled cDNA probe with metaphase chromosomes spread on a glass slide; b) Hybridization sites are revealed by grains (arrows) created by exposure of a photographic emulsion by radioactive decay from the hybridized cDNA probe. The example shows a cDNA probe for rRNA sequences hybridized to a D- and G-group chromosome. Although hybridization has occurred to both chromatids of each chromosome, decay from only one probe in each case was recorded by the film.

Restriction Mapping

The discovery of restriction enzymes and their modes of action opened a new area of linkage analysis. Polymorphisms of restriction sites are more extensive than those of marker genes. Each chromosome contains a unique array of sites for any given enzyme permitting their use for mapping purposes. This variation leads to the production of different-sized fragments when corresponding regions of homologous chromosomes of two individuals are digested with a particular restriction enzyme or combination of restriction enzymes. RFLPs (restriction fragment length polymorphisms) can be used as markers for linkage analysis. Use of known genes or of probes for noncoding DNA sequences in combination with RFLPs can provide a genetic map of each chromosome (Fig. 9-14). Chromosome 14 is shown in the figure with a sampling of genes that have been mapped to the chromosome regions spanned by the vertical bars. The two right-hand sets of vertical bars encompass chromosome regions containing RFLPs identified with autonomous probes (probes that recognize specific noncoding DNA sequences). Each RFLP begins with the letter, D, followed by the chromosome number, and some other label that refers to the order of discovery or some other attribute of the RFLP. Many more RFLPs have been mapped to chromosome 14 but have not been localized to the extent of those pictured in Fig.

9-14. The Human Genome Project, a major project involving several laboratories in the United States and others worldwide, has as its goal the sequencing of the entire human genome. One aspect of this program is the blanketing of the 22 autosomes and the sex chromosomes with RFLPs. The RFLPs will provide a "net" which will enable the location of many genes that have not yet been mapped, the identification of yet undiscovered loci, and the identification of **QTLs**. QTLs (quantitative trait linkages) refer to genes or blocks of genes that provide a major contribution to quantitative traits like height, weight, intelligence, behavior, heart disease, and respiratory disease. It is hoped that the discovery of such regions will permit the identification of genes contributing to these traits, elucidation of the functions of the proteins they encode, and development of strategies for correcting problems associated with mutations at these loci.

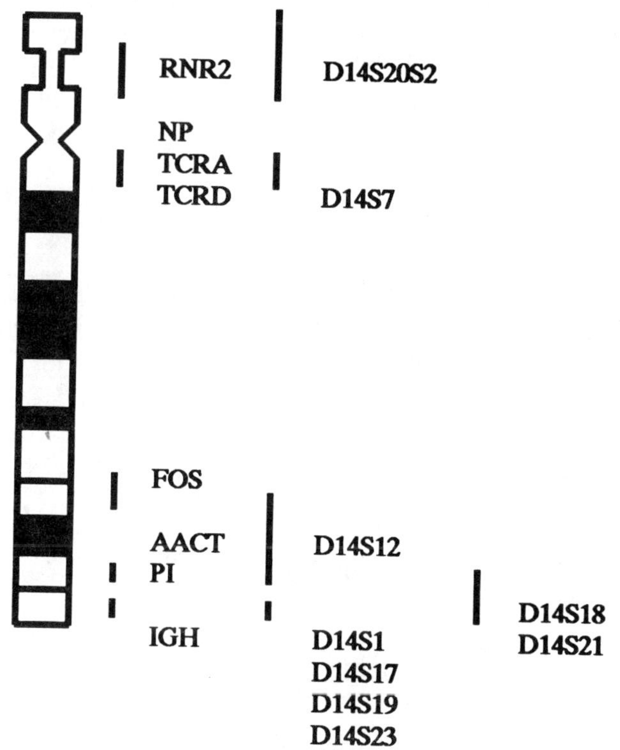

Fig. 9-14. Genetic map of Chromosome 14. Genes that have been localized to chromosome bands are listed adjacent to the chromosome. RFLPs that have been localized to narrow subregions are presented in the center and right-hand panels. Several other genes and many additional RFLPs have been mapped to chromosome 14 but have not been localized to subregions. RNR2 = ribosomal RNA gene cluster 2; NP = nucleoside phosphorylase; TCRA = t-cell receptor gene cluster α; TCRD = t-cell receptor gene cluster δ; FOS = proto-oncogene; AACT = α-antichymotrypsin; PI = α-antitrypsin; IGH = immunoglobulin heavy chain gene cluster.

Nucleic Acid Sequencing

Until recently, this method of linkage analysis was limited to relatively small segments of DNA. However, emerging technologies are permitting the sequencing of relatively long DNA regions. The average human chromosome contains about 100 megabases. Current technology permits the construction of YACS (yeast artificial chromosomes) that can accommodate 0.3 megabases. The DNA in the YACS can be amplified, further subdivided and sequenced. The order of bases in the original YAC segment can be established by inspecting the smaller fragments. The sequences of the fragments derived from the YAC are compared and overlaps identified. With a computer-assisted method, the fragments are placed in their original order in the YAC chromosome which corresponds to a 0.3mb region of the human chromosome from which they were derived. The amount of effort required for sequencing even one chromosome is staggering, since at least 300 YACS would have to be prepared, sequenced, and then ordered in their proper configuration to provided the original chromosome sequence. Current technology permits 1 person using an automatic sequencing machine to sequence about 7000 bases a week. Within three years it is anticipated that this output will reach 50,000 bases per week. At this rate, allowing for two weeks vacation, one technician could sequence about 2.5% of an average-sized human chromosome per year. This calculation, although somewhat simplified, indicates the scope of the problem faced by the project. Hopefully additional technological advances will be made in the generation and evaluation of linkage data. Many associated with the project are convinced that the benefits derived from the project will outway the costs and intensive effort required.

Some regions of the human genome have been extensively sequenced. The entire sequences of the β-like globin region and most of the α-like globin region are known. Several other chromosome regions, including the HLA region on chromosome 6, the immunoglobulin gene regions, and the t-cell receptor regions, are being intensively investigated. Most of the regions span more than 50-100kb of DNA.

APPLICATION OF LINKAGE ANALYSES TO MEDICAL DIAGNOSIS

Presymptomatic and prenatal testing for genetic disorders is facilitated by exploiting linkage of the locus involved to nearby gene or RFLP markers. For example, the **F8C** gene encodes coagulation factor VIIIc and has been mapped to Xq28. Mutations at this locus are responsible for the X-linked

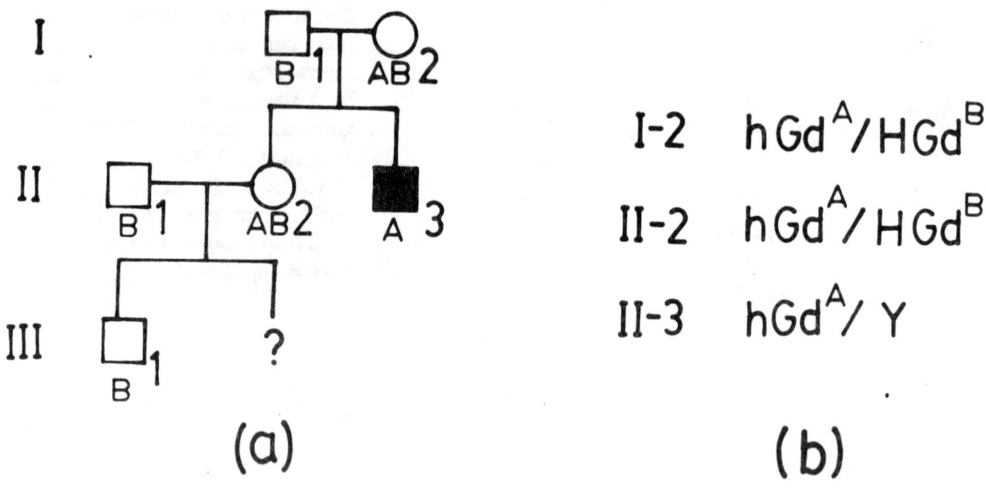

Fig. 9-15. Application of linkage analysis to detection of the hemophilia allele in an unborn child. a) Pedigree: shaded symbol indicates the presence of hemophilia; G6PD phenotypes are listed below the pedigree symbols. b) Chromosomal arrangements of G6PD (Gd) and hemophilia (h) alleles.

recessive disease, Hemophilia A. The **F8C** gene is not expressed in amniotic fluid cells; however, it is within 5cM of the **G6PD** locus which has also been mapped to Xq28. The latter gene is expressed in amniotic fluid cells. Fig. 9-15 presents a couple at risk for a hemophiliac child. The brother of II-2 has Hemophilia A and expresses the G6PD-A phenotype. Since he has only one X chromosome, this indicates that the mutation responsible for hemophilia is cis to the **G6PD^A** allele. II-2 has the G6PD-AB phenotype. Her chromosome carrying the **G6PD^A** allele is also carrying the hemophilia allele. Amniotic fluid cells derived from her fetus (?) would be karyotyped to establish sex, and tested for their G6PD phenotype. If the child is male and proves to be G6PD-B, he has a 95% likelihood of possessing the normal **F8C** allele. On the other hand, if the fetus is male and tests G6PD-A, there is a 95% risk that he will develop Hemophilia A in the future.

An example of the use of an RFLP to assess risk for a disease is presented in Fig. 9-16. The function of the gene responsible for Huntington's disease (HD) is unknown, nor has a candidate gene been identified. However, the HD gene has been mapped to the short arm of chromosome 4 and is situated near an RFLP identified by the probe G8(pK082) which hybridizes with noncoding DNA near the HD gene. Two HindIII restriction sites (*) were found to vary among different persons. If the

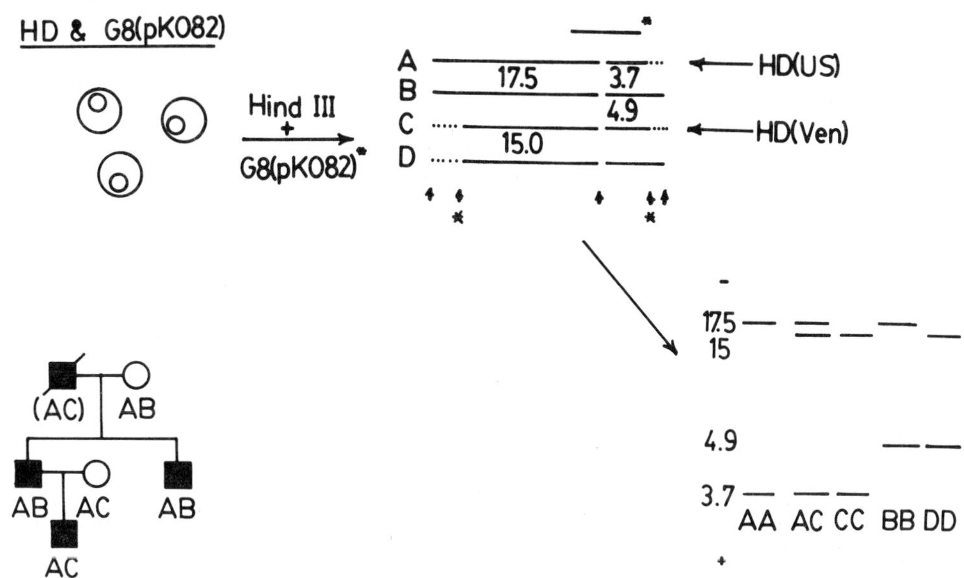

Fig. 9-16. Linkage of the Huntington's disease gene with a HindIII RFLP. Treatment of DNA with HindIII and probing with G8(pK082) produces four chromosome types (upper left). Five HindIII sites occur in the region (arrows) and two of these are polymorphic (*). The sizes (in kb) of the fragments generated by HindIII digestion of the region are indicated below the appropriate fragments. Representative genotypes are shown at the lower right, and cosegregation of the HD allele and pattern A (3.7kb + 17.5kb fragment) is illustrated at the lower left. US and VEN indicate the pattern most commonly associated with HD in the United States and Venezuela, respectively (13).

left-hand site was absent, 17.5 and 3.7kb fragments recognized by the G8(pK082) probe were produced. If both sites were missing, 17.5 an 4.9kb fragments would be generated, and so on. A Venezuelan and U.S. kindred were studied using this polymorphism. The A pattern segregated with the HD gene in the U.S. kindred, while the HD allele segregated with the C pattern in the Venezuelan kindred. These trends indicated that the HD gene was tightly linked (within 2cM) to the HindIII RFLP. This association can be used for predictive testing. Suppose we go back in time to a point where the man in the third generation of the pedigree in Fig. 9-16 has not yet developed symptoms of Huntington's disease. DNA from peripheral leukocytes would be obtained from his father, mother, paternal uncle and himself and tested for their HindIII status. From the pedigree, it can be seen that the HD allele is on the chromosome producing the A pattern. The man is AC and received the C chromosome from his mother. Therefore, he received the A chromosome from his father and has a 98% risk of contracting Huntington's disease in the future.

Suppose the man discussed above has not yet developed the disease and has married a woman with the BB chromosome combination, who is clinically normal, and who has no family history of HD. Suppose the woman is pregnant and the couple want to know the probability that the fetus will eventually develop HD, but the man would rather not know his genotype and associated risk. **Exclusion testing** would be used in this situation. Amniotic fluid or chorionic villus cells from the pregnancy would be obtained and assessed for their HindIII phenotype. If the cells prove to be BC, the couple could be told that their fetus' risk for HD can be excluded with 98% certainty. On the other hand, if the child proves to be AB, they would be told that risk for HD cannot be excluded. In neither case is the man informed of his genotype, although it has been established.

In some cases digestion of a chromosome in the vicinity of a gene with a single enzyme will

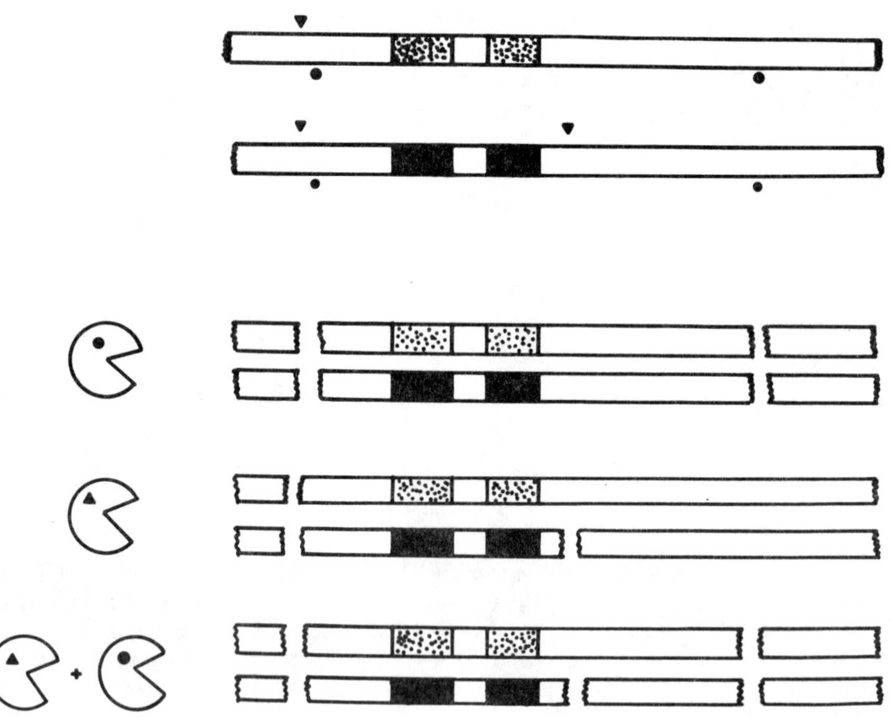

Fig. 9-17. Double digest. Top: Restriction sites for two enzymes (circles and triangles) on homologous chromosomes, one of which carries a mutant allele. Each gene consists of two exons. The exons of the mutant allele are stippled. Center: treatment of these chromosomes with one enzyme (circle) produces similar fragments; however, use of both enzymes discriminates between the chromosomes carrying the mutant and normal alleles (bottom).

not distinguish between the homologues carrying the normal and mutant alleles (Fig. 9-17). In these instances a "double digest" may be used to infer the presence of the mutant or normal allele in an individual. Treatment of DNA from the individual whose chromosomes are shown in Fig. 9-17 with both restriction enzymes will result in the abnormal allele associating with a larger DNA fragment than that of the normal allele. If two parents are each heterozygous for this abnormal gene and have the chromosome types displayed in the figure, presymptomatic testing can be done. If only the large fragment is found in the child, there is a very high probability that he or she will contract the disease in the future. If large and small fragments are observed, that person is most likely heterozygous for the abnormal allele. Two small fragments would be interpreted as a normal homozygote. These inferences are dependent upon tight linkage of the restriction sites that are being exploited and the gene, since recombination between the right-hand triangle and the gene would transfer the abnormal allele to the small fragment and lead to a misinterpretation.

PARTIAL SEX LINKAGE

The term, **partial sex linkage**, has been used to describe the linkage behaviors of genes that are located within the pairing regions of the X and Y chromosomes. The tips of the X and Y chromosomes associate during Prophase I of male meiosis. Recombination between the paired segments would transfer a X-linked gene to the Y chromosome and vice versa, leading to "**pseudoautosomal**" inheritance of traits determined by these genes. Although the X-linked genes within or close to the pairing region (pseudoautosomal region) of the X chromosome will appear to follow the rules of X-linked inheritance much of the time, X-Y recombination moving an allele to the Y chromosome will result in male-to-male transmission. Conversely, a Y-linked gene located within or near the pseudoautosomal region of the Y chromosome will occasionally be inherited by daughters following transfer of the allele to the X chromosome of their fathers by X-Y recombination.

The pseudoautosomal regions of the human X and Y chromosomes and neighboring areas are displayed in Fig. 9-18. There is a high degree of sequence homology between the X and Y pseudoautosomal regions, and recombination occurs frequently throughout the region, ranging from 50% at the telomeric end to about 2% at the centromeric end. The **MIC2** genes encode a cell surface antigen that is detectable using special antibodies (monoclonal antibodies 12E7, F21, and O13). Two loci, **MIC2X** and **MIC2Y**, have been identified that have very similar sequences and which are located at the centromeric border of the pseudoautosomal regions of the X and Y chromosomes, respectively. Alleles at these loci are interchanged by X-Y recombination in about two percent of children; however, they continue to display typical X-linked or Y-linked inheritance patterns in the remaining 98% of the next generation.

Occasional recombinations include loci neighboring the pseudoautosomal region. For example, approximately 1/20,000 offspring are sex-reversed 46,XX males or 46,XY females following a recombination in their fathers which transferred the **TDF** region from the Y to the X chromosome. A similar fraction of children inherited a recombined X chromosome lacking the **STS** locus or a Y chromosome with a functional **STS** locus. The latter gene is also of interest with respect to a number of repetitive sequences that flank the **STS** gene. Mispairing during meiosis in females followed by recombination between the mispaired sequences can lead to deletion of all or part of the **STS** gene. More than 90% of males who have steroid sulfatase deficiency have inherited one of these deleted X chromosomes from their mothers. Steroid sulfatase participates in hydrolysis of sulfated androgen stored in the placenta. Following removal of the sulfate group, the androgen is converted to estrogen, signaling the expectant mother that her child is ready to be born. A large proportion of steroid sulfatase-deficient male infants must be delivered by cesarean section, and these same boys later develop an X-linked form of ichthyosis, a skin disease. Steroid sulfatase also acts as a cholesterol sulfate sulfatase, and deficiency of the enzyme results in the accumulation of cholesterol sulfate in the skin, producing the disease.

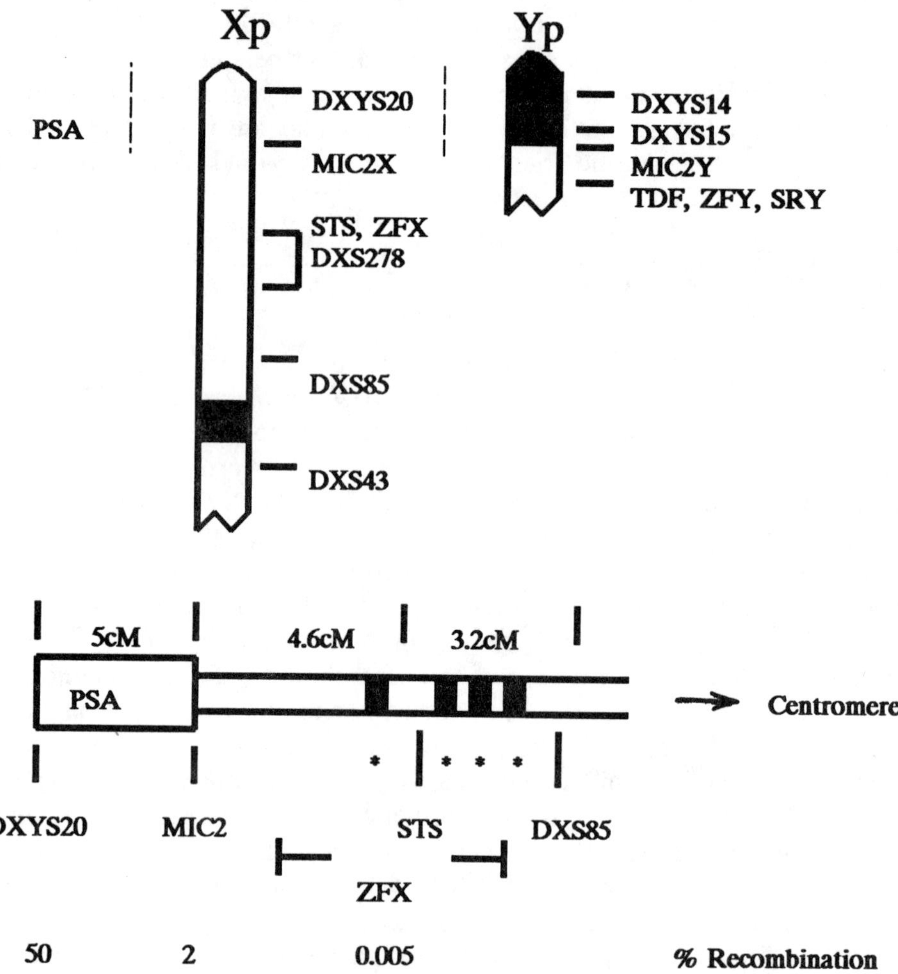

Fig. 9-18. Pseudoautosomal regions (PSA) of the X and Y chromosomes and adjacent loci. Top: PSA regions are indicated by a vertical dashed line to the left of each chromosome. DXYS14, DXYS15, and DXYS20 designate RFLPs that are freely exchanged between the X and Y chromosomes. DXS43, DXS85, and DXS278 represent RFLPs that are specific for the X chromosome. MIC2X, MIC2Y, STS, ZFX, ZFY, and SRY are protein-encoding loci located within or near the PSA regions. TDF may or may not be identical to SRY. Bottom: Enlargement of the PSA region and the adjacent area of the X chromosome. Estimated distances in centiMorgans are presented above the enlarged regions, and percent of observed recombination is listed below. The asterisks represent the repetitive sequences flanking the STS locus which are responsible for the DXS278 polymorphism. The position of ZFX with respect to STS is unknown, but the gene has been localized to the area enclosed by the interval indicated.

Distribution of Genes Encoding Enzymes of Two Representative Pathways

TCA Cycle

Locus	Enzyme	Location
CS	Citrate synthase	12p11-qter
ACO1	Aconitase(cytosolic)	9p22-q32
ACO2	(mitochondrial)	22q11.2-13.31
IDH1	Isocitrate dehydrogenase	
	(cytosolic)	2q32-qter
IDH2	(mitochondrial)	15q21-qter
SDH	Succinate	
	dehydrogenase	1p22.1-qter
FH	Fumarate hydratase	1q42.1
MDH1	Malate dehydrogenase	
	(cytosolic)	2p23
MDH2	(mitochondrial)	7p13-q22

Pentose Pathway

Locus	Enzyme	Location
G6PD	Glucose-6-PO_4-dehydrogenase	Xq28
PGD	Phosphogluconate dehydrogenase	1p36.2-36.13

Selected Gene Clusters

Myosin Heavy Chain

Locus	Enzyme	Location
MYH1	(Adult skeletal muscle)	17pter-p11
MYH2	(" " ")	17p13.1
MYH3	Embryonic	17pter-p11
MYH4	(Adult skeletal muscle)	17pter-p11

Immunoglobulins

Locus	Enzyme	Location
IGH	Immunoglobulin heavy chains	14q32.33
IGK	Immunoglobulin kappa chains	2p12
IGL	Immunoglobulin lambda chains	22q11.1-q11.2

Others

Locus	Enzyme	Location
HB(α-like)	α-like globins	16p133
HB(β-like)	β-like globins	11p155
HLA	Histocompatibility antigens	6p21.3
H1F2, H3F2, H4F2	Histone families	1q21
H1F3	H1 histone family 3	6p12-q21
H1F4	H1 histone family 4	12q11-q21
TCRA/TCRD	T-cell receptors(α/δ)	14q11.2
TCRB	T-cell receptor β	7q35
TCRG	T-cell receptor γ	7p15

ORGANIZATION OF THE HUMAN GENOME

Prokaryotic genes encoding enzymes catalyzing reactions in the same metabolic pathway are generally clustered; however, this type of gene organization is rarely observed in higher eukaryotes (Table 9-8). Human enzymes participating in the TCA cycle tend to be dispersed over many chromosomes. Furthermore, some of these enzymes occur in cytosolic and mitochondrial forms. The two forms of a given enzyme are also encoded by genes on different chromosomes. Two enzymes participating in the oxidative portion of the pentose pathway (hexose monophosphate shunt) are also located on different chromosomes.

Although most genes do not appear to occur in closely linked groups, exceptions are known. In each case, the genes appear to have arisen by tandem duplication of an ancestral gene and subsequent divergence of gene structure following accumulation of mutations in the duplicate products. The α-like and β-like globin gene clusters have been discussed previously. Notice how the genes share a common function (oxygen transport), but have become developmentally restricted with respect to their time of expression. The myosin heavy chain genes provide additional examples of closely linked duplicate genes. The myosin heavy chain genes presented in Table 9-8 are all expressed in adult skeltal muscle or during the embryonic period. **MYH5**, which encodes another adult skeletal muscle heavy chain, and **MYH6**, encoding a cardiac muscle myosin heavy chain, have been mapped to chromosome 7cen-q11.2 and chromosome 14, respectively. Several other examples of clustered gene families are entered in Table 9-8. In each case, the proteins encoded by the genes share a common function.

The human gene map is expanding at a rapid rate, and recent compilations may be found in **Cytogenetics and Cell Genetics** (a journal) or in the most recent edition of **Genetic Maps** which is published by Cold Spring Harbor Laboratory Press.

EVOLUTION OF THE MAMMALIAN GENOME

Some conservation of linkage groups has been observed among a variety of vertebrates. One striking example is provided by X-linked genes of placental mammals. S. Ohno proposed that the X chromosome of mammals is conserved, such that X-linked genes in one mammalian species would also be X-linked in other mammalian species (14). A comparison of the human and mouse X and Y chromosomes is provided in Fig. 9-19. All eight murine X-linked loci that are presented in Fig. 9-19 are also X-linked in humans; however, the order of the loci on the X chromosomes of the two species is somewhat different. The murine **Stsx** locus is also within the pseudoautosomal region of the murine X chromosome, while the human **STSX** locus is situated outside of the human X pseudoautosomal region. This pattern of conservation of X-linked loci but not the relative positions of these X-linked genes is generally observed among placental mammals.

Both human and murine Y chromosomes also possess similar genes. Both chromosomes possess steroid sulfatase loci, testis-determining regions, loci encoding Y-linked zinc finger proteins, and a locus encoding a Y-linked histocompatibility antigen (not shown in figure). It is interesting to note that the human **STSY** gene appears to have been moved away from the human Y pseudoautosomal region by a pericentric inversion; whereas, the murine locus is located within the murine Y pseudo-autosomal region. The human **STSY** locus has also incurred several mutations that have relegated the locus to pseudogene status. As a consequence of the latter event, steroid sulfatase activities tend to be lower in human males than in females, while male and female mice have comparable steroid sulfatase activities.

The human X chromosome is partitioned into two different linkage groups. Loci on the short arm tend to be specialized genes, while those on the long arm are usually housekeeping genes. Studies have revealed that the loci on the long arm of the human X chromosome are also X-linked in non-placental mammals. However, genes on the human X short arm tend to be autosomal in monotremes (egg-laying mammals).

Some conserved autosomal gene linkages have also been reported. For example, **TPI1** (triose phosphate isomerase) and **GAPD** (glyceraldehyde phosphate dehydrogenase) are linked in the mouse, cat, dog, and a variety of primates, including humans. **LDHB**, **PEPB** (peptidase B), **TPI1**, and **GAPD** are all linked in primates. Comparative linkage studies of mice and humans suggest that the average length of conserved autosomal linkages approximates 8cM; that is, genes that are within 8cM on murine chromosomes are also linked in humans (15). This evolutionary conservation of linkage groups has been exploited to map genes in different mammalian species.

Isozyme-encoding genes can be used to trace chromosome evolution in mammals. **MDH1** and **ACP1** (acid phosphatase 1, cytosolic) are located on chromosome 12, and **IDH1** has been mapped to chromosome 13 of chimpanzees. Both of these chromosomes are acrocentric. Human **MDH1** and **ACP1** have been localized to chromosome 2p, and human **IDH1** has been mapped to 2q. These relationships suggest that human chromosome 2 was formed by centric fusion of the two ancestral chromosomes corresponding to chimpanzee chromosomes 12 and 13 following divergence of the two primate lines.

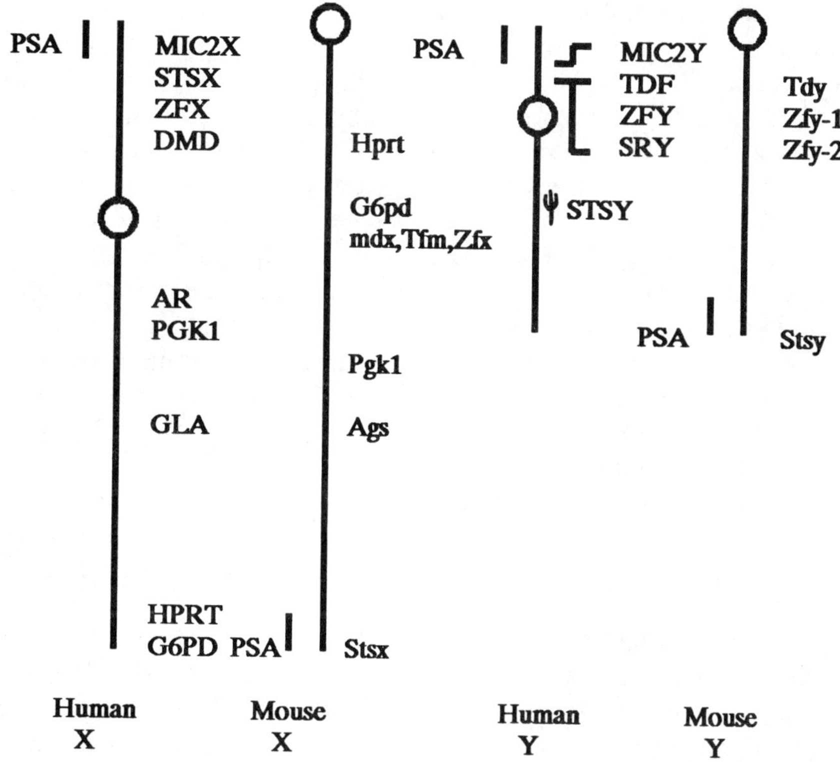

Fig. 9-19. Comparative linkages on the human and murine sex chromosomes. The pseudoautosomal regions of the four chromosomes are indicated by PSA and a vertical bar. Loci are presented at the right of each chromosome. In some cases the gene symbols for corresponding loci differ: DMD = mdx, AR = Tfm, GLA = Ags, and TDF = Tdy. Note that the human STSY is a pseudogene.

SUMMARY

Genes located on the same chromosome are said to be linked. This physical association prevents random assortment of alleles among gametes of a double heterozygote. That is, an **AaBb** individual will not produce **AB, Ab, aB,** and **ab** gametes in equal frequency. For example, if **A** and **B** are on one chromosome, and **a** and **b** are on the homologous chromosome (cis array), the **AB** and **ab** gametes will be most common. The closer the two loci are on the chromosome, the greater the departure from random assortment.

Exchange of linked genes between homologous chromosomes is called crossing over or recombination and occurs during Prophase I of meiosis when homologous chromosomes are associated in a tetrad. The frequency of recombination between linked genes is generally directly proportional to the distance between them. Genes can be ordered on a chromosome using their recombination frequencies, and interlocal distances are generally expressed as percent recombination.

Demonstration of linkage among human loci has proved difficult. Linkage can be inferred from family studies, somatic cell hybrids, deletion mapping, nucleic acid hybridization and restriction mapping. The LOD method is used for determination of linkage using informative pedigrees. LOD is an abbreviation for logarithm of the odds in favor of linkage and is symbolized by Z. Z equals the logarithm of the ratio of the probability of observing a family or series of families given linkage of a certain strength to the probability of observing that family or those families assuming independence of the two loci being considered. Z is calculated for recombination frequencies ranging from 0 (same locus) to 0.5 (independent loci) and achieves a maximum at an appropriate recombination frequency. Since Z is a logarithm, LOD scores from a large number of informative families can be easily combined. Z values for corresponding recombination frequencies for all families in the sample are summed. The antilog of the summed Z scores represents the odds in favor of linkage at that particular recombination frequency. Conventionally, a summed Z score ≥ 3 (1000:1 odds) is interpreted to favor linkage.

Somatic cell hybrids between human cells and those of other species generally segregate human chromosomes. Association between persistence of a human biochemical marker and a particular human chromosome implies that a major gene influencing that trait is located on the chromosome involved. This method can also be used to demonstrate syntenic (same chromosome) relationships between genes influencing two human traits. If both traits are simultaneously retained or lost by clones of hybrid cells, the traits are inferred to be determined by syntenic genes.

Deletion mapping has been used to assign genes to specific chromosome regions. Simultaneous loss of a trait and a specific chromosome region implies that a major gene responsible for that trait is located on the missing segment. This method may be used in combination with somatic cell hybridization or with newer molecular procedures, such as chromosome walking, to localize genes to small regions of chromosomes and to isolate them for characterization.

Nucleic acid hybridization uses radiolabelled cDNA probes or DNA probes coupled to fluorescent ligands. Metaphase chromososomes are denatured and allowed to reanneal in the presence of the probe. The probe will hybridize with the complementary chromosome sequence and can be located by autoradiography or fluorescent microscopy. Although this method was originally more efficient for locating tandemly repeated genes, recent innovations have increased the efficiency of in situ hybridization so that single copy genes can also be found.

Restriction mapping involves saturation of the human chromosomes with RFLPs. Association between segregation of a RFLP with segregation of an allele determining a human trait suggests linkage, and chromosomal assignments can often be made.

Linkage studies have been used to identify persons at risk for genetic diseases. If an individual carries a gene that is known to be close to a second gene responsible for a human disease, then that person has most likely also inherited the gene responsible for the disease. This method is useful for prenatal diagnosis and for detection of persons at risk for late onset genetic disorders. Use of known linkages of RFLPs with disease-causing genes for presymptomatic diagnosis has also proved successful.

Partial sex linkage refers to loci that are situated within the pairing regions of the X and Y chromosomes. Recombination during male meiosis can result in the transfer of an X-linked gene within or near the pairing region to the Y chromosome and vice versa. This recombination occasionally leads to transmission of a "X-linked" gene from father to son and a "Y-linked" gene from father to daughter. The pairing regions of the X and Y chromosomes are designated the pseudoautosomal regions.

Most genes encoding enzymes participating in the same metabolic pathway tend to be located on nonhomologous chromosomes. Some clustering of genes with related functions has been observed. These clusters appear to have arisen following one or more tandem duplications of an ancestral locus and subsequent divergence of the gene sequences following accumulation of mutations in the duplicate products. Although the function of the duplicated genes may be similar, their expression may have been spatially (different tissues) or temporally (different developmental stages) isolated.

X-linked genes in one placental mammal tend to be X-linked in other placental mammals, although their relative orders on the X chromosome may vary among species. Autosomal genes also may occur in conserved linkage groups. Studies of certain linkage groups have revealed patterns of chromosome evolution.

PROBLEMS

1. Traits C and D are segregating through the following kindred. Comment on the genetics of these traits.

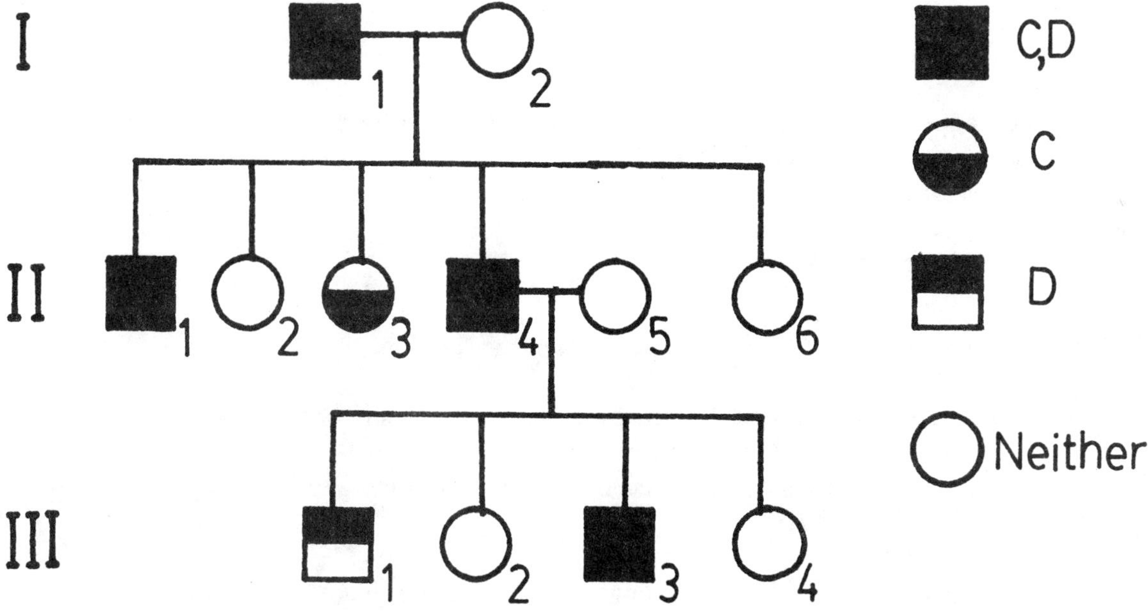

2. Three dominant traits (E, F and G) appear to be determined by linked genes. A large population was screened for these three traits to determine chromosomal types. Assume the population is at equilibrium for the alleles at each locus. The frequencies of the respective chromosomes were:

EFG	0.08	efG	0.38
EFg	0.41	efg	0.02
EfG	0.03	Efg	0.006
eFg	0.07	eFG	0.004

What were the ancestral chromosome types? Identify the single crossovers. What are the apparent map distances between E, F, and G?

3. Trait A is a fatal dominant disease (20 years after onset) with 100% penetrance. Onset of Trait A occurs after 30 years of age. The locus determining trait A is closely linked to the **ABO** locus A 20-year-old woman who is blood type A has recently married and desires to have children. Her mother is type O, normal; and her father has trait A and is type A. Her paternal grandmother was O, normal; and her paternal grandfather had trait A and was type A. The woman's husband is O, normal. How would you counsel this woman and her husband?

4. Consider the following data:

θ	Z_T
0	--
0.1	-0.25
0.2	+0.62
0.3	+1.42
0.4	+2.10
0.5	+4.41

These Z scores apply to segregation of codominant alleles at each of two enzyme-encoding loci through several informative families. Z_T represents the summed Z score for each family for each value of θ. What would you conclude regarding the possible linkage of these loci?

GLOSSARY OF TERMS

Clone colony formed from a single cell by repeated mitotic division.

Coincidence percent of theoretical double crossovers that are actually observed.

Coupling (cis) two dominant alleles are situated on one chromosome and two recessive alleles on the homologous chromosome (**AB/ab**)

Crossover (recombinant) an individual containing a combination of linked genes which differs from that occurring in his/her parents; site of potential genetic exchange between two homologous chromosomes.

Heterokaryon cell containing two genetically distinct nuclei.

Interference (positive) one crossover reduces the probability that a second crossover will occur nearby.

Linkage group array of genes on the same chromosome.

Linked genes that are located within measurable distance on the same chromosome.

Map unit (centimorgan (cM)) unit of distance between two linked genes; often expressed as percent recombination.

Nonrecombinant (noncrossover) an individual who possesses the same array of linked genes occurring in her/his parents.

Partial sex linkage (pseudoautosomal) genes that are located within the pairing regions of the X and Y chromosomes.

Recombination the physical exchange of genes between homologous chromosomes.

Repulsion (trans) one dominant allele and one recessive allele are located on one chromosome, and their respective recessive and dominant alleles are situated on the homologous chromosome (**Ab/aB**).

Restriction mapping determining the relative positions of genes and noncoding polymorphic sequences (RFLPs) in the human genome.

Quantitative trait linkages (QTLs) use of RFLPs and other marker loci to locate chromosome regions that contribute major genetic components to quantitative traits (height, weight, intelligence, behavior, etc.).

Synkaryon cell containing a single nucleus having complete sets of chromosomes from which it has been derived by cell fusion.

Syntenic genes located on the same chromosome whose relative positions are unknown.

BIBLIOGRAPHY

1. Sutton, WS. 1902. On the morphology of the chromosome group in Brachystola magna. Biol Bull 4:24-39.

2. Boveri, T. 1902. Uber mehpolige Mitosen als Mittle zur des Zellkerns. Verh phys med Gesselsch Wurzburg 35:67-90.

3. Sturtevant, AH. 1913. The linear arrangement of six sex-linked factors in Drosophila, as shown by their mode of association. J Exptl Zool 14:43-59.

4. Haldane, JBS and CAB Smith. 1947. A new estimate of the linkage between the genes for colour-blindness and haemophilia. Ann Eugen 14:10-31.

5. Smith, CAB. 1953. The detection of linkage in human genetics. J Roy Stat Soc B 15:153-192.

6. Morton, NE. 1955. The inheritance of human birth weight. Ann Hum Genet 20:125-134.

6. Morton, NE. 1955. The inheritance of human birth weight. Ann Hum Genet 20:125-134.

7. Morton, NE. 1956a. Sequential tests for the detection of linkage. Am J Hum Genet 7:277-318.

8. Morton, NE. 1956b. The detection and estimation of linkage between the genes for elliptocytosis and the Rh blood type. Am J Hum Genet 8:80-96.

9. Morton, NE. 1957. Further scoring types in sequential linkage tests with a critical review of autosomal and partial sex linkage in man. Am J Hum Genet 9:55-75.

10. Morton, NE. Segregation and linkage, in <u>Methodology</u> <u>in</u> <u>Human</u> <u>Genetics</u>, (W.J. Burdette, ed). San Francisco: Holden-Day, 1962.

11. Boyd, Y, V Buckle, S Holt, <u>et al</u>. 1986. Muscular dystrophy in girls with X;autosome translocations. J Med Genet 23:484-490.

12. Ricciuti, FC and FH Ruddle. 1973. Assignment of three gene loci (**PGK, HGPRT, G6PD**) to the long arm of the X chromosome by somatic cell genetics. Genetics 74:661-678.

13. Gusella, JF, NS Wexler, PM Conneally, <u>et al</u>. 1983. A polymorphic DNA marker genetically linked to Huntington's disease. Nature 306:234-238.

14. Ohno, S. Sex chromosomes and sex-linked genes, in <u>Monographs</u> <u>on</u> <u>Endocrinology</u> 1, (Labhart, A, T Mann and LT Samuels, eds). Heidelberg: Springer-Verlag, 1967, pp. 58-69.

15. Nadeau, JH and BA Taylor. 1984. Lengths of chromosome segments conserved since divergence of man and mouse. Proc Natl Acad Sci USA 81:814-818.

ADDITIONAL LEARNING RESOURCES

1. Bollag, RJ, AS Waldman. 1989. Homologous recombination in mammalian cells. Annu Rev Genet 23:199-225.

2. <u>Genetic Maps</u>, 5th ed, (SJ O'Brien, ed). Cold Spring Harbor: Cold Spring Harbor Laboratory Press, 1990.

3. O'Brien, SJ, HN Seuanez and JE Womack. 1988. Mammalian genome organization. Annu Rev Genet 22:323-351.

4. Watson, JD. 1990. The human genome project: Past, present, and future. Science 248:44-51.

5. White, R and J-M Lalouel. 1987. Investigation of genetic linkage in human families. Adv Hum Genet 16:121-228.

6. White, R and J-M Lalouel. 1988. Sets of linked genetic markers for human chromosomes. Annu Rev Genet 22:259-279.

Chapter 10

Genetics and Immunity

OVERVIEW OF THE IMMUNE SYSTEM

Defense against infection is a complicated phenomenon involving a number of systems and has both cell-mediated and humoral components. The particular elements of our defense arsenal that are mobilized to combat infection are dependent upon the type of insult and the route of invasion. An overview of our immune system is provided in Fig. 10-1. An **antigen** is defined as any substance capable

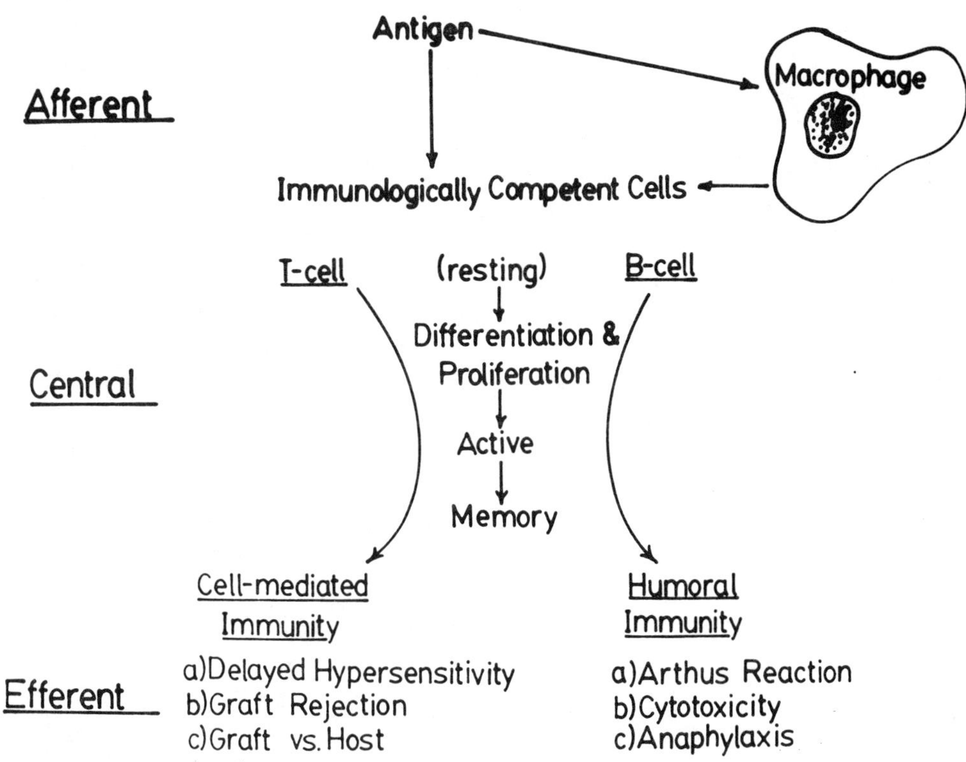

Fig. 10-1. Afferent, central and efferent arms of the immune system.

of stimulating an immune response and, specifically, as any substance capable of stimulating the production of antibodies. **Antibodies** (or **immunoglobulins (Ig)**) are proteins that combine with antigens in a specific way to neutralize them and to promote their destruction. An antigen may directly stimulate immunologically competent cells (such as B-cells) or may be ingested by phagocytosis and processed by macrophages and other cells. The processed antigen is then presented to T-cells that inspect the processed antigen. If the antigen fragment is judged "self", no immune response will be mounted; however, an antigen judged to be foreign will elicit a potent immune response.

Immunologically competent cells comprise the central arm of the immune system. These cells have been genetically programmed to respond to specific antigenic stimuli. Our lymphoid tissues contain two general types of immunologically competent cells: T-cells and B-cells. The **T-cells** are involved with the generation of **cell-mediated** immunity. This component of the immune system is directed toward **intracellular** antigens including viruses that have infected cells and antigens that are components of our own cells or components of tissues that may have been transplanted to a patient. Cells appear to continually present information to T-cell components which, in turn, monitor this information. When something is detected that is unusual (virus infection, cancer antigen), the inspecting T-cell component either sends out an alarm to other cells in the immune system or directly attacks and destroys the abnormal cell. This process will be explored more extensively in a subsequent section.

B-cells and their derivatives, the **plasma cells**, participate in humoral immunity which is directed against **extracellular** antigens. When B-cells are stimulated by antigen or receive signals from certain T-cells, they differentiate into plasma cells which divide rapidly to form clones that synthesize and secrete copious quantities of antibodies directed against the offending extracellular antigen. These antibodies participate in cell killing, neutralization of extracellular antigens, in the Arthus reaction, and contribute to anaphylactic reactions. An Arthus reaction results from interactions between antigen and specific antibodies in tissues and includes localized swelling and inflammation at the site of the interactions. An individual experiencing anaphylaxis has often been exposed to the offending antigen on previous occasions. Subsequent exposure produces a series of physiological responses which include muscular and vessel changes and varying degrees of shock.

A summary of cell-mediated and humoral immunity is presented in Table 10-1. Cell-mediated immunity utilizes a receptor molecule, called the T-cell receptor, which occurs in two general forms. The α/β T-cell receptor is located on T-cells that possess either a CD4 or a CD8 surface antigen closely situated with the T-cell receptor. Cells that carry CD4 are referred to as "helper" T-cells, while those which are CD8+ are called "killer" or "cytotoxic" T-cells. Both of these T-cells monitor peptide fragments that are presented to them by histocompatibility antigens (MHC = HLA in humans). These HLA molecules offer the fragment to the T-cell receptor for inspection. Depending upon the type of fragment presented and the type of cell doing the monitoring: 1) no response will occur; 2) B-cells will be activated; 3) macrophages will be activated; or 4) the presenting cell will be killed. Further details of this type of immunity will be presented below. The second type of T-cell receptor, the $\gamma\delta$-T-cell receptor, is located on intra-epithelial lymphocytes that are found in the intestinal epithelium. This system is poorly characterized and will not be discussed further.

Humoral immunity culminates in the destruction of bacteria, extracellular viruses, parasitic worms, or toxins by specific antibodies secreted by clones of plasma cells. This aspect of the immune response may be divided into primary and secondary stages. The primary response is triggered by direct contact of a bacterial antigen with an IgM receptor or by response of appropriate B-cells to chemical signals released by T-cell helpers that have encountered extracellular protein fragments. Either encounter causes rapid proliferation and differentiation of the responding B-cells into clones of plasma cells that secrete IgM directed against the offending antigen. Some of these cells "switch" to the production and secretion of IgG, IgE, or IgA. This switch constitutes the beginning of the secondary immune response. IgG antibodies are directed against the same antigen that elicited the primary IgM response, but tend to be more efficient in clearing the foreign antigen from the system. IgE antibodies are involved with removal of allergens and the killing of parasitic worms, while IgA molecules tend to be found in

Table 10-1. Summary of Cell-mediated and Humoral Immunity.

Receptor	Recognition Factor	Ligand	Accessory System	Function
I. Cell-mediated immunity (intracellular antigen).				
CD3 T-cell (α/β)	CD4+ T-cell	Extracellular Protein Fragment Class II MHC	B-Lymphocytes	B-cell Activation
	CD4+ T-cell	Fragment from Ingested Bacterium	Macrophage	Macrophage Activation
	CD8+ T-cell	Fragment from Cellular or Viral Protein Class I MHC		Target cell Killing
CD3 T-cell (γ/δ)	CD8+ IEL*	?	?	Cell Killing?

*IEL = Intra-epithelial lymphocyte (intestinal epithelium)

II. Humoral immunity (extracellular antigen).				
Surface Igs	IgM	Bacterial	Complement	Lysis
	IgG	Viruses/toxins	"+" Phagocytes	Lysis Engulfment Neutralization
	IgE	Antigens/Allergens Parasitic Worms	Mast Cells Eosinophils	Vascular Changes Killing
	IgA	Viruses/toxins Bacterial adhesions		Neutralization Block Adherence

secretions, including those of the intestine. Intestinal IgA retards infection of intestinal cells by gut bacteria and neutralizes toxins that may occur in foods and other ingested substances. IgM lysis of bacteria is facilitated by a protein system known as complement. Mast cells release chemicals that contribute to the localized swelling and feeling of lung congestion associated with allergic reactions. A more extensive discussion of the humoral immune system will be presented in a later section of this chapter.

Differentiation of the Immune System

Immunocompetent lymphoid cells develop from primitive stem cells occurring in the bone marrow (Fig. 10-2) and in fetal liver. These stem cells appear to be descendants of a common precursor cell (see below) and are heterogeneous in composition. The stem cells migrate to other sites during immunodifferentiation. A proportion of these cells, whose descendants are subsequently involved with tissue-mediated immunity, migrates to the thymus where they interact with the thymus epithelium, differentiating into T-cells. Not only do these cells develop T-cell receptors and other T-cell specific antigens on their surfaces, but they also undergo a preliminary selection such that self-reactive T-cells are eliminated or otherwise neutralized so they don't mount responses to our own tissue antigens. These T-cells (thymus-derived lymphocytes) undergo a second migration, seeding the paracortical regions of lymph nodes and other lymphoid tissues.

Other primitive stem cells migrate from the bone marrow to seed the germinal centers of lymph nodes and other lymphoid tissues. These cells are called B-cells (bone marrow-derived). B-cells in the lymph nodes, spleen, Peyer's patches, and other lymphoid tissues have IgM receptors and are

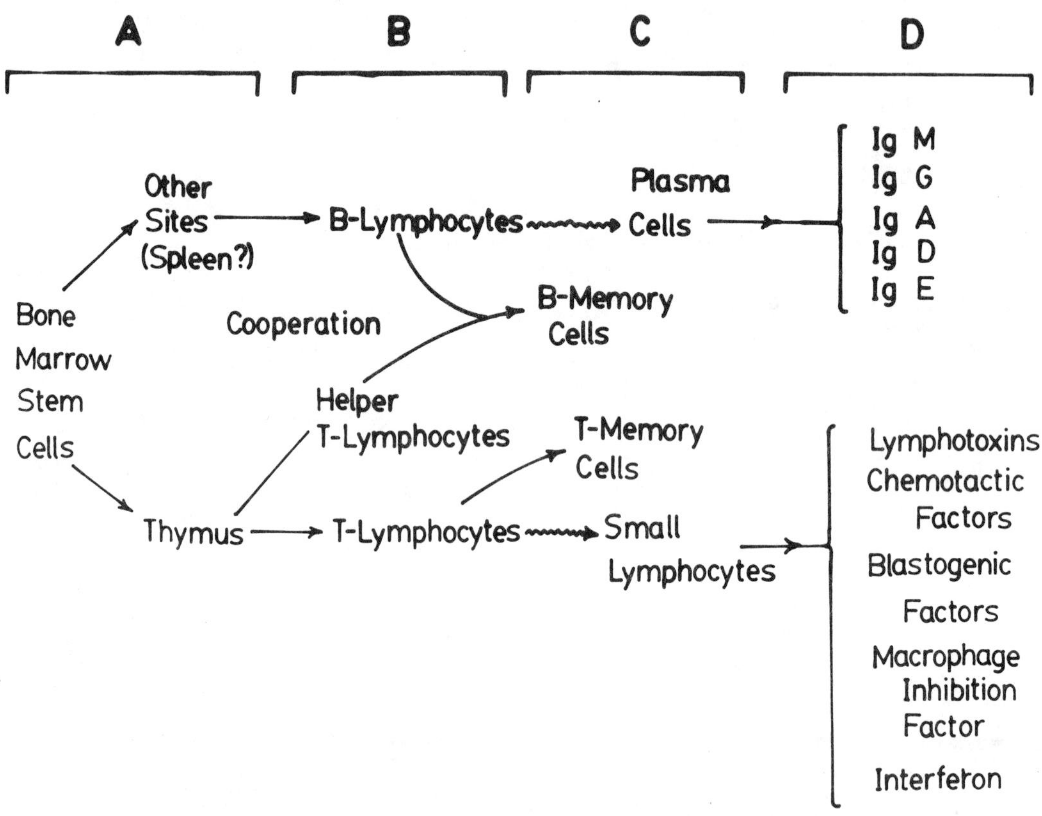

Fig. 10-2. Differentiation of the immune system. A) Cell maturation; B) antigen encounter and lymphocyte activation (lymph nodes); C) differentiation and proliferation; D) effector molecules.

responsive to antigen. Note that B-cell differentiation up to their arrival in the lymph nodes has been accomplished without antigen exposure. These immunocompetent B-cells are called virgin B-cells. Exposure of B- and T-cells to foreign antigen or fragments thereof induces their multiplication and differentiation into plasma cells or small lymphocytes, respectively. The plasma cell clones produce antibodies specific for the inducing antigen. Small lymphocytes produce a variety of small molecules including substances that kill foreign lymphocytes (lymphotoxins), substances that attract other lymphoid cells and promote their interaction (chemotactic factors), chemicals that promote division of lymphoid cells (blastogenic factors), substances that localize macrophages in the region of inflammation (macrophage inhibition factors), and antiviral substances (interferons).

Both T- and B-cell populations possess a memory component. Apparently not all B- and T-cells exposed to antigen take immediate action against the antigen. For example, a proportion of B-cells that have been stimulated by antigen multiply but do not secrete antibodies. These "memory" cells remain in lymph nodes for long periods of time, and, when a second exposure to the same or closely related antigen occurs, they are rapidly mobilized to produce an immune reaction that is more potent than the one that occurred following the initial antigen encounter.

T-cell helpers can facilitate B-cell response to antigens and are apparently essential for proper B-cell response. Helper T-cells are activated by interaction with a peptide fragment derived from a foreign antigen that is presented by another cell, such as a macrophage. B-cells internalize antigens following their binding to Ig receptors on their surfaces. The internalized antigen is degraded to peptides that are recycled to the surface in combination with a class II HLA molecule. The activated helper T-cell interacts with the peptide fragment-HLA complex, and a series of events are initiated which culminate in the release of a signaling molecule, known as an interleukin. This interleukin binds to a receptor on the B-cell surface and stimulates the B-cell to release antibodies against the antigen.

Three rare genetic disorders have been described that appear to result from abnormal maturation of the immune system (Table 10-2). The most severe disorder is an autosomal recessive syndrome, hereditary thymic aplasia (Swiss type agammaglobulinemia). Patients with this disease appear to have a genetic lesion affecting the differentiation of a bone marrow stem cell into precursors of T- and B-cells. Patients lack both cell-mediated and humoral immunity. This total immunodeficiency is reflected in extreme susceptibility to bacterial, viral, and fungal infections, and tolerance of unmatched tissue grafts. Patients die during infancy unless reared in germ-free environments. Occasionally such children are able to find a histocompatible donor for a bone marrow transplant. The stem cells from the donor tissue repopulate the patient's marrow and are able to correct the deficiency. X-linked hypo-gammaglobulinemia (Bruton type) is associated with a lack of circulating antibodies; however, cell-mediated immunity appears intact. The defect in B-cell maturation renders these patients susceptible to certain bacteria including streptococcus, staphylococcus and pneumococcus. These patients appear to be able to combat fungal and viral infections; however, since antibodies are required for a portion of antiviral defense, there may be some increased susceptibility to viral diseases. Bruton's individuals appear to exhibit normal tissue graft rejection. Infusions of gammaglobulins, the fraction of blood plasma that contains antibodies, provide temporary protection against bacterial infections. The DiGeorge syndrome (thymic hypoplasia) encompasses a group of clinically similar disorders, many of which are not inherited. The DiGeorge syndrome usually is found in patients who are heterozygous for a deletion of chromosome 22q11. This deletion presumably renders these individuals hemizygous for a recessive allele that impairs development of the cell-mediated immune system. DiGeorge persons have problems with viral and fungal

Table 10-2. Immune Deficiency Diseases.

Type	Cell Population	
	T-cell	B-cell
Hereditary Thymic Aplasia	-	-
X-linked Hypogammaglobulinemia	+	-
DiGeorge Syndrome	-	+

infections, display increased graft tolerance, and have somewhat impaired humoral immunity as a consequence of lack of T-cell helpers. Bone marrow transplants from compatible donors may correct the disease.

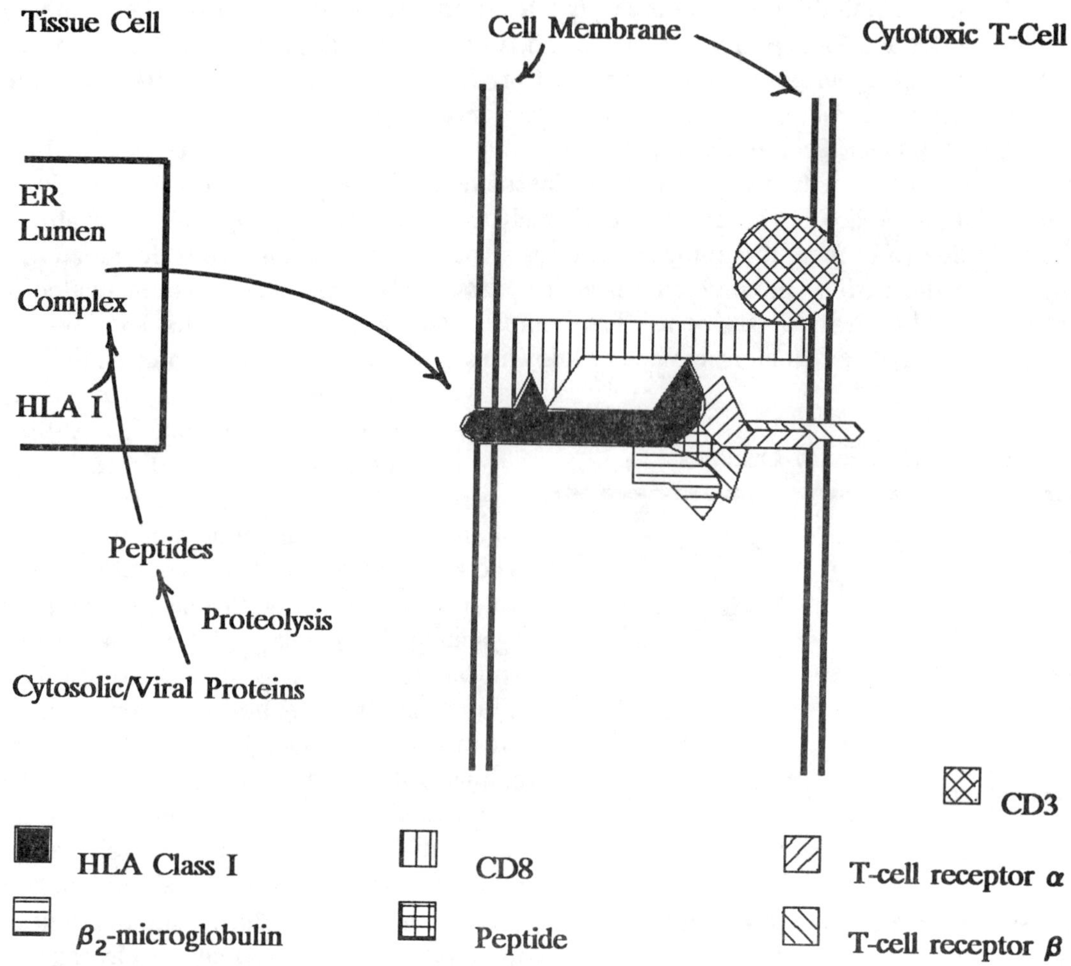

Fig. 10-3. Recognition of Intracellular Antigen. Viral or endogenous intracellular antigens are partially degraded to peptides which enter the endoplasmic reticulum (ER). A peptide is bound by an HLA class I complex (HLA class I polypeptide + β_2-microglobulin), and the complex translocates to the cell membrane. A CD8+CD3+ cytotoxic T-cell "inspects" the peptide fragment in the context of its presenting class I molecule. If the peptide is judged "foreign", the cytotoxic T-cell destroys the presenting cell.

CELL-MEDIATED IMMUNITY

Dr. R.A. Fisher defined the concept of a **supergene** as a block of genetic information that is preserved by evolutionary forces (1). The major histocompatibility complex (**MHC**) provides an excellent example of such a supergene or complex locus. The genes within the locus are very tightly linked and are only rarely separable by recombination. The MHC was first described in the mouse, and similar MHC regions have been detected in a large number of different species, including humans. The human MHC region is located on the short arm of chromosome 6, and certain genes (**HLA** (human lymphocyte antigen)) within the region encode the amino acid sequences of cell surface antigens (histocompatibility antigens) that function in the presentation of peptides for inspection by the T-cell receptor of T-cells. These genes are highly polymorphic, with as many as forty codominant alleles per locus identified in various human populations. Class I HLA antigens which are widely distributed on many cell types are encoded by the **HLA-A, HLA-B** and **HLA-C** loci. The class II region contains at least six different loci, three of which (**HLA-DP, HLA-DR** and **HLA-DQ**) are known to encode histocompatibility antigens found on the surfaces of B-cells, macrophages and certain other immune and endothelial cells. These histocompatibility antigens are involved with presentation of peptides for inspection by the T-cell receptors of regulatory T-cells such as helper T-cells.

Presentation of Intracellular Antigens by Class I HLA Molecules

Viruses infect cells by injecting their nucleic acids (DNA or RNA) into the cytosol of cells. These nucleic acids then utilize the infected cell's replicational and transcriptional machinery for the formation of new virus particles. Human cells appear to break down cytosolic proteins (both endogenous and viral) to their constituent peptides which are 8 to 25 amino acids in length. These peptides enter the endoplasmic reticulum where they form a complex with class I HLA molecules (Fig. 10-3). A class I HLA molecule consists of a polypeptide encoded by a **HLA-A, HLA-B** or **HLA-C** gene and a smaller polypeptide, β_2-microglobulin, specified by the **B2M** locus on chromosome 15. The complex translocates to the cell membrane, where it "presents" the peptide to a cytotoxic T-cell. At least three different polypeptides interact with the Class I - peptide complex. The CD8 protein binds in a specific fashion with a site on the HLA component of the presenting complex. This binding most likely explains why cytotoxic T-cells, and not T-cell helpers, interact with Class I HLA complexes. The T-cell receptor also interacts in a specific fashion with the HLA polypeptide, with β_2-microglobulin and with the peptide. The T-cell receptor contains both α and β polypeptides, with the β polypeptide containing a cytoplasmic signaling domain. The T-cell receptor "inspects" the peptide within the context of its surrounding class I presenting molecule. If there is a "good fit" of the five components of the interacting complex, a chain of events is activated in the cytotoxic T-cell which culminates in the destruction of the T-cell. By destroying a virus-infected tissue cell before the virus can complete its reproductive cycle, the spread of the viral infection can be attenuated. If the peptide is derived from an endogenous cytosolic protein, the fit of the inspecting T-cell receptor, the presenting class I HLA complex and the peptide will be imprecise, and no killing response will be activated. Self-reactive cytotoxic T-cells are either eliminated or inactivated by other means during early development.

Presentation of Internalized Extracellular Anigens

Class II HLA molecules present peptides derived from extracellular antigens that have been phagocytized and degraded in lysosomes by macrophages, B-cells or certain other immune cells (Fig. 10-4). For example, a macrophage may phagocytize a bacterium. The bacterium is incorporated into an endosome and degraded by lysosomal enzymes in an acidified endosomal compartment of the cell. The class II receptor complex, consisting of both α and β chains that are encoded by the **HLA-DP, HLA-DQ,** or **HLA-DR** loci, is processed in the ER and complexed with a polypeptide known as the invariant chain. This polypeptide occupies the cleft in the Class II molecule while it is in the ER and prevents the class II HLA complex from binding with cytosolic peptide fragments. The invariant chain also participates in the trageting of the class II HLA molecule to the endosomal compartment where the invariant chain disengages and is degraded. The class II HLA molecule binds a bacterial peptide

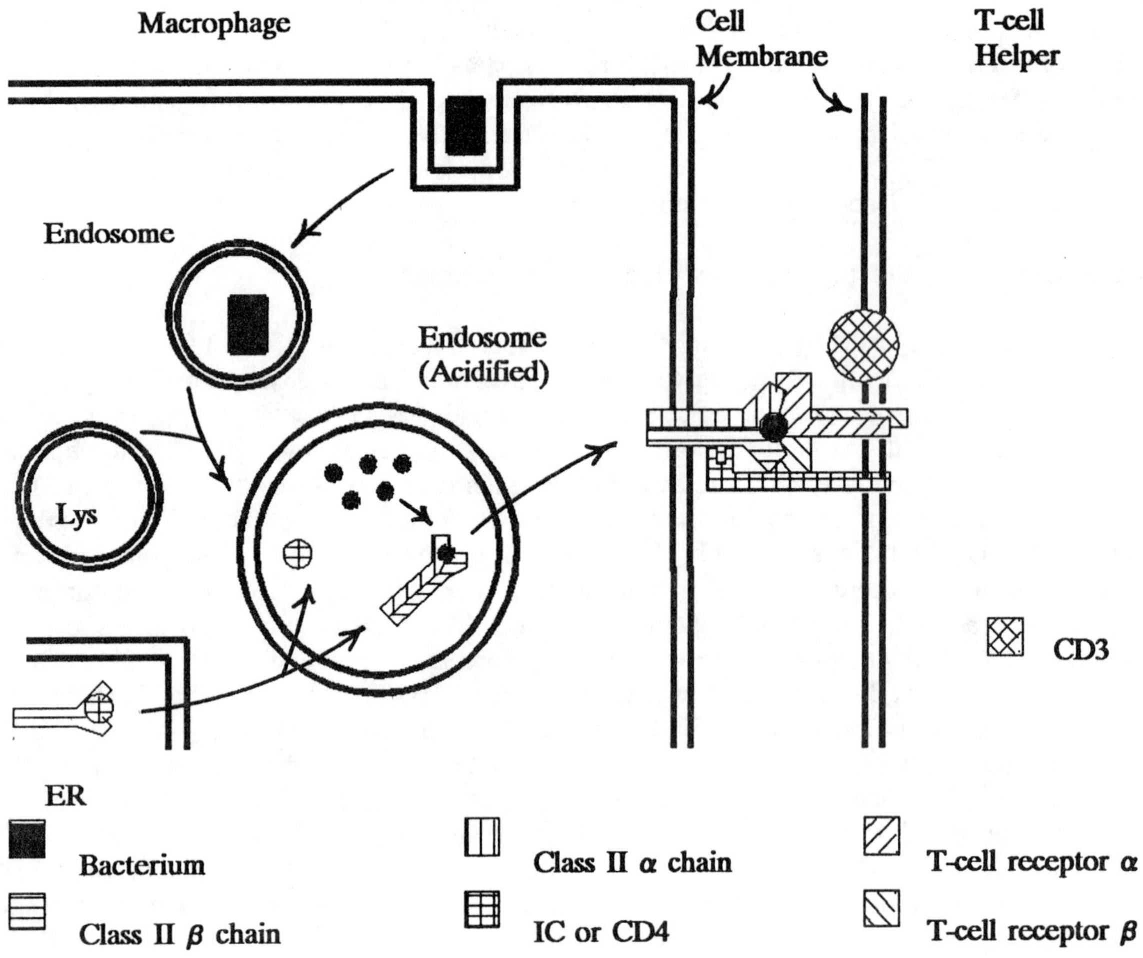

Fig. 10-4. Recognition of an extracellular processed antigen. A bacterium is phagocytized by a macrophage. The endosome containing the bacterium fuses with a lysosome (LYS), and the lysosomal enzymes degrade the bacterial components. Class II HLA molecules are complexed with an invariant chain (IC) in the ER which prevents them from binding endogenous or viral peptides. The IC-HLA complex enters the acidified endosome, where the IC dissociates from the HLA αβ dimer and is degraded. The HLA molecule binds a peptide fragment from the bacterium and translocates to the macrophage cell membrane where it presents the peptide to the T-cell receptor of a helper T-cell. Note how the CD4 antigen facilitates binding of the T-cell receptor to the HLA molecule.

and translocates to the macrophage cell membrane where it presents the peptide for inspection by the T-cell receptor of a CD4+CD3+ helper T-cell. The CD4 surface antigen of the T-cell helper insures that the T-cell helper will bind with class II HLA molecules and not class I HLA molecules. If there is an appropriate fit among the peptide, the α and β HLA chains, and the α and β chains of the T-cell receptor, the cytoplasmic domain of the T-cell receptor β-chain will signal the release of interleukins which will activate other immune cells. Self-reactive T-cell helpers are either eliminated or otherwise incapacitated during early development.

Cooperation Between T-cell Helpers and B-cells

T-cell helpers facilitate the synthesis and secretion of antibodies by B-cells (Fig. 10-5). Antigen

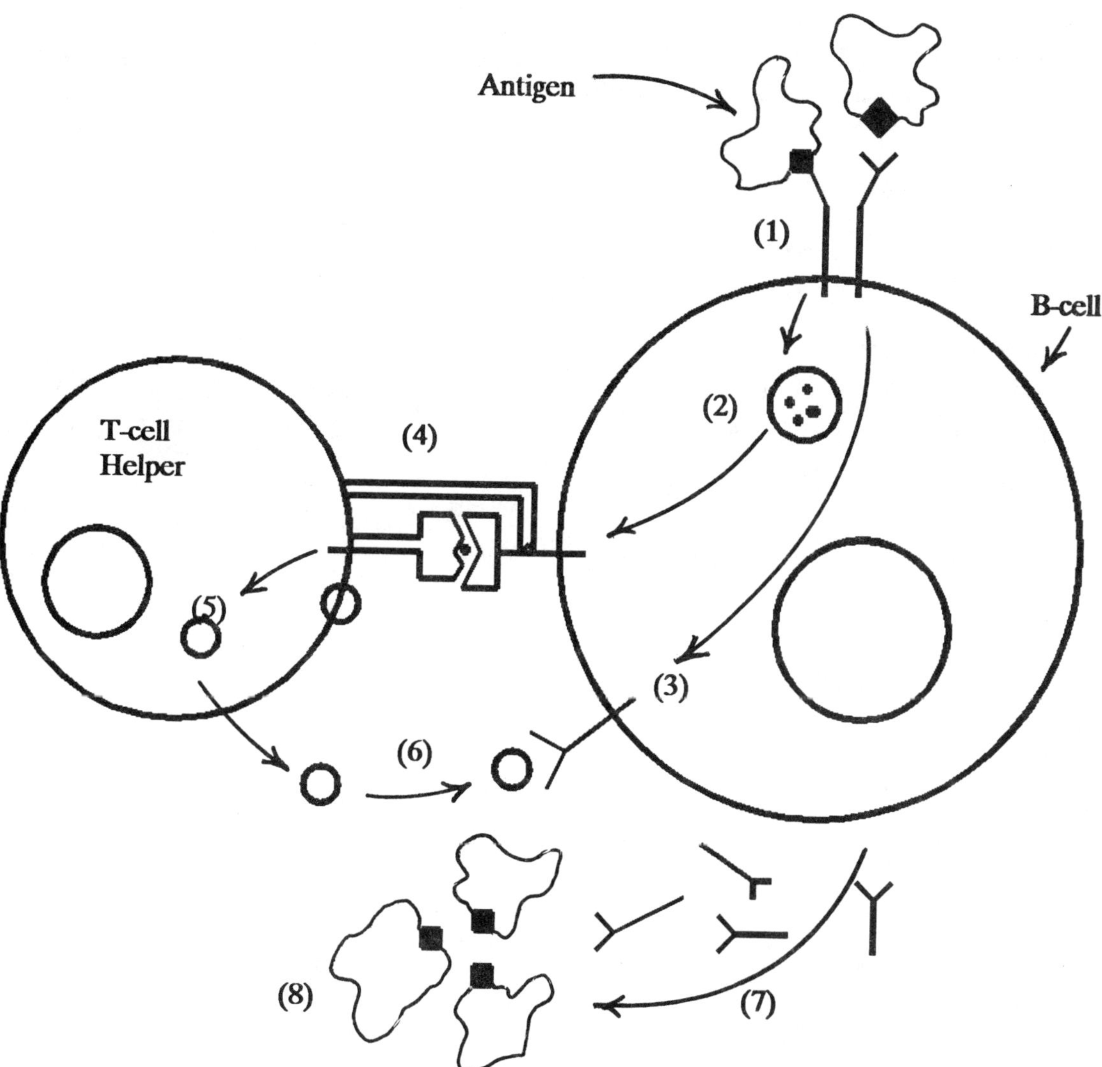

Fig. 10-5. T-cell facilitation of antibody production and release by B-cells. Binding of antigen by Ig receptors (1) results in the internalization and degradation of the antigen (2) and production of interleukin receptors (3). Peptides from the antigen are recycled to the surface of the B-cell complexed with class II HLA molecules. T-cell helpers that have been activated by presentation of the same or very similar peptides by other immune cells, such as macrophages, interact with the peptide fragment (4) and release interleukins (5) which bind to the interleukin receptors on the B-cell membrane (6). This binding promotes synthesis and release of antibodies (7) which attack the antigen (8).

is encountered by specific Ig receptors on the B-cell surface, internalized and degraded in acidified endosomes. Interaction of the antigen with the Ig receptor signals the production of interleukin receptors that are inserted into the B-cell membrane. Peptides from the antigen are recycled to the B-cell surface bound to class II HLA molecules. The same or a very similar antigen may have been presented to a T-cell helper by a macrophage or other immune cell. The activated T-cell helper now interacts with the peptide presented by the B-cell and releases interleukins which stimulate the B-cell to synthesize and secrete specific antibody that attacks the antigen. This example illustrates the interrelationships between cell-mediated and humoral immunity and explains why DiGeorge patients exhibit partial compromise of their humoral system by a mutation that interferes with T-cell development. It also underscores the central importance of T-cell helpers in the immune response. The AIDS (acquired immunodeficiency syndrome) virus incapacitates T-cell helpers and causes a loss of both cell-mediated and humoral immunity. Consequently patients are vulnerable to a variety of infections and are susceptible to certain cancers that are rare in most other people.

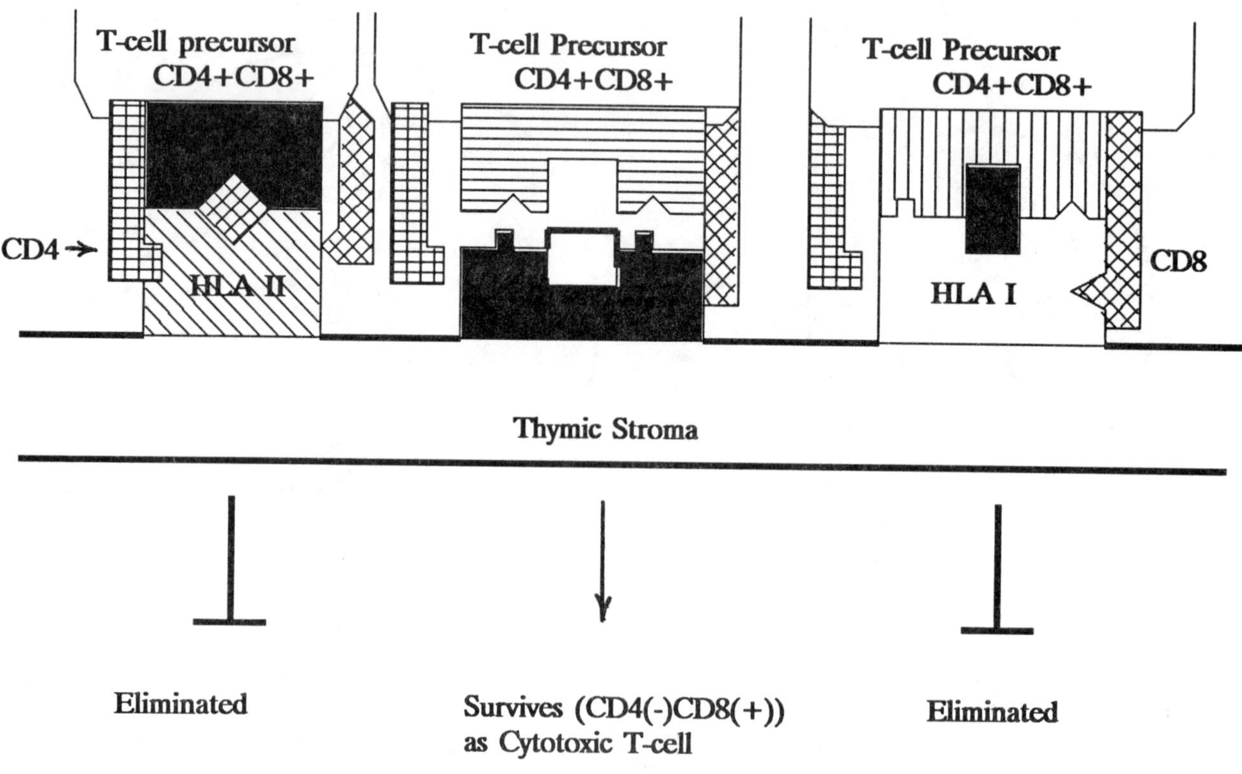

Fig. 10-6. Thymic elimination of self-reactive T-cells. A subset of three T-cell precursors, all of which possess CD4 and CD8 antigens are diplayed. The left-hand cell encounters a cell in the thymic stroma that presents a self-antigen peptide with a class II antigen. The fit between the T-cell receptor on the T-cell precursor, the antigenic peptide, and the HLA molecule is good, and the T-cell precursor is killed by an unknown mechanism. The right-hand cell encounters a self-peptide presented by a HLA class I antigen. A good fit is obtained, and this T-cell precursor is also eliminated. The central cell encounters a self-peptide presented by a HLA I bearing stromal cell; however, a poor fitt occurs, and the cell survives as a cytotoxic T-cell line that is not self-reactive.

Development of Tolerance

Immunological tolerance of self-antigens occurs both in the thymus and subsequently at other sites. The mechanisms proposed for tolerance include elimination of self-reactive clones, anergy of immune cells, and suppression of self-reactive clones. A model for elimination of self-reactive clones is presented in Fig. 10-6. T-cell precursors bearing both CD4 and CD8 antigens interact with thymic stromal cells that present self-antigens by means of a class I or class II HLA molecule. If there is a close fit between the HLA class I or II antigen, the peptide and the T-cell receptor of the T-cell precursor, the T-cell precursor is destroyed, eliminating cell clones that are self-reactive. If the fit is poor, the pre-T cell continues to differentiate, and depending upon the HLA type of the stromal cell with which it interacted, develops into a cytotoxic T-cell or helper T-cell directed against non-self antigenic peptides. Anergy refers to the functional inactivation of self-reactive T-cell clones. Although the manner in which the inactivation occurs has not been elucidated, improper peptide presentation may be responsible. A third way in which self-reactive T-cells are prevented from attacking our own tissues involves immunological suppression. Suppressor T-cells appear to counterbalance the effects of helper T-cells in certain situations. The mechanism by which this suppression occurs is unknown. Unlike clonal elimination, both clonal anergy and clonal suppression may be reversed. For example, a clone escaping suppression may contribute to symptoms of autoimmunity. A number of autoimmune disorders are known, and risk for many of these problems appears to increase as a person ages. Reviews of cell-mediated immunity, immune tolerance, and autoimmunity have recently been published (2-4).

Genetics of Cell-mediated Immunity

Cell-mediated immunity requires appropriate interaction between the presenting molecules (HLA antigens) and the monitoring molecules (T-cell receptors). Genes encoding HLA antigens are located on the short arm of chromosome 6 (Fig. 10-7). Class II loci are clustered on the centromeric side of the MHC region, while the class I loci are on the telomeric end. Class III loci (**C2, C4A, C4B,** and **BF**) are centrally located in the cluster. The class III genes encode proteins that participate in the complement cascade which culminates in the production of proteins that assist IgM with cell lysis. Several other genes encoding complement components have been mapped to other chromosomes.

HLA antigens can be freed from the surface of cells by treatment with nonionic detergents and precipitated by anti-HLA antibodies. After separation from the antibodies, the HLA antigens can be further subdivided by electrophoresis, recovered from the gel, and characterized. The structures of the class I and class II HLA antigens are quite different (Fig. 10-8). A class I antigen contains two chains, a heavy chain (MW approximately 45,000) that is determined by the **HLA-A, HLA-B,** or **HLA-C** locus and a light chain (MW about 12,000) encoded by the **B2M** locus on chromosome 15. The heavy chains consist of three domains encompassing amino acid residues 1-90, 91-180 and 181-271 (5). The majority of the molecule projects into the extracellular space, with the α_1 and α_2 domains forming a groove resembling a "catchers mit" which holds the peptide fragment. The β_2M and the α_3 domain of the heavy chain form the ramainder of the extracellular molecule. The CD8 antigen of the cytotoxic T-cell binds to the α_3 domain. The HLA molecule is anchored in the cell membrane by the HLA-encoded heavy chain. The structure of the class I molecule (β_2M light chain + heavy chain) resembles half of a typical antibody molecule, and structural similarities between domains of HLA antigens and immuno-globulins have been identified, suggesting that genes encoding HLA molecules and immunoglobulins may be descendants of the same ancestral genes. Class II HLA molecules consist of two chains (MW about 34,000 and 29,000, respectively). A typical HLA-DR molecule possesses an α chain and a β chain. The α_1 and β_1 domains form a groove which holds the antigenic peptide. Unlike Class I molecules, both the

α and β chains are encoded by the genes within the class II region, and both chains span the cell membrane.

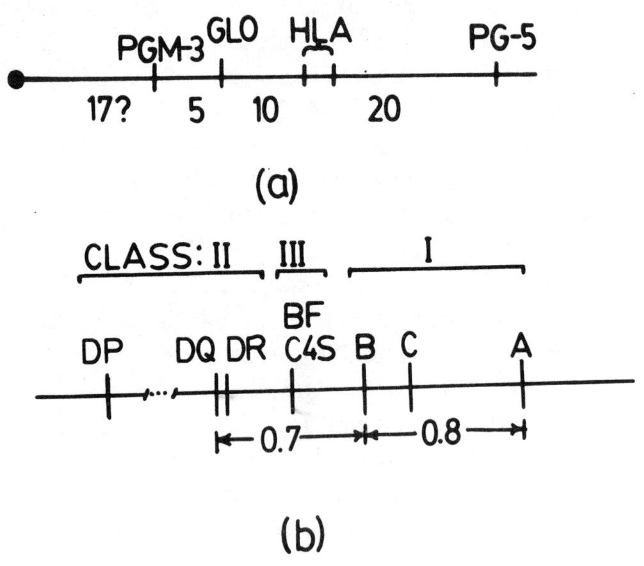

Fig. 10-7. Structure of the human MHC region. a) short arm of chromosome 6. The centromere is at the left, and approximate crossover distances are presented below the chromosome arm. PGM-3 = phosphoglucomutase-3; GLO = glyoxylase; and PG-5 = pepsinogen-5. b) Enlargement of the 6p21.3 region containing the HLA genes. C4S and BF are only two of several complement genes located in the class III region. Only a subset of class II genes is indicated in the figure. A-D = HLA loci.

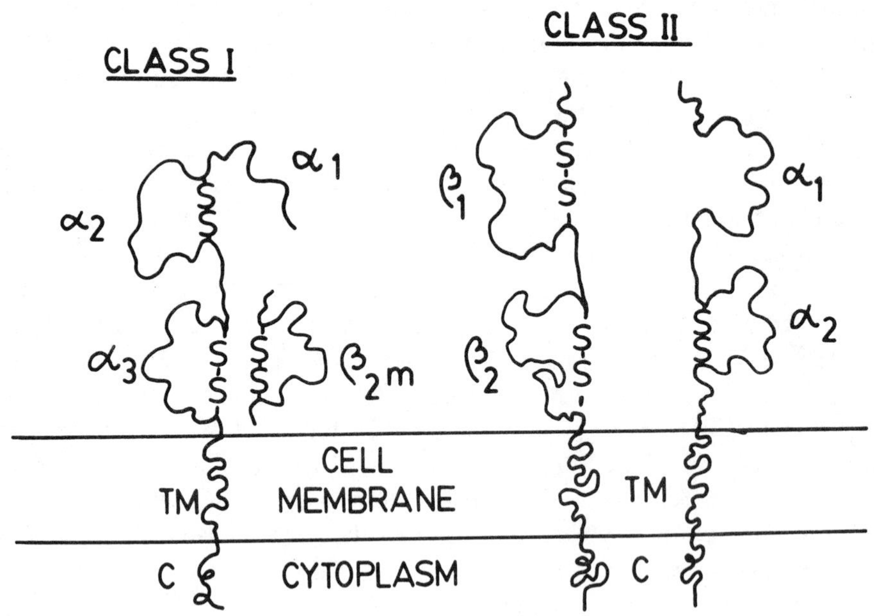

Fig. 10-8. Structures of class I and II HLA antigens. The heavy chain of class I and both chains of class II antigens contain well defined extracellular, transmembrane (TM) and cytoplasmic (C) domains.

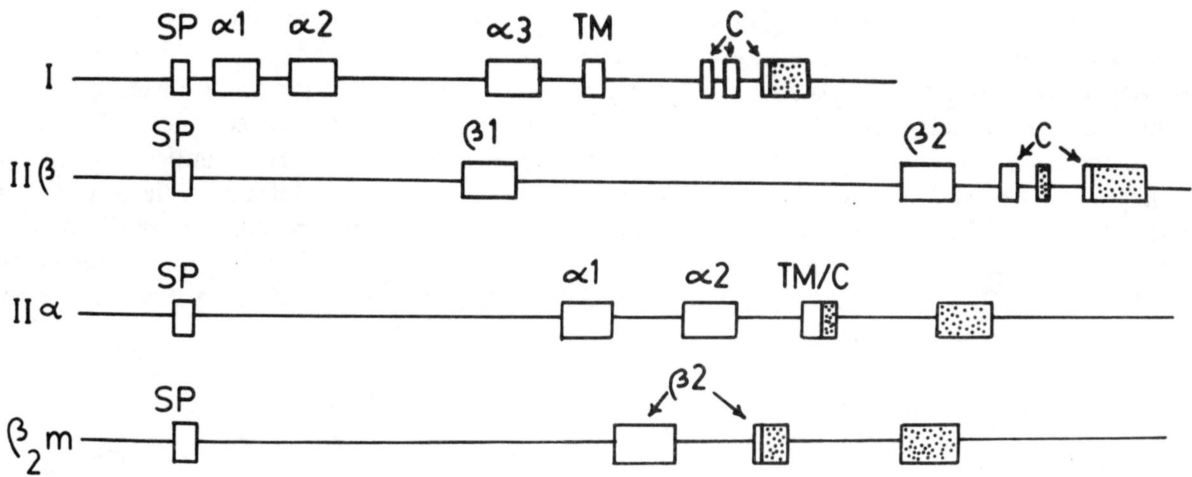

Fig. 10-9. Structures of representative HLA genes. Class I, class II and β_2-microglobulin genes are labelled I, IIα, IIβ, and β_2m, respectively. SP = signal peptide; TM = transmembrane segment; C = cytoplasmic segment; and exons specifying untranslated regions of the mRNA are designated by speckled boxes (redrawn from (6)).

Structures of three HLA antigen genes and of the β_2-microglubulin gene are presented in Fig. 10-9. Each gene consists of multiple exons, with each exon specifying a distinct domain of antigen structure. The similar gene structures suggest they may have evolved by a series of gene duplications from a common ancestral gene. Although the β_2-microglobulin gene is invariant, each of the HLA loci possesses multiple alleles, with as many as 40 alleles occurring in some populations. Most of the antigen variability occurs within the α_1 and β_1 domains, the regions contacting the peptide and the T-cell receptor. The class II α and β chains are encoded by pairs of genes (Fig. 10-10). At least three functional doublets have been found within the class II region, and more may be discovered. There appears to be less allelic variability among the class II loci than among class I loci.

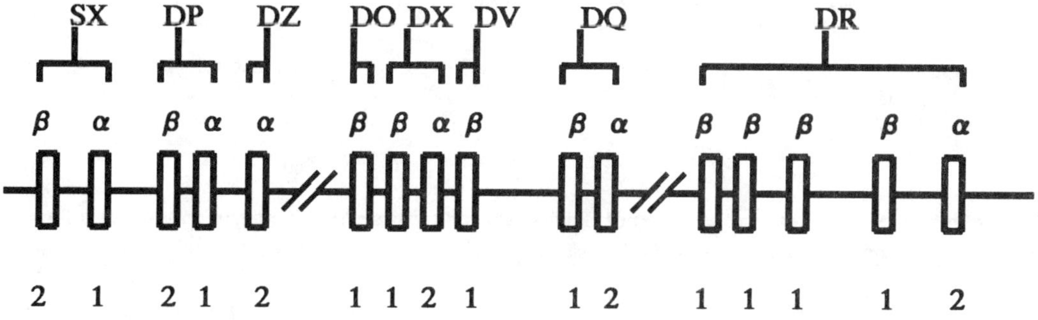

Fig. 10-10. Structure of the class II region. Each box represents a class II α or β gene as diagrammed in Fig. 10-9. The DP, DQ and DR loci are known to be expressed in protein. The number of DR β genes varies among haplotypes. 1 = sequence is transcribed from left to right. 2 = sequence of complementary strand is used and transcribed from right to left.

HLA antigens are usually defined by one of three general tests (Fig. 10-11). Tissue typing centers maintain banks of specific antisera directed against HLA antigens. Lymphocytes of both the potential donor and recipient possess HLA antigens that are used for typing. The lymphocytes of the potential donor (or recipient) are treated in parallel and in the presence of complement with specific antisera for each known antigen and observed for their response. If the HLA antigen is present, the specific antibodies and complement combine with it and lyse the cell. If the antigen is absent, no reaction occurs. The surface antigen array can be scored for both the potential donor and recipient. Chances of a successful transplant are considerably enhanced by matching donor with recipient with respect to their HLA types. HLA-A, HLA-B, and HLA-C antigens are tested in this manner and are said to be serum-defined antigens.

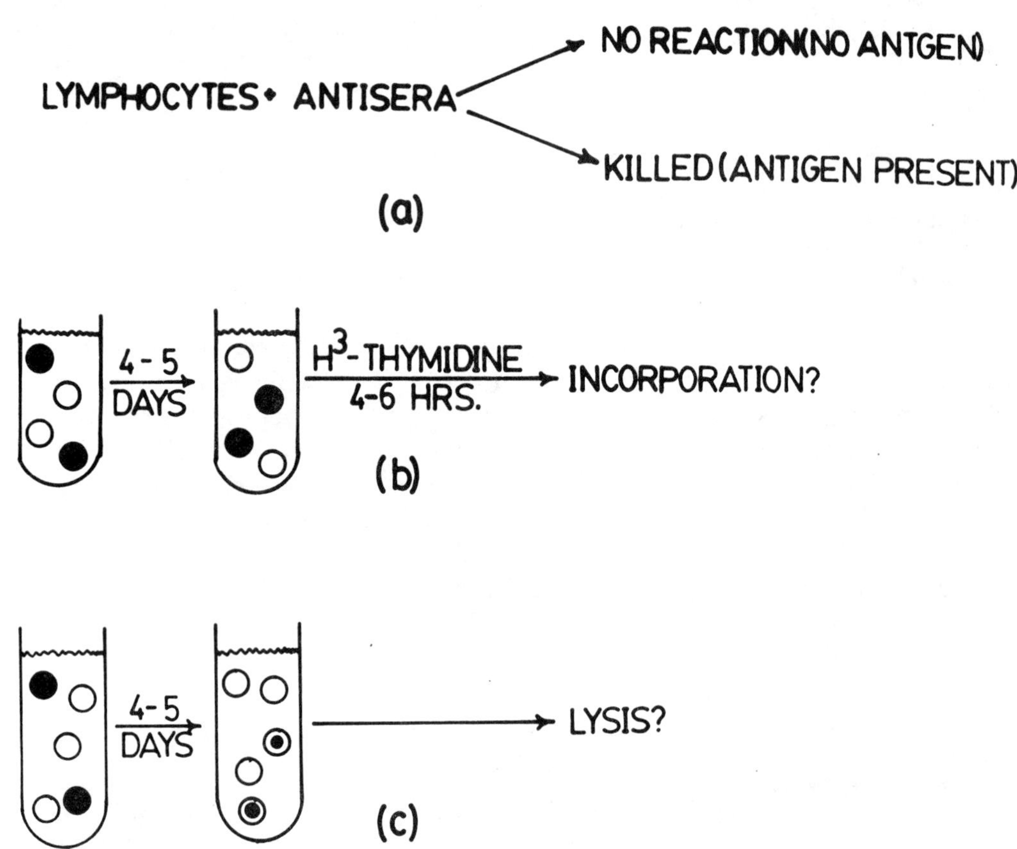

Fig. 10-11. Definition of HLA antigens. a) Serum defined antigens; b) mixed lymphocyte culture; c) cell-mediated lysis. Open circles symbolize recipient lymphocytes; shaded circles represent donor cells and circles with a dot are target cells.

Two lymphocyte-defined test systems are used to assess potential success of tissue transplants. The **mixed lymphocyte culture (MLC)** test provides information regarding class II HLA compatibility and is believed to provide an indication of early stages of transplant rejection if donor and recipient prove incompatible. When lymphocytes of the potential donor and recipient differ with respect to their class II histocompatibility antigens, cultures consisting of mixtures of their lymphocytes will produce blastogenic factors that stimulate donor and recipient cell DNA replication and cell division. In a two-way MLC the donor and recipient lymphocytes are cultured together for 4-5 days, tritiated thymidine is added, and incorporation of the labelled thymidine into cellular DNA is monitored. In a one-way test, either the donor or recipient cells are killed by radiation or other mutagen prior to culture, and the viable cells are monitored for DNA replication. Uptake of label by the cells is interpreted as an indication of class II incompatibility. If no uptake of thymidine occurs, the donor and recipient are judged to be compatible. Another test, **cell-mediated lympholysis (CML)**, detects cytotoxic T lymphocytes (CTLs). Presence of CTLs in tissue transplants leads to their rejection. A one-way MLC (killed donor cells) is set up and allowed to incubate for 4-5 days. The viable recipient cells are then harvested and introduced to a culture flask containing "target" donor cells containing radioactive chromium. If recipient lymphocytes have been stimulated to differentiate into CTLs, the recipient CTLs will attack the donor target cells, releasing the radioactive chromium into the culture medium. If the donor and recipient are compatible, no chromium release will occur.

     RFLP analysis has recently been added to the assays used to type HLA alleles. The RFLP haplotypes of the HLA region are established as described in the preceding chapter and used to determine the arrays of alleles present in the potential donor and recipient. Presence of the same RFLP haplotypes in donor and recipient implies compatibility for the respective HLA antigens. Matching of HLA antigens between donor and recipient greatly enhances the success of a tissue transplant. Mismatch of these antigens results in rejection of the transplant within a few weeks of surgery. In cases where bone marrow tranplants are attempted in HLA-incompatible people, a graft vs. host reaction may occur where the CTLs derived from the transplanted marrow tissue attack the recipient's tissues.

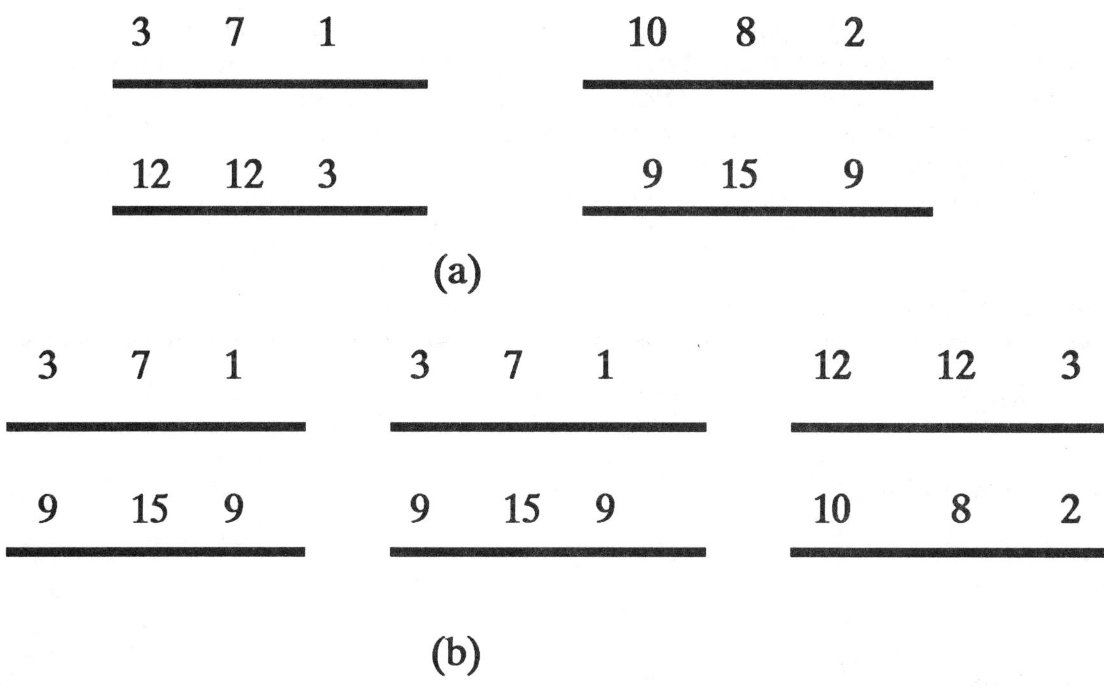

Fig. 10-12. Inheritance of HLA haplotypes. a) Parents; b) Siblings. The chromosomal order of the loci are (left to right): B, C and A.

The HLA loci are closely linked, so arrays of alleles on a chromosome tend to be inherited as blocks. The combination of HLA alleles occurring on the same chromosome is called a **haplotype**. The inheritance of HLA antigens is outlined in Fig. 11-12. Note that each haplotype has been inherited intact. Two of the siblings share all six of their alleles and would have a better chance of tolerating transplants from one another than from either of their parents or the third sibling.

Linkage disequilibrium refers to the occurrence of specific haplotypes in considerably higher frequencies than predicted by chance. For example, the **HLA-A1** allele occurs with a frequency of 0.17, and the **HLA-B8** allele has a frequency of 0.11. The chance occurrence of these two alleles on the same chromosome is estimated by the product of their separate frequencies (0.02). However, the observed frequency of the **HLA-A1, HLA-B8** haplotype is seven- to twenty-fold more common, depending upon the population studied. The reason for this association and others is unknown. One explanation for such an association would be that the two alleles are closely linked to a third allele that carries a selective advantage. The selected allele, as well as the closely linked HLA alleles, become frequent in the population as long as the force responsible for the selective advantage is present. A number of HLA haplotypes have been observed to have relatively high frequencies among certain patients. Many of these individuals are afflicted with autoimmune disorders such as disseminated lupus erythematosis, active chronic hepatitis, Grave's disease, myasthenia gravis, and a subtype of juvenile-onset diabetes mellitus (JOD). The haplotype **HLA-A1, HLA-B8, HLA-DR3** occurs very frequently among caucasians with these disorders. Another haplotype occurring frequently in JOD patients is **HLA-B18, HLA-DR3**. The B4 Coxsackie virus has been isolated from a number of patients with the autoimmune form of JOD. Viral infection of the pancreatic islet cells may have triggered an autoimmune response against the insulin-producing β-cells which is directed toward viral antigens. If the **HLA-DR3** product is ineffective against the virus, viral infection would persist, and more extensive destruction of islet cells would occur. JOD would tend to develop in persons having the DR3 antigen more often than in persons who lack DR3. An alternative explanation for this disease association proposes that there is another locus responsible for JOD that is closely linked to **HLA-DR3**. The HLA-DR3 allele would then occur commonly in JOD as a result of the tight linkage, rather than as a consequence of a direct role in the cause of JOD. A number of other disease associations between HLA alleles and autoimmune diseases are presented in Table 10-3. The striking association between **HLA-B27** and ankylosing spondylitis is used when counseling

Table 10-3. Associations of Certain HLA Antigens and Disease (7).

Disease	HLA Antigen	Incidence (%)		Relative Risk
		Patients	Controls	
Ankylosing Spondylitis	B27	90	7	141
Reiter's Disease	B27	76	6	47
Psoriasis	B17	29	8	5
Coeliac Disease	B8	78	24	10
Myasthenia Gravis	B8	52	24	5

patients regarding risk for the disease. Ankylosing spondylitis is characterized by chronic inflammation of joints and the spinal column. The inflammation gradually leads to fusion of the vertebrae. Approximately one in five persons who have the **HLA-B27** allele have ankylosing spondylitis whose symptoms generally appear among young adults and affect primarily young men (male:female ratio = 9). The reason for this strong disease association is unknown but may be related to an interaction between the HLA-B27 antigen and <u>Klebsiella</u> infection.

T-cell Receptors

T-cell receptors occur in two types, $\alpha\beta$ and $\gamma\delta$. The respective receptor chains are displayed in Fig. 10-13. Each chain can be subdivided into several domains. The leader sequence targets the chain to the endoplasmic reticulum and is then cleaved from the molecule. The variable domain contains at least three complementarity determining regions (CDR) which are involved with receptor-HLA-peptide contacts. The β and δ chains possess diversity segments which also contribute to variability of the respective receptors. The joining segment connects the variable domain with the constant domain of the α and γ chains, and the diversity segment with the constant domain of the β and δ chains. The interface between the V-J, V-D, D-D, and D-J domains exhibit extensive amino acid variability and contribute to the flexibility of the T-cell receptor system and its ability to recognize a diverse range of peptides and HLA molecules. Relatively little sequence variation exists among the constant domains of α T-cell receptor chains, those of β chains, those of γ chains or those of δ chains, respectively. The hinge region of each chain provides a point of rotation for the molecules and participates in disulfide bond formation that helps link different chains of the receptor. Both the β and γ chains possess cytoplasmic domains that participate in signal transduction from the receptor to signaling pathways within the T-cell.

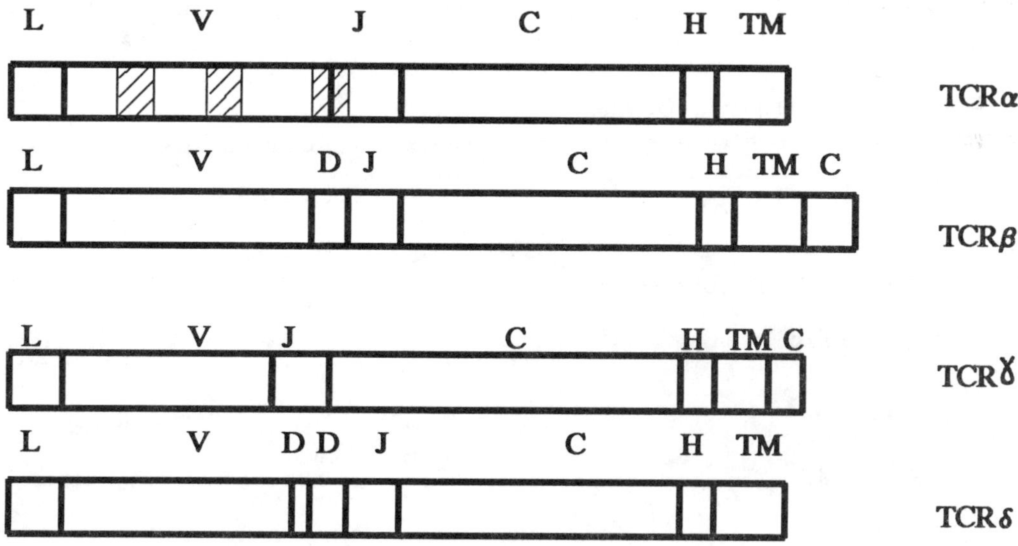

Fig. 10-13. T-cell receptor chains. The $\alpha\beta$ T-cell receptor is found on cytotoxic T-cells and T-cell helpers. The diagonally hatched regions represent the complementarity determining regions (CDRs) which contact the peptide and HLA molecules. Similar areas occur on the remaining three chains. L = leader sequence; V = variable sequence; D = diversity domain; J = joining sequence; C = constant domain; H = hinge region; TM = transmembrane domain; C = cytoplasmic domain. (8).

The human T-cell receptor genes have been mapped to three different chromosomes. The **TCRB** locus is situated in band 7q35, while the **TCRG** locus encoding the γ chain has been assigned to 7p15. The α and δ chains are encoded by the **TCRA** and **TCRD** loci in band 14q11.2. All of these loci are complex. **TCRG** appears to have the simplest structure; however, the gene segments encoding the V, J, and C domains of the receptor are separated and intermixed on the chromosome. DNA rearrangements during T-cell differentiation bring specific combinations of V, J, and C gene segments together to form the mature γ chain gene. The mechanism involved in this chromosome rearrangemnt is apparently similar to that employed for the formation of antibody genes and will be discussed later in this chapter. The **TCRB** locus is more complex than **TCRG** and contains multiple V, D, J, and one or a few C segments. DNA recombination is again responsible for the construction of mature β chain genes during T-cell defferentiation. The **TCRA** and **TCRD** loci are situated in closely linked regions of chromosome 14, and some of the V_α and V_δ DNA segments may be intermixed on the chromosome. This close proximity and intermixture of DNA elements must present some interesting problems for the recombination complex during construction of the respective genes!

BLOOD GROUP ANTIGENS

Human erythrocytes possess an array of surface antigens that are determined by multiple allelic systems. More than 250 blood group antigens have been described that are determined by more than 35 loci.

The first blood group system, the ABO system, was discovered by Karl Landsteiner (9) in 1900. This blood group plays a major role in transfusion complications and tissue rejection, especially with reference to kidney transplants. The genotypes and phenotypes characteristic of the ABO system are presented in Table 10-4. The **ABO** locus on chromosome 9 has three alleles which contribute to four phenotypes. Examination of the table reveals that presence of the A and B antigens is dominant to their

Table 10-4. The ABO Blood Group System.

Type	Genotype	Phenotype RBC Antigens	Phenotype Serum Antibodies
A	$ABO^A ABO^A$ or $ABO^A ABO^O$	A	anti-B
B	$ABO^B ABO^B$ or $ABO^B ABO^O$	B	anti-A
AB	$ABO^A ABO^B$	A + B	neither antibody
O	$ABO^O ABO^O$	H	anti-A + anti-B

absence, and that type A and type B are codominant with respect to each other. Blood type A persons possess antibodies against type B in their serum, and type B individuals have anti-A. Type AB has neither antibody, and, conversely, serum from O individuals contains both antibodies.

A and B antigens occur on red cell surfaces as glycolipids, with an extracellular highly branched carbohydrate moiety connected to the lipid portion which anchors the antigens in the cell membrane (Fig. 10-14. The antigenic qualities of the ABO substances are contributed by the penultimate monosaccharides of the carbohydrate chains. Blood type A, B, and O are associated with N-acetylgalactosamine, galactose, and fucose, respectively. These determinants are displayed in Fig. 10-15. Two other antigens, Lewis "a" or Lewis "b", occur in about 80% of the population and are also associated with sugars that are located near the ends of the same carbohydrate chains. The **ABO** alleles encode the structures of **glycosyl transferases** that catalyze the addition of either N-acetylgalactosamine or galactose to the ends of the carbohydrate chain. Type O individuals produce ABO antigens that have a terminal fucose residue. This antigen is designated H antigen.

Lewis antigenic determinants possess either one internal fucose (Lewis "a") or two fucose residues, one corresponding to the fucose of H antigen and one corresponding to the fucose of Lewis "a" antigen. When both fucose residues are present, the antigen is called Lewis "b".

About 80% of European-derived populations possess ABO substances in secretions (saliva, intestinal secretions, ovarian cyst fluid). Unlike the erythrocyte antigens, the soluble blood group substances are large glycoproteins. These antigens have a protein core from which radiate large numbers of oligosaccharide chains. The antigenic determinants of these substances are very similar to those that occur in the erythrocyte antigens. The penultimate sugars determine the antigenicity of these substances, and secreted A and B antigens also have N-acetylgalactosamine and galactose, respectively, at the ends of their carbohydrate chains. Lewis "a" and Lewis "b" determinants also occur in secretions.

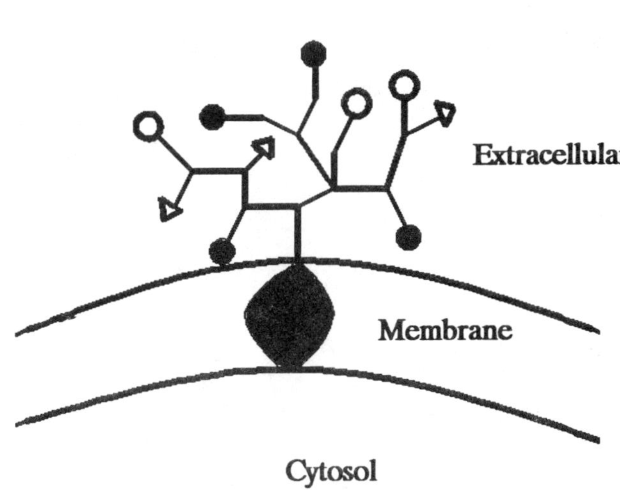

Fig. 10-14. Erythrocyte ABO antigen. The lipid moiety anchors the antigen in the membrane, and the monosaccharides (circles and triangles) at the ends of the carbohydrate chains serve as the antigenic determinants of the molecule.

Epistasis and ABO Phenotypes

Levine et al. (10) described a family in Bombay, India that exhibited an unusual inheritance pattern for the erythrocyte ABO antigens (Fig. 10-16). II-6 produced an AB daughter, indicating that she must have been heterozygous for the **ABO^B** allele, but did not express the B antigen. II-3, II-4, and II-6 all lacked A, B, and H antigens on their erythrocytes. Furthermore, all three persons possessed anti-A, anti-B, and anti-H in their sera. This phenotype was referred to as the Bombay phenotype and represents an example of **epistasis** where alleles at one locus block the expression of genes at a second

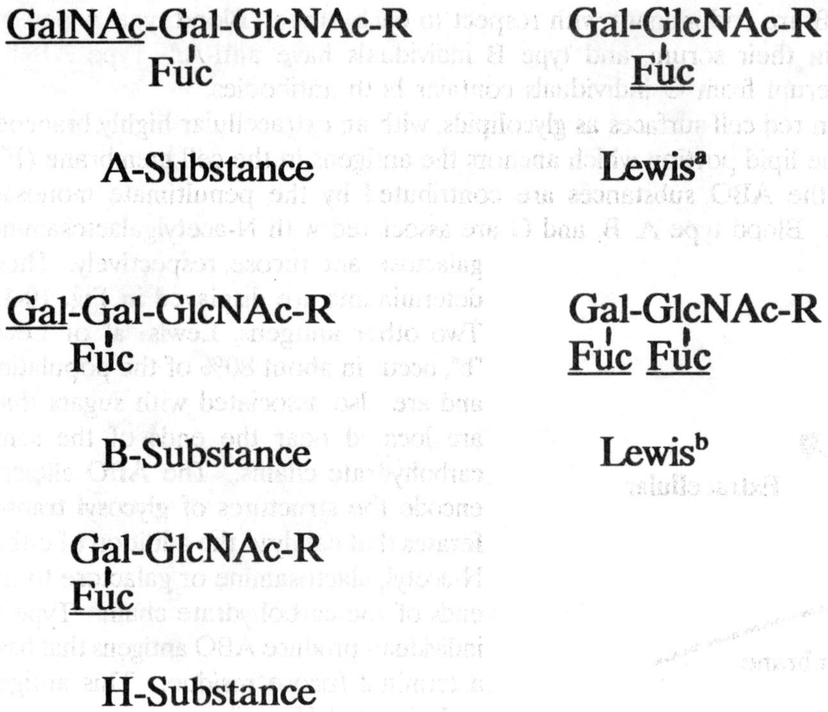

GalNAc-Gal-GlcNAc-R
Fuc
A-Substance
Gal-GlcNAc-R
Fuc
Lewis^a
Gal-Gal-GlcNAc-R
Fuc
B-Substance
Gal-GlcNAc-R
Fuc Fuc
Lewis^b
Gal-GlcNAc-R
Fuc
H-Substance

Fig. 10-15. Partial structures of A, B, H, and Lewis antigens. (GalNAc = N-acetylgalactosamine; Gal = galactose; Fuc = fucose; and R = remainder of molecule.

I
1
2
II
1
2
3
4
5
6
7
III
1
2
AB
A
B
O
Bombay

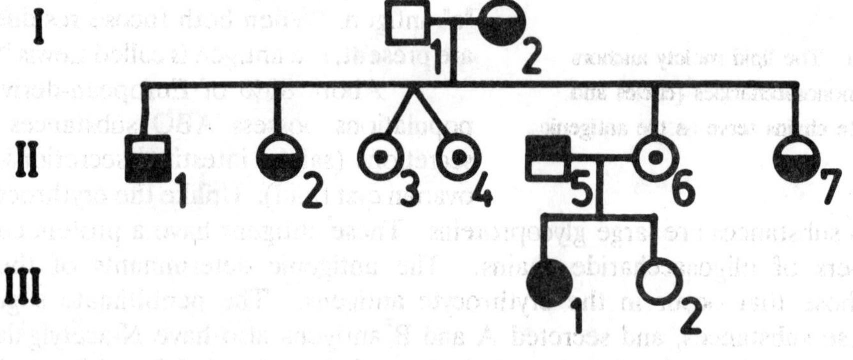

Fig. 10-16. Occurrence of Bombay phenotype in a pedigree (modified from Levine et al. (10)).

Table 10-5. Genotypes of Potential Children of I-1 and I-2 (Fig. 10-16).

Genotype	Phenotype	Proportion
I-1: ABO^0ABO^0Hh	O	--
I-2: $ABO^BABO^?Hh$	B	--
Children		
ABO^BABO^0H-	B	3/8
$ABO^?ABO^0H-$	? = B or O	3/8
ABO^BABO^0hh	Bombay	1/8
$ABO^?ABO^0hh$	Bombay	1/8

locus. Interpretation of this pedigree reveals that I-1 and I-2 were both heterozygous at the H locus (**Hh**), and that II-3, II-4, and II-6 were all **hh**. The possible genotypes of the children conceived by I-2 are provided in Table 10-5. One quarter of the children would be anticipated to lack B and H antigens and to have anti-A, anti-B, and anti-H. The ABO genotype of I-2 is unknown; however, the occurrence of the ABO^B allele in at least four of her children suggests that she may have been homozygous for this allele.

The **Secretor** locus determines the presence or absence of ABO antigens in secretions. Persons who are **sese** lack these secreted substances and may possess Lewis[a] in their saliva. People who have at least one **Se** and one **Le** allele not only are capable of secreting ABO antigens, but also secrete Lewis[b] antigen. Watkins (11) and Oriol et al. (12,13) have developed a biochemical pathway for the production of erythrocyte and secreted blood

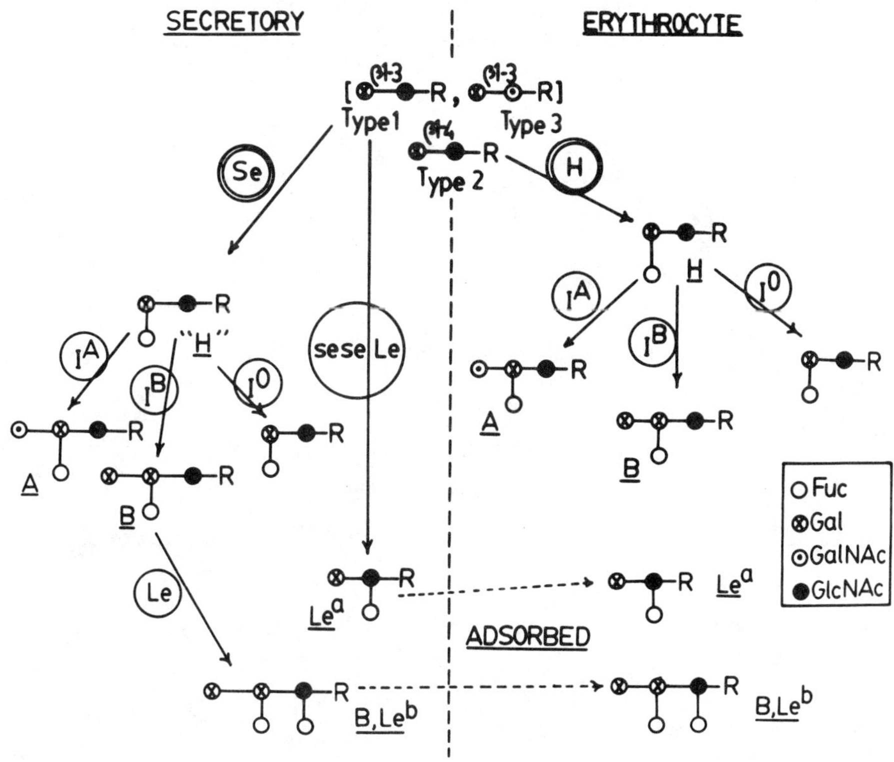

Fig. 10-17. Formation of A, B, H, Le[a], and Le[b]. The Type 1 and 3 precursors flow into the secretory pathway, and the Type 2 precursor enters the erythrocyte pathway. The appropriate precursor is converted to H (RBC) or an H-like substance (secretions) by glycosyltransferases encoded by the H and Se loci, respectively. H and H-like antigens are then converted to A or B by the glycosyltransferases encoded by the ABO^A or ABO^B alleles, respectively. The Lewis glycosyltransferase can add fucose to either the original Type 1 or 3 precursor (in sese persons) or to A, B, or H-like antigens produced by secretors. The resulting Le[a] or Le[b] antigenic determinants are then adsorbed to erythrocytes.

group substances which is summarized in Fig. 10-17. Several precursor-like materials have been isolated from a variety of different sources and characterized. These substances may be apportioned among two groups, depending upon the type of bond linking the last two monosaccharides in the carbohydrate chain. Precursors that possess a β1-3 linkage appear to flow into the secretory pathway where they serve as a substrate for the fucosyltransferase specified by the **Se** locus. The product of this reaction closely resembles the H antigen, differing with respect to the β1-3 bond and with respect to the fact that the oligosaccharide chains are linked to protein. This H-like material serves as a substrate for the glycosyltransferases specified by the **ABO** locus, forming the secreted A and B antigens. The **Le** locus also encodes a fucosyltransferase; however, this enzyme adds fucose at an internal position on the chain. The Le-encoded enzyme displays some flexibility in terms of the type of substrates it will accept. In **sese** persons, the precursors temporarily accumulate as a consequence of the metabolic block produced by lack of the enzyme encoded by the Se allele. These precursors are then converted to Le^a by the Le-encoded enzyme. When an individual's genotype includes at least one **Se** and one **Le** allele, antigens flow through the pathway, with the Le-encoded enzyme acting last in the sequence to form Le^b. Antigens containing Le^a or Le^b determinants are not made by erythrocytes, but are adsorbed to erythrocyte surfaces following their manufacture in the secretory compartment. Notice how the recessive mutation, **se**, acts in an epistatic manner in homzygous state to block expression of A and B antigens in the secretions of individuals possessing at least one **ABO^A** and/or one **ABO^B** allele. It is also important to emphasize that the Le^b antigen represents the joint effects of the actions of the Se and Le loci.

Precursors containing a β1-4 bond linking the last two sugars in their oligosaccharide chains preferentially enter the erythrocyte pathway, serving as substrates for the fucosyltransferase encoded by the **H** locus. The product, H substance, serves as the substrate for the glycosyltransferases encoded by

Table 10-6. Phenotypes of Selected Allelic Combinations at the ABO, H, Le, and Se Loci.

Genotype	Phenotype						
	Secretions			Erythrocytes			
	AB	Le^a	Le^b	AB	H	Le^a	Le^b
ABO^{A}ABO^{B}HhLeleSese	+	-	+	+	-	-	+
ABO^{O}ABO^{O}HhLeleSese	-	-	+	-	+	-	+
ABO^{A}ABO^{B}hhLeleSese	+	-	+	-	-	-	+ (Bombay)
ABO^{A}ABO^{B}HhleleSese	+	-	-	+	-	-	-
ABO^{A}ABO^{B}HhLelesese	-	+	-	+	-	+	- (nonsecretor)
ABO^{A}ABO^{B}hhLelesese	-	+	-	-	-	+	- (Bombay/ nonsecretor)
ABO^{A}ABO^{B}hhlelesese	-	-	-	-	-	-	-

* (+) = antigen present; (-) = antigen weak or absent.

ABO^A and **ABO^B**. Homozygosity for **h** results in a lack of the fucosyltransferase activity necessary for the formation of the substrate for the **ABO** genes, producing the Bombay phenotype. Genotypes illustrating the interactions among these four loci are presented in Table 10-6. Notice how metabolites (blood group antigens) tend to flow to the end of the respective pathways when all of the necessary enzymes are present, and that H antigen and Le^a antigen only accumulate when the **ABO**- or **Se**-encoded activities, respectively, are absent.

Considerable progress has been made with respect to chacterization of the enzymes encoded by these blood group genes and the mutations that compromise their action. Yamamoto et al. (14) have sequenced the three **ABO** alleles. Four base substitutions appear to account for the different catalytic activities of the enzymes encoded by the **ABO^A** and **ABO^B** alleles. The **ABO^O** allele contains a single base deletion, which results in a frameshift and loss of the enzyme. The **Le, Se,** and **H** genes all encode fucosyltransferases and have been mapped to the long arm of chromosome 19. The three alleles apparently arose by duplications of an ancestral gene and retain similar but distinct functions.

The Rhesus (Rh) Blood Group

There has been considerable controversy during the past forty years regarding genetic determination of the Rh antigen. Two general phenotypes are observed in human populations: Rh(+) (about 85% of most populations) and Rh(-). Unlike the ABO system, Rh(-) persons usually do not have anti-Rh antibodies; however, such antibodies can be induced by transfusion of an Rh(-) individual with Rh(+) blood or the occurrence of an Rh(+) pregnancy in an Rh(-) woman. In the latter case, the Rh(+) fetal blood cells cross the placenta, enter the maternal circulation and sensitize her B-cells to synthesize and secrete anti-Rh. The first Rh(+) infant is born before the levels of anti-Rh become critical; however, a second Rh(+) pregnancy is at risk for fetal damage caused by anti-Rh antibodies that cross the placenta. The collection of problems is called erythroblastosis fetalis and occurs in about one out of every six unprotected pregnancies experienced by Rh(-) women. Some natural protection is obtained from a "double" incompatibility such as the marriage of a blood type O(-) woman to a type A(+) man. If the fetus proves A(+), the fetal cells entering the mother's circulation will be destroyed by her anti-A antibodies before they can sensitize her to produce anti-Rh. A more reliable protection can be obtained by administering Rhogam, an anti-Rh preparation, to an Rh(-) woman during each of her pregnancies. The Rhogam destroys the fetal cells entering her circulation before sensitization can occur.

The Rh system was initially defined by Landsteiner, Wiener, and others in terms of a single locus with two alleles with Rh(+) dominant to Rh(-). While this explanation still serves us well in most clinical counseling situations, the actual genetics of the system is more complex. Fisher and Race proposed an alternative theory in which three tightly linked loci, C, D, and E, and their respective alleles determined the large number of Rh phenotypes that were soon discovered. Wiener and his associates later proposed a two-locus, multiple-allelic system to explain the inheritance of the complicated phenotypes that were observed. Biochemical studies of Rh(+) cells have revealed the presence of an integral membrane protein with a relative molecular weight of 32,000 (15), and that Rh(+) and Rh(-) cells contain membrane proteins that differ with respect to portions of their amino acid sequences, or that there are a group of very closely related proteins that possess the respective Rh specificities. Further work indicates that there may be at least three distinct and closely related Rh polypeptides that have a common N-terminal sequence. Furthermore, proteins (M_r 37,000 - 47,000) that possess MN blood group specificity (these proteins are encoded by the **LW** locus which is closely linked to the **H-Se-Le** gene cluster on chromosome 19) are closely associated with the Rh antigens in the red cell membrane. The C-termini of the **RH**- and **LW**-encoded proteins are both oriented to the exterior of the cell and presumably contain the antigenic determinants recognized by anti-Rh and anti-LW sera (16). Current data indicate that the differences between both the respective Rh antigens and the LW antigens are related to amino acid sequence and not carbohydrate.

Blood Groups Genetics and Medicolegal Applications

The inheritance of several blood groups is presented in Table 10-7. Since these systems are polymorphic, they can be used to assess parentage and as evidence in criminal proceedings. Suppose parentage of an infant is being disputed following an alleged interchange of two infants in a nursery. One of the infants has subsequently contracted a serious autosomal recessive disease, and the couple who took the child home are claiming the infant is not theirs. The couple and the infant are typed:

Table 10-7. Genetics of Several Medically Relevant Blood Group Systems.

Blood Group	Inheritance	Dominance Relations
ABO	Autosomal (9q)	A=B>O
Rh	Autosomal (1p)	Rh(+)>Rh(-)
Le	Autosomal (19q)	(+)>(-)
H	Autosomal (19q)	(+)>(-)
Se	Autosomal (19q)	Secretor> Nonsecretor
GYPA	Autosomal (4q)	M=N
Xg^a	X-linked	(+)>(-)

= codominant with; > dominant to

	ABO	Rh	Le	GYPA	Xg
Baby boy	O	-	b	MN	+
Woman	A	+	b	M	-
Man	B	+	b	N	+

There is agreement for the first four systems (assuming heterozygosity for the **ABO** and **Rh** systems in the man and woman); however, the woman is excluded as the mother on the basis of the Xg^a phenotype. Note that parentage can be excluded on the basis of blood types, but not proven by such data. There is always a possibility that a match can be demonstrated between two people by chance, particularly when only five loci are used. In practice, more than twenty loci would be tested to assess the relationship between individuals or between a biological specimen obtained at a crime scene and the suspect. The judge or jury must decide the case on the basis of an interpretation of the test results provided by an expert.

More recently DNA analyses have been used in a similar manner to demonstrate relationship, because the degree of polymorphism is much higher, and the DNA "fingerprints" of a person are almost unique to that individual. PCR can be used to amplify DNA from a single hair taken from the victim or from a small speck of blood, and the DNA is digested with one or more restriction enzymes. The resulting fragments are separated on a gel and probed with a cDNA that recognizes the core sequence of a DNA that displays polymorphic repeat differences among homologous chromosomes of different people (Fig. 10-18). The Southern blot of the specimen is then compared with that of the victim and that of the suspect. When the test is conducted appropriately, when the DNA is undegraded (this can be checked by running undigested DNA on the gel - native DNA stays at the origin), and when the patterns are correctly interpreted (care must be taken not to misinterpret variant patterns that are caused by gel distortion) a high degree of confidence can be achieved with respect to obtaining a conviction or acquittal of the suspect. Enforcement agencies are currently training forensic laboratory technicians and establishing centers for DNA analyses. It is anticipated that DNA analyses will become a standard component of evidence used during criminal proceedings in the near future.

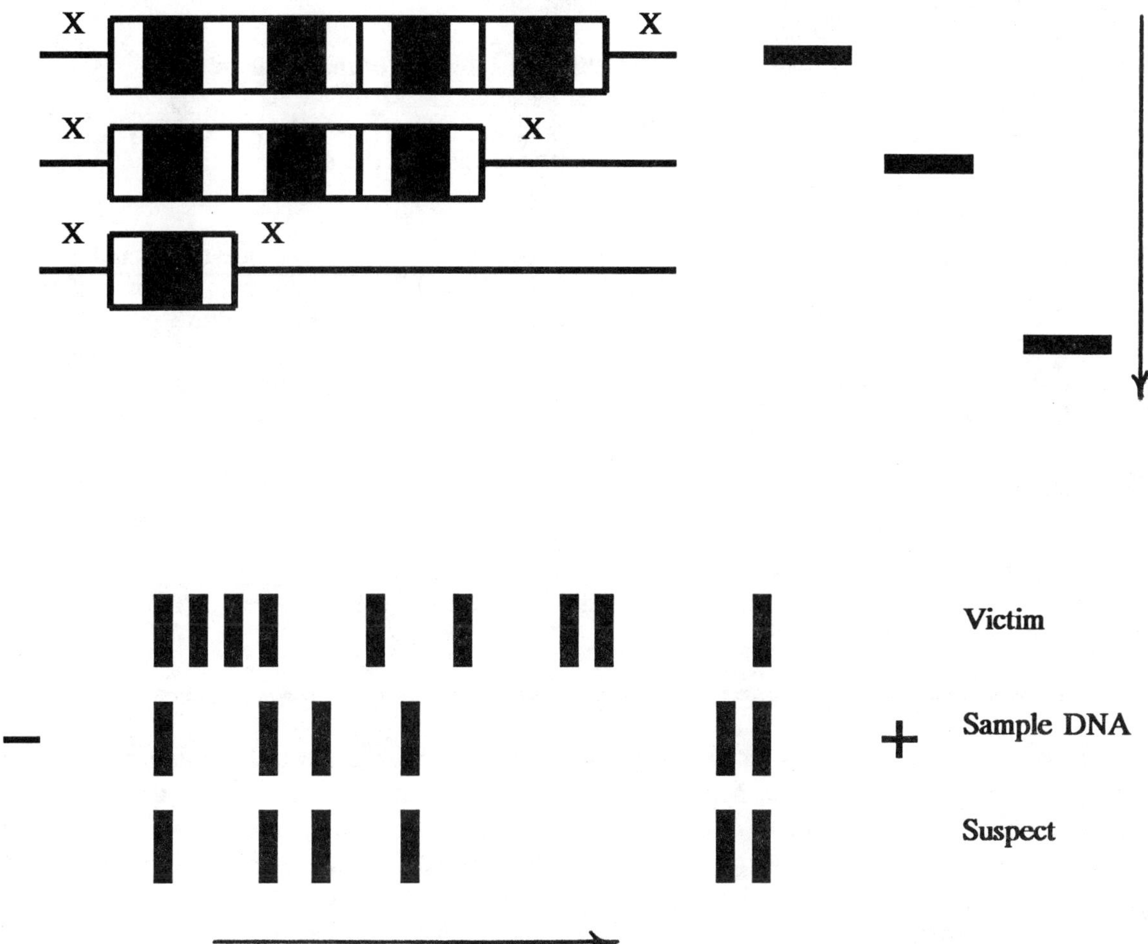

Fig. 10-18. DNA fingerprinting. Top: Three homologous chromosomes from a population that differ with respect to the number of repetitive elements between two restriction sites (X). When the chromosomes are digested with the restriction enzyme, and the fragments are separated by electrophoresis and probed with a DNA sequence complementary to the core sequence of each element (shaded box), three different bands would be observed, with the fragment containing the single element migrating farthest in the gel. Arrow indicates the direction of migration. Bottom: DNA fingerprints from the victim of a crime, a biological specimen, and the suspect. The pattern from the suspect closely matches that of the specimen, implicating the suspect as the perpetrator of the crime. Arrow indicates the direction of migration.

GENES AND ANTIBODIES

Antibodies occur in five general classes (Table 10-8) which vary with respect to their subunit composition, distribution and function. Antibodies typically contain two light chains and two heavy chains (Fig. 10-19). The antibody pictured in the figure is IgG, a predominant antibody involved with the secondary response. Secreted IgM (Fig. 10-20) is released during the primary response, is much larger, and consists of five IgM units connected by double bonds and a joining protein. Antibody chains are characteristically divided into domains (Fig. 10-21). These domains share similar amino acid sequences,

Table 10-8. Antibody Structure and Function.

Class	Composition	Distribution	Function
IgA	$(\lambda_2\alpha_2)_n$* $(\kappa_2\alpha_2)_n$	Secretions	Prevents infection of intestinal cells by gut bacteria.
IgD	$(\lambda_2\delta_2)$ $(\kappa_2\delta_2)$	Serum	May participate in the differentiation of B-cells.
IgE	$(\lambda_2\epsilon_2)$ $(\kappa_2\epsilon_2)$	Serum	Contains reagin activity; fixes mast cells and promotes their release of histamine; increased in asthma and eczema; combats parasitic infections.
IgG	$(\lambda_2\gamma_2)$ $(\kappa_2\gamma_2)$	Serum Extravascular spaces	Neutralizes bacterial toxins; assists phagocytosis of microorganisms; produces cell lysis with complement.
IgM	$(\lambda_2\mu_2)_n$ $(\kappa_2\mu_2)_n$	Serum	Agglutinates and lyses bacteria; receptor for virgin B-cells.

* n = 2 or 3 for IgA; n = 1 for receptor IgM and 5 for secreted IgM. Secreted IgA and IgM complexes contain a joining protein.

suggesting that they originated by a series of duplications of an ancestral sequence. Two of these domains occur in light chains, while heavy chains contain four (α,γ) or five (μ,ϵ) domains. Both light and heavy chains possess a variable domain near the N-terminal end, portions of which participate in antigen-antibody contacts. Each variable region exhibits three highly variable zones (**complementarity determining regions, cdr**) which display a high degree of amino acid variation, which represent antigen contact sites, and which are interspersed with more conservative areas known as **framework regions** (Fig. 10-22). **Allotypic** variation among antibody chains is also observed (Table 10-9). This variation occurs as a consequence of amino acid substitutions in the **constant region** of antibody chains. For example, The **Km1** allotype has leucine at position 191 of the kappa light chain, while **Km2** contains valine at that position. Although these polymorphic substitutions do not affect the function of the antibody, they have proved valuable for population studies and also led to the discovery of a fundamental property of antibody-producing cells: **allelic exclusion**. In B-cells or plasma cells that are heterozygous for an allotypic marker (**Km1/Km2**), only one of the two possible chain types are synthesized.

Genetics of Antibody Expression

A typical individual develops the potential to produce from 10^6 to 10^7 different antibody specificities. Furthermore, the antibody armamentarium is in place before encounter with antigens occurs. The diversity of antibodies produced by a person is obtained by several mechanisms: 1) presence of multiple germline elements; 2) DNA rearrangement to place the elements into a functional gene construct; 3) junctional diversity; 4) combinatorial variability; and 5) antigen-driven somatic mutation. Further flexibility is obtained by the production of polyclonal antibodies during the immune response.

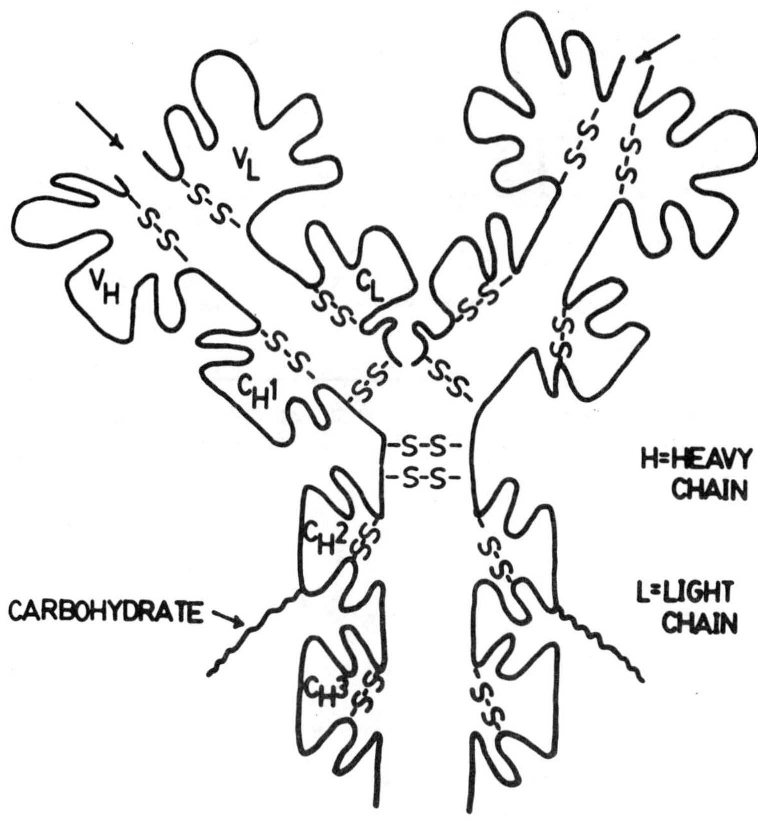

Fig. 10-19. Structure of a typical antibody from the IgG class. Arrows indicate antigen binding sites. Notice that regions of heavy and light chains (V_H and V_L, C_L and C_{H1}, C_{H2}, and C_{H3}) form domains or compact globular zones within the antibody molecule. Each antibody molecule can bind two antigens, and each molecule has flexibility in the "hinge region" (place where the heavy chains are joined by the paired disulfide bonds).

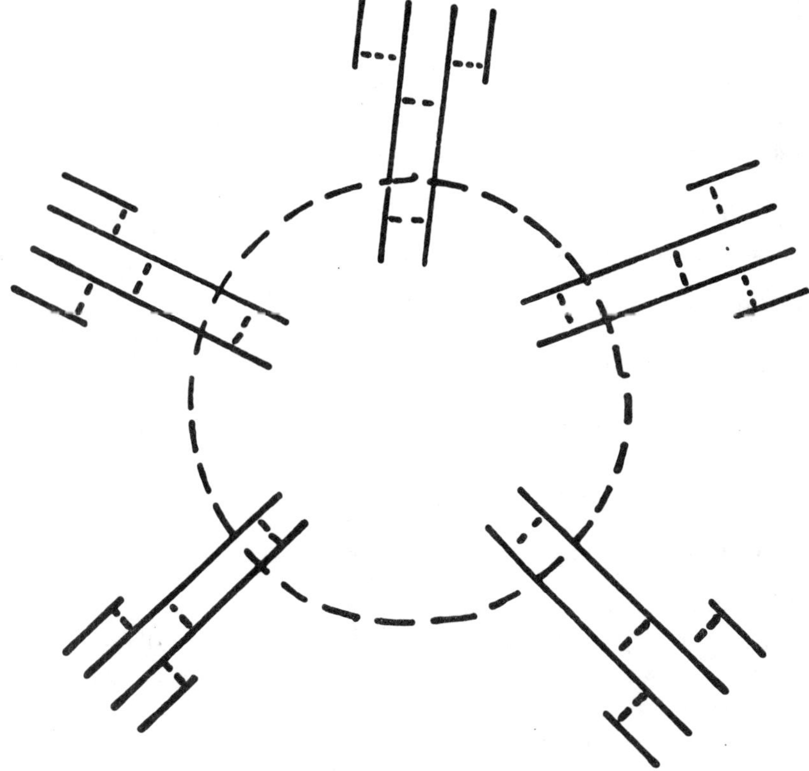

Fig. 10-20. Schematic diagram of an IgM molecule. Notice how the antibody combining regions extend toward the periphery of the molecule. Dashed lines represent disulfide bridges. A joining protein also links the IgM components.

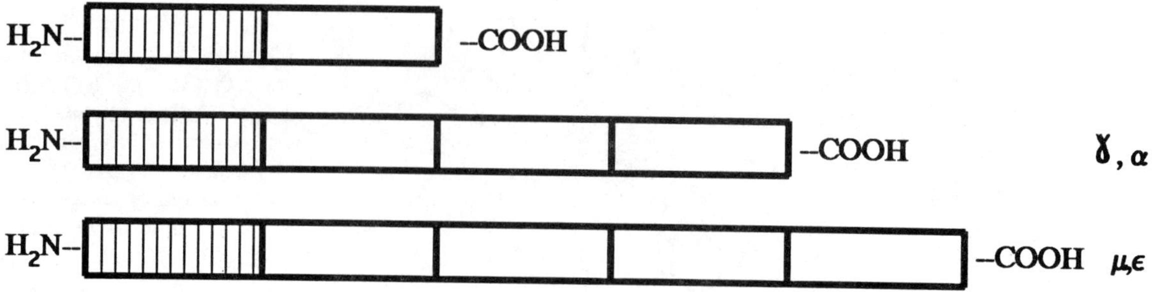

Fig. 10-21. Homology regions (domains) occurring in light and heavy chains of immunoglobulin molecules. Each domain encompasses approximately 110 amino acids. The variable domains are shaded, and the constant domains are represented by open boxes.

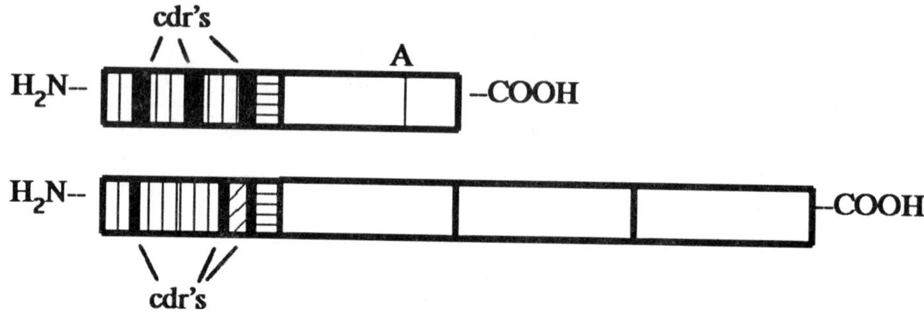

Fig. 10-22. Light and heavy chain of IgG. Variable (V) segment framework regions, the diversity (D) segment, and the joining (J) segment are represented by verticle, diagonal or horizontal hatching, respectively. cdr's occur one or at two locations in the variable segment and at the V-D and D-J interfaces and are indicated by thick vertical lines. A = kappa chain allotypic site.

Table 10-9. Allotypic Immunoglobulin Variants.

Locus	Chain	# of Alleles
Gm	Gamma	>24
Am	Alpha	>2
Km	Kappa	>3

Each of these aspects of immune function will be discussed below.

Antibody gene segments are clustered in three chromosome regions. The kappa light chain variable (V), joining (J) and constant (C) segments are located on chromosome 2 (Fig. 10-23). The corresponding lambda light chain elements are found on chromosome 22, and the heavy chain segments are situated on chromosome 14. The lambda region appears to contain greater numbers of elements than the kappa region. Multiple heavy chain constant region genes are present on chromosome 14, and all share the same V, D, and J segments. The germline configuration of the gene segments on all three chromosomes are clustered according to type. The V, D, J and C clusters tend to be widely separated from one another. Each chromosome region contains an enhancer element between the J cluster and the C segments. This enhancer element is activated in the B-cell lineage, resulting in a high level of transcription of appropriately assembled antibody gene constructs.

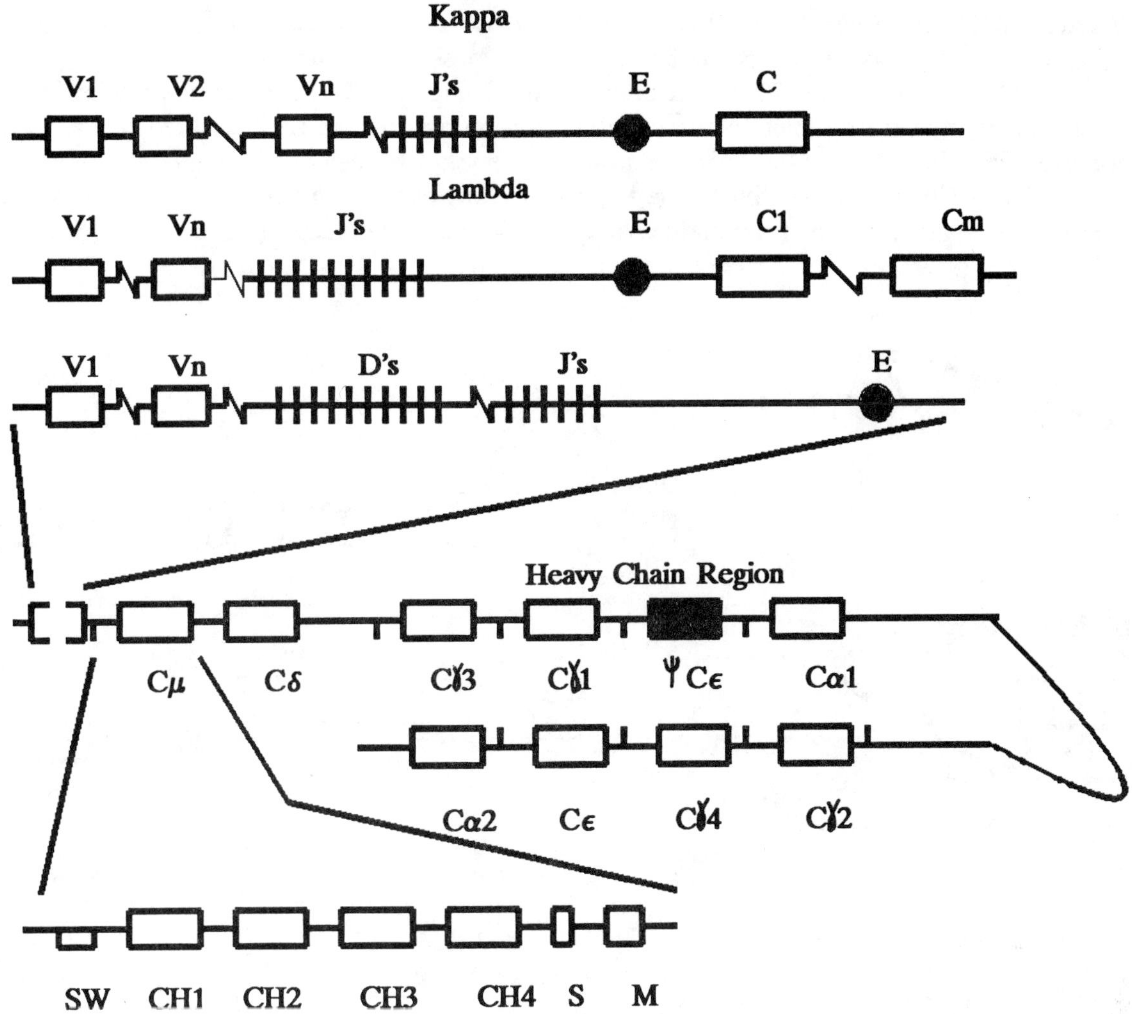

Fig. 10-23. Linkage groups of immunoglobulin gene elements. The elements are displayed in their germline configuration. Kappa cluster (chromosome 2): V = variable segment; J = joining segment; E = enhancer; C = constant region segment. For the kappa V segments, n is about 50-300. It is unknown how many of these V segments are pseudogenes. Lambda cluster: n approximates 200 and m is about 50. At least six of the C segments are functional. Heavy chain region: The region containing the V, diversity (D) and J segments is enlarged above the chromosome, and the Cμ region is enlarged below the chromosome. The number of heavy chain variable segments is believed to range from 500 to 1000. The C regions have several exons and are arranged in a doublet (Cμ-Cδ) and two repeated quartets. The Cε region in the first quartet is not expressed. All C regions excepting Cδ are preceded by a switch sequence (SW). The Cμ region possesses six exons. The S exon encodes a C-terminal segment that appears in the secreted form of IgM, while the M exon encodes a hydrophobic sequence that anchors the receptor form of IgM in the membrane. All C region genes possess S and M exons.

Functional antibody heavy and light chain genes are constructed by DNA recombination within the same chromosome. A gene element is selected from each cluster and joined to form the construct by a recombinase complex. Kappa gene construction is illustrated in Fig. 10-24. A V segment is brought into close proximity with a J segment by the recombinase complex. Each V and J segment is flanked by highly conserved short sequences that are separated by larger unconserved stretches of DNA. For example, the downstream sequence of a V segment contains a highly conserved 7 nucleotide sequence (CACGGTG) that is separated from a second conserved sequence of 9 nucleotides ($AC(A)_5CC$) by 11-12 base pairs. The 9 nucleotide sequence is separated from a highly conserved 9 nucleotide sequence

(**GG(T)$_5$GT**) that is upstream from a J segment by the DNA to be excised during recombination. This conserved 9 base sequence is separated from a highly conserved 7 nucleotide sequence (**CACTGTG**) adjacent to the J segment by 23 nucleotides.   The complementarity of the stretches of 7 and 9 nucleotides suggests that they pair to form a stem and loop structure as illustrated in the figure. The recombinase nicks the DNA strand adjacent to each segment and ligates the segments together to form a VJ pair. This VJ pair is connected to the kappa constant region segment by an intron containing the kappa enhancer element, forming the mature kappa chain gene. The action of the recombinase produces

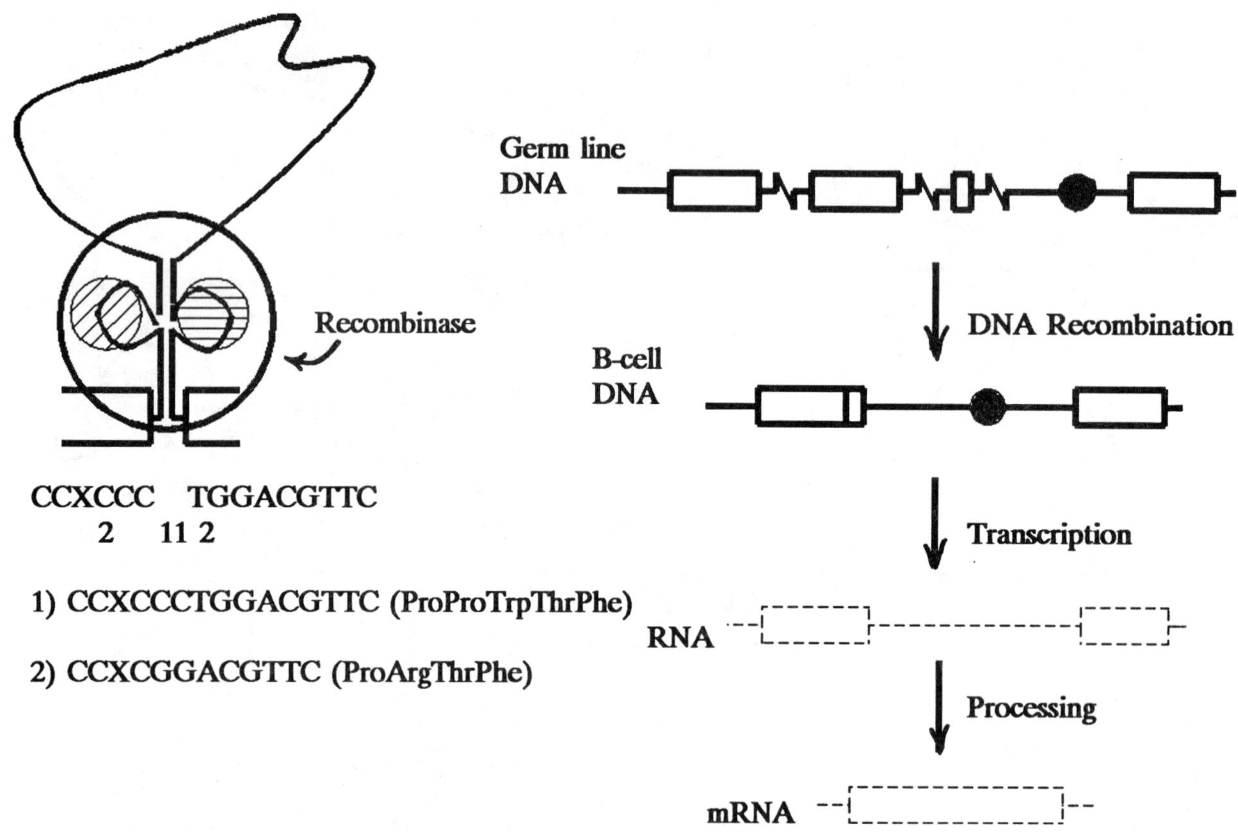

Fig. 10-24.   Generation of a mature kappa chain gene by DNA recombination. Left: The stem and loop structure between a V segment and a J segment is stabilized by DNA binding proteins (hatched circles), and the recombinase cleaves the DNA at the intron-exon boundaries, forming the VJ segment. If cleavage occurs at site 1, the junction sequence would be as indicated (1). Cleavage at site 2 would result in the deletion of three bases and the second sequence. This "junctional diversity" contributes to the hypervariability at the cdr site formed at the V-J junction. Right: Overall view of the process. DNA recombination forms a VJ doublet from particular V and J segments and moves the doublet closer to the C segment. The mature gene consists of the VJ exon and the C exon separated by an intron containing the kappa enhancer element. This gene is transcribed into a nascent RNA molecule containing the intron and subsequently processed to form the mature mRNA.

an in phase junction about one-third of the time. Therefore there is a 67% chance that a functional kappa chain gene will be formed from one of the two homologous chromosomes. Some of the in phase constructs display gain or loss of aminoacids at the V-J junction. This **junctional diversity** contributes to one of the cdr sites in the variable portion of the kappa chain. This example illustrates two of the sources of antibody diversity. If there are about 100 functional kappa V segments, 6 J segments, and 1 C segment, a total of 600 different kappa chain genes could be formed by DNA recombination. Furthermore, V-J junctional diversity increases variability at the third cdr site of the kappa chain variable region. Lambda chain gene elements are rearranged in similar fashion; however, there are more lambda gene segments. If there are 200 functional lambda V elements, 10 J segments, and at least six functional C segments, about 12,000 different lamda chain gene constructs are possible. Junctional diversity is also superimposed upon the variability created by lambda element recombination. Estimates of both the kappa and lambda functional genes may be somewhat high, since preferential use of V and J segments that are close together, rather than widely separated elements, for recombination has been demonstrated.

   Recombination among heavy chain DNA segments follows a pattern similar to that for the light chain gene elements. D-J recombination appears to occur first, followed by recombination between a DJ segment and a heavy chain V element. The first functional heavy chain gene construct consists of a particular VDJ complex and an intron (containing the heavy chain enhancer element) that connects

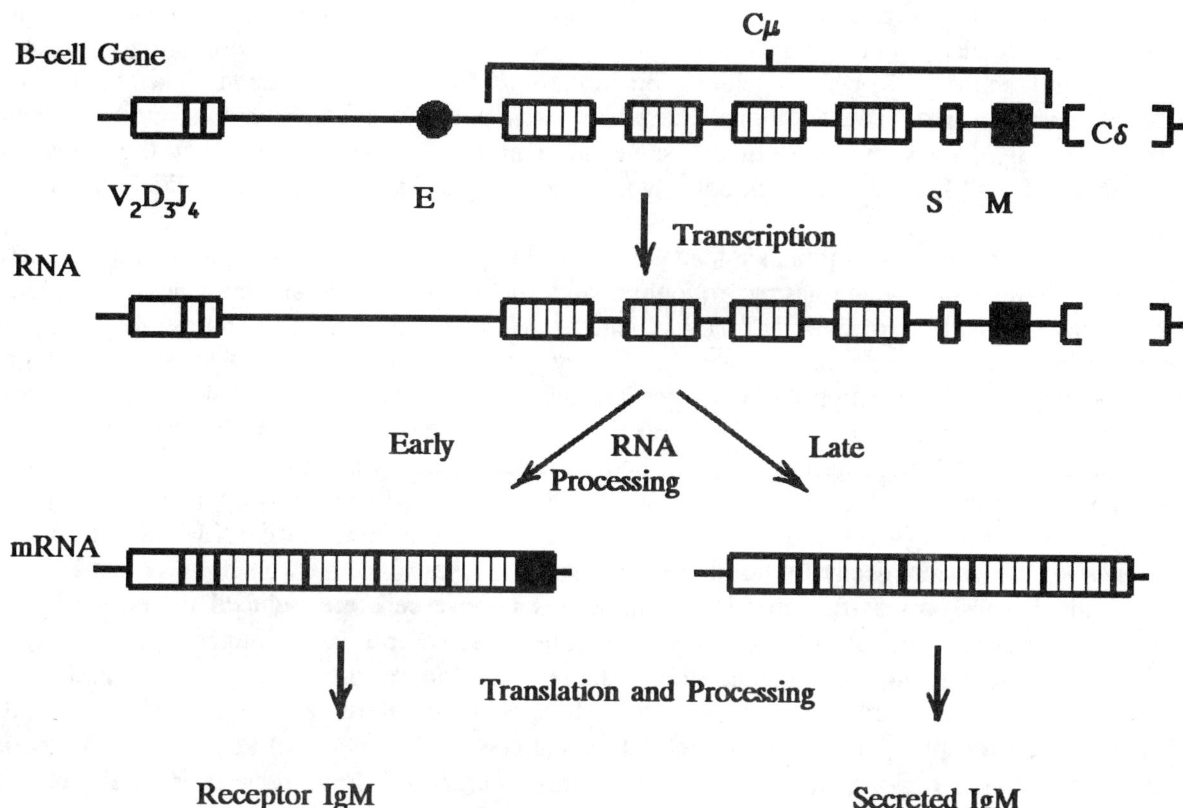

Fig. 10-25. Processing of heavy chain RNA transcripts. The functional heavy chain gene occurring in virgin B-cells (B-cells unexposed to antigen) consists of a VDJ element connected via an intron to both the Cμ (expanded to show exons) and Cδ regions. This gene is transcribed into a large nascent RNA containing both the Cμ and Cδ constant regions. Processing in the virgin B-cell follows the early path forming an mRNA containing the M exon. This mRNA is translated into a heavy chain that becomes part of receptor IgM. Following antigen exposure, the late RNA processing path is followed, producing mRNA containing the S exon. This mRNA is processed into an IgM heavy chain that appears in secreted IgM. Alternative processing pathways may form receptor IgD or secreted IgD from the same nascent RNA molecule.

the VDJ complex to the μ constant region and δ constant region exons (Fig. 10-25). Sequencing studies have revealed relatively close association between the μ and δ constant region exon clusters: J_H - 6.5kb - $C\mu$ - 4.5kb - $C\delta$ - >50kb - remainder of constant region exons (kb - 1000 base pairs). All V, D, and J segments are flanked by conserved sequences that contribute to stem and loop structures which are excised during recombination. Junctional diversity at V-D and D-J junctions form the second and third cdr sites within the variable domain of the heavy chain polypeptide. If there are 500 heavy chain V segments, 10 D segments, 10 J segments, and nine constant region exon clusters, a total of 450,000 different heavy chain genes are possible. **Combinatorial** variability refers to the total antibody molecule and is determined by which particular light chain and which heavy chain associate to form the antibody. An individual has the capability of producing 600 x 450,000 or 27 x 10^7 different kappa-containing antibodies and 12,000 x 450,000 or 540 x 10^7 distinct lambda-containing antibodies during his or her lifetime.

The close proximity of the $C\mu$ and $C\delta$ exon clusters leads to their cotranscription into the same nascent mRNA molecule. This large RNA enters one of four processing pathways. In virgin B-cells, cells unexposed to antigen, the RNA is processed into a mRNA containing the VDJ and all $C\mu$ exons except the S exon (path 1, Fig. 10-25). This mRNA will be translated into a μ heavy chain containing a hydrophobic membrane sequence which is assembled into receptor IgM. Following antigen exposure, the processing path shifts to (2) which excludes the M exon and includes the S exon. This chain will be assembled into secreted IgM. Two additional RNA processing paths form heavy chains found in receptor IgD and secreted IgD, respectively, from the same large nascent RNA. Notice that the **same VDJ complex is shared** by all four of these antibody types, imparting the **same antigen specificity** to all four antibodies.

Each variable domain contains a leader sequence (signal peptide) which targets the respective heavy or light chain to the endoplasmic reticulum and the Golgi processing pathway. This leader sequence is cleaved from the polypeptide following entry into the ER lumen. The chains are glycosylated and assembled into the mature antibody molecules as they pass through the Golgi. Antibodies containing heavy chains possessing the hydrophobic sequence are inserted into the membrane with their N-termini exposed to the cell exterior where they serve as Ig receptors, participating in the internalization of antigen for processing and presentation to T-cell helpers. Antibodies containing the secretory sequence are secreted and combine with extracellular antigen. Both secretory IgM and IgA complex with a joining protein and other IgM or IgA molecules, respectively, to form aggregates. Five IgM and two or three IgA molecules are characteristically found in these large complexes.

Virgin B-cells possess receptor IgM, and some of these cells are induced to secrete IgM by exposure to appropriate antigen (antigen that contains a structural component (**epitope**) that is structurally complementary to the antigen-binding domains of the receptor IgM). This fundamental difference between HLA, T-cell receptor, and antibody specificities must be emphasized. **Both HLA molecules and T-cell receptor molecules interact with small peptide fragments of an antigen. Antibodies recognize larger antigen regions, and three dimensional structure of these regions is important with respect to antigen-antibody interactions.** The primary humoral immune response culminates in the secretion of IgM complexes. Antigen exposure of these B-cells also induces a proportion of them to undergo a **class switch**, wherein the heavy chain class of the receptor and secreted antibody changes from IgM to IgG, IgA, or IgE. In each case the second antibody type possesses **the same VDJ region and recognizes and binds to the same antigen recognized by the original IgM.** The mechanism for class switching is illustrated in Fig. 10-26. The $C\mu$, C , Cϵ and Cα exon clusters are preceded by short switch sequences. After antigen exposure, the chromosome region forms a loop in which the switch sequence preceding the $C\mu$ exon cluster pairs with the switch sequence preceding another constant region exon cluster. A recombinase cleaves the DNA in the paired region and ligates the resulting molecules such that a new functional gene construct containing the **same** VDJ complex is ligated to another constant region exon cluster, and a ring containing the intervening DNA is formed. The ring of DNA is degraded; therefore, once a class switch has occurred in a clone of cells, that clone cannot return to production of IgM. Furthermore, a specific clone of antibody-synthesizing cells can undergo a class switch more than once.

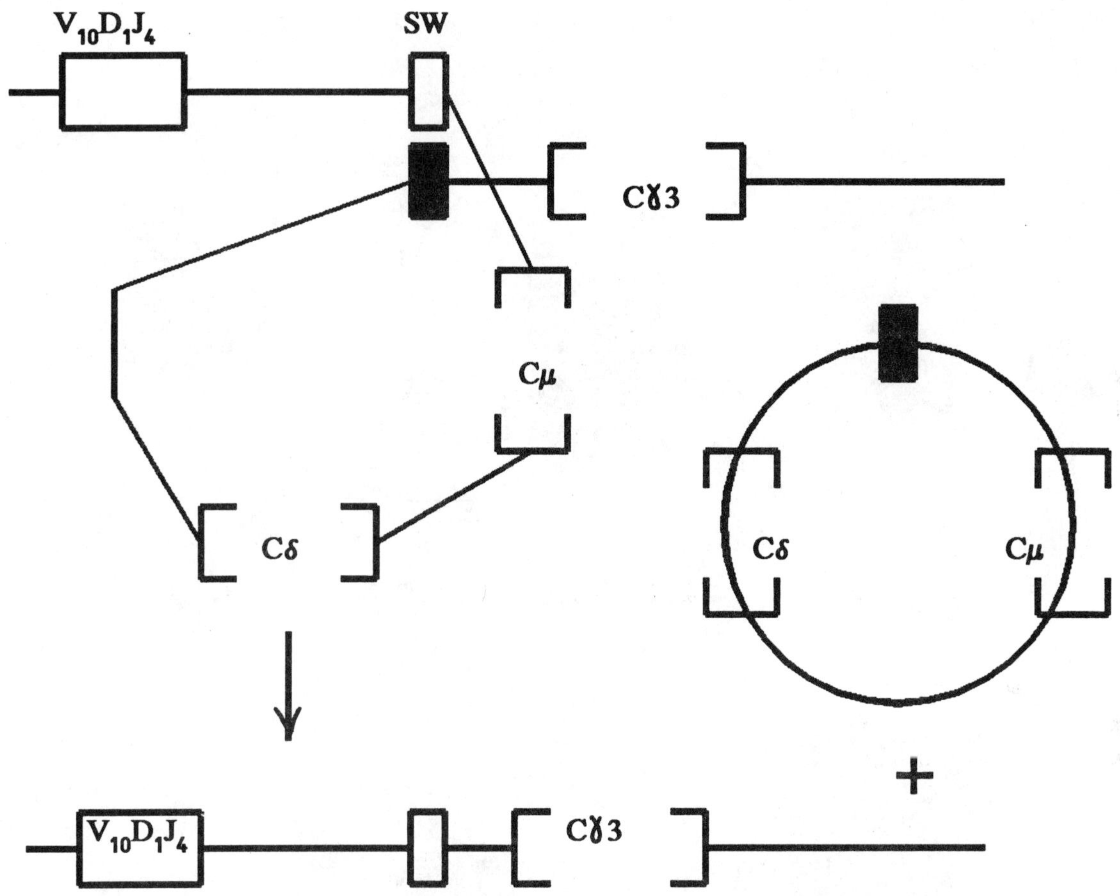

Fig. 10-26. Class switch mechanism. Switch regions preceding the prior expressed constant region exon cluster pairs with the switch region preceding the constant region exon cluster that will be expressed, forming a loop. The variable domain is transferred by recombination to a point proximal to the new constant region exon cluster, and a loop of intervening DNA is formed. The loop is degraded. The new antibody has the same antigen specificity as the old antibody.

Antigen exposure also initiates a relatively high rate of somatic mutation (1/1000) within the framework regions of the heavy chain variable domains. The mechanism by which this occurs is unknown; however, clones of cells begin to produce antibodies of the same heavy chain class that have differing affinities for the antigen. The clones having the highest antigen affinities (greatest efficiency of antigen-antibody interaction) are positively selected, such that they begin to replace the other clones of cells producing antibody against that antigen. Thus, antigen driven **somatic mutation** further enhances the ability of the immune system to combat foreign antigens.

Lymph nodes and other lymphoid tissues have been seeded with millions of B-cells, each with a different antigen specificity determined by the antigen-binding sites of the variable domains of its receptor IgM. A group of these B-cells may be directed toward the same antigen, but to different epitopes on the surface of that antigen (Fig. 10-27). Therefore, an antigen may stimulate several different B-cells, inducing them to synthesize and secrete a population of antibodies each directed against a different site of the antigen. The ability of the individual to eliminate or neutralize an antigen is determined by the particular array of antibodies he or she is able to produce. This aspect of the humoral immune response is referred to as the **polyclonal** immune response.

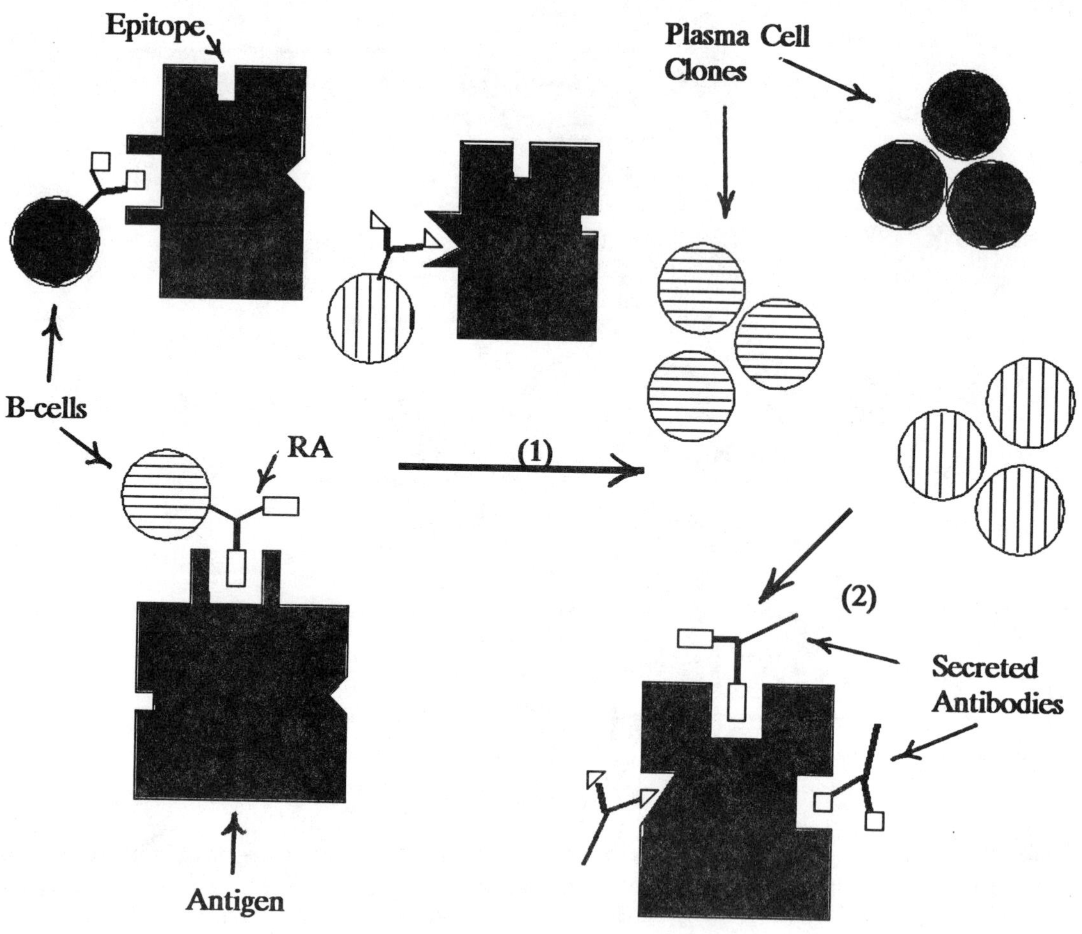

Fig. 10-27. Polyclonal humoral immune response. An antigen (enlarged) containing several epitopes interacts with the receptor antibodies on the surfaces of several B-cells or memory cells that have binding sites complementary to the antigen epitopes. A series of reactions occurs (1), culminating in the proliferation of the stimulated cells into plasma cell clones which secrete (2) antibodies directed against the respective epitopes of the antigen.

Differentiation of the B-cell lineage occurs by a series of steps (Fig. 10-28). Pre-pro B-cells have their antibody gene clusters in the germline configuration. Heavy chain D-J rearrangement commences in the pro B-cell, and the pre B-cell contains μ heavy chains, indicating the completion of the initial stages of heavy chain gene rearrangement. Light chain gene rearrangement is completed by entry to the virgin B-cell stage, and IgM receptors appear on virgin B-cell surfaces. All of these DNA rearrangements and the appearance of IgM surface receptors occur prior to exposure to antigen. When a virgin B-cell encounters an antigen for which it has been programmed to react through its IgM receptor cdrs, the cell begins to proliferate. Some of the descendant cells form a clone of IgM-secreting cells, while others undergo a class switch. During the transition, cells can be identified which possess both IgM and IgG, IgE, or IgA receptors on their surfaces. Completion of the switch is signified by the presence of only IgG, IgE or IgA surface receptors. A second encounter with the same antigen or a very closely related antigen induces a portion of the switched cells to proliferate into plasma cell clones and secrete the corresponding antibody. Others retain their receptor antibody and persist as memory cells.

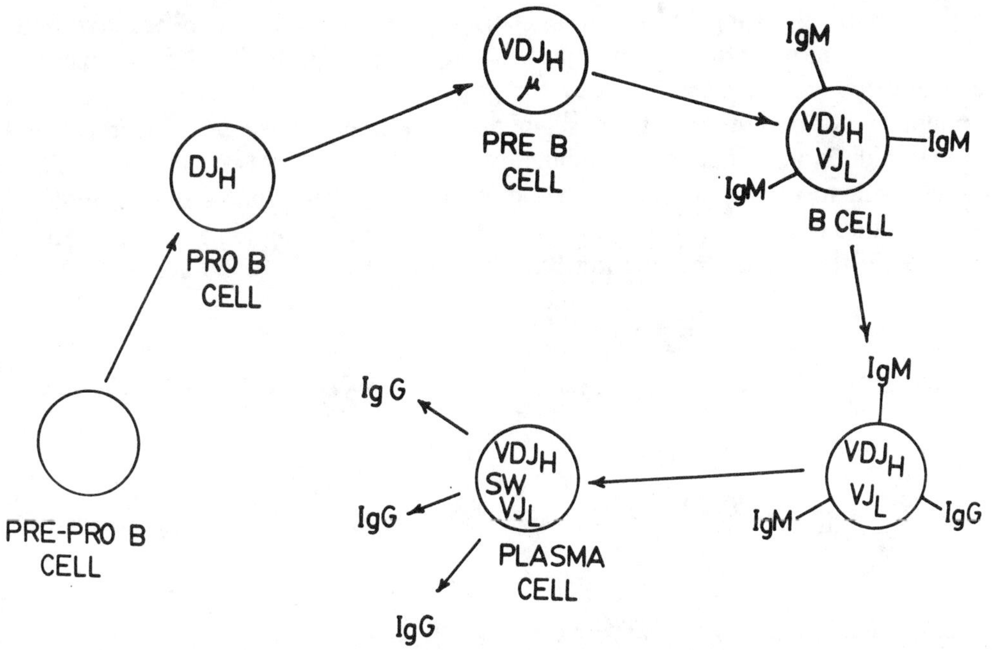

Fig. 10-28. Differentiation of IgG-secreting plasma cells. The transition from virgin B-cells with surface IgM to cells expressing heavy chains encoded by downstream gene segments passes through intermediate stages when two receptor types may be present. Note how D-J joining precedes V-D-J recombination (modified from (17)).

Allelic exclusion, where antibody genes on only one of two homologous chromosomes is expressed in a given cell or plasma cell clone, is apparently related to the stochastic nature of the rearrangement process. Both chromosome 14s begin recombining their heavy chain elements at about the same time. About a third of the pre B-cells fail to achieve an in frame functional gene construct, and these cells die out. In the remaining cells, one homolog achieves a functional gene construct before the other, and the μ chain derived from its gene appears to act as a signal to stop rearrangement of the partner chromosome. Furthermore, only the homolog successfully determining a μ chain can udergo the class switch. This explains why only one of each pair of heavy chain alleles is expressed in a specific cell or cell clone. The μ chain appears to act directly or indirectly as a signal to commence kappa region recombination.   If one chromosome 2 achieves an in frame construct, the kappa chain stops rearrangement of the homologous chromosome, such that only one of two kappa chain alleles is expressed. This cell and its progeny will express antibodies consisting of two heavy chains and two kappa chains. Approximately one-third of the cells initiating kappa region rearrangement fail to construct a functional kappa gene. When this occurs, the lambda regions on chromosome 22 begin recombination. About two-thirds of the cells initiating lambda rearrangement express antibodied containing two heavy chains and two lambda chains. Again, only one of two lambda regions is expressed because the lambda chain encoded by one allele suppresses recombination of the homologous chromosome. Those cells failing to make a functional lambda gene do not express antibody and are eliminated.

Further studies (summarized in (18)) indicate that class switching may occur in two or more steps. Loci that are relatively close together (Cμ - Cδ, Cγ3 - Cγ1, etc.) are initially cotranscribed into the same nascent RNA. Differential processing of this transcript produces mRNAs encoding μ and δ or γ1 and γ3 in the same cell and is responsible for the appearance of two different receptor antibodies on the surface of the same cell during the transition period. As the class switch nears completion, the DNA recombination event removing the earlier expressed segments from the DNA is completed, and only one type of surface receptor occurs on these cells.

The tandem arrays of antibody gene segments with similar base sequences is conducive to mispairing during meiotic division. Unequal recombination between segments on misaligned chromosomes can generate products with too many or too few gene segments. This process undoubtedly led to the current multiplicity of antibody gene segments and is apparently continuing (Fig. 10-29). Approximately 5% of chromosome 14s in the human population have an additional Cγ2 exon cluster (19).

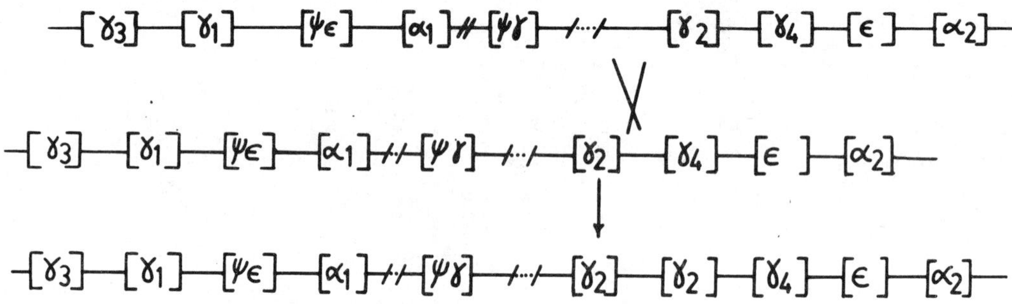

Fig. 10-29. Production of chromosomes containing a duplicated gamma(2) exon cluster. Only the recombinant chromosome carrying the duplicated region is shown (19).

SUMMARY

The immune system is composed of two major divisions: cell-mediated and humoral immunity. Humoral immunity is contributed by B-cells that differentiate from bone marrow stem cells and seed the lymph nodes and other lymphoid tissues. When these cells are challenged by the appropriate antibody they form clones of plasma cells which produce and secrete large amounts of antibody directed against the offending antigen. Antibodies synthesized by B-cells and their plasma cell derivatives fall into five classes: IgM, IgG, IgD, IgA, and IgE.

Cell-mediated immunity involves T-cells and their products. T-cells develop from bone marrow stem cells that have migrated to the thymus where selection against self-reactive T-cells occurs. The mature T-cells subsequently migrate to lymph nodes and other lymphoid organs. Descendants of T-cells and their products are involved with graft rejection, antiviral activities, cell killing, and other aspects of cell-mediated immunity. The T-cell and B-cell arms of the immune system are interconnected. For example, T-cell helpers enhance antibody production by B-cells.

Patients with hereditary thymic aplasia lack both functional T-cells and B-cells. These patients tolerate tissue transplants from unmatched donors and are highly susceptible to viral, bacterial, and fungal diseases. Bruton's hypogammaglobulinemia is associated with a loss of humoral immunity, and patients are susceptible to bacterial infections and have a reduced ability to fight viruses. Other aspects of their immune system are largely intact. The DiGeorge syndrome is often caused by deletions of chromosome 22. These persons have impaired cell-mediated immunity and, as a consequence of deficient T-cell helpers, have defective but not totally absent antibody production and secretion.

The human histocompatibility (HLA) antigens play a major role in cell-mediated immunity. The HLA genes encode molecules that present viral, phagocytized bacterial, and endogenous intracellular antigens in peptide form to T-cell killers and T-cell helpers. T-cell receptors of these cell types inspect the peptide within the context of the presenting HLA molecule and judge the fragment self or nonself. When the fragment is fudged foreign by a T-cell killer, it attacks and destroys the presenting cell. When the fragment is judged foreign by the helper T-cell, the T-cell helper releases interleukins that recruit other T-cell derivatives to attack the presenting cell or stimulate B-cells to produce antibodies directed against the antigen from which the peptide was derived. Autoimmunity occurs as a consequence of an escape of self-reactive T-cells from suppression or anergy. Autoimmune forms of diabetes are frequently associated with HLA-DR3 and HLA-DR4 alleles.   Other disease associations have been reported, including the association of HLA-B27 with ankylosing spondylitis. Linkage disequilibrium refers to the unusually frequent occurrence of certain HLA haplotypes in a population. Some of these associations are disease-related.

Erythrocyte surface antigens are determined by more than 35 loci, many of which have several alleles.   Blood group genes either determine the structure of enzymes involved with processing of oligosaccharides attached to the antigens (**ABO, H, LE, SE**) or encode the protein portion of these antigens (**RH and LW**). Persons with at least one functional secretor gene also have glycoproteins in their secretions that contain blood group specificities. Both erythrocyte and secretory ABO and Lewis antigens are synthesized by a series of metabolic steps. Epistasis occurs when an enzyme required for the synthesis of a subtrate required for a subsequent step in the pathway is deficient. In this way **hh** blocks expression of the erythrocyte ABO antigens, and **sese** blocks expression of secretory ABO antigens.

Alleles occurring at two or more tightly linked loci encode the polypeptides occurring in Rh antigens. Erythroblastosis fetalis may occur when an Rh(-) woman experiences two or more pregnancies involving an Rh(+) fetus. Administration of Rhogam, an anti-Rh preparation, with each pregnancy prevents sensitization of the Rh(-) mother by her Rh(+) fetus and prevents signs of the disease. Some natural protection against erythroblastosis fetalis has also been observed. An O,Rh(-) woman has a lower risk when married to an A or B, Rh(+) man. The anti-A and anti-B antibodies in the woman's circulation destroy fetal A, Rh(+) and B, Rh(+) erythrocytes entering her circulation before they can induce production of anti-Rh.

Antibodies generally consist of two light and two heavy chains. Each light chain contains antigen-binding N-terminal hypervariable regions, while the remainder of the molecule is relatively invariant. Heavy chains also have an N-terminal region whose antigen contacting sites are hypervariable and have a larger constant region consisting of several domains. People are capable of forming more than 10^7 different antibodies during their lifetimes. Multiple germ line antibody-encoding gene segments are clustered on three different chromosomes. The heavy chain cluster on chromosome 14 contains V, D, and J segments that rearrange during B-cell differentiation to produce a complex encoding the variable domain of the heavy chain. Heavy chain constant region domains are encoded by exon clusters that are dowstream from the V, D, and J segments. One chromosome 14 region is expressed in each antibody-synthesizing cell. This expression of only one heavy chain allele per cell is called allelic exclusion. The kappa and lambda light chains are also determined by multiple copies of V, J and C elements that are clustered on chromosomes 2 and 22, respectively. Recombination following heavy chain rearrangement results in the production of one functional light chain gene per antibody-producing cell. Furthermore, either the kappa chain or the lambda chain is synthesized in a specific cell, not both. During the humoral immune response, the heavy chain class of an antibody chain can change. This class switch is completed by a recombination event which moves the VDJ segment closer to the heavy chain ultimately expressed and deletion of the intervening DNA from the chromosome. As a consequence of this process, the antigen specificity of the antibody remains the same. Other factors contributing to antibody diversity (in addition to multiple germline segments and DNA recombination) include junctional diversity (variable joining of V and J, V and D, and D and J segments), combinatorial variability (which light chain type is paired with which heavy chain type in the antibody), and somatic mutation (an antigen-driven hypermutation in framework regions of the variable domains of heavy chains). Antibodies are polyclonal, and the ability of an individual to mount a humoral response against an antigen is determined by the particular array of antibodies secreted against that antigen. Antibodies occur as membrane-anchored receptors and as secreted forms. Targeting of antibody molecules to the membrane or to the cell exterior is accomplished at the RNA splicing level.

PROBLEMS

1. Hereditary thymic aplasia appears to involve a lesion affecting both cell-mediated and humoral immunity. How might this occur?

2. Why do our cells possess HLA antigens?

3. A patient has the class I HLA haplotypes: 10, 7, 1/ 8, 15, 9 (note the order is HLA-B, -C, -A). Two potential organ donors are 10, 7, 1/ 6, 12, 3 and 10, 7, 1/ 8, 12, 9 , respectively. Whose tissue would the patient more readily tolerate?

4. Predict the phenotype of an individual who is: **$ABO^A ABO^B$, Lele, Hh, sese.**

5. Define linkage disequilibrium and provide an example.

6. Draw a mature antibody light chain-encoding gene. Where would you expect to find this type of gene?

7. A boy who lacks ABO antigens on his erythrocytes has a father who is type AB. Assuming the boy is biologically related to his father, how might this occur? How would you verify your explanation?

8. An O, Rh(-) woman married an A, Rh(+) man and has four normal children. The woman has never received Rhogam. How would you explain these normal pregnancies?

9. Two infants have inadvertently been interchanged in the nursery. In order to determine parentage, blood typing was done. One infant, a male, has the phenotype: A, MN, Rh(-), Xg(a)+, secretor, Le^b. The parents claiming parentage of this child are: husband - A, M, Rh(+), Xg(a)+, nonsecretor, Le^a; wife - B, N, Rh(+), Xg(a)-, secretor, Le^b. What is your conclusion?

10. Is it possible for an antibody to possess both Gm and Am allotypes?

11. What is the meaning of the statement: a single antibody-producing cell clone produces a single antibody specificity?

12. What do current data suggest regarding genetic determination of immunoglobulin structure?

GLOSSARY OF TERMS

Allelic exclusion expression of one of two allotypic alleles in a heterozygote.

Allotypic variation allelic variation among antibody chains determined by amino acid substitutions at one or a few sites in the constant region domain of the same light chain or of the same antibody heavy chain class.

Antibody protein which specifically combines with an antigen, contributing to its destruction or neutralization.

Antigen substance capable of eliciting an immune response.

B-cell (bone marrow-derived) lymphocyte capable of differentiating into a plasma cell and secreting antibodies.

Cell-mediated immunity an immune reaction involving the direct participation of immune cells in the rejection of foreign tissue, killing of virus-infected cells, etc.

Cell-mediated lympholysis lysis of histo-incompatible lymphocyte targets by sensitized T-cell killers.

Class switch a change of the class of an antibody produced by a cell or cell clone during an immune response.

Complex locus (supergene) chromosomal region containing a cluster of tightly linked loci that are only rarely separated by recombination.

DNA fingerprinting identification of a sample on the basis of its pattern of DNA fragments.

Epistasis genes at one locus block the expression of genes at a second locus.

Epitope the three-dimensional region of an antigen recognized by an antibody.

Haplotype chromosomal array of HLA genes that are inherited as a unit.

Histocompatibility antigens surface antigens that present self/nonself peptides to the T-cell receptor.

HLA human lymphocyte antigens - genes encoding human histocompatibility antigens.

Junctional diversity variation generated by imprecise joining of V, D, and J segments.

Linkage disequilibrium unusually frequent occurrence of a haplotype.

Mixed lymphocyte culture test coculture of lymphocytes from a potential donor and recipient to see if they are histocompatible.

Plasma cell antibody-secreting cell.

Somatic mutation heritable DNA changes in somatic cells.

T-cell thymus-derived lymphocyte capable of producing or interacting with cells which participate in cell-mediated immunity.

T-cell receptor surface receptor of T-cells that interprets the self/nonself nature of a peptide presented by a HLA molecule.

Tolerance lack of an immune reaction to an antigen, usually because the antigen is of "self" origin.

BIBLIOGRAPHY

1. Fisher, R A. 1930. <u>Genetical Theory of Natural Selection</u>. 2nd ed., New York: Dover. 1958.

2. Blackman, M, J Kappler and P Marrack. 1990. The role of the T cell receptor in positive and negative selection of developing T cells. Science 248:1335-1341.

3. Ramsdell, F and B J Fowlkes. 1990. Clonal deletion versus clonal anergy: the role of the thymus in inducing self tolerance. Science 248:1342-1348.

4. Sinha, A A, M T Lopez and H O McDevitt. 1990. Autoimmune diseases: the failure of self tolerance. Science 248:1380-1388.

5. Pleogh, H L, H T Orr and J L Strominger. 1981. Major histocompatibility antigens: the human (HLA-A, -B, -C) and murine (H-2K, H-2D) class I molecules. Cell 24:287-299.

6. Auffray, C and J L Strominger. 1986. Molecular genetics of the human major histocompatibility complex. Adv Hum Genet 15:197-247.

7. Bach, F H. 1976. Genetics of transplantation: the major histocompatibility complex. Annu Rev Genet 10:319-339.

8. Davis, M M and P J Bjorkman. 1988. T-cell antigen receptor genes and T-cell recognition. Nature 334:395-402.

9. Landsteiner, K. 1900. Zur Kenntnis der antifermentative, lytischen, und agglutinerenden Wirkungen des Blutserums und der Lymphe. Zentralblatt. Bakt. Parasit. 27:357-362.

10. Levine, P, F Robinson, M Celano, et al. 1955. Gene interaction resulting in suppression of blood group substance B. Blood 10:1100-1108.

11. Watkins, W M. 1966. Blood group substances. Science 152:172-181.

12. Oriol, R, J Danilous and B R Harkins. 1981. A new genetic model proposing that the Se gene is a structural gene closely linked to the h gene. Am J Hum Genet 33:421-431.

13. Oriol, R, J LePendu, R S Sparkes, et al. 1981. Insights into the expression of ABH and Lewis antigens through human bone marrow transplantation. Am J Hum Genet 33:551-560.

14. Yamamoto, F, H Clausen, T White, et al. 1990. Molecular genetic basis of the histo-blood group ABO system. Nature 345:229-233.

15. Saboori, A M, B L Smith and P Agre. 1988. Polymorphism in the M_r 32,000 Rh protein purified from Rh(D)-positive and -negative erythrocytes. Proc Natl Acad Sci USA 85:4042-4045.

16. Bloy, C, P Hermand, D Blanchard, et al. 1990. Surface orientation and antigen properties of Rh and LW polypeptides of the human erythrocyte membrane. J Biol Chem 265:21482-21487.

17. Honjo, T and S Habu. 1985. Origin of immune diversity: genetic variation and selection. Annu Rev Biochem 54:803-830.

18. Shimizu, A and T Honjo. 1984. Immunoglobulin class switching. Cell 36:801-803.

19. Bech-Hansen, N T and D W Cox. 1986. Duplication of the human immunoglobulin heavy chain gamma(2) gene. Am J Hum Genet 38:67-74.

ADDITIONAL LEARNING RESCOURCES

1. Bell, J I, J A Todd and H O McDevitt. 1989. The molecular basis of HLA-disease association. Adv Hum Genet 18:1-41.

2. Bjorkman, P J and P Parham. 1990. Structure, function, and diversity of class I major histocompatibility complex molecules. Annu Rev Biochem 59:253-288.

3. Blackwell, T K and F W Alt. 1989. Mechanism and developmental program of immunoglobulin gene rearrangement in mammals. Annu Rev Genet 23:605-636.

4. Davis, D R, E A Padlan and S Sheriff. 1990. Antibody-antigen complexes. Annu Rev Biochem 59:439-473.

5. Kappes, D and J L Strominger. 1988. Human class II histocompatibility complex genes and proteins. Annu Rev Biochem 57:991-1028.

Chapter 11

Genetic-Environmental Interactions

and Multifactorial Traits

The importance of both heredity and environment in the determination of human traits was recognized by Sir Francis Galton during the nineteenth century. Galton and his student, Karl Pearson, were responsible for the early development of biometrics, the branch of biology that deals with measurement and analysis of quantitative phenomena. Just as Mendel's work had laid the foundation for the study of qualitative characters, the work of Galton and Pearson provided the basis for the study of quantitative human characteristics.

Previous discussions have centered around the inheritance of **discontinuous** (qualitative) traits, where the overlap of phenotypes is minimal or entirely lacking. However, there are many traits such as height, weight and skin color that lack distinctive, classifiable phenotypes. Variation within these **quantitative** traits is continuous with extensive overlapping of phenotypes. As a consequence, populations are generally defined in terms of an "average type", and the variation around this mean or average is estimated. Quantitative traits are **multifactorial** and are the result of the interaction of a complex genotype with an equally complex environment. The genetic contribution to continuously variable traits consists of three components: 1) the effects of multiple alleles at one or a few major loci, 2) the effects of alleles at each of many loci with individually small effect but which may have a significant additive effect upon a phenotype, and 3) the interaction effect of alleles at major loci functioning in a genetic background of a specific set of alleles at additive loci. The expression of major and minor genes is strongly influenced by the environment. For example, two individuals who have very similar gene arrays for weight may attain quite different adult weights as a consequence of dietary, occupational, lifestyle, and other environmental differences. It is worthwhile to view human traits as occupying a continuum. Environmental disorders represent one extreme of this continuum in which a uniform genotype is interacting with a heterogeneous environment. Cancer caused by vinyl chloride or asbestos may represent this type of problem. There may be little population variation with respect to genetic factors that detoxify these carcinogens, and development of the cancer is largely a result of the degree of exposure. "Pure" genetic traits occupy the opposite extreme of this continuum. There is considerable genetic variation with respect to genes contributing to these traits, and the environment tends to be fairly uniform. Therefore, development of the genetic condition is largely determined by the genotype of the person. Most human normal and abnormal traits fall somewhere between these extremes. Both the genotype and environment are variable, and occurrence of a particular problem depends upon both the genotype and its interaction with particular environmental variables. Examples of abnormal **genetic-environmental** traits include simple birth defects, epilepsy, mental retardation, diabetes mellitus, and coronary artery disease. It is important to remember that quantitative traits have **both** genetic and environmental components, and that the extent of involvement of each, as well as their interaction, will determine the phenotype of the individual.

CHARACTERISTICS OF MULTIFACTORIAL TRAITS

Consider height as an example of a multifactorial trait displaying continuous variation within a human population. The average male height may approximate 5'10" $\pm$ 3". The first number is the **mean** and is obtained by summing the heights of all males and dividing by the number of males measured:

1) $\qquad$ Mean = $\dfrac{\text{total height}}{\text{number of males}}$ $\qquad$ or $\qquad$ $\overline{X} = \dfrac{\Sigma x}{n}$

where $\overline{X}$ is the mean, x = individual heights, and n = the number of males. The second number is the **standard deviation.** The standard deviation is an estimate of the dispersion or variation of the population around the mean and can be calculated:

2) $\qquad$ $S = \sqrt{\dfrac{\Sigma(x-\overline{X})^2}{n-1}}$

which can be rearranged to facilitate use of a calculator:

3) $\qquad$ $S = \sqrt{\dfrac{\Sigma(x)^2 - (\overline{X})^2/n}{n-1}}$

where S is the standard deviation. If the population follows a normal distribution relative to height (e.g., is distributed according to a bell-shaped curve), two thirds of the population will fall within one standard deviation of the mean (between 5'7" and 6'1"), while 95% will fall within two standard deviations of the mean (5'4" to 6'4"). Another term, **variance,** is also used to describe variation of continuous traits. The variance is equal to the square of the standard deviation.

Variation of continuous traits differs among populations, such that populations may have different means, different variances or both (Fig. 11-1). The magnitude of the mean and the dispersal of the population around that mean are the resultant of environmental forces, genetic constitution of the population, and the interaction between these variables. The position occupied by a member of the population within that distribution (e.g., the height of that person) represents the net effect of the interaction of his genotype with the series of environments he has experienced during his prenatal, perinatal, and postnatal lifetime.

Sources of Phenotypic Variation

The phenotypic variation of a population can be expressed as

4) $\qquad$ $S_P^2 = S_E^2 + S_G^2 + S_{GE}^2$

where S_P^2 = the phenotypic variance, S_E^2 = that portion of the phenotypic variance due to environment, S_G^2 = that portion of the phenotypic variance due to genetic factors, and S_{GE}^2 = that portion of the phenotypic variance due to the interaction between environment and genotype. The genetic variance may be further subdivided into the additive genetic variance (S_A^2) and the variance due to dominance (S_D^2). The additive genetic variance estimates variation caused by differences between homozygotes $(A_1A_1B_2B_2C_1C_1$ vs. $A_2A_2B_3B_3C_2C_2)$ in the population, while the dominance variance arises from effects of alleles in heterozygotes and is a measure of the deviation of a heterozygous (A_1A_2) phenotype from the mean of the respective homozygous (A_1A_1, A_2A_2) phenotypes. Equation (4) becomes:

$$5) \qquad S_P^2 = S_E^2 + S_D^2 + S_A^2 + S_{GE}^2$$

Although this partitioning of the phenotypic variance may appear quite straightforward, study of its components has proved difficult.

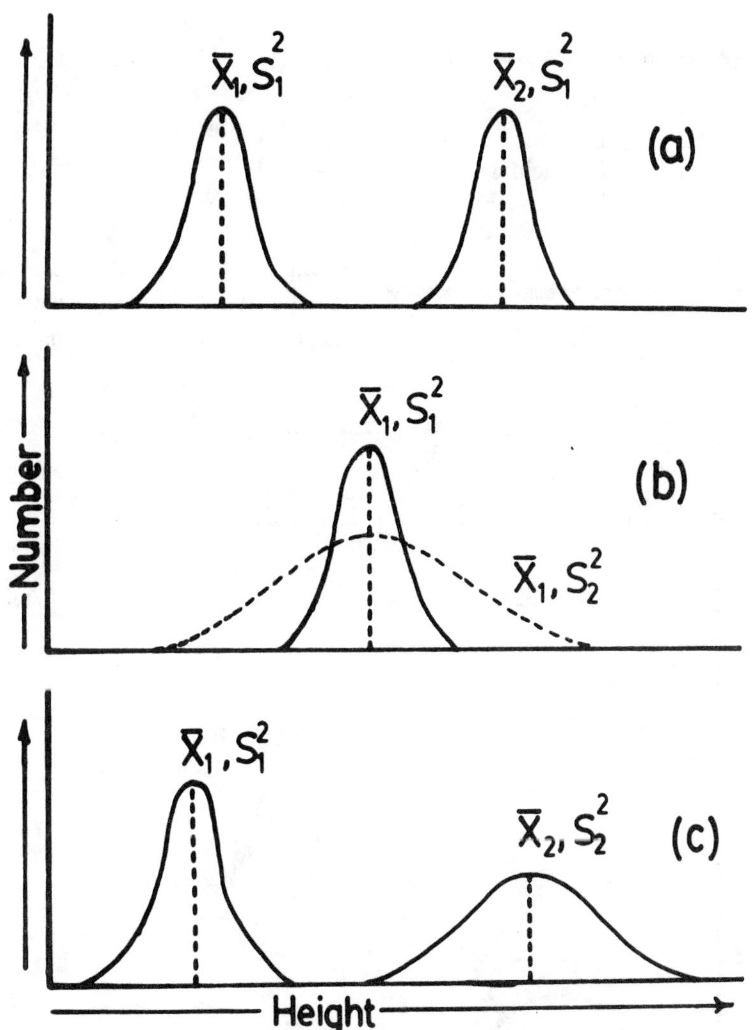

Fig. 11-1. Distributions of populations with respect to a continuous phenotype. a) Different means, comparable variances; b) Same means, different variances; c) Different means and variances.

Environmental Factors

Environmental factors which contribute to phenotypic variation may be experienced by an individual both before and after birth. The age of the mother may influence the probability of a normal pregnancy outcome. Older mothers tend to run a higher risk for children with birth defects and other complications. Intrauterine exposure to Rubella (German measles virus), drugs, and maternally derived metabolites can also influence fetal development. For example, maternal PKU and maternal diabetes mellitus have both been found to cause problems if they are not properly controlled. Postnatally, the varying socioeconomic conditions to which an individual is exposed during her/his lifetime will affect the phenotypic expression of his genotype.

Known environmental agents that affect human development are listed in Table 11-1. The relative effects of these agents vary with respect to the stage of development of the individual. Most of them were mentioned as potential mutagens in Chapter 6. Interactions among two or more of these factors and an individual's genotype are also possible. For example, a person may be homozygous for a mutation that markedly reduces levels of a drug detoxifying enzyme. If this drug is prescribed to this person, adverse and even life-threatening effects may occur; whereas, most people who receive the drug do not experience complications.

Table 11-1.   Environmental Factors Influencing Human Development.

Infection	Trauma	Drugs
Noise	Diet	Emotion
Insecticides	Pollutants	Herbicides
Temperature	Radiation	Maternal-Fetal
Maternal Age	Paternal Age	Incompatibility

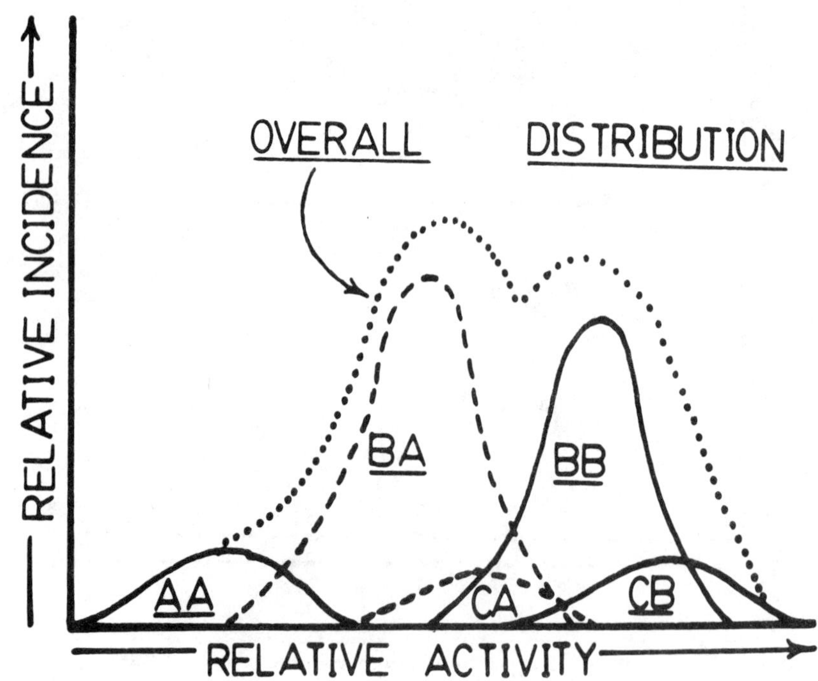

Fig. 11-2. Red cell acid phosphatase activities. Cumulative activity contributions of the six genotypes produce a continuous activity distribution. CC is not shown, since these homozygotes occur relatively infrequently (about 1/600). A = ACP^A, B = ACP^B and C = ACP^C (modified from Harris (1)).

Genetic sources of human variation include the effects of major loci, such as the **PAH** locus or the globin loci, and the additive effects of numerous minor loci, each of which makes a relatively small contribution to the overall phenotype. The concerted effects of both types of genes generate a continuous distribution of phenotypes.

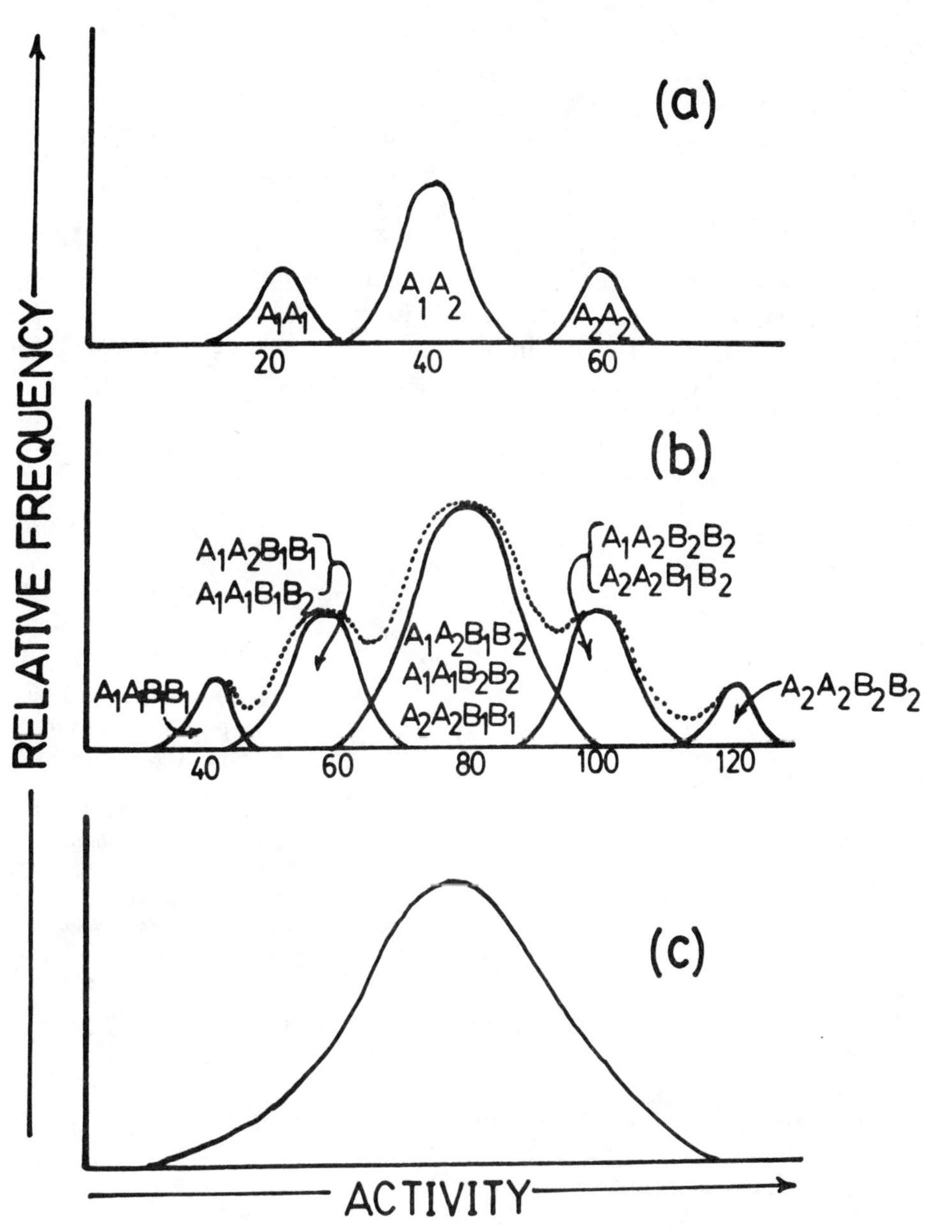

Fig. 11-3. Distributions of a phenotype determined by two alleles with additive effects at one (a), two (b) and several loci (c).

Genetic heterogeneity due to multiple alleles at specific major loci has been discussed previously. Polymorphisms involving multiple alleles can generate a continuous distribution (Fig. 11-2). Red cell acid phosphatase is encoded by the **ACP1** locus (chromosome 2p25). Three common alleles

(ACP1^A, ACP1^B, ACP1^C) are relatively frequent among European and European-derived populations. These alleles occur in three homozygous and three heterozygous combinations that can be distinguished on the bases of their respective electrophoretic patterns. The acid phosphatase activities associated with these six genotypes also vary, and overlap among genotypes is extensive. The cumulative contributions of these six genotypes generate a continuous distribution of acid phosphatase activities in these populations. Serum cholinesterase activities are also continuously distributed in human populations, and this activity variation is largely determined by multiple genotypic combinations of several alleles at a single locus (**CHE1**).

The contributions of multiple additive loci, each with two or more alleles, provides another source of continuous phenotypic variation (Fig. 11-3). The effect of one locus (**A**) and those of two loci are compared. Each locus possesses two alleles that have additive effects. The alleles A_1 and B_1 contribute ten units to the total phenotype, while the alleles A_2 and B_2 add 30 units. For simplicity, the alleles at each locus occur with a frequency of 0.5 in the population. Overlap among the two homozygous and the heterozygous genotypes in Fig. 11-3a is minimal; however, when the additive effects of alleles at a second locus contribute to those of the first, the distributions begin to merge. As the number of loci influencing a trait increase, the distribution becomes bell-shaped (Normal distribution; Fig. 11-3c). Most continuously distributed traits that have genetic components appear to be determined by five or more additive loci, and the actual numbers of loci appear to be very large for complex traits like height, hearing, vision, intelligence and behavior. Certain of the loci may have a greater impact than others. The example of the **PAH** locus and intelligence has been cited earlier.

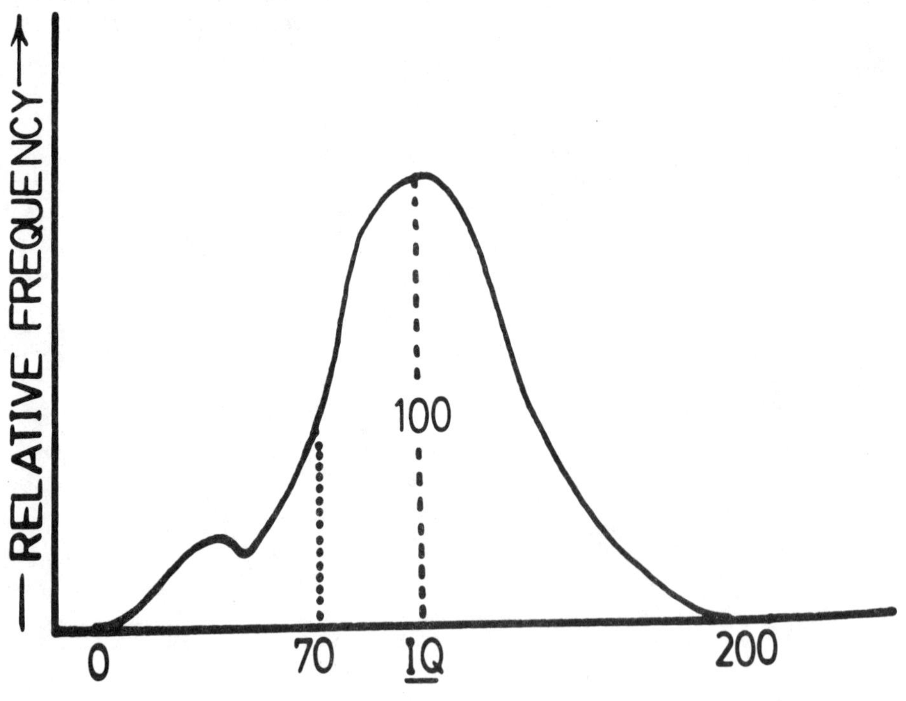

Fig. 11-4. I.Q. distribution of a typical human population. The mean I.Q. for this group is 100. Persons with I.Q.s below 70 tend to be mentally handicapped. The "shoulder" at the low end of the I.Q. distribution includes a special group of mentally handicapped individuals who may have mendelian or chromosomal syndromes or who have mental retardation due to major environmental causes.

Thresholds, Penetrance and Variable Expression of Quantitative Traits

Although quantitative characteristics may be continuously distributed, some appear to be discontinuous with respect to some aspect of their phenotypic expression. The I.Q. distribution for a typical human population is presented in Fig. 11-4. I.Q. scores are determined by tests designed to measure specific aptitudes that are believed to contribute to "intelligence" as defined by Western cultural standards. Performance on these tests is influenced by the genetic makeup of the individual and by a number of environmental variables such as the sociocultural experience and age of the person and the manner in which the test is administered.

When intelligence is estimated by I.Q., it appears to be a continuously varying quantitative trait. However, there exists a point (**threshold**) beyond which persons with lower scores tend to experience problems coping with society and require some degree of support. An I.Q. of 70 has been arbitrarily selected as an approximation of the threshold between "normal" intelligence and mental retardation. Individuals who have inherited a favorable genotype and who are reared in a favorable environment will have an I.Q. that exceeds the threshold and will display normal intelligence. Those who inherit a "deficient" set of genes and/or are exposed to an unfavorable environment will fall below the threshold and are likely to be classified as mentally handicapped. Application of this threshold to the continuous variable, I.Q. score, serves to divide the population into two discontinuous groups: normal and mentally retarded. However, the threshold is "fuzzy", since inspection of persons near the cutoff point (I.Q. = 70 ± 10) will reveal that their classification in one group or the other may prove difficult.

Two distinct populations of mentally handicapped people appear to exist. The majority possess I.Q.s between 50 and 70 and seem to represent the low "tail" of the intelligence distribution. Genetic factors contributing to this mildly to moderately retarded group are believed to involve numerous loci which act in cumulative fashion to limit the potential of individuals who have inherited them. The limits are rather broad, and the particular level of attainment within these limits is heavily influenced by a variety of environmental factors including quality of diet, type of health care, degree of reinforcement in the home, access to educational programs, and other sociocultural factors. The second distribution is largely composed of persons with an I.Q. below 50 and includes cases caused by severe infections, trauma to the central nervous system, single gene disorders, chromosomal anomalies, and unusually large numbers of additive genes predisposing toward mental retardation. Most people in this group are unable to function independently in society, requiring assistance with their financial affairs and other aspects of their lives. Although I.Q. has been used as an estimator of intellectual potential, it is not a self-sufficient predictor of an individual's ultimate achievement in society. Persons with similar I.Q.s may attain quite different degrees of independence. Motivation, presence or absence of associated physical handicaps, and developmental environment must also be considered when judging the potential of a mentally handicapped person and planning management programs.

Attempts have been made to estimate the relative contributions of genetics and environment to mental retardation. Evidence for genetic factors can be revealed by comparing the incidence of mental retardation among relatives of mentally handicapped persons with that among relatives of an age-, race- and sex-matched control group. Pauls and Anderson (2) evaluated genetic factors involved with mental retardation by screening relatives of 289 moderately to severly retarded patients with I.Q.s of 50 or below (Table 11-2) for other cases of mental retardation. The incidence of mental retardation among these relatives was compared with that among 6,000 relatives of an appropriate group of controls. The incidence of mental retardation among relatives of the controls approximated 1.6%. The incidence of mental retardation among the patients' relatives varied with the degree of relationship (note: civil law defines "degree of relationship" in terms of the number of steps that separate the two individuals being considered from one another in a pedigree) and with the sex of the index case. The incidence of mental retardation among first degree relatives (parent-child) was nine-fold higher than that among control relatives. Less than a four-fold increase among second degree relatives (grandparent-grandchild, brother-

Table 11-2. Incidence of Mental Retardation Among Relatives of 289 Retarded Patients (2).

Sex of Index Case	Relationship		
	1st Degree	2nd Degree	3rd Degree
Male	11.0*	5.4	3.3
Female	17.0	6.6	4.0

* % Retarded

sister) and less than a three-fold increase among third-degree relatives (uncle-niece or nephew, aunt-niece or nephew) were observed. These data are most compatible with a multifactorial model including participation of multiple additive loci in the etiology of mental retardation. First degree relatives share one-half of their genes in common; therefore, the chance that a particular combination of genes predis-

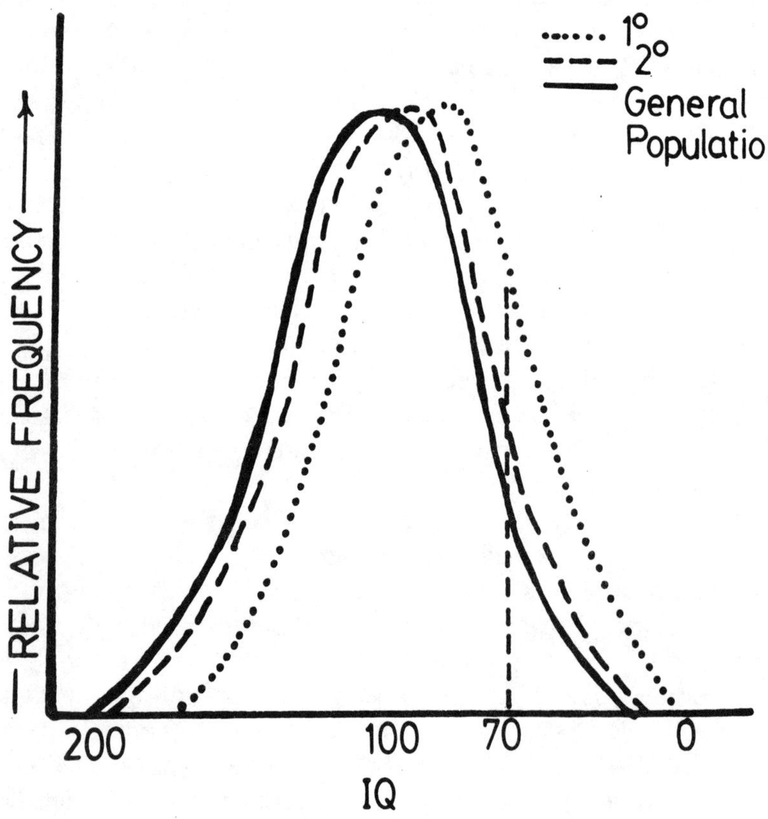

Fig. 11-5. Effect of degree of relationship upon relative frequency of mental retardation. Close relatives would be expected to share more genes predisposing toward mental retardation, and the incidence of retardation would be higher in this group than among more distant relatives or unrelated persons.

posing to mental retardation would occur in a first degree relative of a mentally retarded person would be anticipated to be considerably higher than the occurrence of the same combination in two unrelated people. This enhanced risk is reflected in Fig. 11-5, where a substantially larger proportion of first degree relatives of a retarded patient falls beyond the threshold compared to unrelated persons. The probability that two second degree relatives (who share one-quarter of their genes in common) would share the same combination of genes at multiple loci would be considerably lower than that for first degree relatives, and the proportion of second degree relatives falling beyond the threshold for mental retardation may not be much greater than that for the general population. Furthermore, the distribution of third degree relatives would closely approximate that of the general population. It can also be argued that close relatives also share more similar environments than more distantly related or unrelated persons and that adverse environmental factors could also account for the patterns discussed above. Definitive proof for genetic influence upon mental retardation and other multifactorial traits requires study of genetically similar individuals reared in dissimilar environments. These investigations will be discussed in a later section.

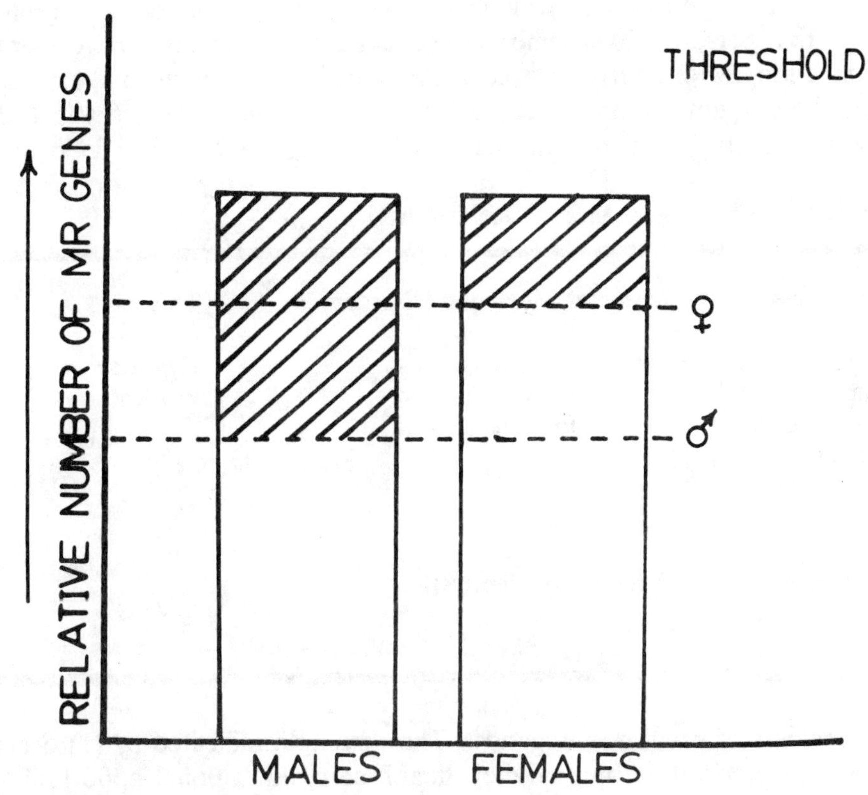

Fig. 11-6. Effect of sex upon threshold traits. Males have a lower threshold for mental retardation and require fewer genes to develop this trait. Therefore, assuming that genes predisposing toward mental handicaps occur equally often in the two sexes, more males than females will be mentally retarded (shaded areas).

The threshold for mental retardation and for other multifactorial traits in general may differ between sexes. This phenomenon is evident in Table 11-2. More relatives of retarded females than of retarded males possess mental handicaps, a trend that is observed at each degree of relationship. This pattern implies that the threshold for mental retardation is lower among males than among females. Therefore, presence of retarded females in a pedigree indicates that an unusually large number of predisposing genes is segregating through the kindred. Male children of a retarded female or brothers of a retarded female would run a higher risk for mental handicaps compared to daughters or sisters of a mentally handicapped female.

Assortative mating is defined as the preferential marriage of persons with similar phenotypes or life styles (positive assortative mating) or the preferential marriage of persons with dissimilar characteristics (negative assortative mating) (3). The former marriage pattern has been erroneously interpreted to suggest that marriage of mentally retarded individuals will gradually lower the average I.Q. of the population and has been advocated as justification for maintaining segregation of mentally handicapped men and women in institutions, rather than permitting them to live in a more natural community setting with appropriate agency support. However, theoretical arguments have demonstrated that positive assortative mating does not appreciably change gene frequencies, but may increase the phenotypic variance of the population by gradually causing a shift of a larger proportion of individuals to both extremes of the distribution. This spreading of the phenotypic distribution assumes fertility is independent of I.Q. In fact, both bright and retarded couples appear to have fewer children than couples of more average intelligence. This reproductive difference tends to counter the phenotypic dispersion promoted by assortative mating.

Although the studies of Pauls and Anderson focused upon the moderately to severely retarded, similar factors are believed to contribute to mild to moderate retardation. As mentioned above, genetic predisposing factors tend to limit the potential of an individual; however, they do not solely determine the specific level of achievement of that person. Environmental agents believed contributory to mental retardation include malnutrition, infection, maternal hypothyroidism, trauma, lead poisoning, and exposure to certain drugs. A sampling of these agents is presented in Table 11-3. These factors affect the **penetrance** of the mental retardation phenotype by altering the threshold (Fig. 11-7). Four individuals

Table 11-3. Environmental Agents Contributing to Mental Retardation.

Fetal Infection	Maternal Drugs	Maternal Metabolic Disorders	Trauma
Rubella	Diphenylhydantoin	Diabetes	Hypoxia
Syphilis	Potassium iodide	Hypothyroidism	Physical
Toxoplasmosis	Propylthiouracil	Phenylketonuria	
	Sulfonamides		

Pollutants	Dietary	Behavioral	
Lead	Malnutrition	Psychosocial deprivation	
Nitrates			

are pictured who possess varying numbers of predisposing genes. The threshold indicated by (1) occurs in a favorable environment that is characterized by optimal diet, health care, educational opportunities and other sociocultural and psychological factors. Only one of the four individuals would develop mental retardation under these cicumstances, largely as a consequence of the presence of an unusually large number of predisposing genes. Threshold (2) is characteristic of a harsh environment that places considerable demands upon a person's "genetic reserve." Alleles of a gene often encode enzymes with different catalytic efficiencies. If genes predisposing toward mental retardation determine the structures of catalytically inefficient enzymes or down regulate normal enzyme activities, harsh environments will accentuate the metabolic deficits that result and will increase the likelihood that mental retardation will penetrate. For example, malnourished children suffer from protein and vitamin deprivation. Not all of these children develop mental retardation, indicating that malnutrition is not a self-sufficient cause. However, children who possess inefficient enzymes require more nutrients to carry out their metabolism, and if malnourished, are more likely to develop mental handicaps than someone who has an average set of enzymes. Under these conditions, three of the four individuals in Fig. 11-7 are likely to become mentally retarded. Children of lower socioeconomic groups are also less likely to have adequate health care, are more likely to be exposed to toxic substances such as lead, and are less likely to receive the educational enrichment enjoyed by children who are members of higher socioeconomic families. These factors further stress an individual's genotype, enhancing the likelihood for developing mental handicaps.

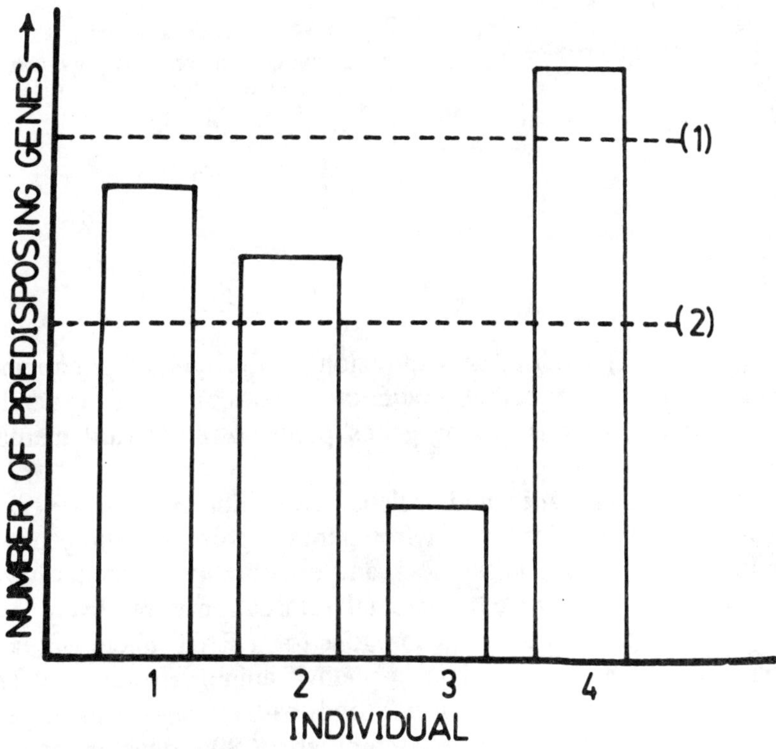

Fig. 11-7. Effect of environment upon expression of mental retardation. 1 = Threshold for mental handicaps in a favorable environment; 2 = threshold for mental retardation in an unfavorable environment.

These interactions between genetic predisposing factors and adverse environmental variables are believed to be at least partially responsible for the greater frequency of mental retardation among lower socioeconomic strata of populations.

Gene expression is also influenced by other genes. The phenomenon of epistasis, described in the preceding chapter, represents one extreme of this type of influence where one genotype can block the phenotypic expression of genes at a second locus. Many genes encode the structures of enzymes that participate in metabolic pathways that are often interconnected. Both major genes and additive genes are subject to influences by this **genetic background**. The effect of the genetic background can be illustrated by examining a small segment of a hypothetical metabolic pathway (Fig. 11-8). Suppose a major locus encodes an enzyme catalyzing reaction (1). Deficiency of this enzyme creates a metabolic block leading to a deficiency of metabolites downstream from the block and accumulation of substrate A. This disruption of the metabolic flow could lead to mental retardation and other problems. Whether or not this phenotype is expressed is influenced by genes encoding enzymes catalyzing reactions (2) and

(3).   If these enzymes have normal activities capable of disposing of **A** and producing **C**, mental retardation may not occur under usual environmental conditions. However, if inefficient enzymes are encoded by mutant alleles at these loci, the probability of mental retardation is enhanced. The loci

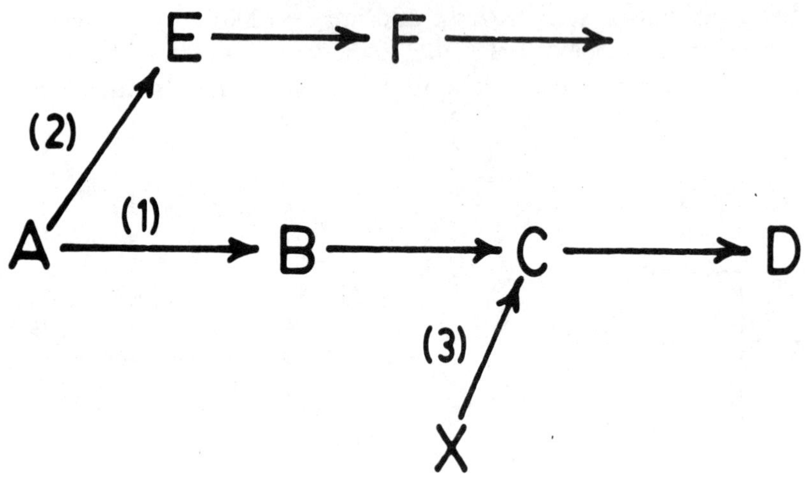

Fig. 11-8. Modifiers affecting expression of a major locus. See text for explanation.

determining enzymes catalyzing reactions (2) and (3) modify the expression of the mutant phenotype characteristic of the enzyme deficiency at step (1) and are called **modifiers.** Extension of this example from one to multiple loci can account for the interactions among genes predisposing toward mental retardation and their genetic background.

As implied by the preceding discussion, penetrance and **variable expression** (varying severity of a phenotype) can be explained by interactions among major loci, environmental factors, and the genetic background (Fig. 11-9). Although major loci (e.g. **PAH**), additive loci, and environmental agents have all been implicated as causes of mental retardation, penetrance of mental retardation is relatively low when only one of the factors is involved. For example, a person homozygous for a PAH^0 allele will not develop mental handicaps if maintained on a corrective diet deficient in phenylalanine. However, when this person ingests foods with average or increased phenylalanine content, about 99% develop mental retardation. The fact that 1% of PKU individuals who are given normal diets do not develop mental retardation suggests that the genetic background also plays a role in the penetrance of mental retardation. Such people may have genes that encode enzymes that enhance the detoxification of phenylalanine or its byproducts. The variable expression of mental retardation may also be explained by Fig. 11-9. Severe to profound mental retardation would be anticipated in cases where all three types of etiological factors overlap (region 4), while overlap of two factors may contribute to mild to moderate mental handicaps.

Mental retardation is **sex-influenced,** that

Table 11-4. Examples of Sex-influenced Traits.

Trait	%Males	%Females
Congenital Dislocation of Hip	15	85
Anencephaly	25	75
Cleft Lip/Palate	67	33
Clubfoot	67	33
Pyloric Stenosis	83	17

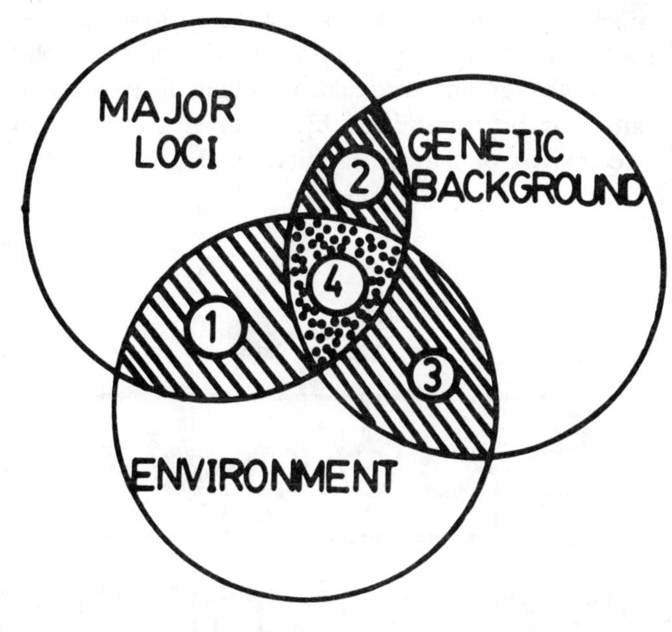

Fig. 11-9. Interactions among etiological factors involved with mental retardation. Overlap of three risk factors engenders the highest risk, while presence of two factors is associated with moderate risk for mental retardation. Severe to moderate mental handicaps may occur when all three risk factors are present; whereas, overlap of two factors may lead to mild to moderate mental retardation.

is, the trait occurs more often in one sex than the other. Other examples of sex-influenced traits are listed in Table 11-4. These sex-influenced conditions appear to involve the effects of different genetic backgrounds, presumably related to the effects of X- and Y-linked genes or the effects of products determined by these genes upon autosomal genes. The excess of females affected with congenital dislocation of the hip appears to be the consequence of a more slanting female pelvis. The excess of females affected with anencephaly may be misleading. The incidence of miscarriage is significantly increased in these families. The more severely affected males may be dying <u>in utero</u>, resulting in the higher miscarriage rate and the relative deficiency of affected males among liveborn infants.

The importance of considering the influence of sex upon the expression of multifactorial traits when counseling prospective parents is illustrated in Table 11-5. The rationale for explaining these data is similar to that provided for mental retardation. Affected females carry a large number of genes predisposing for pyloric stenosis. Since the male theshold is considerably lower, sons of these women run the highest risk. Conversely,

Table 11-5. Risks for Pyloric Stenosis to Sons and Daughters of Affected Men and Women (4).

Affected Parent	Percentage Affected	
	Sons	Daughters
Mother	23	11
Father	6	2

daughters of affected males experience the lowest risk for pyloric stenosis.

Sex-limited traits represent an extreme in which traits are observed in one sex and not the other. Uterine anomalies in women and testicular problems in men are examples of such traits. Genes responsible for such traits may be transmitted by either sex. Sex limited traits may be inherited as dominants (Fig. 11-10), recessives, or may be multifactorial.

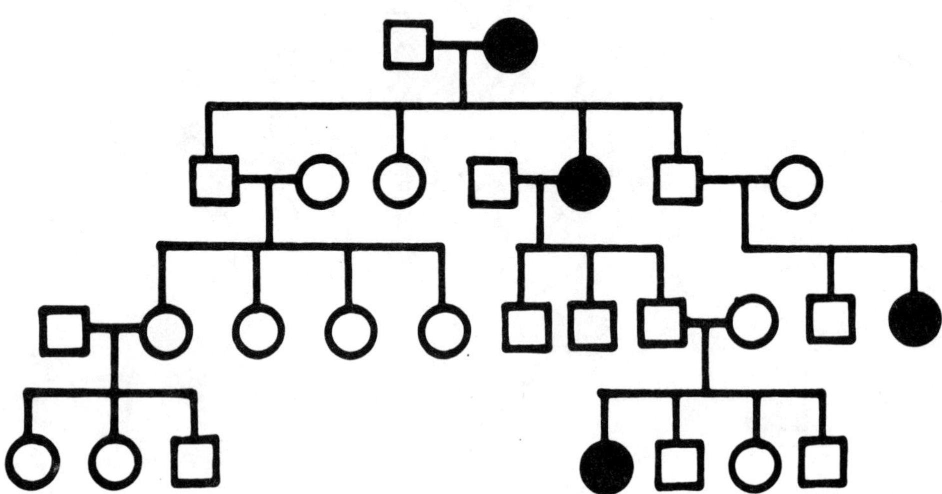

Fig. 11-10.   Inheritance of a sex-limited autosomal dominant trait. Notice how the trait was transmitted through two asymptomatic males to their daughters. About half of the women at risk for the trait developed the problem.

Frequencies of mutlifactorial traits often vary among races and different geographical regions. For example, myocardial infarction occurs more frequently among Black than among Caucasian Americans. Anencephaly and spina bifida occur more frequently among Caucasians from Wales and Ireland than among other groups. Polynesians exhibit a relatively high frequency of severe clubfoot. If these interracial differences are partially determined by additive genetic differences and behave as threshold characters, racial admixture should produce groups who exhibit intermediate frequencies of these traits. The incidence of clubfoot among Asian Americans in Hawaii is quite low (0.4/1000), while native Hawaiians have a high incidence (6.8/1000) (5). The incidence of clubfoot among newborn infants of mixed parentage approximated 3.7/1000.

INDICATORS OF MULTIFACTORIAL INHERITANCE

Frequency of Traits Among First Degree Relatives

When the additive genetic component of a trait is sufficiently high, the frequency of the trait among first degree relatives approximates the square root of the population frequency of the trait (6). An average of 14% of first degree relatives of mentally retarded persons were mentally retarded. This value was close to the square root of the observed control frequency (1.6%). Cleft lip with or without cleft palate occurs in about 1/1000 newborn. The frequency of this trait among first degree relatives of such patients averages 3.5%, which approximates the square root of the newborn frequency. This relationship has been demonstrated to hold for a number of other traits including congenital heart defects, congenital dislocation of the hip, pyloric stenosis, clubfoot, and anencephaly-spina bifida. The square root estimate of risk is often used to forecast risk when no other observational data are available for that purpose, such as in the case of rare birth defects.

Exponential Decline with Decreasing Degree of Relationship

An autosomal dominant disorder displays a stepwise decline with decreasing degree of relationship (Table 11-6). However, as the number of loci determining the trait increases, the decline becomes more precipitous, approaching an exponential decline at about four loci. In each case, the

Table 11-6. Incidence of a Trait Among Relatives Assuming Additive Inheritance and Complete Penetrance.

	Number of Loci				
	1	2	3	4	5
General Population	X	X	X	X	X
1° Relatives	500X	250X	125X	62X	31X
2° Relatives	250X	62X	16X	4X	X
3° Relatives	125X	16X	2X	X	X

X = 1/1000.

affected index case was assumed to be heterozygous at each locus involved. The chance that a child, grandchild or great grandchild would inherit the allele responsible for the trait in column 1 is 1/2, 1/4 and 1/8, respectively. These risks are 500-, 250- and 125-fold higher than for the general population. If additive effects of alleles at two loci, **A** and **B**, are required for development of the trait, the genotype of the index case would be **AaBb**. The probabilities that both the **A** and **B** alleles would be inherited by a child, grandchild or great grandchild become 1/4, 1/16 and 1/64, respectively. These risks are 250-, 62- and 16-fold higher than that of the general population. The remaining entries in the table were calculated in similar fashion. Data collected for several multifactorial traits (Table 11-7) appear to fit a four to five locus model. In each case there is a rapid decline in risk with decreasing degree of relationship. This pattern is not readily explained by single locus models without making a number of qualifying assumptions relevant to penetrance, gene frequencies and other factors.

Increased Risk with Increasing Numbers of Affected Children

Once parents have one or two children affected by a mendelian disorder, the risk to the next child plateaus at a fixed value (1/2, 1/4 or 1/4 for autosomal dominant, autosomal recessive, and X-linked

Table 11-7. Risk to Relatives for the Same Malformation Occurring in the Index Case (4).

Malformation	General Population	1°	2°	3°
		Relative Risk		
Cleft Lip/Palate	1/1000	35X	7X	3X
Congenital Dislocation of Hip	1/1000	40X	4X	1.5X
Pyloric Stenosis	2/1000	20X	5X	2X
Clubfoot	1/1000	20X	5X	2X
Anencephaly-Spina Bifida	2/1000	8X	--	2X

recessive conditions, respectively). In each case, the genotypes of the parents are revealed by birth of the affected child. However, in the case of multifactorial traits that have a **polygenic** component (many loci with additive effects), the genotypes of the parents are never completely known. Threshold theory predicts that parents carrying a relatively large number of predisposing genes should tend to have more affected children than parents who carry fewer predisposing genes. Therefore, repetition of a defect in a family can be used as an estimator of the "gene loading" of the parents. The more affected children, the greater the number of predisposing genes, and the higher the risk to the next child (Table 11-8). This trend has been observed for a number of disorders. These data may also be partially accounted for

Table 11-8. Effect of the Number of Prior Affected Children upon Risk to a Subsequent Child (8,9).

	Number of Prior Affected Children			
	0	1	2	3
Anencephaly-Spina Bifida	1/500	1/20	1/10	1/4
Cleft Lip/Palate	1/1000	1/25	9/100	---

by heterogeneity of causes (7). For example, both anencephaly and cleft lip/palate occur both as "pure" traits in which other systems are unaffected and as components of syndromes. Many of these syndromes are caused by single gene defects, or may be precipitated by unbalanced translocations inherited from a balanced translocation parent. The high risk in these families is superimposed upon the polygenic-environmental type of birth defect. It is important to eliminate these syndromes by thorough clinical evaluation and appropriate laboratory studies before using the risks listed in Table 11-8.

Severity of Malformation and Gene Loading

Threshold theory predicts a direct correlation between numbers of predisposing genes and severity of defect. This prediction is upheld by observation. Cleft lip/palate may occur in either unilateral or bilateral forms, and may affect both the soft and hard palate or the soft palate only. If bilateral clefting of both lip and palate are associated with the presence of a large number of predisposing genes, and unilateral cleft lip signifies the presence of relatively few predisposing genes, then there should be a higher incidence of clefts among relatives of persons with bilateral clefting of the lip and palate. 5.6% of siblings of index cases with bilateral clefting of both lip and palate developed clefting, while 2.6% if siblings of patients with unilateral cleft lip had clefts (10). These trends support the use of severity of a multifactorial birth defect as an indicator of gene loading. These **empiric risks** as well as those data presented in previous sections of this chapter may be used when counseling families relevant to recurrence of multifactorial birth defects.

Twins and Concordance of Multifactorial Traits

Monozygous twins (derived from the same fertilized egg) share all of their genes in common; however, dizygous twins are no more genetically similar than siblings who are products of different pregnancies. If a trait is caused by a dominant or recessive mutation, if a trait is caused by chromosomal nondisjunction, or if a trait is caused by an unbalanced chromosomal rearrangement, then monozygous twins should both have the abnormality about 100% of the time (i.e. they are **concordant** for the trait or syndrome). About 50% of dizygous twins will share a fully penetrant dominant trait, and approximately one in sixteen dizygous twins will be concordant for an autosomal recessive trait. Concordance among dizygous twin pairs for chromosomal abnormalities is usually very low. Multifactorial traits occupy an intermediate position. Concordance rates among monozygous twins for birth defects is generally quite low, indicating a major role for environmental factors in their etiology (Table 11-9). However, dizygous concordance for the same traits is much lower and comparable to those of siblings in general. The latter trend indicates that there is a significant polygenic component in each of the birth defects listed in the table. These data are not strictly comparable, since they were taken from a large

Table 11-9. Concordance Rates for Common Birth Defects (7,10,11).

	Monozygous Twins	Dizygous Twins	Non-twin Siblings
Cleft Lip/Palate	38%	8%	4%
Pyloric Stenosis	22%	4%	3%
Congenital Dislocation of Hip	35%	5%	4%
Clubfoot	33%	3%	3%
Anencephaly-Spina Bifida	21%	--	5%

number of different studies that were performed in somewhat different ways. Furthermore, environments of monozygous twins may be more similar than those of dizygous twins or siblings who are derived from different pregnancies. If environments of monozygous twins are more similar than those of dizygous twins, the effect of this bias on concordance will simulate what is expected for genetic factors. In spite of these limitations, the twin studies described above provide additional evidence for a polygenic contribution to multifactorial traits.

ESTIMATING GENETIC AND ENVIRONMENTAL INFLUENCES UPON COMPLEX TRAITS

Sir Francis Galton, a cousin of Charles Darwin, first proposed twin studies as a means for distinguishing the relative roles of heredity and environment in the determination of a trait. Estimation of genetic and environmental influences upon complex traits requires design of research protocols in which one of these components is maintained constant between groups being compared, while the other is varied to study its effects. This type of research is difficult to perform on human subjects.

The relative effect of heredity upon a trait can be estimated by observing the occurrence of complex traits in siblings and dizygotic and monozygotic twins who have been raised together and comparing the incidences of these traits in these groups with those in siblings and twins who have been reared separately. The rationale behind these studies is as follows. If a complex trait is largely environmental, the frequency of the trait among biological relatives should be largely influenced by the environment in which they have been reared. Therefore, the incidence of the trait should be higher among siblings and twins reared together than among those who have been separated from each other early in life and raised in different homes. On the other hand, if a trait has a major genetic component, the incidence of the trait among adopted siblings and twins should be more similar to that among their biological siblings and cotwins who have stayed with their parents than the incidence among their adoptive siblings in their new home. Monozygotic and dizygotic twins have been used most often for these studies, since they provide a direct estimation of the role of genetics. However, numbers of twins who have been separated from birth or early in life are small.

Many more investigations have used monozygous and dizygous twins who have been reared with their cotwins and biological siblings and parents.The study group is ascertained by selecting twin pairs in which at least one twin is affected by the trait. The frequency of twin pairs in which both twins are affected (**concordance**) is then calculated:

$$5) \qquad C = \frac{\underline{\text{Number of twin pairs in which both are affected}}}{\text{Total number of twin pairs}} \quad \text{x } 100$$

For monozygous twins, concordance values approaching 100% suggest a strong genetic component, while those near 0 indicate a strong environmental component.

There are several limitations to use of concordance. Environments of monozygous twins tend to be more similar than those of dizygous twins. This bias would tend to inflate the role of genetics in the etiology of a trait. A second bias in favor of genetics was also present in many of the early twin studies. Scientists often ascertained their samples by searching hospitals, mental institutions, homes for the mentally retarded, and other special care facilities for twins with a particular problem. This mode of ascertainment would naturally detect more severe cases, and milder forms of the traits would be missed. More severe forms of a disease or birth defect often have larger genetic components than milder forms, and concordance rates among institutionalized cases tend to be much higher than those of unbiased series.

The use of concordance and the bias created by screening institutions for cases of a multifactorial disorder can be illustrated by examining schizophrenia. Schizophrenia afflicts about 1% of the population and most commonly appears during late adolescence or early adulthood. The disease follows an intermittent course with periods of psychotic behaviors interspersed with periods of partial remission. Schizophrenics tend to suffer auditory and, less commonly, visual or tactile hallucinations. Delusions are another hallmark of schizophenic psychosis and usually involve a perception of some outside force or person directing the patient's thoughts or actions. There seems to be a blunting of emotion, typically preventing the patient from experiencing social feelings like empathy. Patients are difficult to motivate, because there is not a true appreciation of rewards. By contrast, feelings of fear and rage may be heightened. Several schizophrenic reactions have been described, including paranoia, simple schizophrenic reaction, hebephrenia, and catatonia. Paranoia usually involves unusual behavior in response to a perceived external threat or danger. There is usually no rational basis for this fear. Furthermore, paranoid patients usually function normally in other aspects of their everyday life. Simple schizophrenic reaction is characterized by a loss of interest in usual activities and moderate withdrawal. Hebephrenia refers to an individual displaying inappropriate behavioral responses such as laughing at sad things and vice versa. Hebephrenics tend to speak in a "word salad", using words and phrases totally out of context. They will grimace in front of mirrors and display other types of bizarre behavior. Catatonic reaction is a severe withdrawal reaction in which the patient appears to be totally out of contact with her/his environment. There is little, if any, communication with others, and patients demonstrate "waxy flexibility". If you raise their arm, they will hold their arm in that position for a long time. Catatonic schizophrenics can be violent as they emerge from this behavioral state. The same person may exhibit two or more behavioral reactions during his/her lifetime, or schizophrenia may present as one reaction type in other individuals. Etiologically, all reactions are viewed as symptoms of the same disease process, with catatonia and hebephrenia being considered as more severe disturbances than paranoia and simple schizophrenia.

The effect of severity upon concordance among monozygotic and dizygotic twins is illustrated by twin studies that were conducted during investigation of genetic factors in schizophrenia. Early studies (12) employed hospitalized twins for their sample. Concordance rates for monozygotic twins in these series generally exceeded 50%, while dizygotic concordance rates averaged 10%. More recent studies (13) utilized twins drawn from the general population. Concordance rates were much lower: monozygous twins = 15.5%; dizygous twins = 4.4%. Although the latter concordance values are much lower, they still support a role for genetics in the etiology of schizophrenia, since concordance for monozygous twins is about four-fold higher than that for dizygous twins.

Adoption studies have also been utilized to explore the effect of genetic factors on liability for schizophrenia. Approximately 15% of children of one schizophrenic and one non-schizophrenic parent ultimately develop schizophrenia. The incidence of schizophrenia among children with the same biological parents, but who were reared apart in foster homes, ranges from 10-17% (14,15). The similar risks experienced by siblings reared in a schizophrenic home and those adopted away support a major genetic component in this disease. The few cases of monozygous and dizygous twins reared apart support the sibling studies.

Heritability

Heritability is defined as the additive genetic contribution to a specific phenotype. The total phenotypic variance was subdivided into genetic and environmental components (equation 4). If monozygous (MZ) twins share the same genotype, then the phenotypic variability among them is largely environmental or

7)
$$S_{MZ}^2 = S_E^2.$$

Dizygous (DZ) twins differ with respect to both genotype and environment. Therefore,

8)
$$S_{DZ}^2 = S_G^2 + S_E^2 = S_G^2 + S_{MZ}^2$$

and

9)
$$S_G^2 = S_{DZ}^2 - S_{MZ}^2.$$

Substituting into equation (4):

$$S_P^2 = S_G^2 + S_E^2 + S_{GE}^2$$

$$S_P^2 = (S_{DZ}^2 - S_{MZ}^2) + S_{MZ}^2 + S_{GE}^2$$

or

10)
$$S_P^2 = S_{DZ}^2 + S_{GE}^2.$$

Genotypic-environmental variance is difficult to measure and is largely ignored in twin studies, such that an approximation of S_P^2 is used:

11)
$$S_P^2 = S_{DZ}^2.$$

That is, the total phenotypic variability of a trait can be approximated by the phenotypic variability among DZ twins. For **discontinuous** traits, the heritability can be calculated :

12)
$$H = (C_{MZ} - C_{DZ}) \div (100 - C_{DZ})$$

where H is the heritability, C_{MZ} is the monozygous twin concordance, and C_{DZ} is the dizygous twin concordance. H assumes values ranging from 0 (minimal additive genetic input) to 1 (largely genetic).
Heritabilities of **continuous** traits may be calculated using the phenotypic variances or correlation coefficients. Since MZ twins are genetically identical, variation between cotwins must reflect environmental influences. Variation of DZ twins is used as an estimator of total or both genetic and environmental effects. Therefore,

$$H = (\text{Total Variation} - \text{Environmental Variation}) \div (\text{Total Variation})$$

or

13)
$$H = (S_{DZ}^2 - S_{MZ}^2) \div S_{DZ}^2.$$

Correlation coefficients estimate trends in covariation of two variables. For example, suppose height and weight of two individuals are being compared. If tall persons tend to weigh more than short individuals, height and weight would be **positively** correlated: as the height increases, weight increases and vice versa. If our sample revealed that tall individuals tend to weigh less than short individuals, height and weight would be negatively correlated. Correlation coefficients (**r**) fall within the range of +1 (positively correlated) to -1 (negatively correlated). When r is close to zero, variation of one trait does not have an appreciable effect upon the variation of the other trait.

MZ twins should display a high degree of inter-twin similarity with respect to traits that are strongly influenced by genetic factors, while DZ twins should be less similar. The heritability equation assumes the form:

14)
$$H = (r_{MZ} - r_{DZ}) \div (1 - r_{DZ})$$

where r_{MZ} and r_{DZ} are the correlation coefficients for MZ and DZ twins, respectively. Note that this equation can be modified to assess the influence of environment (E) upon a trait:

15)
$$E = (r_{MZT} - r_{MZA}) \div (1 - r_{MZA})$$

where r_{MZT} and r_{MZA} are the correlation coefficients for monozygous twins raised together or apart, respectively. If the trait has a major environmental component, MZ twins raised together would be more likely to encounter the environmental factor than twins raised apart. Therefore, r_{MZT} would approach 1, and r_{MZA} would be about 0, making E approximate 1.

Interpretation of heritability estimates must be undertaken with care. Down's syndrome generally has a relatively low recurrence rate. Therefore, although Down's syndrome is genetic in the sense that it is caused by a chromosomal imbalance, the most common type is usually not inherited. By contrast, the heritability of Down's syndrome is greater than 0.9, suggesting a strong additive genetic component. The reason for this discrepancy arises from the biology of Down's syndrome and monozygotic twinning. Since MZ twins originate from the same fertilized egg, approximately 100% of MZ twins will be concordant for primary trisomy 21. On the other hand, the nodisjunctional event leading to the formation of a disomic gamete would be unlikely to occur twice during the formation of the ova that contribute to DZ twins, and the DZ concordance rate would approximate 0. These two trends are responsible for the high heritability value for Down's syndrome. Two other biases that affect concordance also affect heritabilities, and these have both been described above. One of these is the tendency for MZ twins to experience more similar environments than DZ twins, and the other is the biased sampling that is often used to collect twin series. Both of these biases inflate the heritability estimate. Finally, heritabilities are environment-bound. That is, heritabilities should be applied to the same population and in the same general environment in which they were determined. A case where heritabilities were misapplied involved I.Q. scores of Black and Caucasian Americans. A group of scientists argued 1) that I.Q. scores of Caucasian Americans were significantly higher than those of Black Americans and 2) that I.Q. as measured by the usual tests are strongly genetically determined (I.Q. > 0.7). The implications of these arguments for federal, state, and local educational programs for minorities are obvious. There were several problems with the interpretations of these scientists. Studies have shown that I.Q. tests are strongly culturally biased. They contain concepts, facts and words that are more likely to be encountered by Caucasian American than by Black American children. It has also been found that the race and attitudes of the person administering the test can influence the outcome by as much as 10 I.Q. points. Finally, the high heritability for intelligence as determined by I.Q. testing that was sighted by these scientists appears to have been based upon fraudulent data. More recent estimates of heritability generally approximate 0.5, indicating that both genetics and environment play a major role in the determination of intellectual potential.

SELECTED EXAMPLES OF MULTIFACTORIAL TRAITS

Schizophrenia and Manic-Depressive Psychosis

During the last several years, a theory for the etiology of mental illness has emerged. This theory proposes that environmental stresses interact with a genetic predisposition to precipitate episodes of mental illness (Fig. 11-11). A subset of the population carries a significant number of genes that

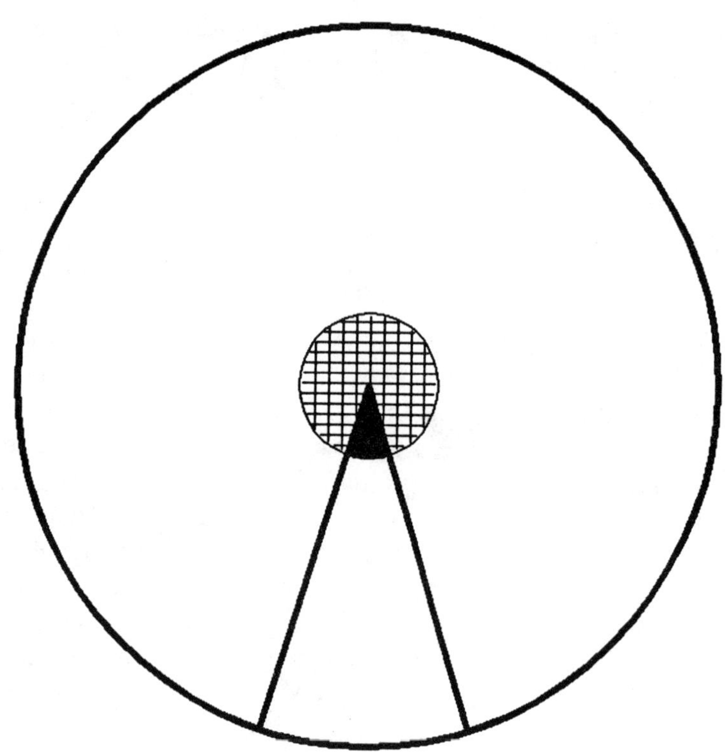

Fig. 11-11. Etiology of mental illness. The large circle encompasses the entire population, and the small circle encloses a group of persons genetically predisposed toward a behavioral reaction. The wedge represents environmental stresses capable of triggering mental illness. People who are genetically predisposed for a behavioral disorder are much more likely to develop symptoms when they encounter the appropriate stresses and are represented by the black shading.

predispose toward behavioral disorders. When this group encounters environmental stresses, they are more likely to develop an abnormal behavioral reaction than someone who lacks the predisposing genes. Schizophrenia was briefly mentioned in a preceding section of this chapter. The incidence of schizophrenia among second degree and third degree relatives (2-3% for both groups) is barely increased over the general population incidence of schizophrenia (1%); however, children of one schizophrenic run a risk of about 15% for the disorder. Children of two schizophrenic parents develop schizophrenia 45-50% of the time (16,17). There is other evidence favoring gene loading for this behavioral disorder (18,19). Recurrence risks for children of patients with more severe schizophrenic reactions are generally higher than those for children of patients with milder disorders. Three-fold higher rates were observed among siblings of catatonic or hebephrenic persons compared to siblings of paranoid and simple schizophrenic patients. Furthermore, concordance among MZ twins who had more severe forms of schizophrenia (unable to work for as long as three years after their original ascertainment) was 80%, while that for MZ twins with milder schizophrenic disorders approximated 27%.

Affective disorders, such as manic-depressive psychosis, have also been demonstrated to be multifactorial. Manic-depressive psychosis is characterized by a gradual change in mood and occurs in bipolar and unipolar forms. Bipolar affective disorder includes individuals who have manic or both manic and depressive episodes and patients with depression who have relatives who have experienced mania. Unipolar disease is considered to be a distinct disease manifested by depression only in the patient and among her/his relatives. Mania involves a positive mood swing characterized by the occurrence of four or more of the following: euphoria, grandiosity, extravagance, racing thoughts, over talkativeness, short attention span, increased motor activity, flight of ideas, impatience with restraint, and excessive plans. Behavior varies among sexes and according to culture. For example, a man may tend to buy several new cars or deliberately pick fights with others during a manic episode, while women may purchase fur coats or other expensive items during their manic episode. Depression presents as loss of energy, loss of appetite, sleep disturbance, loss of interst in usual things, retardation, loss of sex interest, and related problems characteristic of a downward mood swing. Manic-depression occurs more frequently in women and is often associated with grief reaction, alcoholism and other emotional disturbances. Most genetic studies have selected cases that were not precipitated by these problems.

The genetics of manic-depressive illness (MD) have been summarized by Stabenau (20). MD is common, affecting 1-2% of the population, with approximately 12% of the children of an MD parent and 9% of the siblings of an MD patient also contracting the illness. Heritabilities for the disorder have generally approximated 0.6.

Data indicate that three distinct polygenic systems predispose toward schizophrenia, bipolar disease, and unipolar depression. Neither affective disorder is increased in frequency in "schizophrenic" families, and the frequency of schizophrenia is not greater among relatives of patients with either affective disorder compared to the general population. All three diseases appear to respond to different drugs: schizophrenia - phenathiazines; bipolar disease - lithium compounds; and unipolar disease - tricyclic drugs. An excellent presentation of the various genetic and environmental aspects of behavior can be found in Kaplan (21).

Diabetes Mellitus

Diabetes mellitus is a disease associated with abnormal glucose utilization and metabolism. The disease is chemically defined as the occurrence of unusually high blood glucose levels, is heterogeneous with respect to cause, and ultimately involves 3-4% of the population. A diabetic is believed to pass through a series of stages during the development of the disease. The prediabetic has normal glucose tolerance and is free of clinical signs of the disease. Identical twins of a diabetic and women who birth unusually large infants fall in this category. Latent diabetes represents an intermediate stage. Glucose tolerance is normal under usual conditions; however, a stress such as pregnancy can lead to abnormal glucose tolerance. Clinical diabetes has both short term and long term effects. The high blood glucose in clinical diabetics exceeds the ability of the kidney reabsorption system. This excessive loss of glucose in the urine triggers sensations of thirst and hunger. Diabetics are unable to utilize carbohydrate as an efficient source of energy; therefore, reliance is placed upon alternative energy sources including fats and proteins. Use of these alternative energy stores results in a loss of weight and the production of ketones and related substances which depress central nervous system activity, resulting in the diabetic coma. Long term effects of diabetes include development of cataracts and retinopathy leading to blindness and vascular changes that result in impotence, enhanced infection, kidney failure, and heart disease. Management of diabetes depends upon the type of problem. Insulin-dependent varieties require insulin injections, while non-insulin dependent types can be treated by oral agents that stimulate insulin production, by dietary changes, and an exercise program.

The genetics of diabetes is complicated. More than thirty chromosomal or mendelian syndromes have been described in which diabetes is one symptom. This indicates that there are a large number of genes that impact upon normal glucose utilization. Several multifactorial types of diabetes are also known (Table 11-10). In addition to polygenic components, numerous environmental factors have also been implicated, including diet, lifestyle, viral diseases, autoimmunity, and drugs and other toxic

Table 11-10. Heterogeneity of Multifactorial Diabetes Mellitus (modified from 22).

	Non-insulin Dependent		Insulin Dependent
	Maturity Onset	Juvenile Onset	Juvenile Onset
Insulin dependence	-	-	+
Ketotic	-	-	+
Obesity	+	?	-
NIDDM increased in family	+	+	-
IDDM increased in family	-	-	+
MZ Concordance (Impaired glucose tolerance)	100%	100%	<50%
Abrupt onset/severe course	-	-	+
Islet cell antibodies	-	-	+
Autoimmune endocrine disease	-	-	+
HLA associations			
DR3/DR4	-	-	+
B8/B15	-	-	+
Incidence:			
Children of diabetic parent	5-10%	50%	(1-5%)
Siblings of diabetic	5-10%	50%	(2-11%)

NIDDM = non-insulin dependent diabetes mellitus; IDDM = Insulin dependent diabetes mellitus substances.

Two general types of multifactorial diabetes are recognized: non-insulin dependent (NIDDM) and insulin dependent diabetes mellitus (IDDM). The latter disease usually appears during adolescence or early adulthood and requires insulin supplementation. Patients are generally thin and ketotic, and the disease requires close monitoring to make sure insulin dose and metabolic needs during growth periods are properly balanced. IDDM is associated with autoimmune symptoms and anti-insulin antibodies, and at least two subtypes of the disease appear to exist. Certain HLA haplotypes occur more frequently among IDDM patients. Concordance among MZ twins for IDDM has generally been reported to be less than 50%, supporting a major environmental component for the disease. Recurrence risks are generally very low, and appear to be influenced by class II HLA haplotypes. Siblings who share two DR3 and/or DR4 alleles have the highest risk (about 11%), while siblings sharing neither of their DR alleles with the patient run a risk approximating 2%. NIDDM is usually a maturity-onset disease, often occurs in obese individuals, and is generally not associated with ketosis. These patients generally have some residual capacity for insulin production and secretion, and management is directed at amplifying insuling delivery to tissues by using oral agents and at reducing the insulin requirement through dietary management and exercise. NIDDM is not HLA-associated and does not appear to involve immune phenomena. However, there appears to be a stronger genetic component, with concordance rates for impaired glucose tolerance for MZ twins approaching 100%. Concordance for the clinical disease among MZ twins approximates 50%, suggesting contribution by environmental factors. Recurrence risks generally average 5-10% both among children and among siblings of NIDDM patients.

Two additional multifactorial NIDDM phenotypes are known. Both have onset during adolescence and are non-progressive. MODY (maturity onset diabetes of youth) resembles its more common adult counterpart in many respects. Its high incidence among relatives and characteristic

Table 11-11. Incidence of Hyperlipidemia Among MI Survivors and the General Population (23).

	Incidence (%) of Hyperlipidemia		
	MI Survivors		General Population
	<60	>60	
A) Monogenic			
Familial Hypercholesterolemia	4.1	0.7	0.1 - 0.2
Familial Hypertriglyceridemia	5.2	2.7	0.2 - 0.3
Combined Hyperlipidemia	11.2	4.1	0.3 - 0.5
B) Multifactorial (Hypercholesterolemia)	5.5	5.5	?
C) Sporadic (Hypertriglyceridemia)	5.8	6.9	?

pedigree patterns are suggestive of autosomal dominant inheritance. About 10% of Black Americans develop a NIDDM similar to MODY; however, the disease is temporary, disappearing after completion of the adolescent growth period.

Coronary Artery Disease

Appreciation for the role of blood lipids in the development of deposits in vessel walls (atherosclerosis), heart attacks and strokes has increased in recent years. Cholesterol and triglycerides appear to be the major lipids involved with pathogenesis of coronary artery disease and myocardial infarction (MI), especially among persons experiencing heart attacks at younger ages (Table 11-11). Monogenic forms of hyperlipidemia (Chapter 8) appear to be largely responsible for the age difference. Cholesterol levels are continuously distributed, and MI behaves as a threshold trait. There is a strong correlation among cholesterol levels of parents, children and siblings and between MZ twins; however, correlation between cholesterol levels of spouses is much lower. Dietary lipid, sedentary lifestyle, obesity, and smoking are known environmental factors that contribute to risk for heart attacks. Reducing cholesterol and saturated fat intake, weight loss, exercise, and avoidance of smoking have been demonstrated to delay or prevent heart attacks. However, the heterogeneous genetic composition of the human population makes evaluation of the effects of these preventive measures on large groups difficult, and contradictory results are not infrequent.

Cancer

Cancer is a highly heterogeneous group of diseases characterized by uncontrolled cellular growth and behaves as a multifactorial disease. Prior sections of this text have discussed the roles of chromosome rearrangements and proto-oncogenes, mutations incapacitating anti-cancer genes, and the arylhydrocarbon hydroxylase system in the etiology of cancer. A summary of several types of genetic

Table 11-12. Representative Cancers According to Type of Genetic Predisposition.

Autosomal Dominant	**Autosomal Recessive**
Neurofibromatosis	Ataxia Telangiectasia
Retinoblastoma	Fanconi's Anemia
Pheochromocytoma	Bloom's Syndrome
Gardner's Syndrome	Xeroderma Pigmentosum
X-linked Recessive	**Multifactorial**
Bruton's Hypogammaglobulinemia	Breast Cancer
Wiskott-Aldrich Syndrome	Lung Cancer
	Prostate Cancer
Chromosomal	Stomach Cancer
Burkitt's Lymphoma	
Chronic Myeloid Leukemia	
Down Syndrome	

predisposition to cancer is presented in Table 11-12. Many adult cancers fall in the multifactorial category; however, breast cancer is one of the better known and more common types. Breast cancer afflicts about 5% of women. Although men generally have a much lower incidence of this cancer, the incidence among Klinefelter's males approximates that of women. The frequency of breast cancer varies among different regions of the world with Asian populations such as the Japanese having very low frequencies and Caucasians in the United States exhibiting relatively high frequencies. However, these differences disappear within one generation following migration of Japanese to the United States. This trend appears to implicate environmental variables such as diet in the etiology of breast cancer (24).

Macklin (25) provided the first evidence in favor of genetic factors contributing to breast cancer. Female relatives of breast cancer patients had a three-fold higher incidence of breast cancer than either female relatives of patients with cancer of other organs or female relatives of persons without a history of cancer. Macklin also noted that nulliparous women (have had no children) displayed the highest risk for breast cancer (26). These observations have been supported and extended by others. The latter studies indicate that: 1) the site specificity of the cancer is similar among close relatives; 2) familial breast cancers tend to appear earlier than sporadic cases; and 3) risk to relatives is highest when the index case has bilateral, premenopausal breast cancer. Kindreds afflicted with bilateral, premenopausal breast cancer often display patterns suggestive of autosomal dominant inheritance, and as many as 10% of familial cases may be transmitted in this manner. The majority of familial cases appear to have a multifactorial etiology. Recent evidence indicates that breast cancer occurs following several mutations in anti-cancer genes and/or abnormal expression of certain proto-oncogenes. If one or more of these mutations involve the germ cell lineage, inherited predisposition would be expected. Dominant forms of breast cancer predisposition could reflect an inherited mutation in a key gene required for normal breast development and function.

Normal Color Vision and Color Blindness

Four visual pigments are found within retinal receptors. One of these, rhodopsin, occurs within cells known as rods and is especially important for black-white vision and plays a major role in vision under dim light conditions. The remaining photopigments occur within cones which are essential for color perception and are most efficient in bright light. Three types of cones populate the retina: red-sensitive, blue-sensitive, and green-sensitive. Each cone type contains a different photpigment which displays a distinct absorption maximum: red photopigment - 560nm, green photopigment - 530nm, and blue photopigment - 420nm. Color discrimination occurs as a consequence of the differential stimulation of these cone types by a particular mixture of light photons reflected by the object being viewed. If all three cone types are functioning normally a mixture of light containing equal quantities of red, green,

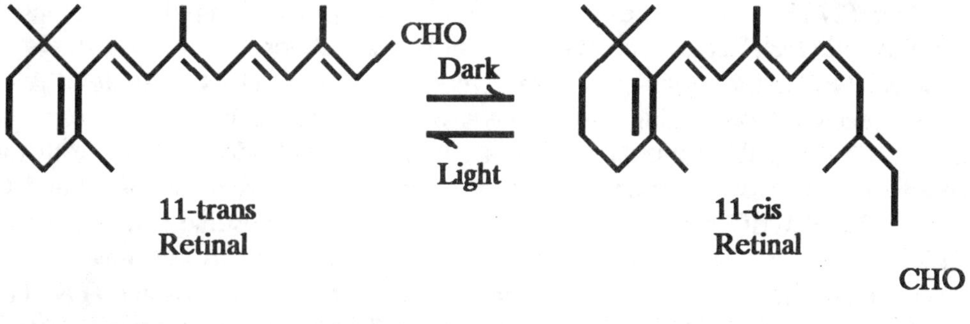

Fig. 11-12. Interconversion of 11-cis and 11-trans retinal.

and blue will be perceived as white.

All four photopigments contain two basic components, 11-trans retinal (derived from vitamin A) and a protein component known as an opsin. The 11-trans and 11-cis retinals are presented in Fig. 11-13. When a particular visual pigment absorbs a photon of light, the retinal moiety isomerizes to the all trans configuration. This change in structure changes the conformation of the attached opsin and sets a complex cascade of enzymatic reactions in motion which ultimately generates a nerve impulse. The nerve impulse travels over the optic nerve network to the visual center in the occipital lobe of the brain. This center monitors the pattern of impulses it receives and interprets the pattern as a color.

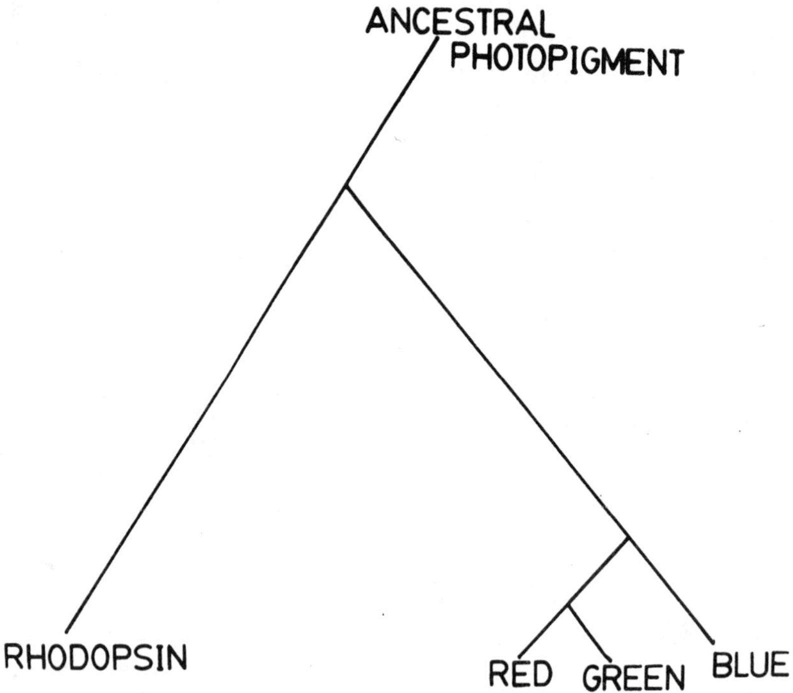

Fig. 11-13. **Evolution of the photopigment genes.**

The opsins occurring in each type of photopigment have different but related structures, suggesting they have evolved from a common ancestral opsin gene (Fig. 11-13). The red and green opsin genes are most similar, followed by the blue opsin and rhodopsin. Molecular analyses of the opsin genes have been published (27,28). The genes encoding rhodopsin (**RHO**) and the blue opsin (**BCP**) are autosomal, while the red (**RCP**) and green (**GCP**) opsin genes are closely linked on the X chromosome. The molecular structures of the opsin genes are presented in Fig. 11-14. All four genes have the specialized type of promoter containing the **TATAA** element and have a downstream **AATAAA** or **AATTAAA** element that signals termination of transcription and polyadenylation. Both the **RHO** and **BCP** genes contain five exons which specify different opsin domains, while the **RCP** and **GCP** possess six exons. **RCP**, **BCP** and **RHO** occur in one copy per haploid chromosome set; however, the number of copies of **GCP** genes vary among X chromosomes, ranging from one to as many as five. The **RCP** and **GCP** genes are linked in a head to tail tandem array on each X chromosome (Fig. 11-15a). The occurrence of one to several **GCP** loci and only one **RCP** gene per chromosome suggests that there must be some type of functional dosage compensation between the **RCP** and **GCP** genes. For example, if there are one **RCP** and three **GCP** loci on an X chromosome, each of the three **GCP** genes may be expressed at about a third of the level of the single **RCP** gene.

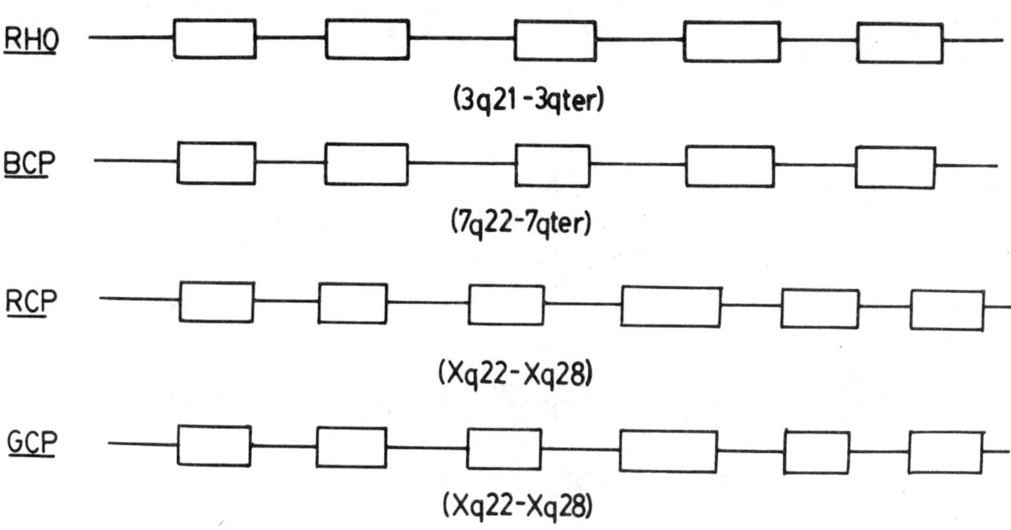

Fig. 11-14. **Structure of the photopigment genes.** Each gene consists of multiple exons and introns. The chromosomal assignments of the genes are listed below the respective loci. RHO = rhodopsin; BCP, RCP, and GCP = blue-, red-, and green-cone pigment genes, respectively. The exons and introns are not drawn to scale.

The occurrence of several genes with similar structure in close proximity is conducive to meiotic mispairing and unequal crossing over leading to a shuffling of genes between X chromosomes and the generation of fusion genes. Both anomalies occur frequently in the **RCP** - **GCP** region of the X chromosome. Sixteen percent of Caucasian American males were found to have **RCP/GCP** fusion genes, half of which appeared to cause anomalous color vision. An even larger proportion of Black American males (21%) had fusion genes; however, the lower incidence of color vision defects among Black males (5%) indicates that only a quarter of these resulted in color vision impairment. Among Japanese Americans, the frequencies of fusion genes and color vision defects is comparable, with both approximating 4% (29). The production of opsin fusion genes is illustrated in Fig. 11-15b. The physiological consequences of a particular fusion gene are apparently determined by the sites of the disruptions of the sequences of the respective genes involved and by the number of intact opsin genes remaining on the rearranged chromosomes. Phenotypic effects of alleles at the opsin loci cover the complete spectrum from normal color sensitivity to total color insensitivity, but, for the sake of discussion, may be apportioned among four rather arbitrary groups: normal sensitivity, moderate reduction of color sensitivity, severe reduction of color sensitivity, and total lack of color sensitivity. Each of these groups can be further subdivided according to the locus involved with the defect (Table 11-13). Anomalous trichromats require more of one of the three primary colors to see white (protanomaly = red weakness; deuteranomaly = green weakness; and tritanomaly = blue weakness). Dichromats can only sense two of the three colors (protanopia = red blindness; deuteranopia = green blindness; and tritanopia = blue blindness). In female compound heterozygotes, a dominance relationship among alleles is usually

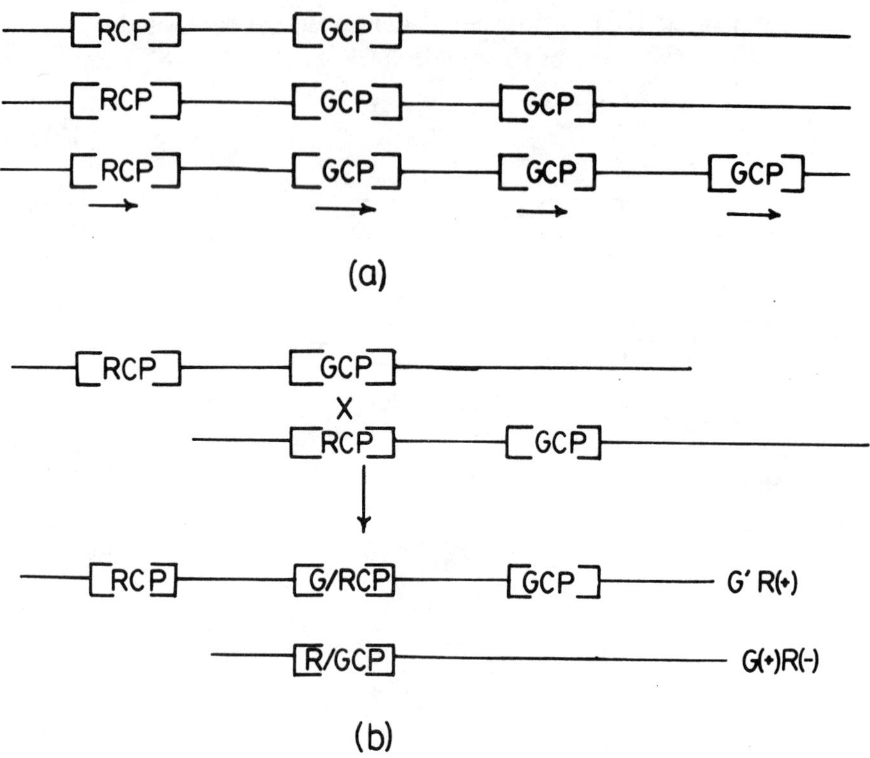

Fig. 11-15. X-chromosome haplotypes of the red- and green-cone pigment genes. a) X-chromosomes from males with normal color vision. b) Intragenic recombination producing a green anomalous trichromat (requires more green to see white (deuteranomaly)) haplotype (G'R(+)) and a red dichromat (cannot sense red (protanopia)) haplotype (G(+)R(-)), respectively. Each gene enclosed within brackets consists of multiple exons and introns. Molecular data do not permit prediction of the color sensitivity phenotype, since the exact breakpoints in the two alleles are often not known, and the effects of residual intact genes must be taken into account.

observed: normal allele > protanomaly allele > protanopic allele (and similarly for other loci). Monochromats are able to sense only one color. For example, blue monochromacy implies both red and green colorblindness. Recombination may also occur between mispaired cone pigment genes, producing chromosomes that lack an **RCP** allele or which have no **GCP** genes. This intergenic recombination can frequently result in total loss of sensitivity to red or green light. In some cases in which the deleted chromosomal material encompasses two or more **RCP** and **GCP** loci, blue monochromacy may result. The multiplicity of **GCP** genes in many haplotypes leads to relatively more rearrangements that involve **GCP** and contributes to an excess of green vision defects such as deuteranomaly.

Mutations in the **RHO** gene occur less often than those affecting **RCP** and **GCP**. Some of these mutations totally compromise an individual's ability to see at night (when cone vision is very poor), and may present as night blindness.

Normal color vision behaves as a multifactorial trait. At least four loci contribute to sensation of light and color. Environmental factors such as the brightness of light and the color qualities of the objects being viewed also influence color perception.

Table 11-13. Anomalous Color Vision.

Anomaly	Genotypes
Anomalous Trichromats	
Protanomaly	BCP(+), RCP', GCP(+) (moderate); BCP(+), RCP", GCP(+) (severe)
Deuteranomaly	BCP(+), RCP(+), GCP' ("); BCP(+), RCP(+), GCP" (")
Tritanomaly	BCP', RCP(+), GCP(+) ("); BCP", RCP(+), GCP(+) (")
Dichromats	
Protanopia	BCP(+), RCP(-), GCP(+)
Deuteranopia	BCP(+), RCP(+), GCP(-)
Tritanopia	BCP(-), RCP(+), GCP(+)
Blue Monochromasia	BCP(+), RCP(-), GCP(-)

*Dominance relations in female heterozygotes: RCP(+) > RCP' > RCP" > RCP(-)
**Note: RCP(-) may refer to deletion of the RCP gene from the chromosome or either a point mutation or fusion gene that has totally compromised red sensitivity.

SUMMARY

Phenotypic variation has three basic components: genotypic, environmental and genotype-environmental interaction. The genetic component of phenotypic variation includes an effect due to major loci superimposed upon a background of minor loci with significant cumulative effects.

Mulitfactorial traits behave as threshold characters. Major and additive genes predispose an individual to these traits by placing him/her closer to or beyond this threshold, increasing the likelihood that certain environmental factors will precipitate the appearance of the clinical problem. Close relatives share more genes in common than unrelated persons. Therefore, first degree relatives of an index case with the multifactorial trait are situated closer to the threshold than members of the general population, and the incidence of the trait is markedly increased among these relatives. The proportion of shared alleles at several to many loci falls rapidly with decreasing degrees of relationship, and risks of second and third degree relatives are barely increased compared to that of the population at large. Thresholds often vary between sexes, and risks to children of the less frequently affected sex are correspondingly increased. Recurrence risks to siblings of an affected child and to children of an affected parent increase with increasing severity of the problem and with increasing numbers of prior born children with the defect.

Heritability is broadly defined as an estimation of the genetic contribution to a trait. Heritability is also defined in a narrower sense as a measure of the additive genetic component of a human trait. Heritability can be estimated by comparing concordance rates for monozygous and dizygous twins or among siblings in general. Continuously varying phenotypes can be compared in a similar manner by utilizing variation or correlation of the quantitative variable. Heritabilities range from 0 (negligible genetic component) to 1 (strong genetic component). Heritabilities should be interpreted with care, since they are subject to a number of biases and cannot be applied to populations possessing different gene pools or living in quite different environments.

PROBLEMS

1. What type of inheritance is responsible for the following trait?

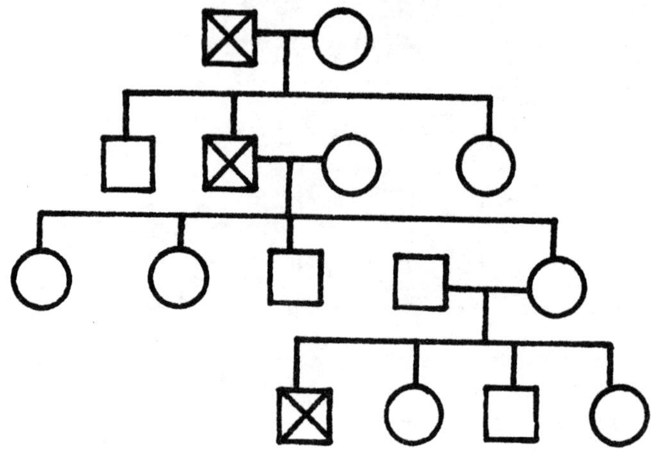

2-7: Match the following terms with their appropriate definitions.

2. variable expressivity

    a. gene that influences expression of a gene at another locus

3. sex-influenced

    b. additive genetic component of a trait

4. heritability

    c. varying manifestations of a trait in genotypes predisposing toward that trait

5. modifier

    d. trait that is expressed more frequently in one sex

6. incomplete penetrance

    e. percent of twin pairs in which both cotwins express a trait

7. concordance

    f. failure of expression of a trait in all individuals of appropriate genotype

8. Why does pattern baldness occur more frequently among men than among women?

9. Fifty percent of MZ twins and 20% of DZ twins were concordant for trait B. Calculate the heritability for this trait.

10. How would you interpret this value?

11. What limitations should be placed upon the interpretation and utilization of heritabilities?

12. Trait Z occurs at the following frequencies among relatives of affected persons:

Relationship	% Affected
$1°$	30
$2°$	15
$3°$	8

What is the most likely cause of this trait or its predisposition?

13. Pyloris stenosis affects approximately 1/500 newborn. A man and woman, both of whom have negative family histories for this disorder and who are clinically normal, have a son with pyloric stenosis. What is the chance that their next child will develop pyloric stenosis?

14. A certain drug is detoxified by a liver microsomal enzyme. The MZ twin variance for rate of clearance of the drug was about 400, while that for DZ twins was 1200. Calculate the heritability.

GLOSSARY OF TERMS

Assortative mating preferential marriage of similar phenotypes (positive) or dissimilar phenotypes (negative)

Concordance the percent of twin pairs in which both cotwins are affected by a trait

Continuous trait (quantitative trait) a trait that displays graded variation and cannot be clearly resolved into discrete phenotypic classes

Correlation coefficient a measure of the covariation of a variable (such as weight) among different pairs of individuals (MZ and DZ twins) or of two different variables (such as height and weight) among different people

Discontinuous trait (qualitative trait) a trait that occurs as distinct phenotypes

Empiric risk risk estimates that are based upon observations of pertinent groups of relatives (i.e.: children of affected parents, siblings of index cases)

Genetic background the genome in which a major gene or set of genes functions

Genetic predisposition persons with a particular genotype are more likely to develop a problem or trait than people with other genotypes

Heritability the additive genetic contribution to a specific phenotype

Mean the average value or norm for a population

Modifiers genes at other loci that influence the expression of a phenotype determined by a major locus

Multifactorial a trait that is caused by the interaction of several factors of genetic and environmental origin

Penetrance occurrence of any aspect of a phenotype characteristic of a genotype in persons of that genotype

Polygenic multiple genes, each with small effect, that contribute to a trait

Sex-influenced a trait occurs more often in one sex than the other

Sex-limited a trait only occurs in one sex

Standard deviation the square root of the variance

Threshold point beyond which people are more likely to develop a trait (usually applied in an abnormal context)

Variable Expression variation of a phenotype once it is expressed

Variance a measure of the dispersion of a population around the mean

BIBLIOGRAPHY

1. Harris, H. <u>The Principles of Human Biochemical Genetics</u>. 2nd ed. New York: American Elsevier Publishing Co., 1975.

2. Pauls, D and VE Anderson. 1972. A genetic analysis of mental retardation. Am J Hum Genet 24:35a.

3. Cavalli-Sforza, LL and WF Bodmer. 1971. <u>The Genetics of Human Populations</u>. San Francisco: WH Freeman and Co., p. 537.

4. Carter, CO. 1965. The inheritance of congenital malformations. Progr Med Genet 4:59-84.

5. Chung, CS. Cited in Smith, DW and JM Aase. 1970. Polygenic inheritance of certain common malformations. Pediatrics 76:653-659.

6. Edwards, JH. 1960. The simulation of Mendelism. Acta Genet Statist Med 19:63-79.

7. Holmes, LB. 1976. Congenital malformations. N Eng J Med 295:204-207.

8. Curtis, E, FC Fraser and D Warburton. 1961. Congenital cleft lip and palate: risk figures for counseling. Am J Dis Child 102:853-857.

9. Carter, CO and JA Roberts. 1967. The risk of recurrence after two children with central nervous system malformations. Lancet i:306-308.

10. Fraser, FC. 1970. The genetics of cleft lip and cleft palate. Am J Hum Genet 22:336-352.

11. Smith, DW and JM Aase. 1970. Polygenic inheritance of certain common malformations. Pediatrics 76:653-659.

12. Kallman, FJ. 1946. The genetic theory of schizophrenia: an analysis of 691 schizophrenic twin index families. Am J Psychiat 103:309-322.

13. Hoffer, A and W Pollin. 1970. Schizophrenia in the NAS-NRC panel of 15,909 veteran twin pairs. Arch Gen Psychiat 23:469-477.

14. Rosenthal, D, PH Wender, SS Kety et al. 1971. The adopted-away offspring of schizophrenics. Am J Psychiat 128:307-311.

15. Heston, L. 1966. Psychiatric disorders in foster home reared children of schizophrenic mothers. Br J Psychiat 112:819-825.

16. Slater, E and VA Cowie. Cited in Gottesman, II and J Shields. 1973. Genetic theorizing and schizophrenia. Br J Psychiat 122:15-30.

17. Heston, LL. 1977. Schizophrenia: genetic factors. Hosp Pract 12 (June): 43-49.

18. Gottesman, II and J Shields. 1971. Schizophrenia: geneticism and environmentalism. Hum Hered 21:517-522.

19. Kallman, FJ. The Genetics of Schizophrenia. New York: Augustin, 1938.

20. Stabenau, JR. 1977. Genetic and other factors in schizophrenia, manic-depressive, and schizo-affective psychoses. J Nerv Ment Dis 164:149-167.

21. Kaplan, AR. 1976. Human Behavior Genetics. Springfield, IL: Charles C Thomas Publishing Co.

22. Goldstein, S and S Podolsky. 1978. The genetics of diabetes mellitus. Med Clin N Am 62:639-654.

23. Goldstein, JL, WR Hazzard, HG Schrott et al. 1973. Hyperlipidemia in coronary heart disease. I. Lipid levels in 500 survivors of myocardial infarction. II. Genetic analyses of lipid levels in 176 families and delineation of a new inherited disorder. III. Evaluation of lipoprotein phenotypes of 156 genetically defined survivors of myocardial infarction. J Clin Invest 52:1533-1577.

24. Buel, P. 1973. Changing incidence of breast cancer in Japanese-American women. J Natl Cancer Inst 51:1479.

25. Macklin, MT. 1959. Comparison of the number of breast-cancer deaths observed in relatives of breast-cancer patients and the number expected on the basis of mortality rates. J Natl Cancer Inst 22:927-951.

26. Macklin, MT. 1959. Relative status of parity and genetic background in producing human breast cancer. J Natl Cancer Inst 23:1179-1189.

27. Nathans, J, D Thomas and DS Hogness. 1986. Molecular genetics of human color vision: the genes encoding blue, green, and red pigments. Science 232:193-202.

28. Nathans, J, TP Piantanida, RL Eddy, <u>et al.</u> 1986. Molecular genetics of inherited variation in human color vision. Science 232:203-210.

29. Jorgensen, A. S Deeb and AG Motulski. 1989. Molecular architecture of color vision pigment genes among Blacks: a new marker gene for African ancestry may be the ancestral X-linked visual pigment gene. Am J Hum Genet 45:A198.

ADDITIONAL LEARNING RESOURCES

1. Devor, EJ and CR Cloniger. 1989. Genetics of alcoholism. Annu Rev Genet 23:19-36.

2. Knudsen, Jr., AG. 1986. Genetics of human cancer. Annu Rev Genet 20:231-251.

3. Manschreck, TC. 1981. Schizophrenic disorders. N Eng J Med 305:1628-1632.

Chapter 12

Genetics and Development

Meiotic division in the female produces an ovum that has a sizeable proportion of cytoplasm. This cytoplasm has rich stores of mitochondria and mRNAs required for the early stages of development of the new organism, during which reliance is placed upon the maternal genome for driving the early developmental events. The human egg, like those of other mammals, is much smaller than eggs of insects such as Drosophila or amphibians like Xenopus. Consequently, less emphasis is placed upon the maternal genome than is the case for amphibians and insects, and the genome of the human embryo is activated much earlier than those of amphibian or Drosophila embryos. Studies of mouse embryos have revealed the appearance of transcripts as early as the two-cell stage, and human embryonic genes are presumably activated during the same period. The orchestration of gene expression during development has fascinated scientists for a long time, and only recently have the tools required to begin the critical molecular analysis of development become available. This chapter will examine some of the methods currently being utilized for study of gene regulation during development and will describe a few systems that have been examined in some detail. An overview of human development will also be presented, as well as a brief discussion of examples of birth defects that occur following failures of normal processes.

HUMAN DEVELOPMENT

The human organism passes through a series of quite different environments that place varying metabolic and physiological demands upon the developing cells. For example, human development is initiated in a free-floating state in the upper reaches of the Fallopian tube. The embryo obtains most of its nutrients from the cytoplasm derived from the egg and from the fluid contained within the Fallopian tube. The first 7 to 8 days of human life are characterized by division of cells without marked increase in size, with all of the cells contained within the confines of the zona pellucida (Fig. 12-1). The early cleavage stages terminate with the formation of a solid mass of 12-16 cells. The embryo continues to divide, forming a blastocyst . An early blastocyst is composed of about 32 cells while a late blastocyst possesses approximately 256 cells and consists of an inner cell mass, a layer of trophectoderm cells, and a well-defined cavity known as the blastocele. Cells of the inner cell mass form the three germ layers of the embryo proper and the extraembryonic membranes, while the trophectoderm cells assist with the penetration of the uterine lining and form the fetal portion of the placenta. The second stage of intra-uterine development, the early implantation stage, occurs from two to three weeks after conception. This period is initiated by the process of gastrulation, forming the three germ layers, and terminates following neurulation which produces the neural tube. The anterior and posterior regions of the neural tube ultimately give rise to the brain and spinal cord, respectively. Embryonic development (4-9 weeks post-conception) begins with morphogenesis when the three major body folds (longitudinal, head and tail folds) are produced and continues with the formation of the **developmental field** complexes essential for organogenesis. The last stage of intra-uterine development is called the fetal period, extends from the ninth through the thirty-eighth week of gestation, and encompasses the time during which growth and maturation of the fetus occur. It is important to stress the importance of the embryonic stage of development. The major body plan is laid down during this time. Intrinsic problems in the embryo's developmental program or interference with normal developmental processes by extrinsic agents during this critical stage will have a major impact upon body regions and specific organs within those regions.

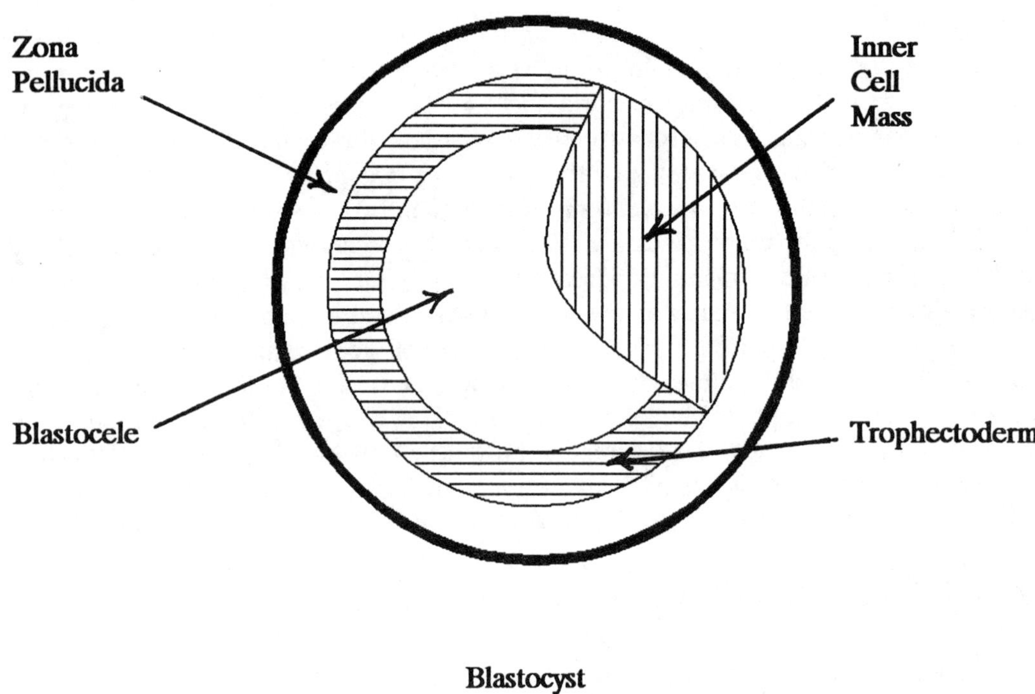

Fig. 12-1. Human blastocyst. A typical blastocyst possesses 32 (early) to 256 (late) cells. The inner cell mass contributes to the three primary germ layers (ectoderm, mesoderm and endoderm) of the embryo and to the extraembryonic membranes (amnion, chorion, allantois and yolk sac). The trophectoderm cells actively participate in the invasion of the uterine lining and establishing the fetal portion of the placenta.

The initiation of maternal-fetal contact promotes a major shift in the manner by which the human organism obtains its nutrients and eliminates its wastes. For example, oxygen is obtained from the maternal blood supply, and fetal hemoglobin must have properties suitable for extracting oxygen from maternal hemoglobin and sequestering it in the fetal circulation. To meet this need embryonic globin genes are inactivated, and fetal globin genes are turned on, resulting in the replacement of embryonic hemoglobins in fetal erythrocytes by fetal hemoglobins. Most likely many other gene systems also shift their patterns of expression during the embryonic-fetal transition. A second major change of environment occurs during birth. Activation of adult β- and δ-globin genes shortly precedes birth, resulting in a gradual replacement of fetal hemoglobin by the two adult hemoglobins and enabling the infant to utilize its lungs for obtaining oxygen and eliminating carbon dioxide. The newborn infant can no longer rely on its mother to carry out intermediary metabolism, and many liver enzyme systems are activated as the infant develops its own metabolic capabilities. The ability of a conceptus to meet each of the major developmental hurdles, as well as many smaller ones, requires the proper timing of activation of new genes and inactivation of genes that are no longer required. The high loss of life during the embryonic and early fetal periods and the occurrence of a substantial number of stillbirths and neonatal deaths most likely are consequences of failures to throw the necessary developmental "switches" at critical transition points in the developmental program.

Monozygous Twinning

Monozygous twins may be formed as a consequence of a separation of the embryo into two roughly equivalent cell masses. About a third of monozygous twins originate during the 2-8 cell stage about one to three days after conception, whereas two-thirds of monozygous twinning occurs during the blastocyst stage. Three to four percent of monozygous twins arise 9-15 days after conception. The latter group has a high incidence of malformations and intra-uterine death of at least one of the cotwins. Congenital abnormalities afflict approximately one-quarter of all twin births, reflecting major problems in apportioning equivalent cell types among both cotwins, effects of intrauterine competition for nutrients, and problems derived from the cramped intrauterine quarters in which the twins develop. Siamese twinning occurs when the cell masses fail to separate entirely, resulting in the cotwins remaining joined in one or more body areas. In some cases, organs are shared, making their separation very difficult.

Congenital Birth Defects

There are two fundamental concepts that should be remembered when considering birth defects: 1) a pattern of anomalies occurring in an infant is most commonly derived from a disturbance of a single developmental field; and 2) an error in the development of an early structure or region will not only influence subsequent development of structures derived from it, but also the differentiation of tissues and organs nearby. The latter phenomenon is referred to as a **sequence**. The birth defect syndrome, holoprosencephaly, illustrates both concepts. Holoprosencephaly affects about 1/5000 newborn; however, the abnormalities characteristic of this syndrome are much more frequent among spontaneous abortices. The developmental error responsible for the syndrome may occur as early as the third week of development and affects a developmental field known as the precordal mesoderm. This developmental field contributes to the mid-facial region and to underlying brain structures. The prosencephalon fails to form normal subregions in all three planes, such that severe cases fail to form cerebral hemispheres, optic and olfactory bulbs, and telencephalon and diencephalon regions of the brain. Underdevelopment of the mid-facial area and the underlying regions causes microcephaly (small brain and skull), hypotelorism (closely spaced eyes or one centrally located eye (cyclopia)), abnormal nose, cleft lip and palate, and severe mental deficiency and other central nervous system abnormalities. A second type of sequence is illustrated by Pierre-Robin syndrome. These fetuses develop an unusually small jaw. As a consequence, there is insufficient room to accommodate the descent of the tongue. Retention of the tongue in the roof of the oral cavity prevents closure of the palatal shelves, and one of the primary features of this syndrome, cleft palate, results.

A **malformation** is defined as a morphological defect of an organ or larger region of the body resulting from an **intrinsically** abnormal developmental process. Although the specific cause of holoprosencephaly, which is categorized as a malformation, is unknown, the syndrome is often associated with trisomy 13 and 18p- syndrome, may be inherited as an autosomal recessive or autosomal dominant, or may occur as a component of other mendelian disorders such as Meckel syndrome.

A second type of birth defect, known as a **disruption**, is often caused by **teratogens** (agents that cause birth defects). A disruption is a morphological defect of an organ or region caused by an **extrinsic** agent that interferes with an intrinsically normal process. About 1/1000 newborn infants are afflicted with fetal alcohol syndrome. These babies have characteristic faces with a flattened central region, short nose, and long upper lip. They are generally small at birth, a sign of intra-uterine growth retardation. Babies with fetal alcohol syndrome have a higher risk for birth defects and later exhibit both growth and mental retardation (1,2). There seems to be no safe dose of alcohol for expectant mothers, and many obstetricians recommend total avoidance of alcoholic beverages during pregnancy. More severe manifestations of fetal alcohol syndrome are observed among infants of older mothers who use alcohol and among infants of severe alcoholics who have a poor nutritional status.

Fetal hydantoin syndrome is a second disruption caused by a known teratogen (3). Diphenylhydantoin and related drugs are often prescribed to persons at risk for convulsions. As many as 10% of fetuses exposed to diphenylhydantoin develop severe forms of the syndrome, while about 30% have milder symptoms. The syndrome is characterized by widely spaced eyes, short nose with a depressed bridge and a "cupids bow" mouth. The distal bones of the fingers and toes and the nails are abnormally formed. These infants are also at risk for a variety of other birth defects. A physician is often faced with a dilemma regarding prescription of these anticonvulsants to a pregnant woman. If she is having periodic seizures, their continuation has been found to be associated with abnormal pregnancy outcome. Current practice appears to involve a trial removal of an expectant mother from anticonvulsants. If no seizures develop, no drugs are prescribed. However, if seizures recur, she will be prescribed the least number and lowest doses of anticonvulsants required for management of her epilepsy.

Retinoids have recently been recognized to have teratogenic effects upon human fetuses. One of these compounds, retinal, was discussed in the previous chapter in relation to visual pigments. Two other retinoids, retinol and retinoic acid, play an important role in normal development of limbs and other structures. However, administration of high doses of retinoids during pregacy can lead to the occurrence of a variety of birth defects. Accutane (isotretinoin), a retinoid analog, is often prescribed for the control of acne and other skin disorders. When given to pregnant women, a 26-fold increase in birth defects was observed among their offspring (4). Birth defects involved the central nervous system, ears, palate, heart and major vessels, thymus, and the eyes. Variation of the particular array of defects and their relative severity among infants was noted. Retinoids are rapidly cleared from tissues, and an intensive educational campaign concerning the teratogenic effects of retinoid analogs has been undertaken to reduce the incidence of these birth defects. The teratogenic effect of retinoids and their analogs has not been elucidated; however, the compounds appear to act upon cephalic neural crest cells. Several additional teratogens are presented in Table 12-1.

A fourth disruption is caused by amniotic bands. Strips of placental connective tissue may grow across a body part, such as a limb or digit, and interrupt the blood supply to the involved tissue. The body part eventually dies and decays, leading to an amputation-like defect.

Two other classes of birth defects include a number of both environmental and inherited problems. **Dysplasias** occur as a consequence of an abnormal organization of cells into morphologically abnormal tissues. Several dysplasias are known which are usually inherited. Ectdodermal dysplasia includes abnormalities of the skin and its components. Patients with anhidrotic ectodermal dysplasia may have sparse hair, lack sweat glands, and have defective teeth. Because of the absence of sweat glands, these persons can experience overheating, particularly upon exercising. Some patients have reduced or absent breasts. Nails of these patients are usually normal. This syndrome is inherited as an X-linked recessive. A second form of ectodermal dysplasia is inherited as an autosomal dominant. This syndrome can be distinguished from the anhidrotic form on the bases of the presence of sweat glands, defective nails, excessively course skin on the palms of the hands and soles of the feet, and hyperpigmented areas of the skin. Neurofibromatosis is a dysplasia in which multiple neurofibromas may develop on the skin. Other features include Lisch nodules in the iris, and the occurrence of coffee-colored (cafe-au-lait) spots on the skin. Some patients are very mildly affected, many presenting only with cafe-au-lait spots.

Table 12-1. A Series of Teratogens Known to Produce Human Birth Defects.

Infectious Agents	Physical Agents	Non-prescription Drugs
Rubella	Radiation	Thalidomide
Cytomegalovirus	Heat	Ethyl alcohol
Herpes simplex		
Syphilis	**Pollutants**	
Toxoplasmosis		
	Organic mercury compounds	

Prescription Drugs

Anti-cancer agents	Anticonvulsants
(aminopterin, methotrexate)	(trimethadione, hydantoin)
Anticoagulants	Endocrine agents
(Warfarin)	(diethylstilbestrol, androgens)
Antibiotics	Psychotropic agents
(Tetracyclines, Streptomycin)	(amphetamines, phenothiazines, tricylic drugs, barbiturates)

Patients may develop other tumors affecting the brain, eyes, and other organs. Some tumors may cause disfigurement and scoliosis. Seizures may occur secondary to the development of a brain tumor. Neurofibromatosis is one of the most common autosomal dominant diseases and has an unusually high mutation rate. A candidate gene, **NF1**, has recently been discovered and is being investigated. Almost all tissues affected by neurofibromatosis have a neural crest cell origin. **Deformations** occur as a consequence of mechanical stress upon the fetus. Causes include breech presentation and deficient amniotic fluid (oligohydramnios).

Many birth defects may involve interaction between an intrinsic developmental problem caused by a polygenic or mendelian predisposition that interacts with one or more, often unknown, environmental agents. Anencephaly-spina bifida seems to be an example of this phenomenon. Anencephaly-spina bifida is caused by a failure of closure of the neural tube during the third to fourth week of gestation. Failure of closure of the anterior end produces anencephaly in which the upper regions of the brain and a portion of the skull fails to form. Failure of closure of the posterior end of the neural tube results in spina bifida. Some infants exhibit both defects. Spina bifida is known to occur in three subtypes (Fig. 12-2). Spina bifida occulta is quite common, with estimates approaching 25%. In these cases, the spinal cord and meninges which surround the cord and contains the cerbrospinal fluid remain below the plane of the back. In many cases of spina bifida occulta, the dorsal arch of the vertebral column is intact, but thin. Spina bifida occulta usually is not associated with problems and is most commonly ascertained when X-rays are taken of that region of the back. Meningocele can create problems if the meninges are not covered by skin. Infection can occur followed by a number of complications. Meningomyelocele is often associated with paralysis of the lower limbs and other problems. Anencephaly-spina bifida often displays an increase with decreasing social class. This trend suggests that there may be environmental factors such as nutrition, disease, and exposure to toxic substances involved with its cause. Recent studies have indicated that recurrence of anencephaly-spina bifida may be preventable by enriching expectant mothers' diets with multivitamin supplements. Programs are now under way in the United Kingdom and the United States to assess this possibility.

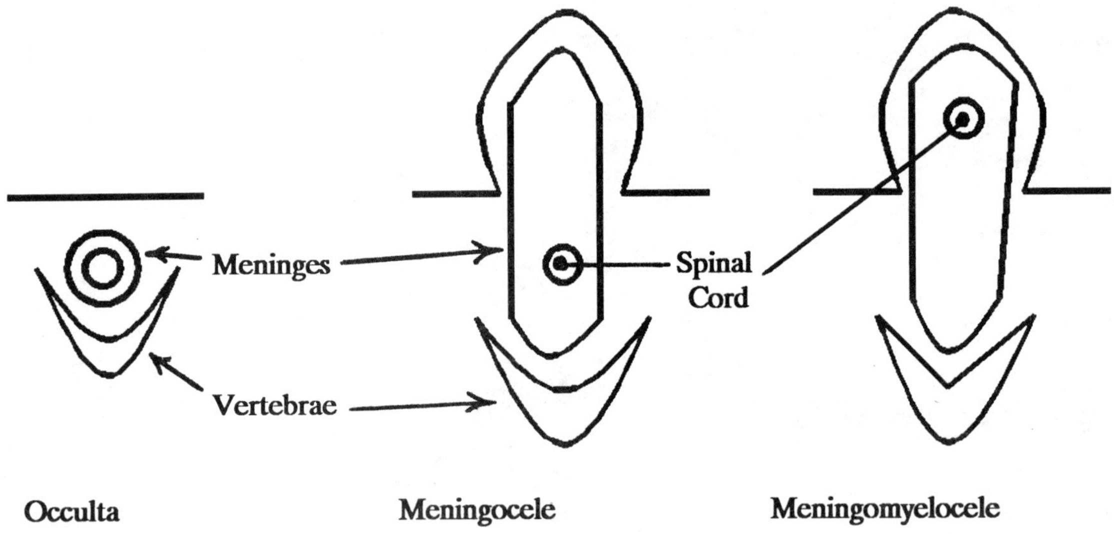

Fig. 12-2. Subtypes of spina bifida. Occulta: both meninges and spinal cord remain below the plane of the back. In many cases, the dorsal arch of the spine is present, but thin. Meningocele: the meninges protrude through the back. Meningomyelocele: both meninges and spinal cord protrude through the back. Risk of paralysis is high in this form of the birth defect.

Maternal-fetal Interactions

Certain maternal factors can impact upon the development of a fetus, increasing risks for growth and mental retardation and birth defects. Uncontrolled maternal diabetes, maternal hyperphenylalaninemia, and Rh incompatibility have been mentioned in previous sections of this text in relation to their teratogenic effects. Prevention of adverse fetal effects caused by maternal diabetes and maternal phenylketonuria is feasible, provided rigorous control of maternal blood glucose or phenylalanine, respectively, is practiced. Erythroblastosis fetalis caused by Rh incompatibility between the mother and fetus can be prevented by administration of Rhogam (anti-Rh) during the pregnancy and parturition. Maternal hypothyroidism is another cause of mental and physical retardation and birth defects. When the mother is maintained in a euthyroid state by thyroid hormone replacement, adverse effects on her children can be prevented. Androgen-secreting adrenal or ovarian tumors can interfere with normal sexual development of a female fetus, leading to masculinization or sexual intermediacy. Precocious sexual development of male fetuses can also be induced by these tumors. Surgical removal of the tumors following their diagnosis will ameliorate their harmful effects upon the baby.

REGULATION OF GENE EXPRESSION DURING DEVELOPMENT

Gene regulatory mechanisms were discussed in Chapters 2 and 3. Regulation occurs at a variety of levels ranging from transcription through processing of the proteins encoded by genes. Other factors such as chromosomal structure also impact upon gene expression, with active genes clustered in relaxed chromosomal domains in expressing cells. This section will describe some of the new approaches being used to explore the orchestration of gene expression during development and will discuss several systems where these methods have been used to probe developmental phenomena.

Molecular Analysis of Development

A) Transgenic Mouse Models

Use of transgenic mice as an experimental model for assessment of the function of specific genes during development has become widespread. The use of transgenic models for the exploration of the developmental regulation of human globin genes was mentioned in chapter 7. Human globin gene constructs containing embryonic, fetal and/or adult globin genes and varying amounts of upstream and downstream sequences. These constructs are incorporated into suitable retroviral vectors, and the recombinant vectors are injected into in vitro-fertilized mouse eggs. The mouse embryos are implanted into female mice that have been hormonally induced to a state of pseudopregnancy and either recovered at specific developmental intervals or allowed to complete development. An example of this type of experiment was reported by Shih et al. (5). A gene construct containing the human ϵ-globin gene, a 13.5kb upstream sequence, and 0.2kb of the flanking downstream sequence was used to make transgenic animals. At approximately 13.5 days after conception, murine circulating erythrocytes consist of both yolk-sac derived nucleated erythroid cells and liver-derived anucleate erythroid cells. The yolk-sac derived cells express embryonic globins, and the liver-derived cells contain adult globins. Transgenic mouse embryos were recovered at 13.5 days gestation, and RNA was extracted from peripheral blood, brain and liver tissue. The RNAs were partially digested with restriction enzymes, separated by electrophoresis, blotted and probed with labelled cDNAs corresponding to the murine embryonic and adult globin genes and the human ϵ-globin gene. These experiments revealed expression of the human ϵ-globin gene construct in peripheral blood cells, but not in liver nor brain. By contrast, the murine adult β-globin gene was expressed in both peripheral blood cells and in liver cells. These results suggested that regulatory elements in the upstream sequence of the ϵ-globin construct were sufficient not only for restriction of expression of the gene to erythroid cells, but also timing its expression to the appropriate developmental period.

Sites within the upstream sequence responsible for these two regulatory functions (spatial and temporal) may be identified by site-directed mutagenesis. The construct is mutagenized, and the sites of the changes are characterized by sequencing. Intact and mutagenized constructs are tested in the transgenic model to determine which mutagenized sequences eliminate the spatial and/or temporal regulation of the expression of the construct.

Another variation of the transgenic model can be used to determine the contribution of a population of cells to the development of body structures. A gene construct is made containing the promoter and upstream flanking sequence of a gene known to display a cell-specific expression pattern coupled to a diphtheria toxin gene. The gene will be activated only in those cells of the transgenic animals that contain the required trans-activating factors. Production of the diphtheria toxin kills the cells. When these transgenic animals are histologically examined, the structures derived from the cells in which the toxin was expressed will be absent.

B) Mouse Embryonic Stem Cells

Cells from the inner cell mass of mammalian blastocysts are pluripotent, that is, they are capable of differentiating into many different cell types. Embryonic stem (ES) cell lines can be produced by removing cells of the inner cell mass from 4½-day mouse blastocysts and culturing them on monolayers of "feeder" cells. The cells are grown on these layers until cell aggregates form, and the aggregates are dispersed into single cells that are used to produce single cell colonies. These ES cell colonies can be implanted into other mouse blastocysts, introduced to pseudopregnant females and allowed to develop into **chimeras**, which are animals containing cell lines of different genetic origins.   ES cells can be obtained as described and transfected with genes of interest. Colonies of the transfected cells are analyzed to determine which cells have incorporated the transfected gene into their genomes. The colonies that have incorporated the gene construct are then used to make chimeric mice.

Chisaka an Capecchi (6) combined this approach with site-specific recombination to demonstrate the role of a homeobox gene in murine development. The term, **homeobox**, refers to a promoter element that is found in developmental genes that were first described in Drosophila. These genes appear to determine positional values of cells forming specific segments of the fruitfly. Genes containing these homeoboxes are referred to as **homeotic** genes. Families of homeotic-like genes have been identified in a wide variety of organisms, including mice and humans. One of the mouse homeotic genes, **Hox-1.5**, is expressed during early embryonic development and is believed to encode a transcription factor that specifies regional information in presomitic mesoderm and ectoderm of 7.5-8.0- day embryos. **Hox-1.5** transcripts are most abundant in neural exctoderm but later can be found in a number of different organs. A special recombination system using the neo^r gene, which determines neomycin resistance, was employed to replace the normal Hox-1.5 genes of ES cells with a defective **Hox-1.5** gene construct. The neo^r system mediated a site-specific recombination between the defective allele and the normal allele on the ES cell chromosome. The recombinant ES cells were microinjected into blastocysts, and the treated blastocysts were implanted in pseudopregnant females. Embryos were recovered at 13.5 and 15.5 days and at term. Embryos and newborn mice who were derived from chimeric blastocysts containin ES cells heterozygous for the defective allele appeared normal; however, those derived from chimeric blastocysts possessing ES cells that were homozygous for the defective **Hox-1.5** allele had defects in the neck region and lacked thymus and parathyroid glands. Other defects were observed in the thyroid, heart and arteries. All of these defective liveborn mice died shortly after birth. The authors pointed out parallels between the problems caused by the defective **Hox-1.5** allele and the symptoms of the human DiGeorge syndrome (see chapter 10). This experimental approach provides a powerful method for demonstrating that the Hox-1.5 gene participates in the regulation of normal development of the affected regions and implies that a similar gene may be involved with development of the corresponding human structures.

C) Cross-specific Experiments

The functional equivalence of two genes from different species could theoretically be tested by performing a reciprocal transfer of each gene to the chromosomal environment of the opposite species. This would be especially interesting with respect to homeotic genes; however, the reciprocal exchange is not always feasible, particularly with reference to human genes. Homeotic genes have been most extensively studied in Drosophila. Experiments have recently been reported (7,8) in which mammalian homeotic genes were introduced into Drosophila. Expression of the respective genes was induced by heat shock, and their effects mimicked those of the Drosophila genes they are believed to resemble. The human **HOX-4.2** gene is structurally homologous to the Drosophila Deformed gene (**Dfd**). The human **HOX-4.2** gene was coupled with a Drosophila heat shock promoter and transfected into Drosophila embryos. The transfected gene was activated by heat shocking the embryos. Effects of the human gene were similar to those produced by the **Dfd** gene, indicating that the human gene can substitute for the Drosophila gene in Drosophila development and implying that the **HOX-4.2** gene may play a similar role in human development. A similar experiment was carried out with a murine **Hox-2.2** gene.

Expression of developmental regulatory genes can be followed at both the transcriptional and translational levels. Transcriptional activity can be monitored by using in situ hybridization. Radioactive or biotinylated cDNA probes possessing sequences identical to those of developmental genes are hybridized with complementary mRNAs in histological sections from whole embryos or embryonic tissues. The vertebrate brain develops in a segmental fashion, with different groups of neurons exiting from different segments. The segments are called rhombomeres, and pairs of rhombomeres give rise to the various cranial nerves. For example, cranial nerves V, VII and IX arise from rhombomeres 2 and 3, 4 and 5, and 6 and 7, respectively. The **Hox-2** gene family consists of nine homologous genes that are tightly linked in the order: 5' -**Hox-2.5** - **Hox-2.4** - **Hox-2.3** - **Hox-2.2** - **Hox-2.1** - **Hox-2.6** - **Hox-2.7** - **Hox-2.8** - **Hox-2-9** -3'. Wilkinson et al. (9) sectioned whole mouse embryos at 8 through 9.5 days of development when the genes of this cluster are prominently expressed. The sections were treated with cDNAs for segments of **Hox-2.1**, **Hox-2.6**, **Hox-2.7**, **Hox-2.8**, and **Hox-2.9** that were labelled with ^{35}S. Specific hybridization patterns for each of the probes were observed. Expression of **Hox-2.6**, **Hox-2.7**, and **Hox-2.8** transcripts were observed only at the posterior boundary of the neural plate in the 8-day embryos, while **Hox-2.1** expression was limited to the boundary between the hindbrain and the spinal cord. This segmental pattern of expression precedes appearance of the rhombomeres, and may indicate that these genes are actually involved with regional specification of the boundaries of these structures. Further studies demonstrated that anterior limits of expression of these genes were different: **Hox-2.6** = boundary between rhombomeres(r) 6 and 7, **Hox-2.7** = r4/r5 boundary, **Hox-2.8** = r2/r3 boundary; however, some overlap of expression regions was observed. These and other studies (summarized in 10,11) have been interpreted to suggest that fates of cells along the anteroposterior axis of the neural tube, mesoderm and neural crest may be determined by the particular patterns of expression of the **Hox** genes which seem to assign a positional marker to these cells. Although this work has largely utilized mouse embryos, the high degree of homology among the human **HOX** and murine **Hox** genes suggests that organization of the human central nervous system and associated structures may be determined in a similar manner.

Morphogens are defined as chemical molecules that are distributed along a concentration gradient and whose distribution influences the pattern formation required for development of body structures. The anteroposterior axis of developing vertebrate limbs is regulated by a specific region, the zone of polarizing activity (ZPA). Experiments in amphibians and birds have demonstrated that the ZPA produces one or more morphogens which diffuse from the ZPA, setting up a concentration gradient. Retinoic acid (RA) has been proposed as a candidate for the morphogen produced by the ZPA. Retinal is converted to retinoic acid in developing limb tissue, and cellular retinoic acid binding proteins (CRABP) have been identified in cells within these regions. Retinoic acid is known to act at the level of transcription, presumably by the mechanism presented in chapter 2 where retinoic acid receptors complexed with retinoic acid bind to specific response elements near genes that are activated by this chemical. Tamura et al. (12) prepared a monoclonal antibody against retinoic acid and treated sections of developing chick limb with the antibody. The antibody demonstrated the presence of retinoic acid throughout mesoderm in the developing region; however, as development proceeded it became progressively localized in distal mesoderm and then ectoderm. This pattern was interpreted to indicate that retinoic acid was playing an active role in limb pattern formation. Many other studies have corroborated these findings; however, recent work (13) suggests that retinoic acid may participate indirectly in pattern formation by inducing the formation of ZPAs which subsequently secrete morphogens that actually establish the pattern which will govern limb differentiation. Regardless of the exact mode of action of retinoic acid, these studies provide a framework for understanding its teratogenic effects on human development when taken in excess by pregnant women. Numerous other studies have been reported which have employed monoclonal and polyclonal antibodies against gene products believed to have roles in differentiation and have augmented the <u>in situ</u> hybridization work in providing both a temporal and spatial view of gene expression during development.

EXAMPLES OF DEVELOPMENTAL SYSTEMS

Three general patterns of gene expression are observed in human tissues. Many genes, known as housekeeping genes, are continuously expressed throughout development in most tissues. Examples include **HPRT** which encodes an enzyme involved with purine biosynthesis and **HEXA** which determines the α-subunit of β-hexosaminidase A. A second developmental pattern of gene expression is illustrated by lactate dehydrogenase (LDH). This tetrameric enzyme contains two different subunits encoded by the **LDHA** and **LDHB** loci, respectively. The two loci are equally expressed in young human fetal tissues, giving rise to a symmetrical isozyme pattern in many tissues (Fig. 12-3). As development progresses, the **LDHA** locus expression declines in heart and kidney; and expression of **LDHB** decreases in liver and skeletal muscle, producing the skewed isozyme patterns observed in the adult. The mechanisms repsonsible for this graded variation in gene expression have not been fully elucidated.

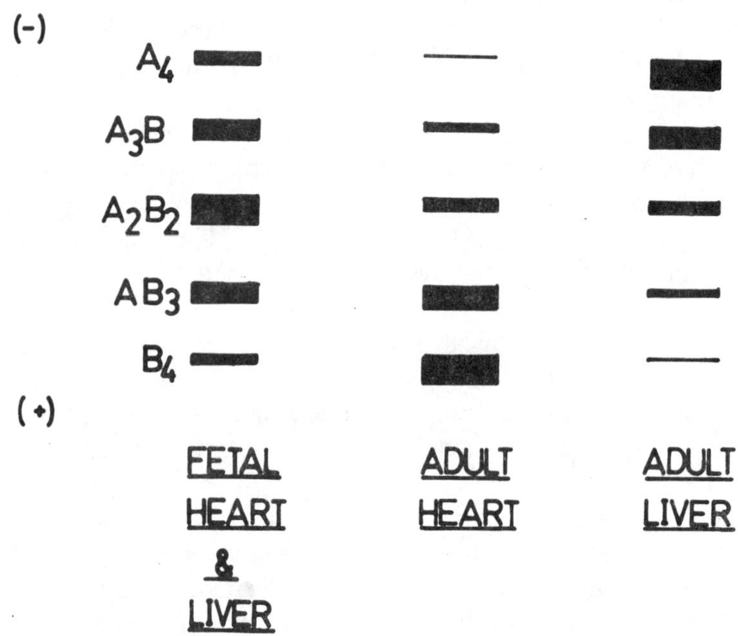

Fig. 12-3. Alteration of the expression of the human LDHA and LDHB genes during differentiation of heart and liver tissues.

A third pattern of gene expression during human development is illustrated by alcohol dehydrogenase. This enzyme, along with aldehyde dehydrogenase, functions in the detoxification of alcohol. Early fetal liver (approximately 20 weeks in gestation) possesses an enzyme that migrates as a single isozyme during electrophoresis. This isozyme is a homodimer consisting of two α chains that are encoded by the **ADH1** locus which is one of five duplicate loci clustered on chromosome 4. Loci from more mature fetuses contain alcohol dehydrogenase activity that is contributed by two isozymes, and newborn liver alcohol dehydrogenase migrates in a symmetrical three-banded pattern (Fig. 12-4). The second band in mature fetuses is a heterodimer, and its appearance signals the activation of the **ADH2** locus which encodes the β subunit. As the expression of **ADH2** increases, $\beta\beta$ homodimers are formed, contributing to the third band observed in newborn liver extracts. A third locus, **ADH3**, is activated later in development, such that a complicated pattern generated by a mixture of homo- and heterodimers containing various combinations of the subunits encoded by the three loci appears. Further diversity is displayed with respect to tissue-specific expression of the three loci. Although all three loci are expressed in hepatic

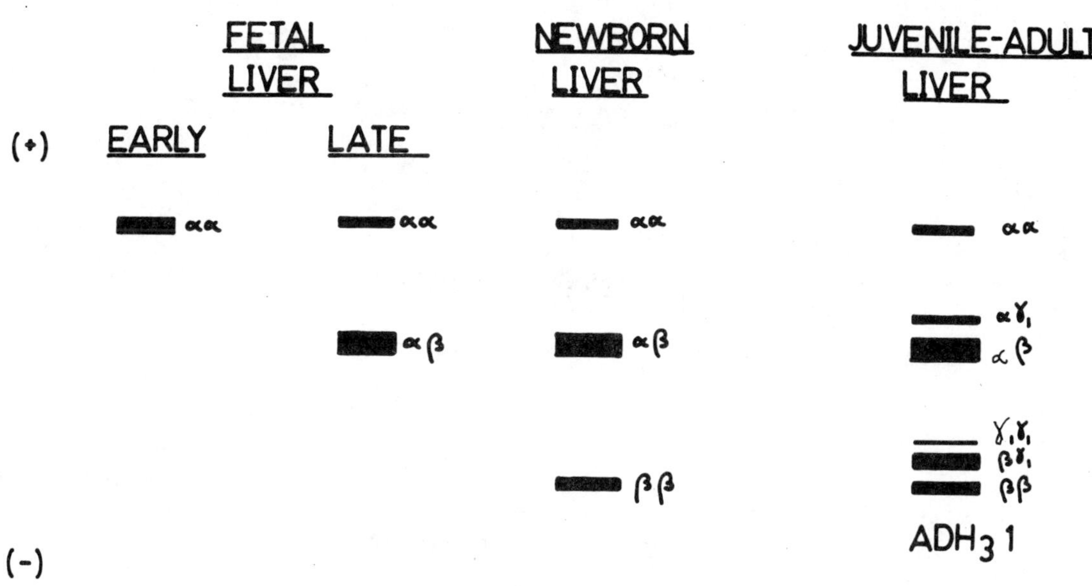

Fig. 12-4. Developmental variation of human liver alcohol dehydrogenase class I isozymes. The ADH2 and ADH3 loci are polymorphic. The ADH$_3$1 phenotype is displayed (modified from Harris (14)).

cells, kidney cells express two (**ADH2** and **ADH3**), and lung (**ADH2**) and stomach cells (**ADH3**) express only one of the three loci. The isozymes determined by these loci tend to preferentially act on short chain alcohols, and there is a high degree of homology among the sequences of the three subunit types. A fourth locus **ADH4** encodes the π subunit which has markedly different properties and is classified as a class II isozyme. **ADH4** is preferentially expressed in liver and kidney. The fifth ADH gene, **ADH5**, encodes the chi chain which is part of an enzyme that displays activity toward long chain alcohols and is categorized as a class III enzyme. This gene is also expressed in liver cells. These examples illustrate the range of developmental variation in gene expression. Regulatory mechanisms responsible for temporal and spatial variation of human gene expression during development are being actively investigated, and several systems will be presented below.

Developmental Regulation of Human Globin Gene Expression

Many vertebrate species synthesize a number of different hemoglobin types during development. Mice produce two different embryonic hemoglobins and either one or two adult hemoglobins, depending upon the strain. Humans produce two embryonic hemoglobins, three fetal hemoglobins, and two adult hemoglobins. The globins are encoded by two clusters of genes situated on chromosome 11 (β-like cluster) and chromosome 16 (α-like cluster) (see chapter 7). One of the hypothesized ancestral mammalian chromosomes is believed to have contained five functional β-like globin loci: 5'-ϵ-γ-η-δ-β-3'. The η locus is expressed as an embryonic globin chain in eutherian mammals, has been lost from the chromosomes of rodents and lagomorphs (rabbits) and may be represented by a pseudogene in humans ($\psi\beta$). The human β-like and α-like gene clusters and their associated regulatory regions are presented in Fig. 12-5. The human locus has recently duplicated forming the $^G\gamma$ and $^A\gamma$ loci, both of which are expressed during fetal development. There are six hypersensitive sites, five of which

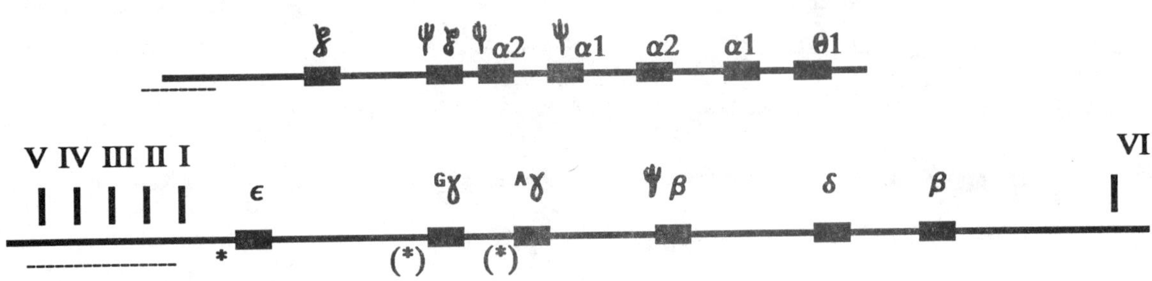

Fig. 12-5. The human α-like (top) and β-like (bottom) globin genes clusters. The Roman numerals and vertical bars indicate the approximate sites of hypersensitive sites in the β-like cluster. The dashed lines denote the locus controlling regions (LCRs). Asterisks represent negative control sites (silencers). The silencer preceding the ϵ locus has been identified and sequenced, while those upstream from the $^G\gamma$ and $^A\gamma$ genes are hypothesized. Hypersensitive sites have also been identified upstream and downstream from the α-like cluster, but are not shown in the figure.

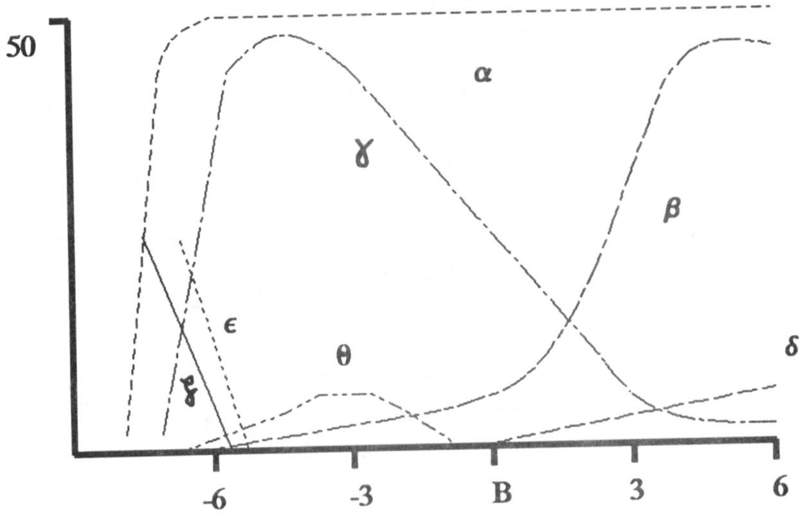

Fig. 12-6. Globin gene switching during human development. Left axis: percent of total globin chains synthesized. Horizontal axis: age in months as measured from birth. During the embryonic period, globin synthesis occurs in nucleated megaloblasts in the yolk sac. During the fetal period (ca. -7 to -6 months through birth), the site of globin synthesis shifts to macrocytes in fetal liver and spleen. Following birth, globin synthesis occurs in normoblasts of the bone marrow.

are located in the locus controlling region (LCR) upstream from the ϵ-globin gene. The LCR has recently been substituted for the term, locus activating region (LAR), described in chapter 7. A number of experiments have defined the general aspects of the regulation of hemoglobin switching that occurs during human development (Fig. 12-6). The figure has been somewhat simplified by representing gamma chains as a single entity. The 0-globin is an α-like globin that is expressed at low levels during the fetal period. The promoter of this gene is unusual, since it has both a **CCAAT** and **ATA** box like a specialized gene and a GC-rich sequence like a housekeeping gene. The GC-rich promoter is actually used, leading to the very low expression. Most fetal hemoglobin contains two alpha and two gamma chains; however, trace amounts of fetal hemoglobin possessing two 0 and two gamma chains are also present.

Transgenic experiments clarifying globin gene regulation were briefly described in chapter 7. Early experiments (15) using constructs containing the human β-globin promoter and the human β-globin gene to transfect mouse embryos resulted in expression of the human β-globin as an adult globin in murine erythrocytes. When similar constructs were prepared using the corresponding human -globin elements, the human -globin gene was expressed as an embryonic globin in yolk-sac derived erythroid cells. When a construct containing all four human gene sequences was transfected into mouse embryos, no gene switching was observed, with both human globins expressed throughout development. These experiments demonstrated that regulatory sites proximal to each globin locus were sufficient for expression of the human globin genes in murine erythroid cells, but that additional elements located elsewhere were essential for globin gene switching. Behringer <u>et al.</u> (16) prepared several gene constructs combining varying numbers of hypersensitive sites I-V and α-, β-, -globin genes. Murine embryos were either injected with constructs containing both hypersensitive sites and globin genes or coinjected with two types of constructs - one containing the five hypersensitive sites and the other containing either the -globin genes or both the - and the β- and δ-globin genes. The constructs containing the hypersensitive sites and a globin gene were expressed at a high level throughout mouse development; however, when the hypersensitive sites were coinjected with the fragment containing the fetal and adult globin genes, the globin genes were not only expressed at a high level, but also at the right time. These experiments were interpreted to indicate that the individual globin genes were competing with one another for an enhancer substance produced in the region of the hypersensitive sites. The region of the upstream hypersensitive sites is now called the **locus controlling region (LCR)**. These authors have proposed a model to explain globin gene switching during development which involves interactions among the LCR, proximal regulatory sites, and trans-acting factors that are present in erythroid cells at specific developmental stages (Fig. 12-7). Three general types of trans-acting factors influence temporal expression of the genes within the β-like globin gene cluster. Each erythropoietic tissue produces up-regulating factors that interact with sites within the proximal promoter regions of the appropriate gene. The array of positive-acting transcription factors produced by yolk sac erythroid cells preferentially binds to the promoter element of the ϵ-globin gene, while those produced in the fetal liver and spleen preferentially interact with -globin promoters, and those synthesized by bone marrow bind to the β-globin gene promoter. The interaction of the proximal promoter-binding factors with the respective genes leads to their low- to moderate-level expression. Each erythropoietic site also produces a combination of down-regulating factors that interacts with the silencer elements of the globin genes that are not expressed in that tissue. For example, yolk sac erythroid cells produce silencer-binding proteins that interact with the silencer elements near the -globin genes, preventing their expression in yolk sac cells. The third class of trans-regulators include enhancer factors that are capable of binding to the promoter elements of downstream globin genes, resulting in high-level transcription. The downstream genes compete for the enhancer substances. During the embryonic period of development, the yolk sac cells synthesize proximal promoter-binding factors that render the ϵ-globin promoter more accessible to the enhancer factors. These cells also produce silencers that retard binding of the enhancer factors to downstream globin genes. The result is preferential expression of the ϵ-globin gene in yolk sac erythroid cells. As erythropoiesis shifts to fetal liver, levels of ϵ-factors decline, and synthesis of γ-factors commences. The hepatic erythroid cells also produce silencer factors that prevent interaction of the enhancer factors with the ϵ-globin gene promoter. Consequently, the γ-globin genes are predominantly expressed in this tissue. The final step in the differentiation of the hemoglobin system is accomplished in the marrow cells. Marrow erythroid cells produce silencer factors that retard interaction of upstream genes with the enhancer factors and positively acting β-factors that promote binding of the enhancer substances to the β-globin promoter. These events result in synthesis of β-globin at a high level in these cells. The α-like globin gene cluster has been less throughly studied; however, similar events are believed to occur during the developmental regulation

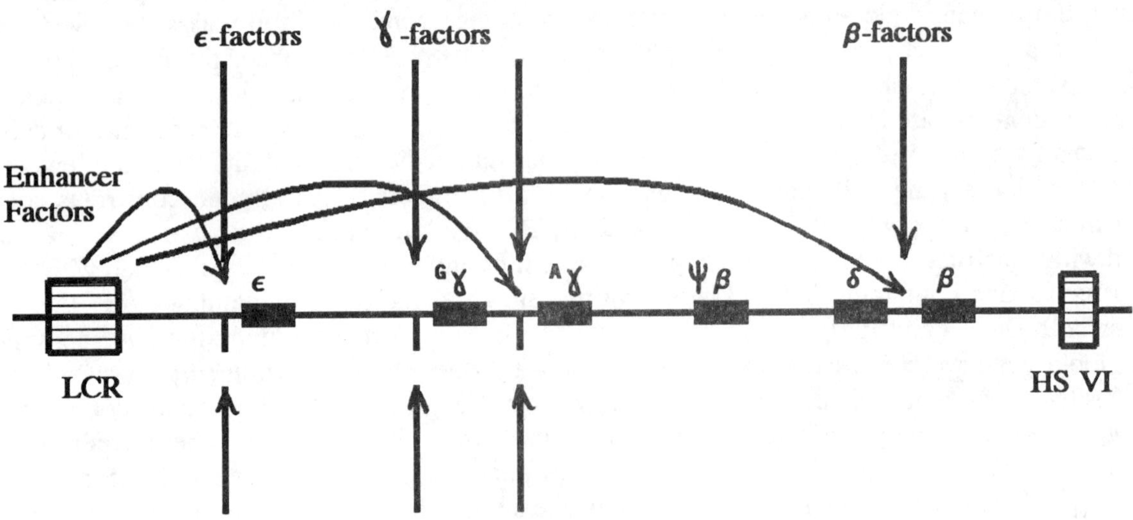

Fig. 12-7. A model of human hemoglobin gene switching. The model illustrates interaction of three types of promoter binding factors. For example, yolk sac cells produce up-regulating ϵ-factors that make the promoter of the ϵ-globin gene accessible to the enhancer factors released by the LCR. Yolk sac cells also produce down-regulating factors that bind to the silencer elements near the γ-globin genes, preventing access of the enhancer factors to downstream promoter elements. The short vertical bars represent known or hypothesized silencer sequences. LCR = locus controlling region; HS VI = hypersensitive site VI.

of this system. Hypersensitive sites occur at comparable positions on chromosome 16, and a region corresponding to the β-cluster LCR has been identified upstream from the ζ-locus.

Evidence supporting the existence of diffusible trans-acting transcriptional factors was provided by experiments utilizing heterospecific heterokaryons (17). Mouse erythroleukemia cells (MEL) express murine adult globins (α- and β-globins) in culture; whereas, human embryonic/fetal erythroid cells (K562) express both embryonic (ϵ- and ζ-globin) and fetal (α- and γ-globin) genes in culture. Heterokaryons containing MEL and K562 nuclei were prepared and allowed to grow for about 24 hours. The RNA produced by these cells was then probed using a cDNA specific for mouse ϵ-globin and for human β-globin. The heterokaryons were found to express both globin types, neither of which was expressed by the cells from which the nuclei were derived. Furthermore, when K562 and MEL cells were cultured together but not fused, neither the mouse ϵ-globin gene nor the human β-globin gene were expressed. These experiments demonstrated that the respective nuclei were producing trans-acting factors that diffused from one nucleus (eg. MEL) to the other in the heterokaryon, activating the gene not normally expressed in the opposite nucleus (eg. K562).

The events involved with hemoglobin switching seem to follow a developmental clock (18). Somatic cell hybrids were formed by fusing MEL cells with human fetal erythroblasts. The latter cells were obtained from young and from older fetuses. Human hemoglobin switching was monitored by using fluorescent monoclonal antibodies prepared against the human γ- and β-globin chains, respectively. Hybrid cells switched from human γ-globin gene to human β-globin gene expression in culture. Furthermore, hybrids formed by fusion of second trimester fetal erythroblasts with MEL cells switched earlier than those derived from fusions of first trimester erythroblasts with MEL cells. The only human chromosome retained in common by the hybrid cells showing the clocklike behavior was chromosome 11, indicating that regulatory elements responsible for this behavior were linked to the β-like globin gene cluster.

Environmental Induction of Human Gene Expression

The human infant is separated from his/her maternal lifeline at birth. This sudden separation from the maternal blood supply triggers a number of events in several tissues and organs that permit the infant to adapt to its new environment. This adaptation involves both cytoplasmic and nuclear events that activate metabolic processes required for execution of new physiolocal processes.

Both glucose-6-phosphatase (G6Pase) and tyrosine aminotransferase (TAT) activities display precipitous increases in hepatic cells following birth (see (19) for a review). Both of these enzymes participate in gluconeogenesis, mobilizing energy stores from liver cells. G6Pase is a microsomal enzyme that hydrolyzes glucose-6-phosphate:

$$\text{Glucose-6-PO}_4 + H_2O \text{ --------------> Glucose} + PO_4.$$

The liberated glucose can than be released from hepatic cells and used as an energy source by other tissues. G6Pase activity displays a modest increase in activity prior to birth in response to fetal thyroxine; however, the postnatal increase is more dramatic and is mediated by the hormone, glucagon.

TAT occurs in both cytosolic and microsomal forms; however, only the cytosolic isozyme displays the rise after birth (20). TAT catalyzes the deamination of tyrosine:

$$\text{Tyrosine} + \alpha\text{-Ketoglutarate --------------> Glutamate} + \text{p-Hydroxyphenylpyruvate.}$$

The p-hydroxyphenylpyruvate enters an oxidative pathway that generates products which can be used as sources of energy. The enhancement of TAT activity is also promoted by glucagon.

Parturition is followed by hypoglycemia as the infant is cut off from the maternal blood supply which supplies nutrients to the fetus. This hypoglycemia triggers a surge in glucagon secretion by the α-cells of the pancreatic islets, and the glucagon stimulates an enzyme, adenyl cyclase, which produces the "second messenger," cyclic adenosine monophosphate (cAMP). The increased cAMP activates a number of protein kinases which catalyze the phosphorylation of several key proteins setting a cascade of events in motion that culminates in the elevation of blood glucose levels. These changes occur very rapidly, reversing the neonatal hypoglycemia that originally set them in motion. The enhancement of expression of the G6Pase and TAT genes occurs primarily at the post-translational level.

Neonatal tyrosinemia is observed in premature infants and among 1/50 to 1/10 infants, depending upon the population studied and the tendency of mothers to breast-feed their babies. In the former case, both TAT and the next enzyme in the pathway, p-hydroxyphenylpyruvate oxidase, display low activities in hepatocytes of premature infants. Placement of the premature babies on low protein diets removes tyrosine stress and restores blood tyrosine to more normal levels. Some full term infants seem to have low levels of TAT and p-hydroxyphenylpyruvate oxidase activities. It has not been determined which enzyme is responsible for the transient tyrosinemia in these infants. Human milk typically contains lower quantities of protein than cow's milk, and breast-fed infants are less likely to have neonatal tyrosinemia. Some controversy over the long term effects of tyrosinemia in the newborn continues. The tyrosinemia eventually corrects itself in most infants, and many of these fail to develop later complications. However, there may be subtypes that may have adverse effects, such as increased risk for mental retardation, later in life. Physicians may encourage breast-feeding of infants with transient tyrosinemia or place them on a low protein (2-3g/day/kg body weight) diet for a period of time.

Genetic Abnormalities of Sex Determination and Sex Differentiation

Determination of the sex of an individual is dependent upon the interactions among

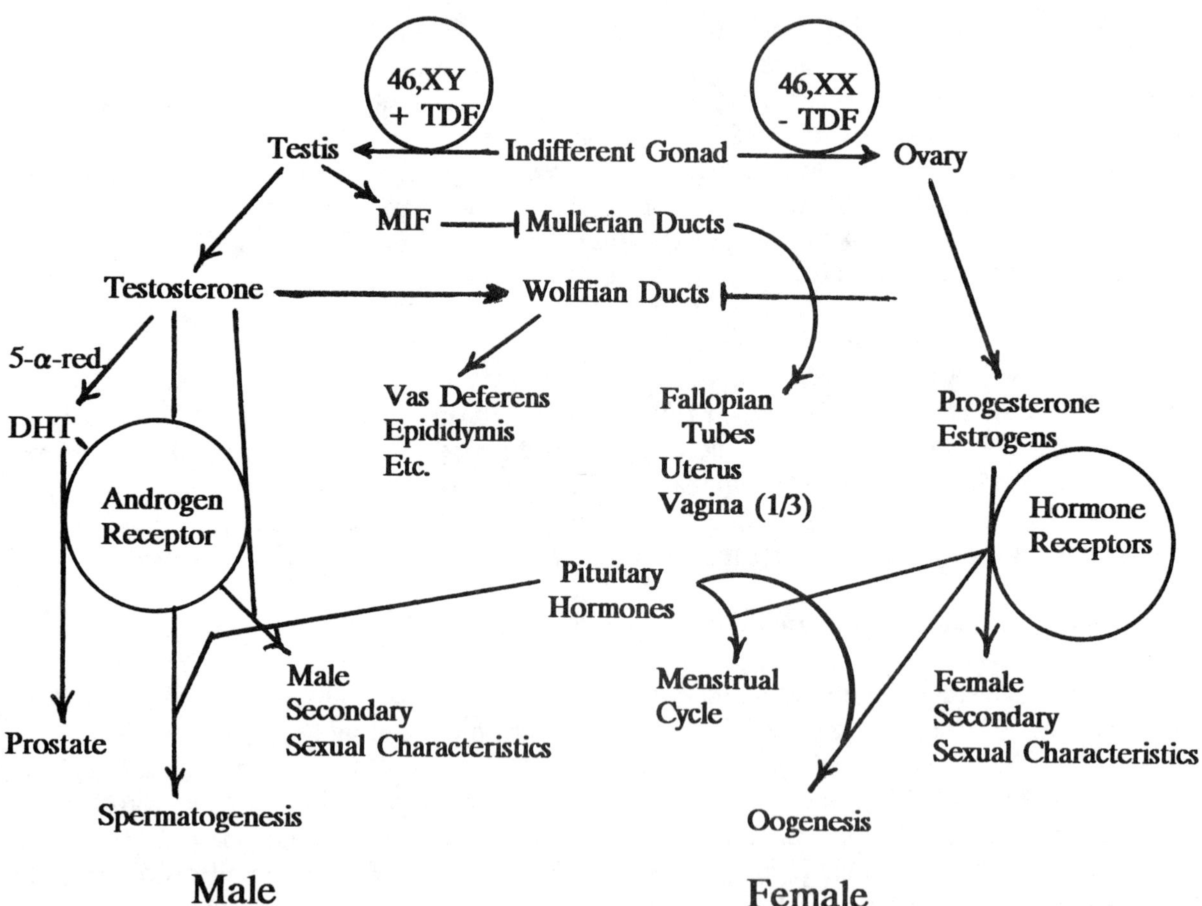

Fig. 12-8. Summary of Sex Determination and Sex Differentiation. See text for description. TDF = testis-determining factor; MIF = Mullerian Inhibitory Factor; 5-α-red = 5-α-reductase; DHT = dihydrotestosterone.

undifferentiated gonadal cells and upon their response to the presence of specific gene products. Determination of human sex and differentiation of the male and female reproductive systems were discussed in chapter 3 and are summarized in Fig. 12-8. Commitment of the indifferent gonad to testicular development requires the presence of one or more products encoded by the testis-determining region on the short arm of the Y chromosome. If there is no testis-determining region, the indifferent gonad differentiates somewhat later into an ovary. This process is believed to require a number of factors encoded by other genes, but none of these have been characterized.

The second phase of sexual development in both sexes requires hormonal stimulation. The fetal testis secretes two homones, testosterone and mullerian inhibitory factor (MIF). Testosterone is required for further development of the testis and Wolffian ducts, and MIF prevents continued differentiation of the Mullerian ducts. The Wolffian ducts form the epididymis, Vas deferens and certain other internal male accessory structures. The fetal ovary does not appear to secrete appreciable amounts of steroid hormones. In the absence of testosterone, the Wolffian ducts degenerate. The Mullerian ducts continue their development into the Fallopian tubes, uterus and upper third of the vagina. Both sexes possess an autosomal gene that encodes 5-α-reductase. This enzyme converts testosterone to dihydrotestosterone (DHT). DHT is essential for the normal formation of the male prostate gland which contributes to the fluid portion of semen. At puberty, hormonal surges occur in both sexes. Testosterone and DHT share the same androgen receptor which is encoded by an X-linked gene and expressed in both sexes. Testosterone and pituitary gonadotropins cooperate to bring about spermatogenesis. Formation of male secondary sexual characteristics is induced by the cooperative actions of testosterone and DHT. The pubertal ovary secretes large amounts of both estrogens and progesterone which stimulate development of female secondary sexual characteristics. These effects of the steroid hormones require mediation by their respective receptor proteins. Both the estrogen receptor and progesterone receptor are determined by autosomal genes that are expressed in both sexes. The ovarian hormones interact with pituitary gonadotropins during the menstrual cycle. Both men and women synthesize androgens, and the androgens produced by women interact with their androgen receptors to promote development of axillary and pubic hair. Androgen levels of women are much lower than those of men as a consequence of regulation of the rate-limiting enzyme required for androgen biosynthesis. Synthesis of androgens normally occurs in the testis and in the adrenal cortex.

Normal spermatogenesis requires haploid dosage of genes on the X chromosome and presence of certain Y-linked genes such as HY. Klinefelter's males possess all of the genes required for normal male sexual development and fertility; however, the double dosage of X-linked genes results in a failure of spermatogenesis, and these males are sterile. Women must have a double dose of genes located in the "critical regions" of the X chromosome (see chapter 3). Turner's (45,X) females have atretic ovaries and fertility of 47,XXX women may be impaired. As the number of X chromosomes increases, both mental and reproductive problems become more severe.

Sexual ambiguity occurs in approximately 1/1000 newborn and has a variety of causes. Diagnostic procedures required for elucidation of the cause of a specific disorder include a thorough physical examination, evaluation of steroid, pituitary, and thyroid hormones, and a chromosome evaluation. Genetic disorders contributing to sexual ambiguity and infertility include single gene, multifactorial, and chromosomal anomalies. Chromosomal abnormalities associated with sexual ambiguity and infertility were described in chapter 4. Several single gene disorders and sex reversals caused by chromosomal rearrangements or recombination between the X and Y chromosomes will be discussed in the following section, emphasizing the contributions these diseases have made to an understanding of normal human sexual development.

About 1/20,000 males possess a 46,XX karyotype, and a similar number of females are 46,XY. 46,XX males may arise as a consequence of X-Y or X-autosome translocations or following recombination between the X and Y chromosomes. In the latter case, exchange of chromosomal material extends sufficiently beyond the pseudoautosomal boundary to include the testis-determining region. The external phenotype of a 46,XX male is usually normal. The recombination event transferred the testis-determining region to the X chromosome; therefore, testes are present. However, there is a double dose of the critical region genes on the X chromosome, and the major portion of the Y chromosome is missing, including **HY** which is presumably necessary for normal male spermatogenesis. Consequently, 46,XX males lack mature sperm and are sterile. 46,XY females often inherit a recombinant Y chromosome from their fathers. The chromosome contains the distal portion of the short arm of the X chromosome and lacks the distal Y short arm, including the testis-determining region. These women may have an external phenotype ranging from a Turner's phenotype to approximately that of a normal woman. The combination of a haploid condition for X-linked genes in the critical region and the presence of some Y-linked genes causes sterility, and the ovaries may degenerate to "streak" gonads consisting largely of connective tissue.

A number of androgen receptor mutations have been described. The **AR** locus has been mapped to the proximal long arm of the X chromosome, and males hemizygous for **AR** mutations display a range of phenotypes that are dependent upon the amount of residual receptor activity of the mutant proteins. Some AR^0 alleles are not expressed, while others encode proteins that have totally lost their ability to activate androgen-responsive genes. Their inability to induce expression of these genes will probably be traced to heterogeneous causes such as inability to bind androgen, failure to bind to the androgen response element, inability to dimerize, or a combination of these problems. Sai et al. (21) reported a family that was segregating for a **G** to **A** transition in the fourth exon which generated a premature stop codon and resulted in a truncated receptor lacking most of its androgen-binding region. This and several other mutations summarized by these authors result in complete androgen insensitivity. 46,XY individuals who are hemizygous for one of these mutations have an essentially normal female appearance, but lack pubic and axillary hair. Testes may be present intra-abdominally, within the inguinal canal, or occasionally in the labia majora. Reason for detection is often associated with their failure to menstruate. Their gender identity is usually female, and they are often counseled and treated as an infertile female. Studies indicate that these women make good mothers, and adoption may be recommended for those who want children. Some surgery may be performed to enlarge the vagina to facilitate sexual intercourse, and the testes may be removed to avoid development of benign tumors. Many mutations encode proteins that retain some receptor function. Persons with these mutations exhibit varying degrees of masculinization, depending upon the amount of residual androgen sensitivity. The mildest manifestations of androgen receptor mutations appear to present as normal-appearing males with low sperm counts and resulting infertility.

The enzyme, 5-α-reductase, is encoded by an autosomal gene, and mutations affecting its activity have effects limited to males. Patients who totally lack enzyme activity are unable to form DHT, a potent androgen derived from testosterone. They tend to have less facial and body hair and smaller external genitalia. These men lack the prostate gland and often have hypospadias. Pubertal patients may escape the teenage problem of acne. Newborn infants with 5-α-reductase deficiency may appear female or may have ambiguous external genitalia. Testes and Wolffian duct structures are present. Many patients have some residual enzyme activity, and the degree of masculinization often parallels the quantity of enzyme activity present. Management depends upon the appearance of the individual and gender identity. In cases where castration was performed early in life, gender identity and rearing pattern is generally female. When no surgery was performed, masculinization (due to testosterone effects and possibly some residual DHT effects) occurs at puberty; and gender identity often shifts from female to male.

Mullerian inhibitory factor (MIF) is essential for curtailing further development of the mullerian duct system into the uterus, Fallopian tube and upper third of the vagina. Cases have been reported where men have been homozygous for mutations that result in a lack of MIF or non-functional MIF. These persons possess rudimentary Fallopian tubes and a uterus. However, their external appearance is typically that of a normal male, and fertility is also normal. The female structures are often unexpectedly discovered during surgery or at autopsy.

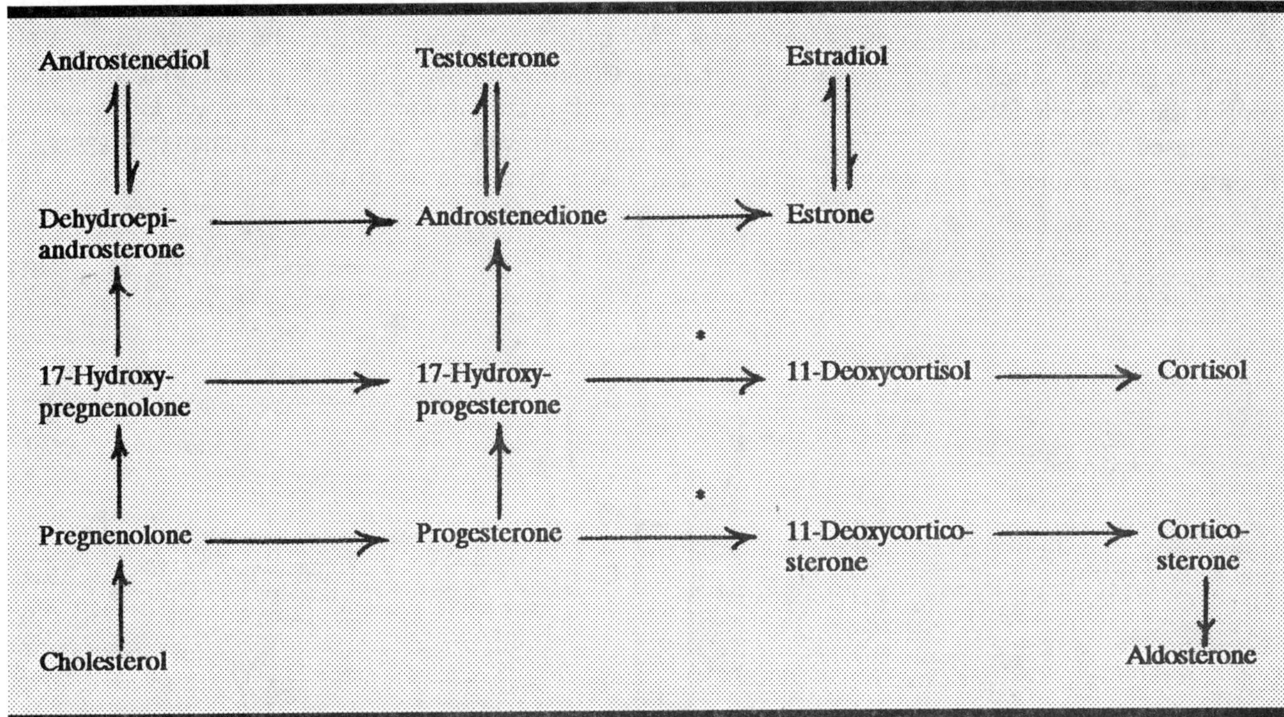

Fig. 12-9. Synthesis of steroids by the adrenal cortex.

21-hydroxylase catalyzes the conversion of progesterone to 11-deoxycorticostcronc and 17-hydroxyprogesterone to 11-deoxycortisol in the adrenal cortex (asterisks; Fig. 12-9). 21-hydroxylase is encoded by a gene, **CYP21**, which is located within the HLA gene cluster on chromosome' 6. About 1/14,000 newborn have 21-hydroxylase deficiency, and have or will develop symptoms of congenital adrenal hyperplasia (CAH). Heterogeneity is observed among these infants. Some have the "simple virilizing" form of CAH in which the preponderant effect is masculinization of females and precocious sexual development in males. A subgroup of CAH patients also experiences moderate to severe salt loss ("salt-wasting" form of CAH). The net effect of 21-hydroxylase deficiency is to cause a diminution of the amounts of aldosterone and cortisol produced by the adrenal cortex and to increase the levels of adrenal androgens. The latter effect occurs as a consequence of the accumulation of 17-hydroxyprogesterone, and entry of this precursor into the androgen biosynthetic pathway. These hormonal imbalances impact upon several homeostatic systems (Fig. 12-10). Aldosterone promotes salt retention by the kidney tubules. As salt loss increases, the renin-angiotensin-adrenal homeostatic mechanism is activated. Renin is released by the kidney which induces the formation of angiotensin II. The latter compound stimulates the release of more aldosterone to promote salt retention. As more salt is retained by the kidney, the renin release declines and aldosterone secretion diminishes. The second homeostatic system is the hypothalamic-pituitary-gonadal axis. A fall in sex hormones stimulates the release of luteinizing hormone releasing hormone (LHRH) by the hypothalamus. LHRH induces the release of luteinizing hormone

(LH) and follicle stimulating hormone (FSH) which stimulate production and secretion of sex steroids by the gonads. As circulating levels of sex steroids rise, they have a negative feedback upon both the hypothalamus and the pituitary, curtailing the release of LHRH, LH, and FSH, and ultimately release of the sex steroids by the gonads. The final homeostatic system is comprised of the hypothalamic-pituitary-adrenal axis. The hypothalamus releases corticotropin-releasing hormone (CRH) which causes the secretion of adrenocorticotropic hormone (ACTH) by the pituitary. ACTH acts on the adrenal cortex to promote synthesis and release of androgens (normally low levels), aldosterone and cortisol. Both cortisol and sex steroid levels negatively regulate CRH secretion. Another effect that should be considered, is the tendency for excessive water loss as salt excretion increases. The water loss results in a decreased blood pressure as a consequence of lowered intravascular volume. This change triggers the release of CRH and ACTH to promote additional aldosterone delivery to the kidneys.

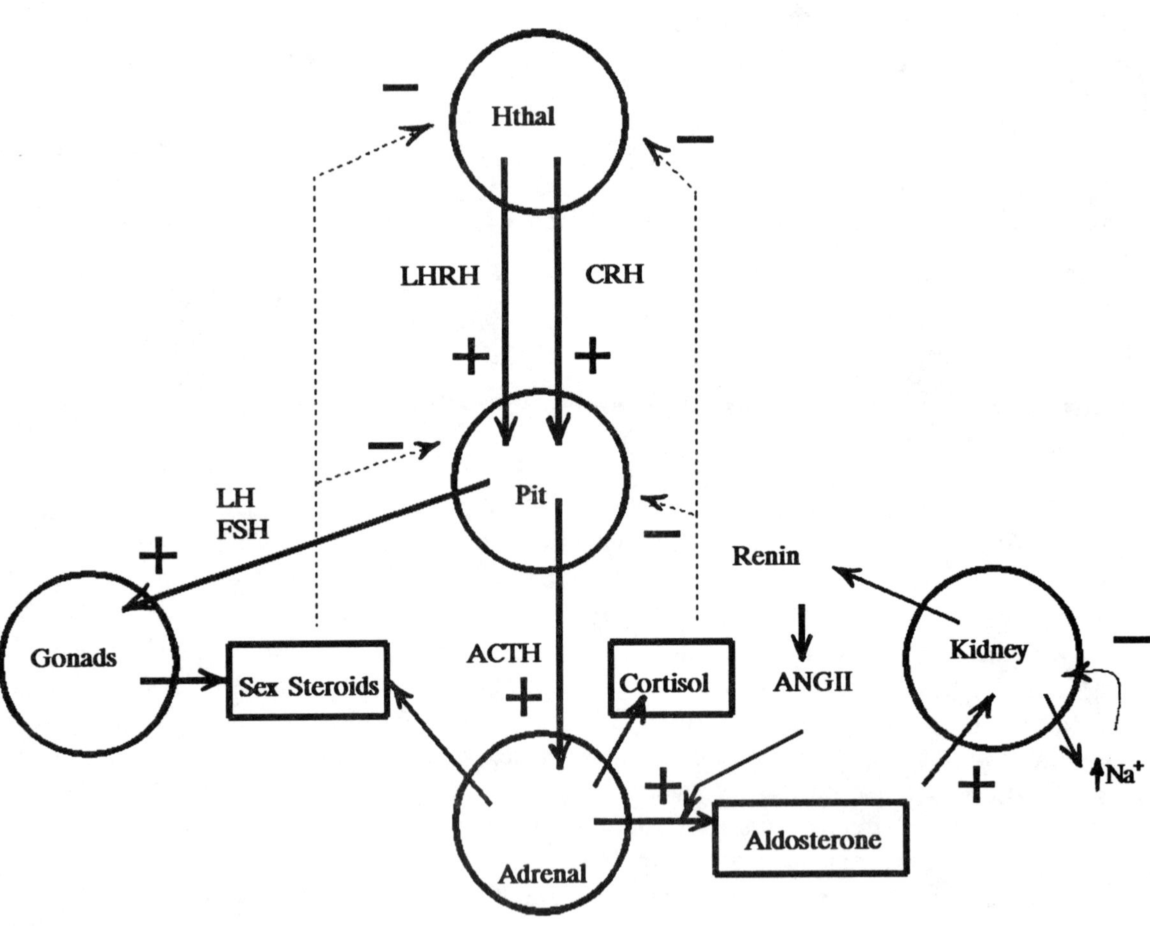

Fig. 12-10. Homeostatic mechanisms. See text for explanation. Hthal = hypothalamus; Pit = pituitary; LHRH = luteinizing hormone releasing hormone; CRH = corticotropin releasing hormone; ACTH = adrenocorticotropic hormone; LH = luteinizing hormone; FSH = follicle stimulating hormone; ANGII = angiotensin II; (+) = stimulates; (-) = inhibits.

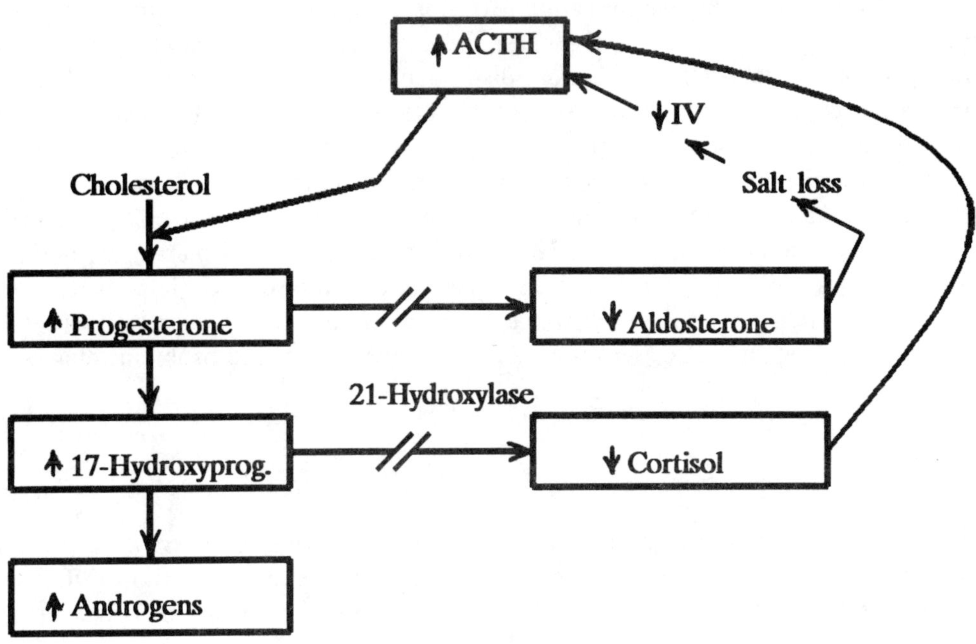

Fig. 12-11. Derangement of homeostatic regulation of adrenocortical steroid hormone biosynthesis in congenital adrenal hyperplasia. **IV** = intravascular volume.

21-hydroxylase deficiency in CAH results in deficiencies of both cortisol and aldosterone. Reduction of aldosterone levels leads to salt loss and decreased intravascular volume. The latter combines with the deficiency of circulating cortisol to cause increased secretion of ACTH. ACTH promotes enhanced steroid biosynthesis, most of which enlarges the androgen pool, causing masculinization. The increased androgen initially causes some down regulation of LHRH, LH and FSH; however, long term effects result in advancement of skeletal aging and precocious puberty in males. Females are masculinized to varying degrees, depending upon the severity of 21-hydroxylase deficiency. They may be born with ambiguous external genitalia or fully masculinized genital structures. The latter may lead to their misclassification as males. Patients with severe deficiencies of 21-hydroxylase activity are at high risk for salt-wasting and adrenal crises that can precipitate death. The average time from birth to diagnosis of CAH in states that lack neonatal screening programs for the disorder is 62 months for boys and 19 months for girls (22); therefore, males run a much higher risk for experiencing the salt-loss adrenal crisis. Because of the skeletal effects of excessive adrenal androgen secretion, CAH adults are shorter than their normal siblings. Following growth, women are infertile, have excessive body and facial hair, and fail to menstruate. Many of these women also experience psychosocial maladjustment. CAH males are also infertile and have an increased risk for testicular tumors.

A family history of CAH permits prenatal diagnosis of fetuses at risk for CAH, and prenatal treatment has proven successful in preventing ambiguous development of the genitalia in females. Furthermore, many states now have neonatal screening programs for early diagnosis of CAH. Cortisol replacement has been attempted to prevent the clinical complications of CAH. This treatment has achieved variable success in ameliorating symptoms of the disease. Lowering of androgen levels to the normal female range often necessitates administration of high levels of glucocorticoids. High levels of these steroids slow growth and development, lead to excessive weight gain and other problems. Prevention of post-pubertal signs of CAH has also been unsuccessful. Combined therapy approaches are now being tried on an experimental basis which utilize physiological doses of hydrocortisone, anti-androgens, and an inhibitor of an enzyme that converts androgens to estrogens. The goal is to maintain patients within normal physiological ranges with respect to their glucocorticoids and sex steroids. These studies have not been conducted long enough to evaluate the long term effects of this strategy.

Infertility is a common problem, involving about 15% of all marriages, and an additional 10% of couples have fewer children than they desire. As the above discussion implies, causes of infertility are heterogeneous, and many cases have significant genetic components. Investigations of the causes of infertility are progressing, and effective treatments for several of these problems may be developed in the near future.

SUMMARY

Normal development requires the timely integration of genetic systems as the human organism passes through a series of different environments. Patterns of gene expression differ among loci. Some loci are active for a relatively short time. For example, a series of globin genes are expressed during embryonic, fetal and adult stages. As each new globin locus is activated, the expression of its developmental predecessor begins to decline. Regulation of these genes occurs primarily at the level of transcription and involves a complex interaction of trans-acting factors with regulatory elements both distal and proximal to the locus that is regulated.

Other loci are constitutively expressed throughout development; however, preferential expression of some of these loci may occur at certain times and in specific tissues. **LDHA** and **LDHB** are expressed in all tissues and are symmetrically active during early fetal development. After that time, **LDHA** expression predominates in liver and skeletal muscle, while preferential expression of **LDHB** occurs in kidney and cardiac muscle.

Environmental agents have significant impact upon developmental processes. Although the maternal environment is generally protective, derangements of maternal metabolism associated with diabetes, hypothyroidism, or phenylketonuria can adversely affect the fetus. Maternal infections such as rubella or cytomegalovirus are associated with malformations. These viruses are most harmful during the first gestational month. A variety of drugs taken by pregnant women have proved harmful to the infants they are carrying. Diphenylhydantoin and alcohol have been implicated as teratogens. Oral progestins and other steroids can alter the sexual differentiation of the fetus. Not all fetuses exposed to teratogenic agents develop complications, suggesting that genetic predisposition for adverse effects may exist. Normal human sexual development requires an appropriate chromosome complement and proper orchestration of autosomal, X-linked and Y-linked genes. These genes determine structures of regulatory proteins, enzymes required for steroid hormone biosynthesis, and hormone receptors. Mutation in the Y-linked sex-determining genes or movement of these genes from the Y to the X chromosome or an autosome can lead to sex reversal. These persons are sterile as a consequence of an inappropriate dosage of X-linked genes in the critical regions and other genetic problems. Mutations affecting the androgen receptor result in pseudohermaphroditism (testicular feminization) in extreme cases, varying degrees of intersexuality, or normal-appearing infertile males with low sperm counts at the other extreme. The severity of sexual maldevelopment appears to parallel the residual receptor activity present in the mutant proteins. Congenital adrenal hyperplasia occurs as a consequence of cortisol and aldosterone deficiency. Patients lack normal 21-hydroxylase activity and overproduce adrenal androgens. Management has not been completely successful, and new approaches directed at control of the cortisol deficiency, and restoration of normal androgen and estrogen effects are being introduced on a clinical trial basis.

PROBLEMS

1. An 18-year-old woman has never menstruated. She has elevated urinary 17-ketosteroids, low serum cortisol, and has a 46,XX karyotype. External genitalia are normal with the exception of an enlarged clitoris. Tests of genital fibroblasts revealed normal androgen receptor function. What is your diagnosis? Provide an explanation for the masculinization in this patient. How would you treat her?

2. Studies of couples in which the husband is A (B or AB), Rh(+) and the wife is O, Rh(-) have shown that there are significantly fewer cases of erythroblastosis fetalis among their children than among those of other Rh-incompatible couples. Why?

3. It has been said that ontogeny (development) recapitulates phylogeny (evolutionary history). During development, structures are formed which resemble those in more primitive forms of life and are subsequently replaced by more advanced structures. If this was also true for proteins, and specifically for hemoglobins, what evolutionary significance would you attribute to the embryonic and fetal hemoglobins?

4. Provide a possible mechanism for activation of genes during development.

5. A young couple has experienced three consecutive abortions between the twelfth and fourteenth gestational week. The husband has a balanced translocation: 46,XY,-4,-6,+t(4;6)(p12;p11). Provide an explanation for the fetal loss in this family.

6. Why do some infants of women who have been taking diphenylhydantoin during pregnancy develop malformations and other problems while others with similar exposure develop normally?

7. Parents of a patient with incomplete testicular feminization have inquired as to whether testosterone therapy would correct their child's sex reversal. How would you answer them?

GLOSSARY OF TERMS

Deformation developmental anomaly produced by the effects of mechanical forces upon the fetus.

Developmental field region of an embryo giving rise to tissues and organs at later stages.

Disruption abnormality caused by interference with an intrinsically normal developmental process by an extrinsic agent.

Dysplasia abnormal organization of cells into tissues.

Malformation a disturbance of a developmental field, usually as a consequence of an intrinsically abnormal process, which causes the abnormal formation of a body part or larger region.

Maternal effect a) trait that is predominantly transmitted by the female parent to her offspring (e.g. mitochondrial genes); b) phenotypic effect imposed upon a fetus by an unsatisfactory uterine environment secondary to an "abnormal" maternal genotype (e.g. the occurrence of mental retardation in a son or daughter of a PKU mother).

Sequence defect in a developmental field results in a cascade of events leading to subsequent abnormalities in tissues and organs derived from or influenced by the embryonic tissue in which the original defect occurred.

Teratogen an agent capable of inducing developmental abnormalities.

BIBLIOGRAPHY

1. James, KL, DW Smith, N Ulleland and AP Streissguth. 1973. Pattern of malformation in offspring of chronic alcoholic mothers. Lancet 1: 1267-1271.

2. Ouellette, EM, HC Rosett, NP Rosman and L Wiener. 1977. Adverse effects on offspring of maternal alcohol abuse during pregnancy. N Eng J Med 297: 528-530.

3. Hanson, JW and DW Smith. 1975. The fetal hydantoin syndrome. J Pediatrics 87: 285-290.

4. Lammer, EJ, DT Chen, RM Hoar et al. 1985. Retinoic acid embryopathy. N Eng J Med 313: 837-841.

5. Shih, DM, RJ Wall and SG Shapiro. 1990. Developmentally regulated and erythroid-specific expression of the human embryonic β-globin gene in transgenic mice. Nucleic Acids Res 18: 5465-5472.

6. Chisaka, O and MR Capecchi. 1991. Regionally restricted developmental defects resulting from targeted disruption of the mouse homeobox gene hox-1.5. Nature 350: 473-479.

7. Malicki, J, K Schughart and W McGinnis. 1990. Mouse Hox-2.2 specifies thoracic segmental identity in Drosophila embryos and larvae. Cell 63: 961-967.

8. McGinnis, N, MA Kuziora and W McGinnis. 1990. Human Hox-4.2 and Drosophila Deformed encode similar regulatory specificities in Drosophila embryos and larvae. Cell 63: 969-976.

9. Wilkinson, DG, S Bhatt, M Cook et al. 1989. Segmental expression of Hox-2 homeobox-containing genes in the developing mouse hindbrain. Nature 341: 405-409.

10. Wright, C and B Hogan. 1991. Another hit for gene targeting. Nature 350: 458-459.

11. Kessel, M and P Gruss. 1991. Murine developmental control genes. Science 249: 374-379.

12. Tamura, K, K Ohsugi and H Ide. 1990. Distribution of retinoids in the chick limb bud: analysis with monoclonal antibody. Dev Biol 140: 20-26.

13. Wanek, N, DM Gardiner, K Muneoka and SV Bryant. 1991. Conversion by retinoic acid of anterior cells in the chick wing bud. Nature 350: 81-83.

14. Harris, H. 1975. <u>The Principles of Human Biochemical Genetics</u>. 2nd ed. New York: American Elsevier Publishing Co., p.76.

15. Kollias, G, N Wrighton, J Hurst and F Grosveld. 1986. Regulated expression of human $^A\gamma$-, β-, and γ/β-globin genes in transgenic mice: manipulations of the developmental expression patterns. Cell 46: 89-94.

16. Behringer, RR, TM Ryan, RD Palmiter <u>et al.</u> 1990. Human γ- to β-globin gene switching in transgenic mice. Genes and Development 4: 380-389.

17. Baron, MH and T Maniatis. 1986. Rapid programming of globin gene expression in transient heterokaryons. Cell 46: 591-602.

18. Papayannopoulou, T, M Brice and G Stamatoyannopoulos. 1986. Analysis of human hemoglobin switching in MEL x human fetal erythroid cell hybrids. Cell 46: 469-476.

19. Greengard, O. 1969. Enzymic differentiation in mammalian liver. Science 163: 891-895.

20. Koler, RD, PJ Van Bellinghen, JH Fellman <u>et al.</u> 1969. Ontogeny of soluble and mitochondrial tyrosine aminotransferases. Science 163: 1348-1350.

21. Sai, T, S Seino, C Chang <u>et al.</u> 1990. An exonic point mutation of the androgen receptor gene in a family with complete androgen insensitivity. Am J Hum Genet 46: 1095-1100.

22. Lebovitz, RM, RM Pauli and R Laxova. 1984. Delayed diagnosis in congenital adrenal hyperplasia: need for newborn screening. Am J Dis Child 138: 571-573.

ADDITIONAL LEARNING RESOURCES

1. Cutler Jr., GB and L Laue. 1990. Congenital adrenal hyperplasia due to 21-hydroxylase deficiency. N Eng J Med 323: 1806-1813.

2. Garbers, DL. 1989. Molecular basis of fertilization. Annu Rev Biochem 58: 719-743.

3. Gehring, WJ and Y Hiromi. 1986. Homeotic genes and the homeobox. Annu Rev Genet 20: 147-173.

4. McLaren, A. 1990. What makes a man a man? Nature 346: 216-217.

5. Phillips III, JA and CL Vnencak-Jones. 1989. Genetics of growth hormone and its disorders. Adv Hum Genet 18: 305-363.

6. Pinsky, L and M Kaufman. 1987. Genetics of steroid receptors and their disorders. Adv Hum Genet 16: 299-472.

7. Smith, M. 1986. Genetics of human alcohol and aldehyde dehydrogenases. Adv Hum Genet 15: 249-290.

Chapter 13

Population Genetics and Human Evolution

Charles Darwin was one of the first to note that the members of a population have inherent variation which could provide the raw material for speciation through natural selection. Studies of variation both within and among species led to the recognition of the mechanisms responsible for maintenance of this variation and have laid the foundations for modern evolutionary theory. Population geneticists are concerned with estimation of genetic variation within and among populations and with development of theories and models which explain how the current genetic structure of a population was achieved and which will permit prediction of the genetic structure of future populations in the same or somewhat different environments.

The genetic structure of a population can be characterized by investigating genetic systems determining phenotypes that are easily resolved by electrophoresis or immunological methods. Enzymes, plasma proteins, blood group antigens and HLA antigens have been used in the past, and nuclear and mitochondrial DNA RFLPs have been added recently. **Polymorphic** loci are those loci that possess two or more alleles, the rarest of which occurs with a frequency of at least 1%. Selection of this value eliminates alleles that are maintained by mutational pressure. **ABO, RH, G6PD**, and the globin loci are examples of polymorphic loci.

HUMAN GENETIC VARIABILITY

Several problems have been encountered during attempts to estimate the average heterozygosity of human populations. Variability of enzymes differs among tissues. Erythrocyte enzymes tend to be more variable than those of solid tissues such as brain. Use of the former enzymes will lead to an overestimate of average heterozygosity, while exclusive utilization of the brain systems will tend to underestimate average heterozygosity. The number of loci sampled will also affect heterozygosity estimates. The more polymorphic loci will be ascertained first, while monomorphic loci will be discovered relatively late during a study (1). Therefore, a reasonable number of loci must be selected without prior knowledge of whether they are polymorphic. The resolving power of the technique used to evaluate phenotypes will also influence the estimate of average heterozygosity. For example, electrophoresis has proved to be quite effective in distinguishing among phenotypes determined by allelic genes. However, examination of the genetic code reveals that only one-third of the possible base substitutions will result in an alteration of the charge of the polypeptide. Racial heterogeneity of a population also influences estimates of average heterozygosity. Certain loci may be polymorphic in one population but not in others. For example β^S and β^C occur at relatively high frequencies among Black Americans, but not among Caucasian Americans. Asian Americans would also be anticipated to have high frequencies of alleles, such as β^D and β^E, that are rare among Caucasian Americans.

Allowing for these limitations, Harris and Hopkinson (2) have estimated human genetic variability based upon a study of 72 different enzymes. Twenty-eight percent of these loci were polymorphic in the population, and the average heterozygosity per locus approximated 6.7%. If only one-third of base substitutions are detectable by electrophoresis, then the "corrected" average heterozygosity approximates 20%. That is, an average of one out of every five loci are heterozygous in a given individual. This estimate agrees well with Lewontin's independent estimate of 16% based upon blood group antigens (1). If there are 100,000 protein-encoding loci in human cells, these estimates suggest that 16,000 to 20,000 are heterozygous.

Such extensive inherent variation provides a population with the genetic flexibility required for meeting the stresses imposed by changing environments. As the species (or population) experiences a changing environment, those genetic combinations that are beneficial increase in frequency through differential suvival and reproduction.

Harris and Hopkinson confined their studies to Caucasians living in England. Extensions of their work have shown that genetic differences are maximal within local populations, and much less extensive between populations or between races.

FACTORS INFLUENCING POPULATION GENE STRUCTURE

Hardy-Weinberg Equilibrium

The concept of the **gene pool** or total number of genes and alleles within a population was first defined by Hardy and Weinberg in 1908 (3,4). Working independently, they designed a mathematical model which provided a method for analyzing populations in a quantitative fashion. Some individuals at that time believed that genes with dominant phenotypic effects would eventually replace those with recessive effects. The Hardy-Weinberg equilibrium provided algebraic proof that demonstrated that this replacement would not occur. In fact, under appropriate conditions, the frequencies of the genotypes reach an equilibrium within one generation and do not change thereafter.

The Hardy-Weinberg model can be used to predict expected gene and genotypic frequencies in subsequent generations. If the gene and genotypic frequencies remain the same from generation to generation, then the population is said to be at **equilibrium** for those alleles. If a change in these frequencies occurs over time, then the population is changing or **evolving**. When studying a natural population, this model provides a means for calculating the expected frequencies for this population under static conditions. Comparisons between the predicted frequencies and the observed frequencies can be used to determine if the population in question is at equilibrium.

According to the predictions of this model, the gene frequencies will stay about the same from generation to generation if the population is large, mates randomly with respect to the traits being examined, experiences minimal migration, and the effects of selective and mutational pressures upon the locus responsible for the traits are negligible. The loci determining the traits should be autosomal, to permit attainment of equal frequencies in both sexes, and, if pairs of traits are being evaluated, the genes determining those phenotypes should be unlinked.

Consider a single locus with two alleles determining codominant traits. The frequency of one allele A^1 is defined as p, and the frequency of the other allele A^2 is q. Since these are the only two alleles occurring in this population,

1) $$p + q = 1.$$

For a diploid organism, the allele frequencies approximate the gametic frequencies, since the haploid gametes will have only one allele for this locus. Hypothetically, p and q could be determined by scoring the gametes as to type of allele they are carrying. Since this is not possible, gene frequencies are usually calculated from the genotypic frequencies that can be observed within a population.

Assuming random mating, gametes carrying A^1 or A^2 will combine in the respective genotypes according to the following predicted frequencies: $A^1A^1 = p^2$; $A^1A^2 = 2pq$; $A^2A^2 = q^2$. Suppose $p =$

Table 13-1. Calculation of Genotypic Frequencies for a Population in Hardy-Weinberg Equilibrium.

| Parental Frequencies: | A^1 frequency = p = 0.8. |
| | A^2 frequency = q = 0.2. |

Female Gametes

		A^1	A^2
		(0.8)	(0.2)
Male	A^1 (0.8)	A^1A^1 (0.64)	A^1A^2 (0.16)
Gametes	A^2 (0.2)	A^1A^2 (0.16)	A^2A^2 (0.04)

Genotypic Frequencies in Next Generation:

$$A^1A^1 \;=\; 0.64 \;=\; p^2$$

$$A^1A^2 \;=\; 0.32 \;=\; 2pq$$

$$A^2A^2 \;=\; 0.04 \;=\; q^2$$

Note: $p^2 + 2pq + q^2 = 0.64 + 0.32 + 0.04 = 1.00$

0.8 and q = 0.2. The genotypic frequencies can be determined as illustrated in Table 13-1. Note that the genotypic frequencies are obtained from the binomial expansion of

$$2) \qquad (p + q)^2$$

and that

$$3) \qquad p^2 + 2pq + q^2 = 1.$$

Therefore, if the genotypic frequencies within a

Table 13-2. Calculation of Allele Frequencies for Codominant Traits.

Phenotype	Genotype	Number
M	$GYPA^MGYPA^M$	360
MN	$GYPA^MGYPA^N$	480
N	$GYPA^NGYPA^N$	160
	Total	1000

$p = [2(360) + 480] \div [2(1000)] = 0.6$

$q = 1.0 - 0.6 = 0.4$

---

population can be determined from observation of the phenotypes within that population, then the gene frequencies can also be calculated. Codominant traits are most appropriate for analyses based upon the Hardy-Weinberg equilibrium because all three genotypes are readily identifiable. Calculation of gene frequencies for two alleles with codominant effects is illustrated in Table 13-2. Notice how the number of $GYPA^MGYPA^M$ homozygotes is doubled to reflect the presence of two $GYPA^M$ alleles in each person of this genotype. The total number of people in the population is also doubled, since each person possesses two alleles at the **GYPA** locus. If the population in Table 13-2 is in Hardy-Weinberg equilibrium, both the gene and genotypic frequencies will remain unchanged generation after generation. Therefore, the genotypic frequencies for the generation illustrated in Table 13-2 could be used to predict the numbers of each phenotype in the next generation. For example, if the population grows to 2000, the number of type M persons would be predicted by

$$p^2(\text{total}) = (0.6)^2(2000) = 720.$$

The relative frequencies of the genotypes A^1A^1, A^1A^2 and A^2A^2 at different gene frequencies are presented in Table 13-3. The heterozygote frequency increases rapidly as the value of p or q increases and reaches a maximum at 0.5, when the gene frequencies are equal. The frequency of the heterozygote exceeds the frequency of the least prevalent homozygote at all values. This trend is most striking at gene frequencies below 0.1. For example, when $q = 0.01$, the heterozygote occurs with a frequency 198-fold higher than that of the A^2A^2 homozygote. This illustrates an important point. Most alleles responsible for a rare trait are carried by heterozgotes.

The example utilized to demonstrate attainment of Hardy-Weinberg equilibrium involved a two-allele system. Equation (3) can be expanded to multiple allelic systems, provided the alleles have codominant effects. The expansion to a three-allele system can be illustrated by the β-globin locus using only the β^A, β^S and β^C alleles.

Table 13-3. Genotypic Frequencies at Varying Allele Frequencies.

Allele Frequencies		Genotypic Frequencies		
A^1	A^2	A^1A^1	A^1A^2	A^2A^2
1	0	1	0	0
.99	.01	.901	.0198	.0001
.8	.2	.64	.32	.04
.5	.5	.25	.50	.25
.2	.8	.04	.32	.64
.1	.9	.01	.18	.81
0	1	0	0	1

Let p = frequency of β^A, q = frequency of β^S and r = frequency of β^C. Then

Table 13-4. Composition of a Population Containing Hemoglobins A_1, S and C.

Phenotype	Frequency	Number
A_1A_1	p^2	3000
A_1S	$2pq$	1000
A_1C	$2pr$	850
SC	$2qr$	50
SS	q^2	50
CC	r^2	50

4) $$(p + q + r)^2 = p^2 + 2pq + q^2 + 2pr + 2qr + r^2 = 1.$$

A three-allele system generates three homozygotes and three heterozygotes. A sample population is entered in Table 13-4. The Hardy-Weinberg frequency of each genotype is listed in the central column, and the observed numbers of each phenotype are entered at the right. The frequency of β^A is calculated:

$$p = [2(3000) + 1000 + 850] \div [2(5000)] = 0.785.$$

The frequency of β^S is:

$$q = [2(50) + 1000 + 50] \div [2(5000)] = 0.115.$$

The frequency of β^C is obtained by subtraction:

$$r = 1 - p - q = 1 - 0.785 - 0.115 = 0.100.$$

A modification of this method can provide the frequencies of two codominant alleles and a recessive allele. The **ABO** locus possesses three alleles, with two having codominant effects (**ABO^A** and **ABO^B**) and one having recessive effects (**ABO^0**). The frequency of type O (**ABO^{0}ABO^0** $= r^2$) can be used to estimate the frequency of the **ABO^0** allele (**r**):

$$r = \sqrt{\text{frequency of type O.}}$$

Tests used for blood typing cannot discriminate between the **ABO^{A}ABO^A** and **ABO^{A}ABO^0** genotypes. The phenotypes, type A and type O, include three genotypes, and, if **p** symbolizes the frequency of **ABO^A**, the frequencies of these two phenotypes and the three genotypes can be given by:

$$\text{Type A} + \text{Type O} = \text{ABO}^A\text{ABO}^A + \text{ABO}^A\text{ABO}^0 + \text{ABO}^0\text{ABO}^0$$

$$\text{Frequency (A + O)} = p^2 + 2pr + r^2 = (p + r)^2$$

$$p = (p + r) - r$$

$$p = \sqrt{\text{Frequency of A + O}} - \sqrt{\text{Frequency of O.}}$$

The frequency of **ABO^B** can be estimated in a similar way:

$$q = \sqrt{\text{Frequency of B + O}} - \sqrt{\text{Frequency of O.}}$$

Suppose a population contains 2000 A, 975 B, 25 AB and 2000 O persons. The frequency of **ABO^0** is approximated by:

$$r = \sqrt{2000 \div 5000} = 0.63.$$

The frequency of **ABO^A** is estimated:

$$p = \sqrt{(2000 + 2000) \div 5000} - 0.63 = 0.26.$$

The frequency of **ABO^B** is calculated by subtraction:

$$q = 1 - 0.26 - 0.63 = 0.11.$$

The error caused by the failure to include **ABO^{A}ABO^B** persons is usually small, if such individuals are infrequent in the population.

For dominant traits, the dominant homozygote cannot be distinguished from the heterozygote at the phenotypic level. The Hardy-Weinberg equilibrium frequencies are often used to estimate the proportion of heterozygotes who carry recessive alleles. The frequency of the recessive allele can be estimated by taking the square root of the disease frequency in the population. This value is doubled to provide an estimate of the population frequency of heterozygotes that can be used for genetic counseling purposes (see chapter 5). For example, recessive albinism occurs with a frequency of about 1/40,000 (q^2) in most populations. The frequency of an albino allele (**q**) is 1/200, and an estimate of the heterozygote frequency (**2q**) is 1/100. The **2q** approximation of the heterozygote frequency is inappropriate for relatively common recessive traits (Table 13-5), because it overestimates the actual heterozygote frequency for these traits. The transition between a good and a bad estimate appears to occur at a trait frequency between 1% and 9%, and the approximation is not recommended for use when the trait occurs more commonly than 1% in the population.

Table 13-5. Comparison of Approximations and Hardy-Weinberg Frequencies of Heterozygotes for Different Values of q.

q^2	q	2q	2pq
1/100,000	0.0032	0.0064	0.0064
1/10,000	0.0100	0.0200	0.0198
1/2000	0.0224	0.0448	0.0438
1/100	0.1000	0.2000	0.1800
9/100	0.3000	0.6000	0.4200
1/4	0.5000	1.0000	0.5000

The Hardy-Weinberg equilibrium assumes that the traits being studied are autosomal, that the population mates at random with respect to these traits, that the population is large, and that effects of migration, mutation and selection are minimal. Many systems, including MN and ABO blood types, seem to fit these requirements, and the Hardy-Weinberg frequencies provide fairly accurate estimates of the respective gene and genotypic frequencies. In other cases, the Hardy-Weinberg estimates poorly approximate actual frequencies and are poor predictors of frequencies in future generations. These exceptions will be discussed in the following sections.

Effect of X-linkage Upon Approach to Hardy-Weinberg Equilibrium

X-linked traits violate a Hardy-Weinberg assumption that allele frequencies are the same for males and females. Since males are hemizygous, (XY) with respect to X-linked loci, males carry one-third of all X-linked alleles in the population, while females possess the other two-thirds. Furthermore, since males have only one allele at each X-linked locus, the frequency of males with an X-linked trait is a direct reflection of the frequency of the allele responsible for that trait.   Suppose Trait Z is inherited as an X-linked recessive and that one allele, TZ^0, is primarily responsible for the occurrence of Trait Z in a population. The frequency of TZ^0 is approximated by the frequency of affected males:

$$\text{Frequency of } TZ^0 = q = \text{Frequency of affected males} = X_M$$

or

5) $$q = X_M$$

This equation is based upon the assumption that TZ^0 occurs with similar frequencies in males and females. From a theoretical standpoint, the male and female frequencies of X-linked alleles are never the same; however, after many generations they will be very close (Fig. 13-1). Since males inherit their

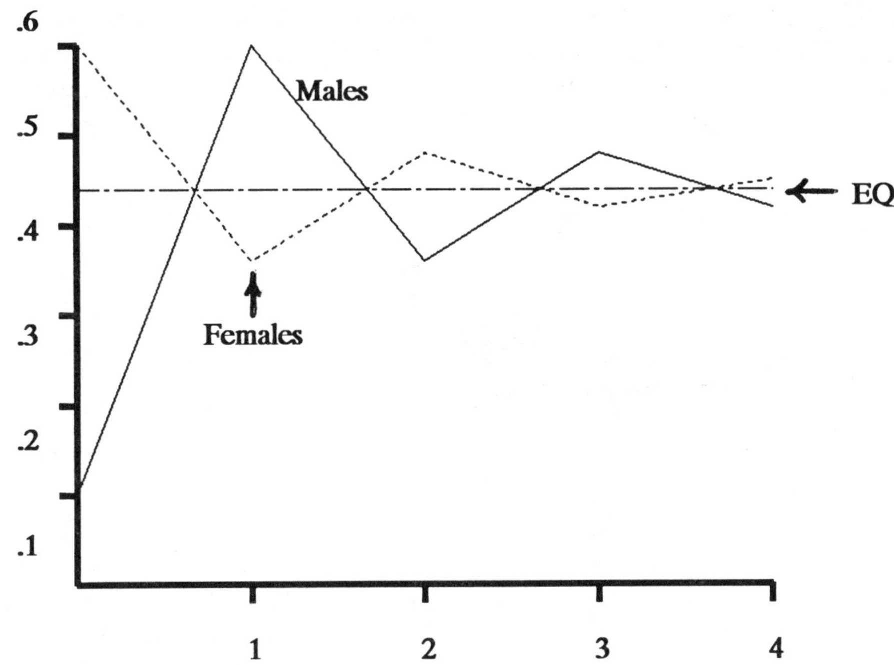

Fig. 13-1. Approach of different male and female X-linked allele frequencies to equilibrium. Males from one population having an allele frequency of 0.1 and a female population with an allele frequency of 0.6 intermarry. After several generations of reproduction among the offspring of the founding populations, the male and female allele frequencies closely approximate an equilibrium value (EQ) of about 0.43. Allele frequencies are presented on the vertical axis, and generations increase along the horixontal axis. Notice how male frequencies equal female frequencies from the preceding generation, while female frequencies approximate the mean of the male and female frequencies of the prior generation.

X chromosomes from their mothers, the male allele frequency is equivalent to the female frequency of the preceding generation. Women receive one X chromosome from their fathers and one from their mothers; therefore, the female allele frequency equals the mean of the male and female allele frequencies of the previous generation. The Figure displays an extreme situation. Even with such large intersexual differences, the allele frequencies of the two sexes rapidly approach equilibrium. For this reason, allele frequencies in both sexes are often simplified to the frequency of affected males.

This example assumed that the trait was caused by only one type of allele at the **TZ** locus. Recent studies of Duchenne muscular dystrophy, Hemophilia A, colorblindness, and other X-linked traits indicate that the respective loci are occupied by many alleles within a population. In these and many other traits, equation (5) represents the frequency of a group of alleles. Hemizygosity for any one of them will result in the trait being studied. Individual frequencies of any specific allele would be expected to be much lower.

Extrinsic and Intrinsic Factors Influencing Gene Pools

The Hardy-Weinberg equilibrium views a population in a somewhat idealized manner. The gene pool contains a large number of alleles whose movement from generation to generation is not restricted by assortative mating. Furthermore, the population is well insulated against perturbations caused by mutation, selection, and migration. Many genetic systems are strongly influenced by combinations of intrinsic and extrinsic pressures that may have significant impact upon the gene pool of populations exposed to these pressures (Fig. 13-2). The sizes of the circles represent the relative sizes

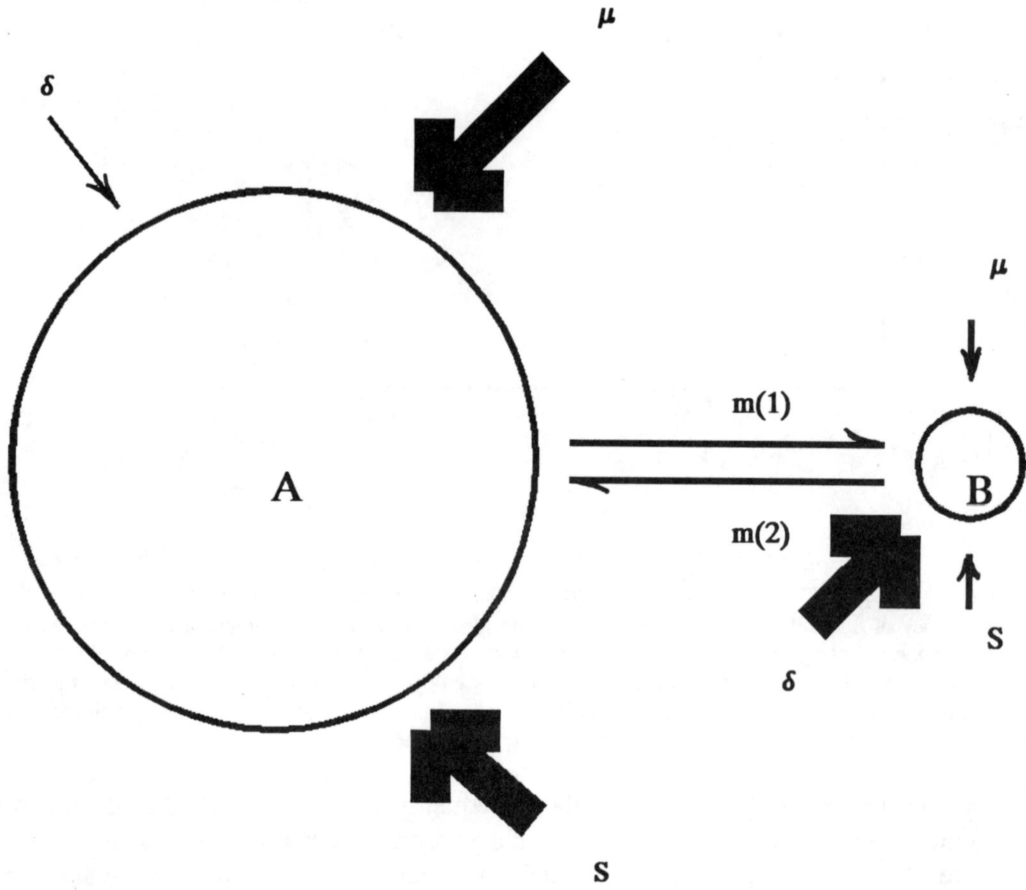

Fig. 13-2. Relative impact of intrinsic and extrinsic forces upon gene pools. Two populations of different sizes are represented by circles. The arrows represent extrinsic (μ = mutation; m(1) and m(2) = migration; s = selection) and intrinsic (δ = drift) pressures acting on each gene pool. The sizes of the arrows indicate the relative importance of each pressure in molding the gene pool of the respective populations.

of the gene pools of two populations. Mutation and immigration introduce new alleles to populations, while emigration and selection tend to decrease the frequencies of alleles in a population. Selection is viewed as acting in a negative manner. That is, a population has generally been living in an environment for a long period of time, and its gene pool generally contains alleles which promote adaptation to that environment. New alleles introduced to the gene pool by mutation or by immigration may not be as adaptive and would tend to be eliminated by selective pressures. Unlike selection which is a directed pressure, drift is random. Drift appears to operate under two situations. Small groups of reproducing individuals (population B) have relatively few targets for mutational and selective pressures. Chance largely determines which alleles are transmitted to the next generation. The second case where drift appears to be important is in relation to alleles which do not appear to play a significant role in the ability of an individual to reproduce. Large populations have many more gene targets for rare events like mutation and for selective pressures of small magnitude. Therefore, mutation and selection have a much greater impact upon larger populations than upon smaller populations. The following sections will discuss each of these factors in greater detail.

Mutation

Mutation (spontaneous or induced) provides one route through which new alleles are introduced into human populations. Although mutation is bidirectional,

$$A \underset{\eta}{\overset{\mu}{\rightleftharpoons}} a$$

the forward mutation rate (μ) is about ten-fold higher than the back mutation rate (η). Therefore, the net effect of mutation is to increase the genetic variability of a population. It is generally estimated that about 10% of human autosomal dominant traits arise in this manner (5), and as many as one-third of Duchenne muscular dystrophy cases represent new mutations (6).

A conservative estimate for the average mutation rate per locus per generation of 1×10^{-6} may be used to illustrate the effect of mutation upon the gene pool. The total mutation rate may be calculated using equation (6):

6) $\quad \mu^{T} =$ (average mutation rate per locus) $\times$ (number of protein-encoding loci)

$$= (1 \times 10^{-6}) \times (1 \times 10^{5}) = 10^{-1}.$$

Based on this total mutation rate, about one in ten people transmit a mutant allele to the next generation. If the number of reproducing individuals worldwide in a given generation approximates 2×10^{9}, then about 2×10^{8} mutant alleles enter the global gene pool of the next generation. Therefore, mutation plays a significant role in diversifying the array of alleles in the global population and has a major impact upon human health and reproductive fitness.

Selection

The number of mutant alleles entering the gene pool of the next generation represents a small sampling of the total number occurring in the previous generation. Human family sizes are small; therefore, the likelihood that a given mutation will be included in a gamete participating in fertilization is very low. Those that are transmitted will persist for varying periods of time. For example, an autosomal dominant mutation that totally prevents reproduction of the person inheriting it (i.e. a genetic lethal) will be eliminated within one generation. By contrast, autosomal mutations with recessive effects are eliminated only in homozygotes and will persist for long periods of time.

The **fitness** of a genotype is measured in terms of the reproductive success of persons with that genotype. The genotype having the largest number of offspring is generally assigned a fitness of 1. All other genotypes have lower fitnesses, and genotypes failing to transmit any genes to the subsequent generation are assigned a fitness of 0. It is important to stress that fitness is measured by relative reproductive success and not necessarily by longevity. Mutations which do not affect reproduction but which shorten longevity from 75 to 55 years do not usually influence fitness. A second point that should be stressed is related to the level at which selection acts, the organism. Therefore, fitness not only is influenced by the particular allele being studied, but also by other linked genes and by the genetic background in which that allele functions.

Fitness is environment bound. For example, hemoglobin AS heterozygotes may tend to produce more children than either $\beta^A\beta^A$ or $\beta^S\beta^S$ homozygotes in areas infested with Falciparum malaria, but have no reproductive advantage in an environment that is malaria-free. In the former case, the fitness of the heterozygote would be 1, while those of the homozygotes would be less than 1. By contrast, the fitness of $\beta^A\beta^S$ heterozygotes in a malaria-free environment would be equivalent to that of $\beta^A\beta^A$ homozygotes. Environments are subject to change over time, and genotypes that may have been favorable in the past may not enjoy that advantage now. For example, certain thalassemia mutations continue to be frequent in the Mediterranean, reflecting their protection against malaria that once infested the region.

The forces occurring in the environment that promote nonrandom differential reproduction of genotypes are called **natural selection**, and the degree to which the environment limits the reproductive performance of a genotype is termed the **selective coefficient**. The relationship between fitness (**W**) and the selection coefficient (**s**) can be formulated:

7) $$W = 1 - s.$$

Selection against recessives can be illustrated using Hurler syndrome, an autosomal recessive. Hurler syndrome is characterized by severe mental retardation, growth failure, and early death. No homozygotes reproduce. The relative fitnesses of the three genotypes are presented in Table 13-6.

Table 13-6. Selection Against the Recessive Homozygote.

	\multicolumn{3}{c}{Genotype}	Total Frequency
	HH	Hh	hh	
Before Selection	p^2	$2pq$	q^2	1
Fitness (W)	1	1	0	
After Selection	p^2	$2pq$	0	$1-q^2$

The fitnesses of the normal homozygote and heterozygote appear to be essentially equivalent and are assigned a value of 1, while the fitness of the Hurler's patients is 0. For comparative purposes, the three genotypes are assumed to originally occur at their Hardy-Weinberg frequencies. After one generation of selection, the frequency of h (q_1) can be calculated:

$$q_1 = pq \div (1 - q^2) ,$$

(note: only 1/2 of the alleles in heterozygotes are h; therefore, pq rather than $2pq$ is used for the frequency of residual h alleles) and the change in q can be determined:

$$q = q_1 - q = [pq \div (1-q^2)] - q = (-pq^2) \div (1-q^2) .$$

The sign of the expression is negative, indicating that the frequency of h will decline under this form of selection. The rate of decline will be very slow, since a large majority of the recessive alleles are carried and transmitted by heterozygotes. The number of generations (n) required to reduce q_0 to q_n is estimated by:

8)
$$n = (1/q_n - 1/q_0) .$$

Hurler's syndrome currently afflicts about 1/60,000 newborn. The Hardy-Weinberg estimate of the current frequency of the Hurler allele is about 1/250 or 0.004. It would require 250 generations to reduce the frequency of the Hurler allele to 0.002! This example illustrates the futility of attempting to reduce the frequency of an infrequent recessive allele by preventing reproduction of affected homozygotes. Most recessive genes that cause human diseases occur with a frequency of 0.01 or less, and many generations would be required to reduce them to one-half of their present frequency. Conversely, relaxation of the selective pressure by medical treatment of affected persons such that they are able to reproduce will only slightly increase the frequencies of rare alleles with recessive effects.

This discussion has simplified the etiology of Hurler syndrome to a single recessive allele. As with many other diseases, Hurler syndrome is most likely caused by either homozygosity for one of several different alleles or compound heterozygosity for any two of these alleles. Selective pressures may vary for different alleles at the same locus and depend upon the relative loss of enzyme activity.

The example of Hurler syndrome can be generalized to other cases where fitness of the affected homozygotes is greater than 0:

9)
$$q = (-spq^2) \div (1-sq^2) .$$

Many genes that have frequencies less than 0.1% are maintained in human populations by recurrent mutation. After many generations, the rate of introduction of new alleles by mutation is balanced by their rate of loss following selection. In cases where selection acts exclusively against the recessive homozygote:

10)
$$\mu = sq^2$$

where μ is the mutation rate, s is the selective pressure and q^2 is the frequency of the recessive trait. Note that s is one for Hurler syndrome and

$$\mu = q^2 .$$

In this case and in others where the recessive homozygote fails to reproduce, the mutation rate equals the frequency of the recessive trait.

Infrequent autosomal dominant traits are also largely maintained by mutation. Consider the trait represented in Table 13-7. Most dominant disorders occur with a frequency of 1/10,000 or less,

Table 13-7. Selection Against a Rare Autosomal Dominant Trait.

	Genotype		Total Frequency
	Aa	**aa**	
Before Selection	$2pq \approx 2p$	$q^2 \approx (1\text{-}2p)$	1
W	1 - s	1	
After Selection	2p - 2ps	1 - 2p	1 - 2ps

permitting elimination of the **AA** homozygote and the term p^2 from consideration. Furthermore, since q is approximately 1, the frequency of **Aa** can be approximated by the term **2p**. The frequency of **aa** is closely approximated by **1 - 2p**. The loss of heterozygotes following selection is about **2ps**, and half of their alleles (**sp**) must be replaced by mutation at equilibrium:

11)
$$\mu = sp \ .$$

This equation provides an indirect estimate of mutation rates for dominant traits and may be compared to the direct method discussed in chapter 6. The indirect method is subject to the same biases experienced when using the direct method for estimating mutation rate. Application of the indirect method is presented in the text box. The relative fitness of achondoplastic individuals is determined by comparing their reproduction with that of their normal siblings. This approach is used to circumvent fertility problems that are unrelated to the mutation.

For dominant lethal traits equation (11) reduces to:

$$\mu = p \ .$$

In other words, each case of the disorder represents a new mutation.

X-linked recessive traits most commonly occur in males. Since males possess only one-third of all X chromosomes and one-third of all X- linked alleles, only one-third of all X-linked alleles are subject to selection. The balance between mutation and selection for X-linked recessive traits assumes the form:

12)
$$\mu = (1/3)sX_{M} \ .$$

When the fitness of affected males is 0 (e.g. Duchenne muscular dystrophy), the mutation rate becomes:

$$\mu = (1/3)X_{M} \ .$$

Let: 1) freq. of Achondroplasia = 1/10,000 = n/N;
2) the frequency of A = p.
3) W_A = the fitness of achondroplastic persons;
4) W_N = the fitness of their normal siblings;
5) W_R = the relative fitness of achondro-
    plastics.

$p = n/2N = 1/2(10,000) = 1/20,000$

If 40 achondroplastic dwarfs had 8 children and 80 of their normal siblings had 160 children,

$$W_R = W_A \div W_N = (8/40) \div (160/80) = 1/10 ,$$

and $s = (1 - W_R) = 9/10$.

The mutation rate is:

$$\mu = sp = (9/10)(1/20,000) = 9/200,000.$$

Indirect Method for Calculating Mutation Rates.

If the frequency of Duchenne muscular dystrophy is 1/3500, then the mutation rate is about 1/10,500. This is one of the highest mutation rates recorded for humans and reflects the large target size of the dystrophin gene. The equation also indicates that one-third of all cases of Duchenne muscular dystrophy appear to arise by new mutations.

The Burden Of Mutations Upon Populations

Most mutations tend to be deleterious, reducing the fitness of individuals possessing genotypes containing the mutant alleles. The total burden upon a population exerted by a deleterious mutation is independent of its fitness (Fig. 13-3). However, the number of generations bearing this burden increases as the fitness of genotypes having the mutation increases.

Balanced Polymorphisms

R.A. Fisher (7) first introduced the

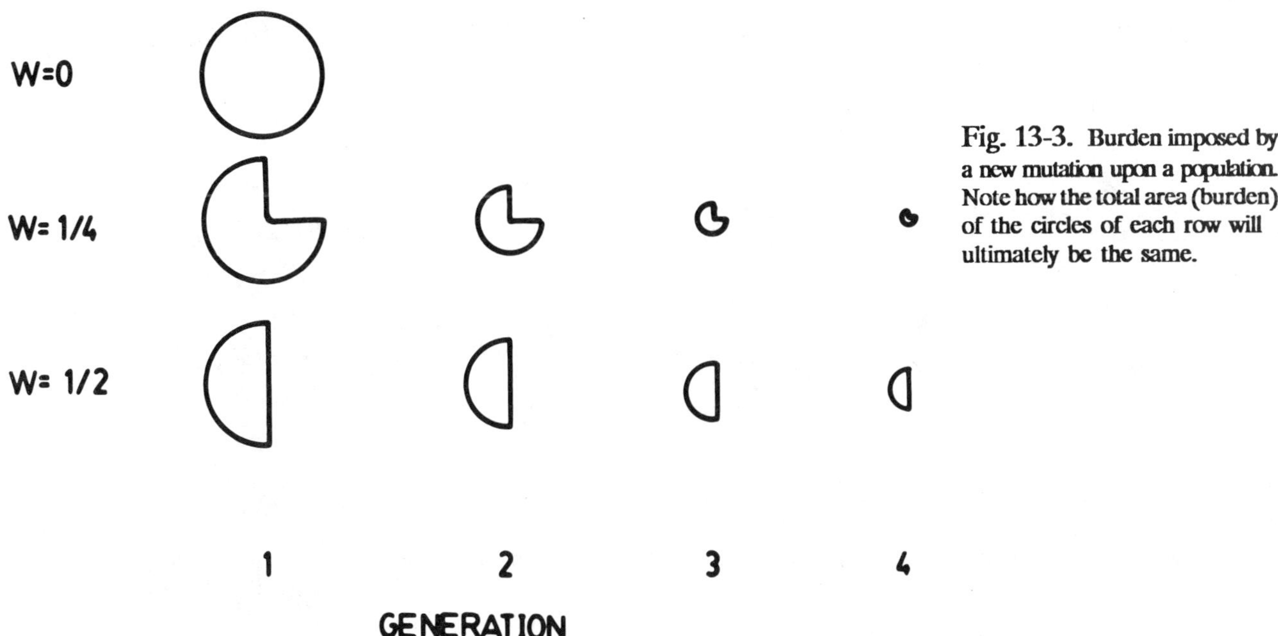

Fig. 13-3. Burden imposed by a new mutation upon a population. Note how the total area (burden) of the circles of each row will ultimately be the same.

concept of **balanced polymorphisms,** and examples of this phenomenon were subsequently found in human populations (8). A balanced polymorphism occurs when the heterozygote enjoys a fitness that is superior to either of the homozygotes, and serves to maintain both alleles in the population at relatively high frequencies.

Certain globin mutations provide an excellent example of a polymorphism maintained by heterozygous advantage. Several alleles occur frequently in Africa (β^S, β^C), the Middle East (β^S) and Southeast Asia (β^D, β^E). Each of these regions have or had infestations of Falciparum malaria, and the occurrence of any one of these alleles with the β^A allele imparts a selective advantage. The β^S allele achieves a frequency as high as 17% in central and west Africa (9). Although $\beta^S\beta^S$ homozygotes often die in these regions before the reproductive period from complications due to sickle cell anemia and $\beta^A\beta^A$ homozygotes often die from malaria, heterozygotes appear to be resistant to infection (10, 11).

The malarial parasite, <u>Plasmodium falciparum</u>, requires both mosquitoes and human erythrocytes to complete its life cycle (Fig. 13-4). When erythrocytes of heterozygotes are deprived of an adequate oxygen supply, HbS crystallizes. This form of hemoglobin may be less readily used by the parasite. A second possible reason for the protective effect of sickle cell trait may be related to the tendency for sickle-shaped cells to be more rapidly removed from the circulation by the spleen and other sites. This

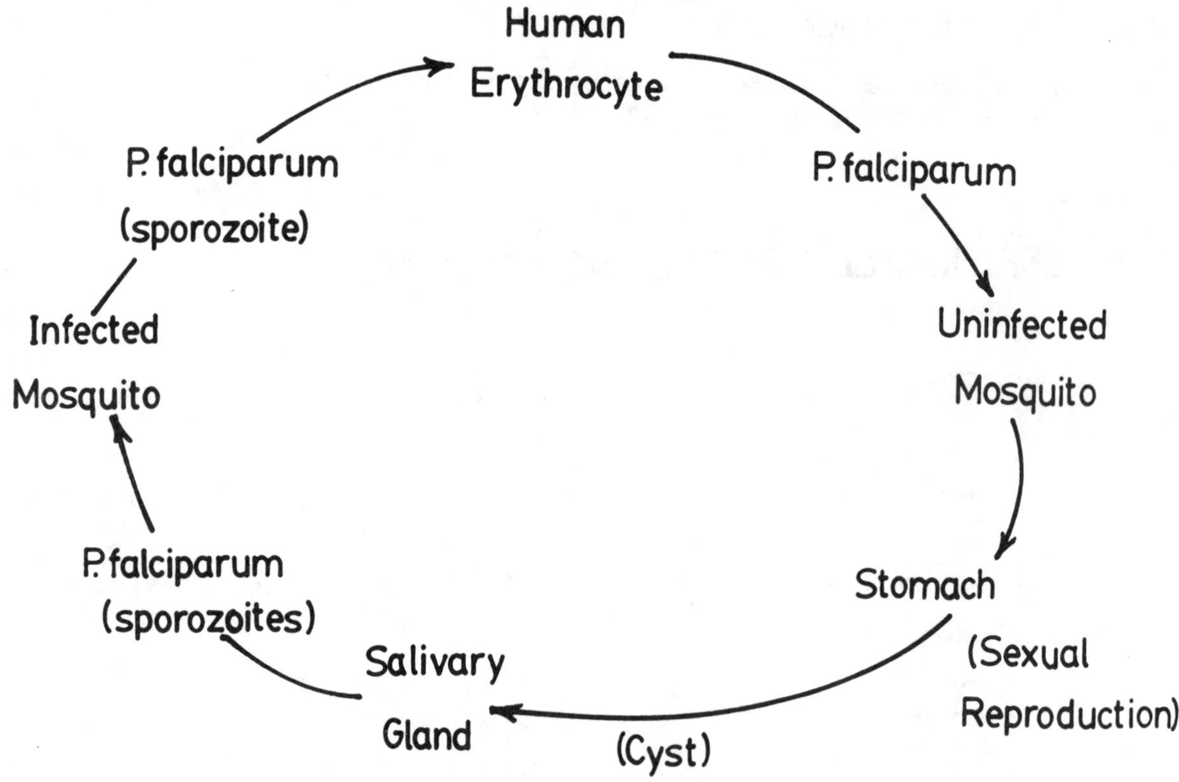

Fig. 13-4. Life cycle of <u>Plasmodium falciparum</u>.

premature destruction may limit multiplication and spread of the parasite by providing it inadequate time to complete the required developmental changes. <u>In vitro</u> studies of the sickling rates of $\beta^A\beta^A$ and $\beta^A\beta^S$ cells that have been infected by the parasite support the latter possibility (12).

These advantages suggest that persons with sickle cell trait tend to survive longer than both homozygotes and may have larger families. Consequently, both alleles are maintained in the population despite the strong selective pressures against the respective homozygotes. The advantage does not have to be very large to account for the present gene frequencies in these regions. Falciparum malaria became a major problem coincident with the development of agriculture. Stagnant pools of water used for irrigation and watering stock animals provided a breeding environment for the mosquito vectors and led to the rapid spread of the disease. Agriculture was developed in east central Africa about 3000 years ago. If the selection coefficient (s) for sickle cell anemia is 0.86, and the selection coefficient for the normal homozygote (t) is 0.12, the current allele frequencies can be accounted for (13).

A general case illustrating the role of heterozygous advantage in maintaining a balanced polymorphism is presented in Table 13-8. The change in the frequency of **a** is given by:

Table 13-8. Balanced Polymorphism Maintained by Heterozygous Advantage.

	Genotype			Total Frequency
	AA	**Aa**	**aa**	
Before Selection	p^2	$2pq$	q^2	1
W	$1-t$	1	$1-s$	
After Selection	$(1-t)p^2$	$2pq$	$(1-s)q^2$	$1-tp^2-sq^2$

$$\Delta q = q_1 - q = [pq + q^2(1-s)] \div [1-tp^2-sq^2] - q.$$

The equation can be simplified to:

13)
$$\Delta q = [pq(tp-sq)] \div [1-tp^2-sq^2] .$$

The numerator of equation (13) will become positive as **p** increases in frequency resulting in an increase in **q**. By contrast, as **q** increases, the numerator becomes negative, and **q** decreases. Consequently, the frequency of **q** will fluctuate around an equilibrium value as long as the selective pressures remain constant (Fig. 13-5). The equilibrium values for **p** and **q** can be predicted by:

14)
$$\hat{p} = s/(s+t) \qquad \text{and}$$

15)
$$\hat{q} = t/(s+t) .$$

Using the values for **s** and **t** given above, the equilibrium value of **q** would be about 0.12.

Large numbers of Black Africans were forced into slavery during the sixteenth through eighteenth century, and many were brought to the American Colonies. These populations carried high frequencies of the β^S allele with them, and the frequencies remained high so long as malaria continued to be a problem. However, eradication of the disease from the United States and elsewhere has resulted in a loss of heterozygous advantage, such that the population dynamics reverted to selection against the sickle cell homozygote only. Consequently, the frequency of β^S has fallen from about 0.13 to approximately 0.04.

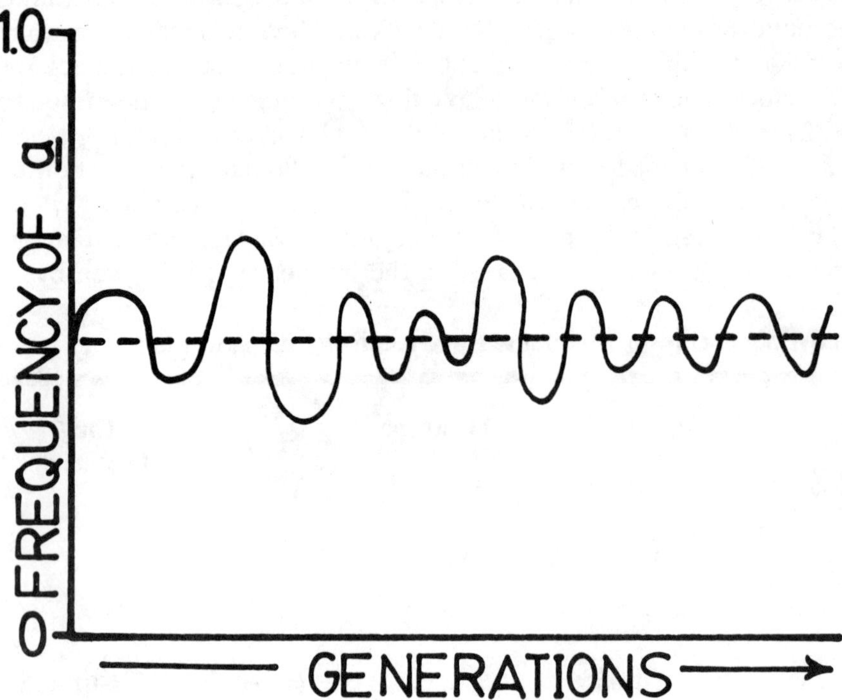

Fig. 13-5. Variation in the frequency of an allele when the heterozygote has a selective advantage.

The relatively high frequencies of other globin alleles and several **G6PD** mutations also exhibit geographic distributions that parallel that of Falciparum malaria, indicating that they also protect against the disease in heterozygous state. In the case of the **G6PD** alleles, it is the heterozygous female who enjoys the selective advantage.

Migration

The movement of individuals into (immigration) or out of (emigration) a population is called **gene flow**. Such exchange of genetic information can occur between two established populations and lead to changes in gene frequencies over time. Therefore, these populations may not be at equilibrium, since the introduction or loss of genes would result in changes in gene frequencies from generation to generation. However, two populations could eventually attain equilibrium if exchange is mutual and fairly constant:

16) $$mp_x = np_y$$

where p_x is the frequency of the allele **A** in population 1; p_y is the frequency of **A** in population 2; **m** is the rate of emigration; and **n** is the rate of immigration.

Migration and subsequent geographic barriers led to isolation of splinter groups and subsequent development of races. Genetic differences among races rarely involve presence or absence of alleles, but rather reflect variation in relative frequencies of alleles. Early migrations involved movements of peoples with more advanced technologies or more aggressive behaviors into regions that were often sparsely populated, or inhabited by populations who had developed less advanced technologies or were less warlike. Such migrations were slow and provided ample opportunity for admixture of genes and development of **clines** of gene frequencies (Fig. 13-6). One example of a cline is provided by the frequency of **ABO**^{B}. The frequency of this allele is highest in central Asia and declines at the rate of about 1% per 400 kilometers until it reaches its lowest frequency in southwestern France and northeastern Spain (13). The cline is believed to parallel the invasion route of Mongol hordes during the centuries following the collapse of the Roman Empire. Another cline is the fall in frequency of the **ABO**^{A} allele extending from southern England to northern Scotland. The Britons were a celtic people who are believed to have been predominantly type O and were invaded by Anglo-Saxons who are thought to have been type A. The celtic population either eventually intermarried with the Anglo-Saxons or retreated to the far northern reaches of the island. Rh(+) blood type is also distributed along a cline extending from eastern Europe to southwestern France and northern Spain. The Basques who inhabit the latter region are considered to be the last remnants of a population who once inhabited all of Europe.

Sociocultural barriers may also impose isolation of gene pools of populations residing in the same geographical region. Allele frequencies of the Duffy and Rh blood groups and the Gm system

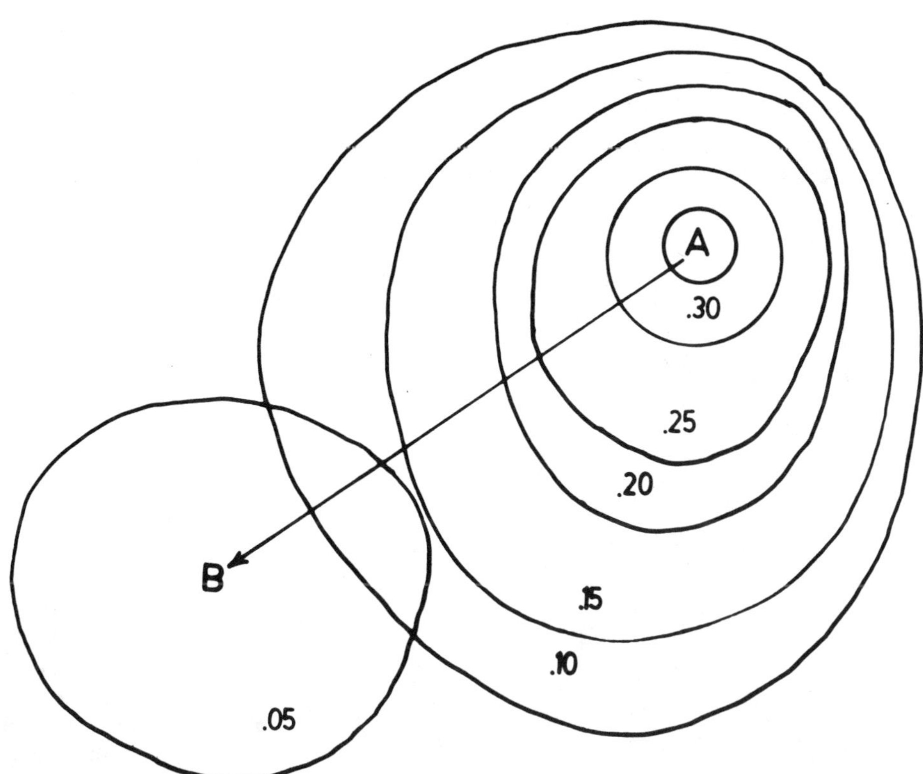

Fig. 13-6. Effect of migration upon gene frequencies. Gradual migration of population A into the territory of population B is paralleled by gradual spread of an allele for A to B, producing a cline or gradient of allele frequencies.

differ markedly among Africans and Caucasians. For example, **FY**^{0} (a Duffy allele) is very frequent among Africans but rare among Caucasians. Conversely, **FY**^{a} and **FY**^{b} are relatively frequent among Caucasian but rare among Black Americans. These alleles have been used to estimate racial admixture in the United States and other countries. Based upon these data, the average proportion of Caucasian ancestry among Black Americans approximates 30%. The percentage is highest in northern cities. Data relevant to the average proportion of Black ancestry among Caucasian Americans are fragmentary, but this proportion is believed to be smaller as the result of historical sociocultural barriers between the two races. In other countries where interracial barriers have been relaxed, complete mixing of gene pools has occurred.

Genetic Drift and Unstable Polymorphisms

Most studies of the genetic structure of populations have covered short periods of time. Studies of this type cannot distinguish between transient polymorphisms and the more stable balanced polymorphisms. Transient polymorphisms may occur in small populations where the number of reproducing individuals is limited and gene frequencies fluctuate randomly from one generation to the next. This random fluctuation of gene frequencies is called **genetic drift**. When drift is operating in a small population over the period of many generations, one of two alleles will eventually be eliminated, while the other becomes fixed (the only allele) in the population. The time required for fixation or elimination of an allele is inversely proportional to population size and can be estimated:

17) $$N_e = 4N_M N_F \div (N_M + N_F) \, ,$$

where N_e is the effective size (an estimate of the number of reproducing individuals) of the population and N_M and N_F are the numbers of males and females, respectively, in the reproductive age group. Sampling of gene frequencies at the time indicated in Fig. 13-7 would reveal the occurrence of two alleles at frequencies that might suggest the presence of a balanced polymorphism maintained by heterozygous advantage. In fact, one of the alleles is progressing toward extinction.

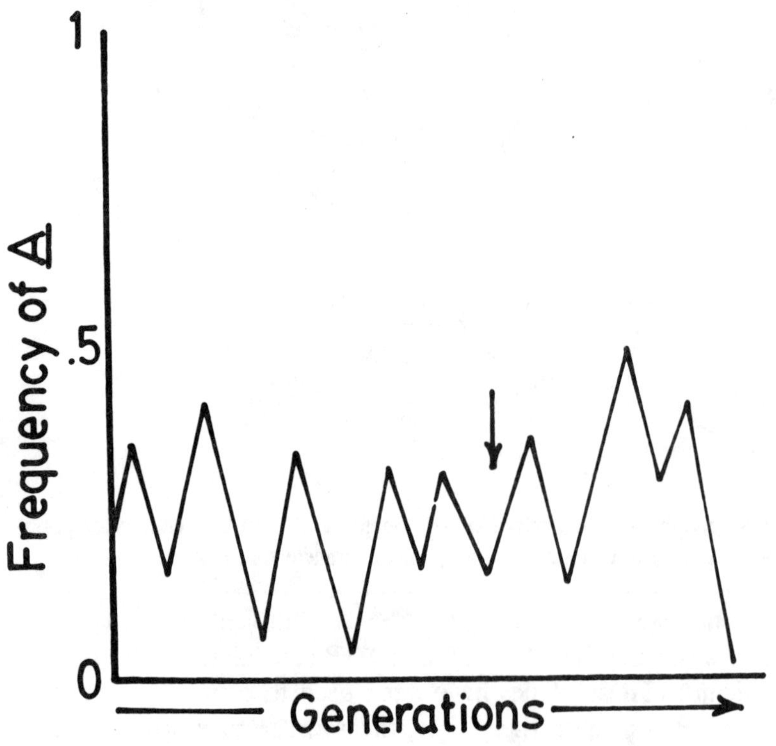

Fig. 13-7. Genetic drift in a small population. The frequencies of two alleles are fluctuating by chance, and $\underline{A}$ is progressing toward extinction. Sampling at the time indicated by the arrow may erroneously suggest that the alleles may be maintained by heterozygous advantage.

Splitting of small populations by emigration of small groups to new geographic regions and reproductive isolation of these peoples from others who live around them creates an opportunity for drift to occur. If one of the "founders" of the new population carries an allele that was rare in the population of origin, drift may lead to its becoming a predominant allele in the group of immigrants. Examples of this **founder effect** include porphyria variegata in the Union of South Africa, limb-girdle muscular dystrophy, Hemophilia B, and pyruvate kinase deficiency among Amish peoples in the United States, and albinism among the San Blas Indians. In each case, the respective diseases can be traced to one or a few founders of the respective groups. Porphyria variegata was apparently brought to South Africa by one Dutch settler. The Dunkers remained reproductively isolated from all non-Caucasians, resulting in the remarkable increase of the allele responsible for the disease by drift. The Amish settlements in the United States are derived from thirty families, and one or a few founders among this group are believed to have carried the rare alleles with them to America. The religious and cultural characteristics of the Amish have maintained a largely closed population through the years, leading to the increase in allele frequencies by genetic drift.

In some cases, it is difficult to distinguish between forces responsible for observed polymorphisms (14,15). Tay-Sachs alleles ($HEXA^0$) occur frequently among certain groups, such as Ashkenazi Jewish and French-Canadian populations. Most scientists tend to favor a founder effect for the unusually high frequencies of these alleles in these populations, and this appears to be the most likely cause in some cases. However, in others, a small heterozygous advantage operating in the past would also explain the relatively high allele frequencies.

Human populations lived in small groups during the pre-agricultural period, and genetic drift most likely played a major role in determining allele frequencies during this important period of human evolution. Furthermore, relatively large groups were decimated by epidemics or other natural catastrophies. The residual small breeding groups created "bottlenecks", and genetic drift most likely had a major impact upon the flow of alleles to subsequent generations.

Inbreeding

Marriage of blood relatives (**inbreeding**) subdivides a larger population into smaller marriage groups. Since two related persons have a greater chance of both being heterozygous for the same recessive allele than two unrelated individuals, recessive disorders will occur more frequently among children from inbred parents. Multifactorial problems that have a polygenic component will also be found more often among children of consanguineous parents, since these parents are more likely to share groups of additive genes that predispose to these problems.

The effect of inbreeding upon the occurrence of recessive alleles is illustrated in Fig. 13-8. III-1 and III-2 are first cousins who share I-1 and I-2 as common ancestors. Suppose I-1 is heterozygous for an allele that causes Hurler syndrome ($IDUA^0$). The chance that III-1 will inherit this allele from I-1 is $1/2 \times 1/2$ or $1/4$. Similarly, there is a $1/4$ chance that III-2 will inherit the same allele. The probability that II-1 and II-2 will produce a child with Hurler syndrome is $1/4 \times 1/4 \times 1/4$ or $1/64$. This risk is about 1000-fold higher than that for two unrelated parents. The more frequent occurrence of diseases among children of consanguineous parents than among children of unrelated persons has been used as an indicator for recessive inheritance of these conditions.

The probability that an inbred individual will be homozygous at a locus that is heterozygous in an ancestor common to both of his/her parents is called the **inbreeding coefficient** which is symbolized by **F**. The inbreeding coefficient may be calculated:

18) $$F = \Sigma m(1/2)^n$$

where **n** is the number of ancestors in the path of relationship from one inbred parent to the other, and where **m** is the number of paths of length **n.** Examples of paths of relationship are given in Fig. 13-8b.

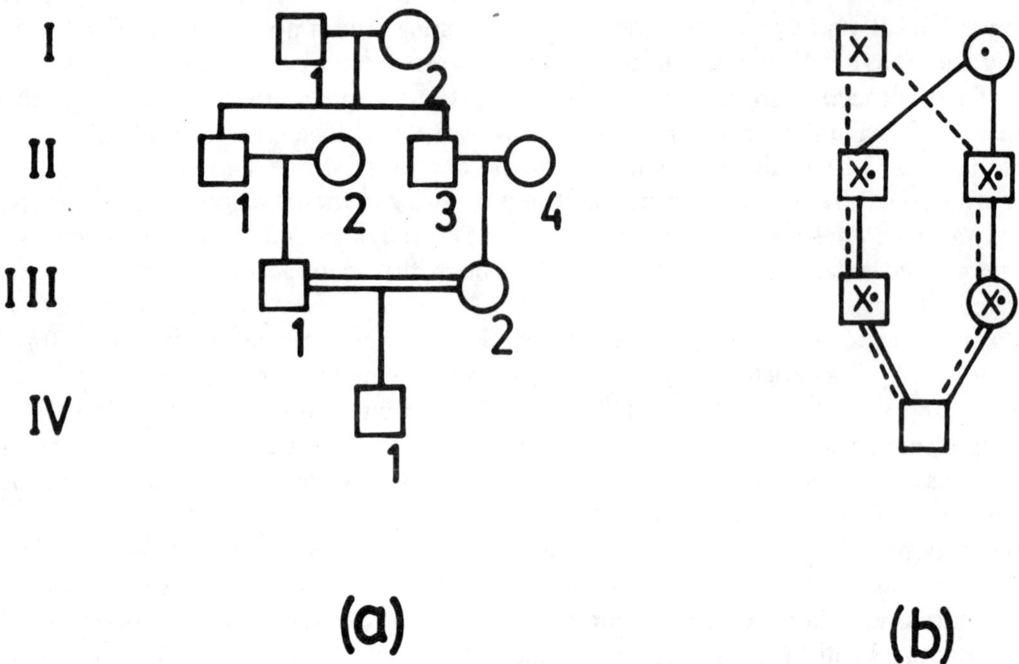

Fig. 13-8. Pedigree illustrating marriage of first cousins (a). The pedigree is redrawn (b) to show paths of relationship.

Five ancestors are included in each of the two paths of relationship (dashed and solid lines). Therefore, the inbreeding coefficient for a child of this couple is:

$$F = 2(1/2)^5 = 1/16.$$

The probability that a locus will be rendered homozygous by descent in a child of first cousins is 1/16. This value may also be interpreted in a more general sense: 1/16 of all heterozygous loci in a common ancestor will become homozygous in a child of first cousins. The inbreeding coefficient for incestuous matings involving parent-child or brother-sister pairings is 1/4, while those for aunt-nephew or uncle-niece marriages and for marriages of second cousins are 1/8 and 1/64, respectively.

There have been a number of cases where inbreeding was practiced to maintain royal lines. For example rulers of ancient Egypt perpetuated their family line by close inbreeding. The father (Ptolemy XIII) of Cleopatra VII was the son of a brother and sister. Cleopatra VII is well known for her relationships with Julius Caesar and Marc Antony. There were five brother-sister and two uncle-niece relationships among Cleopatra's ancestry that contributed to the royal line.

Inbreeding occurs frequently in many regions of the world today, particularly in the sub-Saharan region of Africa and in Asia (16). Marriages involving second cousins or closer relatives account for 20-55% of the total marriages in Muslim communities in these areas, with first cousin marriages being the most common. Custom favors marriages of men with their father's brother's daughter. Hindu communities in southern India also have high incidences of consanguineous marriages (20-45%), where the most frequent marriage involves the marraige of a man to his mother's brother's daughter. Somewhat lower frequencies of marriage of relatives occur among Buddhists, Christians, Jews and other sects living in the same regions. Reasons cited as favoring first cousin marriages include better relationships between the bride and her husband's family, maintenance of family property, less danger of hidden medical or other problems emerging after marriage, and no or a smaller dowry, where dowries are customarily expected as part of the marriage contract.

The negative aspect of inbreeding presents as increased risk for congenital birth defects and recessive traits. Children derived from incestuous marriages (parent-offspring or brother-sister) have an inbreeding coefficient of 1/4. Deaths and other defects display a 30% increase among these children compared to those born to unrelated parents (17). A number of studies have addressed occurrence of early fetal loss, stillbirths, congenital defects, and other problems among children of consanguineous marriages on a larger scale. In many cases the studies were not properly controlled or otherwise gave equivocal results. More recent studies have concentrated on larger numbers of people and have also studied noninbred populations living in the same communities more thoroughly. A large Brazilian study (18,19) involved 5719 people and included a number of different controls, including cousin controls and sib controls. This study estimated enhanced risks due to inbreeding for morbidity and precocious mortality of 32%, 18%, 9%, 5% and 2.5% for pregnancies from incestuous, uncle-niece and aunt-nephew, first cousin, first cousin once removed (only one common ancestor), and second cousin matings. An investigation of 2752 families in Beirut, Lebanon (20) revealed that 25% of all marriages were consanguineous, with about 57% of these involving first cousins. This study found that the proportion of dead among children ever born was higher among the consanguineous families than among nonconsanguineous families; however, differences disappeared when the sampling was adjusted for socioeconomic status, duration of marriage and religious affiliation.

Perhaps the largest investigation encompassed more than 111,000 mothers in Bangalore and Mysore, India over a ten year period (16). The sample included Muslims, Hindus, and Christians, and the corresponding frequencies of consanguineous marriages were 23.7%, 33.5%, and 18.6%, respectively. The coefficient of inbreeding for each group ($\alpha = \Sigma p_i F_i$; where p_i is the frequency of consanguineous marriage of type i which has an inbreeding coefficient of F_i) was 0.0333, 0.0160 and 0.0173, respectively. Several trends were noted. 1) Age at marriage and birth of first child decreased as the degree of consanguinity increased. 2) Infant and early childhood deaths were increased in families of uncle-niece and first cousin marriages. 3) Many of the losses appeared to be clustered in a subset of inbred families. 4) An associated study revealed a wide range of recessive disorders in these inbred communities. Overall, this study indicated that consanguineous families tend to have more children, but to suffer higher postnatal mortality. Furthermore, the excess mortality compared to nonconsanguineous families appears to be contributed by a minority of families.

Some have proposed that inbreeding practiced over many generations will eventually eliminate harmful recessive alleles. However, the large study described in the preceding paragraph suggests that this may not be the case. The tendency for inbred couples to have larger families to "compensate" for the lost children they experience would tend to maintain deleterious alleles in the heterozygous state. That is, for every homozygous child who dies, two heterozygous children would be born, resulting in perpetuation of the abnormal alleles.

Genetic Load

The preceding discussion suggests that human populations possess a significant number of alleles which, when rendered homozygous, would cause death or other human suffering. This burden of lethal and detrimental alleles is called our **genetic load.** Two major forces that contribute to genetic load are mutation and heterozygous advantage. The former tends to produce alleles that have phenotypic effects which detract from the adaptive norm of the population. Since mutation is infrequent, the vast majority of these alleles are carried in heterozygous state and are usually revealed only when two heterozygotes conceive a homozygous recessive child. Heterozygous advantage which maintains balanced polymorphisms also imposes a load on the population, since the respective homozygotes are at a selective disadvantage compared to the heterozygotes. Marriage of optimally fit heterozygotes results in the segregation of the respective alleles and formation of less fit homozygotes.

POPULATION GENETICS AND MEDICINE

The medical community must be aware of the genetic structure of the population that it serves. Most populations, particularly in large urban areas, are heterogeneous and include racial and ethnic subgroups whose gene pools may be quite different from those of the most common sector of the population. These differences often involve genes contributing to recessive diseases that may occur infrequently in the general population, but at relatively high frequencies in ethnic and other subgroups within this population. Examples include sickle cell anemia, thalassemias and G6PD deficiency among Black Americans, Tay-Sachs and certain other diseases among Ashkenazi Jews, thalassemias and G6PD deficiency among persons of Mediterranean or Asian ancestry, and Hemoglobin D and E diseases among Asian Americans. Effective medical management requires proper awareness, treatment and counseling of families at risk for these diseases.

HUMAN EVOLUTION

Investigations of human evolution have relied upon two general types of information. Anthropologists usually utilize fossil evidence to reconstruct human ancestry. Measurements of fossilized skeletal fragments have been used to demonstrate relationships among human and nonhuman primates and among primates and other mammals. Two major limitations of this approach are convergent evolution and the lack of knowledge of how many gene changes are required to produce the observed differences. Two evolutionarily distant species may develop similar skeletal features as a consequence of similar environmental selection mechanisms. For example, both whales and fish have fins; however, this similarity does not mean that a whale is more closely related to fish than to mammals. On the other hand, two species may have quite different skeletal elements. However, without knowing how many gene changes were required to generate the differences, it is difficult to assess the relationship of the two species based upon this trait. The second type of information is based upon molecular evolution. Proteins and, more recently, DNA have been used to construct evolutionary relationships among different species. Both types of molecular evidence are based upon the same assumption. Species that have proteins or genes with very similar amino acid or nucleotide sequences are more closely related than are two species whose respective sequences are very different. Comparison of molecular information with the fossil record can lead to the development of "molecular clocks" that can be used to determine when ancestors of the two species being compared diverged from one another along the evolutionary path. This approach is somewhat circular, since the setting of the molecular clock utilizes fossil evidence to develop a time frame for the accumulation of the number of sequence differences between the two species being compared. A second limitation of molecular evidence is the tendency to assume that the clock runs at a more or less constant rate. This assumption may be incorrect in at least some cases, and would lead to errors in estimating the time of divergence of two species. A third limitation of molecular evidence is related to genetic-environmental associations. Two species or races may appear more similar due to selective advantages imparted by certain alleles in specific environments. Examples include certain HLA alleles, Gm alleles, and the example of selective pressure favoring retention of the β^S allele. Another problem with molecular data involves the placement of the root and branches of the evolutionary trees that are developed. It is often assumed that once two groups have diverged, they remain reproductively isolated. There are probably many exceptions, most of which are difficult to conclusively demonstrate today. An illustration of this problem is presented in Fig. 13-9. Three

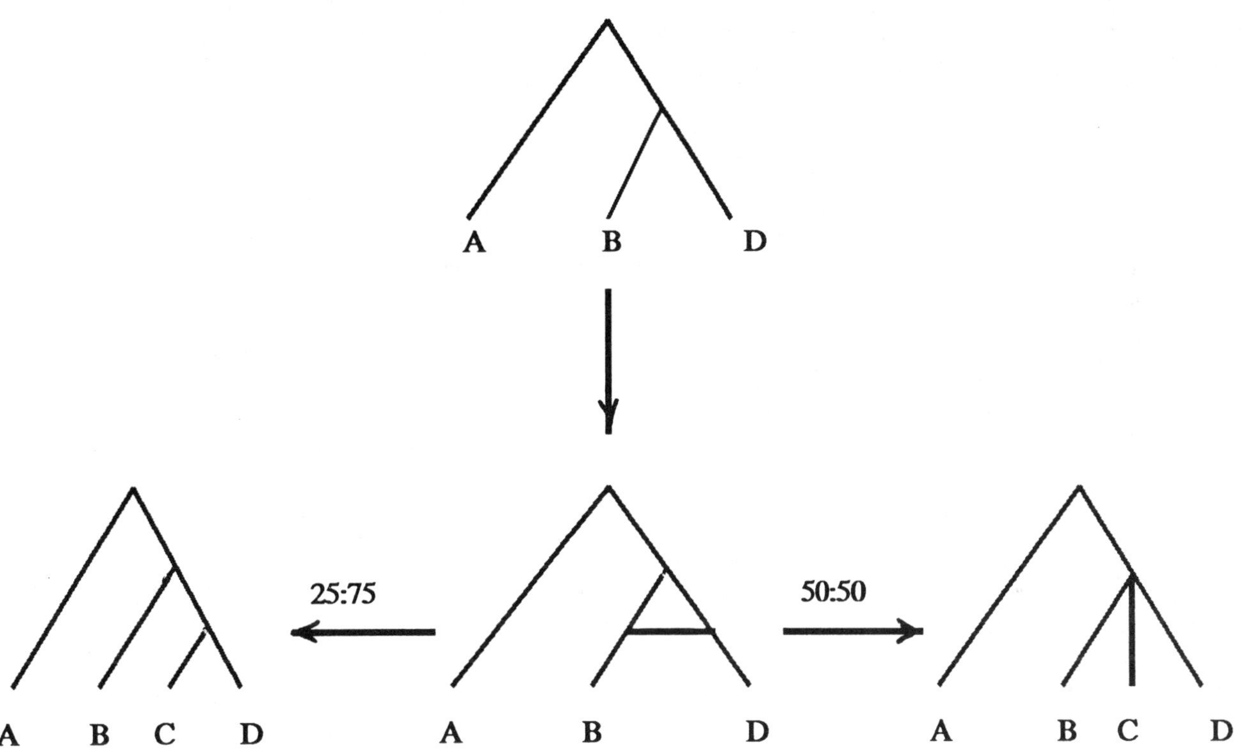

Fig. 13-9. Effect of racial admixture upon inferred evolutionary relatedness. Divergence of three races without admixture is presented at the top of the figure. A fourth racial group formed by admixture of two of the original races is shown in the lower portion of the figure. Depending upon the ratio of gene exchange between races B and D, race C may appear to have evolved very recently and be more closely related to race D, or may appear to have diverged at about the same time as races B and D. The horizontal line between races B and D in the lower central tree represents the time of admixture.

races are displayed at the top of the figure, and racial admixture is indicated in the lower center. Race C may appear to have diverged very recently from race D or to have diverged much earlier, depending upon the relative amount of gene exchange during the period of racial admixture.

Despite these limitations, both anthropological and molecular data have been very useful in piecing together relationships among current human races, among humans and other primates, and among primates and other mammalian groups. The evolutionary tree most consistent with all of the available information may most closely approximate what actually occurred. However, individual types of data may provide conflicting stories. Comparisons of African, European and Asian populations based upon Gm, HLA and blood group polymorphisms indicate that Europeans are more closely related to Africans than to Asians; however, when enzyme polymorphisms are used for these same comparisons, Europeans appear more closely related to Asians.

Two general theories for emergence of modern humans have been proposed. One promotes a single origin in Africa, and the other several origins in different regions of the world (Fig. 13-10; (21,22)). This dispute is far from settled; however, the bulk of the genetic data favor a sub-Saharan origin for modern humans (Homo sapiens sapiens). Several early hominids inhabited this African region about four million years ago (mya) to 2.5mya. Striking changes were taking place in the skull bones during this period, some of which resemble structures in H. sapiens. Australopithecus afarensis is believed to be the ancestral species for three other hominids: Homo habilis, A. boisei, and A. robustus. The latter two species are believed to be tangential to the main evolutionary path leading to H. sapiens sapiens. Two or more species of H. habilis may have coexisted with A. boisei. Evidence from Olduvai

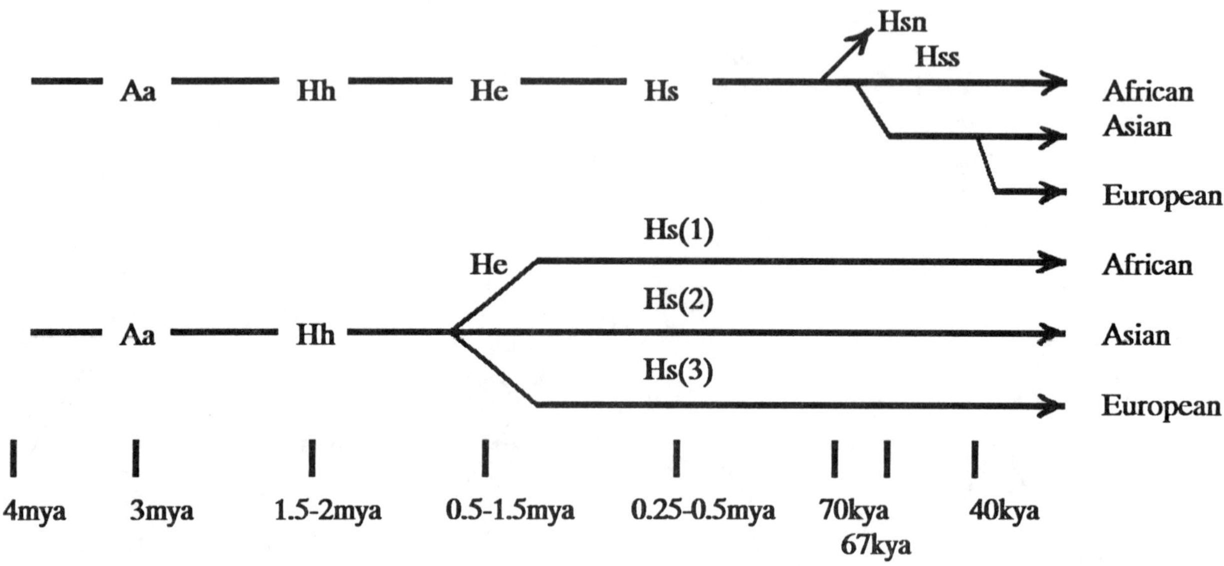

Fig. 13-10. Human evolution. Two schemes of evolution are displayed. The top model favors a direct line of evolution culminating in a single origin of modern humans (Homo sapiens sapiens = Hss). According to this model, Australopithecus afarensis (Aa), Homo habilis (Hh), Homo erectus (He), and archaic Homo sapiens (Hs) are in the main stream of the evolutionary path, and Homo sapiens neanderthalensis (Hsn) evolved away from the main path. Genetic data suggest that Hss radiated out of sub-Saharan Africa and into Eurasia, producing a split of non-African from African about 70,000 years ago. Divergence of European from Asian races occurred 30-40kya. The alternative theory proposes a multiple origin for Hs and Hss. This model favors radiation of He from Africa to different global regions, isolation of the migrant populations and independent evolution into Hs, Hss and the current African, European and Asian races. Mya = million years ago; kya = thousand years ago.

Gorge suggests that H. habilis used primitive pebble tools. About 1.6mya, Homo erectus fossils begin to appear in the African digs. Similar skeletons, much younger in age, have been found in China and Indonesia, suggesting that this hominid was distributed over vast regions of Africa, Asia and perhaps Europe. Their migrations out of Africa apparently occurred approximately 1mya. These hominids are quite similar to H. sapiens sapiens with respect to many of their skeletal features; however, the spinal canal was much smaller in diameter, and they had a smaller brain size. Their cervical and upper thoracic vertebrae were more apelike than human, and their rib cage was conical like apes and not barrel-shaped like humans. H. erectus evolved into archaic Homo sapiens, most likely in several global regions. Nei and Livshits (23) employed 84 protein-encoding loci, 33 blood group loci, 8 HLA and immunoglobulin loci, and 61 nuclear DNA RFLPs to decipher the origin of these three human races. Their analysis favored an African origin of Homo sapiens sapiens. This result has been corroborated by studies of mitochondrial DNA (mtDNA) RFLPs (24). mtDNA is exclusively of maternal origin, and mtDNA analyses indicate one or a few "Eves" appeared in Africa about 200 thousand years ago (kya). These studies and more recent work indicate that H. sapiens sapiens migrated out of Africa about 60-70kya and into Eurasia. A second branching occurred between 30-40kya, with the European and Asian races arising from the descendants of the split populations. Theoretically, Y-chromosomal DNA RFLPs should be able to provide evidence confirming or refuting the proposed African origin. Y chromosomes are limited to males, and RFLPs involving sites outside the pseudoautosomal region could be used to trace male ancestry. These studies are just beginning and have proved more difficult than anticipated (25). Preliminary data indicates that "Adam" may have originated in Africa as well.

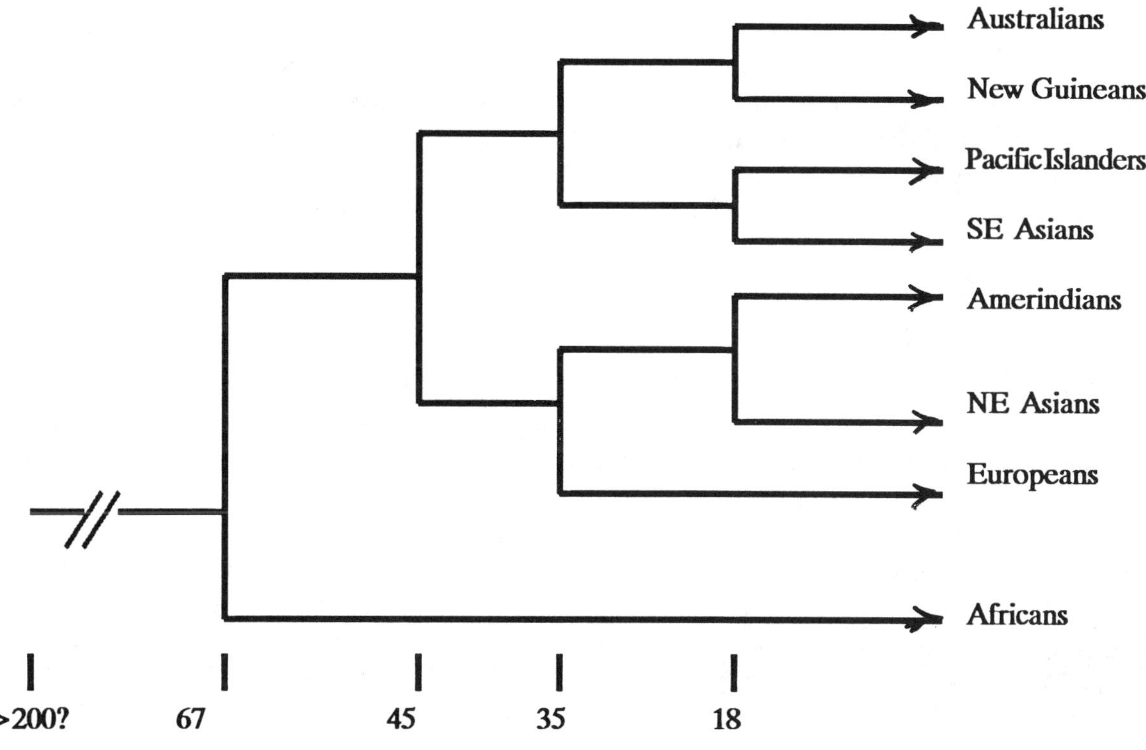

Fig. 13-11. Proposed scheme for the evolution of current human races. The number of races is in dispute. This model accepts eight discrete racial groups. The deepest branch occurred about 65-70kya, separating Africans from non-Africans. Ancestors of southeast Asian and northeast Asian and races derived therefrom occurred approximately 45kya. Ancestors of Europeans branched from those of northeast Asians about 35kya, and Amerindian forebears migrated across the Bering land bridge about 18kya. An approximate time scale is listed at the bottom of the figure.

Examination of the preceding evolutionary scheme reveals that most scientists appear to agree on many points of human evolution. The major difference of opinion centers on the end of the path. For example, at least two radiations out of Africa appear to have taken place: one involving H. erectus and the other H. sapiens sapiens. Many believe at least two of the H. erectus groups evolved to H. sapiens. As stated above, the African branch evolved into H. sapiens sapiens which has given rise to the present human races (Fig. 13-11). H. erectus in Europe and western Asia also continued to evolve into a species that closely resembled H. sapiens sapiens. Homo neanderthalensis was quite different from H. sapiens sapiens; however, there were also a number of similarites, both anatomically and culturally. Neandertals buried their dead, had similar "tool kits", had comparable brain volumes and may have been capable of speech. Neandertals and modern humans lived close together for thousands of years and even alternated at the same sites. However, there is no evidence supporting forms of life that were intermediate to the two subspecies, supporting reproductive isolation. This continues to be puzzling, since the differences between the two subspecies do not appear to be that great, and subspecies of apes often interbreed in zones of contact. Furthermore, humans have historically incorporated women and children from groups with which they were in conflict into their own groups. One school of thought proposes that modern humans emerging from Africa overran the indigenous populations of H. sapiens and H. neanderthalensis, replacing the latter in Europe about 35kya.

A proposed evolutionary scheme for eight modern races is displayed in Fig. 13-11. The evolutionay pattern is based upon a large body of genetic data composed of mtDNA and nuclear DNA RFLPs and supporting protein and antigen polymorphism data. Two major branchings appear to have occurred, one separating Africans from non-Africans and the other separating forebears of Europeans and northeast Asians from southeast Asians and derived races. Northeast Asians crossed the Bering land bridge about 18kya, giving rise to Amerindians who subsequently settled throughout the Americas.

The story of human evolution is still very incomplete, and the theories described in this section must be supported by additional data, both molecular and anthropological.

SUMMARY

Populations of intermarrying individuals share a gene pool defined by the types and frequencies of alleles occurring within that population. Gene frequencies will remain constant from one generation to the next, provided perturbations by mutation, migration, selection and drift are minimal. Genotypic frequencies under these conditions will approximate the Hardy-Weinberg equilibrium frequencies of p^2, $2pq$ and q^2 for a two allele system.

Mutation introduces new alleles into a population, and is balanced by selection at equilibrium: rate of gain of new alleles equals the rate of loss of those alleles. Selection acts at the organismic level and restricts reproduction of less fit genotypes. Selection against recessive homozygotes tends to be inefficient, particularly with respect to rare recessives, because most recessive alleles are carried by heterozygotes. Balanced polymorphisms in which at least two alleles occur with a frequency >0.01 are maintained by heterozygous advantage. Falciparum malaria is responsible for balanced polymorphisms involving globin alleles and G6PD alleles. Selection is environment-bound. For example, β^S is maintained at high frequencies in malarial environments, but not in malaria-free regions.

Migration from a large population to a sparsely populated region, or to an area where the immigrants are reproductively isolated from the surrounding population, produces evolutionary bottle-necks. The migrant or founding groups consist of few reproducing individuals and provide an opportunity for chance variation (genetic drift) to operate. Genetic drfit is believed responsible for the occurrence of certain alleles at high frequencies in small isolates, while those same alleles are rare in most populations. One or a few of the founders of the isolate presumably possessed the alleles in question, and genetic drift subsequently led to the marked increase in the frequencies of these alleles among their descendants.

Slow migration of populations followed by intermarriage with the indigenous populations promotes generation of clines of gene frequencies in which there are graded decreases in frequency of specific alleles along the migratory path. A cline of the ABO^B allele extends from maximal frequencies in central Asia to minimal levels in southwestern France and northeastern Spain.

Inbreeding promotes homozygosity for recessive alleles and the sharing of polygenes by inbred persons. A significant proportion of rare recessive diseases occurs among children of consanguineous parents, and multifactorial problems are also increased in this group. The inbreeding coefficient, F, is the probability that an inbred individual will be homozygous at a locus that was heterozygous in an ancestor shared by his parents. Effects of inbreeding upon the reproduction and children of highly inbred populations have been difficult to measure. However, these people appear to suffer a higher rate of early prenatal and childhood loss. Furthermore, recessive conditions occur frequently in these communities. Congenital birth defects are also increased among children of inbred parents in these areas. Many of these problems appear to concentrate in a relatively small number of families.

Human evolution has been an area of great interest for many years, and a resurgence of research has occurred using molecular methods to infer past history from present distributions of markers. Enzyme and protein polymorphisms and nuclear and mtDNA RFLPs favor two radiations of hominids out of Africa. Modern humans seem to have originated in sub-Saharan Africa and spread to Asia and Europe, and eventually to all global regions. As groups became isolated, they evolved into our present array of races. An earlier radiation of <u>H. Erectus</u>, regional isolation and subsequent genetic divergence led to evolution of other hominids, including a closely related subspecies of <u>H. sapiens</u> which arose in Europe and western Asia. <u>H. neanderthalensis</u> apparently lived in the same locales with <u>H. sapiens sapiens</u>, but the two species did not interbreed. Modern humans replaced neandertals in Europe about 35kya.

PROBLEMS

1. The frequency of phenylketonuria among Caucasian newborn in the United States is about 1/8000. Calculate the frequency of the heterozygote. What have recent molecular studies of PKU revealed about this estimate?

2. Traits **A** and **A'** are inherited as autosomal codominants. 1600 **AA**, 800 **AA'** and 600 **A'A'** individuals were identified among 3000 persons in a random mating population. Calculate the frequencies of the two alleles.

3. If this population grows to a size of 4000 in the following generation, how many heterozygotes will there be? If the population size remains the same (4000), how many heterozygotes will occur in the third generation?

4. An autosomal dominant disease occurs with a frequency of 1/20,000. Studies of families segregating for this disease have indicated that patients with the problem tend to have fewer children than do their normal siblings: 1000 patients had 200 children and 2000 of their normal siblings had 4000 children. Calculate the mutation rate that would be required to maintain the frequency of this allele at equilibrium.

5. Two American Indian tribes live on opposite sides of a mountain range and have not intermarried. Both tribes are small with the breeding size of the population averaging 100 persons during the last four generations. Typing of these tribes revealed that one tribe was 95% A and 5% O, while the other was 100% O. Amerindians are generally type O. Tribal records indicate that French-Canadian trappers frequently visited the area about 200 years ago and often selected wives from among the Indians. How would you account for the different blood type frequencies in these two tribes?

6. Calculate the inbreeding coefficient for an aunt-nephew marriage.

7. Calculate the frequency of Duchenne muscular dystrophy, if the aggregate mutation rate to Duchenne alleles is 1/25,000, and the fitness of Duchenne patients is 0. Duchenne muscular dystrophy is inherited as an X-linked recessive.

8. If there are 200 second cousin marriages, 100 first cousin marriages and 25 uncle-niece marriages among 100,000 marriages in a population, calculate the average inbreeding coefficient, α, for this population.

9. The grandmother-in-common of a man and woman who are first cousins birthed a child with phenylketonuria during a previous marriage. What is the probability that these first cousins will have a child at risk for PKU? Phenylketonuria is an autosomal recessive trait.

10. A community in northern Illinois was blood typed and found to include 2000 A, 1600 O, 400 B and 100 AB individuals. Calculate the frequencies of ABO^A, ABO^B and ABO^O.

11. Which of the following diseases appear to be maintained by mutation?

Disease	Frequency
Hurler's Syndrome	1/60,000
Tay-Sachs Disease	1/3600*
Sickle Cell Anemia	1/500**
Cystic Fibrosis	1/2100***

*Ashkenazi Jewish Americans; ** Black Americans; ***Caucasian Americans.

GLOSSARY OF TERMS

Balanced polymorphism a polymorphic system maintained by selective advantage of the heterozygote.

Cline gradual change in allele frequences over geographic distance.

Consanguineous blood-related.

Equilibrium a balanced state in which the proportions of two or more components of a system remain relatively constant through time.

Evolution change in the composition of the gene pool of a species through time.

Fitness a measure of the reproductive success of a genotype.

Founder effect most or all of the current alleles of a specific type can be traced to one or a few persons who originally established the population.

Gene flow movement of genes from one gene pool to another.

Genetic drift random fluctuation of gene frequencies.

Genetic load the concealed allelic variation of a population that is maintained by mutation, heterozygous advantage or other pressures and which is capable of potential harm to an individual if these alleles are rendered homozygous.

Inbreeding coefficient the probability that an inbred person will be homozygous at a locus that is heterozygous in ancestors shared by his/her parents.

Natural selection environmental forces that promote retention of favorable genotypes and loss of unfavorable genotypes.

Polymorphism a genetic system in which two or more alleles occur within a population at frequencies exceeding 1%.

Population a representative sampling of a species occupying a geographic area.

Selection coefficient the degree to which the environment limits the reproduction of a genotype.

BIBLIOGRAPHY

1. Lewontin, RC. 1967. An estimate of average heterozygosity in man. Am J Hum Genet 19:681-685.

2. Harris, H. 1972. Average heterozygosity per locus in man: an estimate based on the incidence of enzyme polymorphisms. Ann Hum Genet 36:9-20.

3. Hardy, GH. 1908. Mendelian proportions in a mixed population. Science 28:49-50.

4. Weinberg, W. 1908. Uber den Nachweis des Vererbung beim Menschen. Jahreschrift Verein Vaterlandische Naturk. Wurtemburg 64:369-382. Reprinted (translation) in Boyer, SH, ed. 1963. <u>Papers on Human Genetics</u>. Englewood Cliffs, NJ: Prentice-Hall, pp.4-15.

5. Neel, JV. 1978. Mutation and diseases in man. Canad J Genet Cytol 20:295-306.

6. Davis, AM and AEH Emery. 1978. Estimation of proportion of new mutants among cases of Duchenne muscular dystrophy. J Med Genet 15:339-345.

7. Fisher, RA. 1922. On the dominance ratio. Proc Roy Soc (Edinburgh) 52:321-341.

8. Haldane, JBS. 1949. Disease and evolution. Ricera Sci 19(Suppl 1):3-10.

9. Dobzhansky, T, FJ Ayala, GL Stebbins and JW Valentine. 1977. _Evolution_. San Francisco: WH Freeman and Co.

10. Motulsky, AG. 1964. Hereditary red cell traits and malaria. Am J Trop Med Hyg 13:147-158.

11. Allison, AC. 1954. Protection afforded by the sickle cell trait against subtertian malarial infection. Brit Med J 1:290-294.

12. Roth, EF, M Friedman and U Ueda _et al._ 1978. Sickling rates of human AS cells infected in vitro with _Plasmodium falciparum_ malaria. Science 202:650-652.

13. Bodmer, WF and LL Cavalli-Sforza. 1976. _Genetics, Evolution and Man_. San Francisco: WH Freeman and Co., pp. 318-319.

14. Chakravarti, A and R Chakraborty. 1978. Elevated frequency of Tay-Sachs disease among Ashkenazic Jews unlikely by genetic drift alone. Am J Hum Genet 30:256-261.

15. Wagener, D, LL Cavalli-Sforza and R Barakat. 1978. Ethnic variation of genetic disease: roles of drift for recessive lethal genes. Am J Hum Genet 30:262-270.

16. Brittles, AH, WM Mason, J Greene and NA Rao. 1991. Reproductive behavior and health in consanguineous marriages. Science 252:789-794.

17. Adams, MS and JV Neel. 1967. Children of incest. Pediatrics 40:55-62.

18. Ruas, PM and N Freire-Maia. 1984. Inbreeding effect on morbidity. Am J Hum Genet 18:381-400.

19. Freire-Maia, N. 1984. Effects of consanguineous marriages on morbidity and precocious mortality: genetic counseling. Am J Hum Genet 18:401-406.

20. Khlat, M. 1988. Consanguineous marriage and reproduction in Beirut, Lebanon. Am J Hum Genet 43:188-196.

21. Stringer, CB and P Andrews. 1988. Genetic and fossil evidence for the origin of modern humans. Science 239:1263-1268.

22. Simons, EL. 1989. Human origins. Science 245:1343-1350.

23. Nei, M and G Livshits. 1989. Genetic relationships of Europeans, Asians and Africans and the origin of modern _Homo sapiens_. Hum Hered 39:276-281.

24. Cann, RL, M Stoneking and AC Wilson. 1987. Mitochondrial DNA and human evolution. Nature 325:31-36.

25. Gibbons, A. 1991. Looking for the father of us all. Science 251:378-380.

ADDITIONAL LEARNING RESOURCES

1. Bodmer, WF and LL Cavalli Sforza. 1976. <u>Genetics, Evolution and Man.</u> San Francisco: WH Freeman and Co.

2. Carson, HL. 1987. The genetic system, the deme, and the origin of species. Annu Rev Genet 21:405-423.

3. Cavalli-Sforza, LL. 1971. <u>The Genetics of Human Populations.</u> San Francisco: WH Freeman and Co.

4. Felsenstein, J. 1988. Phylogenies from molecular sequences: inference and reliability. Annu Rev Genet 22:521-565.

5. Goodman, M, BF Koop, J Czelusniak <u>et al.</u> 1989. Molecular phylogeny of the family of apes and humans. Genome 31:316-335.

Chapter 14

Genetics and Medicine

This chapter will focus upon the impact of human genetics upon medicine, the family and society. Medicine has made many advances in the areas of treatment and prevention of infectious diseases. Eradication of many infectious diseases has resulted in the emergence of genetic diseases as a major health problem. Genetic diseases differ in several respects from their infectious counterparts. Infectious diseases often have acute manifestations. By contrast most genetic diseases involve chronic illness that is protracted over long periods of time and which often requires extensive involvement by specialists and other members of the health care community. Costs for medical treatment and other supportive services during the lifetime of an individual can be staggering and place a considerable burden on family resources, community and governmental agencies, and the health insurance industry. Psychological problems frequently arise in families of persons with genetic handicaps. Not only care of existing children with genetic problems must be dealt with, but also the threat of recurrence of the disorder must be coped with. In cases of adult-onset disease, parents are not only confronted with their own risk for the disorder, but also with the burden of knowing that they may have transmitted the gene responsible for the disease to one or more of their children.

Mendelian problems, particularly those characterized by autosomal dominant or X-linked dominant or recessive inheritance, involve members of the extended family. Doctors are used to dealing with individuals or, at most, mother and child. They are unaccustomed to following a disease through families and rarely have had to confront multiple members of a kindred with regard to management and prevention of a health problem. Furthermore, new advances in the area of medical genetics have created ethical dilemmas. How aggressive should a physician be when caring for problems in a severely handicapped child who will most likely die from a genetic disease? How much information regarding an unborn child should be told to the expectant couple? Should a doctor aggressively seek out other members of a kindred and tell them they are at risk for a serious dominant disorder? How should a health professional behave when his or her moral values conflict with those of a patient or family? These and many other dilemmas face modern medicine today, and many have not been adequately resolved.

New genetic and reproductive technologies have opened a Pandora's box of problems related to rights of individuals, families and society. Legal and ethical dimensions of these problems have not been adequately explored. The following sections will discuss medical aspects of human genetics and the moral and ethical dilemmas faced by individuals, families and society with regard to treatment and prevention of genetic diseases.

GENETIC COUNSELING

The following definition of **genetic counseling** was drafted by the Workshop on Genetic Counseling sponsored by the National Genetic Foundation, Inc. in December of 1972:

"Genetic counseling is a communication process which deals with the human problems associated with the occurrence, or the risk of recurrence, of a genetic

disorder in a family. This process involves an attempt by one or more appropriately trained persons to help the individual or family (1) comprehend the
medical facts, including the diagnosis, the probable course of the disorder, and
the available management; (2) appreciate the way heredity contributes to the
disorder, and the risk of recurrence in specific relatives; (3) understand the
options for dealing with the risk of recurrence; (4) choose the course of action
which seems appropriate to them in view of their risk and their family goals
and act in accordance with that decision; and (5) make the best possible adjustment to the disorder in an affected family member and/or to the risk of recurrence of that disorder (1)."

Goals of genetic counseling include: 1) assistance with the diagnosis and treatment of the genetic disease,
2) helping families cope with the physical, financial, and emotional burden of genetic disease, and
3) prevention of genetic disease.

Effective genetic counseling requires correct diagnosis of the clinical disorder to permit
accurate assessment of the risk of recurrence and determination of the appropriate treatment. Validation
of the diagnosis requires obtaining a thorough pregnancy and family history, conducting a physical
examination of the patient and close relatives where appropriate, and may require laboratory tests such
as chromosome analyses or metabolic studies. In spite of the efforts of highly trained personnel, a
precise diagnosis cannot be achieved in about 40% of the cases presenting to a typical genetics clinic.
Based upon the results of these evaluations, a pattern of inheritance may be inferred and risks
determined for specific family members.

The psychological climate of the patient or parents must be assessed to determine their ability
to comprehend and to accept the information that needs to be discussed and to establish the best way
in which to communicate this information. Parents transcend five psychological stages of adjustment to
knowledge of the presence of a severe disability in themselves or in their child: shock, denial, sadness
and anger, equilibrium, and reorganization of their life plans. People who are in the early stages of
coping with a problem are not ready for counseling, and, if counseled during this time, may deny that
they were told specific information by the counselor. Prior to meeting with the couple or individual, the
counselor must explore the availability of specialized care centers and support services in the community
and pursue sources of financial assistance that may be required. Identification of parent support groups
has proved to be invaluable in helping persons adjust to the presence of a genetic problem in their
family. These support groups are often composed of parents who have had children with the problem
and who can provide the new parents with the special insight and support that they need to cope with
their handicapped child.

When these preparations are completed, the counselor meets with the persons involved and
explains the disease and its implications. This is often the most important aspect of counseling, because
most parents have not been adequately informed about the nature of the disease and its management.
They want to know what to expect from their affected children and what can be done to help their
children achieve their maximal potential. Risks of recurrence are discussed in a manner that can be
comprehended by the counselees, and possible alternatives for avoiding future affected children are
explained. Care of the handicapped child is explained at great length, because parents often play an
important role in the early therapy for their child. These parental activities may include exercising their
infant's or child's limbs to develop muscle tone and early educational activities to ready the child for the
school period. When the counselor detects emotional problems in the family, he or she may make a
referral to a social worker, psychiatrist, or clergy, where appropriate. Contacts between the parents and
a support group in the community or region will be initiated by the counselor.

It is recommended that families be contacted at intervals to determine whether they have been able to gain access to required supportive services, to assess comprehension of information given during the counseling session, and to appraise their adjustment to the demands presented by the genetic disorder. The counselor may also inform the family of recent progress relative to the disorder present in their child, especially with regard to management. The extent of follow-up and follow-along of clients by counseling services varies considerably, and this important service needs to be improved.

These diverse counseling processes often require participation of a team of appropriately trained and certified health professionals including a clinical geneticist, genetic counselor, a social worker, and, where necessary, a clinical psychologist or psychiatrist, clinical nutritionist, and public health nurse. All of these persons may be involved with a genetics clinic at a medical center; however, counseling also occurs at outreach sites with limited staff and in the offices of primary care physicians.

During the past few years, a number of career opportunities have appeared in the area of clinical genetics and genetic counseling. Clinical geneticists must have the M.D. and usually train in pediatrics, obstetrics and gynecology, or internal medicine. They must enter a certified training program in clinical genetics and pass the appropriate certification examination administered by the American Board of Medical Genetics. Genetic counselors generally complete a two-year certified masters program in human genetics and genetic counseling. Prospective counselors must also be certified by the American Board of Medical Gentics. Directors of cytogenetic, biochemical genetic, or molecular genetic diagnostic laboratories may have either an M.D. with two to three years training in a certified program, a Ph.D. from a certified program, or hold both degrees. All three specialists must pass the respective certification examination administered by the American Board of Medical Genetics. Cytogenetic technologists must meet certification requirements of the Association of Cytogenetic Technologists. A list of training programs and other information regarding careers in human and medical genetics may be obtained from: Administrative Office, American Society of Human Genetics, 9650 Rockville Pike, Bethesda, MD 20814.

Training and Certification of Persons Participating in Medical Genetics

Most genetic counselors provide informational counseling. Their main purpose is to provide a family with all of the information and support they need to make decisions regarding family planning, to get them started with the adjustment process, and to assist them with caring for their child. These counselors avoid influencing reproductive decision-making by their clients. An increasing number of counselors are beginning to provide information-guidance counseling. This practice was initiated when studies revealed that parents appeared to be avoiding natural pregnancy more than seemed warranted by the recurrence risk and burden associated with the problems present in their existing children. An alternative form of counseling was developed in which some guidance regarding reproduction was provided along with the information during the counseling process (2). Several aspects of the genetic defect and the family were considered before formulating advice. These included the burden the problem placed upon the child, parents and family, and the availability of treatment for the disorder. The magnitude of the risk of recurrence and the possibility and acceptance of prenatal diagnosis were weighed by the counseling team. The family was evaluated with respect to their ability to care for a child with the handicap, and maternal risk during a complicated pregnancy was considered. The recommendations listed in Table 14-1 were made based upon these six considerations. This counseling approach most likely encouraged more couples in the first three advice classifications to plan additional pregnancies, and pregnancy outcome appeared to approximate the predicted risk figures. Couples receiving advice categories IV and V were strongly deterred from planned pregnancies, and those pregnancies that

Table 14-1. Outcomes of Information-Guidance Genetic Counseling (2).

Advice Offered	Parental Attitudes	Outcome
I. No problem, pregnancy recommended	I-II. 95% planned pregnancies	I-III. 4% of babies were abnormal
II. Pregnancy recommended after some preparation		
III. Pregnancy recommended with prenatal diagnosis	III. 90% planned pregnancies	
IV. Major problem: pregnancy should be carefully considered	IV. 60% were deterred	IV. 19.4% of babies were abnormal
V. Pregnancy not recommended	V. 62% were deterred	V. 43.5% of babies were abnormal

occurred were associated with a high incidence of problems with genetic components. These investigators concluded that information-guidance counseling was superior to informational counseling with respect to reproductive behavior of the counselees.

Factors Influencing Reproductive Decision-Making

Reproductive decision-making is influenced by a large number of factors including preconceived notions of risk, existing children, available alternatives, magnitude of risk and burden of the disease. Two major factors are risk and burden. Generally speaking, couples tend to be deterred from attempting pregnancies by recurrence risks exceeding 10%. Risks below 5% do not usually discourage parents from having their own children. However, preconceived notions of risk associated with a problem may cause couples to behave in unexpected ways. For example, a couple who believed their recurrence risk was 100% prior to receiving counseling may be relieved by a risk of 25% and decide to have additional children. On the other hand, a couple who believes they will not have a child with a disorder may be deterred by a risk of recurrence approximating 5%. Couples who desperately want a child and who cannot adopt a child or utilize any of the alternatives to natural pregnancy may elect to attempt a pregnancy when the risk exceeds 10%.

Burden proves to be as important as risk in influencing reproductive decisions. Diseases like cystic fibrosis and sickle cell anemia are associated with a large burden for the family, since considerable support is required (medical, familial, and financial) for a relatively long period. By contrast, anencephaly, although a more severe problem, is considered a low burden disease, because the affected infant either dies prenatally or only survives a few hours. Studies have demonstrated that patients with genetic diseases tend to be hospitalized more frequently and for longer periods than other patients. Such chronic and expensive illnesses make it difficult for parents to obtain insurance coverage for their affected children, and more than 30% pay their own bills (3). In spite of this financial burden, some families have not been deterred from having additional children. In one case, the family welcomed the birth of a second child, because the first affected child had drawn their family unit closer together while caring for him.

Availability of acceptable alternatives (Fig. 14-1) to natural pregnancy can influence family planning. Men and women at high risk for a genetic problem associated with a large burden often elect to practice contraception to avoid natural pregnancy. Methods used ranged from the rhythm method to oral contraceptives. The latter may not be an approved form of birth control for some people who are followers of certain religions. Sterilization of the wife or husband may be requested to avoid pregnancy; however, sterilization is forbidden by some faiths. Many couples at high risk for children with moderate to severe genetic handicaps may elect to adopt children. However, the waiting lists for couples seeking such children are long, and some couples may be found unsuitable for placement of adoptive children. For example, a parent afflicted with a genetic problem that will be fatal prior to completion of the child's rearing period will find adoption difficult. This may be the case for a parent at risk for Huntington's disease or who has Marfan's syndrome. Access to prenatal diagnosis may encourage many parents at risk to attempt natural pregnancy. However, certain religions find these procedures unacceptable, since there is a chance that abortion of a fetus with an abnormal genotype may be performed. Prenatal diagnostic procedures will be discussed in a subsequent section of this chapter. Artificial insemination using sperm from an anonymous donor is becoming a common means of avoiding risk. This procedure may be utilized by couples when they are at risk for having a child with an autosomal recessive disorder, a dominant disease present in the father, or when the father is a balanced carrier for a chromosome rearrangement. The sperm donor must be screened for potential chromosomal rearrangements, and a thorough family history should be taken to exclude those who have relatives with heritable genetic diseases. However, even these precautions will not exclude the possibility that the sperm may be carrying a recessive allele or a new mutation; and the parents must be informed of these possibilities. Gamete intra-fallopian transfer (GIFT) may be used to avoid transmission of a dominant allele, X-linked recessive allele, or chromosomal rearrangement from mother to child. A donor egg may be placed in the upper region of the woman's Fallopian tube prior to intercourse with her husband. The egg donor must be screened for possible genetic problems in a manner similar to that employed for sperm donors.

A)   Contraception   Prenatal Diagnosis
       Sterilization   GIFT
       Adoption   ZIFT
       AID   Surrogate

B)   In Vitro Fertilization (IVF)

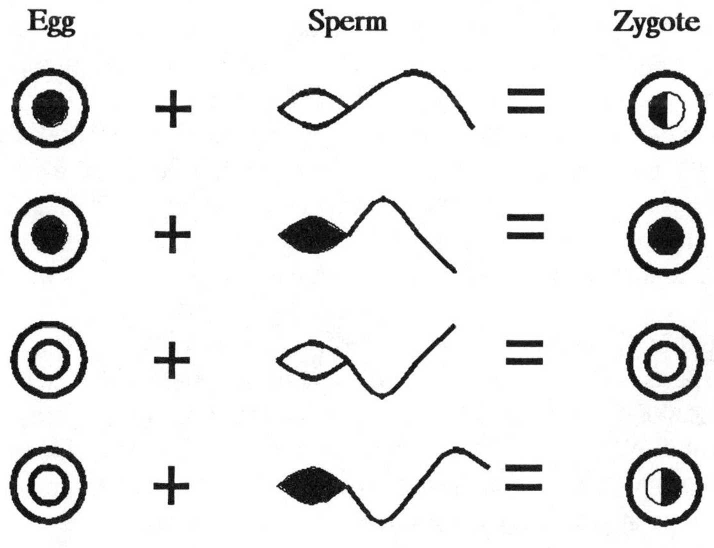

Egg   Sperm   Zygote

Fig. 14-1. Alternatives to natural pregnancy. A) AID = artificial insemination by donor; GIFT = gamete intra-fallopian transfer; ZIFT = zygote intra-fallopian transfer. B) Several possible combinations of gametes are shown. Shaded = genes derived from wife or/and husband.

In _vitro_ fertilization (IVF) using a donor egg and the husband's sperm may also be used to avoid transmission of a dominant allele, X-linked recessive allele, or a chromosomal rearrangement from mother to child. The woman is treated hormonally to induce ovulation of several eggs, and the ova are collected surgically. The ova are incubated with capacitated sperm to promote _in vitro_ fertilization. The fertilized eggs are allowed to undergo a few cleavages and several human embryos are implanted in the uterus. Several embryos are employed because this appears to increase the frequency of successful pregnancies. As indicated in the figure, four different parental combinations are possible, ranging from the case where the gametes are biologically related to the couple (used in cases of infertility) to the opposite extreme, where neither gamete is biologically related to the couple. The frequency of successful fertilizations appears to be improved by freezing sperm prior to use. This approach also permits testing the donor for AIDS prior to _in vitro_ fertilization. Donor sperm utilized for artificial insemination is also frozen prior to use to permit AIDS testing. One possible complication of IVF is the possibility of the implantation of multiple fetuses and the associated risks multiple conceptions present.

ZIFT, zygote intra-fallopian transfer, is a variation of IVF, where the fertilized egg is introduced to the Fallopian tube, and pregnancy is allowed to proceed. This method has a somewhat lower success rate, since only one zygote is used. ZIFT is often used in combination with first polar body diagnosis (see below).

Surrogate mothers may be sought by couples who may be unable to use any of the preceding reproductive methods, are unsuccessful with obtaining an adoptive child, and who desperately want a baby. A contract may be arranged with another woman and her husband to conceive and bear a child for the couple. A variation of this procedure involves artificial insemination of the surrogate mother with sperm from the adoptive husband. Problems with surrogates include cases where the surrogate mother decides she wants to keep the child and responsibility for a child who is born with handicaps. A number of these disputes have appeared in the courts. Cases where another couple contracts to conceive and bear the child for a couple at risk for having a child with a genetic problem appear to fit the "adoptive" model from a legal perspective. Some of these cases have been settled for the surrogate and some for the adoptive parents, depending upon the specific nature of the contract and of the case at hand. The situation where the adoptive husband supplies the sperm for impregnation of the surrogate has been troublesome, and the legal aspects of these cases are still being clarified.

Alternative reproductive options open to a couple at risk for a child with a genetic disorder are often very limited. Due to cultural, religious, and/or financial considerations, rhythm may prove to be the only option acceptable to them. Since this method has a significant failure rate, there is considerable anxiety regarding a possible unwanted pregnancy. Failure of the rhythm method that culminates in another child with a problem may force a couple to behave in a manner contrary to their beliefs. Many people do not have the financial resources required for access to prenatal diagnosis or some of the other reproductive technologies, and insurance carriers rarely provide adequate coverage for these services. As a consequence, these technologies tend to be unavailable to a large sector of the population who needs them.

A number of couples may be unable to come to a clear decision regarding their family planning. They may attempt to diffuse the responsibility for reproductive decision-making to others, such as their physician or peers. They may be more concerned about what others think about their potential choices and are unable to accept the responsibility for them. Women who are caught in this dilemma may practice "reproductive roulette (4)." Although they may be using a reliable means of birth control, they practice the method in an erratic fashion, such that they eventually become pregnant.

Family Problems Associated with Genetic Disease

One of the goals of the genetic counselor is to help the family adjust to their handicapped child. The counselor attempts to present a picture of the child as a person who needs the nurturing of the family, and who has certain capabilities, the extent of which are often unknown. Abilities of mentally handicapped children are variable, may be limited by associated physical handicaps, and may not be appreciated until the child is five or older. A pediatrician may often tell parents that a child is the only one that can provide the clues that suggest his/her potential, and considerable patience is required of the parents during the interval between birth and when these capabilities are known.

Families afflicted with genetic diseases experience a number of psychological problems precipitated by the occurrence of a genetic disorder and the threat of recurrence of that disorder. These problems include both those that arise as a consequence of the presence of the affected person in the family and the implications of the genetic cause of the disease for parents and other relatives, and those that are associated with alternatives for prevention of recurrence of the disorder, one or more of which may be contrary to the moral beliefs of the family involved (5). Several of these problems will be discussed here and following the section dealing with prenatal diagnosis.

Autosomal recessive diseases most commonly occur in families in which both parents are clinically normal. Tay-Sachs disease cannot be treated effectively, and the impact of the deteriorating child upon parents and siblings is staggering from both psychological and economical perspectives. Parents often feel guilt and shame for contributing to the disease afflicting their baby. One or both parents may insist upon concealment of the diagnosis from both friends and their normal children. Many years later they may find themselves in a dilemma caused by their desire to suppress information regarding the deceased individual and the feeling that they want to spare their normal offspring from the suffering associated with potential recurrence of the disorder in their families. If the latter desire is sufficiently strong, they may be motivated to seek help from a physician or genetic counselor. In the case of Tay-Sachs disease, carrier detection and pregnancy monitoring are possible.

Cystic fibrosis (CF), another autosomal recessive disease is associated with a longer average lifespan and a variable phenotype. Many of the psychological ramifications of CF are similar to those of Tay-Sachs disease; however, CF is more difficult to diagnose prenatally, and the variable phenotype presents problems for prognosis. The ΔF508 allele probe makes it possible to diagnose carrier status in appropriate CF families and to prenatally detect fetuses heterozygous or homozygous for this allele. Since the ΔF508 allele is often associated with more severe forms of CF, some prognostic statements can be made relevant to a homozygote detected in utero. Complications are introduced by the presence of a large number of other alleles that can cause CF in homozygous form or as one of the abnormal alleles in genetic compounds. Suitable probes for detection of these genes have not been developed for large scale use in a clinical setting. These uncertainties tend to compound the difficulty parents face with regard to continuation or termination of a pregnancy involving a fetus with an abnormal test result.

X-linked recessive and dominant alleles, autosomal dominant alleles, and balanced chromosomal rearrangements tend to be associated with high recurrence risk and introduce two additional dimensions to psychological complications experienced by couples with an affected child: 1) the perceived "blame" is largely concentrated upon one of the two parents, and 2) more extended portions of the kindred run a risk for occurrence of the respective disorders in their families. The former problem may precipitate divorce, while the latter presents a number of ethical and legal problems for physicians and counselors. These difficulties are illustrated by a case history described in reference 5. A couple was referred for genetic counseling following birth of a Down syndrome infant. During the compilation of the family history, the husband was found to be at risk for Huntington's disease. The child had the primary trisomic form of Down syndrome, suggesting low recurrence risk. However, Huntington's disease is highly penetrant, and the husband faced a 50% chance for developing the neurological deterioration characteristic of the disease. Furthermore, the burden of the disease is large, and about 5% of men and women afflicted with the disease attempt suicide. Anxiety among those who know they are at risk for Huntington's disease is also substantial, precipitating a variety of behavioral disorders. The counselor's ethical and legal obligations in this situation might be questioned. The couple was unaware of the familial nature of Huntington's disease. Is the counselor obligated to inform them of this risk, consequently causing great anxiety? The courts hold both physicians and counselors liable for relating this information to people at direct risk, since birth of a person with the Huntington's allele could be grounds for a malpractice suit under the Tort for Wrongful Birth (see below). The couple was informed of the nature of the disease, and the husband was referred for neurological evaluation and subsequently has required psychiatric care. In this case, the recurrence risk for Down syndrome was low, and the course of action chosen by the counselor may be judged as correct. Suppose the child had been found to have a translocation form of Down syndrome, and one of the parents was a balanced carrier of the translocation? If the couple elected for one of them to be sterilized, would withholding of information concerning Huntington's disease now be acceptable? This decision is less clear. What is the counselor's responsibility toward other paternal relatives? In this case the counselor also contacted the husband's siblings and referred them to a genetic counselor in their regions. Suppose the husband had refused permission to divulge information about his (and their) risk for Huntington's disease? Now we have a conflict between the siblings' right to know and the husband's right for confidentiality. Furthermore, the siblings are receiving unpleasant news and may resent the counselor telling them about this problem. The availability of the RFLP test for the Huntington's disease allele has facilitated presymptomatic testing. However, the motivation for consenting to the test is counterbalanced by the fear of knowing an abnormal outcome. The latter problem can now be partially alleviated by exclusion testing (see chapter 9).

Some parents are able to cope with a severely handicapped child knowing that the child will most likely die within a year of birth. Examples include Trisomies 13 and 18 and other chromosomal disorders. However, improved medical care of these infants has led to longer average lifespans without an improvement of the quality of life they experience. Furthermore, medical costs associated with these cases are very high and often surpass the ceilings that are in force on health insurance policies. These parents love their handicapped children and do everything possible to help them; however, the increased burden associated with prolonged survival leads to emotional and financial crises. Another aspect of this difficulty is related to the tendency of trisomic conditions to appear relatively late in families. Parents of these children are genuinely concerned about being able to provide the support they will continue to require during adolescence and adulthood.

These are a few of the problems faced by families and their genetic counselors. Additional difficulties arise from: 1) appraising the couple of previously unknown aspects of a hereditary disease which are of a life-threatening nature, 2) dealing with problems associated with an "accidental" pregnancy subsequent to the initial consultation, 3) the guilt feelings associated with inherited diseases, and 4) the unusually strong fears, stresses and hostilities rampant in some families with severly handicapped children. These difficulties accent the need for a team approach to genetic counseling and the importance of follow-up and follow-along to help families cope with rearing their child.

The success story observed in relation with Down syndrome (chapter 4) provides an example of what can be done to improve the quality of life experienced by persons with genetic handicaps. Hopefully, others may benefit from similar approaches in the future.

GENETIC SCREENING

The number of genetic diseases is large and is steadily increasing; therefore, it is unlikely that any one individual would be able to diagnose all or even a major proportion of this group of conditions. One of the most common questions asked by health professionals and other agency personnel is: "Who should be referred for genetic counseling?"

Indications for referral have been discussed in various sections of this text. Many chromosomal and Mendelian disorders present as syndromes, and one common type of patient referred for genetic workup is one who has **a collection of traits that are each individually rare.** For example, the occurrence of mental retardation in combination with three or more physical anomalies may suggest presence of a chromosome abnormality. **Occurrence of two or more children with a congenital birth defect** in the same family also suggests involvement of a genetic problem. Cleft lip with or without cleft palate (CL(P)) is found in one per thousand newborn. Chance occurrence of two CL(P) patients in the same family would be rare (1×10^{-6}); therefore, either the parents carry a large number of factors predisposing to this birth defect, or both children were subjected to the same or similar environmental insults. If review of the pregnancy history fails to reveal an environmental etiology, referral for genetic counseling seems appropriate. A third basis for referral is the **occurrence of multiple spontaneous abortions,** particularly if fetal loss occurs at about the same time during the first trimester. **Recognition or suspicion of a known genetic disorder** in a person is also due cause for referral. Persons who are **close relatives** of a person with a genetic disease or **who are members of a racial, ethnic or age group** that has increased frequencies of certain traits would also benefit from genetic counseling. An example of the latter group is comprised of women who are over age 35. These women have an increased risk for birthing infants with trisomic conditions and/or birth defects.

Genetic screening programs attempt to identify individuals who are at risk for genetic disease or who are at risk for having children with these diseases. Goals of genetic screening programs include: 1) identification of persons at risk for having children with genetic problems and supplying them with necessary educational materials that will assist them with responsible family planning (sickle cell anemia, Tay-Sachs disease); 2) identification of infants at high risk for contracting inherited diseases for the purpose of treatment and prevention; and 3) provision of a total picture of the variability of a phenotype associated with specific genotypes. Metabolic or chromosomal variants may be discovered in persons who have various clinical problems. However, association of these problems with the variant may prove coincidental. Histidinemia was initially discovered in a mentally retarded patient and was believed to be responsible for the retardation. However, large scale screening and follow-up of persons with histidinemia soon revealed that most do not become mentally retarded, suggesting clinical heterogeneity of the metabolic variant (6).

Effective and appropriate screening programs should be accessible by the entire population and should not employ compulsion to obtain compliance. It is desirable to involve the community in which the screening will take place in the planning and execution of the program. The benefits and limitations of the screening program should be carefully explained to those seeking access, and unrealistic promises should not be made to the participants. Tests employed for screening should be economical but accurate and should provide information that will effectively discriminate between heterozygotes and homozygotes. For example, an efficient screening program for sickle cell anemia should accurately distinguish between anemia and trait. Results should be accurately interpreted in a manner that is comprehensable, and supportive counseling should be available when needed. Informed consent should be sought from persons seeking the test or from their parents or legal guardians, and the testing procedure must be practiced according to Health and Human Services guidelines for protection of human subjects. Results of tests must be dislosed to the person tested or his legal representative or physician. Furthermore, the confidentiality of test results must be adequately protected to ensure the right of privacy of tested individuals and their families. Violations of one or more of these principles have occurred in the past and have led to unfortunate consequences. More stringent safeguards promise to make currently operational and planned programs more effective and appropriate.

An efficient screening program should involve a test that is sufficiently sensitive to permit identification of the person at risk and yet avoid an unacceptable rate of "false positives." Two examples of false positives are observed in neonatal screening for PKU and in maternal serum α-fetoprotein screening (MSAFP). Many states and other countries have employed the Guthrie test as an initial screening procedure for PKU. Benign hyperphenylalaninemia presents as a false positive in this system. Alleles responsible for the clinically benign condition are about as prevalent among the newborn as those that cause PKU. Discrimination between them requires additional testing. The Guthrie test continues to be used because it is inexpensive, easily applied to large numbers of samples, and identifies essentially all infants who are at risk for PKU, permitting their confirmation and placement on a low phenylalanine diet. Open neural tube defects release α-fetoprotein (a protein of unknown function that is encoded by a gene tightly linked to the albumin gene) into the amniotic fluid, and lesser quantities appear in maternal serum. MSAFP screening has been proposed as a means for detecting pregnancies at risk for neural tube defects. Problems began to appear with respect to a relatively high frequency of false positives with concomitant anxiety in women who were notified of these results. A false positive may be caused by miscalculation of the length of gestational age of the fetus, multiple gestations (twins, triplets, etc.), kidney disease, and fetal death. The American Society of Human Genetics proposed guidelines for introduction of this test for screening which included: 1) physician/health professional education, 2) performance of the tests by qualified laboratories, 3) provision of adequate follow-up for abnormal results, and 4) adequate patient education regarding the interpretation of the test (7). The policy statement stipulated that testing should be done between 16 and 18 weeks of gestation, and that a positive test should be confirmed by a second evaluation of MSAFP. If the result was confirmed, the expectant couple/woman should receive counseling and a level II ultrasound evaluation should be performed (note: level II ultrasonography provides a detailed evaluation of the fetus). If the latter indicates that a neural tube defect may be present, then amniocentesis for α-fetoprotein measurement, acetylcholinesterase evaluation and possible chromosomal evaluation should be performed. The policy statement also stipulated that MSAFP screening should be voluntary. MSAFP testing has proved to be a good screening test, but it has its limitations. It has an acknowledged false positive rate which is dealt with by the protocol outlined above. Like α-fetoprotein testing of amniotic fluid, MSAFP testing will also miss some fetuses with neural tube defects ("false negatives"), particularly those with smaller lesions and those which are covered by membranes and skin. Provided MSAFP screening is viewed as a screening test and not as a diagnostic test, its use can be beneficial in many cases. Critics of the program have stressed the anxiety that a positive test creates for an expectant couple, a considerable majority of whom are not at risk for having an infant with a neural tube defect.

In order for a screening program to be instituted, it must demonstrate an appropriate cost-benefit ratio. PKU testing costs about five dollars per test or five million dollars to screen the entire newborn population in the United States. About 90 infants at risk for PKU will be identified and treated by this screening program. If there was not a screening program, a high proportion of these infants would become brain damaged. A conservative estimate of the cost to the government to care for 90 PKU persons throughout their lifetime approximates eighteen million dollars. By contrast, about 16,000 newborn infants are born each year who have major chromosome abnormalities. It would cost more than one billion dollars per year to test all newborn for chromosome abnormalities, an investment of about $86,000 per case identified. It costs approximately eight thousand dollars to care for a mentally handicapped individual per year. The cost-benefit ratio is not appropriate for large scale cytogenetic screening. Furthermore, most severely affected infants die shortly after birth, and most sex chromosome errors that would be encountered do not require institutionalization. Cytogenetic problems cannot be prevented by means other than abortion, and abortion of infants possessing sex chromosome errors seems inappropriate. For these reasons, mass neonatal or prenatal screening for chromosome anomalies is not likely to occur.

Availability of a voluntary screening program does not necessarily imply that it will be extensively utilized (8). A planned screening program for identification of Tay-Sachs carriers in a large metropolitan area concentrated upon 30,000 eligible subjects. Public media were used to advertise the program, and pamphlets and information packets were distributed. Physicians and rabbis serving the community were contacted, and the screening program was explained by letter or verbally. Students were ascertained by a variety of routes including posted notices, personal letters, and through religious groups. Compliance with the screening program averaged 9.8% and ranged from a high of 27.6% among graduating high school seniors to a low of 8.1% among the general Jewish community. The study demonstrated a problem with perception of the relevance of heredity to specific individuals, not only among the potential clients of the program, but also among the physicians and clergy who helped the advocates of the program.

Conflicts between screening persons at risk for a hereditary disorder and confidentiality of government records can lead to distressing situations. A French research group was studying the segregation of manic-depressive psychosis and came across a large French kindred that was afflicted with a form of juvenile-onset glaucoma (9). If individuals at risk are identified and monitored for the condition, early treatment can prevent the blindness that eventually occurs. The autosomal dominant mutation apparently arose in a couple who died in 1495 and currently involves an estimated 30,000 families. The researchers know the individuals who are at risk; however, a government agency has forbidden release of names of those at risk to physicians serving the communities in which they reside because of privacy laws that prohibit release of such computerized data. As a consequence, many French children are destined to become blind from a preventable condition.

Neonatal Screening

Several large newborn screening programs are currently conducted in many states and foreign countries. These programs screen all newborn for the following conditions: PKU, galactosemia, certain hemoglobinopathies, hypothyroidism, congenital adrenal hyperplasia, and biotinidase deficiency.
The justification for each of the programs is that early identification and treatment can prevent or ameliorate many of the adverse symptoms of the diseases. Mothers of the newborn infants sign a consent form granting permission to test their infants, so, at least on the surface, the programs appear voluntary. However, it is unclear what would transpire if a mother refused to approve the testing. These large screening programs also have problems with educational and interpretational aspects of the testing. In some cases follow-up of positive test results is difficult, and cases of the diseases may occur as a consequence of failure to relocate the infant involved. Pamphlets are printed and distributed by public health departments, and pediatricians and public health nurses appear to be carrying out most of the follow-along activities. In general, these screening programs have had a positive effect. The treatments have varying degrees of success. PKU is largely preventable with this approach; however, patients must apparently be maintained on the diet throughout life. Treatment of both galactosemia and congenital adrenal hyperplasia prevents some but not all of the symptoms, and more research is necessary to develop more successful approaches. Provision of more aggressive medical treatment for infections contracted by children with sickle cell anemia and generally upgrading their medical treatment appears to improve their quality of life. These programs are generally economical and effectively managed, although some

patients may be missed.

Prenatal Diagnosis

Prenatal screening for genetic defects is done on a voluntary and selective basis. MSAFP screening was described above. Clients for prenatal diagnosis are physician referred on the basis of one of the criteria listed in Table 14-2. In general, the risk of the problem must significantly exceed the risk of the procedure before referral is made for testing. The major complication of prenatal diagnosis is loss of the fetus as a consequence of fetal trauma or damage to the placenta. Risks are very low for amniocentesis (generally <0.5%) and somewhat higher for chorionic villus sampling (<1%). A number of prenatal diagnostic procedures are listed in Table 14-3. Ultrasonography and amniocentesis are most widely performed, while chorionic villus sampling in still limited to specialized centers. First polar body diagnosis is still classified as an experimental procedure.

First polar body diagnosis can be used to detect the presence of abnormal chromosomes or certain abnormal alleles. Maternally-derived alleles separate from paternally-derived alleles during anaphase I, and this division is not completed until ovulation. A woman is hormonally induced to ovulate, and the ova and their attached polar bodies are collected surgically. If the couple is at risk for a chromosome error, the chromosomal complement of the polar body is determined. If the polar body contains the abnormal chromosome or abnormal doses of a chromosome, the ovum is inferred to have a normal chromosome complement. If the woman is a carrier for a testable allele responsible for a metabolic disease or other anomaly, the DNA of the polar body is amplified by the polymerase chain reaction. The amplified DNA is partially digested, electrophoresed and probed with a cDNA specific for the allele in question or for an RFLP known to be associated with the abnormal allele. If the abnormal allele is found in the polar body, then its normal counterpart is inferred to be present in the ovum. In both cases, the ovum is fertilized in vitro with the husbands sperm and implanted in the woman's Fallopian tube or uterus. If the ovum is found to contain a chromosome anomaly or an abnormal allele, respectively, it is not fertilized. This procedure holds promise, since abortion of a defective fetus need not be performed to prevent a disease. Two limitations of the procedure center around the relatively poor pregnancy success rate due to implantation of a single zygote, and the perception of the reproductive technology as being abnormal by certain religious groups. The loss of zygotes is also perceived as undesirable by these groups.

Chorionic villus sampling (CVS) is generally performed between the 9th and 12th weeks of gestation (trans-cervical method) or as late as the 13th week (trans-abdominal method). Chorionic villi are projections of fetal material, some of which contribute to the fetal portion of the placenta. The

Table 14-2. Indications for Prenatal Diagnosis.

High Risk:

Mother a carrier for an X-linked disease
Parent with a testable dominant disorder
Both parents carriers of a testable recessive disorder
Parent a carrier of a balanced chromosomal
    rearrangement

Intermediate Risk:

Maternal age >40
Parent mosaic for chromosome error
Parent or prior child with detectable birth defect
    (neural tube defect, etc.)

Low Risk:

Maternal age 35-40
Previous child with primary trisomy

Table 14-3. Methods for Prenatal Detection of Genetic Disease.

First Polar Body Diagnosis:

1) Chromosome Errors
2) DNA Analysis

Chorionic Villus Sampling (CVS):

1) Chromosome Errors
2) Enzyme Assays
3) DNA Analysis

Fetoscopy (largely replaced by ultrasound):

1) Limb and Other Surface Defects

Peri-umbilical Blood Sampling:

1) Blood-specific Disorders
2) Resolution of CVS-Amnio Discrepancies

Amniocentesis (Amnio):

1) α-fetoprotein
2) Chromosome Errors
3) Enzyme Assays
4) DNA Analysis

Ultrasound-Level I (General) and II (High Resol.):

1) Neural Tube Defects
2) Sexual maldevelopment
3) Cystic Hygromas (surgically correctable)
4) Limb Malformations
5) Heart Malformations
6) Used for Guidance During CVS and Amnio

remaining chorionic villi decay later in the pregnancy. Since the cells of the villi share the same genetic material with the fetus, they can be used for diagnostic purposes. A small amount of material is obtained with guidance of ultrasonography, and the cells are teased apart and studied directly or cultured for a short time. Cells from chorionic villi have a high division rate, and results can be communicated to a physician and the parents in a matter of a few days. Many chromosome errors and enzyme deficiencies can be detected by CVS. DNA studies can be performed either on direct preparations or on cultured cells to detect the disease-causing alleles, provided the required probes are available. Advantages of this approach include the high cell division rate of the villus cells and the timeliness of the procedure. The high division rate permits rapid communication of results. Furthermore, the study is done at a time when the mother has not yet experienced fetal movements, and maternal-fetal bonding is not as strong as it would be at later stages of pregnancy. Disadvantages include a somewhat higher loss rate due to the procedure, the inability to diagnose the presence of neural tube defects at this stage, and the potential decision to abort a defective fetus. A number of chromosomal abnormalities will be detected by CVS that do not reach term. These include trisomy 16 (the most common human trisomy), triploidy and tetraploidy, and several other trisomies and monosomies that may occur as uniform disorders or in mosaic condition. Several discrepancies have been encountered during CVS. These include mosaic patterns and tetraploidy in villus cells which were not present at amniocentesis or in the liveborn fetus. Guidelines for use of CVS now require confirmation of mosaicism and ploidy anomalies by amniocentesis.

Amniocentesis (Amnio; Fig. 14-2) is performed during the 14th-16th week of pregnancy. The position of the fetus is determined by ultrasound, and amniotic fluid is obtained trans-abdominally. The fluid is tested for α-fetoprotein levels, and the amniocytes which are derived from a number of fetal tissues (skin, respiratory tract, genitourinary tract, and digestive tract) are cultured and used for DNA studies, chromosomal evaluation or enzyme determinations. These cells have a slower division rate, and the time between obtaining the sample and reporting of results is considerably longer than for CVS. Furthermore, if a second amnio must be performed consequent to culture failure, problems may be encountered if an abortion is desired, since few places perform abortions after 20 weeks of gestation.

Maternal-fetal bonding is stronger at the time when amnio is performed, and parents find decisions regarding possible termination of pregnancy more difficult than at earlier times.

Results from both CVS and amnio have been found to be highly accurate, exceeding 99%. Possible sources of error include contamination of a specimen by maternal cells and the problem of mosaicism. Multiple cultures are established from each sample. This practice reduces the likelihood for mistakes being made due to sampling of maternal cells or overgrowth of a culture by unusual cell types. DNA markers could be used to differentiate between maternal and fetal cells if necessary. Presence of a paternally-derived marker in the cell cultures would confirm their fetal origin. Cases where there is a discrepancy between CVS and amnio cultures with regard to mosaicism are usually interpreted to indicate that the fetus is normal in this respect. Resolution of the discrepancy could be achieved by peri-umbilical blood sampling and cytogenetic analysis of fetal white blood cells. This technique must be done with care, because trauma to the placenta or fetus during the sampling procedure can precipitate abortion.

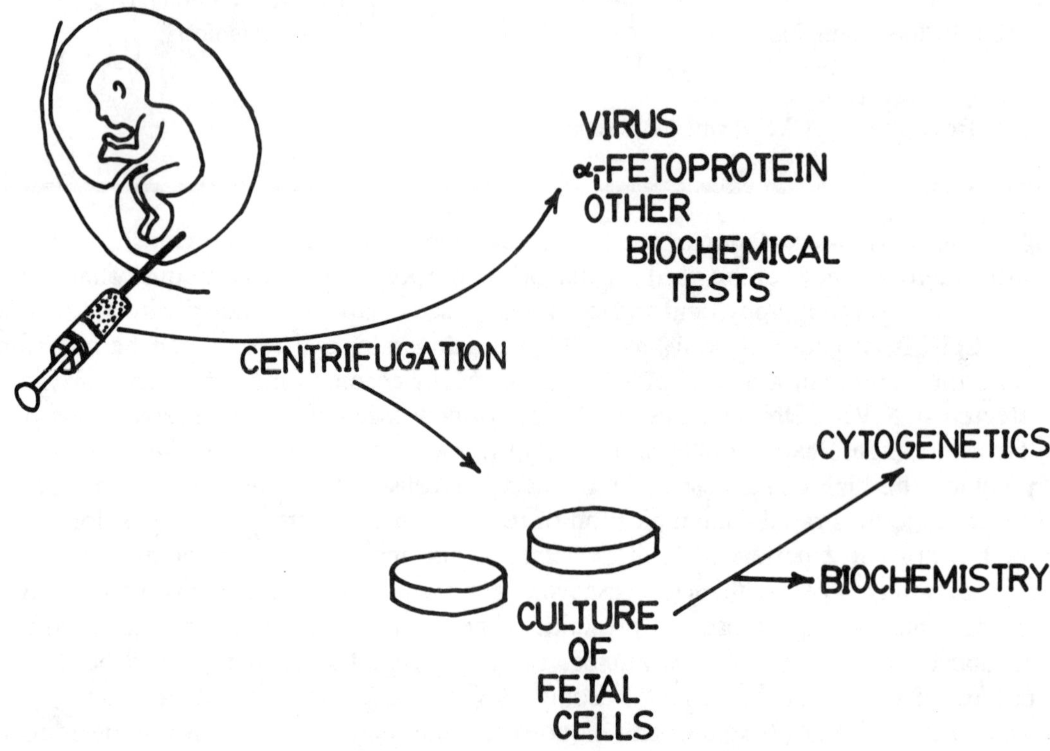

Fig. 14-2. Amniocentesis and prenatal diagnosis.

Fetoscopy utilizes a fiberoptic system inserted into the uterus to directly visualize the fetus. This procedure can detect surface abnormalities and limb defects; however, it is an invasive procedure and is associated with some risk for fetal loss. Fetoscopy has essentially been replaced by high resolution (level II) ultrasound for this purpose. Ultrasound is used to scan the fetus for structural defects involving the central nervous system, external genitalia, skeleton and heart. Cystic hygromas, which are tumor-like growths derived from the lymphatic system and filled with lymph, can also be detected by this means and later surgically removed before they cause major problems. Ultrasound has been proved harmless to both mother and fetus and is an invaluable resource for the obstetrician.

The development of DNA probes for genes predisposing to late-onset genetic diseases has opened the possibility for presymptomatic testing of children and adults. Huntington's disease (HD) is determined by a gene that has been mapped to the short arm of chromosome 4, although its precise location is uncertain. A DNA probe recognizing an RFLP that maps close to the HD locus has been shown to be useful in identifying people who are at risk for the disease. The segregation of HD and the RFLP is followed in each pedigree to establish the relationship between the RFLP type and the HD allele (chapter 9). This relationship can then be used in a predictive manner to assign genotypes to children and young adults who have not yet developed signs of the disease. Recombination between the RFLP marker and the HD locus is less than 2%; therefore, 98% certainty of the allele assignment can be attained in appropriate pedigrees. HD appears to be determined by a single locus, permitting general use of the RFLP method.

A number of screening programs for HD have been organized in the United States, Canada and other countries. These programs are voluntary, and exclusion testing is offered when requested. Two major reasons appear to motivate persons at risk for HD to seek testing. Those who would like to have children may want to know their genotype to assist them with their family planning. Secondly, some prefer not to live with uncertainty. If the test proves negative, they are relieved to learn they are not carrying the HD allele. On the other hand, if they test positive, they claim they are better able to plan their lives. Many who have a positive test indicating they most likely carry the HD allele say they will live their lives a little faster to enable them to accomplish the things they want to do before the disease begins to affect their abilities.

A form of amyotrophic lateral sclerosis (Lou Gehrig's disease), a subtype of Alzheimer's disease and one type of polycystic kidney disease are determined by autosomal dominant genes that have been mapped to specific chromosome regions. RFLPs are known that are situated near each locus and which can be used to identify individuals carrying the respective abnormal alleles. The Human Genome Project has,as one of its goals, the identification of chromosome regions that contribute a major predisposition to common adult-onset disorders. Molecular analysis of these regions may reveal key genes participating in the pathogenesis of these problems. Furthermore, early identification of persons carrying these predisposing genes may permit early treatment and prevention of the respective disorders.

GENETIC ENGINEERING

Most approaches to treatment of genetic disease have focused upon modification of the disease phenotype or **phenogenics**. Surgical correction of birth defects, dietary modification, vitamin supplementation, and hormone replacement are typical examples of more traditional phenogenic methods. Enzyme replacement therapy has also been attempted. Gaucher's disease, a sphingolipid storage disorder which occurs in several subtypes, is caused by glucocerebrosidase deficiency. Accumulation of gluco-cerebroside in macrophage lysosomes is observed in Type I Gaucher's disease. Infiltration of the spleen, liver and bone marrow by these macrophages leads to the characteristic features of the disease. Involve-ment of the nervous system occurs in a minority of patients with this form of the disease. A recent report (10) described visceral, hematological and skeletal improvement in a number of patients who were infused at biweekly intervals with glucocerbrosidase containing a signal that targeted the enzyme to macrophages. The enzyme was synthesized by a biotechnology company. All twelve patients participating in the study showed some improvement; however, the degree of improvement and the systems displaying changes varied among patients. Enzyme replacement therapy has the disadvantage that the enzyme must be infused at regular intervals and, if performed over a long period of time, would prove to be expensive.

Genetic intervention by manipulation of the genetic material (**genetic engineering**) also holds considerable promise for correction of human genetic diseases. The striking advances during the last few years make it increasingly likely that genetic engineering will be used in the near future to cure genetic diseases. This approach to therapy for genetic diseases has the advantage that it would provide a more permanent correction of the problem, making it unnecessary for the patient to follow a special diet or receive periodic injections of hormones or enzymes.

Studies of human gene structure have indicated that the critical cis-acting elements regulating gene expression are closely linked to the structural genes they control (Fig. 14-3). These sites include enhancers, hormone response elements, sites that interact with factors modulating the spatial expression of genes, and other promoter elements that govern the accuracy and amplitude of gene transcription. Temporal and spatial regulation of globin gene expression was discussed in chapter 12. The technology

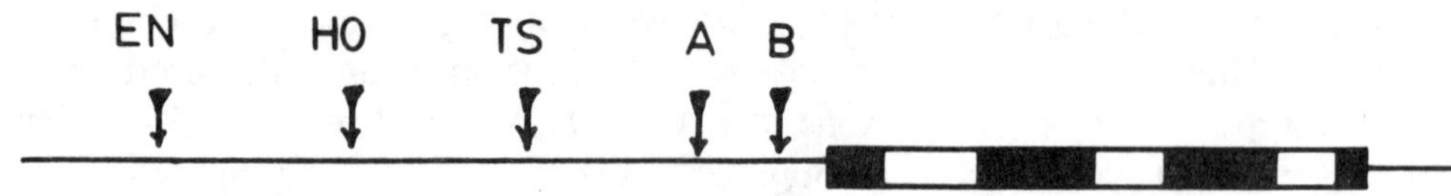

Fig. 14-3. Upstream controlling elements of a hypothetical gene. EN = enhancer element; HO = hormone response element; TS = tissue-specific factor response element; A and B = other proximal promoter elements, such as the TATA and CCAAT boxes or GC-rich promoter regions.

is now available to construct genes consisting of the required promoter regions and protein-encoding regions, to insert these genes into vectors, and to transfect cells. Limitations of these methods that have been difficult to correct are: 1) only a minority of transfected zygotes develop into mature individuals; 2) only a minority of these transgenic animals incorporate the transfected gene into their chromosomes; 3) only a minority of the incorporated genes are regulated in the appropriate manner; and 4) very few of the transgenic animals transmitted the transgene to their progeny. An early experiment of this type (11) is presented in Figures 14-4 and 14-5. Female mice were hormonally induced to superovulate (Fig. 14-4), and the ova were fertilized in vitro. The male pronuclei were injected with many copies of the genetically-engineered gene, and the embryos were implanted into other female mice who were hormonally prepared for pregnancy (pseudopregnant mice). These animals carried the embryos to term, and the animals were screened for expression of the transgene. The gene construct used for this experiment (Fig. 14-5) contained the promoter from the mouse metallothionein gene spliced to a hybrid gene consisting of a portion of the mouse metallothionein-encoding region and the rat growth hormone gene. The murine promoter contained the information believed necessary to target expression of the gene to hepatic cells (the normal site of synthesis of metallothionein, a metal-binding protein) and to make the gene construct responsive to regulation by metals such as zinc. Twenty-one of 170 implanted embryos developed to term. Seven of these animals revealed the presence of at least one copy of the transgene in their DNA. Six of the seven transgenic animals responded to zinc stimulation by synthesizing and releasing growth hormone from their hepatocytes. One of these animals grew to about twice normal size; however, actual expression of the transgene varied considerably among different transgenic mice. One transgenic mouse died as a consequence of overexpression of the transgene, and only one mouse transmitted the transgene to its progeny.

There are two major objections to this type of approach to genetic engineering of human cells: 1) the manner of reproduction that would be required (in vitro fertilization) is objectionable to many;

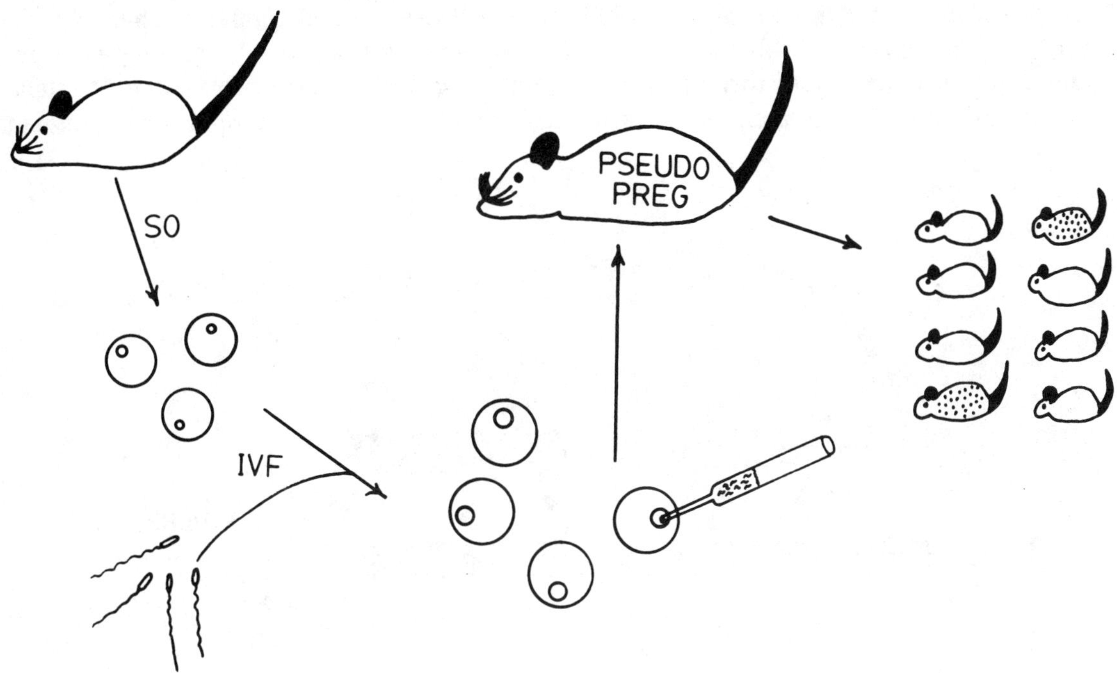

Fig. 14-4. Production of transgenic mice. The stippled pups represent animals who have incorporated the fusion gene into their DNA. SO = superovulated; IVF = *in vitro* fertilization; pseudopreg = pseudopregnant.

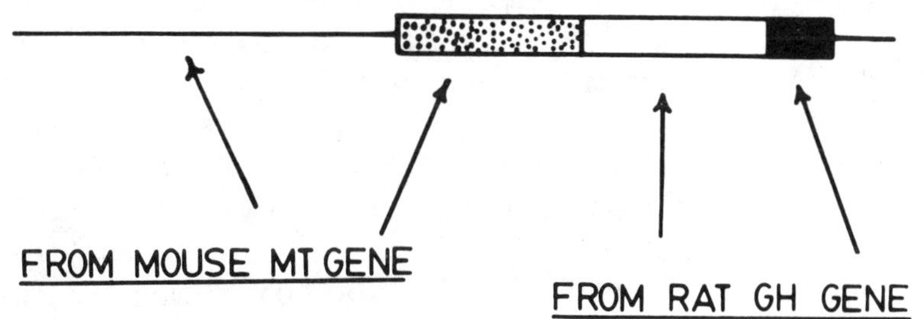

DEVELOPED   21/170
INCORPORATED   7/21
Zn ↑ GH RELEASE 6/7
GERMLINE/PROGENY 1/7

Fig. 14-5. Construction of a metallothionein-growth hormone fusion gene. MT = metallothionein; GH = growth hormone (11).

and 2) the death of many human embryos would be viewed in a manner similar to abortion by a number of different religious groups. Although transgenic technology has improved markedly since the above experiment was performed, these two criticisms remain major problems for proponents of human embryo manipulation; and it is very unlikely that this particular approach will be supported for humans.

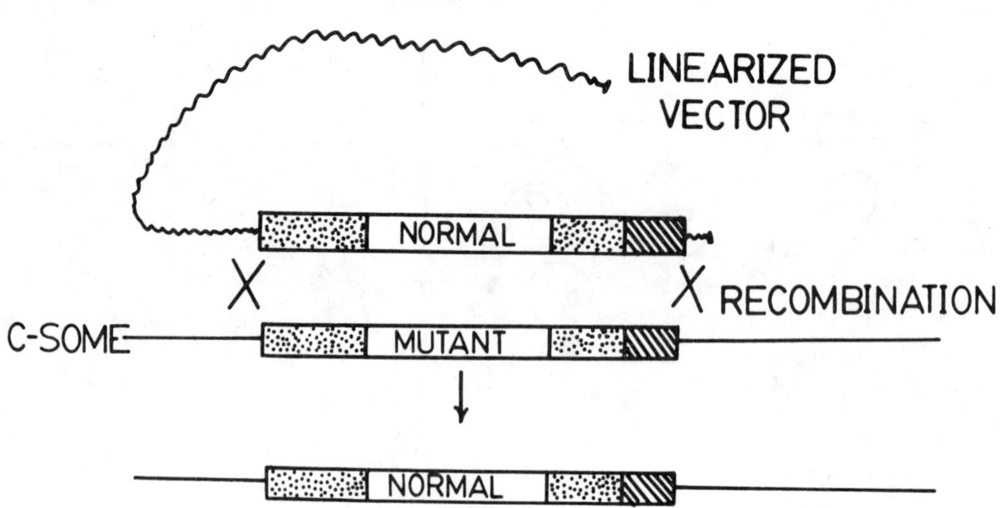

Fig. 14-6. Site-specific integration of a donor gene into a host chromosome. C-some = host chromosome.

Genetic engineering of somatic cells is feasible and supported by most groups, since it is viewed as an extension of more traditional forms of medical treatment and is therapeutic for the patient. An example of this approach is presented in Figures 14-6 and 14-7. A gene construct is prepared containing the required regulatory elements and spliced into a retrovirus vector that contains elements within it that target the gene to the appropriate chromosomal site (Fig. 14-6). **Site-specific recombination** transfers the recombinant gene from the linearized vector to the natural location of the gene in the host chromosome and results in the replacement of the defective allele by a "normal" allele. An example of site-specific recombination was presented in chapter 12 in relation to homeobox genes. By placing the gene in its natural chromosomal environment, it is believed that regulation of gene expression will be comparable to that of endogenous genes. A theoretical example of this method utilizing marrow cells is illustrated in Fig. 14-7. Marrow cells are obtained from a patient suffering from severe combined immunodeficiency caused by adenosine deaminase deficiency. The marrow cells are treated with a retrovirus vector containing the normal adenosine deaminase gene including its promoter region, and the transgene is assimilated at the adenosine deaminase locus of the host chromosomes by site-specific recombination. The transgenic cells are reintroduced into the blood stream and reseed the marrow. Proliferation and further differentiation of these cells is made possible by the expression of the adenosine deaminase gene, and the recipient develops a functional immune system.

This method has worked in cultured human cells and in a variety of animal model systems. One limitation is related to the low frequency of site-specific integrations; however, researchers are working to improve the efficiency of the process. In the mean time, housekeeping genes, such as the adenosine deaminase gene, do not need to be expressed at high levels. Therefore, the initial treatments will most likely utilize retrovirus vectors that randomly integrate into host chromosomes. The retroviruses have been genetically engineered to prevent their inherent pathogenicity.   Transfer of a gene into human cells _in_ _vivo_ has recently been reported (12). Patients with advanced malignant melanoma were treated with tumor-infiltrating lymphocytes (TILs) transfected with a bacterial neomycin gene introduced with a retroviral vector. The lymphocytes infiltrated the tumors, and the neomycin gene was transcribed and translated. Lymphocytes containing the gene were observed in the circulation of four of the five treated patients for as long as two months after treatment. No harmful side effects of the experiment were observed. Although this experiment was not designed to cure the patients of their melanoma, it clearly demonstrated that this type of somatic genetic engineering could be used for curative purposes and introduced a new way to potentially combat cancer. The strategy would be to transfect the gene for tumor necrosis factor into TILs and introduce the genetically-engineered TILs to patients with solid tumors. The TILs would infiltrate the tumors and express the tumor necrosis factor gene. The factor would then shrink the tumors (13). Experiments using somatic cell transfection methods to treat adenosine deaminase deficiency are close at hand (14). Use of retroviruses to introduce the transgene requires cell division for incorporation of the gene into the host chromosome. Recent work with adenovirus shows promise for targeting genes to and achieving expression of those genes in nondividing lung cells (15). This approach has tremendous potential for pulmonary diseases such as α_1-antitrypsin deficiency and cystic fibrosis. Similar approaches may be possible for other cell types.

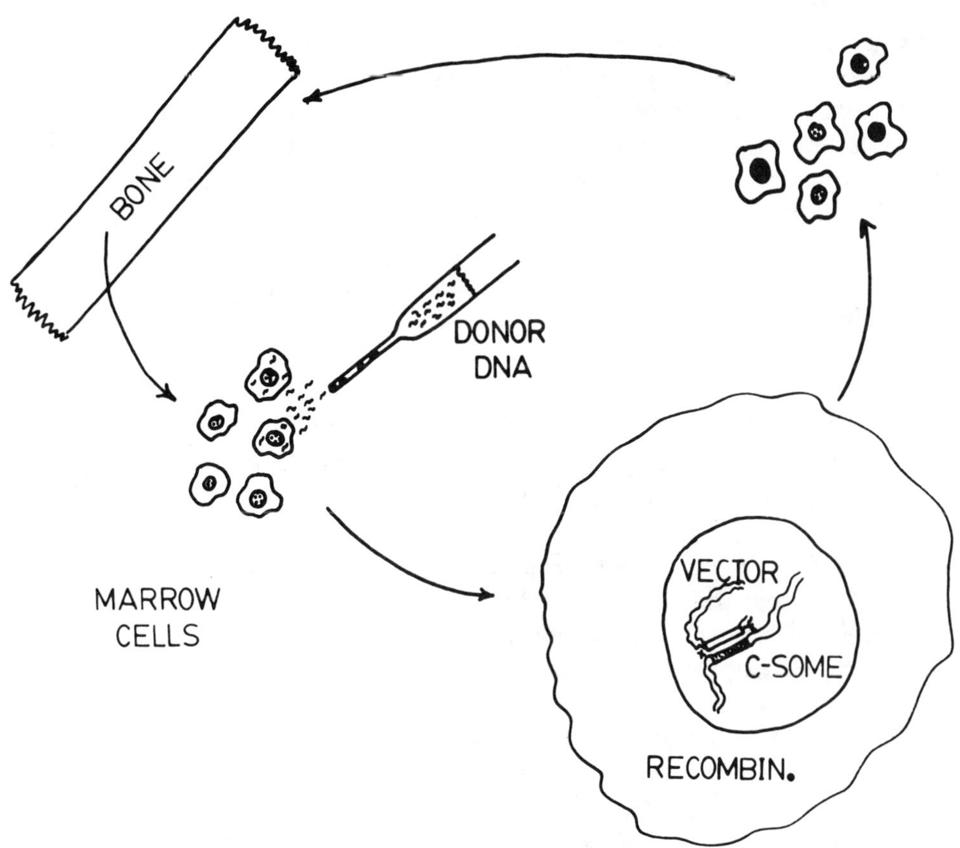

Fig. 14-7. **Somatic correction of a mutant allele by site-specific recombination. See text for description.**

ETHICAL ISSUES IN MEDICAL GENETICS

Prenatal Diagnosis and Pregnancy Termination

Parents' decisions regarding continuation of a pregnancy are voluntary and are influenced by the severity of the defect, its effects upon the family in terms of emotional, financial and sociocultural factors, and the availability of treatment for the disorder. Couples may elect to abort a fetus afflicted with a severe problem that precludes any semblance of a normal life. In other cases, such as Down syndrome, a significant proportion of parents may opt to continue the pregnancy in the hope that the child will have a relatively moderate form of the suyndrome and will eventually be able to lead a semi-independent life. Other parents, particularly those who have experienced a severely handicapped Down syndrome child, may not be able to accept a second person with the problem. Still other diseases are effectively treated or are not seriously incapacitating, and parents will decide to have the child. Surveys of prenatal diagnoses indicate that less than 5% of referred cases involve a child with the disease in question. Therefore, it appears that prenatal diagnosis relieves anxiety in most cases and encourages couples to have their own children.

Many individuals view family planning as a highly personal activity of the couple involved and view the government's role as guaranteeing the right of free choice to such couples. Persons electing abortion view their decision as a responsible act for the purpose of sparing themselves and existing or future normal siblings in their family from the burden of rearing severely handicapped children. These choices are very difficult to make, and abortion of a child often creates marital stress, depression, and other emotional problems. The situation is compounded by the **Tort of Wrongful Life** in which children born to parents who underwent prenatal diagnosis have sued their parents, claiming they were responsible for inflicting the suffering and disabilities associated with their disease upon them. These cases have largely been decided in favor of the parents, since the court has ruled that any life is better than not to have been born.

The moral and ethical controversy concerning abortion for genetic reasons continues today. Proponents of abortion for genetic cause cite one or more of the following reasons as justification for termination of pregnancy (5):

1. Prevention of suffering by the child. An individual has a right to normal physical and mental health. Failure to terminate a pregnancy involving a severely defective fetus or one at risk for future severe handicaps is a violation of this right (see Tort of Wrongful Life).

2. Welfare and rights of siblings. Handicapped children potentially deprive normal siblings of needed parenting. Siblings have a right to receive this care. Associated with this right is the "replacement" concept: abortion of the defective or potentially defective fetus permits subsequent birth of a normal child.

3. Parental welfare and right to healthy children. Pressures of a difficult economy and burgeoning world population have combined to limit the average family size. Parents should not be forced to assume the financial and emotional burdens associated with severely handicapped children. Women have a right to control their own reproduction. If a family is destined to be small, parents have the right to normal children.

4. Parental responsibility to the community. Handicapped children and adults require special, often costly, assistance from the community and state. These resources are limited, and the

community has the right to promote parental decisions that would limit the burden related to special needs of the handicapped.

5. Consideration of the gene pool of the species. Survival and possible reproduction of genetically defective persons will ultimately lower the fitness of the species. We are interfering with nature when we permit birth of defective infants and provide care which permits their eventual reproduction. Prenatal diagnosis and selective abortion of defective fetuses will permit utilization of limited resources for other human needs.

The first four reasons are worthy of careful consideration and present the "free choice" side of the debate. The fifth argument is weakened by the fact that recessive diseases account for a significant proportion of handicapped persons, and most recessive alleles are present in heterozygotes. Couples at risk for having children with recessive problems will conceive two heterozygotes for every recessive homozygote produced. Therefore, selective abortion will not appreciably improve the fitness of the gene pool.

The "free choice" position emphasizes the rights of parents, of siblings and of society. The opposite side of the debate focuses upon the rights of the individual, namely the fetus. Proponents of the "right to life" position view prenatal monitoring as a "search and destroy" activity. They consider couples' decisions to abort affected fetuses as selfish and irresponsible acts of murder and interference with the basic right to life of the fetus. They are seeking legislation to prevent abortion and are demanding that private and governmental agencies withhold financial support from programs that include prenatal diagnosis and abortion among their services. Proponents of the right to life position present the following arguments (5):

1. The state has the obligation to protect members of its society. The fetus is a member of this society and is entitled to equal protection under the laws of the state. Termination of the life of a defective fetus to permit birth of a normal fetus violates the fundamental principle of equal protection under the law.

2. Value of the person. Human worth cannot be judged by a cost/benefit ratio. Handicapped individuals should be accepted as full members of society. It is often argued that the whole individual should be considered. A person with a physical handicap may have an excellent mind and creative talent that would benefit society. Society would lose this benefit if the individual were aborted because of his physical deformity.

3. Moral responsibility. Society has the obligation to strive for moral virtue and growth. Society should reaffirm the fact that the end does not justify morally wrong means used to achieve that end. Justice should be cushioned by compassion, care and concern for the handicapped.

4. The power of society has limits. The fundamental rights of married couples to reproduce and raise children according to their moral and religious convictions should not be infringed upon by society. Society exists to promote the good of its individuals and families.

Each of these arguments also has considerable merit, and parents, physicians and other health professionals experience considerable difficulty in arriving at an ethical position that is consistent with their personal beliefs and yet addresses the dilemma they face in an appropriate way. Most couples seeking prenatal diagnosis approach the process in two steps. They tend to view prenatal monitoring as an extension of normal health care and enter the process anticipating a normal outcome. For many there is no need to progress any further with facing a potential moral conflict. However, some couples will have an abnormal test result, and they must now make a very difficult choice. They are caught in a dilemma in which they do not feel they can provide the care for a severely handicapped individual, and yet find abortion objectionable consequent to their moral upbringing. It is not unusual for parents to act contrary to their moral beliefs when confronted with this problem, and they experience strong feelings of guilt and may experience depression. Studies have demonstrated that people may express strong positions regarding abortion for genetic reasons; however, when actually placed in a position where they are the ones caught in the dilemma, often behave in a manner opposite to that which they so strongly support.

Physicians and counselors also face difficult ethical decisions, particularly when their moral beliefs differ from those of their patients/clients. Medical standards of care require physicians to: 1) appraise couples of increased risk for genetic disorders where appropriate and 2) inform them of the availability of prenatal monitoring. In cases where patients or clients request prenatal diagnosis and where the physican or counselor does not approve of abortion, they must refer their patients or clients to another health professional who will do so. If either health professional fails to inform persons of their increased risk for having a child with a genetic problem or does not inform them of the availability of prenatal diagnosis, and a child with that problem is born, they are susceptible to a malpractice suit based upon the **Tort of Wrongful Birth**. This suit is filed by parents against their physician, other health professionals, and/or medical institutions. The courts have generally decided in favor of the plaintiffs.

Opponents of abortion for genetic cause point out the dangers of establishing criteria for who should survive and who should be aborted. However, their arguments are considered extreme by supporters of free choice. They point out that most fetuses who are aborted because of genetic problems are so severely handicapped that they could not possibly contribute to society. However, this position is counterbalanced by the opponents of abortion who point out the variable severity of genetic defects, who cite the achievements of persons with Down syndrome, and who cite the normal male fetuses who may be aborted in cases where their mother is a carrier of an X-linked recessive allele that cannot be detected by prenatal testing. The availability of RFLPs and increased use of level II ultrasonography has reduced the number of situations where a normal male fetus is at risk of being aborted. The availability of probes for prenatal detection of fetuses carrying alleles responsible for adult-onset disorders provides an additional dimension to the abortion problem. In these instances, the fetus would develop into a person who has a strong likelihood of leading a productive life for many years prior to onset of symptoms. It is obvious that both sides of the abortion controversy are cogent when applied to genetic diseases, and a simple resolution does not appear possible. Most health care professionals would prefer to have a cure for the various disorders that would make abortion unnecessary. However, a particularly distressing trend has surfaced in which some agencies funding medical research are reluctant to commit funds to research projects investigating treatments for genetic disorders that can be prenatally detected, and affected fetuses aborted.

In many regions of the world where there is strong governmental pressure to limit family size or where sociocultural pressures strongly favor birth of one sex over the other, prenatal diagnosis is used to determine the sex of the fetus, and the wrong-sexed fetuses are aborted. Some couples are also motivated to seek prenatal diagnosis for sex determination in this country. Although some clinics will perform this service, most will not. Can geneticists legally and ethically withhold information regarding sex of a fetus until about 20 weeks of gestation when such information would be strictly used for sex selection by clients? The conflict here is occurring between the health professional who does not want to have a part in the destruction of a healthy fetus and prospective parents who want to terminate the pregnancy because the fetus is of the "wrong sex". There does not appear to be agreement among medical ethicists regarding this issue. Perhaps a central concern is the meaning of parenthood. Are children the "property" of their parents, or are the parents merely acting as "stewards" of their children, preparing them for life as an adult? A person's approach to prenatal gender monitoring would be quite different from these two viewpoints. There are at least four models that could be used to address ethical behavior of the health professional relative to prenatal gender selection. The consumer model proposes that "anything goes", and that the role of the geneticist is to provide clients with all of the available information and allow the clients to follow the course of action most appropriate for them. A second approach would be to adopt a contract model. The geneticist and clients would negotiate a mutually agreeable operational protocol prior to undertaking prenatal diagnosis. If the clients were dissatisfied with the withholding of information regarding sex of the fetus until it is too late to perform an abortion, they could go elsewhere for this service. The medical professional model has the geneticist setting the standards and behaving accordingly. If the geneticist decides that sex is not a disease and refuses to release requisite information to clients, so be it. Most health professionals dislike being placed in this position where they must set the moral standards for their clients. The final model is the state regulatory model. Laws are passed which prohibit abortion for reason of gender selection. This model is also objectionable to many health professionals.

Another dilemma that arises during prenatal diagnosis concerns the presence of a serious genetic defect in one member of a multiple gestation but not the others. For example, a woman at risk for having a child with Duchenne muscular dystrophy requests amniocentesis and RFLP testing. The woman is informed that she is carrying different-sexed twins, that the female fetus is most likely a carrier, and that there is a 98% chance that the male fetus will contract muscular dystrophy. Although reports have appeared describing selective abortion of one fetus and subsequent normal birth of the other, this procedure is difficult and frequently results in the abortion of both twins. What should the woman do? In this particular case, the mother carried both twins to term, and muscular dystrophy was confirmed in the boy. This example can be extended to multiple gestations in general. For example, in vitro fertilization is frequently followed by implantation of several human embryos to improve the probability that one will survive to term. Suppose a woman learns she is carrying quintuplets? The odds of all five fetuses surviving to term are small. Furthermore, if one or more should die in utero the decay of fetal remains can result in death of the others. What should be done? Should "lifeboat ethics" be practiced by aborting all but one fetus to insure its survival and normal development? Opinions regarding ethical behavior in this case vary considerably.

Ethical Issues Related to Fetal Research and Use of Fetal Organs

A number of states have passed laws prohibiting research involving fetal tissues, and several others are in the process of such legislation. Proponents of these laws equate the fetus to a mature human, stipulating that the fetus and all of its organs must receive appropriate burial or cremation. Opponents of the laws argue that fetal research is absolutely necessary to understand normal human development, to determine how birth defects occur, and to have any hope for developing cures for these problems. Furthermore, fetuses aborted for genetic reasons may also be used as a source for tissues and organs for human transplants. A number of test cases are in the courts which challenge the constitutionality of these laws.

Several scenarios might be developed relevant to use of fetal tissues for transplant or research purposes. Each of the following introduces an additional dimension to the ethical implications of the disposal of fetal tissues. Case 1) A woman is carrying a fetus with a genetic problem and elects to have an abortion. She expresses a desire to have the tissues of her fetus used for research purposes.

Case 2) The woman designates the type of research for which the fetal tissues are to be used or the person to whom the transplant should be given. Case 3) The woman elects to terminate the pregnancy for the purpose of using the fetal tissues for research or for transplants. Case 4) The woman becomes pregnant so that she can terminate the fetus for provision of tissues for transplants or for research. Case 5) She demands payment for the fetal tissues. Case 6) Cell lines are established from the fetal tissues, and a fundamental discovery of commercial value is made. The woman demands to receive some of the profits. An example of case (4) has already occurred. A woman deliberately became pregnant so that her fetus could provide a bone marrow transplant for her daughter who had contracted a severe leukemia. The ethical and legal implications presented by these six scenarios are far from resolved. However, several laws exist which prohibit some activities, and a number of guidelines have been suggested for these cases.

At least twenty-five states have laws forbidding the use of tissues from living fetuses for research purposes. Several states, as mentioned above, have forbidden all fetal research. One of these (Lousiana) has been declared unconstitutional, and another (Illinois) is currently in litigation. Many feel that the decision to abort a fetus should be separate from the decision regarding disposal of fetal tissues, and that a woman's rights end with abortion. This issue centers around who "owns" the fetus. Although an adult can designate his or her organs and tissues for transplants, considerable controversy exists regarding whether a pregnant woman can ethically make this decision for her fetus. Many believe she does not have this right. Others feel that the mother may be able to assume the role of "legal guardian" of the fetus and may designate the fate of its tissues, provided use is for a laudable purpose and non-profit. Adults cannot sell their organs, nor can the organs or tissues be transported across state lines. These limitations also appear appropriate for application to fetal organs and tissues.

Studies with mice indicate that intra-uterine transplants are a potential method for correcting immunological deficiency diseases and certain other problems, such as thalassemias. There appears to be an "immunological window" during which an animal will tolerate tissue grafts from a histo-incompatible donor. The graft is ultimately recognized as "self" by the recipient. Although a similar window of immunological tolerance appears to exist for primates, attempts to perform intra-uterine transplants between fetal Rhesus monkeys have largely failed. Premature labor and fetal loss were the most common complications. More recent primate research has met with some success. Hematopoietic stem cells were obtained from fetal liver and transplanted into the marrow space of other primate fetuses of similar gestational age. The transplants were tolerated by the recipients, and two of three animals expressed the phenotypes of the donor cells long after the grafts. The success of this work opens up a new means for treating thalassemias, sickle cell anemia, severe combined immunodeficiency, and other diseases that are correctable by marrow transplants. Current major ethical limitations of this approach include the tendency for the donor to die during the operation and the high rate of loss of recipients of the tissue.

Anencephalic fetuses have also been considered as a possible source of organs for transplants. None of these infants live more than a few hours after birth, and many die before birth. Although the central nervous system is malformed, other organs are largely intact and fully functional. Such transplants are not being performed at present; however, guidelines are being discussed. Surgeons would like to obtain organs for transplants that are in the best possible physiological condition; however, current views stipulate that the life of the anencephalic fetus cannot be prematurely terminated nor deliberately prolonged by artificial means for this purpose. Since more than 95% of women carrying anencephalic fetuses elect to terminate the pregnancy in the first trimester, transplants using anencephalic organs may not be a major problem.

Medical Management of Severely Handicapped Persons

When an infant presents with unusually severe physical handicaps, some question arises as to how aggressive a medical team should be in trying to correct the problems that are present. In the case of hopelessly ill adults, competent patients have the right to refuse life-sustaining treatment. However, infants are not able to make these choices on their own, and decision-making becomes the responsibility of parents or legal guardians. Suppose an infant is born with severe manifestations of trisomy 13, and is being maintained on a life support system. Prognosis for this patient is very poor, and the infant will not live more than a few days. The infant exists in a largely vegetative state with little awareness of others or its environment. The management strategy in this example would be to make the infant as comfortable as possible and to fully explain the nature of the infant's problems to the parents who have the responsibility for determining whether or not to continue artificial support. It would be advantageous to ask the parents if they would like to consult with clergy regarding this important decision. In this particular case, death of the infant is certain without artificial support. It does not appear appropriate to attempt surgery to try to remedy some of the problems. Furthermore, the infant is too immature and weak to withstand the stress of surgery. Many parents elect to withdraw life support, and prepare themselves for the eventual death of the infant.

Dr. John Fletcher (16) describes a second case where a decision must be made regarding performing essential surgery on a newborn Down syndrome infant. The infant has duodenal stenosis, a blockage of the esophagus, and is having difficulty feeding. The parents are experiencing major emotional problems with regard to coping with the occurrence of Down syndrome in their child and with attitudes of the husband's family. The medical team is split regarding whether the surgery should be performed, or whether the child should be allowed to die. The parents seem to be inclined toward the latter course of action. A Down syndrome parents support group had apparently heard of the case and was pressuring the hospital for information, and the hospital was concerned about adverse publicity. Dr. Fletcher, a clergyman trained in genetics and medical ethics, has a strong interest in the moral issues associated with clinical research, genetic counseling, and prenatal diagnosis. He was requested by the chief physician to work with the parents and to help resolve the differences among the medical staff. Before the parents could make any decision regarding their baby, they had to resolve their own problems and those placed upon them by their families. The latter was taken care of by inviting the husband's parents to the hospital so they could be fully informed of the nature of the baby's syndrome and the immediate medical emergency. Some of the medical staff felt the parents were attempting to abandon their infant, while others were sympathetic with the parents. The central issue seemed to be whether mental retardation was a self-sufficient reason for withholding treatment from the baby. Dr. Fletcher called the entire medical group together along with the hospital administrator and encouraged them to verbalize their feelings. A geneticisit met with the grandparents and helped them resolve their misconceptions regarding Down syndrome. The baby was maintained by mouth feeding for about a week, when the parents finally resolved to authorize the surgery. Dr. Fletcher states:

"My own moral position with respect to this particular medical condition was greatly clarified by this case. Mental retardation associated with Down syndrome is not profound enough in itself (and uncomplicated by any other physical problems) to warrant nontreatment. The child's life need not be overwhelmed by the disease, especially if the child has good care in a concerned family. Parents should be persuaded to accept this particular damaged child as an acceptable substitute for their lost, ideal child. In the event of their refusal to consent to the operation, medical authorities should seek legal remedies and persuade the parents to forego their parental responsibilities. Others are willing to adopt retarded children and substitute for the lost parents. There are many other conditions more serious than Down syndrome, however, that can warrant decisions not to bring treatment or discontinue a trial of treatment." (16, p.93)

This story is commonly encountered in many forms in medical genetics and illustrates problems faced by parents and by medical personnel regarding management of handicapped persons. Part of the problem seems to be related to the tendency for some people to perceive a handicapped individual as an object rather than a person.

Ethical Issues Associated with Genetic Data Banks

Genetic networks have been established in several locations in the United States and throughout the world that collect genetic data on families who present for diagnosis and counseling for genetic defects. One of the purposes for maintaining these computerized repositories of family data is its value in counseling families at risk for hereditary disorders. "DNA banks" are also being established for the purpose of providing a resource for both applied and basic genetic studies. An example of the former application involves a dominantly inherited disease with late onset. Blood is collected from members of a pedigree that is segregating for the disorder, and DNA is obtained from white blood cells and stored for future use. When members of the kindred appear for counseling, DNA from the stored samples can be used for RFLP or other DNA analyses which will provide valuable information regarding presence or absence of the disease allele in persons at risk. A third repository of genetic information is being assembled by the Human Genome Project. Large data banks required for mapping the human genome are being developed in the United States and Europe. One possible outcome of the Human Genome Project is that DNA from a person will be able to be studied to yield a prediction of risk for many of the known genetic diseases, including certain common multifactorial problems.

Many people are genuinely concerned about the maintenance of the confidentiality of these records and possible uses of the information contained therein. For example, business corporations and insurance carriers would benefit from knowing who is at risk for debilitating diseases that will impair a person's ability to work and which will involve substantial medical costs. Leakage of computerized data to the wrong persons may result in a loss of employment and insurance coverage by persons at risk for these problems. Precedent for this type of difficulty may be seen in the early sickle cell screening programs that were conducted several years ago. Misinterpretation of the meaning of sickle cell trait led to a loss of employment and/or insurance coverage by many individuals with sickle cell trait who were identified by the screening program. Considerable effort is being devoted to the development of systems that will maintain confidentiality of genetic data bases that will prevent recurrence of the sickle cell trait fiasco. The example of juvenile glaucoma in France which was discussed earlier in this chapter illustrates an adverse effect of strict confidentiality laws. In this particular case release of the information to appropriate parties would spare many children from blindness and ultimately reduce associated costs for both insurance carriers and the French government. The resolution of this dilemma is uncertain.

Ethical Issues Associated with Genetic Engineering

New reproductive technologies and transfection technology have raised new ethical questions. A number of these procedures are unacceptable to certain religious groups from two perspectives: 1) there is potential for loss of human embryos (that is, they are not therapeutic) and 2) many of the reproductive technologies are unnatural. The imprecision of most transfection procedures also presents a major problem for manipulation of the DNA of human embryos. For these reasons, the major emphasis in genetic engineering of human cells is currently confined to somatic cells, and it is unlikely that human embryos will be involved in the foreseeable future.

Another concern relevant to human genetic engineering is the possibility that adoption of this form of treatment may occur before all implications of its use are fully appreciated. Guidelines for use of somatic cell genetic engineering (transfection) have been established (Table 14-4). Genetic manipulation of human somatic cells will not be attempted unless other, more traditional approaches to therapy have been demonstrated to be unsuccessful. Furthermore, the genetics and biochemistry of the lesion responsible for the disease must be understood at the molecular level. This requirement eliminates many common diseases such as diabetes and the majority of coronary artery disease from consideration for this approach. The requirement for the candidate disease to be life-threatening or seriously incapacitating precludes "cosmetic" uses of the procedures.

One of the first applications of somatic cell gene manipulation will involve cells obtained from the bone marrow as described earlier in this chapter. Three categories of disease (Table 14-4) may be amenable to this approach. Obvious candidates include the hemoglobinopathies, thalassemias, and Bruton's hypogammaglobulinemia. A second category includes diseases that are caused by mutations in genes that are generally expressed but appear to have their major impact upon marrow cells and their derivatives. Adenosine deaminase deficiency is an excellent example of this type of problem and is the leading candidate for somatic cell transfection. Gaucher's disease is another example of disease that may respond to this approach, since it largely involves macrophages. The final category of disease is a questionable group of disorders that are composed of generalized sympotoms but which may respond to stem cell transfection. Lesch-Nyhan syndrome (HPRT deficiency) is an X-linked recessive disease involving self-mutilation, mental retardation, and other symptoms that may fall in this category. Metachromatic leukodystrophy (arylsulfatase A deficiency) is another questionable member of this group of diseases.

Recent progress with the use of adenovirus as a vector for targeting normal human genes to lung alveolar epithelial cells represents a major breakthrough (15). Adenovirus carrying the human gene encoding α_1-antitrypsin was placed in the tracheas of rats. The virus infected the rat alveolar epithelial cells, where the human gene was expressed. Human α_1-antitrypsin appeared in the secretions covering lung tissue. Although this experiment is an exciting step forward and removed the need for cell division required for retroviral insertion into chromosomes, it must be evaluated further to determine whether the adenovirus can cause complications.

Table 14-4. Criteria for Application of Genetic Engineering Technology to Human Subjects.

1. The disease must result from a specific biochemical lesion that is well understood.

2. The structure of the normal allele and its regulatory elements must be known.

3. The disease must be life-threatening or severely incapacitating.

4. The disease must be untreatable by other means.

5. For bone marrow stem cell engineering:
   a) major site of gene expression is in hematopoietic or lymphoid cells
   b) expression of the gene is generalized but disease symptoms are limited to hematopoietic or lymphoid cells
   c) possible: expression of gene is generalized and symptoms of disease are generalized, but disease responds to marrow manipulation

---

MEDICAL GENETICS AND SOCIETY

The most recent International Congress of Genetics (Toronto, Canada, 1988) included a symposium addressing the past, present and future impact of medical genetics upon society and the effects of past societal pressures for eugenic programs (17-21). **Eugenics** was initially defined by Sir Frances Galton in 1883 as "the study of the agencies under social control that may improve or impair the racial qualities of future generations, either physically or mentally." The term was redefined by the American Charles Davenport as "the science of the improvement of the human race by better breeding (19)." During the early twentieth century, Mendel's work was rediscovered and a strong interest in human genetics rapidly developed. Many geneticists strongly believed that many human social problems were genetically determined and that controlling the reproduction of persons with various social ills including "feeblemindedness", epilepsy, criminal behavior, alcoholism, pauperism and prostitution would improve the gene pool of the population and lead to the disappearance of these problems.

There had been a substantial effort during the 1880s and 1890s to attempt to address social problems by the traditional means of charity and through activities of social and religious organization who worked with the poor and disadvantaged. The economy of the United States worsened during this period and immigration was occurring at a high rate, placing additional stresses upon limited resources. The more affluent and powerful members of American society were having fewer children than those at the opposite end of the socioeconomic ladder, and the "survival of the fittest" philosophy that had influenced society through the mid- to late 1800s no longer appeared to be working.

The time was ripe for a new science such as Mendelism to restore efficiency and vitality to the American population. The eugenics movement became very strong and gained considerable political influence during the first twenty years of the twentieth century. Many prominent American geneticists and other biologists joined the movement and publicly promoted its goals. This loose-knit movement had two fundamental beliefs: 1) human evolution must be carefully planned and controlled; and 2) social traits are inherited, and "bad" genes must be eliminated from and "good" genes favored in the gene pool. As a consequence of the activities of these scientists and their political allies, a number of laws were passed which prohibited intermarriage of and sexual relations between members of different races and of people believed to be "unfit." Manditory sterilization laws were passed in a number of states which specified sterilization for people who were "feebleminded", alcoholics, insane, or who had been incarcerated for criminal behavior. Political pressures were placed upon Congress to pass immigration laws which favored peoples from Northern and Western Europe and discriminated against all others. The purpose of the immigration law of 1924 was to protect the American gene pool from influx of genes from Eastern and Southern Europe which were going to make the American population shorter, darker and more "mercurial."

These activities of their colleagues were so offensive to many geneticists, that scientific support for the eugenics movement quickly declined. These feelings were slow to heal, and geneticists actively avoided any political activities and shunned public statements or positions on sociogenetic issues until very recently. The eugenics movement began to change, making room for the effect of environment upon human traits, but continued to stress the importance of genetics and environment in social behavior and supported birth control measures in the form of population control movements after World War II. Many of the laws passed during the 1920s remained in effect until the 1960s and 1970s, and the immigration law of 1924 was not modified until 1962.

Strong eugenics movements were also present in England, Scandinavia (20), Germany (21) and Russia (18), and culminated in the social programs practiced by Nazi Germany. The eugenics movement in Germany began in the 1890s, and the establishing of eugenics as a scientific discipline called race hygiene occurred about 1905. Race hygienists joined forces with social hygienists in Germany to lay the groundwork for programs that would be placed into operation 20-30 years later. Following World War I, there was major concern over the declining birth rate and the need for numbers of people to assist economic recovery from the war. The stress placed upon quantity and not quality of the German people caused a temporary decline in the popularity of race hygienists. Race hygienists began to propose the adoption of marriage permits and sterilization of persons deemed hereditarily defective, and the popularity of these proposals gradually increased. This eugenic movement was facilitated by difficult economic times in Germany during the late 1920s, and its proposals were accepted as a solution for the social problems facing Germany at that time. In 1933, the Nazi government approved a draft of a sterilization law that had been developed before they took power. However, one fundamental change was incorporated into the legislation - sterilization became compulsory. In 1935 a law protecting German blood and honor and a law protecting hereditary health were passed. The law protecting German blood and honor prohibited marriages between Jews and Germans as well as marriages between Germans and all other races. The second law required health certificates for marriages. This law was apparently never enforced. Race biology was established as a discipline at German universities with the goals of providing a unified scientific basis for a racial policy (protection of German blood) and a eugenic policy (protection of German heredity). Eventually the anthropological aspects of the discipline became separated from its human genetic aspects and evolved into a responsibility of the SS by 1940. It is interesting that the term, human genetics, was first introduced in Germany in 1940 by a German scientist. More than 350,000 sterilizations were conducted by the Nazi regime. Indications for most of the sterilizations included hereditary deafness and blindness, epilepsy, Huntington's disease, mental retardation and presence of severe birth defects. Americans were horrified by the manner in which the Nazis practiced eugenics; however, it should be pointed out that, not only did the Germans obtain their eugenic ideas from the Americans, but that more than half of the sterilizations performed for eugenic reasons in the United States occurred after strong anti-Nazi criticism surfaced in this country.

The lesson learned from the history of eugenics described above seems to be that extensive legislation restricting individual freedom of choice in reproductive matters such as choice of marital partner, numbers of children, and pregnancy monitoring, seems unwarranted and can have counterproductive effects. This appears to be particularly true when the potential social impact of new technologies or new knowledge is not fully comprehended. Areas of conflict among the rights of society, the family, and the individuals still exist. Resources are limited, and many are concerned that problems may surface again. For example, what potential pressures may be brought against a couple who decide to carry a physically and mentally handicapped child into the world after they have been alerted to the presence of the handicaps by prenatal diagnosis? Is it possible that government policies may be established in the future that will mandate prenatal diagnosis for couples at risk and abortion of imperfect fetuses? Many geneticists as well as others favor protection of the right of free choice and desire to maintain the autonomy of the family.

SUMMARY

Genetic counseling involves a number of activities including diagnosis, determination of risk, communication and interpretation of risk and the ramifications of the genetic disorders, and assistance with management of the patient with genetic handicaps. Diagnosis of genetic disease requires careful clinical evaluation, laboratory studies and investigation of the medical and family histories. Determination of risk of occurrence or recurrence requires a precise medical diagnosis, and depending upon the information available, is calculated using prior or posterior probabilities. Studies have indicated that counseling is best provided by one individual supported by a number of specialists. Couples' decisions based upon information gained from genetic counseling appear to be primarily influenced by the burden of the disease and its risk of occurrence or recurrence. Alternatives to having their own children include adoption, artificial insemination by donor, pregnancy monitoring, and in vitro fertilization. A variety of procedures are available for evaluationg the morphological, biochemical, cytogenetic and molecular genetic state of the fetus. Decisions to continue or terminate a pregnancy complicated by presence of a genetic disorder is influenced by the severity and burden of the disease, available treatment and a number of sociocultural and religious factors.

Genetic screening is designed to detect individuals affected with genetic disease and carriers who are at risk for having a child with a genetic problem. These screening programs are most efficient when they employ an accurate and economical screening test and are applied to selected populations who are at an increased risk for the disorder. Counseling should be available for persons identified by the screening program.

Treatment and prevention of genetic diseases have concentrated upon phenotypic modification. Dietary modification, vitamin enrichment, drug therapy, hormone replacement, and surgery have been employed to correct the phenotype of certain diseases and malformations. Enzyme replacement is in an experimental stage. Genetic manipulation of somatic cells is now being investigated and clinical trials will commence shortly. This type of intervention involves transfection of bone marrow or other cells with retroviruses or other viruses carrying the normal allele and its controlling elements. Future use of site-specific-recombination may increase the success rate because this procedure places the transgene in its natural chromosomal environment, promoting its appropriate regulation in vivo. Somatic cell gene manipulation will correct the problem in the patient but not in his or her children. Strong sociocultural objections to human embryo manipulation will preclude use of this genetic technique in the foreseeable future.

Many geneticists are reluctant to become involved in political activities related to applications of medical genetics. This posture has resulted from unfortunate past experiences with eugenics movements in the United States and elswhere. These events led to discriminatory practices against and suffering by large numbers of people and have left lasting scars on human history. Current events that impact upon society include prenatal monitoring and selective abortion, new reproductive technologies, genetic screening, and provision of care for handicapped persons. Many believe that preservation of free choice in reproductive decisions, the autonomy of the family, and the right of handicapped individuals to have access to care and services that will provide them with the opportunity for an independent lifestyle are goals that are worth guaranteeing by law.

PROBLEMS

Review appropriate sections of this text when answering the following questions. You will find the index to be a helpful guide to locating the material.

1. A man is carrying a 14/21 balanced translocation in 100% of his peripheral lymphocytes. What is his theoretical risk for a Down syndrome child? Assume his wife has a normal karyotype and is under age 35. How does the theoretical risk compare with the observed risk?

2. Determine the probability that a man with hemophilia A (factor VIII deficiency) and a woman known to be a carrier for hemophilia B (factor IX deficiency) will produce a child with hemophilia. Both diseases are inherited as X-linked recessives.

3. If this couple has two consecutive normal sons, what is the probability that a third son will develop hemophilia?

Match the following diseases with an appropriate screening test capable of detecting them.

4. Sickle cell anemia		a. Chromosome studies
5. Huntington's disease		b. RFLP analysis
6. Edward's syndrome		c. Chromatography
7. Succinylcholine sensitivity		d. Dibucaine test
8. Phenylketonuria		e. Electrophoresis
9. Turner syndrome		f. Guthrie test
10. Homocystinuria		g. Sickledex test

11. Screening for 47,XYY has been severely criticized. Why?

12. Would prenatal diagnosis be useful for identification of any of the following disorders?

a. Tay-Sachs disease		d. Galactosemia
b. Huntington's disease		e. X-linked muscular dystrophy (woman heterozygous)
c. Cystic fibrosis		f. Phenylketonuria

GLOSSARY OF TERMS

Amniocentesis extraction of amniotic fluid and suspended fetal cells for diagnostic purposes.

Chorionic villus sampling surgical removal of chorionic villi for genetic evaluation.

First polar body diagnosis cytogenetic or DNA testing of first polar body for abnormalities.

Eugenics improvement of the fitness of the gene pool through selective marriage and reproduction of individuals.

Genetic counseling provision of information concerning burden, risk and treatment of genetic disorders to permit parents to make informed decisions.

Genetic engineering manipulation of DNA to correct genetic disorders.

Genetic screening testing large numbers of individuals for presence of genotypes predisposing to diseases or malformations or for carrier status.

Site-specific recombination   the mechanism used to target a transgene to its natural site in the chromosome.

Surrogate mother woman who agrees to bear another couple's child.

Tort of Wrongful Birth legal suit brought by parents against a health professional for medical malpractice leading to the birth of a genetically defective infant.

Tort of Wrongful Life legal suit brought by a handicapped individual against his/her parents and health professionals for allowing her/his birth and suffering as a consequence of the handicaps that are present.

Transfection the process by which foreign DNA is deliberately introduced to a cell.

Transgenic an organism or cell containing one or more foreign genes.

BIBLIOGRAPHY

1. Fraser, FC. 1974. Genetic counseling. Am J Hum Genet 26: 636-659.

2. Czeizel, A, J Metneki and M Osztovics. 1981. Evaluation of information-guidance genetic counseling. J Med Genet 18: 91-98.

3. Hall, JG, EK Powers, RT McIlvaine and VH Ean. 1978. The frequency and financial burden of genetic disease in a pediatric hospital. Am J Med Genet 1: 417-436.

4. Lippman-Hand, A and FC Fraser. 1979. Genetic counseling - the postcounseling period: II. Making reproductive choices. Am J Med Genet 4: 73-87.

5. Atkinson, GM and AS Moraczewski. 1980. Genetic Counseling, the Church and the Law. St. Louis: the Pope John XXIII Medical-Moral Research and Education Center.

6. Levy, HL, VE Shih and PM Madigan. 1974. Routine screening for histidinemia. N Eng J Med 292: 1214-1219.

7. Garver, KL, J Davis, F Greenberg et al. 1987. American Society of Human Genetics policy statement for maternal serum alpha-fetoprotein screening programs and quality control for laboratories performing maternal serum and amniotic fluid alpha-fetoprotein assays. Am J Hum Genet 40: 75-82.

8. Beck, E. S Blaichman, CR Scriver and CL Clow. 1974. Advocacy and compliance in genetic screening. N Eng J Med 291: 1166-1170.

9. Dorozynski, A. 1991. Privacy rules blindside French glaucoma effort. Science 252: 369-370.

10. Barton, NW, RO Brady, JM Dambrosia, et al. 1991. Replacement therapy for inherited enzyme deficiency - macrophage-targeted glucocerebrosidase for Gaucher's disease. N Eng J Med 324: 1464-1470.

11. Palmiter, RD, RL Brinster, RE Hammer, et al. 1982. Dramatic growth of mice that develop from eggs microinjected with metallothionein-growth hormone fusion genes. Nature 300: 611-615.

12. Rosenberg, SA, P Abersold, K Cornetta et al. 1990. Gene transfer into humans - immunotherapy of patients with advanced melanoma, using tumor-infiltrating lymphocytes modified by retroviral gene transduction. N Eng J Med 323: 570-578.

13. Cullitan, BJ. 1990. Gene therapy: into the home stretch. Science 249: 974-976.

14. Cullitan, BJ. 1990. ADA gene therapy enters the competition. Science 252: 975.

15. Hoffman, M. 1991. New vector delivers genes to lung cells. Science 252: 374.

16. Fletcher, JC. 1982. <u>Coping with Genetic Disorders</u>. San Francisco: Harper & Row, pp. 86-93.

17. Motulski, AG. 1989. Societal problems in human and medical genetics. Genome 31: 870-875.

18. Adams, MB. 1989. The politics of human heredity in the USSR, 1920-1940. Genome 31: 879-884.

19. Allen, GE. 1989. Eugenics and American social history, 1880-1950. Genome 31: 885-889.

20. Roll-Hansen, N. 1989. Eugenic sterilization: a preliminary comparison of the Scandinavian experience to that of Germany. Genome 31: 890-895.

21. Weingart, P. 1989. Politics of heredity - Germany 1900-1940, a brief overview. Genome 31: 896 -897.

ADDITIONAL LEARNING RESOURCES

1. Andrews, LB. 1987. <u>Medical Genetics: A Legal Frontier</u>*. Chicago: American Bar Foundation.

* Copies of this book (NTIS #PB 88-100045) may be obtained from: National Technical Information Service, U.S. Dept. Commerce, Springfield, VA 22161.

Appendix I

Answers to Problems

Chapter 1:

1. See Fig. 1-1 and associated reading material.

2. The molecular charge of hemoglobin reflects the sum of the charges of the amino acids located on its surface. The charges of these amino acids vary according to the conditions of the molecular environment (see Fig. 1-3), altering the net charge of hemoglobin and its electrophoretic mobility.

3. Cells utilize five methods to adapt their biochemistry to the available energy supply: 1) use of exogenous energy sources when these are available; 2) use of stored energy when exogenous sources are depleted (carbohydrate then fats then protein); 3) employment of enzymes to catalyze reactions; 4) breaking the energy requirement for synthesis of molecules into small increments through use of metabolic pathways; and 5) shutting down metabolic pathways that consume large amounts of energy when energy supplies are low.

4. Low enzyme avtivity can result from: 1) deficient enzyme protein or 2) defective enzyme function. The latter problem may be caused by abnormal interaction with substrate, with coenzyme or other cofactors.

5. The cell membrane helps the cell maintain its integrity by regulating entry and exit of molecules.

6. The gene often determines the structure of an enzyme that is of central importance in cell metabolism. Alteration of the enzyme's catalytic function will deplete or enhance the quantity of a chemical that has a variety of metabolic fates, leading to the observed multiple gene effects.

7. Amino acids, proteins, triglycerides and other fats, nucleic acids, complex carbohydrates, etc.

8. Compounds E and G which are proximal to the block should accumulate, while the end products L, K and N should be deficient, since they are dependent upon the formation of compounds distal to the block.

9. Nucleoli actively form RNA species that are integral parts of ribosomes. Ribosomes are required for protein synthesis. These individuals would not be able to manufacture cellular proteins and would die before detection.

10. Separate pathways for synthesis and degradation of molecules provide a better opportunity for efficient control of their metabolism. Both pathways may be subject to coordinate control (one turned off while the other is functioning) or each may be regulated independently of the other.

Chapter 2:

1. mRNA codons: 5'-UUCAGCUUU-3'
   amino acids: -Phe-Ser-Phe-

2. j	4. k	6. a	8. b	10. h
3. i	5. e	7. g	9. d	11. c

12. f

13. An appropriate codon sequence for the normal N-terminal sequence would be: 5'-UUUCGAAGC-3'. The corresponding DNA sequence is: 3'-AAAGCTTCG-5'. The mutant DNA sequence could be obtained by deleting one of the first three A residues and inserting a G between the T and C of the last triplet.

14. Consider the structural gene:

        ----Exon 1-------Exon 2------Exon 3------sol Exon----mem Exon----

This gene is transcribed into a nascent RNA that is subject to two processing pathways. The formation of the mRNA encoding the membrane form is produced by excising the hydrophilic exon (sol) and splicing in the hydrophobic exon (mem). The reverse of these two events will generate the soluble form of the protein.

15. Mutations have accumulated in the δ-globin promoter or downstream from the δ-globin gene that reduce the frequency of its transcription or which slow the processing of its RNA to mRNA.

16. The deficient activity may be due to decreased transcription of the gene, defective processing of its RNA, deficient translation, defective posttranslational processing, or enhanced clearance of the enzyme from the cell. Alternatively, the deficient activity may be caused by a structural alteration of the protein which alters its function.

17.
      DNA: 3'-GCATTACCC-5'
      mRNA: 5'-CGUAAUGGG-3'
      tRNA: GCAUUACCC
      aa: ArgAsnGly

18.   a

19. Proteins are synthesized from the N-terminal to the C-terminal end. The 3' half of the gene determines the N-terminal half of the polypeptide, and the 5' end of the gene encodes the C-terminal half of the polypeptide. Therefore, the δ-globin gene must be located on the 3' side of the β-globin locus (i.e. upstream from the β-globin locus).

20. There are several ways that this could be done. One scenario is as follows. Purify superzymase to homogeneity and prepare a polyclonal antibody against the purified protein. There are several human

X-chromosome specific libraries available. Gain access to an expression library derived from the X chromosome and screen for the superzymase gene. Extract the DNA insert from the positive clones, confirm their identity with your protein, and sequence the ones that contain superzymase sequences. Alternative: Purify superzymase and sequence the N- and C-terminal ends. Prepare oligonucleotides encoding those sequences and use them as primers for PCR of genomic DNA or of X-chromosomal DNA. Sequence the amplified DNA.

21. The protein appears to consist of two identical subunits. The mother's gene encodes the α subunit, and her enzyme has the molecular formula: $\alpha\alpha$. Her husband is homozygous for a mutant allele that encodes an α' subunit with a more positive charge. His isozyme has the formula: $\alpha'\alpha'$. Their children are heterozygous and produce both types of subunits which associate in three different combinations: $\alpha\alpha$, $\alpha\alpha'$ and $\alpha'\alpha'$ to generate the three-banded pattern.

22. Amplify the promoter and label. Run a gel shift assay using the promoter mixed with the protein and compare its migration with that of promoter DNA that has not been exposed to the protein.

Chapter 3:

1. Sequences of adjacent labelled nucleosomes would be anticipated to alternate with runs of adjacent unlabelled nucleosomes.

2. b

3. When runs of similar base sequences occur in close proximity in the chromosome, mispairing may occur during prophase I of meiosis. If recombination occurs within the mispaired region, the exchange of DNA is nonreciprocal; and one of the products will lack sequences present in the original chromosome. This mechanism is apparently responsible for the STS deletions. The recombination apparently occurred in a maternal ancestor of the STS-deficient boys.

4. Trypsin does extract some DNA from interband (light-staining) regions; however, it most likely accomplishes most of its effects through partial digestion of nonhistone chromosomal proteins and consesequent alteration of their conformational relationships with DNA. Chromomeric DNA is highly compact, and its proteins would be expected to be resistant to trypsin. By contrast the more relaxed euchromatic chromatin is susceptible to trypsin attack. The altered nonhistone proteins may cover phosphate groups in the DNA backbone, denying access to Giemsa dyes. Therefore, chromomeric DNA would stain dark; and euchromatic areas would stain poorly with this technique.

5. c. Marrow reticulocytes are the only one of the adult cell types listed that normally express β-globin. The nucleosomes in the β-globin gene would contain HMG-14 and/or17 and have a relaxed conformation permitting access of DNase I to the β-globin nucleosomal DNA. The nucleosomes of the β-globin genes of the other cell types would be more compact and resistant to DNase I digestion.

6. This person would lack critical testis-determining DNA and would develop into an infertile female, since two doses of the critical regions of the X chromosome are essential for normal female fertility.

7. The woman is not the mother of the boy, because he lacks her **G6PD^B** allele.

8. a,d   9. c

10. The patchwork distribution of fluorescence is caused by inactivation of the X chromosome bearing
    the normal dystrophin allele in some cells and not others.

11. All of the meiotic products derived from an anaphase I nondisjunctional event would be abnormal.
    Half of the meiotic products formed following anapahase II nondisjunction would contain normal
    chromsome complements.

12. Presence of both a Y chromosome and the paternal **G6PD**^{A} allele indicates that he is the result of
    a nondisjunction of his father's sex chromosomes during anaphase I of spermatogenesis.

Chapter 4:

1. Patient (1) would have the more severe clinical phenotype. The normal cell line occurring in
    patient (2) would partially compensate for the abnormal cell line. The degree of compensation
    approximates the relative frequencies of the two cell lines.

2. X-inactivation tends to reduce the imbalance of X-linked alleles. Y chromosomes carry few genes
    that are critical for normal development.

3. e

4.

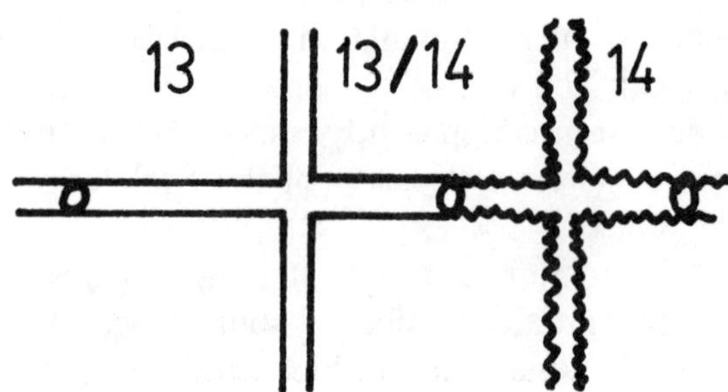

5. a) alternate: 13,14 and 13/14 gametes
    b) adjacent: 13,13/14 and 14 gametes or 13 and 13/14,14 gametes

6. Repetition of Down syndrome usually indicates parental mosaicism or occurrence of a balanced
    translocation in one parent. Chromosome studies for both parents are indicated.

7. Turner syndrome

8. f   10. g   12. a   14. d

9. e   11. c   13. b   15. j

16. Testicular differentiation is determined by the Tdf region which is normally Y-linked. These exceptional males may have been conceived as 47,XXY males. Nondisjunction or anaphase lag may produce a 46,XX cell line and a Y-bearing cell line. The cell line containing the Y chromosome may have been lost following sex determination, or may be confined to gonadal tissues. Two other causes are possibly more likely: 1) a recombination event in a male ancestor moved the Tdf region to the distal short arm of the X chromosome, or 2) a translocation moved a small piece of the Y containing the Tdf region to another chromosome.

17. The average frequency of trisomic cells is 20%. If 20% of the germinal cells are also trisomic, the frequency of disomic gametes approximates 10%. Therefore, the theoretical risk for Edward's syndrome is about 10%. The actual risk is lower because of a high rate of prenatal lethality.

Chapter 5:

1. Autosomal recessive inheritance.   2. Autosomal dominant inheritance.

3.

Prior P(Carrier)	1/2	Prior P(noncarrier)	1/2
Likelihood normal	2/5		1
Joint probability	1/5		1/2

Posterior P(Carrier) = (1/5) ÷ (1/5 + 1/2) = 2/7

P(Child with Trait) = (2/7)(1/2)(6/10) = 12/140 = 3/35.

4. $P = 2/3 \times 1/30 \times 1/4 = 1/180$.

5. No. One child of the doubly affected parent failed to inherit either hemoglobin variant. This indicates that the hemoglobin variants are determined by nonallelic genes.

6. X-linked recessive.

7. $P = 1/2 \times 1/2 \times 1/2 = 1/8$.

8.

P(Carrier)	1/2	P(normal)	1/2
Likelihood normal son	1/2		1
Joint Probability	1/4		1/2

Posterior P(Carrier) = (1/4) ÷ (1/4 + 1/2) = 1/3

P(affected son) = 1/3 × 1/2 × 1/2 = 1/12.

9. a) All daughters of affected males will inherit the trait (barring X-inactivation of defective allele)
    b) 1/2 of the sons and 1/2 of the daughters of an affected female will inherit the trait
    c) No male to male transmission of the trait will occur
    d) For common X-linked dominant traits that do not limit fertility of affected males, more females than males will have the trait

10. Heterozygous females would display varying degrees of defect ranging from normal to severely affected. Most of these women would exhibit a degree of severity intermediate between these extremes and most likely milder than that observed in affected males.

11. You will miss both fragile-X males and females who have normal intelligence. This excluded group is very important in understanding why some fragile-X carriers fail to express mental retardation and other features of the Martin-Bell syndrome.

12. X-linked dominant. Look at rows two and three of the table.

Chapter 6:

1. The mutant protein occurred as the result of a frameshift produced by an insertion (or deletion) in codon 36 and a deletion (or insertion) in codon 51.

2. The δ-globin locus is closely linked to and upstream from the β-globin locus : 3'- δ-globin - β-globin -5'.

3. Production of mutations requires cell division. Stem cells giving rise to the white cell population retain a high division rate and are more likely to accumulate mutations than are more slowly dividing or nondividing cells of other tissues. Mutagens are carcinogens. Therefore, accumulation of mutations triggers neoplastic changes in white cell precursors.

4. Organisms have evolved over long periods of time to a state where they contain a genotype that is optimal for the existing environment. Since mutation is random and since most regions of proteins cannot tolerate alteration without changing their functional properties, mutations will generallly produce alleles that encode functionally inferior proteins.

5. Mutations account for about 10% of autosomal dominant diseases and about one-third of the cases of Duchenne muscular dystrophy. The role of mutation in contributing to cases of autosomal dominant and X-linked recessive problems is inversely related to the ability of the respective genotypes to reproduce. Mutations are also responsible for the development of cancers.

6. The woman. Most of the gametic cells in a woman are at the highly sensitive oocyte stage. Even though these cells possess effective repair systems, a significant fraction of damage would persist and potentially could be transmitted to her children. On the other hand, the radiosensitive gametic cells of the male are cleared from the seminiferous tubules after about 6 months. The spermatogonia that remain are radioresistant. The relatively small amount of damage they incur is largely corrected, and the risk of transmission would be significantly less. However, as the man enters his fifties, the accumulation of this damage is sufficiently large to lead to an observable paternal age effect with respect to autosomal dominant and perhaps X-linked recessive mutations.

7. d 9. d 11. b

8. c 10. e

1. The enzyme is a homodimer, and the respective parents are homozygous for alternative alleles at the locus encoding the subunit. The children are heterozygous at this locus, and their isozyme patterns contain bands corresponding to each of their parents' isozymes and an intermediate band that is a heterodimer consisting of each parental subunit type.

2. The patient's father must be herterozygous for a β^0 allele. The patient received a β^S allele from her mother and a β^0 allele from her father. Therefore, she can only synthesize hemoglobin S.

3. If the amino acid is similar to the one it replaces, no significant change would be anticipated. If the site is one that is intimately involved with interactions with the heme iron, methemoglobinemia may occur if the amino acid is sufficiently different from the one it has replaced. If the amino acid occurs at a site participating in the shaping of the heme pocket and is sufficiently different from the one it is replacing, it may promote instability of the molecule. Modification of a site participating in the interaction of globin subunits may promote unusual oxygen retention and polycythemia.

4. The embryo and fetus are "protected" by their respective embryonic and fetal hemoglobins which lack β chains. Adult hemoglobin synthesis does not replace fetal hemoglobin synthesis until the fifth or sixth month after birth. δ-globin mutations have little effect upon persons because it is part of a hemoglobin that normally accounts for less than 3% of the hemoglobin synthesized by adults.

5. This boy has inherited a mutant allele that has sustained a base substitution in an exon. This substitution results in an amino acid change that causes G6PD to be unstable, resulting in the observed deficient activity. The activity is higher in reticulocytes because they can partially compensate by synthesizing new enzyme.

6. a) Different structural loci may contribute to the various isozyme patterns in different tissues. b) The same gene may be expressed in both tissues, but the polypeptides may be subject to different post-translational processing or different rates of clearance from the respective cells.

7. The β-thalassemia mutation in the left branch of the pedigree is segregating with haplotype 3, and the mutation in the right branch is segregating with haplotype 2. The fetus has most likely inherited both mutations and will develop β-thalassemia.

Chapter 8:

1. The patient and his brother are homozygous for an autosomal allele (or genetic compounds for different mutations in this gene) which encodes a defective enzyme A or in some way limits the quantity of the enzyme. The functional block at the step catalyzed by enzyme A results in accumulation of substrate or byproduct, producing the granules and accounting for the major symptoms of this disease by drastically interfering with cellular processes and/or by causing cell death. The parents and normal sister are heterozygous for a defective allele.

2. Assuming that suitable probes are available, the amount of mRNA could be estimated by northern blotting. The amount of enzyme protein could be assessed by western blotting. If the structure of the normal allele is known, PCR could be used to amplify the abnormal allele(s), and the sequence

compared with that of the normal allele to identify the particular lesion(s) responsible for the disease.

3. Individuals who possess a genotype that renders them more susceptible to a disease are said to be genetically predisposed. Examples include PKU (mental retardation), G6PD deficiency (anemia), α_1-antitrypsin deficiency (emphysema, cirrhosis) and pseudocholinesterase deficiency (succinylcholine sensitivity).

4. c   6. b

5. d   7. a

8. A enzyme defect such as that occurring in phenylketonuria results in overflow aminoaciduria or loss of related ketoacids. Both blood and urinary levels of the substrate or related metabolites will be elevated. In the case of a transport defect, the metabolite would either be diminished or normal in the blood and elevated in the urine. Your suspicions based upon these trends may be verified by testing the tissue activity of the appropraite enzyme.

9. CRM(+). The patient would be less likely to develop antibodies against infused enzyme that may be used in treatment of the disorder.

10. Cells from your patient did not complement with those from patients B through D. This indicates that these four cell types share a defect at the same metabolic step. Check the clinical features of your patient, family history and other data you have and compare them with those of B through D to see if you can make a diagnosis.

Chapter 9:

1. Traits C and D are autosomal dominant traits determined by linked but nonallelic genes. The probable chromosome arrangement in I-1 is **CD/cd** (cis alignment). II-3 and III-1 are recombinants.

2. Ancestral chromosome types: **EFg, efG**
    Single crossovers: **EFG, EfG, eFg, efg**
    Map distances: **E-F** = 0.11; **F-G** = 0.11; **E-G** = 0.22.

3. Trait A and the **ABO^A** allele are segregating in cis combination. Since this woman has the **ABO^A** gene, the probability she will develop trait A after age 30 is very high (approximately 100%). She should be counseled with respect to this probability, the effect of the trait upon her ability to raise a family and available treatment. Prenatal monitoring is possible, since the A blood type is detectable during the first trimester.

4. The two traits are most likely determined by unlinked loci or are determined by loci that are widely separated on the same chromosome.

Chapter 10:

1. These patients may have a defect affecting development of a stem cell common to both cell populations. Alternatively, differentiation of B-cells into plasma cells requires a T-cell helper.

Therefore, a defect in a T-cell precursor may affect both cell-mediated and humoral immunity.

2. HLA antigens appear to have evolved as part of a system required for recognition of normal cells and for detection of cells invaded by viruses. B-cells also use HLA antigens to present foreign antigens to T-cell helpers which subsequently stimulate B-cells to differentiated into plasma cells that synthesize and secrete antibodies directed against those anitgens.

3. That of the second donor, since only one incompatibility exists.

4. Erythrocytes: AB, LE^a Secretions: Le^a

5. Linkage disequilibrium refers to the unexpectedly high frequency of certain combinations of closely linked genes. For example, HLA-A1 and HLA-B8 occur on the same chromosome at a frequency of 0.13 to 0.40, whereas the product of their separate frequencies is about 0.02.

6.   -----V$_{10}$J$_3$--------[Enhancer]------C$_1$------

7. The parents of the boy are most likely **Hh**, and the boy is **hh**. Test for H-antigen. If the boy lacks the antigen, your hypothesis is correct.

8. a) If the husband is **Rr**, about half of their children would be Rh(-). b) Maternal anti-A will destroy fetal A,(+) fetal cells entering her circulation and prevent sensitization.

9. The infant most likely does not belong to this couple. The man cannot transmit an **Xg(a)+** allele to his son under normal circumstances. Note: **Xg(a)** is not inactivated.

10. No. An antibody contains only one heavy chain type.

11. A given antibody-synthesizing cell may produce two different antibody types (e.g., IgM and IgG); however, these antibodies share specificity for the same antigen.

12. The germline contains a sufficient number of variable, diversity, junctional and constant region segments to account for a large portion of the diversity of antibodies we synthesize. Recombination among these elements, junctional approximation, combinatorial association and somatic mutation considerably enhance our antibody repertoire.

Chapter 11:

1. Sex-limited autosomal dominant.

2. c   4. b   6. f

3. d   5. a   7. e

8. Pattern baldness appears to be determined by two or more autosomal loci. However, expression is dependent upon androgen levels. Women rarely possess sufficient androgen to develop baldness when an average number of predisposing genes are present.

9. H = (50 - 20) ÷ (100 - 20) = 0.38.

10. The heritability is quite low, indicating that environmental factors play a major role in the etiology
of trait B. Higher concordance among monozygous twins suggests that genetic factors are also
involved.

11. Heritability estimates derived from twin studies are based upon the assumption that environments
experienced by monozygous and dizygous twins are similar. This is probably not the case.
Environmental agents contributing to an abnormality at the level of or prior to formation of the
zygote will tend to display greater concordance among MZ than DZ twins. Heritability is environment bound and cannot be applied to populations in dissimilar environments.

12. An autosomal dominant with 60% penetrance would fit these data.

13. Observation suggests a risk of 6% for sons and 2% for daughters.

14. H = 0.67.

Chapter 12:

1. The phenotype appears to be characteristic of adrenalgenital syndrome. Decreased cortisol synthesis
disturbs the feedback control of ACTH release. Excessive ACTH release enhances synthesis of
adrenal androgens which produce the masculinization. Cortisol replacement and possibly administration of anti-androgens may promote greater feminization. Surgical enlargement of the vagina, if
indicated, may facilitate intercourse. The patient should be counseled regarding the cause of her
problem and the possibility of infertility. Siblings of the patient should be checked for the disease.

2. The anti-A and anti-B of the O,Rh(-) mothers destroy fetal A (B or AB), Rh(+) cells before they
can stimulate anti-RH production.

3. Fetal hemoglobin may be related to a functional adult hemoglobin of a relatively recent human
ancestor. Embryonic hemoglobins may be related to the functional adult hemoglobins of more distant
ancestors.

4. Consult the section: **Regulation of Gene Expression during Development.**

5. The husband is forming unbalanced gametes:

Gametes:

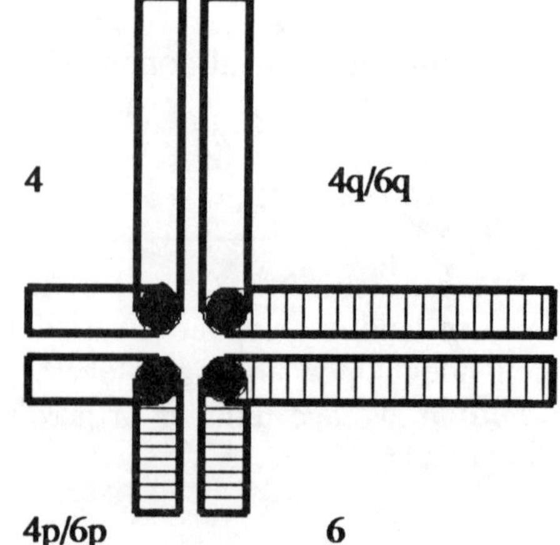

Adjacent:	4, 4q/6q	(duplication, deficient)
	4p/6p, 6	(duplication, deficient)
	4, 4p/6p	(duplication, deficient)
	4q/6q, 6	(duplication, deficient)
Alternate:	4, 6	(normal)
	4p/6p, 4q/6q	(balanced)

Two-thirds of their children would theoretically contain unbalanced gene combinations. These unbalanced combinations would most likely terminate during the embryonic or fetal period.

6. Affected infants may be genetically predisposed to the teratogenic effects of diphenylhydantoin.

7. Their child lacks androgen receptors and will not respond to testosterone. Many counselors would contend that this case has been mishandled. Most cases of testicular feminization present as normal-appearing females with delayed menstruation and/or infertility. The patients have been raised as women, and knowledge of sex reversal can produce considerable psychological harm and marital discord. It appears more appropriate to counsel on the basis of an infertile female. Abdominal testes may be removed to reduce cancer risk, and vaginal enlargement may be indicated. The patient should be advised of sterility, and suggestions may be made regarding possible adoption. Checks for other affected siblings may be recommended.

Chapter 13:

1. $q^2 = 1/8,000$; $q \approx 1/90$; $2pq \approx 2q = 1/45$.

2. Frequency of $\mathbf{A} = p = (3200 + 800) \div 6000 = 0.67$; $q = 1 - 0.67 = 0.33$.

3. Number of heterozygotes $= 2pqN = 2(0.67)(0.33)(4000) = 1769$. The third generation will also contain about 1769 heterozygotes.

4. First calculate the relative fitness of the patients compared to their normal siblings: $W_r = (200/1000) \div (4000/2000) = 1/10$. Now calculate the frequency of the allele with dominant effects: $p = n/2N = 1/2(20,000) = 1/40,000$. Now calculate the mutation rate: $\mu = sp = (1 - W_r)p = (9/10)(1/40,000) = 9/400,000 = 2.25 \times 10^{-5}$.

5. The $\mathbf{ABO^A}$ allele was most likely introduced to one of the tribes by a French-Canadian trapper. Subsequent drift led to near fixation of this allele.

6. $F = 2(1/2)^4 = 1/8$.

7. $\mu = (1/3)sX$. Solve for X: $X = 3\mu/s$. $s = 1\text{-}0 = 1$. Therefore, $X = s\mu = 3(1/25,000) = 3/25,000$.

8. The inbreeding coefficients for these marriages are:

second cousin:	$F = 2(1/2)^7 = 1/64$
first cousin:	$F = 2(1/2)^5 = 1/16$
uncle-niece:	$F = 2(1/2)^4 = 1/8$

$\alpha = \Sigma p_i F_i = (200/100,000)(1/64) + (100/100,000)(1/16) + (25/100,000)(1/8) = 1.25 \times 10^{-4}$.

9. $r = $ freq. $\mathbf{ABO^O} = \sqrt{(1600/4100)} = 0.625$; $p = $ freq. $\mathbf{ABO^A} = \sqrt{(2000 + 1600)/4100} - r = 0.312$;

$q = $ freq. $\mathbf{ABO^B} = 1 - 0.625 - 0.312 = 0.063$.

10. Hurler's syndrome.

Chapter 14:

1. 33%. The apparent risk is much lower, approximating 5%.

2. Approximately one-half of their sons will develop hemophilia B.

3. 1/2

4. b,e,g   7. d   10. c

5. b   8. b,c,f

6. a   9. a

11. XYY exhibits a statistically increased association with criminal behavior. However, the majority of
    XYY men do not display this behavior. Concern has been expressed regarding the possibility that
    early identification of XYY fetuses and notification of parents of the karyotype and its implications
    would alter rearing patterns of these individuals leading to actual development of criminal behavior.

12. Prenatal diagnosis can now detect all of these diseases with reasonable certainty in appropriate
    families using biochemical or molecular tests (RFLPs). In the latter case, the families would have
    to be studied carefully to determine whether the RFLP marker is cis or trans with respect to the
    disease-causing allele. Furthermore, the probability of recombination between the RFLP marker and
    the abnormal allele would have to be considered when giving advice regarding risk.